Proceedings of GeoShanghai 2018 International
Conference: Advances in Soil Dynamics
and Foundation Engineering

Tong Qiu · Binod Tiwari
Zhen Zhang
Editors

Proceedings of GeoShanghai 2018 International Conference: Advances in Soil Dynamics and Foundation Engineering

 Springer

Editors
Tong Qiu
The Pennsylvania State University
University Park, PA
USA

Zhen Zhang
Tongji University Shanghai
Shanghai
China

Binod Tiwari
Civil and Environmental Engineering
California State University, Fullerton
Fullerton, CA
USA

ISBN 978-981-13-0130-8   ISBN 978-981-13-0131-5   (eBook)
https://doi.org/10.1007/978-981-13-0131-5

Library of Congress Control Number: 2018939621

Printed on acid-free paper

This Springer imprint is published by the registered company Springer Nature Singapore Pte Ltd.
part of Springer Nature
The registered company address is: 152 Beach Road, #21-01/04 Gateway East, Singapore 189721, Singapore

Preface

The 4th GeoShanghai International Conference was held on May 27–30, 2018, in Shanghai, China. GeoShanghai is a series of international conferences on geotechnical engineering held in Shanghai every four years. The conference was inaugurated in 2006 and was successfully held in 2010 and 2014, with more than 1200 participants in total. The conference offers a platform of sharing recent developments of the state-of-the-art and state-of-the-practice in geotechnical and geoenvironmental engineering. It has been organized by Tongji University in cooperation with the ASCE Geo-Institute, Transportation Research Board, and other cooperating organizations.

The proceedings of the 4th GeoShanghai International Conference include eight volumes of over 560 papers; all were peer-reviewed by at least two reviewers. The proceedings include Volumes 1: Fundamentals of Soil Behavior edited by Dr. Annan Zhou, Dr. Junliang Tao, Dr. Xiaoqiang Gu, and Dr. Liangbo Hu; Volume 2: Multi-physics Processes in Soil Mechanics and Advances in Geotechnical Testing edited by Dr. Liangbo Hu, Dr. Xiaoqiang Gu, Dr. Junliang Tao, and Dr. Annan Zhou; Volume 3: Rock Mechanics and Rock Engineering edited by Dr. Lianyang Zhang, Dr. Bruno Goncalves da Silva, and Dr. Cheng Zhao; Volume 4: Transportation Geotechnics and Pavement Engineering edited by Dr. Xianming Shi, Dr. Zhen Liu, and Dr. Jenny Liu; Volume 5: Tunneling and Underground Construction edited by Dr. Dongmei Zhang and Dr. Xin Huang; Volume 6: Advances in Soil Dynamics and Foundation Engineering edited by Dr. Tong Qiu, Dr. Binod Tiwari, and Dr. Zhen Zhang; Volume 7: Geoenvironment and Geohazards edited by Dr. Arvin Farid and Dr. Hongxin Chen; and Volume 8: Ground Improvement and Geosynthetics edited by Dr. Lin Li, Dr. Bora Cetin, and Dr. Xiaoming Yang. The proceedings also include six keynote papers presented at the conference, including "Tensile Strains in Geomembrane Landfill Liners" by Prof. Kerry Rowe, "Constitutive Modeling of the Cyclic Loading Response of Low Plasticity Fine-Grained Soils" by Prof. Ross Boulanger, "Induced Seismicity and Permeability Evolution in Gas Shales, CO_2 Storage and Deep Geothermal Energy" by Prof. Derek Elsworth, "Effects of Tunneling on Underground Infrastructures" by Prof. Maosong Huang, "Geotechnical Data Visualization and Modeling of Civil

Infrastructure Projects" by Prof. Anand Puppala, and "Probabilistic Assessment and Mapping of Liquefaction Hazard: from Site-specific Analysis to Regional Mapping" by Prof. Hsein Juang. The Technical Committee Chairs, Prof. Wenqi Ding and Prof. Xiong Zhang, the Conference General Secretary, Dr. Xiaoqiang Gu, the 20 editors of the 8 volumes and 422 reviewers, and all the authors contributed to the value and quality of the publications.

The Conference Organizing Committee thanks the members of the host organizations, Tongji University, Chinese Institution of Soil Mechanics and Geotechnical Engineering, and Shanghai Society of Civil Engineering, for their hard work and the members of International Advisory Committee, Conference Steering Committee, Technical Committee, Organizing Committee, and Local Organizing Committee for their strong support. We hope the proceedings will be valuable references to the geotechnical engineering community.

Shijin Feng
Conference Chair
Ming Xiao
Conference Co-chair

Organization

International Advisory Committee

Herve di Benedetto	University of Lyon, France
Antonio Bobet	Purdue University, USA
Jean-Louis Briaud	Texas A&M University, USA
Patrick Fox	Penn State University, USA
Edward Kavazanjian	Arizona State University, USA
Dov Leshchinsky	University of Illinois, USA
Wenhao Liang	China Railway Construction Corporation Limited, China
Robert L. Lytton	Texas A&M University, USA
Louay Mohammad	Louisiana State University, USA
Manfred Partle	KTH Royal Institute of Technology, Switzerland
Anand Puppala	University of Texas at Arlington, USA
Mark Randolph	University of Western Australia, Australia
Kenneth H. Stokoe	University of Texas at Austin, USA
Gioacchino (Cino) Viggiani	Université Joseph Fourier, France
Dennis T. Bergado	Asian Institute of Technology, Thailand
Malcolm Bolton	Cambridge University, UK
Yunmin Chen	Zhejiang University, China
Zuyu Chen	Tsinghua University, China
Jincai Gu	PLA, China
Yaoru Lu	Tongji University, China
Herbert Mang	Vienna University of Technology, Austria
Paul Mayne	Georgia Institute of Technology, USA
Stan Pietruszczak	McMaster University, Canada
Tom Papagiannakis	Washington State University, USA
Jun Sun	Tongji University, China

Scott Sloan   University of Newcastle, Australia
Hywel R. Thomas   Cardiff University, UK
Atsashi Yashima   Gifu University, Japan

Conference Steering Committee

Jie Han   University of Kansas, USA
Baoshan Huang   University of Tennessee, USA
Maosong Huang   Tongji University, China
Yongsheng Li   Tongji University, China
Linbin Wang   Virginia Tech, USA
Lianyang Zhang   University of Arizona, USA
Hehua Zhu   Tongji University, China

Technical Committee

Wenqi Ding (Chair)   Tongji University, China
Charles Aubeny   Texas A&M University, USA
Rifat Bulut   Oklahoma State University, USA
Geoff Chao   Asian Institute of Technology, Thailand
Jian Chu   Nanyang Technological University, Singapore
Eric Drumm   University of Tennessee, USA
Wen Deng   Missouri University of Science and Technology,
                                        USA
Arvin Farid   Boise State University, Idaho, USA
Xiaoming Huang   Southeast University, China
Woody Ju   University of California, Los Angeles, USA
Ben Leshchinsky   Oregon State University, Oregon, USA
Robert Liang   University of Dayton, Ohio, USA
Hoe I. Ling   Columbia University, USA
Guowei Ma   Hebei University of Technology, China
Roger W. Meier   University of Memphis, USA
Catherine O'Sullivan   Imperial College London, UK
Massimo Losa   University of Pisa, Italy
Angel Palomino   University of Tennessee, USA
Krishna Reddy   University of Illinois at Chicago, USA
Zhenyu Yin   Tongji University, China
ZhongqiYue   University of Hong Kong, China
Jianfu Shao   Université des Sciences et Technologies
                                        de Lille 1, France
Jonathan Stewart   University of California, Los Angeles, USA

Wei Wu   University of Natural Resources
                                    and Life Sciences, Austria
Jianhua Yin   The Hong Kong Polytechnic University, China
Guoping Zhang   University of Massachusetts, USA
Jianmin Zhang   Tsinghua University, China
Xiong Zhang (Co-chair)   Missouri University of Science and Technology,
                                    USA
Yun Bai   Tongji University, China
Jinchun Chai   Saga University, Japan
Cheng Chen   San Francisco State University, USA
Shengli Chen   Louisiana State University, USA
Yujun Cui   École Nationale des Ponts et Chaussees (ENPC),
                                    France
Mohammed Gabr   North Carolina State University, USA
Haiying Huang   Georgia Institute of Technology, USA
Laureano R. Hoyos   University of Texas at Arlington, USA
Liangbo Hu   University of Toledo, USA
Yang Hong   University of Oklahoma, USA
Minjing Jiang   Tongji University, China
Richard Kim   North Carolina State University, USA
Juanyu Liu   University of Alaska Fairbanks, USA
Matthew Mauldon   Virginia Tech., USA
Jianming Ling   Tongji University, China
Jorge Prozzi   University of Texas at Austin, USA
Daichao Sheng   University of Newcastle, Australia
Joseph Wartman   University of Washington, USA
Zhong Wu   Louisiana State University, USA
Dimitrios Zekkos   University of Michigan, USA
Feng Zhang   Nagoya Institute of Technology, Japan
Limin Zhang   Hong Kong University of Science
                                    and Technology, China
Zhongjie Zhang   Louisiana State University, USA
Annan Zhou   RMIT University, Australia
Fengshou Zhang   Tongji University, China

Organizing Committee

Shijin Feng (Chair)   Tongji University, China
Xiaojqiang Gu   Tongji University, China
  (Secretary General)
Wenqi Ding   Tongji University, China
Xiongyao Xie   Tongji University, China

Yujun Cui	École Nationale des Ponts et Chaussees (ENPC), France
Daichao Sheng	University of Newcastle, Australia
Kenichi Soga	University of California, Berkeley, USA
Weidong Wang	Shanghai Xian Dai Architectural Design (Group) Co., Ltd., China
Feng Zhang	Nagoya Institute of Technology, Japan
Yong Yuan	Tongji University, China
Weimin Ye	Tongji University, China
Ming Xiao (Co-chair)	Penn State University, USA
Yu Huang	Tongji University, China
Xiaojun Li	Tongji University, China
Xiong Zhang	Missouri University of Science and Technology, USA
Guenther Meschke	Ruhr-Universität Bochum, Germany
Erol Tutumluer	University of Illinois, Urbana—Champaign, USA
Jianming Zhang	Tsinghua University, China
Jianming Ling	Tongji University, China
Guowei Ma	Hebei University of Technology, Australia
Hongwei Huang	Tongji University, China

Local Organizing Committee

Shijin Feng (Chair)	Tongji University, China
Zixin Zhang	Tongji University, China
Jiangu Qian	Tongji University, China
Jianfeng Chen	Tongji University, China
Bao Chen	Tongji University, China
Yongchang Cai	Tongji University, China
Qianwei Xu	Tongji University, China
Qingzhao Zhang	Tongji University, China
Zhongyin Guo	Tongji University, China
Xin Huang	Tongji University, China
Fang Liu	Tongji University, China
Xiaoying Zhuang	Tongji University, China
Zhenming Shi	Tongji University, China
Zhiguo Yan	Tongji University, China
Dongming Zhang	Tongji University, China
Jie Zhang	Tongji University, China
Zhiyan Zhou	Tongji University, China
Xiaoqiang Gu (Secretary)	Tongji University, China
Lin Cong	Tongji University, China
Hongduo Zhao	Tongji University, China

Fayun Liang	Tongji University, China
Bin Ye	Tongji University, China
Zhen Zhang	Tongji University, China
Yong Tan	Tongji University, China
Liping Xu	Tongji University, China
Mengxi Zhang	Tongji University, China
Haitao Yu	Tongji University, China
Xian Liu	Tongji University, China
Shuilong Shen	Tongji University, China
Dongmei Zhang	Tongji University, China
Cheng Zhao	Tongji University, China
Hongxin Chen	Tongji University, China
Xilin Lu	Tongji University, China
Jie Zhou	Tongji University, China

Contents

Shafts and Deep Foundations

Offshore Geotechnics

About the Editors

Tong Qiu received his Ph.D. in Civil Engineering from the University of California, Los Angeles. His research interests include soil dynamics, landslides, and numerical analysis of geosystems. He is currently an Associate Professor in the Department of Civil and Environmental Engineering, The Pennsylvania State University, University Park.

Binod Tiwari received his Ph.D. from Niigata University, Japan. He got his Ph.D. in 2003 in Geotechnical Engineering and is expert in the field of slope stability and stabilization, landslide mitigation, applied GIS, soil behavior and characterization, and geotechnical earthquake engineering. He is currently a Professor in the Department of Civil and Environmental Engineering, California State University, Fullerton.

Zhen Zhang received his Ph.D. in Geotechnical Engineering from Tongji University. His research interests include ground improvement, geosynthetic reinforcement, and roadway engineering. He is currently an Associate Professor in the Department of Geotechnical Engineering, Tongji University.

Probabilistic Assessment and Mapping of Liquefaction Hazard: From Site-Specific Analysis to Regional Mapping

C. Hsein Juang[✉], Qiushi Chen, Mengfen Shen, and Chaofeng Wang

Glenn Department of Civil Engineering, Clemson University,
Clemson, SC 29634, USA
hsein@clemson.edu

Abstract. Probabilistic methods have been increasingly used in liquefaction hazard assessments for purposes of considering the substantial uncertainties in both the liquefaction case histories and the model development process, and for the risk assessment and performance-based earthquake engineering. In this paper, a review on the probabilistic methods of site-specific liquefaction assessment, including logistic regression, the Bayesian method and various performance-based methods, is first undertaken. Another important topic in the liquefaction hazard assessment is to understand its spatial extent, leading to mapping of liquefaction hazard over a region. The regional liquefaction hazard maps are being employed as planning tools and provide guidance to assess the need for site-specific evaluations. The second focus of this paper details a review of methods for regional liquefaction hazard mapping, including geology-based, geotechnical data-driven and geostatistical methods as well as multiscale methods. The review of the site-specific probabilistic methods and regional mapping methods involves a discussion of their formulations, key assumptions, advantages and applications in liquefaction assessment. The challenges and the need for further research in these areas are also mentioned.

Keywords: Liquefaction · Probabilistic assessment · Geostatistics
Random field models

1 Introduction

Infrastructures such as buildings, bridges and underground utility lines are often threatened by ground deformation caused by soil liquefaction during an earthquake. Site-specific assessment of liquefaction hazard involves ground shaking hazard analysis and estimation of liquefaction resistance of soil deposits, where various in-situ tests such as the cone penetration test (CPT), the standard penetration test (SPT) and the shear wave velocity test (Vs) are commonly utilized to estimate liquefaction resistance or to classify levels of liquefaction severity at individual locations. Of various methods, the cyclic stress-based simplified procedure pioneered by Seed and Idriss is the most widely used method for liquefaction evaluation (e.g., Seed and Idriss 1971, 1982; Seed 1979; Seed et al. 1985). A summary report by Youd et al. (1978) reviewed and recommended simplified procedures for the most commonly used in-situ tests.

© Springer Nature Singapore Pte Ltd. 2018
T. Qiu et al. (Eds.): GSIC 2018, *Proceedings of GeoShanghai 2018 International Conference: Advances in Soil Dynamics and Foundation Engineering*, pp. 1–16, 2018.
https://doi.org/10.1007/978-981-13-0131-5_1

In this simplified procedure, the evaluation of liquefaction involves the use of two variables: (1) the seismic demand on a soil layer placed by a given earthquake, expressed in terms of the cyclic stress ratio (CSR); and (2) the capacity of soil to resist liquefaction, expressed in terms of the cyclic resistance ratio (CRR). In a deterministic analysis, the factor of safety (F_S) of a soil against liquefaction is computed as the ratio of CRR to CSR, and the computed F_S is used to determine if the soil liquefies for a given set of conditions in an earthquake.

There are several significant limitations of the deterministic methods in terms of both model development and practical application. These arose because of the uncertainty in deriving a boundary curve that separates liquefied from non-liquefied cases, uncertainties in assessed soil properties and ground motion parameters, and incompatibility between the deterministic approach for liquefaction evaluation and the performance-based design paradigm in the earthquake engineering profession (Juang et al. 2017). Consequently, probabilistic methods for assessing liquefaction potential and its impact have been developed (e.g., Boulanger and Idriss 2012; Cetin et al. 2004; Juang et al. 2001, 2002, 2004, 2005, 2006, 2009, 2010, 2012a, b, 2013, 2017; Kayen et al. 2013; Khoshnevisan et al. 2010; Ku et al. 2012).

The site-specific assessments provide estimation of the liquefaction potential of soil deposits or the liquefaction-induced damages at individual locations. To assess the consequence of liquefaction, it is oftentimes necessary to understand the potential spatial extent of liquefaction, leading to maps of liquefaction hazard over a region. The continuing evolution of research in this area has also resulted in a concurrent evolution of the liquefaction hazard maps from the descriptive and qualitative to the quantitative and probabilistic. These maps are of use to government agencies as planning tools, to insurance companies for seismic loss estimation, and as a guide for site-specific evaluations as needed. The recent trends in regional liquefaction mapping involve integrations of multiple sources of information (e.g. surficial geologic, geotechnical and hydrological data) via the use of advanced geostatistics tools and multiscale random field models for an accurate assessment of liquefaction hazard over a vast area.

Given the continued growth in the development of complex methodologies for the probabilistic assessment and mapping of liquefaction hazards, the objective of this paper is to provide a systematic review on the state-of-the-art in both site-specific probabilistic assessment and regional mapping of liquefaction hazards, including limitations and advantages, key assumptions involved, and applications. The challenges and the needs for further research are also discussed.

2 Site-Specific Probabilistic Assessment of Liquefaction Potential and Effects

The early research in probabilistic assessment of liquefaction focuses mainly on predicting the occurrence and potential of liquefaction. When a large number of liquefaction case histories are available, the prediction of liquefaction may be viewed as a binary classification problem in which the event falls into either the liquefaction or the non-liquefaction realm. Probabilistic methods such as discriminant analysis and logistic regression deal with the uncertainties of the boundary separation between the liquefied and the non-liquefied cases. Subsequent developments (e.g., Bayesian methods) have been used to address the uncertainties in the case histories. Recent efforts in probabilistic assessment have focused on the study of both the liquefaction potential and its effects (e.g., lateral spread, settlement, and building damages) and adopted the performance-based approach widely used in the earthquake engineering community. The availability of the site-specific probabilistic assessment methods provides the basis for regional liquefaction hazard mapping. The assumptions adopted in the site-specific probabilistic assessment may also be implicitly followed in the regional hazard mapping. Therefore, it is essential to understand the site-specific probabilistic assessment methods. In this section, the selected probabilistic assessment methods of liquefaction, e.g., logistic regression, Bayesian methods, and performance-based assessment methods, are briefly discussed. The reader is referred to Juang et al. (2017) for a more detailed description and comparison of these methods.

2.1 Logistic Regression

The problem of determining whether a soil liquefies for a given earthquake loading may be viewed as predicting the response of a binary system, which can be solved using the logistic regression (e.g., Liao et al. 1988; Toprak et al. 1999; Juang et al. 2002, 2003, 2006; Lai et al. 2006). Let $x_1, x_2, \ldots, x_r$ denote explanatory variables in a regression model. In the logistic regression model, the liquefaction potential is measured by the probability of liquefaction, P_L, under the assumption that $\ln[P_L/(1 - P_L)]$ is a linear function of explanatory variables as follows (e.g., Liao et al. 1988):

$$\log\left(\frac{P_L}{1 - P_L}\right) = \theta_0 + \theta_1 x_1 + \theta_2 x_2 + \cdots + \theta_r x_r \tag{1}$$

where $\boldsymbol{\theta} = \{\theta_1, \theta_2, \ldots, \theta_r\}$ are regression coefficients. Let $\mathbf{D}$ denote the calibration database. The likelihood function of the calibration database can be written as

$$l(\boldsymbol{\theta}|D) = \prod_{i=1}^{N_L} \frac{1}{1 + \exp\{-[\theta_0 + \theta_1 x_1 + \theta_2 x_2 + \cdots + \theta_r x_r]\}}$$
$$\times \prod_{i=1}^{N_{NL}} \frac{1}{1 + \exp\{-[\theta_0 + \theta_1 x_1 + \theta_2 x_2 + \cdots + \theta_r x_r]\}} \tag{2}$$

where N_L and N_{NL} is the number of liquefied cases and non-liquefied cases, respectively (Juang et al. 2015). The regression coefficients $\boldsymbol{\theta}$ can be readily estimated by maximizing the above likelihood function using the maximum likelihood principle (Juang et al. 2015). In practice, the calibration database often contains more liquefied cases than non-liquefied cases because of the greater interest in liquefied sites in the post-liquefaction survey (Cetin et al. 2002; Oommen et al. 2010). It is also possible to modify the likelihood function to consider choice-based sampling bias in the calibration database (Juang et al. 2017).

The use of logistic regression however prevents a direct consideration of the uncertainties in the case histories. Therefore, it is most suitable to analyze case histories with a similar level of uncertainties in the soil properties and ground motion parameters. The logistic regression is a member of the generalized linear regression family. Although it may not always produce the most accurate results, the logistic regression is relatively easy to use. The reader is referred to Juang et al. (2015) for further details on the generalized linear regression family.

2.2 Bayesian Methods

The reliability method has also been used to assess the liquefaction probability, which requires a thorough knowledge of the model uncertainty associated with a deterministic model. In the absence of such knowledge, the failure probability computed by the structural reliability method is only considered "notional." Therefore, there is a need to reestablish the β-P_L relationship. As illustrated in Juang et al. (2000), the β-P_L can be calibrated with real performances of soils observed during past earthquakes. Let $f(\beta|L)$ and $f(\beta|NL)$ denote the probability density functions (PDFs) of the reliability index for the liquefied and non-liquefied cases, respectively, calculated based on a structural reliability method without considering the model uncertainty. Let $P(L)$ and $P(NL)$ denote the prior probabilities of liquefaction and non-liquefaction, respectively, when analyzing the liquefaction potential for a future case. Through the use of the Bayes' theorem, the β-P_L relationship for a future case is thus established by the following equation (Juang et al. 2000):

$$P_L = P(L|\beta) = \frac{f(\beta|L)P(L)}{f(\beta|L)P(L) + f(\beta|NL)P(NL)} \tag{3}$$

This equation is known as the reliability index mapping method in the literature. Here, the β-P_L relationship is established based on a comparison with the actual distributions of β of the liquefied cases and non-liquefied cases, thus encompassing the model uncertainty. Therefore, the effect of the model uncertainty on the assessment of the liquefaction potential is automatically considered although not explicitly characterized. It is also possible to use the same procedure to establish the relationship between the factor of safety for a deterministic method and the liquefaction probability based on previous case histories. One noticeable advantage of the above Bayesian mapping function is that it is affected by the sampling bias in the database, i.e., the bias from the greater ratio of liquefied to non-liquefied cases in the database than that in the real world.

The Bayesian methods have also recently been used to explicitly characterize the model uncertainty associated with geotechnical models (e.g., Zhang et al. 2009; Ku et al. 2012). These methods have also been suggested for developing new liquefaction models with an explicit consideration of both parameter and model uncertainties (e.g., Cetin et al. 2002; Moss et al. 2006; Boulanger and Idriss 2015). For a more detailed discussion on application of Bayesian method for liquefaction potential assessment, the reader is referred to Juang et al. (2017), and for a discussion on the use of the Bayesian method to solve different types of geotechnical problems, the reader is referred to Juang and Zhang (2017). The probability of liquefaction obtained by semi-empirical models discussed previously is a conditional probability for a given set of seismic parameters, i.e., the peak horizontal acceleration at the ground surface (a_{max}) and the moment magnitude of the earthquake (M_w). For performance-based earthquake engineering (PBEE), it is necessary to consider contributions from all possible ground motion hazard levels (Kramer 2008; Ekstrom and Franke 2016). As illustrated in Kramer and Mayfield (2007), it is possible to conduct performance-based assessment for liquefaction problems following the general framework suggested by the Pacific Earthquake Engineering Research (PEER) Center (e.g., Cornell and Krawinkler 2000). Nevertheless, Juang et al. (2008) suggested an intuitive method for performance-based analysis of liquefaction potential based on the total probability theory. Let $P(L|M_w = M_{wi}, a_{max} = a_{maxj})$ denote the conditional probability of liquefaction given a set of seismic parameters a_{max} and M_w, and $P(M_w = M_{wi}, a_{max} = a_{maxj})$ denote the joint probability of $M_w = M_{wi}$ and $a_{max} = a_{maxj}$ during the given exposure time T. Based on the total probability theorem, the probability of liquefaction during a given exposure time T can be computed as follows:

$$P_{L,A} = \sum_{j=1}^{N_{a_{max}}} \sum_{i=1}^{N_{M_w}} P\left[L|\left(M_w = M_{wi}, a_{max} = a_{maxj}\right)\right] P\left(M_w = M_{wi}, a_{max} = a_{maxj}\right) \quad (4)$$

where $N_{a_{max}}$ and N_{M_w} are the numbers of possible values for a_{max} and M_w, respectively. This method is also useful in assessing the exceedance probability of the liquefaction induced settlements and lateral spreading during a given exposure period (e.g., Lu et al. 2009; Liu et al. 2016). While the above performance-based framework is well established, the vulnerability function is still necessary for assessing the possible damage from soil liquefaction during a given exposure time. Such vulnerability functions, however, are remain largely absent.

3 Regional Mapping of Liquefaction Hazard

The probabilistic site-specific assessment methods reviewed in the previous section are used to provide an estimation of liquefaction potential of soil deposits or the liquefaction-induced damages at the individual locations. Assessing the consequence of liquefaction often requires elucidating the potential spatial extent of the liquefaction, leading to mapping of liquefaction hazard over a region. The regional liquefaction hazard maps are being employed as planning tools and provide guidance to assess the need for site-specific evaluations. In this section, current methods for regional liquefaction hazard mapping are reviewed.

3.1 Geology-Based Mapping of Liquefaction Susceptibility and Potential

Many of the early liquefaction evaluation and regional hazard mapping projects focus on the occurrence of liquefaction and conventionally adopted approaches that relate surficial geology to liquefaction susceptibility and potential (Youd and Perkins 1978; Knudsen et al. 2000; Witter et al. 2006). The recent trend involves the use of supplementary data such as hydrological and geotechnical data (Holzer et al. 2006a, b;

Brankman and Baise 2008; Hayati and Andrus 2008; Heidari and Andrus 2010, 2012) or geomorphological data (Papathanassiou et al. 2017) along with the surficial geology when characterizing and mapping the liquefaction hazard. In most of the early studies, liquefaction hazard level (quantified in terms of either liquefaction susceptibility or liquefaction potential) is assumed to be constant within each surficial geologic unit. One commonly used approach, as shown in Fig. 1, involves calculating the cumulative frequency distributions (CFDs) of the liquefaction potential index LPI (Iwasaki et al. 1982; Li et al. 2006) of surficial geologic units, and then assigning a constant liquefaction hazard level to each geologic unit based on the calculated CFDs (Holzer et al. 2006a, b).

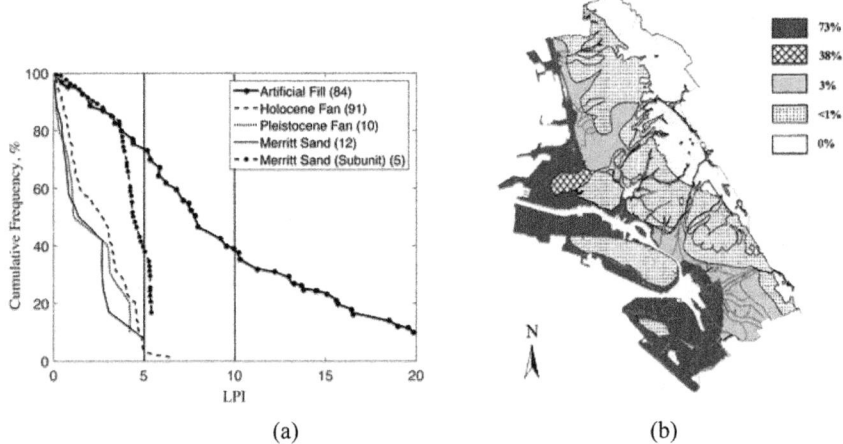

Fig. 1. Liquefaction hazard mapping of the Alameda county site in California for a hypothetical magnitude 7.1 earthquake event (modified from Holzer et al. 2006a): (a) Cumulative frequency of LPI by geologic units; (b) Percentage of area predicted to liquefy based on LPI.

The assumption of a constant liquefaction hazard level, though convenient for mapping, ignores the inherent spatial variability of soil properties and therefore may limit the accuracy of the generated liquefaction hazard maps. Moreover, one of the main challenges when combining surficial geology information with geotechnical data is a representation of the spatial variability at two vastly different scales. These limitations and challenges motivate further research in this area, where more sophisticated geostatistics approaches and multiscale methods have been proposed for regional liquefaction hazard mapping.

3.2 Geostatistics and Random Field Model-Based Liquefaction Hazard Mapping

To address the challenges and limitation of early regional liquefaction studies, recent research has involved the use of geostatistical methods to interpolate site-specific data over an extended region (e.g., Liu and Chen 2006; Baise et al. 2006; Lenz and Baise

2007; Baker and Faber 2008; Vivek and Raychowdhury 2014; Liu et al. 2016; Juang et al. 2017). The use of geostatistical methods permits the incorporation of geotechnical data and the explicit consideration of spatial correlations of soil properties or lique-faction potential. The hypothesis behind the geostatistical methods is the existence of spatial correlations in soil properties that are present in either the measured geotech-nical data or the estimated liquefaction potentials. For instance, using datasets acquired from nine liquefaction hazard mapping projects, Baise and Lenz (2006) found that densely spaced CPT or V_s data exhibit spatial correlations with the correlation distances on the order of several kilometers. Similarly, Juang and co-workers (Chen et al. 2016b; Wang et al. 2017a) determined the existence of spatial correlations in both the CPT data and the calculated LPI values using field data from the 2011 Christchurch earthquakes, as illustrated in Fig. 2.

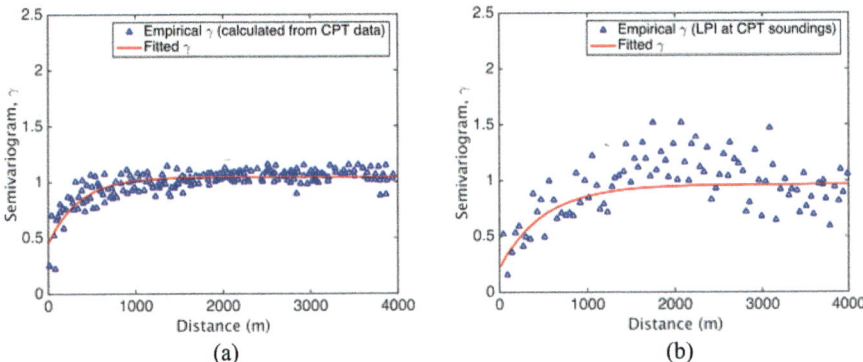

Fig. 2. Empirical and fitted semivariogram of Christchurch data (modified from Wang et al. 2017a): (a) tip resistance of CPT; (b) calculated LPI values.

To model the spatial structure of soil properties, a form of covariance known as the semivariogram $\gamma(\mathbf{h})$, is commonly adopted in the geostatistics field. The semivariogram $\gamma(\mathbf{h})$ is equal to half the variance of the difference of two random variables separated by a distance $\mathbf{h}$

$$\gamma(\mathbf{h}) = \frac{1}{2}\mathrm{Var}[Z(\mathbf{u}) - Z(\mathbf{u}+\mathbf{h})] \qquad (5)$$

where $Z(\mathbf{u})$ is the distribution of the Gaussian random variable at location $\mathbf{u}$. It is possible to define a vector $\mathbf{h}$ to account for both the separation distance and orientation. The semivariogram is related to other commonly used measures to quantify spatial correlation, i.e., the covariance $\mathrm{COV}(\mathbf{h})$ and the correlation $\rho(\mathbf{h})$. In geostatistics, the semivariogram is often used in lieu of covariance as it requires second-order station-arity of only the increments, which is a weaker requirement than the second-order stationarity of the variable itself.

The local soil property and the averaged index approaches are two commonly used to account for the spatial correlation of soil properties in the liquefaction evaluation procedure, as described and compared in Juang et al. (2017) and Wang et al. (2017a). These two approaches are shown in Fig. 3 where the local soil property approach can be further divided into two-dimension (2D) and three-dimension (3D) approaches depending on the treatment of the 3D spatial correlation.

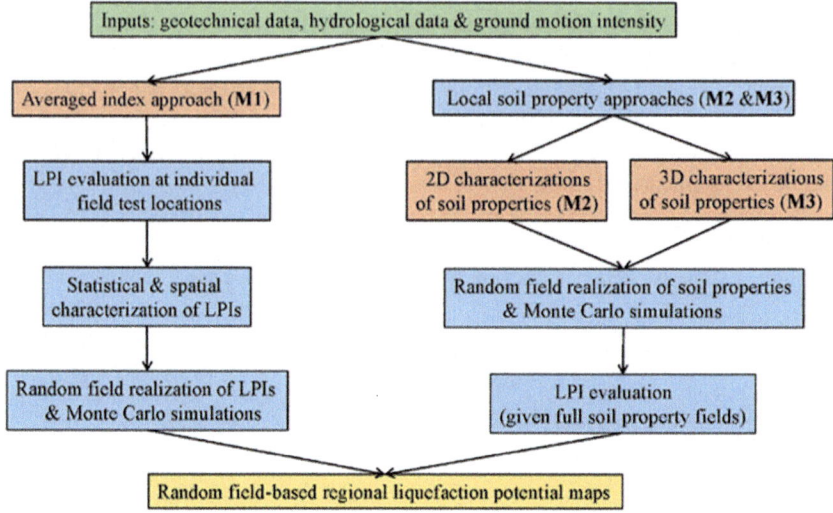

Fig. 3. Approaches used to account for spatial variabilities in regional liquefaction mapping (modified from Juang et al. 2017). Here, the LPI is used to quantify liquefaction hazard.

Juang and co-workers recently extended the previously described spatial correlation to consider the multiscale nature of soil variability (Chen et al. 2012; Chen et al. 2016a, b). It is possible to use this extension to generate adaptively a higher resolution random field around the regions of high interest. The novelty of this multiscale extension is to represent consistently fine and coarse scale random fields while maintaining the appropriate spatial correlation structures across scales.

In the multiscale framework, two scales of interest are considered, i.e., scale 1 for coarse scale and scale 2 for fine scale. The coarse scale is defined as the average of all fine-scale points within its area (a coarse element), mathematically written as (Chen et al. 2012; Chen et al. 2016a)

$$Z_{1(a)} = \frac{1}{N} \sum_{i=1}^{N} Z_{2i(a)} \tag{6}$$

where N is the number of fine-scale points within a coarse-scale area (element a); the subscripts '1' and '2' indicating the coarse and fine scale, respectively; and 'a' indicating a coarse-scale element. Using such a relationship, the expression of the variances and spatial correlations of coarse-scale quantities of interest can be explicitly derived, as detailed in Chen et al. (2016a) and briefly summarized as follows.

If the variance of fine-scale values is unity, the coarse-scale variance $\sigma_{Z1(a)}$ can be computed as

$$\sigma_{Z1}^2 = \frac{1}{N^2} \sum_{i=1}^{N} \sum_{j=1}^{N} \rho_{Z2i,Z2j} \cdot \sigma_{Z2i} \cdot \sigma_{Z2j} \tag{7}$$

The correlation between all considered scales can be calculated by expanding the definition of covariance and rearranging terms of the covariance definition such that

$$\rho_{Zi,Zj} = \frac{COV\left[Z_i, Z_j\right]}{\sigma_{Zi}\sigma_{Zj}} \tag{8}$$

where Z_i and Z_j are the two elements within the random field at any scale with means μ_{Zi} and μ_{Zj}, respectively. Using Eq. (7), the correlation between elements at different scales can be derived (Chen et al. 2016a)

$$\rho_{Z1(a),Z1(b)} = \frac{\sum_{i=1}^{N} \sum_{j=1}^{N} \rho_{Z2i(a),Z2j(b)}}{\sqrt{\sum_{i=1}^{N} \sum_{j=1}^{N} \rho_{Z2i(a),Z2j(a)}} \sqrt{\sum_{i=1}^{N} \sum_{j=1}^{N} \rho_{Z2i(b),Z2j(b)}}} \tag{9}$$

$$\rho_{Z2,Z1(a)} = \frac{\sum_{i=1}^{N} \rho_{Z2,Z2i(a)}}{\sqrt{\sum_{i=1}^{N} \sum_{j=1}^{N} \rho_{Z2i(a),Z2j(a)}}} \tag{10}$$

where $\rho_{Z1(a),Z1(b)}$ is the correlation between two coarse-scale (scale 1) elements a and b; and $\rho_{Z2,Z1(a)}$ is the correlation between a fine-scale element (scale 2) and a coarse-scale element a. A comparison of the spatial correlation (ρ) versus the normalized scalar distance measure between all combinations of scales is shown in Fig. 4. It can be seen that averaging of the fine-scale points effectively increases the correlation for a given distance relative to the fine scale, i.e., the coarse-to-coarse scale correlation ($\rho_{Z1,Z1}$) is always larger than the fine-to-fine scale correlation ($\rho_{Z2,Z2}$).

The multiscale random field models have been applied to map liquefaction hazard in terms of LPI (Chen et al. 2016a; Wang et al. 2017a) and settlement (Chen et al. 2016b). The methods provide regional liquefaction maps while at the same time allow generation of high-resolution information in critical areas, which can be used to perform small-scale liquefaction hazard analysis.

3.3 Integration of Geotechnical and Geological Data

Current geostatistical methods consider the spatial variability of the mapped properties (e.g., soil property indices, liquefaction potential index) derived from geotechnical

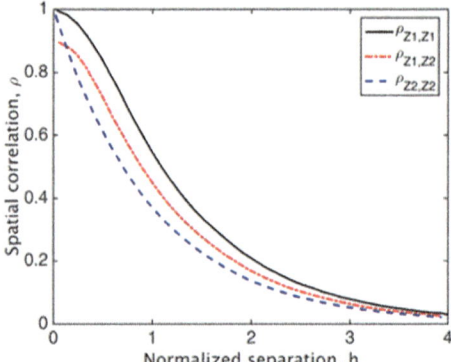

Fig. 4. Exponential correlation versus normalized distance between scales for a given realization of liquefaction potential index (modified from Chen et al. 2012).

data. However, the spatial structure of the mapped properties could vary within and across different geologic units. The geologic data provides information on large-scale (regional scale) material heterogeneity, which is not considered in current geostatistical method-based regional liquefaction hazard mapping studies. A hybrid geotechnical and geologic data-based framework has been developed to address this challenge (Wang and Chen 2017; Wang et al. 2017b). Using the liquefaction potential index (LPI) as a measure of liquefaction hazard, Figure illustrates the newly developed framework. The resulting liquefaction hazard map is generated taking into account two types of LPI data. The first type, termed the primary data, is evaluated using geotechnical data-based LPI model, e.g., the Robertson and Wride (1998) CPT-based liquefaction model implemented within the Iwasaki et al. (1982) LPI framework. The second type, termed the secondary data, is obtained based on geologic information such as a surficial geologic unit map. The role of the secondary data is to constrain and/or improve the primary data-based LPI map so that the final map conforms to the large-scale geologic boundaries in the region studied (Fig. 5).

The maps of the regions of liquefaction hazard within the Alameda County site in California are illustrated in Fig. 6. Most of the observed liquefied areas (along the coastline and in the artificial fill unit) are predicted to have high LPI values as indicated by the predicted LPI map. However, when no secondary data is incorporated (Fig. 6b), the resulting LPI map may be incorrect in some areas. For example, a lack of geologic constraint yields an incorrect prediction of the bedrock unit with a relatively high LPI value, which is in direct contradiction to the well-established concept that bedrock is not prone to liquefaction. Another example is the Merritt sand unit (highlighted by a solid star). No liquefaction manifestation was observed and previous studies confirmed that this unit has low liquefaction potential (Holzer et al. 2006a). Without proper geologic constraint, the LPI map in Fig. 6b again incorrectly predicts high LPI values. The integration of secondary geologic data through the proposed framework, in which these misclassifications are mitigated, results in LPI map with improved accuracy as shown in Fig. 6c.

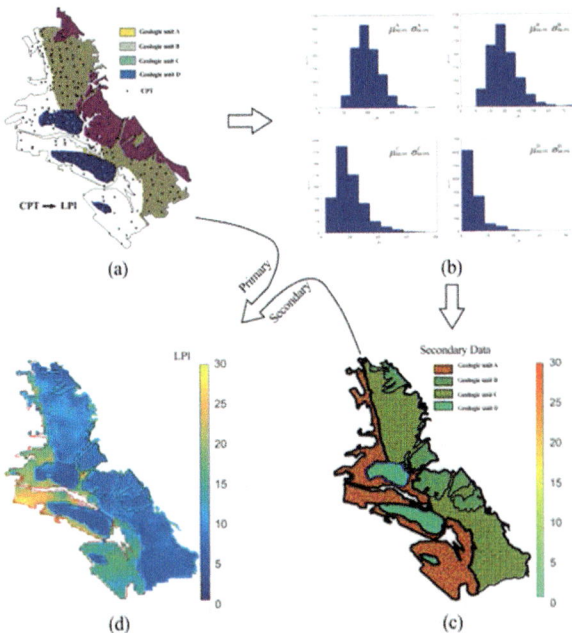

Fig. 5. A hybrid geotechnical and geologic data-based framework for regional liquefaction hazard mapping (modified from Wang et al. 2017b).

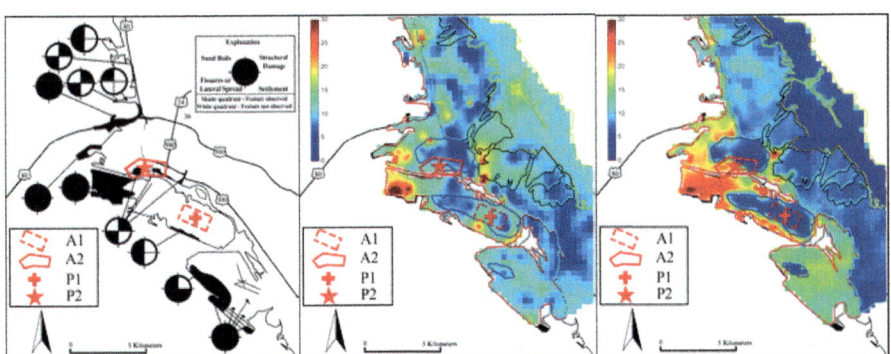

Fig. 6. Maps of the liquefaction hazards at the Alameda County site in California: (a) Observed liquefaction after the 1989 Loma Prieta earthquake; (b) an LPI map without secondary geologic data; (c) an LPI map with hybrid data generated with this proposed framework (modified from Wang et al. 2017b; Wang and Chen 2017).

4 Summary, Challenges and Future Research

The cyclic stress-based simplified procedure has evolved over the past few decades into a widely used practical tool for the site-specific assessment of liquefaction potential and effect. In this paper, a systematic review of both probabilistic site-specific methods for liquefaction potential evaluation and regional liquefaction hazard mapping is presented. While significant progress has been made in these areas, key challenges are identified that deserve further research:

Probabilistic assessment of liquefaction hazard. The availability of the case histories and how they are analyzed currently limits the efficacy of the present generation of site-specific liquefaction potential assessment methods.

As illustrated in detail in Juang et al. (2017), the following concepts may be the subject of a concerted and focused research effort: (1) the collection of calibration case histories in the high CSR region, (2) an elucidation of the disparity between different probabilistic liquefaction analysis models, (3) a reduction of the model uncertainty through a better understanding of the liquefaction phenomena, and (4) a characterization of the inter-region variability of the liquefaction potential models.

Mapping of liquefaction hazard in case of insufficient data. The use of Geostatistics often requires a substantial amount of in-situ test data, which when limited may greatly hinder the application of geostatistics models. Consequently, there is an urgent need to develop a new generation of models for use in liquefaction hazard mapping with explicit consideration of the amount of in-situ test data. In this regard, the use of multiple sources of information may be help in improving current liquefaction hazard schemes, specifically in terms of (i) a robust and consistent integration of heterogeneous sources of data (e.g., geotechnical, geologic, geomorphology, hydrologic), (ii) a better understanding of the multiscale spatial variability of different data sources, and (iii) a collection of calibration case histories for use in regional liquefaction hazard maps.

Vulnerability of buildings to liquefaction induced hazard. The implementation of performance-based design to mitigate the liquefaction hazard requires a thorough understanding of the relationship between the liquefaction-induced ground deformation, and its effect on the degree of building damage, otherwise known as the vulnerability function. This expected building vulnerability will depend upon both the type and the extent of the ground deformation, and the type of superstructure. Given the current paucity of building vulnerability studies, the creation of a set of building-specific vulnerability functions would be most valuable for the evolution of performance-based liquefaction hazard assessments.

Acknowledgements. The authors would like to acknowledge the financial support provided by the U.S. Geological Survey (Grant Nos. G17AP00044 & 07HQGR0053) and National Science Foundation (Grant Nos. CMS-9612116, CMS-0085143 & CMS-0218365) for their projects on soil liquefaction. The opinions expressed in this paper do not necessarily reflect the view and policies of the National Science Foundation and the U.S. Geological Survey. The first author also wishes to acknowledge the contributions of the following individuals to his studies of liquefaction in the past 20 years: Dr. Ron Andrus (Clemson University), Dr. Jie Zhang (Tongji University), Dr. Chieh-Seng Ku (I-Shou University), and Dr. Jin-Hung Hwang (National Central University); former Ph.D. students, Jinxia Chen, Mihai Chiru-Danzer, Tao Jiang, Peter Haiming Yuan, Susan Hui Yang, David Kun Li, Sunny Ye Fang, Zhe Luo, Lei Wang, Wenping Gong, and Sara Khoshnevisan; and former visiting Ph.D. students, Jesse Chen and Chieh Lu.

References

Baise, L.G., Lenz, J.A.: Guidelines for regional liquefaction hazard mapping. Final technical report for U.S. Geological Survey Award No. 05HQGR0103, U.S. Geological Survey, Reston, VA (2006)

Baker, J.W., Faber, M.H.: Liquefaction risk assessment using geostatistics to account for soil spatial variability. J. Geotech. Geoenviron. Eng. **134**(1), 14–23 (2008)

Boulanger, R.W., Idriss, I.M.: Probabilistic standard penetration test–based liquefaction–triggering procedure. J. Geotech. Geoenviron. Eng. **138**(10), 1185–1195 (2012)

Boulanger, R.W., Idriss, I.M.: Magnitude scaling factors in liquefaction triggering procedures. Soil Dyn. Earthq. Eng. **79**, 296–303 (2015)

Brankman, C.M., Baise, L.G.: Liquefaction susceptibility mapping in Boston, Massachusetts. Environ. Eng. Geosci. **14**(1), 1–16 (2008)

Cetin, K.O., Der Kiureghian, A., Seed, R.B.: Probabilistic models for the initiation of seismic soil liquefaction. Struct. Saf. **24**, 67–82 (2002)

Cetin, K.O., Seed, R.B., Der Kiureghian, A., Tokimatsu, K., Harder Jr., L.F., Kayen, R.E., Moss, R.E.: Standard penetration test-based probabilistic and deterministic assessment of seismic soil liquefaction potential. J. Geotech. Geoenviron. Eng. **130**(12), 1314–1340 (2004)

Chen, Q., Seifried, A., Andrade, J.E., Baker, J.W.: Characterization of random fields and their impact on the mechanics of geosystems at multiple scales. Int. J. Numer. Anal. Meth. Geomech. **36**(2), 140–165 (2012)

Chen, Q., Wang, C., Hsein Juang, C.: CPT-based evaluation of liquefaction potential accounting for soil spatial variability at multiple scales. J. Geotech. Geoenviron. Eng. **142**(2), 04015077 (2016a)

Chen, Q., Wang, C., Juang, C.H.: Probabilistic and spatial assessment of liquefaction-induced settlements through multiscale random field models. Eng. Geol. **211**, 135–149 (2016b)

Cornell, C.A., Krawinkler, H.: Progress and challenges in seismic performance assessment. PEER News **3**(2), 1–3 (2000)

Ekstrom, L., Franke, K.: Simplified procedure for the performance-based prediction of lateral spread displacements. J. Geotech. Geoenviron. Eng. 04016028 (2016). https://doi.org/10.1061/(ASCE)GT.1943-5606.0001440

Hayati, H., Andrus, R.D.: Liquefaction potential map of Charleston, South Carolina based on the 1986 earthquake. J. Geotech. Geoenviron. Eng. **134**(6), 815–828 (2008). https://doi.org/10.1061/(ASCE)1090-0241

Heidari, T., Andrus, R.D.: Mapping liquefaction potential of aged soil deposits in Mount Pleasant, South Carolina. Eng. Geol. **112**, 1–12 (2010). https://doi.org/10.1016/j.enggeo.2010.02.001

Heidari, T., Andrus, R.D.: Liquefaction potential assessment of Pleistocene beach sands near Charleston, South Carolina. J. Geotech. Geoenviron. Eng. **138**, 1196–1208 (2012). https://doi.org/10.1061/(ASCE)GT.1943-5606.0000686

Holzer, T.L., Bennett, M.J., Noce, T.E., Padovani, A.C., Tinsley III, J.C.: Liquefaction hazard mapping with LPI in the greater Oakland, California, area. Earthq. Spectra **22**(3), 693–708 (2006a)

Holzer, T.L., Blair, J.L., Noce, T.E., Bennett, M.J.: Predicted liquefaction of east bay fills during a repeat of the 1906 San Francisco earthquake. Earthq. Spectra **22**(S2), S261–S277 (2006b)

Iwasaki, T., et al.: Microzonation for soil liquefaction potential using simplified methods. In: Proceedings of the 3rd International Conference on Microzonation, Seattle. vol. 3 (1982)

Juang, C.H., Chen, C.J., Rosowsky, D.V., Tang, W.H.: CPT-based liquefaction analysis, Part 2: reliability for design. Géotechnique **50**(5), 593–599 (2000)

Juang, C.H., Chen, C.J., Jiang, T.: A probabilistic framework for liquefaction potential by shear wave velocity. J. Geotech. Geoenviron. Eng. ASCE **127**(8), 670–678 (2001)

Juang, C.H., Jiang, T., Andrus, R.D.: Assessing probability-based methods for liquefaction evaluation. J. Geotech. Geoenviron. Eng. ASCE **128**(7), 580–589 (2002)

Juang, C.H., Yuan, H., Lee, D.H., Lin, P.S.: Simplified cone penetration test-based method for evaluation liquefaction resistance of soils. J. Geotech. Geoenviron. Eng. **129**(1), 66–80 (2003)

Juang, C.H., Yang, S.H., Yuan, H., Khor, E.H.: Characterization of the uncertainty of the Robertson and Wride model for liquefaction potential. Soil Dyn. Earthq. Eng. **24**(9–10), 771–780 (2004)

Juang, C.H., Yang, S.H., Yuan, H.: On the model uncertainty of shear wave velocity-based methods for liquefaction potential evaluation. J. Geotech. Geoenviron. Eng. **131**(10), 1274–1282 (2005)

Juang, C.H., Fang, S.Y., Khor, E.H.: First order reliability method for probabilistic liquefaction triggering analysis using CPT. J. Geotech. Geoenviron. Eng. **132**(3), 337–350 (2006)

Juang, C.H., Fang, S.Y., Tang, W.H., Khor, E.H., Kung, G.T.C., Zhang, J.: Evaluating model uncertainty of an SPT-based simplified method for reliability analysis for probability of liquefaction. Soils Found. **49**(2009), 147–152 (2009)

Juang, C.H., Ou, C.Y., Lu, C.C., Luo, Z.: Probabilistic framework for assessing liquefaction hazard at a given site in a specified exposure time using Standard Penetration Test. Can. Geotech. J. **47**(2010), 674–687 (2010)

Juang, C.H., Ching, J., Ku, C.S., Hsieh, Y.H.: Unified CPTu-based probabilistic model for assessing probability of liquefaction of sand and clay. Géotechnique **62**(10), 877–892 (2012a)

Juang, C.H., Ching, J., Luo, Z., Ku, C.S.: New models for probability of liquefaction using standard penetration tests based on an updated database of case histories. Eng. Geol. **133** (2012), 85–93 (2012b)

Juang, C.H., Ching, J., Luo, Z.: Assessing SPT-Based probabilistic models for liquefaction potential evaluation: a ten-year update. Georisk Assess. Manag. Risk Eng. Syst. Geohazards 7 (3), 137–150 (2013)

Juang, C.H., Khoshnevisan, S., Zhang, J.: Maximum likelihood principle and its application in soil liquefaction assessment. Phoon, K.K., Ching, J. (eds.) Risk and Reliability in Geotechnical Engineering, pp. 181–219. CRC Press, Boca Raton (2015)

Juang, C.H., Zhang, J.: Bayesian methods in geotechnical engineering—a practical guide. In: Geotechnical Safety and Reliability: Honoring Wilson H. Tang (GSP 286), ASCE, Reston (2017)

Juang, C.H., Zhang, J., Khoshnevisan, S., Gong, W.P.: Probabilistic methods for assessing soil liquefaction potential and effect. In: Proceedings of Georisk 2017 (Keynote) (2017)

Kayen, R., Moss, R.E.S., Thompson, E.M., Seed, R.B., Cetin, K.O., Kiureghian, A.D., Tokimatsu, K.: Shear-wave velocity–based probabilistic and deterministic assessment of seismic soil liquefaction potential. J. Geotech. Geoenviron. Eng. **139**(3), 407–419 (2013)

Khoshnevisan, S., Juang, C.H., Zhou, Y.G., Gong, W.: Probabilistic assessment of liquefaction-induced lateral spreads using CPT - Focusing on the 2010–2011 Christchurch events. Eng. Geol. **192**, 113–128 (2015)

Kramer, S.L.: Performance-based earthquake engineering: opportunities and implications for geotechnical engineering practice. In: Proceedings of the Geotechnical Earthquake Engineering and Soil Dynamics IV, ASCE, Reston, VA, pp. 1–32 (2008)

Kramer, S.L., Mayfield, R.T.: Return period of soil liquefaction. J. Geotech. Geoenviron. Eng. **133**(7), 802–813 (2007)

Ku, C.S., Juang, C.H., Chang, C.W., Ching, J.: Probabilistic version of the Robertson and Wride method for liquefaction evaluation: development and application. Can. Geotech. J. **49**(1), 27–44 (2012)

Knudsen, K.L., et al.: Preliminary maps of Quaternary deposits and liquefaction susceptibility, nine-county San Francisco Bay Region, California: A digital database (2000)

Lai, S.Y., Chang, W.J., Lin, P.S.: Logistic regression model for evaluating soil liquefaction probability using CPT data. J. Geotech. Geoenviron. Eng. **132**(6), 694–704 (2006)

Lenz, J.A., Baise, L.G.: Spatial variability of liquefaction potential in regional mapping using CPT and SPT data. Soil Dyn. Earthq. Eng. **27**(7), 690–702 (2007)

Liao, S.S.C., Veneziano, D., Whitman, R.V.: Regression model for evaluating liquefaction probability. J. Geotech. Eng. **114**(4), 389–410 (1988)

Liu, C.N., Chen, C.H.: Mapping liquefaction potential considering spatial correlations of CPT measurements. J. Geotech. Geoenviron. Eng. **132**(9), 1178–1187 (2006)

Liu, F., Li, Z., Jiang, M., Frattini, P., Crosta, G.: Quantitative liquefaction-induced lateral spread hazard mapping. Eng. Geol. **207**, 36–47 (2016)

Lu, C.C., Hwang, J.H., Juang, C.H., Ku, C.S., Luo, Z.: Framework for assessing probability of exceeding a specified liquefaction-induced settlement at a given site in a given exposure time. Eng. Geol. **108**, 24–35 (2009)

Moss, R.E., Seed, R.B., Kayen, R.E., Stewart, J.P., Der Kiureghian, A., Cetin, K.O.: CPT-based probabilistic and deterministic assessment of in situ seismic soil liquefaction potential. J. Geotech. Geoenviron. Eng. **132**(8), 1032–1051 (2006)

Oommen, T., Baise, L.G., Vogel, R.: Validation and application of empirical liquefaction models. J. Geotech. Geoenviron. Eng. **136**(12), 1618–1633 (2010)

Papathanassiou, G., et al.: Liquefaction susceptibility map of Greece. Bull. Geol. Soc. Greece **43**(3), 1383–1392 (2017)

Robertson, P.K., Wride, C.E.: Evaluating cyclic liquefaction potential using the cone penetration test. Can. Geotech. J. **35**(3), 442–459 (1998)

Seed, H.B.: Soil liquefaction and cyclic mobility evaluation for level ground during earthquakes. J. Geotech. Eng. **105**(2), 201–255 (1979)

Seed, H.B., Idriss, I.M.: Simplified procedure for evaluating soil liquefaction potential. J. Geotech. Eng. **97**(9), 1249–1273 (1971)

Seed, H.B., Idriss, I.M.: Ground Motions and Soil Liquefaction During Earthquakes. Earthquake Engineering Research Institute, Oakland (1982)

Seed, H.B., Tokimatsu, K., Harder, L.F., Chung, R.M.: The influence of SPT procedures in soil liquefaction resistance evaluations. J. Geotech. Eng. 1425–1445 (1985). https://doi.org/10.1061/(ASCE)0733-9410(1985)111:12(1425)

Toprak, S., Holzer, T.L., Bennett, M.J., Tinsley, J.C.I.: CPT- and SPT-based probabilistic assessment of liquefaction potential. In: Proceedings of the 7th US - Japan Workshop on Earthquake Resistant Design of Lifeline Facilities and Countermeasures against Soil Liquefaction, State University of New York at Buffalo, New York, pp. 69–86 (1999)

Vivek, B., Raychowdhury, P.: Probabilistic and spatial liquefaction analysis using CPT data: a case study for Alameda County site. Nat. Hazards **71**(3), 1715–1732 (2014)

Youd, T.L., Perkins, D.M.: Mapping liquefaction-induced ground failure potential. J. Soil Mech. Found. Div. **104**(4), 433–446 (1978)

Youd, T.L., Idriss, I.M., et al.: Liquefaction resistance of soils: summary report from the 1996 NCEER and 1998 NCEER/NSF workshops on evaluation of liquefaction resistance of soils. J. Geotech. Geoenviron. Eng. **127**(4) (2001)

Wang, C., Chen, Q.: A hybrid geotechnical and geologic data-based framework for multiscale regional liquefaction hazard mapping. Géotechnique (2017). https://doi.org/10.1680/jgeot.17.P.074

Wang, C., Chen, Q., Shen, M., Juang, C.H.: On the spatial variability of CPT-based geotechnical parameters for liquefaction potential evaluation. Soil Dyn. Earthq. Eng. **95**, 153–166 (2017a). https://doi.org/10.1016/j.soildyn.2017.02.001

Wang, C., Chen, Q., Juang, C.H.,: Regional liquefaction mapping accounting for multiscale spatial variability of soil parameters with geological constraints. In: Proceedings of the Geotechnical Risk - from Theory to Practice, Denver, CO (2017b)

Witter, R.C., et al.: Maps of Quaternary deposits and liquefaction susceptibility in the central San Francisco Bay region, California, No. 2006-1037. Geological Survey (US) (2006)

Zhang, J., Zhang, L.M., Tang, W.H.: Bayesian framework for characterizing geotechnical model uncertainty. J. Geotech. Geoenviron. Eng. **135**(7), 932–940 (2009)

Soil Dynamics and Earthquake Engineering

Effect of Initial Relative Density on Post-cyclic Stress-Strain Response of Liquefied Samples

Bashar Ismael[1(✉)], Domenico Lombardi[1],
and Subhamoy Bhattacharya[2]

[1] School of Mechanical, Aerospace and Civil Engineering,
University of Manchester, Manchester, UK
`Bashar.Ismael@manchester.ac.uk`
[2] Department of Civil and Environmental Engineering, University of Surrey,
Guildford, UK

Abstract. This paper presents results from multi-stage cyclic triaxial tests carried out on samples prepared at different relative densities to investigate the post-cyclic response of the liquefied soil. The experimental results show that the stress-strain behaviour of liquefied samples is strain-hardening, thus it increases in strength and stiffness upon shearing. This strain-hardening response can be represented by means of a simplified stress-strain relationship defined by three parameters: (i) take-off shear strain γ_{to}, (ii) shear modulus; G_1 at the beginning of shearing stage for $\gamma < \gamma_{to}$; (iii) shear modulus G_2 at large strains $\gamma > \gamma_{to}$. The results show that these parameters are a function of the initial relative density of the samples, whereby higher densities result in lower γ_{to} but higher G_1 and G_2.

Keywords: Liquefaction · Post-cyclic response · Multi-stages triaxial test
Undrained soil behavior

1 Introduction

Understanding the behaviour of liquefied soil under shearing (i.e. post-liquefaction response) is important to model and predict the response of foundations during liquefaction phenomena. One of the pioneering works on the post-liquefaction response of sand was published by (Seed 1979). According to his research, when the liquefied soil was monotonically sheared, it started to mobilise increasing strength and stiffness due to a gradual decrease in EPWP. This unusual strain hardening behaviour for soils undergoing large strain was associated with the tendency of the liquefied soil to dilate upon shearing, and was confirmed by other researchers (Yoshida et al. 1994; Kiku and Tsujino 1996; Yasuda et al. 1999). Vaid and Thomas (1995) concluded that despite the contractive behaviour of loosely-prepared samples, upon liquefaction samples show a dilative behaviour. A comparison between the undrained monotonic stress-strain response observed prior to and after the onset of liquefaction is illustrated in Fig. 1.

Several attempts have been made to characterise the undrained post-liquefaction behaviour by investigating the stress–strain behaviour of liquefied soils. To the best of author's knowledge, the first proposed post-liquefaction stress-strain curve was proposed by Thomas (1992). This consisted of three segments: a first, starting from zero

© Springer Nature Singapore Pte Ltd. 2018
T. Qiu et al. (Eds.): GSIC 2018, *Proceedings of GeoShanghai 2018 International Conference: Advances in Soil Dynamics and Foundation Engineering*, pp. 19–26, 2018.
https://doi.org/10.1007/978-981-13-0131-5_2

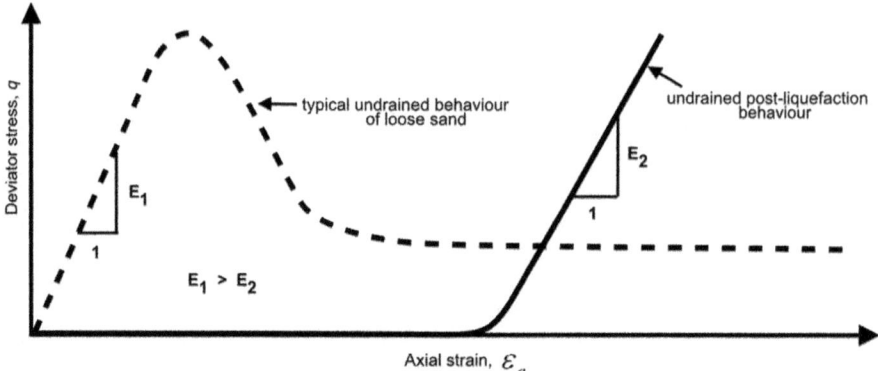

Fig. 1. Comparison between pre and post liquefaction behaviours of sand subjected to undrained monotonic loading

strain to the value of axial strain required to mobilise 5 kPa deviator stress, characterised by small stiffness and strength; a second segment characterised by a parabolic curve with increasing strength and stiffness with increasing strain levels. A final segment characterised by a linear relationship between stress and strain whose slope (stiffness) was a function of the confining pressure and relative density. Yasuda et al. (1995) proposed a different post-liquefaction behaviour in which the stress-strain curve consisted of two linear functions with increasing slope to simulate the strain-hardening behaviour. Dash (2010), Lombardi (2014), Lombardi and Bhattacharya (2016), Dash et al. (2017) introduced a simplified stress-strain curve to simulate the post-liquefaction behaviour, requiring three parameters, which can be obtained from cyclic triaxial test. The first parameter is called take-off shear strain γ_{to}, representing the value of shear strain corresponding to 1 kPa mobilized shear strength of the liquefied soil when it is sheared monotonically under undrained condition. The second parameter is the shear modulus at the beginning of shearing stage, which following the definition of take-off shear strain was computed as $1/\gamma_{to}$. The third parameter was the shear modulus at large strain for $\gamma > \gamma_{to}$, named critical state shear modulus G_2. The schematic interpretation of the three aforementioned models is illustrated in Fig. 2.

This paper presents the results obtained from multi-stages cyclic triaxial tests, whereby the samples were initially liquefied by application of cyclic loading and subsequently sheared monotonically to investigate its post-liquefaction behaviour. The results are used to calibrate the three parameters introduced by Dash (2010), Lombardi and Bhattacharya (2016), Lombardi et al. (2017), Dash et al. (2017) to define the simplified stress-strain curve for liquefied soils. The results show that these parameters are a function of the initial relative density of the samples, whereby higher densities result in lower γ_{to} but higher G_1 and G_2.

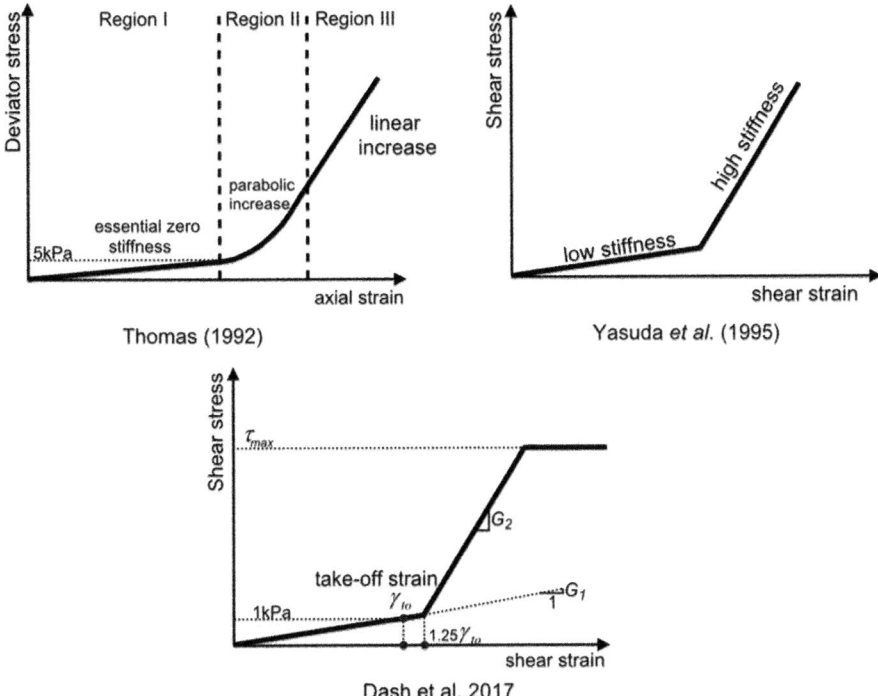

Fig. 2. Post-liquefaction stress-strain models available in literature.

2 Material

All the tests were conducted on a fine-grained silica sand with angular particles, known with the commercial name Redhill 110 sand. The grain size distribution curve for this sand is plotted corresponding to two liquefaction boundaries as suggested by the Japanese Seismic code for Harbour Structures (Ministry of Transport Japan 1999) as shown in Fig. 3. The sand index properties are presented in Table 1. The maximum and minimum densities were estimated according to the (BS: 1377: part 4: 4.2) and (BS: 1377: part 4: 4.4) respectively. The coefficient of uniformity for this sand is found to be 2.25 indicating that this material is poorly graded.

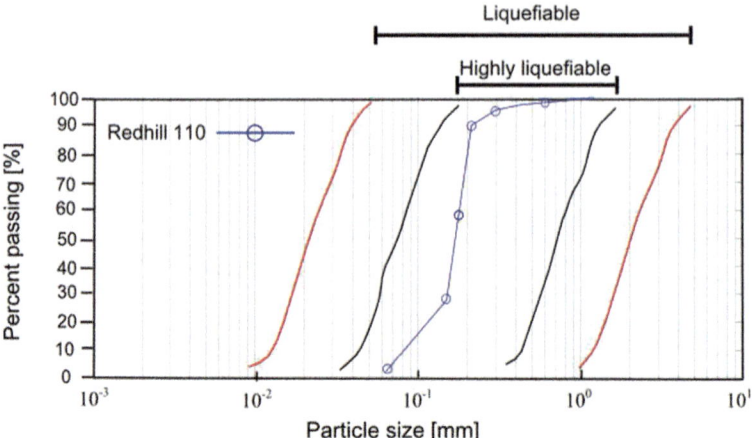

Fig. 3. Grain size distribution for Redhill 110 sand corresponding to two liquefaction boundaries as recommended by the Japanese Seismic code for Harbour Structures.

Table 1. Redhill sand index properties

Specific gravity	Minimum density (g/cm³)	Maximum density (g/cm³)	Maximum void ratio	Minimum void ratio	D_{60}	D_{10}	C_U
2.65	1.302	1.648	1.035	0.608	0.18	0.08	2.25

3 Specimen Preparation Method and Multi-stage Soil Element Test

All tests were conducted using a cyclic triaxial apparatus available at the Surrey Advanced Geotechnical Engineering laboratory (SAGE). Samples were prepared using dry pluviation method because of the wide range of densities could be prepared by adjusting the height of fall from the funnel and nozzle diameter (Miura and Toki 1982). Targeted relative densities were selected taking into account the typical relative densities achieved in engineering practice which ranges from (40–80)% (Shamoto et al. 1997). Some tapping on the mould perimeter was required to prepare very dense sample (i.e. 80% Dr). All tested samples where consolidated under 100 kPa confining pressure before the application of the cyclic loading.

To study the post-liquefaction behaviour of the Redhill sand, the soil was liquefied first by applying a cyclic loading at constant deviator stress and 0.1 Hz frequency so that the viscosity effect was reduced to a minimum value (Lombardi et al. 2014). The cyclic loading was ceased when the effective stress reduced to zero (i.e. initial liquefaction) and the double amplitude axial strain reached 5%. It is worth mentioning that for loose samples the zero effective stress condition is called liquefaction by Seed and Lee (1966) whereas for denser samples the zero confining pressure is momentarily approached and liquefaction is defined by 5% double amplitude axial strain according to Ishihara (1993). Then the samples were subjected to monotonic undrained loading at 0.1 axial strains rate to investigate the post-liquefaction resistance for the tested samples without releasing the generated excess Pore Water Pressure due to the cyclic loading (i.e. the drainage valve was kept close). This represents the short shaking period where the water drainage is prohibited.

4 Post-liquefaction Stress-Strain Relation

To examine the effect of initial relative density on the post liquefaction response, samples prepared at different relative densities and cyclically loaded up to 5% double amplitude axial strain were tested. The post liquefaction responses for various relative densities are depicted in Figs. 4 and 5. The results reveal that the three parameters required for the construction of the simplified stress-strain relation (γ_{to}, G_1 and G_2) are largely sensitive to samples' relative density. Specifically, from the results plotted in Figs. 4 and 5 it can be noted that all samples deformed without mobilisation any resistance to the applied

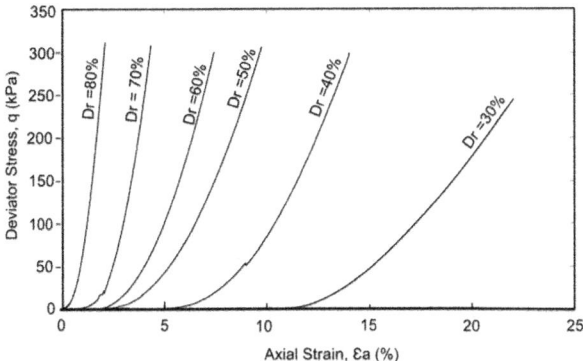

Fig. 4. Post-liquefaction stress-strain responses for different relative densities.

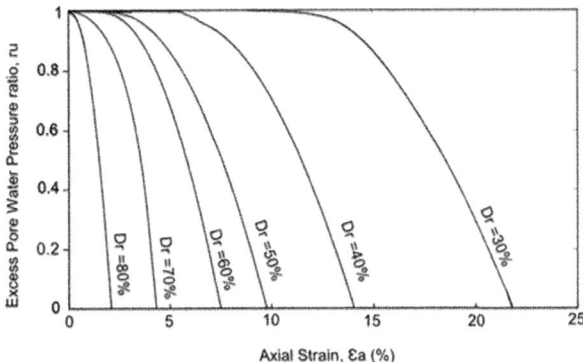

Fig. 5. Post-liquefaction EPWP, ru – axial strain, Ɛa for different relative densities.

monotonic loading up to a certain value of strains beyond which the strain-hardening behaviour developed due to tendency of liquefied soil to dilate.

Using these data, the three parameters required for the construction of the afore-mentioned stress-strain curve are presented as a function of the samples relative density. The results confirm the high sensitivity of these parameters to the degree of packing, whereby a change in relative density from 30% to 80% causes a significant reduction in the threshold strain from 16.95% to 0.15% (see Fig. 6). Equation 1 is herein proposed to estimate the threshold strain as a function of the initial relative density.

$$\gamma_{to} = 258.84e^{-0.084.Dr} \tag{1}$$

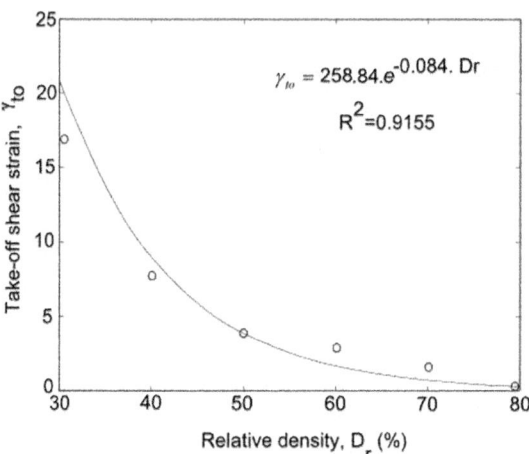

Fig. 6. Threshold shear strain versus axial strain at 5% preceding axial strain.

The post-liquefied shear modulus, G_1 versus relative density is presented in the semi-log plot in Fig. 7. It can be seen that G_1 increases with increasing relative density according to the relationship given by Eq. 2.

$$G_1 = 0.3863e^{0.0842.D_r} \tag{2}$$

Figure 8 plots the post-liquefaction shear modulus, G_2 versus initial relative density. The results confirm the increase in G_2 with increasing relative density following the expression given by Eq. 3.

$$G_2 = 282.8e^{0.0429.D_r} \tag{3}$$

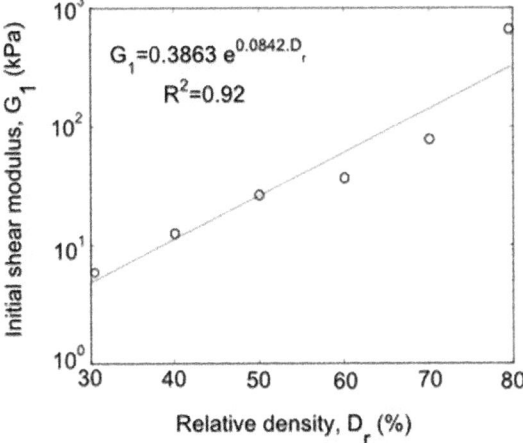

Fig. 7. Initial Shear modulus G_1 versus relative density.

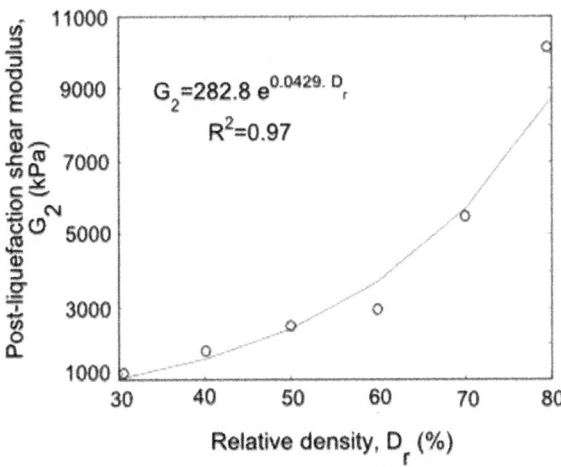

Fig. 8. Post-liquefied shear modulus G_2 versus relative density.

5 Conclusion

Using multi-stage triaxial tests, the cyclic and post-liquefaction undrained monotonic behaviour of redhill sand were examined. Multi-stage triaxial test refers to two loading conditions. Firstly, the cyclic loading was applied to liquefy the samples followed by static loading-without dissipation the generated EPWP developed during the cyclic loading- to investigate its post-liquefaction response. The post-liquefaction response for the tested samples show a strain hardening behaviour where the initial stiffness is practically zero and increase rapidly after exceeding a specific strain value named threshold strain. This is related to the tendency of liquefied soil to dilate under shearing.

As a result, the generated EPWP from the cyclic loading dissipates and the effective stress – which is linked to both strength and stiffness, increases with increasing strain. The results are finally used to calibrate three equations for the derivation of three parameters, i.e., γ_{to} G1, G2, used for the construction of the simplified stress-strain curve for liquefied sands.

References

Dash, S.: Lateral pile soil interaction in liquefiable soils. Ph.D. thesis, Oxford University (2010)

Dash, S., Rouholamin, M., Lombardi, D., Bhattacharya, S.: A practical method for construction of p-y curves for liquefiable soils. Soil Dyn. Earthq. Eng. **97**, 478–481 (2017)

Ishihara, K.: Liquefaction and flow failure during earthquakes. Geotechnique **43**(3), 351–451 (1993)

Kiku, H., Tsujino, S.: Post liquefaction characteristic of sand. In: Proceedings 11th World Conference on Earthquake Engineering (1996)

Lombardi, D., Bhattacharya, S.: Evaluation of seismic performance of pile-supported models in liquefiable soils. Earthq. Eng. Struct. Dynam. **45**, 1019–1038 (2016)

Lombardi, D.: Dynamics of pile-supported structures in seismically liquefiable soils. Ph.D. thesis, Bristol University (2014)

Lombardi, D., Bhattacharya, S., Hyodo, M., Kaneko, T.: Undrained behaviour of two silica sands and practical implications for modelling SSI in liquefiable soils. Soil Dyn. Earthq. Eng. **66**, 293–304 (2014)

Lombardi, D., Dash, S.R., Bhattacharya, S., Ibraim, E., Muirwood, D., Taylor, C.A.: Construction of simplified design p–y curves for liquefied soils. Géotechnique **67**, 216–227 (2017)

Ministry of Transport Japan: Design standard for port and harbour facilities and commentaries (1999)

Miura, S., Toki, S.: A sample preparation method and its effect on static and cyclic deformation-strength properties of sand. Jpn. Soc. Soil Mech. Found. Eng. **22**(1), 61–77 (1982)

Seed, H.B., Lee, K.L.: Liquefaction of saturated sands during cyclic loading. J. Soil Mech. Found. Div. **92**(6), 105–134 (1966)

Seed, H.B.: Considerations in the earthquake-resistant design of earth and rockfill dams. Geotechnique **29**(3), 215–263 (1979)

Shamoto, Y., Zhang, J.-M., Goto, S.: Mechanism of large post-liquefaction deformation in saturated sand. Soils Found. **37**(2), 71–80 (1997)

Thomas, J.: Static, cyclic and post liquefaction undrained behaviour of Fraser River sand. Retrospective Theses and Dissertations, 1919–2007 (1992)

Vaid, Y., Thomas, J.: Liquefaction and postliquefaction behavior of sand. J. Geotech. Eng. **121**(2), 163–173 (1995)

Yasuda, S., Yoshida, N., Masuda, T.: Stress-strain relationship of liquefaction sands. Earthq. Geotech. Eng., 811–816 (1995)

Yasuda, S., Yoshida, N., Kiku, H., Adachi, K., Gose, S.: A simplified method to evaluate liquefaction-induced deformation. In: Seco e Pinto, P. (ed.) Proceedings of the Earthquake Geotechnical Engineering, Balkema, Rotterdam, vol. 2, pp. 555–566 (1999)

Yoshida, S.Y., Masanori, K., Masuda, T., Finn, W.L.: Behavior of sand after liquefaction. Technical report. **94**(26), 181 (1994)

Seismic Response of a Cross Transfer Subway Station in Soft Soil Subjected to Different Ground Motion Directions

Weifeng Wu[1,2], Shiping Ge[1,2], Wenqi Ding[1,2(✉)], and Yong Yuan[1,2]

[1] Department of Geotechnical Engineering, College of Civil Engineering,
Tongji University, Shanghai, China
dingwq@tongji.edu.cn
[2] Key Laboratory of Geotechnical and Underground Engineering of Ministry
of Education, Tongji University, Shanghai, China

Abstract. Cross transfer station plays an important role in the urban transit system, but few studies focus on its seismic performance. A 3-D numerical model of a typical cross transfer station structure-soil system is conducted based on the finite element code ABAQUS to study the seismic response under different ground motion directions. The Shanghai artificial seismic waves are chosen to be input. The structure and soil deformation are studied respectively. The structure deformation contains the relative displacement between the end well and intersection of the station structure, and the non-uniform story drifts of structural sections. The results shows the relative relation between the soil deformation and structure-soil stiffness.

Keywords: Seismic performance · Cross transfer station structure
Numerical analysis

1 Introduction

Urban subway system is an important part of the modern urban passenger transport system. With the development of the subway network, operational needs have led to the construction of two-line or multi-line subway transfer station (Ge 2000a, b). In recent years, the number of transfer stations in Shanghai has increased rapidly. In 2005, Shanghai had five subway lines with 15 subway transfer stations. As of the end of 2015, it became 13 lines and 51 subway transfer stations. The cross transfer station is one of the most common transfer station type, the passengers can transfer directly through the stairs between the platforms, which means a short transfer distance.

Most of the cross transfer stations are built in recent years, and often located in urban central areas. According to the important role it plays in the urban traffic system, it is commonly believed that once the transfer station is damaged or even collapses during the earthquake, it can cause significant impact and loss. After the serious damage to the Daikai station occurred during the 1995 Hanshin earthquake in Japan, there was increasing research into the seismic performance of single rectangular subway stations (Yang et al. 2003; Chen et al. 2015; Chen et al. 2016). Because of the

© Springer Nature Singapore Pte Ltd. 2018
T. Qiu et al. (Eds.): GSIC 2018, *Proceedings of GeoShanghai 2018 International Conference:*
Advances in Soil Dynamics and Foundation Engineering, pp. 27–35, 2018.
https://doi.org/10.1007/978-981-13-0131-5_3

conformity of the cross-section, most of the published work studied the subway station seismic response based on the plane strain conditions. But to those transfer station structures that normally have abrupt structural stiffness change at the connection position, it's necessary to use three-dimensional analysis. A few studies involve the seismic performance of cross transfer station structures, but were only for some special and complicated cross structure forms.

This present work describes a 3-D numerical model of a typical two-story to three-story cross transfer station based on the general purposed finite element code ABAQUS/CAE. A series of time history analyses were performed to investigate the seismic response characteristics of the transfer station structure in soft soil.

2 Project Background

The typical two-story to three-story cross transfer station chosen for our model is located in Shanghai, China. The underground water depth of the site is 0.70–2.90 m. According to the Code for Seismic Design of buildings, the site belong to a weak type of Grade IV with layered soil. The designed peak ground acceleration (PGA) is 0.1 g. As shown in Fig. 1, along the Y direction is a three-span, two-story station structure, whose first layer is the station hall, while the second layer is the island type platform. Along the X direction is a three-span, three-story station structure. The first layer and the second layer in the three-story station are continued with the two-story station, and the third layer is another island platform. In order to prevent the leakage and differential settlement of the subway station structures (Ge 2000a, b), no disconnected or flexible joints are in the structure. The cut and cover method is used to construct the station structure The station lining wall and the diaphragm wall are connected by embedded parts which ensure that the lining wall and the diaphragm wall can bear the loads together. The two-story and three-story station have a total length of 211.4 m, including a 180 m long standard section and two 15.7 m long end wells. Central columns of both stations are of the same size as 1.4 m long and 0.7 m wide.

3 Numerical Model Setup

3.1 Numerical Model

A detailed 3D numerical model of the structure and surrounding soil was established in the finite element code ABAQUS. The geometry is described on the basis of an X-Y-Z orthogonal coordinate system with the X axis along the three-story station longitudinal axis, the Y axis along the two-story station longitudinal axis and the Z axis along the vertical direction, as shown in Fig. 1. The effects of the diaphragm wall and the non-structural components like stairs are not taken account into the model.

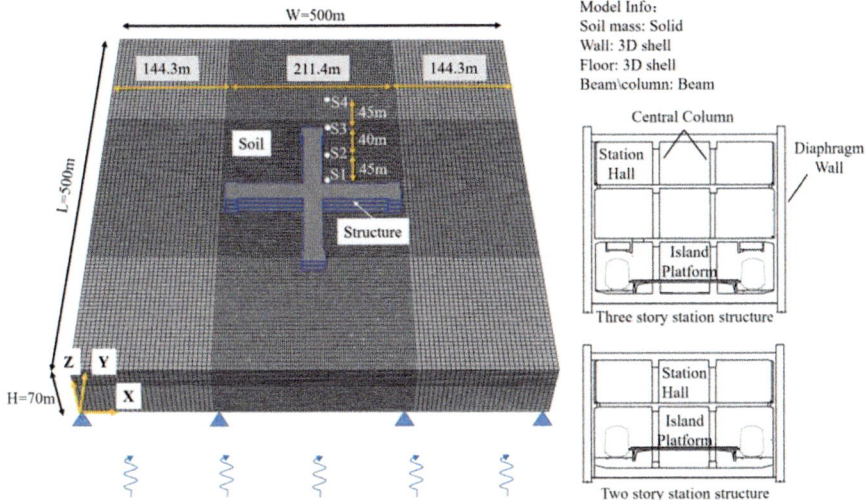

Fig. 1. Numerical analysis model.

3.2 Boundary Conditions

In the dynamic analysis, the seismic wave will be reflected on the lateral discretized boundary surface, which will return energy to the model. In this study, a free boundary condition was imposed at the lateral boundaries. A number of different-sized preliminary numerical models were tested to get the influence scope of boundary effect. Finally, the distance between the boundaries and the station is determined to ensure that the free-field conditions were recovered in the area between the structure and the boundaries. According to the previous studies (Wang 2007), the soil was modeled down to a 70 m depth from the ground surface.

3.3 Seismic Input

The ground motions imposed at the bottom of the model for the numerical analyses are the Shanghai artificial seismic waves which are based on the seismic risk analysis of Shanghai, and have the duration of 40.94 s. The vertical displacement of the model element nodes at the bottom boundary was constrained, and the ground motion are applied in the direction of X axis and Y axis (Fig. 1), respectively, to study the seismic response of cross transfer station under different directions of ground motion.

Since the peak ground acceleration in the horizontal direction should be 0.1 g to meet the intensity of a moderate earthquake with which a 10% exceedance probability in 50 years, the peak value of the input ground motion in the X and Y direction at the 70 m depth were adjusted to 0.069 g based on the previous numerical test. Figure 2 presents the acceleration time histories and Fourier spectra of the input motion.

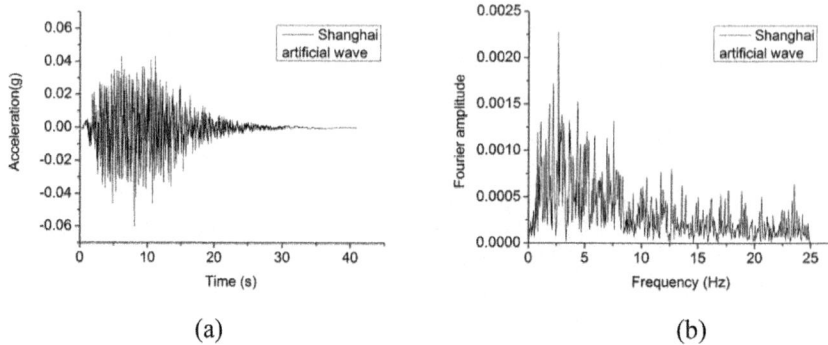

Fig. 2. Shanghai artificial seismic wave: (a) acceleration time histories; (a) Fourier spectrum

3.4   Properties of Materials

In the numerical model, the soil continuum was represented by 3D solid (brick) elements. Based on the engineering geological exploration, the site soil could be roughly divided into 9 layers. 4 boreholes at the site were drilled to investigate the soil shear wave velocity by single-hole method, then, the average shear wave velocity of each layer was gotten. According to the geotechnical investigation report, the specific soil properties are shown in Table 1, and the Poisson's ratio is taken as 0.35. The elastic model which considered material damping as Rayleigh type was applied for the soil mass using the method. The beams and columns of the station structure were modeled by the beam element (B31). The wall and the floor of the station were simulated by the 3D shell element. An elastic model was adopted to stimulate the concrete. Perfect bonding is assumed at the soil-structure interface.

Table 1. Soil parameters of every layer

Layer	Soil type	Depth (m)	Density (ton/m^3)	Shear wave velocity (m/s)	Dynamic shear modulus (MPa)
1	Artificial fill	0–1.38	1.765	127	27.90
2	Silty clay	1.38–2.18	1.888	126	29.37
3	Mucky silty clay	2.18–3.98	1.745	125	26.72
4	Clayey silt	3.98–5.68	1.878	125	28.75
5	Mucky silty clay	5.68–16.48	1.745	124	26.29
6	Clay	16.48–26.28	1.776	192	64.14
7	Sandy silt	26.28–34.48	1.980	255	122.90
8	Silty fine sands	34.48–65.89	1.929	322	195.96
9	Silty clay and silty sand interbedded soil	65.89–73.690	1.867	373	254.61

4   Numerical Results and Discussion

4.1   Structure Displacement

Under the seismic input along Y axis or X axis station (Fig. 3), it could be found that the displacement of the cross transfer station structure is mainly on the excitation orientation of the acceleration and is much smaller on the other 2 directions, less than 1%. Figure 3 shows the lateral displacement diagram of the structure at 3.72 s of the input motion, the scale factor is 1000 in the figure.

As shown in Fig. 3(a), for the input motion along Y direction, the three-story station structure mainly has the rigid body displacement along Y direction, and the displacement of the two-story station includes two types, (1) the relative displacement between the end well and the intersection of the station structure, (2) the story drifts of the structural sections which are diverse at different sections. This kind of displacement is probably due to the different constraint imposed on the structure by ground soil. Figure 3(b) shows the other case, for input motion along X direction. The two-story structure mainly occurs the displacement along Y axis, compared to the former, the three-story station also have different story drifts of the cross sections, but the relative displacement between the end well and intersection seems to be smaller.

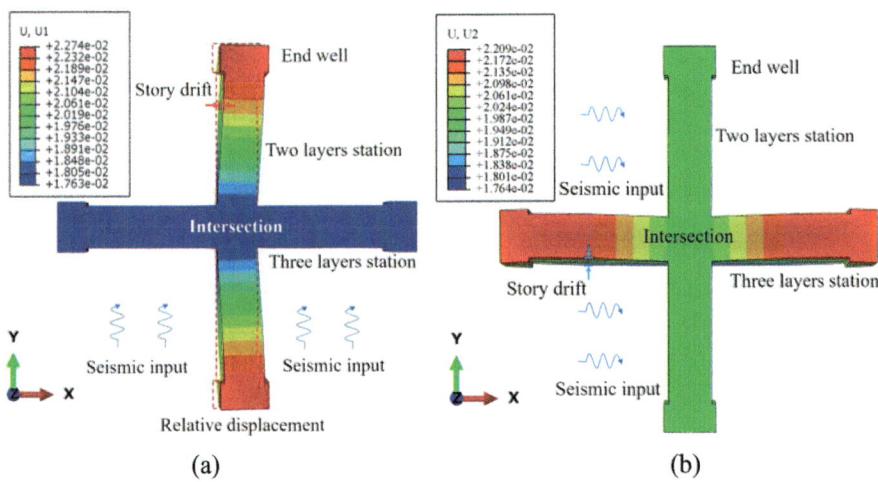

(a)   (b)

Fig. 3. Horizontal displacement of transfer station structure: (a) under the input motion along Y direction; (b) under the input motion along X direction. (unit: m, scale factor: 1000)

4.2   Soil Deformation Response Within the Height of Structure

To investigate how the soil deformation influenced by the structure, several soil test holes were set along the two-story station structure-soil surface in advance. As shown in Fig. 1, the test holes S1, S2, S3 are located near the cross joint, quartile and end of the two-story station, respectively, while the S4 is 45 m away from the S3 along Y-axis into the soil. The soil deformation response reflected by the S4 can be seen as a far-field response when the ground motion is input along Y-axis, since it can hardly influenced by the structure. The soil deformation within the height of structure is what we cared most, thus the relative soil displacement between two positions corresponding to the top and bottom of the structure was gotten. From Fig. 4, we can find that the peak soil relative displacement at S4 is bigger than others. So, it can be assumed that the structure is stiffer than the ground. Compared to the structure deformation (Fig. 3), the soil deformation near the structure is strongly affected by the structure. Because of the materials of soil and structure are elastic, there is no residual deformation after shaking. However, the plastic behavior should be considered in the future to study great earthquakes.

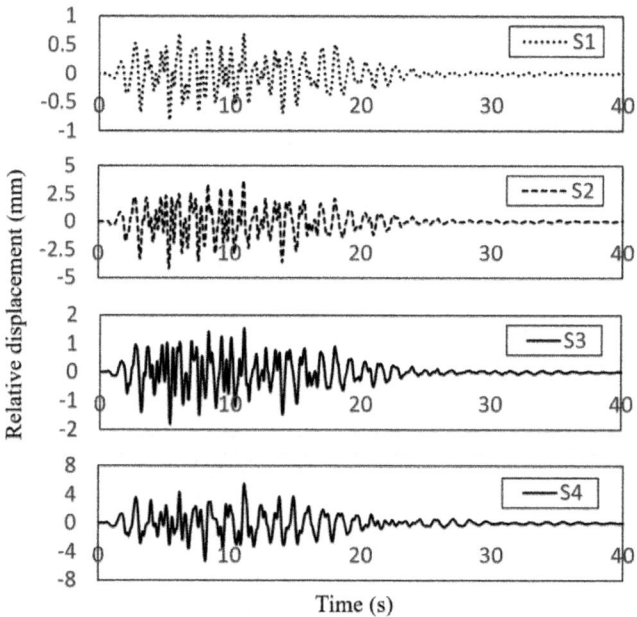

Fig. 4. Time histories of soil relative displacement between the top and bottom positions of the structure (mm)

4.3   Column Drifts

Central columns are vulnerable components of a subway station. Figure 5 shows the variation curves of central column drifts in different floors of the station structure along the motion input direction, and the horizontal axis represent the position of central columns while the vertical axis means the drift deformation. As shown in Fig. 5(a), under the input motion along Y direction, the column drifts are inconsistency which means the plane strain assumption is not applicable for this type of structures. Overall, the standard sections between the intersection and the end wells has the greater column drifts, followed the end wells, the column drifts in the intersection are the smallest probably due to the enhanced structural stiffness. The data also shows a distinct characteristic that the column drifts in the second floor are larger than the first floor in the same cross section position. Under the moderate earthquake shaking (PGA = 0.1 g), the maximum column drift of the first floor and the second floor are 1.81 mm and 2.36 mm, respectively, in which the cross section is located 65.77 m from the center of the intersection. The minimum column drift of the first floor and the second floor are 0.37 mm and 0.45 mm, respectively, in the intersection. Researchers have found that drift is related to the relative stiffness ratio of soil to structure (Tsinidis 2017; Ulgen et al. 2015). It seems the three-story station structure provide the extra stiffness for the two-story station in the intersection part, so the structure stiffness here is much stronger than the other part of the station structure. As the soil stiffness has not changed much along the station, the uneven structure stiffness may be the main reason for the varying column drifts.

Figure 5(b) shows the column drifts of the three-story station structure under the input motion along X direction. It can be found that the two curves of the column drifts in the first and second floor of three-story station structure are generally as same as what shows in Fig. 5(a). It is interesting that the column drifts of the second floor in the three-story structure are little larger than what in the two-story station, possibly because of the racking impetus effect of the third floor in the three-story station. The columns drifts in third floor are different with those in first and second floor. The third floor column drift in the intersection is the largest, followed the standard sections, and the drift in the end wells is the smallest. That's maybe because of the traction of the two-story station structure, the third floor columns in the intersection part have to bear more racking loads. Meanwhile, since the third floor do not has the stiffness support from two-story station structure, the stiffness of the intersection part and the standard sections are almost the same actually. That may be the reason why column drifts in the intersection of the third floor become obviously larger.

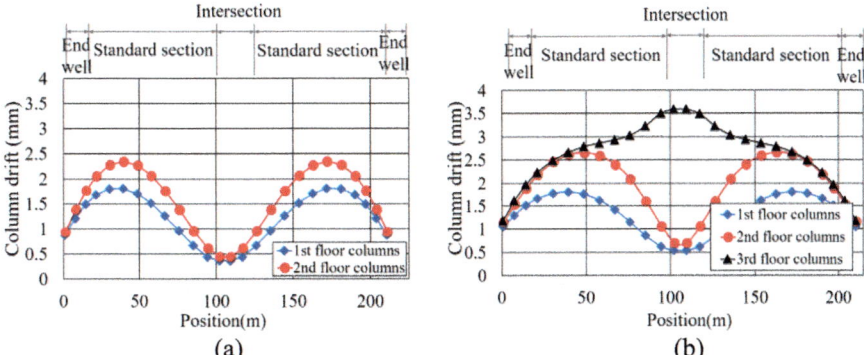

Fig. 5. Columns story drift of different sections under the ground motions (mm): (a) the 2-story station sections under the input motion along Y direction; (b) the 3-story station sections under the input motion along X direction.

5 Conclusions

In this study, a numerical model of a typical two-story to three-story cross transfer station was presented, and time history analyses of the station structure under the input motions of different direction were carried out. The main conclusions are as follow:

1. Under different directions of the seismic input motion, the cross-transfer station presents different deformation, including the relative displacement of the end wells to the intersection of the station structure and the story drifts of the cross section.
2. The soil deformation near the structure is strongly influenced by the structure. The far-field soil relative displacement within the structure height is larger than those near the structure, which shows the structure is stiffer than the soil.
3. Under the ground motion inputs along Y direction, the column drifts are not uniform in different cross sections, and are symmetry about the station center. The standard sections between the intersection and the end wells have the greater column drifts, followed the end wells, and then those in the intersection. The structure stiffness difference may be the main reason for the varying column drifts. The similar situation is presented at the upper two layer of the three-story station when the ground motion is input along X direction. But the columns drifts in the third floor distribute differently with those in the first and second floor, which have the largest column drift in the intersection part.

Acknowledgement. This work was sponsored by the National Key R&D Program of China (No. 2017YFC0806004), Seismic safety control technology study of cross transfer station structure, and the Study of key technology in the construction of large span subway station in the soft soil.

References

Chen, G., Su, C., Xi, Z., et al.: Shaking-table tests and numerical simulations on a subway structure in soft soil. Soil Dyn. Earthq. Eng. **76**, 13–28 (2015)
Chen, Z., Chen, W., Li, Y., et al.: Shaking table test of a multi-story subway station under pulse-like ground motions. Soil Dyn. Earthq. Eng. **82**, 111–122 (2016)
Ge, S.: Development trend of subway transfer hub domestic and overseas. Urban Rail Traffic **1**, 6–8 (2000a)
Ge, S.: Study on Longitudinal Deformation Prediction and Countermeasure of Soft Subway Station Structure. Tongji University (2000b)
Tsinidis, G.: Response characteristics of rectangular tunnels in soft soil subjected to transversal ground shaking. Tunn. Undergr. Space Technol. **62**(1), 1–22 (2017)
Ulgen, D., Saglam, S., Ozkan, M.Y.: Dynamic response of a flexible rectangular underground structure in sand: centrifuge modeling. Bull. Earthq. Eng. **13**(9), 2547–2566 (2015)

Wang, G.: Study on Theory and Method of 3D Seismic Response Calculation of Soft Subway Station Structure. Tongji University (2007)

Yang, L., Yang, C., Ji, Q., et al.: Shaking table test and calculation method of earthquake response in subway station. J. Tongji Univ. (Nat. Sci. Ed.) **31**(10), 1135–1140 (2003)

A Study on the Natural Periods of Soil Site Based on Ground Motion Data from KiK-Net in Japan

Yu Miao[✉] and Su-Yang Wang

School of Civil Engineering and Mechanics, Huazhong University of Science and Technology, Wuhan 430074 China
miaoyu@hust.edu.cn

Abstract. To gain insight into the soil site amplification, utilizing ground motion data from KiK-net in Japan during January 2009 to June 2014, and employing the surface-to-borehole spectral ratio method, the natural periods of soil site are studied. The main results are as follows. (1) We compared the correlationships between fundamental period and different influencing factors based on 249 KiK-net stations, including soil depth, soil stiffness, surficial sediment stiffness and the combination of soil depth and soil stiffness. The highest correlation coefficient (greater than 0.85) was obtained by the combination of soil depth and soil stiffness. And there is a medium correlation between fundamental period and surficial stiffness, which indicates the limitations of the time-averaged velocity to 30 m (Vs30). (2) We built empirical relationships between high-order natural frequencies and fundamental frequency based on 32 KiK-net stations, and the results fit well with the previous theory. Using the empirical relationships, the natural frequencies of the first 8th or even higher orders can be estimated from the fundamental frequency.

Keywords: Site effect · Fundamental period · Natural frequency

1 Introduction

It has been long recognized that ground motions can be dramatically affected by local site conditions. Previous research has shown that shallow sediments can significantly amplify the ground motion because the wave impedance of soft sediments is generally much lower than that of the bedrock. Due to the well-known effect of resonance, the largest amplifications of the ground motions usually occur at the natural frequencies of soil site [1–3]. It is necessary to consider the natural frequencies of soil site in site selection or seismic design of buildings to mitigate or avoid earthquake damage caused by resonance between buildings and site.

The fundamental period of site, often referred to simply as site period, defined as the reciprocal of the lowest frequency of nature frequencies, is a key parameter for site classification [4–8] and ground-motion prediction equations (GMPE) [5]. Comparing with Vs30, some researchers suggested that the fundamental period of site is a better proxy for site amplification [5, 7]. A major reason is that the fundamental period

© Springer Nature Singapore Pte Ltd. 2018
T. Qiu et al. (Eds.): GSIC 2018, *Proceedings of GeoShanghai 2018 International Conference: Advances in Soil Dynamics and Foundation Engineering*, pp. 36–43, 2018.
https://doi.org/10.1007/978-981-13-0131-5_4

accounts for both stiffness and depth of the soil sediments, while Vs30 only accounts for the stiffness of top 30 m.

While most previous studies focused on the fundamental frequency, a few studies have also investigated the high-order natural frequencies [1–3]. Zhao [1, 2] applied modal analysis technique to estimate equivalent modal parameters of soft-soil layers on an elastic half-space. From those studies, it can be found that the ratios of high-order natural frequencies to fundamental frequency fit an arithmetic progression with common difference between $1 \sim 2$. Based on ground motion data at Treasure Island, Wang [3] found the ratios of the high-order natural frequencies to fundamental frequency are smaller than corresponding values for a single layer with a constant shear wave velocity model.

The purpose of this study is to estimate the nature periods of soil site by using surface-to-borehole spectral ratio method with ground motion data in Japan, to compare the correlationships between fundamental period and different influencing factors, and to discuss relationships between high-order natural frequencies and fundamental frequency.

2 Data

The Kiban-Kyoshin Network (KiK-Net) is composed of ~ 700 vertical arrays with an uphole/downhole pair of strong-motion seismometers. These arrays were built by the National Research Institute for Earth Science and Disaster Resilience (NIED) after the great Hanshin Awaji earthquake disaster. The velocity profiles and geotechnical cross sections for most KiK-Net stations are available on the website (http://www.kyoshin.bosai.go.jp). These velocity profiles are obtained from downhole PD logging measurements. A majority of KiK-Net stations are located in valleys in hilly terrain with shallow sediments over rock [9–11].

There are about a million accelerograms (more than 160,000 events, and each event includes two sets of three-component accelerograms) recorded between January 2009 and June 2014 by all 697 KiK-Net sites during 5007 earthquakes. The time interval and the total length of those accelerograms are, respectively, 0.01 s and 300 s (the total length includes 15 s trigger delay).

Recent studies suggested that the nonlinear thresholds for soil sites are generally between 20 to 80 gal [10, 12–14]. To estimate the linear site response, we select accelerograms according to the following criteria: (1) the peak ground accelerations (PGA) are between 5 and 20 Gal, (2) the signal-to-noise ratios (SNR) are higher than 5. The SNR can be calculated by Eq. 1.

$$\text{SNR} = 20 * \log_{10}(Sig/Noise) \tag{1}$$

in which *Sig* and *Noise* are the amplitudes of the signal and noise between 0.1 to 20 Hz, respectively.

For all 697 sites in KiK-net, 54 sites were eliminated firstly because of null or zero entries for the velocity profiles. Out of the rest 643 sites, we picked out 425 sites which have more than 5 sets of recordings after selecting accelerograms for further study.

3 Method

3.1 Method for Estimating Site Response

Ground motion is the result of a complex system composed of three physical processes, including the source rupture process (source effect), wave propagation in the crust (path effect) and the site response (site effect). According to Boore [15], the Fourier spectrum $A(M_0, R, f)$ of the observed ground motion can be expressed as a product of the source effect (S), path effect (P), site effect (G) and instrument effect (I) in frequency domain.

$$A(M_0, R, f) = S(M_0, f)P(R, f)G(f)I(f) \qquad (2)$$

in which M_0 is earthquake moment, R is hypocentral distance, f is frequency.

Surface-to-borehole spectral ratio method (SBSR) is an alternative to the standard spectral ratio (SSR) method [16], where the reference site is at the bottom of the borehole directly, rather than at a bedrock surface near the site. However, destructive interference between the up-going incident waves and down-going reflected waves from the surface at specific frequencies can produce a notch in the FAS (Fourier Amplitude Spectrum) of the borehole recording [17]. This destructive interference (depth effect) may produce pseudo resonances in the spectral amplification estimates. In this study, we corrected the depth effect by multiplying spectral ratio by the coherence (γ^2) between the signals at surface and downhole. The coherence can be calculated by Eq. 3.

$$\gamma^2 = \frac{|S_{12}(f)|^2}{S_{11}(f)S_{22}(f)} \qquad (3)$$

in which $S_{11}(f)$ and $S_{22}(f)$ are the power spectral densities of the accelerograms recorded at the surface and downhole, respectively, and $S_{12}(f)$ is the cross-power spectral density function.

The main steps of the SBSR method in this study are as follows: (1) apply non-causal, band-pass $(0.1 \sim 20$ Hz$)$, 4-order Butterworth filters to the accelerograms, (2) calculate the FAS of the surface and downhole recordings, and smooth the FAS by the Parzen window with bandwidth of 0.4 Hz, (3) calculate the spectral ratio and the coherence, and multiply the spectral ratio by the coherence to estimate site response.

3.2 Method for Identifying Nature Frequencies

As soil column is a multiple-degree-of-freedom system, different vibration modes of soil site are excited when seismic wave propagates into soil layers from bedrock [3]. And "peaks" of the site response are used to determine the nature frequencies corresponding to the excited vibration modes. However, in some cases, it is difficult to identify those peaks with confidence because of the irregularity of site response, especially for sites with volcanic or interbedded layers.

Cadet et al. [18] has defined the peak of fundamental frequency by two conditions: (1) It must correspond to a specific local maximum. The local maximum was practically considered fulfilled whenever the site amplification factor exhibits values at least 10% lower than the amplitude at f0 within the frequency ranges $[0.5f_0, f_0]$ and $[f_0, 2f_0]$; (2) Its amplitude should be larger than 2 in a statistical sense. On the basis of the definition, we proposed two similar conditions for selecting the peaks for nature frequencies as follows, and then identified the nature frequencies manually. (1) Those peaks must correspond to clear local maximums. We discarded nearly half of sites because the site response is too jagged, unlike a smooth curve with well-defined peaks at the nature frequencies in theory. (2) Their amplitudes should be larger than 20% of the largest amplification factor between $0.1 \sim 20$ Hz.

4   Results

4.1   Correlationships Between T_0 and Different Influencing Factors

Utilizing ground motion data from KiK-net in Japan during January 2009 to June 2014, and employing the surface-to-borehole spectral ratio method, the fundamental periods for 249 sites are identified from the site response curves. To have a better understanding of the site amplification, we compared the correlationships between fundamental period and different influencing factors, including soil depth (H), soil stiffness (V_{SH}), surficial sediment stiffness (T_{Vs20} & T_{Vs30}) and the combination of soil depth and soil stiffness (T_S & T_M). Those influencing factors can be calculated by Eqs. $4 \sim 9$.

$$H = \sum h_i \tag{4}$$

$$V_{SH} = \frac{H}{\sum \frac{h_i}{v_{Si}}} \tag{5}$$

$$T_{Vs30} = \frac{4 \times 30}{Vs30} = \frac{120}{Vs30} \tag{6}$$

$$T_{Vs20} = \frac{4 \times 20}{Vs20} = \frac{80}{Vs20} \tag{7}$$

$$T_S = \frac{4H}{V_{SH}} \tag{8}$$

$$T_M = \sqrt{\sum_{i=1}^{N} (\frac{4h_i}{v_{si}})^2 \frac{2H_i}{h_i}} \tag{9}$$

in which h_i and v_{si} are the thickness and shear-wave velocity for the *i-th* soil layer (shear-wave velocity lower than 760 m/s), respectively. Vs30 and Vs20 are the time-averaged shear-wave velocity at top 30 m and 20 m, respectively. T_S and T_M are fundamental frequencies derived from singer-layer model and multiple-layer model, respectively.

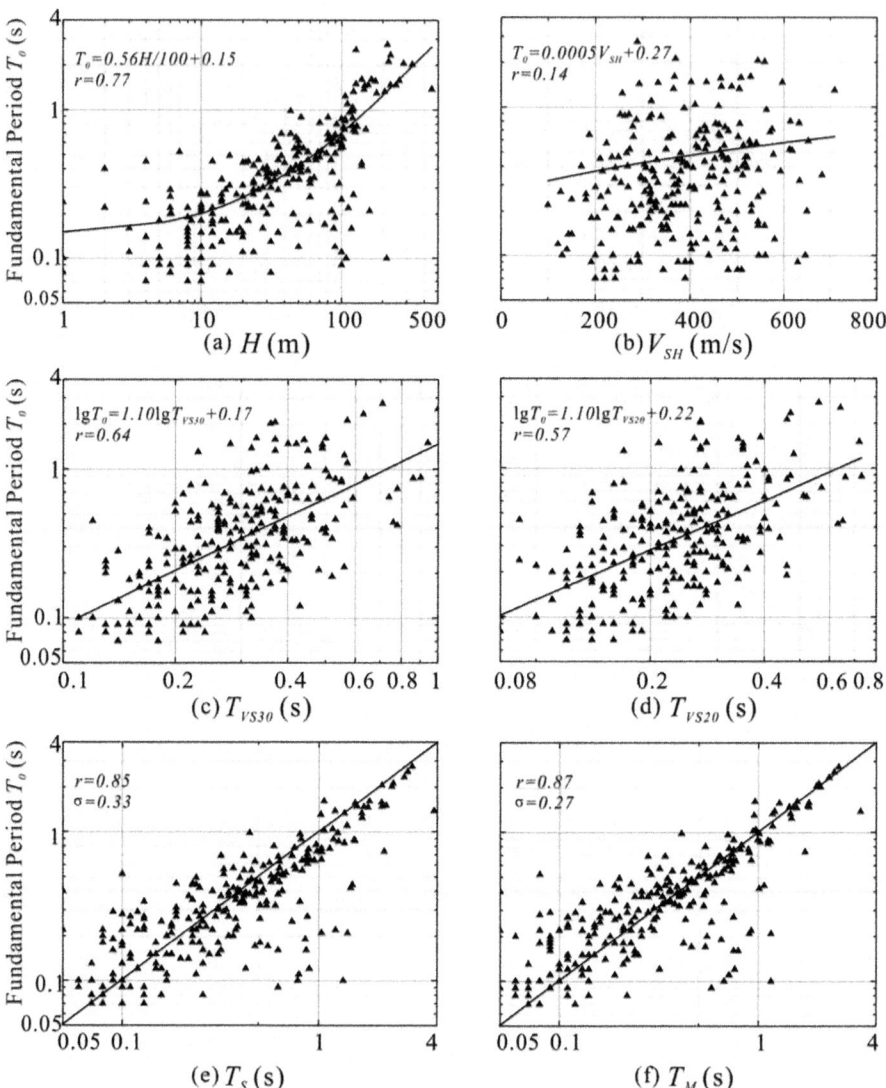

Fig. 1. Correlationships between T_0 and different influencing factors, including (a) soil depth, (b) soil stiffness, (c) and (d) surficial sediment stiffness, (e) and (f) the combination of soil depth and soil stiffness.

The correlation coefficients derived from different influencing factors are compared in Fig. 2. From Fig. 2, it can be easily found that (1) the highest correlation coefficient (greater than 0.85) was obtained by the combination of soil depth and soil stiffness;

(2) soil depth is much more strongly related to the fundamental period than soil stiffness; (3) there is a medium correlation between fundamental period and surficial stiffness, which indicates the limitations of Vs30.

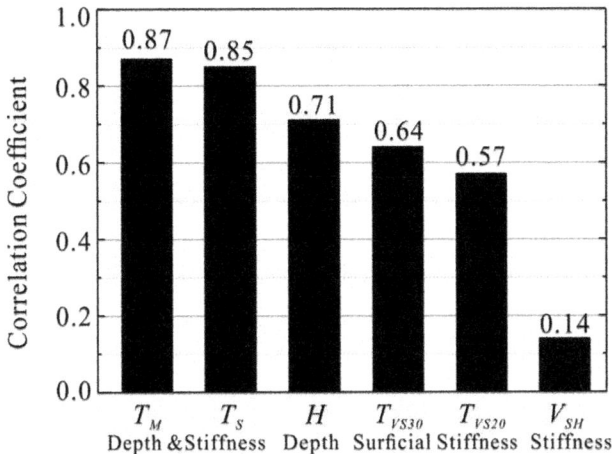

Fig. 2. Comparison of correlation coefficients derived from different influencing factors

4.2   Relationships Between High-Order Natural Frequencies and f_0

Among the 249 KiK-net stations, there are 32 stations which at least the first three order natural frequencies can be identified from the site response curves between 0.1 to 20 Hz. Those stations are selected to discuss the relationships between high-order natural frequencies and fundamental frequency. In theory, the frequency ratios of the high-order natural frequencies to fundamental frequency fit an arithmetic progression with common difference between $1 \sim 2$ [1, 2].

We calculated the frequency ratios for the 32 stations, and then fit the mean ratios and order of vibration modes by a regression form of $y = k(n-1) + 1$ (k is the regression parameter and n is the order of vibration modes, see in Fig. 3). The empirical relationships between high-order natural frequencies and fundamental frequency are listed as Eq. 10.

$$y = 1.45(n - 1) + 1 \qquad (10)$$

in which y is the ratio of high-order frequency to fundamental frequency.

It can be found that the empirical relationships fit well with the previous theory. Using the empirical relationships, the natural frequencies of the first 8th or even higher orders can be estimated from the fundamental frequency.

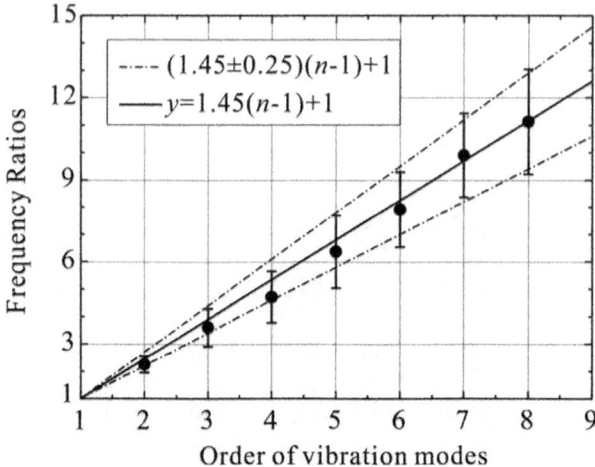

Fig. 3. Empirical relationships between frequency ratios and order of vibration modes

5 Conclusions

In this study, utilizing ground motion data from KiK-net in Japan, and employing the surface-to-borehole spectral ratio method, the natural periods of soil site are studied. We compared the correlationships between fundamental period and different influencing factors based on 249 KiK-net stations, and built empirical relationships between high-order natural frequencies and fundamental frequency based on 32 KiK-net stations. The main conclusions are as follows.

(1) The highest correlation coefficient between fundamental period and different influencing factors was obtained by the combination of soil depth and soil stiffness. And there is a medium correlation between fundamental period and surficial stiffness, which indicates the limitations of Vs30.
(2) The empirical relationships between high-order natural frequencies and fundamental frequency fit well with the previous theory. Using the empirical relationships, the natural frequencies of the first 8th or even higher orders can be estimated from the fundamental frequency.

References

1. Zhao, J.X.: Estimating modal parameters for a simple soft-soil site having a linear distribution of shear wave velocity with depth. Earthq. Eng. Struct. Dyn. **25**(2), 163–178 (1996)
2. Zhao, J.X.: Modal analysis of soft-soil sites including radiation damping. Earthq. Eng. Struct. Dyn. **26**(1), 93–113 (1997)

3. Wang, H.Y.: Study on variation of soil site amplification with depth: a case at treasure island geotechnical array, San Francisco bay. Chin. J. Geophys. **57**(5), 1498–1509 (2014). (in Chinese)

4. Zhao, J.X.: An empirical site-classification method for strong-motion stations in japan using h/v response spectral ratio. Bull. Seismol. Soc. Am. **96**(3), 914–925 (2006)

5. Zhao, J.X., Xu, H.: A comparison of vs30 and site period as site-effect parameters in response spectral ground-motion prediction equations. Bull. Seismol. Soc. Am. **103**(1), 1–18 (2013)

6. Zhao, J.X., Hu, J., Jiang, F., et al.: Nonlinear site models derived from 1d analyses for ground-motion prediction equations using site class as the site parameter. Bull. Seismol. Soc. Am. **105**(4), 2010–2022 (2015)

7. Luzi, L., Puglia, R., Pacor, F., et al.: Proposal for a soil classification based on parameters alternative or complementary to vs30. Bull. Earthq. Eng. **9**(6), 1877–1898 (2011)

8. Pitilakis, K., Riga, E., Anastasiadis, A.: New code site classification, amplification factors and normalized response spectra based on a worldwide ground-motion database. Bull. Earthq. Eng. **11**(4), 925–966 (2013)

9. Boore, D.M., Thompson, E.M., Cadet, H.: Regional correlations of vs30 and velocities averaged over depths less than and greater than 30 meters. Bull. Seismol. Soc. Am. **101**(6), 3046–3059 (2011)

10. Regnier, J., Cadet, H., Bonilla, L.F., et al.: Assessing nonlinear behavior of soils in seismic site response: statistical analysis on KiK-net strong-motion data. Bull. Seismol. Soc. Am. **103**(3), 1750–1770 (2013)

11. Wang, S.Y., Wang, H.Y.: Site-dependent shear-wave velocity equations versus depth in California and Japan. Soil Dyn. Earthq. Eng. **88**, 8–14 (2016)

12. Wu, C., Peng, Z., Ben-Zion, Y.: Refined thresholds for non-linear ground motion and temporal changes of site response associated with medium-size earthquakes. Geophys. J. Int. **182**(3), 1567–1576 (2010)

13. Rubinstein, J.L.: Nonlinear site response in medium magnitude earthquakes near Parkfield, California. Bull. Seismol. Soc. Am. **101**(1), 275–286 (2011)

14. Ghofrani, H., Atkinson, G.M., Goda, K.: Implications of the 2011 M9.0 Tohoku japan earthquake for the treatment of site effects in large earthquakes. Bull. Earthq. Eng. **11**(1), 171–203 (2013)

15. Boore, D.M.: Simulation of ground motion using the stochastic method. Pure. appl. Geophys. **160**(3–4), 635–676 (2003)

16. Borcherdt, R.D.: Effects of local geology on ground motion near San Francisco bay. Bull. Seismol. Soc. Am. **60**(1), 29–61 (1970)

17. Steidl, J.H., Tumarkin, A.G., Archuleta, R.J.: What is a reference site? Bull. Seismol. Soc. Am. **86**(6), 1733–1748 (1996)

18. Cadet, H., Bard, P.Y., Rodriguez-Marek, A.: Site effect assessment using KiK-net data: part 1. a simple correction procedure for surface/downhole spectral ratios. Bull. Earthq. Eng. **10**(2), 421–448 (2012)

Multiscale Regional Liquefaction Hazard Mapping Accounting for Heterogeneous Data Sources

Chaofeng Wang[1], Qiushi Chen[1(✉)], C. Hsein Juang[1], and Fang Liu[2,3]

[1] Glenn Department of Civil Engineering,
Clemson University, Clemson, SC 29634, USA
qiushi@clemson.edu
[2] State Key Laboratory of Disaster Reduction in Civil Engineering,
Tongji University, Shanghai 200092, China
[3] Department of Geotechnical Engineering, Tongji University,
Shanghai 200092, China

Abstract. In this work, a multiscale random field-based method is presented to integrate heterogeneous data sources, i.e., geologic data at the regional scale and geotechnical data at the site-specific scale, for regional liquefaction hazard mapping. The multiscale random field model accounts for spatial variability of soil parameters over multiple length scales. Uncertainties and spatial variability of soil parameters are appropriately accounted for in the process. The proposed method is then applied to an earthquake-prone region for liquefaction hazard mapping. It is found that the spatial correlation structure of the calculated liquefaction potential index (LPI) is prominent in the study region. The influence of geologic data on the generated liquefaction hazard map is significant. As the weight of the geologic data increases, the geologic boundaries become more distinguishable in the generated map. With an appropriately selected or calibrated Markov-Bayes coefficient, the geotechnical and geologic data could be properly accounted in the mapping process.

Keywords: Liquefaction · Random field models · Geostatistics
Cone penetration test · Geologic constraint

1 Introduction

Soil liquefaction is regarded as one of the major damages during earthquakes and it leave infrastructure such as bridges, buildings and underground utility lines threatened. It is possible to predict liquefaction hazard based on ground shaking hazard analysis and the knowledge of the resistance of soil deposits, the latter of which can be obtained with various in-situ tests such as the cone penetration test (CPT), the standard penetration test (SPT) and the shear wave velocity test, have been proved effective for estimating liquefaction resistance at individual locations [1–6]. It is, however, more complicated when mapping liquefaction hazard over an extended region. Classical approaches often rely on surficial geologic data and knowledge of past liquefactions [7]. Although efficient and simple to apply, in these approaches, liquefaction hazard

© Springer Nature Singapore Pte Ltd. 2018
T. Qiu et al. (Eds.): GSIC 2018, *Proceedings of GeoShanghai 2018 International Conference: Advances in Soil Dynamics and Foundation Engineering*, pp. 44–52, 2018.
https://doi.org/10.1007/978-981-13-0131-5_5

level is assumed to be constant within each surficial geologic unit, which ignores the inherent spatial variability of soil properties and therefore limits the accuracy and the applicability of the generated liquefaction hazard maps.

In this paper, a novel integrated framework is developed for regional liquefaction assessment accounting for the spatial variation of liquefaction potential across different scales and hybrid data from different sources.

2 Methodology

In this study, liquefaction hazard is quantified and mapped in terms of the liquefaction potential index (LPI) defined by Iwasaki et al. [8] as

$$\text{LPI} = \int_0^{20} w(z)F_L dz \tag{1}$$

where z is the soil depth in meters; $w(z) = 10 - 0.5z$; F_L is a function of factor of safety (FS)

$$F_L = \begin{cases} 0 & \text{FS} \geq 1.2 \\ 1 - \text{FS} & \text{FS} \leq 0.95 \\ 2 \times 10^6 e^{-18.427\text{FS}} & 0.95 < \text{FS} < 1.2 \end{cases} \tag{2}$$

Two types of LPI data are taken into account when generating the liquefaction hazard map. The first type is termed the primary data, which is evaluated using a CPT-based liquefaction model [2]. At the beginning, CPT data are collected within the study area upon which the primary LPI are calculated. The second type data is termed the secondary data, which is obtained based on secondary information such as a surficial geologic map. Then, the distribution of primary LPIs within each geologic unit are characterized (Table 1), upon which the secondary LPI data within each geologic unit are generated based on the characterized distributions of primary LPIs. The role of the secondary data is to constrain and improve the primary data-based LPI map such that the final map conforms to the large-scale geologic boundaries in the analysis region.

Table 1. Characterization of the LPI values within each geological unit

	Qhbm	Qhf	Qhff	Qhfy	Qhl	Qhly	Qpf
Mean	8.3	4.7	6.2	7.6	5.3	8.8	2
Standard deviation	5.8	4.1	4.6	4.1	4.9	7.3	2.5

Based on the primary CPT-based LPI data and the secondary geology-based LPI data, multiscale random field models are developed to generate realizations of LPIs across the region of interest. Coupled with Monte Carlo simulations, uncertainties associated with the generated liquefaction hazard maps can also be obtained.

3 Data Characterization

3.1 Primary Data

The investigated region is an earthquake-prone area, the Santa Clara Valley. Its map of geology is shown in Fig. 1.

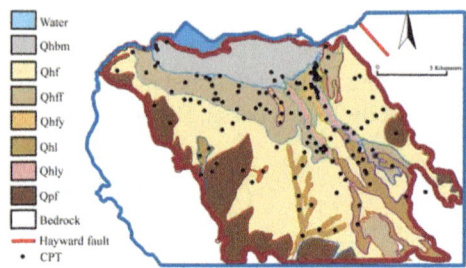

Fig. 1. Map of the Santa Clara Valley, surficial geology and locations of 162 CPT soundings (black dots). The surficial geology map is generated based on Holzer et al. [9].

In this work, the classical CPT-based liquefaction model by Robertson and Wride [2] is adopted to calculate the liquefaction potential of a soil layer, where two variables, i.e., the cyclic stress ratio (CSR) and the cyclic resistance ratio (CRR), are evaluated. Details of this classical liquefaction model and the calculation of LPI have been summarized in a previous study [10] and are not repeated here. In this study, a hypothetical earthquake event (a rupture of the Hayward Fault) with Mw = 7.1 and amax = 0.5 g is assumed. The primary LPI values are calculated at all 162 CPT soundings. The histogram of the calculated 162 LPI values is plotted in Fig. 2(a). To assess the spatial correlations of the primary LPI data, the empirical semivariogram $\hat{\gamma}(h)$ is calculated as [11]

$$\hat{\gamma}(\boldsymbol{h}) = \frac{1}{2N(\boldsymbol{h})} \sum_{\alpha=1}^{N(\boldsymbol{h})} [z(\boldsymbol{u}_\alpha) - z(\boldsymbol{u}_\alpha + \boldsymbol{h})]^2 \tag{3}$$

where $N(\boldsymbol{h})$ is the number of pairs of data z located a vector $\boldsymbol{h}$ apart (i.e., a lag bin $\boldsymbol{h}$). Figure 2(b) shows the calculated empirical semivariogram based on LPIs at 162 CPT soundings. Given the sample semivariogram, a weighted least square method by Cressie [12] is implemented to fit an analytical semivariogram model, shown as solid line in the plot.

3.2 Secondary Data

It is very important to consult surficial geology when assessing regional liquefaction hazard. Surficial geology provides broader area coverage and information on large-

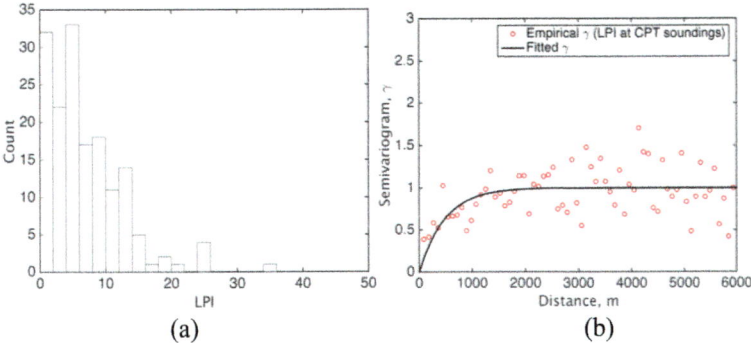

Fig. 2. Primary LPI data at 162 CPT soundings in the Santa Clara Valley. The fitted model semivariogram is an exponential model $\gamma = 1 - e^{\left(-\frac{h}{480}\right)}$ with h being the scalar distance. (The R^2 and the root-mean-square error (RMSE) of the fit are 0.2163 and 0.2532, respectively.)

scale material heterogeneity. For the project area, the surficial geology information is used to generate secondary LPI following the procedure stated in methodology section. The generated results are shown in Fig. 3 for all geologic units in the study region. Such secondary LPI values will be incorporated into the multiscale random field models described in the next section.

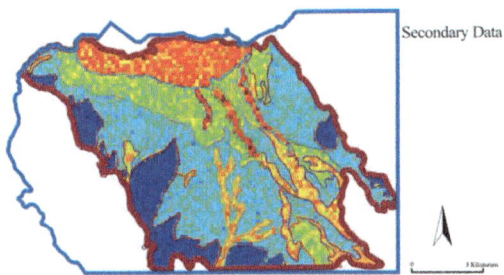

Fig. 3. Generation of secondary data within each geologic unit.

3.3 Spatial Correlation

It is required to know the spatial correlations between data point before a random field simulation. The spatial correlation of the primary LPI data has been shown in Fig. 2(b). The left two type of correlations, i.e., the correlation for secondary variable and the correlation across primary and secondary variable, are still unknown.

Usually, direct calculation of the secondary and cross-covariances can be challenging due to lack of data. In this work, one simplified approach is adopted based on the Markov-Bayes hypothesis described by Goovaerts [11] to derive the secondary and cross-covariances by calibrating them to the primary covariance as [11, 13]

$$COV_s(\boldsymbol{h}) = \begin{cases} |B| \cdot COV_p(\boldsymbol{h}) \text{ for } \boldsymbol{h} = 0 \\ B^2 \cdot COV_p(\boldsymbol{h}) \text{ for } \boldsymbol{h} > 0 \end{cases}$$
$$COV_{ps}(\boldsymbol{h}) = B \cdot COV_p(\boldsymbol{h}) \tag{4}$$

where B is the Markov-Bayes coefficient; COV_p is the covariance for the primary variable, COV_s is the covariance for the secondary variable and $COV_{ps}(\boldsymbol{h})$ is the covariance between the primary and the secondary variables. The Markov-Bayes coefficient B affects the relative importance of primary data and secondary data and this effect will be illustrated in the application section.

4 Multiscale Sequential Simulation

A conditional sequential simulation algorithm is implemented in this work to generate random field realizations of LPIs. The following derivation is for variables having Gaussian distributions. A normal score mapping technique has been used in the present study to transform variables between Gaussian distributions and non-Gaussian distributions in order to take advantage of the desirable Gaussian properties in describing spatial correlations [14]. This algorithm integrates and preserves multiple sources of known data (e.g., primary and secondary LPI data). In this algorithm, the realization of a random variable Zn is represented by a joint distribution as follows

$$\begin{bmatrix} Z_n \\ \boldsymbol{Z}_p \\ \boldsymbol{Z}_s \end{bmatrix} \sim N \left(\begin{bmatrix} \mu_n \\ \boldsymbol{\mu}_p \\ \boldsymbol{\mu}_s \end{bmatrix}, \begin{bmatrix} \sigma_n^2 & \boldsymbol{\Sigma}_{np} & \boldsymbol{\Sigma}_{ns} \\ \boldsymbol{\Sigma}_{pn} & \boldsymbol{\Sigma}_{pp} & \boldsymbol{\Sigma}_{ps} \\ \boldsymbol{\Sigma}_{sn} & \boldsymbol{\Sigma}_{sp} & \boldsymbol{\Sigma}_{ss} \end{bmatrix} \right) \tag{5}$$

Where $\sim N(\boldsymbol{\mu}, \boldsymbol{\Sigma})$ denotes the vector of random variables following a joint normal distribution with the mean vector $\boldsymbol{\mu}$ and the covariance matrix $\boldsymbol{\Sigma}$; Z_n is the random variable to be generated with the expected value μ_n; $\boldsymbol{Z}_p$ is the vector of previously generated or known primary random variables with the vector of expected values $\boldsymbol{\mu}_p$; $\boldsymbol{Z}_s$ is a vector of secondary random variables with the vector of expected values $\boldsymbol{\mu}_s$; σ_n is the standard deviation of Z_n; $\boldsymbol{\Sigma}$ is the covariance matrix with subscripts "n", "p" and "s" denoting "next", "previous primary" and "secondary", respectively. The individual terms in the covariance matrix is defined as

$$COV[Z_i, Z_j] = \rho_{Z_i, Z_j} \sigma_{Z_i} \sigma_{Z_j} \tag{6}$$

where ρ_{Z_i, Z_j} is the correlation between two elements Z_i and Z_j within the random field at any scale with a standard deviation of σ_{Z_i} and σ_{Z_j}, respectively. Given the joint distribution (5), the distribution of the random variable Z_n, conditional upon all previously simulated and known primary and secondary data, could be written by a univariate normal distribution with the updated mean and variance. Then the value of a random variable Z_n at an unsampled location is drawn from the above joint distribution. Once generated, Z_n becomes a data point in the vector $\boldsymbol{Z}_p$ to be conditioned upon by all subsequent simulations. This process is repeated by following a random path to each unknown location until all the values in the field have been simulated, i.e., a map of the primary variable for the region of interest is generated. More details of this procedure could be found in [10].

The authors extended the sequential random process to account for spatial correlations across different scales. This could be done by defining two new correlations:

$$\rho_{Z_I^c, Z_{II}^c} = \frac{\sum_{i=1}^{N} \sum_{k=1}^{N} \rho_{z_{i(I)}^f, z_{k(II)}^f}}{\sqrt{\sum_{i=1}^{N} \sum_{j=1}^{N} \rho_{z_{i(I)}^f, z_{j(I)}^f}} \sqrt{\sum_{i=1}^{N} \sum_{j=1}^{N} \rho_{z_{i(II)}^f, z_{j(II)}^f}}} \tag{7}$$

$$\rho_{Z^f, Z_I^c} = \frac{\sum_{i=1}^{N} \rho_{z^f, \rho_{z_{j(I)}^f}}}{\sqrt{\sum_{i=1}^{N} \sum_{j=1}^{N} \rho_{z_{i(I)}^f, z_{j(I)}^f}}} \tag{8}$$

where $\rho_{Z_I^c, Z_{II}^c}$ is the correlation between two coarse scale grids I and II to be simulated; ρ_{Z^f, Z_I^c} is the correlation between a fine grid and a coarse scale grid I, $\rho_{z_{i(I)}^f, z_{k(II)}^f}$ is the correlation between a fine scale grid i and a fine scale grid k, which belong to two different coarse scale grids I and II, respectively. In the right-hand-side of Eqs. 7 and 8, $\rho = 1 - \gamma(\boldsymbol{h})$, where γ is the fitted semivariogram model in Fig. 2(b).

Given the multiscale spatial dependence specified by (7) and (8) and an inferred or assumed probability distribution of the random variable, the conditional sequential simulation algorithm is employed to generate random field realizations of variables of interest. Development of the multiscale correlations is elaborated in details in [10].

5 Application

5.1 Random Field Realization

While both primary and secondary LPI data obtained from CPT and geologic information are prepared, random field realizations of LPIs across the Santa Clara Valley are performed. The region is firstly discretized by a relatively coarse grid and an example realization of LPIs on the coarse grid is shown in Fig. 4(a). Upon the coarse scale realization, the grid can be adaptively refined in subregions of higher interest while maintaining consistent spatial correlations across scales. As an example, a subregion marked by the red box in Fig. 4(b) is selected for refinement. This subregion is identified as an area of high liquefaction potential by Holzer [9] and is evidenced by the concentration of higher LPI values.

5.2 Monte Carlo Simulation

For regional liquefaction hazard mapping, the multiscale random field models are coupled with Monte Carlo simulations to obtain expected liquefaction hazard across the region and to perform probabilistic analysis on quantities of interest.

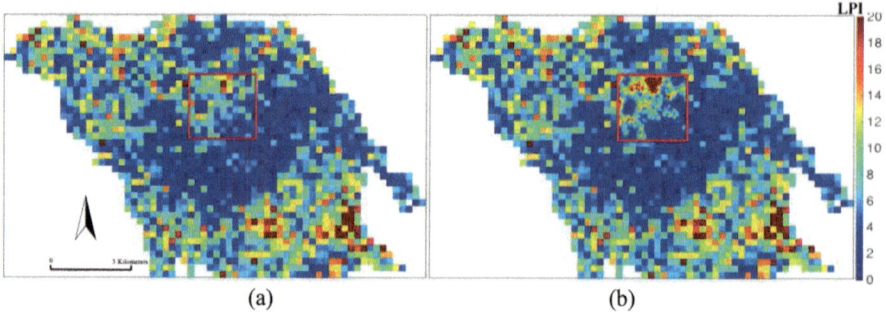

Fig. 4. Example realizations of LPIs across the Santa Clara Valley site for a hypothetical earthquake event (M_w = 7.1, a_{max} = 0.5 g) and the Markov-Bayes coefficient B = 0: (a) Single scale; (b) Multiscale.

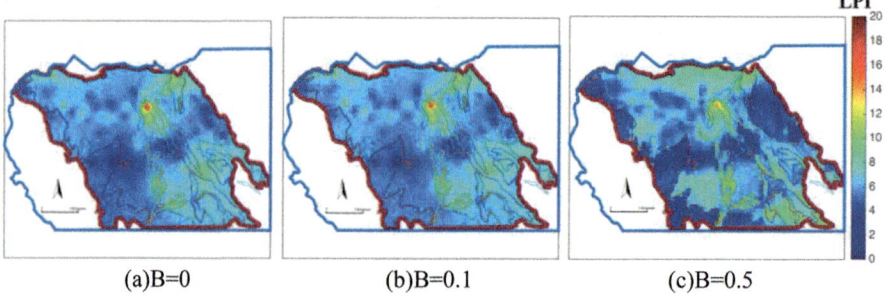

Fig. 5. Maps of expected LPI values for three cases of Markov-Bayes coefficient B. B = 0 is the case without secondary LPI data. Each map is obtained by averaging results from 1,000 Monte Carlo simulations.

Figure 5 shows maps of expected LPI values for three cases of Markov-Bayes coefficient B. Each of the three maps is obtained by averaging results from 1,000 Monte Carlo simulations. The Markov-Bayes coefficient B is essentially a "scaling" factor between the primary covariance and the secondary covariance matrices. The larger the coefficient B, the stronger influence the secondary data has on the generated LPI maps. In this work, the secondary LPI data come from geologic information. Therefore, as the value of B increases, the geologic boundaries become more distinguishable in the resulting LPI maps as evidenced in Fig. 5(b)-(c). When no secondary data is incorporated, i.e., B = 0, no geologic constraint is applied to the LPI map. Such case is shown in Fig. 5(a) and the resulting LPI map might be inaccurate. When there is sufficient of real-word data points, Markov-Bayes coefficient B could be calibrated to reflect the true correlation between variables.

$$B = E(\Pr ob\{Z_s \leq z\}|Z_p \leq z) - E\big(\Pr ob\{Z_s \leq z\}|Z_p > z\big) \qquad (9)$$

where E is the expectation operator; Zs is the secondary variable (e.g. the geological data-based LPI value) and $Prob\{Z_s \leq z\}$ is the probability of Zs less than or equal to a threshold value z (e.g. a given LPI threshold value); Zp is the primary variable (e.g. the geotechnical data-based LPI value). In this paper, the secondary data is randomly generated pseudo LPI, therefore it is hard to calibrate. Among the above three cases, B = 0.5 is most reasonable since it not only shows spatial variations of LPI but also, more importantly, keep consistent with the geologic boundaries.

6 Conclusion

The multiscale random field-based method is presented to integrate heterogeneous data sources, i.e., geologic data at the regional scale and geotechnical data at the site-specific scale, for regional liquefaction hazard mapping. Uncertainties and spatial variability of soil parameters are appropriately accounted for in the process. The geologic data are incorporated in the process as secondary data and essentially constraint geologic boundaries in the generated map. Both geotechnical and geologic data are integrated into multiscale random field models through a conditional sequential simulation technique. The proposed framework is applied to map liquefaction hazard at the Santa Clara Valley site in California. It is found that the spatial correlation structure of the calculated liquefaction potential index (LPI) is prominent in the study region. The influence of geologic data on the generated liquefaction hazard map is significant. As the weight of the geologic data increases, the geologic boundaries become more distinguishable in the generated map. With an appropriately selected or calibrated Markov-Bayes coefficient, the geotechnical and geologic data could be properly accounted for in the mapping process.

Acknowledgement. This work is supported by the U.S. Geological Survey (Grant No. G17AP00044) and the Program of Shanghai Academic Research Leader (Grant No. 17XD1403700).

References

1. Seed, H.B., Idriss, I.M.: Simplified procedure for evaluating soil liquefaction potential. J. Soil Mech. Found. Div. **97**(9), 1249–1273 (1971)
2. Robertson, P.K., Wride, C.E.: Evaluating cyclic liquefaction potential using the cone penetration test. Can. Geotech. J. **35**(3), 442–459 (1998)
3. Youd, T.L., Idriss, I.M.: Liquefaction resistance of soils: summary report from the 1996 NCEER and 1998 NCEER/NSF workshops on evaluation of liquefaction resistance of soils. J. Geotech. Geoenviron. Eng. **127**(10), 817–833 (2001)
4. Cetin, K.O., Seed, R.B., Der Kiureghian, A., Tokimatsu, K., Harder Jr., L.F., Kayen, R.E., Moss, R.E.: Standard penetration test-based probabilistic and deterministic assessment of seismic soil liquefaction potential. J. Geotech. Geoenviron. Eng. **130**(12), 1314–1340 (2004)

5. Idriss, I.M., Boulanger, R.W.: Soil liquefaction during earthquakes. Earthquake Engineering Research Institute (2008)
6. Shen, M., Chen, Q., Zhang, J., Gong, W., Juang, C.H.: Predicting liquefaction probability based on shear wave velocity: an update. Bull. Eng. Geol. Env. **75**(3), 1199–1214 (2016)
7. Youd, T.L., Perkins, D.M.: Mapping of liquefaction severity index. J. Geotech. Eng. **113**(11), 1374–1392 (1987)
8. Iwasaki, T., Tatsuoka, F., et al.: A practical method for assessing soil liquefaction potential based on case studies at various sites in Japan. In: Proceedings of 2nd International Conference on Microzonation, pp. 885–896 (1978)
9. Holzer, T.L., Noce, T.E., Bennett, M.J.: Scenario liquefaction hazard maps of Santa Clara Valley, northern California. Bull. Seismol. Soc. Am. **99**(1), 367–381 (2009)
10. Wang, C., Chen, Q.: A hybrid geotechnical and geologic data-based framework for multiscale regional liquefaction hazard mapping. Geotechnique, 1–12 (2017)
11. Goovaerts, P.: Geostatistics for Natural Resources Evaluation. Oxford University Press, New York (1997)
12. Cressie, N.: Fitting variogram models by weighted least squares. J. Int. Assoc. Math. Geol. **17**(5), 563–586 (1985)
13. Deutsch, C.V., Journel, A.G.: Geostatistical Software Library and User's Guide. Oxford University Press, New York (1998)
14. Chen, Q., Seifried, A., Andrade, J.E., Baker, J.W.: Characterization of random fields and their impact on the mechanics of geosystems at multiple scales. Int. J. Numer. Anal. Meth. Geomech. **36**(2), 140–165 (2012)

Dynamic Response of Soil Around the Tunnel Under Subway Vibration Loading

Zhong-Liang Zhang[1] and Zhen-Dong Cui[1,2(⊠)]

[1] State Key Laboratory for Geomechanics and Deep Underground Engineering,
School of Mechanics and Civil Engineering,
China University of Mining and Technology,
Xuzhou 221116, Jiangsu, People's Republic of China
czdjiaozuo@163.com
[2] Fujian Research Center for Tunneling and Urban Underground
Space Engineering, Huaqiao University,
Xiamen 361021, Fujian, People's Republic of China

Abstract. With the development of urban rail transit construction, the land subsidence, deformation even failure of the subway tunnel and other environmental issues induced by subway vibration have been paid more attention by researchers and engineers recently. The dynamic response of soil under the vibration load is of significance for the safe operation of the subway. In this study, the numerical simulations of soil-tunnel model considering the factors of the model damping and the viscoelastic boundary were conducted. The dynamic response of soil under subway vibration is symmetrically distributed along the vertical centerline of tunnel. The longer the distance to the vibration, the less the effect is. The energy of vibration, which is mainly propagated along the vertical direction and causes the uneven displacement by extruding the soil beside the tunnel, attenuates quickly when spreading in soil. The range of acceleration response of the soil is only about 5 m around the tunnel. The acceleration of soil around the tunnel is superimposed by a series of harmonics with different frequencies and amplitudes. The variations of the amplitude with frequency can be obtained from the acceleration curves by FFT. The dominant frequency of soil under subway vibration is mainly between 5–25 Hz.

Keywords: Subway tunnel · Dynamic response · Subway vibration
Viscoelastic boundary

1 Introduction

The subway has been considered as a significant way to solve the urban traffic problems. However, the vibration load may result in degradation or even failure of soil foundations and structures [1]. Besides, the normal use of precision instrument and human health can be affected by the vibration load [2]. In recent years, the dynamic response of soil around tunnel has become one of the most important subjects. The dynamic response of ordinary soil around the tunnel under vibration load has been studied by many researchers [3–5]. However, in general, current researches are mainly concentrated on theory analysis and test study. In this study, the governing equations of

© Springer Nature Singapore Pte Ltd. 2018
T. Qiu et al. (Eds.): GSIC 2018, *Proceedings of GeoShanghai 2018 International Conference: Advances in Soil Dynamics and Foundation Engineering*, pp. 53–61, 2018.
https://doi.org/10.1007/978-981-13-0131-5_6

the tunnel-soil dynamic interaction were derived and the solving method was given. The dynamic response of soil around the tunnel under the subway vibration load were analyzed with the nonlinear finite element software ADINA. The results of the numerical analysis can offer a reference to the design and construction of subway tunnel in soft soil areas.

2   Tunnel-Soil Dynamic Interaction

In order to simplify the establishment of model equations, the model in this paper consist of three parts: the tunnel, the soil and their interface. It is based on the following hypotheses: (1) the strain of the soil and tunnel is small enough so that elastic model can be applied; (2) the properties of the model remain unchanged in the longitudinal direction, that is the model can be treated as a plane strain problem; (3) the interface of the tunnel and the soil satisfies the deformation compatibility.

The relationship between the dynamic strain ϵ and the perturbation displacements $\mathbf{u}$ can be expressed as follows:

$$\epsilon = \frac{1}{2}\left[\nabla\mathbf{u} + (\nabla\mathbf{u})^{\mathrm{T}}\right] \tag{1}$$

According to Hooke's law, the dynamic stress σ can be calculated by

$$\sigma = \lambda(\nabla \cdot \mathbf{u})e + 2\mu\epsilon \tag{2}$$

where λ and μ the Lame constants and e is the 3×3 unit matrix.

For the soil, the governing equation can be written as:

$$\rho_s \frac{\partial^2 \mathbf{u}_s}{\partial t^2} = (\lambda + \mu)\nabla(\nabla \cdot \mathbf{u}_s) + \mu\nabla^2\mathbf{u}_s + \mathbf{f}_s \tag{3}$$

For the tunnel, the governing equation can be written as:

$$\rho_t \frac{\partial^2 \mathbf{u}_t}{\partial t^2} = (\lambda + \mu)\nabla(\nabla \cdot \mathbf{u}_t) + \mu\nabla^2\mathbf{u}_t + \mathbf{f}_t \tag{4}$$

The boundary conditions of their interface can be written as:

$$\mathbf{u}_s = \mathbf{u}_t \qquad \sigma_s = \sigma_t \tag{5}$$

As the form of Eqs. (3) and (4) are the same, only Eq. (3) is solved following. The displacement is transformed by Laplace transformation for the time t and Fourier transformation for x, y, z direction, that is:

$$\tilde{u}\left(s, \omega_x, \omega_y, \omega_z\right) = \int_0^{+\infty} \int_{-\infty}^{+\infty} \int_{-\infty}^{+\infty} u(t, x, y, z)e^{-st}e^{-i\omega_x x}e^{-i\omega_y y}e^{-i\omega_z z}dt\,dx\,dy\,dz \tag{6}$$

Equation (3) can be transformed as:

$$\rho_s s^2 \tilde{u}_x = -(\lambda+\mu)\omega_x\left(\omega_x\tilde{u}_x, \omega_y\tilde{u}_y, \omega_z\tilde{u}_z\right) - \mu\left(\omega_x^2+\omega_y^2+\omega_z^2\right)\tilde{u}_x + \tilde{f}_{sx}$$
$$\rho_s s^2 \tilde{u}_y = -(\lambda+\mu)\omega_y\left(\omega_x\tilde{u}_x, \omega_y\tilde{u}_y, \omega_z\tilde{u}_z\right) - \mu\left(\omega_x^2+\omega_y^2+\omega_z^2\right)\tilde{u}_x + \tilde{f}_{sy} \quad (7)$$
$$\rho_s s^2 \tilde{u}_z = -(\lambda+\mu)\omega_z\left(\omega_x\tilde{u}_x, \omega_y\tilde{u}_y, \omega_z\tilde{u}_z\right) - \mu\left(\omega_x^2+\omega_y^2+\omega_z^2\right)\tilde{u}_z + \tilde{f}_{sz}$$

Define $U = \left(\tilde{u}_x, \tilde{u}_y, \tilde{u}_z\right)^{\mathrm{T}}, \Omega = \left(\omega_x, \omega_y, \omega_z\right)^{\mathrm{T}}, F = \left(\tilde{f}_{sx}, \tilde{f}_{sy}, \tilde{f}_{sz}\right)^{\mathrm{T}}$ as three vectors, Eq. (7) can be simplified as

$$\rho s^2 U = -(\lambda+\mu)\Omega\Omega^{\mathrm{T}}U - \mu\Omega^{\mathrm{T}}U + F \quad (8)$$

Then U can be calculated as follows:

$$U = \left[\rho s^2 e_3 + (\lambda+\mu)\Omega\Omega^{\mathrm{T}} + \mu\Omega^{\mathrm{T}}\Omega\right] - F \quad (9)$$

Therefore, the displacement fields in time-space domain can be obtained by inverse Fourier transformation and inverse Laplace transformation with the contact conditions considered. Then the strain fields and stress fields can be obtained.

3 Numerical Model

3.1 Model Size and Material Properties

The numerical model in this paper is based on Subway Line 8 in Shanghai, 20 m in depth from the center of the tunnel to the ground surface. The characteristics of the tunnel are assumed to be the same, therefore a plane strain problem is considered and only one interception is chosen, as shown in Fig. 1. The external and inner radiuses are 3.1 and 2.75 m, respectively. To simplify the model, the grey silty clay of layer No. 5 is assumed around the tunnel. A, B, C and D in the model are taken as the reference points to investigate the differences of soil dynamic response around the tunnel. The Mohr-Coulomb plastic constructive model is adopted for the silty clay, while the tunnel is considered as an elastomer. The parameters are summarized in Table 1.

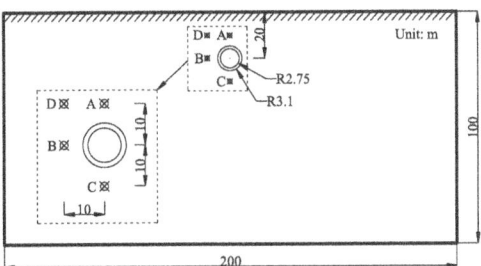

Fig. 1. Model size and the reference points.

3.2 Vibration Load and Boundary Conditions

The vibration load of the subway can be regarded as the superposition of the static load and the dynamic load [6, 7]. Moreover, the vibration load of the train occurs mainly in the three ranges of frequency based on the previous tests of the Center of British Railway [8]. So the function of the vibration load can be written as:

$$F(t) = F_0 + F_1 \sin \omega_1 t + F_2 \sin \omega_2 t + F_3 \sin \omega_3 t \tag{10}$$

where F_0 is the static load of the wheel, which the value is 160 kN. F_1, F_2 and F_3 are the amplitudes of the vibration load which correspond to the three ranges of frequency, which the values are 0.548 kN, 1.566 kN and 5.012 kN, respectively. ω_1, ω_2 and ω_3 are 14.45 rad/s, 72.26 rad/s and 289.03 rad/s, respectively. The curve of the vibration load is shown in Fig. 2.

Table 1. Material properties in the model.

Parameters	Soil	Tunnel
Density (kg·m^{-3})	1810	2500
Elastic modulus (MPa)	13.8	35000
Poisson ratio	0.4	0.2
Friction angle (°)	30	–
Cohesive strength (kPa)	8	–

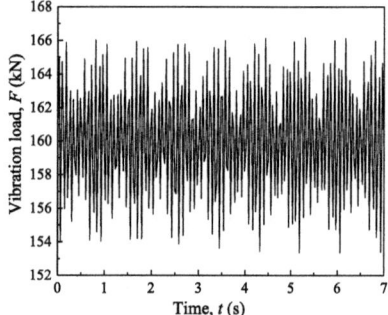

Fig. 2. The curve of the vibration load.

In order to prevent the vibration waves from being reflected to the soil dynamic response analysis, the consistent artificial viscoelastic boundaries have been used at the bottom or side of the model [9]. Besides, the Rayleigh damping was adopted to quantify the damping property of the dielectric material in this model [10]. The analysis of the vibration mode was carried on to obtain the circular frequencies of the model. Assuming the damping ratio as 0.05, the Rayleigh damping coefficients α and β can be calculated as 0.09213 and 0.02695, respectively. Since the natural vibration period is 5.5 s, the step length was chosen as 0.01 s for the step length should smaller than one tenth of the natural vibration period in the dynamic problems [11].

4 Results and Discussion

The vertical displacement of the silty clay after vibration in the vertical direction is illustrated in Fig. 3. The displacement of the soil is symmetrically distributed along the vertical centerline of tunnel. But the direction of the vertical displacement is inconsistent. The area of larger displacement is mainly concentrated on the lower half of the

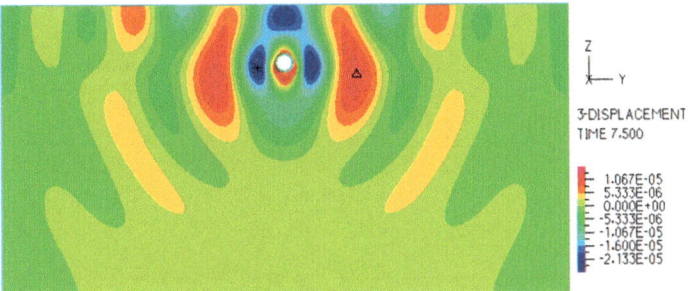

Fig. 3. Vertical displacement of the silty clay after vibration.

tunnel and the position about two times the diameter from the tunnel. The uneven displacements caused by the vibration load spread out with the loops of the vibration cycle. Figure 4 depicts the attenuations of the acceleration after the vibration. The longer the distance to the vibration, the less the effect is. It should be mentioned that the wave of vibration attenuates quickly when spreading in soil [12].

Fig. 4. Acceleration of the silty clay after vibration.

Figure 5 illustrates that the vertical displacements of the reference points. It can be seen that the displacement curves of the soil are fluctuating with the vibration. Compared with Points A, B and C, Point D has smaller displacement, which means that the longer the distance to the vibration, the less the effect is.

The amplitude-frequency curves of acceleration can be obtained after Fast Fourier Transform (FFT) of the time history curves, as shown in Fig. 6. The acceleration of soil around the tunnel is superimposed by a series of harmonics with different frequencies and amplitudes. The dominant frequency of soil under subway vibration is mainly between 5–25 Hz, which is accordance with the monitoring results in literature [13].

The vertical ground displacements with distance to the left side of the model is shown in Fig. 7. The displacement of the soil is symmetrically distributed along the vertical centerline of tunnel. It is obvious that the displacement is fluctuating and the maximum of the displacement gradually attenuates with distance to the tunnel on both

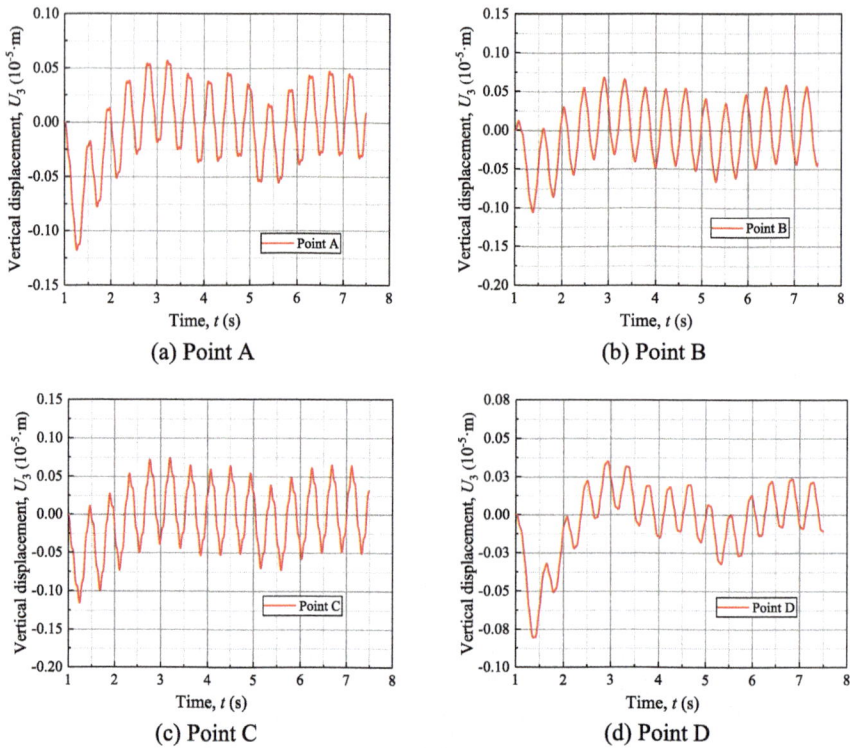

Fig. 5. Vertical displacement curves of the silty clay during vibration.

sides. The reason is that the energy of vibration, which is mainly propagated along the vertical direction and causes the uneven displacement by extruding the soil beside the tunnel, attenuated quickly when spreading in soil. In this model, the viscoelastic boundaries and the adopted Rayleigh damping absorbed most of the vibration energy. The wave propagation of the vibration load is hardly reflected to the soil dynamic response analysis. The energy spread radially to the soil from the bottom of the tunnel and the uneven ground displacement is resulted from the superposition of continuous energy.

The variation of the acceleration with the depth to the bottom of the tunnel is illustrated in Fig. 8. It can be seen that the change of the soil acceleration is quite obvious in the range 0–5 m. When the depth to the bottom of the tunnel is over than 20 m, the acceleration in the vertical direction can be ignored. The soil near the bottom of the tunnel directly affects the deformation of the tunnel. If the vibration acceleration is too large, the soil will be more prone to be damaged in the long term cyclic loads, and the local stability of the tunnel will be affected.

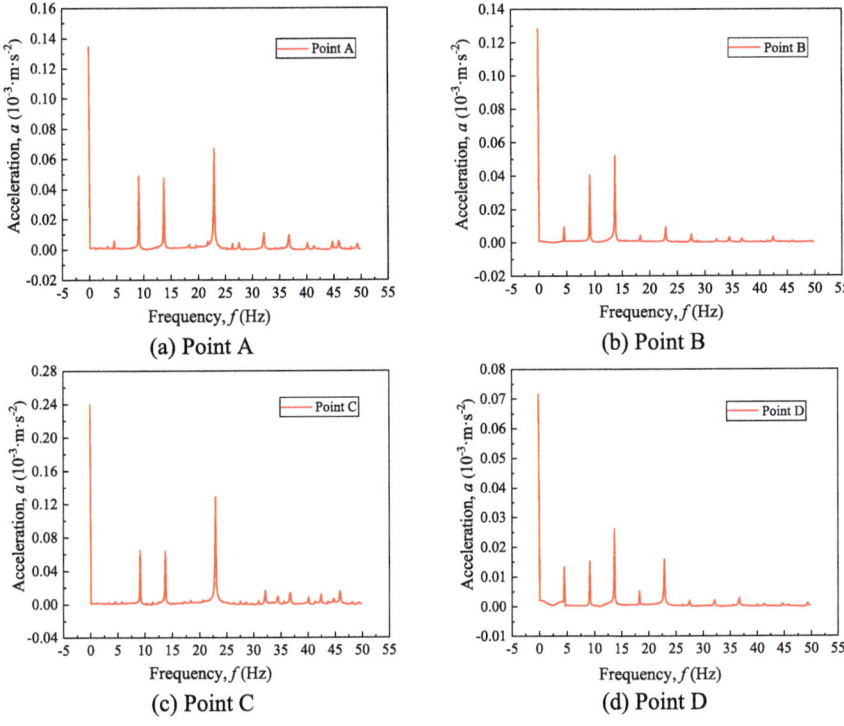

Fig. 6. Amplitude-frequency curves of acceleration.

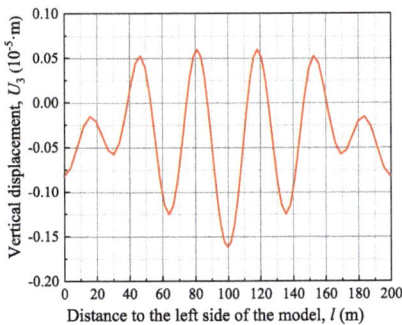

Fig. 7. Variation of the ground displacement with distance to the left side of the model.

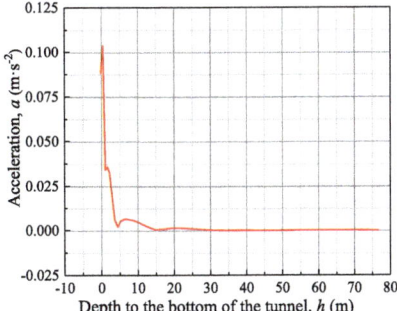

Fig. 8. Variation of the acceleration with depth to the bottom of the tunnel.

5 Conclusions

In this paper, the governing equations of the soil-tunnel model were derived and the solving method was given. The numerical simulations of soil dynamic response were conducted. The main conclusions are as follows.

(1) The dynamic response of soil under subway vibration is symmetrically distributed along the vertical centerline of tunnel. The longer the distance to the vibration, the less the effect is. The energy of vibration, which is mainly propagated along the vertical direction and causes the uneven displacement by extruding the soil beside the tunnel, attenuated quickly when spreading in soil.
(2) The range of acceleration response of the soil is only about 5 m around the tunnel. The variations of the amplitude with frequency can be obtained from the acceleration curves by FFT. The acceleration of soil around the tunnel is superimposed by a series of harmonics with different frequencies and amplitudes. The dominant frequency of soil under subway vibration is mainly between 5–25 Hz.

Acknowledgements. This work presented in this paper was by National key research and development program (2017YFC1500702) and the research grant (16FTUE03) from Fujian Research Center for Tunneling and Urban Underground Space Engineering (Huaqiao University).

References

1. Wang, X., Yang, P., Wang, H., et al.: Experimental study on effects of freezing and thawing on mechanical properties of clay. Chin. J. Geotech. Eng. **31**(11), 1768–1772 (2009)
2. Ren, X.W., Tang, Y.Q., Xu, Y.Q., et al.: Study on dynamic response of saturated soft clay under the subway vibration loading I: instantaneous dynamic response. Environ. Earth Sci. **64**(7), 1875–1883 (2011)
3. Tang, Y.Q., Sun, K., Zheng, X.Z., et al.: The deformation characteristics of saturated mucky clay under subway vehicle loads in Guangzhou. Environ. Earth Sci. **75**(5), 1–10 (2016)
4. Xiao, J.H., Juang, C.H., Wei, K., et al.: Effects of principal stress rotation on the cumulative deformation of normally consolidated soft clay under subway traffic loading. J. Geotech. Geoenviron. Eng. **140**(4), 04013046 (2013)
5. Yuan, Z.H., Xu, C.J., Cai, Y.Q., et al.: Dynamic response of a tunnel buried in a saturated poroelastic soil layer to a moving point load. Soil Dyn. Earthq. Eng. **77**, 348–359 (2015)
6. Li, S.: Study on subway-induced environmental vibration and floating floor isolation. Tongji University, Shanghai (2008)
7. Liu, W., Xia, H., Guo, W.: Study of vibration effects of underground trains on surrounding environments. Chin. J. Rock Mechan. Eng. **15**(1), 586–593 (1996)
8. Jenkins, H.H., Stephenson, J.E., Clayton, G.A., et al.: The effect of track and vehicle parameters on wheel/rail vertical dynamic forces. Railway Eng. J. **3**(1), 2–16 (1974)
9. Liu, J., Gu, Y., Du, Y.: Consistent viscous-spring artificial boundaries and viscous-spring boundary elements. Chin. J. Geotech. Eng. **28**(9), 1070–1075 (2006)
10. Shu, F., Qian, Z.: Analysis on the dynamic response of asphalt pavement under moving load. J. Transp. Eng. Inf. **5**(3), 90–95 (2007)

11. Zhai, J.: Test and analysis of vibration propagation caused by the subway train. Tongji University, Shanghai (2007)
12. He, P.P., Cui, Z.D.: Dynamic response of a thawing soil around the tunnel under the vibration load of subway. Environ. Earth Sci. **73**(5), 2473–2482 (2015)
13. Shen, Y.: Study on the propagation laws of subway-induced vibration and isolation or reduction methods of building vibration. Tongji University, Shanghai (2007)

Liquefaction Characteristics of Four Ya-An Low-Plastic Silty Sands with Presence of Initial Static Shear Stress

Xiao Wei[1], Yi Guo[1], Jun Yang[1(✉)], and Chang-Bao Guo[2,3]

[1] Department of Civil Engineering, The University of Hong Kong,
Hong Kong, China
junyang@hku.hk
[2] Key Laboratory of Neotectonic Movement & Geohazard,
Ministry of Land and Resources, Beijing 100081, China
[3] Institute of Geomechanics, Chinese Academy of Geological Sciences,
Beijing 100081, China

Abstract. Following the construction of Sichuan-Tibet railway, further development of infrastructures can be anticipated in the cities or counties along the railway. However, these cities or counties, including Ya-An, are located in mountain areas with relatively high seismicity, making the initial static shear stress to be an important factor in soil liquefaction assessment. In this study, four low-plastic sands, which liquefied in the 2013 Lu-Shan earthquake, were extracted for investigation. A series of cyclic triaxial tests were performed to evaluate the liquefaction resistance ratios (*CRR*) of the four sands with different initial static shear stress ratios (α). The failure patterns of the soils were identified to be either cyclic mobility or plastic strain accumulation, depending on the cyclic shear stress amplitude and the initial static shear stress, i.e. the stress reversal conditions. The effects of initial static shear stress on the liquefaction resistance were found either beneficial or detrimental depending on the initial static shear stress level. The threshold α concept (α_{th}) proposed by Yang and Sze [6] may be used to characterize such effects for natural sands containing a substantial amount of plastic fines. The liquefaction resistance first increases with increasing α if $\alpha < \alpha_{th}$, and then decreases with further increase of α if $\alpha > \alpha_{th}$. The α_{th} can be estimated using the no-reversal line representing $\alpha = CRR$.

Keywords: Liquefaction · Low-plastic sand · Undrained cyclic triaxial test
Initial static shear stress · Liquefaction resistance

1 Introduction

The Sichuan-Tibet Railway, connecting Chengdu (Sichuan) and Lhasa (Tibet) via Ya-An and Nyingchi, will serve as a vital artery in the Southwest of China. One of the sections joining Chengdu and Ya-An is located very close to the southern section of Long-Men-Shan Tectonic Belt (which induced the 2008 Wenchuan earthquake and the 2013 Lu-Shan earthquake [1]), and thus may be vulnerable to geohazards induced by

© Springer Nature Singapore Pte Ltd. 2018
T. Qiu et al. (Eds.): GSIC 2018, *Proceedings of GeoShanghai 2018 International Conference: Advances in Soil Dynamics and Foundation Engineering*, pp. 62–69, 2018.
https://doi.org/10.1007/978-981-13-0131-5_7

seismic activities. Liquefaction of sands has been reported on several sites in Ya-An during the 2013 Lu-Shan earthquake (e.g. Hong et al. [2], Liu and Huang [3]). Thus, it is of great importance to investigate the cyclic behavior of the local soils due to earthquake events.

After the 1964 Niigata earthquake, liquefaction of soils has become a hot issue attracting tremendous efforts from the academia and the industry. Liquefaction assessment can be achieved by in-situ tests (e.g. SPT, CPT, seismic wave methods) or laboratory tests (e.g. cyclic triaxial test, cyclic simple shear test). A lot of laboratory database have been accumulated, covering a variety of factors that may affect the liquefaction resistance of soils, such as gradational properties, packing density, confining pressure, initial static shear stress, and fabric [e.g. 4–9]. Most of these investigations are based on standard sands with very well-controlled soil properties. There seems to be a lack of database using natural sands containing a substantial amount of plastic fines.

Among the aforementioned factors, the initial static shear stress is the shear stress on the horizontal plane of a soil element subjected to static loading due to sloping ground condition or surcharge. It is an important factor in liquefaction analysis for large projects involving tailing dams, embankments and slopes [5–7], and thus may be of particular interest in this study due to the mountainous terrain in and around Ya-An City.

The present study investigates the cyclic behavior of four Ya-An sands, which are vulnerable to liquefaction. And the effects of the initial static shear stress are also considered.

2   Testing Program

2.1   Testing Materials

Liquefied Soils. The liquefied sites were reported by Han et al. [10] and Zhang et al. [11]. Figure 1 presents these sites with cracks and sand boils. Sampling pits were excavated, and the soil profiles were recorded for these liquefied sites. Four liquefied layers were identified and summarized in Table 1. The soils were classified as silty fine sands to fine sands by the engineers performing ground investigation. Then the four liquefied soils were extracted from the liquefied layers for undrained cyclic triaxial tests.

Fig. 1. Liquefaction manifestation (a) in Jian-Gan-Lin Village [10]: sand boil and through cracks; (b, c) in Shuang-He Village: sand-boils [11], and sand-boil through a crack [11].

Table 1. Sampling locations.

Sample ID	Sampling locations	Depth
SC1	Long-Quan Village, Yu-Cheng District, Ya-An City	−2.1 m
SC2	Lin-Xia Village, Shuang-Shi Town, Lu-Shan County, Ya-An City	−3.0 m
SC3	Lin-Xia Village, Shuang-Shi Town, Lu-Shan County, Ya-An City	−1.2 m
SC4	Jian-Gan-Lin Village, Yu-Cheng District, Ya-An City	−0.6 m

Soil Properties. The particle size distributions of the four soils were obtained following the British Standard BS1377–2:1990, and are presented in Fig. 2. The mass contents of the fines (<0.063 mm) and the clay-sized particles (<0.002 mm) are summarized in Table 2 together with other basic properties of the soils. The liquid limit (*LL*) and plastic index (*PI*) were measured for the fines only. This is partly because the plastic limit cannot be measured confidently by rolling the natural sands into shreds with diameter of 3 mm. All the four sands are considered vulnerable to liquefaction.

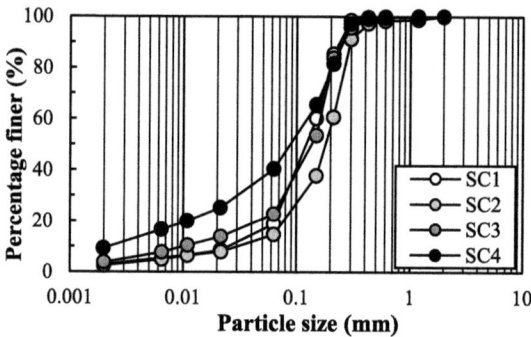

Fig. 2. Particle size distribution of the four tested soils

2.2   Testing Program

A series of undrained cyclic triaxial tests were performed under various initial states. Moist tamping method was adopted to reconstitute the specimens. All specimens were saturated by percolation of CO_2, circulation of de-aired water, and then by increasing back pressure. Given that liquefaction resistance is sensitive to the degree of saturation (e.g. Yang [12]), the condition of full saturation was considered when the B-value is greater than 0.98. The specimens were anisotropically consolidated to different initial static shear stress ratios ($\alpha = 0 - 0.25$), defined as follows for triaxial conditions.

$$\alpha = \frac{\tau_s}{\sigma'_{nc}} = \frac{q_s}{2\sigma'_{nc}} \tag{1}$$

where, $\sigma'_{nc} = (\sigma'_{1c} + \sigma'_{3c})/2$ and $\tau_s = q_s/2 = (\sigma'_{1c} - \sigma'_{3c})/2$ are the post-consolidation effective normal stress and the static shear stress on the maximum shear stress plane,

Table 2. Properties of the soil samples.

ID	Natural water content	G_s	D_{50} mm	D_{10} mm	C_u	C_c	FC* %	CC[†] %	LL[‡] %	PI[‡]
SC1	20.5%	2.67	0.129	0.027	5.5	1.8	18.9	2.9	38	20
SC2	28.9%	2.63	0.183	0.035	6.1	2.0	14.7	2.5	40	15
SC3	27.0%	2.65	0.140	0.010	16.1	4.2	22.7	3.7	42	17
SC4	23.7%	2.68	0.096	0.002	55.6	3.9	40.4	9.2	40	21

*: Content of fines (<0.063 mm); †: Content of clay-sized particles (<0.002 mm); ‡: Measured for fines.

respectively. Then cyclic deviatoric stresses are applied. The cyclic loading magnitude is represented by the cyclic stress ratio, $CSR = \tau_{cyc}/\sigma'_{nc} = q_{cyc}/2\sigma'_{nc}$, where τ_{cyc} is the amplitude of the cyclic shear stress on the maximum shear stress plane, and q_{cyc} is the amplitude of the cyclic deviatoric stress. The post-consolidation void ratio, e_c was determined by measuring the water content after testing. Three replicable specimens were used to obtain the relationship between the applied CSR (ranges from 0.1 to 0.2) and the number of cycles to failure (N_l). CRR is defined as the CSR causing failure in 10 cycles.

3   Test Results

3.1   Failure Patterns

Cyclic Behavior without Initial Static Shear Stress ($\alpha = 0$). The typical stress-strain curves, excess pore water pressure build-up and the stress paths of the four soils are presented in Fig. 3, showing that all the four soils failed in the pattern of cyclic mobility [9]. The axial strain develops cyclically, showing gradually increased double amplitude (DA). The excess pore water pressure also increases and finally reach a state called initial liquefaction, at which its value equals to the initial effective lateral pressure (σ'_{3c}), i.e. effective stress equals zero. During subsequent loading cycles, transient zero-effective-stress states and large deformation of the sample can be observed when the deviatoric stress reverses its direction.

Cyclic Behavior with Initial Static Shear Stress ($\alpha \neq 0$). The stress-strain relationship, excess pore water pressure build-up and stress paths of SC1 subjected to $\alpha = 0.1$ are presented in Fig. 4(a). Since α is relatively low and $CSR > \alpha$, the specimen is subjected to reversed stress cycles. The failure pattern is identified to be cyclic mobility. The axial strain development is more obvious on the compression side.

When α is relatively higher (e.g. $\alpha = 0.25$) and $CSR < \alpha$, the deviatoric stress will not reverse its direction (Fig. 4(b)). The specimen exhibits a different failure pattern known as plastic strain accumulation [9]. The irrecoverable strain accumulates in each cycle and leads to large deformation of the specimen. The excess pore water pressure

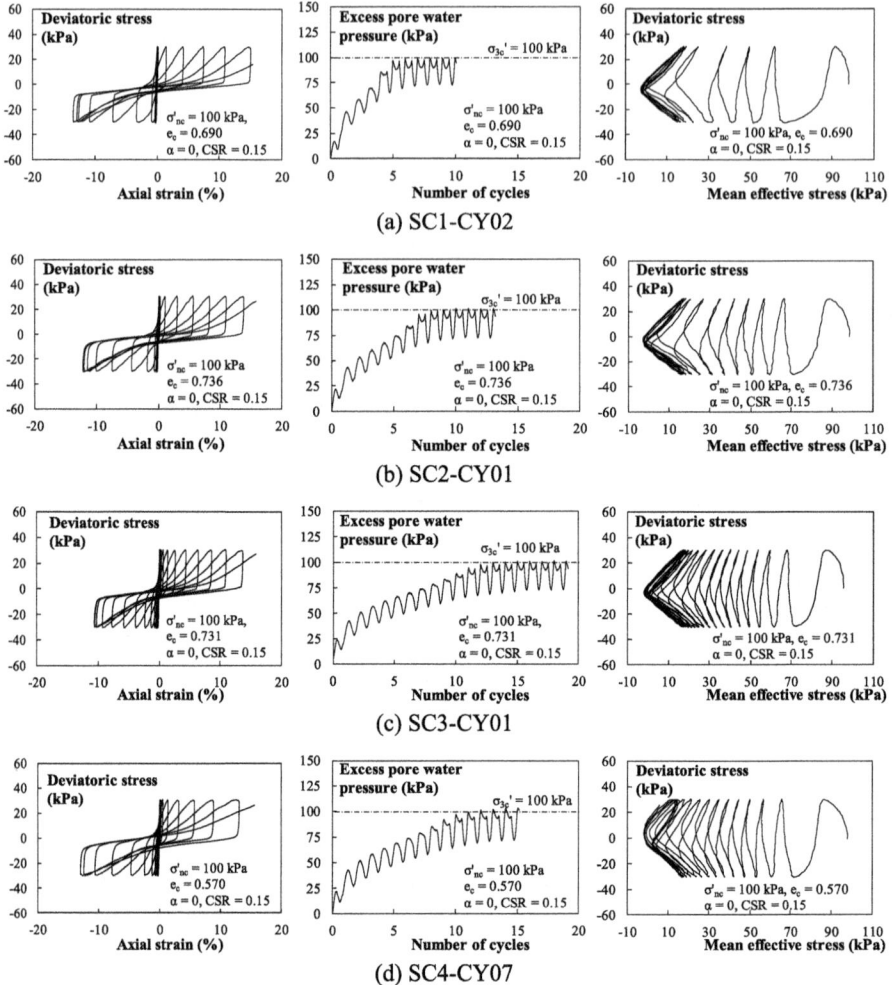

Fig. 3. Typical cyclic response of SC1–SC4 under $\alpha = 0$

increases cyclically and becomes steady at a value usually lower than σ'_{3c}. The effective stress does not reach zero as shown in the figure.

The behavior of SC4 under $\alpha = 0.1$ is similar to that of SC1 under $\alpha = 0.1$ (Fig. 5).

3.2   Cyclic Resistance

If the specimen exhibits cyclic mobility, the failure is defined when the DA of axial strain reaches 5%. If the specimens exhibit plastic strain accumulation, the failure is defined when the axial strain reaches 5%. The number of cycles to failure and the corresponding *CSR* are plotted in Fig. 6, for the four materials.

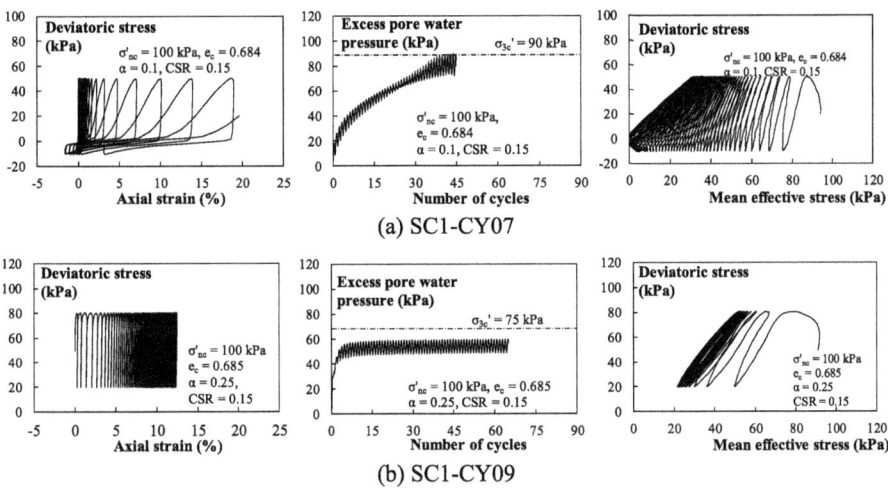

(a) SC1-CY07

(b) SC1-CY09

Fig. 4. Typical cyclic response of SC1 under $\alpha \neq 0$

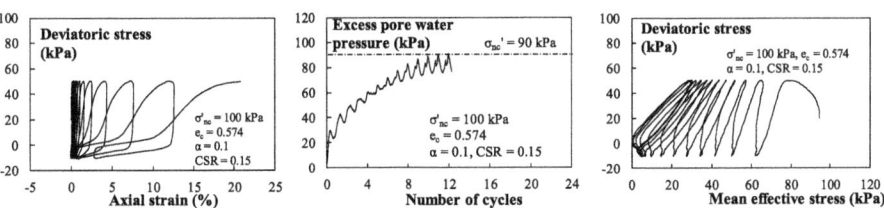

Fig. 5. Typical cyclic response of SC4 under $\alpha \neq 0$: SC4–CY13, $\alpha = 0.1$

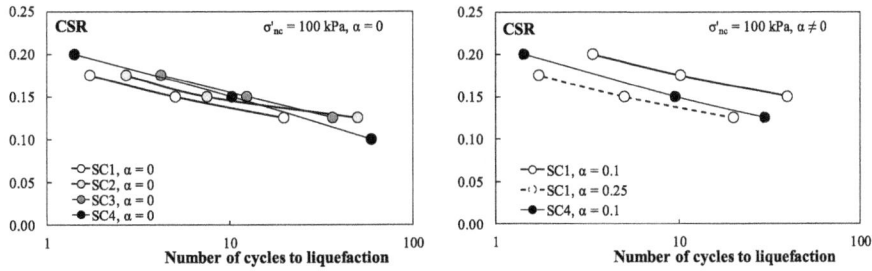

Fig. 6. CSR-N_l relationships of SC1–SC4 under $\alpha = 0$ (left); CSR-N_l relationship of SC1 and SC4 under $\alpha \neq 0$ (right)

4   Effects of Initial Static Shear Stress

The effects of α on the CRR of sands can be either beneficial or detrimental, depending on the initial state of the soils and the initial static shear stress levels (e.g. Yang and Sze [6, 7], Wei and Yang [5]). Yang and Sze [6] proposed the concept of threshold α (α_{th}). Under triaxial conditions, CRR firstly increases with increasing α until α_{th} is reached, whereas a further increase in α leads to decreased CRR. The threshold can be approximately estimated by the intersection of the CRR-α curve and the no-reversal line representing $CRR = \alpha$ (Fig. 7).

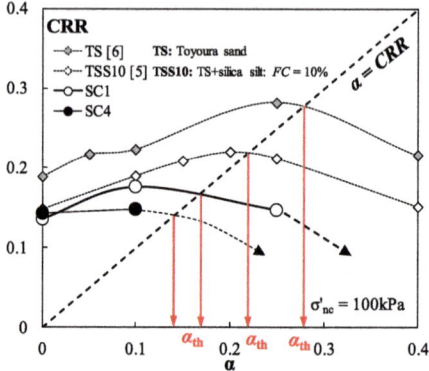

Fig. 7. Comparison of tested data of SC1 and SC4 with literature data

The CRR-α curves of the tested materials are presented in Fig. 7, showing that SC1 and SC4 exhibit similar behaviors with those of standard sands. When α increases from 0 to 0.1, CRR of both materials increases. But the increment of SC1 is larger than that of the SC4. This may be because SC1 contains fewer fines than SC4. When α increases from 0.1 to 0.25, the CRR-α curve of SC1 crosses the no-reversal line leading to a decrease of CRR. Further decrease of CRR may be anticipated if α increases from 0.25 to 0.4 (as shown by the dashed line in Fig. 7). For SC4, the further increase of α from 0.1 to 0.25 may lead to a decrease of CRR because the CRR-α will cross the no-reversal line.

5   Conclusions

The liquefaction behavior and cyclic resistance of four natural sands, which liquefied during the 2013 Lu-Shan earthquake, are investigated with consideration of the effects of initial static shear stress. The major findings are summarized as follows.

(1) When $\alpha = 0$, specimens of the four soils failed in the pattern of cyclic mobility. When $\alpha \neq 0$, the failure pattern depends on the reversal condition of the cyclic stress. When the cyclic stress is reversed, the specimen exhibits cyclic mobility, otherwise exhibits plastic strain accumulation.

(2) When the cyclic stress is reversed, *CRR* of SC1 and SC4 increases with increasing α. When the stress is not reversed, *CRR* decreases with increasing α. The no-reversal line may serve as a boundary separating the beneficial and the detrimental effects of α.

(3) The concept of threshold α is originally developed from test results of standard sand. The present investigation shows that it may be also applicable to natural sand. Future work may be promising to further extend this concept to other natural soils.

Acknowledgement. Financial support provided by the Research Grants Council of Hong Kong (No. 17250316) is gratefully acknowledged.

References

1. Xu, T.-W., Chen, G.-H., Xu, G.-H., Cheng, J., Tan, T.-B., Shu, A.-L., Wen, H.-Z.: Tectonic origin of the Lushan earthquake and its connection with the Wenchuan earthquake. Earth Sci. Front. **20**(3), 11–20 (2013). [in Chinese]
2. Hong, H., Xu, H., Song, F., Zhang, X., Li, G.: Discussion on seismo-geological hazards induced by 2013 Lushan MS 7.0 earthquake and its seismogenic fault. Acta Seismol. Sin. **35** (5), 738–748 (2013). [in Chinese]
3. Liu, Y., Huang, R.: Seismic liquefaction and related damage to structures during the 2013 Lushan Mw6.6 earthquake in China. Disaster Adv. **6**(10), 55–64 (2013)
4. Vaid, Y.P., Fisher, J.M., Kuerbis, R.H., et al.: Particle gradation and liquefaction. J. Geotech. Eng. **116**(4), 698–703 (1990)
5. Wei, X., Yang, J.: The effects of initial static shear stress on liquefaction resistance of silty sand. In: Proceedings of 6th International Conference on Earthquake Geotechnical Engineering, Reference No. 145. Christchurch, New Zealand (2015)
6. Yang, J., Sze, H.Y.: Cyclic behaviour and resistance of saturated sand under non-symmetrical loading conditions. Géotechnique **61**(1), 59–73 (2011)
7. Yang, J., Sze, H.Y.: Cyclic strength of sand under sustained shear stress. J. Geotech. Geoenvironmental Eng. **137**(12), 1275–1285 (2011)
8. Vaid, Y.P., Sivathayalan, S.: Static and cyclic liquefaction potential of Fraser Delta sand in simple shear and triaxial tests. Can. Geotech. J. **33**(2), 281–289 (1996)
9. Sze, H., Yang, J.: Failure modes of sand in undrained cyclic loading: impact of sample preparation. J. Geotech. Geoenvironmental Eng. **140**(1), 152–169 (2014)
10. Han, Z.-J., Ren, Z.-K., Wang, H., et al.: Records of the earthquake-induced hazards in Jian-Gan-Lin Village, Shang-Li Town, Yu-Cheng District, Ya-An City. Report published by Institute of Geology, China Earthquake Administration (2013). [in Chinese]
11. Zhang, Y., Dong, S., Hou, C., Guo, C., et al.: Geohazards Induced by the Lushan Ms7.0 Earthquake in Sichuan Province, Southwest China: Typical examples, types and distributional characteristics. Acta Geol. Sin. **87**(3), 646–657 (2013)
12. Yang, J.: Liquefaction resistance of sand in relation to P-wave velocity. Géotechnique **52**(4), 295–298 (2002)

Influence of Soil-Pile-Structure-Fluid Interaction on Seismic Behavior of a Liquid Storage Tank

Rulin Zhang[1]([⊠]), Zhiwei Zhang[1], and Huaifeng Wang[2]

[1] College of Pipeline and Civil Engineering,
China University of Petroleum (East China), Qingdao 266580, China
zhangrulin@upc.edu.cn
[2] School of Civil Engineering and Architecture,
Xiamen University of Technology, Xiamen 361000, China

Abstract. The seismic response of the tank-fluid system considering soil-structure interaction is studied. A three-dimensional (3-D) finite element method (FEM) model of soil-pile-structure-fluid interaction system is built based on ANSYS, in which, the structure and fluid interaction is considered by Lagrangian fluid FEM approximation. An artificial earthquake and Qian'an earthquake are selected as load inputs for the model. The results show that, after considering the interaction between soil and structure, the dynamic characteristics of the tank changes, and produce the whole vibration shape of soil-tank. There are differences in the spectral components of the two seismic waves. The prominent frequency of the artificial earthquake is closer to the frequency of liquid-tank coupled vibration, and causes a larger horizontal displacement of the tank wall, which shows obvious elephant-foot deformation near the bottom of the tank. The stress response of the tank wall under the excitation of two seismic waves is small. The conclusion may provide some reference for the seismic design of the tank considering the effect of soil-structure interaction.

Keywords: Soil-structure interaction · Seismic response
Elephant-foot deformation · Storage tank · Dynamic characteristics analysis

1 Introduction

Storage tanks are very important components, principally in water supply, nuclear plants, refineries and petrochemical facilities. The importance goes beyond its economic cost because the effects of a failure are not limited to the risk of human lives and equipment's in the proximity, but also can lead to serious consequences on the environment [1]. Some tanks were severely damaged and some failed with disastrous consequences, which revealed their vulnerability in previous earthquakes [2, 3]. Many storage tanks are located in areas of high seismicity, and easy to be threatened by earthquakes. So it is very important to study the seismic behavior of tanks and ensure the operational reliability of storage tanks when earthquakes occur.

Some works have been developed in an attempt to obtain the seismic performance of tanks [4–6]. However, most studies are only limited to the analysis of the influence

© Springer Nature Singapore Pte Ltd. 2018
T. Qiu et al. (Eds.): GSIC 2018, *Proceedings of GeoShanghai 2018 International Conference: Advances in Soil Dynamics and Foundation Engineering*, pp. 70–77, 2018.
https://doi.org/10.1007/978-981-13-0131-5_8

of the tank-liquid system, and the interaction effect between the soil and the tank is not considered, which is not consistent with the actual situation of tanks. In fact, the interaction between soil and structure may change the dynamic characteristics of the tank-fluid system, which will produce some impact to the dynamic response of the tank-fluid system. Therefore, in order to reflect the actual state of the tank as much as possible, it should establish the whole model of the tank including the foundation and soil. The problem of the interaction between soil and structure is also a hot topic in the field of civil engineering in these years. In this study, a 3-D model of soil-pile-tank-liquid system is established based on ANSYS, and the seismic response of the tank is discussed under the inputs of two kinds of seismic waves, which includes the displacements, stress of the tank wall, sloshing of liquid, the base shear, and so on.

2 The Dynamic Equation of the Soil - Tank - Liquid System

According to structural dynamic theory, with the fluid displacement as the unknown, the finite element scheme of the fluid can be established based on displacement. To carry out the calculation of standard finite element format, the fluid, the tank and the soil all adapt the calculating mode of displacement in this study. The standard finite element method based on the displacement is used for the fluid, the tank and the soil. In this way, the dynamic equation of the soil-tank-liquid system under earthquake loads can be obtained,

$$([m_t] + [m_l] + [m_s])\{\ddot{u}\} + ([c_t] + [c_l] + [c_s])\{\dot{u}\} + ([k_t] + [k_l] + [k_s])\{u\}$$
$$= ([m_t] + [m_l] + [m_s])\{a_g\} \tag{1}$$

where $\{\ddot{u}\}$, $\{\dot{u}\}$ and $\{u\}$ are the acceleration vector, velocity vector and displacement vector of nodes, $[m_t]$, $[m_l]$ and $[m_s]$ are the mass matrix of tank, fluid and soil, $[c_t]$, $[c_l]$ and $[c_s]$ are the damping matrix of tank, fluid and soil, $[k_t]$, $[k_l]$ and $[k_s]$ are the stiffness matrix of tank, fluid and soil, respectively, $\{a_g\}$ is the seismic acceleration.

In this study, Rayleigh damping is used for the damping matrix $[C]$,

$$[C] = \alpha[M] + \beta[K] \tag{2}$$

where α and β are the constant values of damping which are proportional to the mass and the stiffness, and they can be obtained by the two specific natural frequency ω_i and ω_j, and the damping ratio ξ_i and ξ_j,

$$\alpha = \frac{2(\xi_j\omega_i - \xi_i\omega_j)\omega_i\omega_j}{\omega_i^2 - \omega_j^2}, \quad \beta = \frac{2(\xi_i\omega_i - \xi_j\omega_j)}{\omega_i^2 - \omega_j^2} \tag{3}$$

3   Description of FEM Model for Soil - Tank - Liquid System

3.1   Geometry and Material Properties of the Tank

The FEM model of the tank in this study mainly includes tank, liquid, piles and soil. The tank uses the material of Q235. In this paper, the inner diameter is 16.8 m, the height is 14.3 m, and the total height is 14.3 m. The height of liquid surface is 10.5 m, and the fluid is almost full of tank. The tank wall is divided into four sections with changing thicknesses, which are 12 mm, 10 mm, 8 mm and 6 mm, respectively from the bottom to the roof of the tank. The thickness of tank vault is 6 mm. The material parameters of tank are as follows, elastic modulus is 206 GPa, Poisson's ratio is 0.3, the density is 7800 kg/m^3, the linear expansion coefficient is 1.0×10^{-5}. There is a reinforcing ring in the middle of the tank height, which is made of the angle bar L100 $\times$ 63 $\times$ 8. For the liquid, the density is 1000 kg/m^3, the elastic modulus is 2.04 GPa and the viscosity coefficient is 0.00113 Ns/m.

3.2   The Parameters of the Soil Site and Boundary Conditions

The tank is located in the soft soil site, and adapts the pile foundation, which is made of 32 reinforced concrete piles. For the piles, the diameter is 0.5 m, the elastic modulus is 32 GPa, the Poisson's ratio is 0.3, the density is 2100 kg/m^3, and the length of each pile is 20 m. The total thickness of the soil layer is 36 m, and the physical and mechanical parameters of all the soil layers are shown in Table 1.

Table 1.   The parameters of soil layers

Number	Soil name	Thickness/m	Density/(kg/m^3)	Shear wave velocity/(m/s)
1	Mucky silty clay	3.2	1810	128
2	Silty clay	5.4	1870	140
3	Mucky clay	10.5	1740	130
4	Silty clay	6.8	1820	310
5	Fine sand	10.1	1930	400

Due to the limitation of the scale of computing model, it can only select some soil in a limited length range for the interaction analysis between soil and structure. The calculation model of soil is usually composed of two areas, one is the core computing area for the dynamic response results, and the other is the extension area that is used to eliminate the influence of the remote artificial boundary. A simple lateral shift boundary can be used as a vertical remote artificial boundary. In order to eliminate the energy of the reflected seismic waves at the artificial boundary, the horizontal extension area of soil should be 5 times larger than the depth of the lateral soil layer outside the core calculation area for each side [7]. In this way, the energy of the reflected seismic waves will not affect the seismic response of the core calculation area. The total horizontal length of soil site is 10 times of the depth of soil layer, which reaches 180 m along the direction of horizontal length.

3.3 The FEM Model of the Tank

For the FEM model of the tank based on the ANSYS software, the tank wall, the bottom plate and the vault of the tank all use Shell181 element. Fluid80 element is used to simulate the incompressible fluid in the tank. The reinforcing ring adapts Beam188 element, and the mesh of the ring is coordinated with that of the tank wall in the same positions. The soil and the foundation use Solid45 element. All the concrete piles use Beam188 element. In order to simplify the calculation of model, it does not consider the contact effect between the soil and the piles. Based on the above parameters, the 3-D FEM model of soil - tank - liquid system is established as shown in Fig. 1.

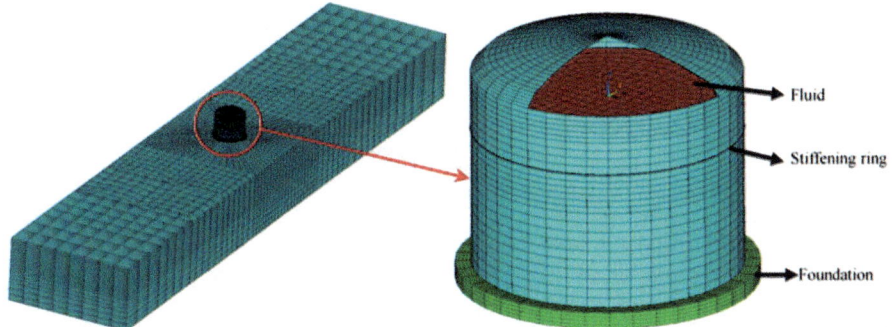

Fig. 1. The FEM model of soil-structure-fluid system

3.4 Selection of Seismic Waves for Dynamic Responses Analysis

To obtain the seismic response of soil-tank-liquid system, the seismic wave should be input from the bedrock. Therefore the input seismic waves should be earthquakes recorded at the bedrock. In this paper, two kinds of seismic waves are selected included a natural recorded earthquake and an artificial synthetic earthquake. The natural recorded earthquake is Qian'an earthquake of north-south direction, which is recorded in the Tangshan earthquake in 1976, and the peak acceleration of the seismic wave appears at about 5 s.

In order to save the calculation time, it only intercepts the first 10 s in the acceleration time-history wave. The artificial synthetic seismic wave is obtained by fitting the standard response spectrum on Class-I site in the Code for Seismic Design of Buildings (GB 50011–2010). According to the standard response spectrum on Class-I site (the characteristic period is 0.2 s) in the Code for Seismic Design of Buildings, the time-history acceleration wave of artificial ground motion can be obtained by Fourier transform, and the peak acceleration is adjusted to 0.4 g, and g is gravitational acceleration.

Usually, the bedrock ground motion is input from the top of the bedrock, and in the ANSYS software it is achieved by applying equivalent external excitation to the soil-tank-liquid system. The input acceleration time-history curves of Qian'an earthquake and the artificial earthquake are shown in Fig. 2 respectively.

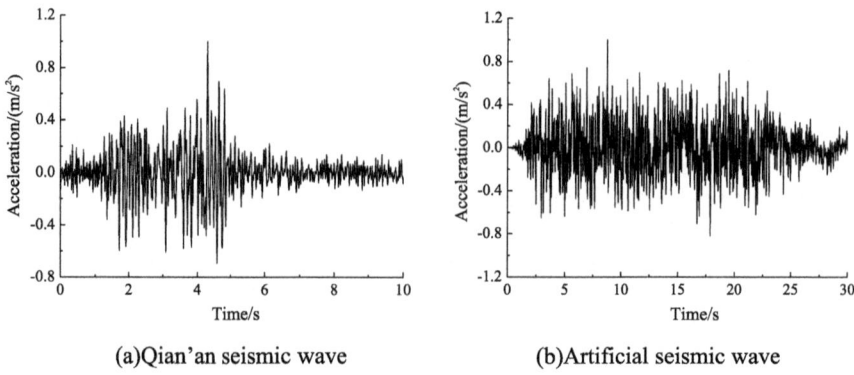

(a)Qian'an seismic wave   (b)Artificial seismic wave

Fig. 2. The acceleration time-history curves of the two input seismic waves

4   Calculation Results and Analysis

4.1   Results of Dynamic Characteristics

For the tank on the rigid foundation, the vibration mode shape is mainly sloshing of liquid and coupled vibration of tank-fluid. According to the author's previous study, if not consider the soil and structure interaction, the first-order frequency of liquid sloshing and tank-fluid coupled vibration frequency are respectively 0.228 Hz and 7.046 Hz. However, after considering the soil and structure interaction, the two corresponding frequencies are 0.226 Hz and 5.089 Hz, respectively. It can be seen that the frequency of the tank-fluid coupled vibration is reduced from 7.046 Hz to 5.089 Hz. It should be noted that, before the occur of tank-fluid coupled vibration frequency, the whole horizontal vibration mode frequency 1.549 Hz (along the x-axis direction) of the soil-tank-liquid system occurs, which is much smaller than the frequency 5.089 Hz of tank-fluid coupled vibration. In this study, the dynamic characteristics results of the pile-foundation model without the tank and liquid is also carried out. The overall horizontal vibration frequency of the model is 1.471 Hz (along the x-axis direction). In this case, the frequency of the whole structure is dominated by the vibration of the pile-foundation, so the fundamental horizontal vibration frequency changes from 1.471 Hz to 1.549 Hz.

Therefore, when the interaction between soil and structure is not considered, the vibration frequency of the whole model can be divided into two zones, one is the low-frequency zone including the liquid sloshing, and the other is the high-frequency zone including the coupled vibration of tank-fluid. However, after considering the interaction of soil and structure, the vibration frequency of the whole model changes into three zones, high, medium and low, i.e., the low-frequency zone of the liquid sloshing, the medium-frequency zone of whole vibration of the soil-tank-liquid system, high-frequency zone of the coupled vibration of tank-liquid. This is also consistent with the conclusion of the previous study [8].

4.2 The Stress Distribution Along the Height of the Tank Wall

Some characteristic points are selected for analysis, and these points are located on the section in the tank along the horizontal seismic wave. The distribution of the hoop stress and the axial stress along the height of the tank wall at the characteristic points are as shown in Fig. 3.

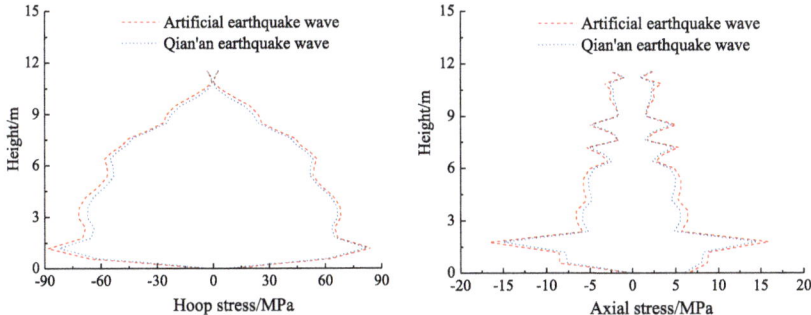

Fig. 3. The stress results distribution along the height of the tank wall

It can be seen from Fig. 3, the stress distributions under the two seismic waves are very close along the height of the tank wall, and the stress values under artificial seismic wave are bigger. The trend of the change of the axial stress and the hoop stress are almost the same. It indicates that the value firstly increases close to the bottom and then reduces near the top. The peak of hoop stress appears in the vicinity of the floor near the bottom plate (1.2 m from the bottom of the tank). Obviously, "elephant-foot" deformation occurs in this location. Note that the place 1.8 m from the bottom of the tank, it appears significant relative shrinkage deformation on the tank wall. The axial stress is larger near the bottom of the tank, while drastically reduces at the upper locations. Moreover, it does not change much at other positions, and the locations of peak axial stress are slightly higher, about 1.8 m from the bottom of the tank. The result of this change is also the reason of the sudden decrease in the horizontal displacement distribution at 1.8 m from the bottom of the tank. That is to say, the "elephant-foot" deformation is due to the smaller hoop tensile stress and the larger axial compressive stress.

4.3 Comparison of Peak Dynamic Response

In order to compare the influence of seismic waves on the seismic response of the tank, the peak results of the tank under the excitation of Qian'an seismic wave and the artificial seismic wave are shown in Table 2. In Table 2, SW-1 means Qian'an earthquake, SW-2 means the artificial earthquake, U_x means the horizontal displacement, S_h means the hoop stress, S_z means the axial stress, D_v means the vertical sloshing displacement of liquid, V means the base shear, M means the overturning moment.

From Table 2, the maximum horizontal displacement under the artificial earthquake is much larger than that under Qian'an earthquake. According to the Fourier spectrum of the earthquake and dynamic characteristics results, the predominant frequency of the artificial earthquake is much closer to the tank-liquid coupled vibration frequency than Qian'an earthquake, and the artificial earthquake also includes more earthquake energy contents near the predominant frequency, thus it leads to a stronger vibration of the tank-liquid system. In fact, the tank-liquid coupled vibration under earthquake loads is the most important factor which causes the horizontal deformation of the tank. The same reason can be explained for the change of base shear and overturning moment under the excitation of two seismic waves in Table 2. The sloshing deformation of the liquid is mainly affected by the convective sloshing vibration of the liquid. The energy of the artificial earthquake is more intensive than that of Qian'an earthquake, and the predominant frequency is closer to the first-order convective sloshing frequency of liquid, which is 0.226 Hz. Thus the sloshing displacement of liquid is much larger than that under Qian'an earthquake.

Table 2. Comparison of the maximum seismic response results

Seismic wave	U_x/mm	S_h/MPa	S_z/MPa	D_v/m	V/(kN)	M/(10^6N·m)
SW-1	7.21	82.22	14.92	0.005	3160.86	15.73
SW-2	19.99	88.15	16.60	0.515	4329.54	20.11

According to Code for Design of Vertical Cylindrical Welded Steel Oil Tanks (GB 50341–2014) [9], the critical stress σ_{cr} of the tank wall under earthquake loads can be calculated as follows,

$$[\sigma_{cr}] = \frac{0.22Et}{D} \tag{4}$$

where E is the elastic modulus of the tank wall, t is the effective thickness of the tank wall of the bottom circle, D is the inner diameter of the tank. From Eq. (4), the allowable critical stress of this tank is 28.77 MPa. It can be seen from Table 2 that, the maximum axial stress is 16.60 MPa under the excitation of two seismic waves, which does not exceed the allowable compressive stress specified in the above code.

From the above descriptions, the seismic response of the tank under the excitation of the artificial earthquake is increased compared with Qian'an earthquake. The horizontal sloshing displacement of the liquid caused by the artificial seismic wave is larger than that under Qian'an earthquake. The growth rate of the hoop stress, and axial stress response is small. Moreover, the base shear and overturning moment at the bottom of the tank become greater under the artificial earthquake, which will make the tank in a more adverse state. Therefore, it also should pay attention to the effect of the spectrum of seismic waves.

5   Conclusion

The current study mainly focuses on the effect of the interaction between soil and structure on the dynamic behavior of the tank. Based on the ANSYS software, the 3-D FEM model of soil-tank-liquid system is established considering the interaction between soil and structure in this study. Using Qian'an earthquake and an artificial earthquake as load inputs, the seismic response results of the tank are analyzed. After considering the soil and structure interaction, the dynamic characteristics of the soil-tank-liquid system is different from that of the tank-liquid system without soil. There exist three obvious frequencies, liquid convective sloshing frequency, the vibration frequency of the whole system and the coupled vibration frequency of tank-liquid. The "elephant-foot" deformation appears near the bottom of the tank, and it is especially much obvious under the excitation of artificial seismic wave. The maximum axial stress does not exceed the allowable stress specified in the code. The base shear and overturning moment resulted by the artificial earthquake is very large, which may make the tank in unfavorable state. The dynamic characteristics of the soil-tank-liquid system change due to the effect of the soil and structure interaction. It is suggested to consider the effect of spectrum characteristic of the seismic waves on the seismic response analysis of soil-tank-liquid system.

Acknowledgments. The research described in this paper was financially supported by the National Natural Science Foundation of China (Grant No. 51408609) and the Fundamental Research Funds for the Central Universities (Grant No. 18CX02081A).

References

1. Curadelli, O.: Equivalent linear stochastic seismic analysis of cylindrical base-isolated liquid storage tanks. J. Constr. Steel Res. **83**(2), 166–176 (2013)
2. Manos, G.C., Clough, R.W.: Tank damage during the May 1983 Coalinga earthquake. Earthquake Eng. Struct. Dynam. **13**(4), 449–466 (1985)
3. Sezen, H., Livaoglu, R., Dogangun, A.: Dynamic analysis and seismic performance evaluation of above-ground liquid-containing tanks. Eng. Struct. **30**(3), 794–803 (2008)
4. Goudarzi, M.A., Sabbagh-Yazdi, S.R.: Numerical investigation on accuracy of mass spring models for cylindrical tanks under seismic excitation. Int. J. Civ. Eng. **7**(3), 190–202 (2009)
5. Bayraktar, A., Sevim, B., Altunışık, A.C., et al.: Effect of the model updating on the earthquake behavior of steel storage tanks. J. Constr. Steel Res. **66**(3), 462–469 (2010)
6. Moslemi, M., Kianoush, M.R.: Parametric study on dynamic behavior of cylindrical ground-supported tanks. Eng. Struct. **42**, 214–230 (2012)
7. Lou, M.L., Pan, D.G., Fan, L.C.: Effect of vertical artificial boundary on seismic response of soil layer. J. Tongji Univ. **31**(7), 757–761 (2003)
8. Liu, S., Weng, D.G., Zhang, R.F.: Seismic response analysis of a large LNG storage tank considering pile-soil interaction in a soft site. J. Vib. Shock **33**(7), 24–30 (2014)
9. GB 50341–2014: Code for design of vertical cylindrical welded steel oil tanks. China Planning Press, Beijing (2014)

Influence of Soil Plugging on Dynamic Responses of Open-Ended Driven Pipe Pile

Chong Zhao, Weizhen Jiang, and Yong Tan[(✉)]

Tongji University, Shanghai, China
tanyong21th@tongji.edu.cn

Abstract. Dynamic response of open-ended pipe pile is complicated by soil-plugging effect. By a series of model tests, this focuses on soil-squeezing effect and characteristics of ground-borne vibration due to pile driving. During experimentally testing, both open-ended piles (control piles) and pipe piles with different fixed soil plug lengths and were driven into soil mass. During pile driving, both earth pressure and vibration acceleration at different pile penetration depths and radial distance from model piles were measured by earth pressure cells and accelerometers. Based on analyzing the instrumentation data, the dynamic responses of driven pipe piles were characterized and the effect of soil plugging on pile dynamic response was investigated. The relevant results will contribute some to the current knowledge of piling driving.

Keywords: Pipe pile · Soil plugging · Dynamic pile response
Earth pressure · Vibration wave

1 Introduction

With rapid urbanization in China in recent years, pile foundations have been widely adopted in practice. Because of their great bearing capacity, good quality and low price, precast piles are preferred by owner, designer and contractor. In many cases, the precast open-ended pipe piles are used, since their installation incurred small soil squeezing effect and encountered smaller penetration resistance. During installation of open-ended pipe pile, soil plug will be formed with soil being squeezed into the piles. Ground-borne vibration due to pile driving will impose adverse effect on environment. One case (Qi et al. 2006) showed that driving pipe piles caused permanent damage to an adjacent building within 40 m distance away. To safeguard environment, it is worthwhile to study effect of soil plugging on ground vibration resulting from driving open-ended pipe piles.

So far, many studies have been contributed to soil plugging. Based on laboratory tests, Paik et al. (2003) pointed out that the behavior of driven piles in sand could be divided into different stages. At initial stage, the inner space of pile was filled by soil plug completely. Then, as the pile went deeper, the ratio of the length of soil plug to pile driven depth was gradually reduced to a stable constant. Via model tests, Miller and Lutenegger (1997) indicated that soil plug in open-ended pipe pile was relevant to the way of pile driving, radial dimension of pile and the properties of soil.

© Springer Nature Singapore Pte Ltd. 2018
T. Qiu et al. (Eds.): GSIC 2018, *Proceedings of GeoShanghai 2018 International Conference: Advances in Soil Dynamics and Foundation Engineering*, pp. 78–85, 2018.
https://doi.org/10.1007/978-981-13-0131-5_9

To date, most of the studies in literature focused on the effect of soil plugging on pile bearing capacity and environmental (e.g., Tan and Lan 2012, Holeyman et al. 2013, Zheng et al. 2006, Zhou et al. 2010, Guo et al. 2013). However, influence of soil plugging on ground-borne vibration hardly received attention. Tao and Cui (2013) proposed a vibration equation for driving open-ended pipe pile, which can consider both soil plugging and squeezing effects; however, the reduction effect of blow energy during driving and the heterogeneity of stress distribution in soil plug were ignored. Therefore, its calculated results would not be accurate. Based on regression analysis of measured data, Wiss (1974) introduced a method focusing on blow energy and maximum soil velocity; however, the fitting of vibration fluctuation at short distance from pile was not considered.

In this study, the entire process of driving open-ended pile was simulated by physical modeling tests, in which dynamic responses of surrounding soil mass were recorded by earth pressure cells and accelerometers. Thus, influence of soil plugging on soil-squeezing behavior and vibration wave propagation can be examined in detail.

2 Test Model and Procedures

2.1 Instruments

The major experimental equipment was an equilateral model box with a dimension of 1 m × 1 m × 1 m. Wang and Sun (2004) proposed that a zone beyond ten pile diameters distance away from model pile can be regarded as an undisturbed zone. To minimize the boundary effect of model box, the model pile was designed to have external and internal diameters of 40 mm and 26 mm, and length of 600 mm, refer to Fig. 1.

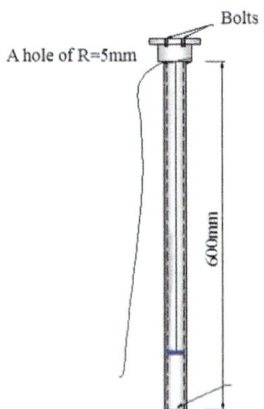

Fig. 1. Configuration of the model pile (Unit: mm)

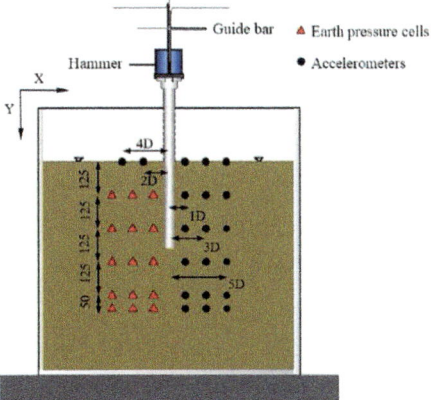

Fig. 2. Schematic diagram of the model box (Unit: mm)

Figure 2 illustrates the loading system of the model test, which consisted of a piercing hammer (weighs = 3.3 kg) and a guide rod. Constant pile driving energy was supplied by a free fall movement of the hammer (the hammer was elevated to a height of 400 mm). Earth pressure cells were installed inside the soil mass to monitor lateral earth pressures surrounding model pile. To prevent reflection of pile-driving induced waves at the boundary between soil mass and the model box, wave-absorbing material (20 mm thick pearl cotton) was affixed tightly to the inner surface of the box.

2.2 Test Procedure

To investigate the influence of soil plugging on lateral earth pressure and ground-borne acceleration during driving, two types of piles were tested for driving: (1) open-ended pipe pile; (2) open-ended pipe pile with different fixed soil plug lengths. During piling, both the earth pressure and ground-borne acceleration were recorded by sensors for analysis.

During driving open-ended model pile, as shown in Fig. 1, a movable shim was placed inside the pipe pile and it could be lifted during soil plugging; a string was fixed on the shim while its free end was left outside. Thus, soil plug length could be determined by referring to the location of the free end of the string. Earth pressure cells and accelerometers were instrumented in the sand at different depths during piling. Figures 3 and 4 were the overlook of the embedded sensors. Below the sand surface, sensors were kept at 1D (60 mm), 3D (140 mm) and 5D (220 mm) distance from the pile; the embedded depths (H) of sensors were 125 mm, 250 mm, 375 mm, 500 mm and 550 mm. Accelerometers were placed on the sand surface at a distance of 1D, 2D, 3D, 4D and 5D from the pile. Because of the limited sensors, it was impossible to place sensors at all the aforementioned locations in one test and then each type of testing was divided into four tests (Table 1). In each of the first three sub tests, earth pressure cells and accelerometers were embedded at a certain depth and kept the same distance to the pile. In the fourth sub test, five accelerometers were regularly placed on sand surface, as shown in Fig. 4. It was found that the radial vibration was more remarkable than those

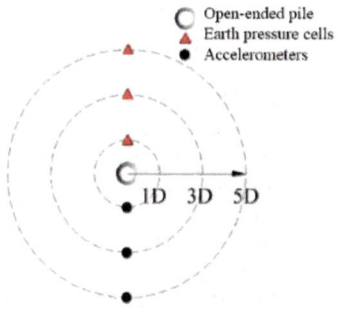

Fig. 3. Arrangement of sensors in soil

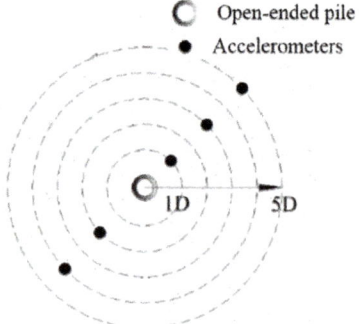

Fig. 4. Arrangement of acceleration sensor on soil surface

Table 1. Experimental items

Stage	Step	Position	Depth	Index
Open-ended piles	Test1	X = 1D	H = 125, 250, 375, 500, 550	Acceleration Earth pressure
	Test2	X = 3D		
	Test3	X = 5D		
	Test4	X = 1~5D	H = 0	Acceleration
Open-ended piles with fixed soil plug length	Test5	X = 1D	H = 125, 250, 375, 500, 550	Acceleration Earth pressure
	Test6	X = 3D		
	Test7	X = 5D		
	Test8	X = 1~5D	H = 0	Acceleration

in vertical and tangential direction and there existed little difference among the values of radial, tangential and vertical surface wave accelerations. Therefore, for the convenience of monitoring, the radial peak ground acceleration (RPGA) was selected as the representative of each vibration in the first three sub tests, and the vertical peak ground acceleration (VPGA) was selected as the representative of each vibration in the fourth sub test.

3   Testing Results of Open-Ended Pile

3.1   Squeezing Effect

Figure 5 shows the lateral earth pressure recorded at five different depths (i.e., H = 125 mm, 250 mm, 375 mm, 500 mm, and 550 mm) and three different lateral distance (i.e., X = 1D, 3D, and 5D). Following pile driving, the lateral earth pressure experienced a rapid increase to the peak, and then slowly decreased to a low value. The maximal value did not come at the moment of the pile reaching the depth of sensors but earlier, which can be called advance effect. During driving, the sandy soil beneath the pile tip was first crushed and then squeezed to the periphery of the pile. The earth pressure increased to the peak as a result of soil squeezing. Afterwards, the pile tip reached the depth of the sensors and the surrounding soil of pile tip was subjected to

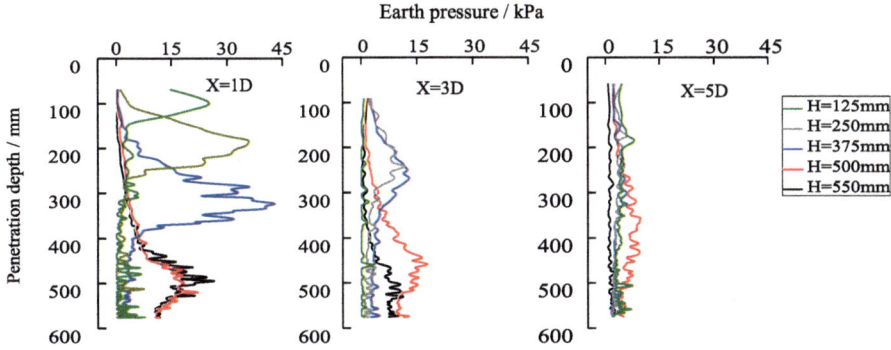

Fig. 5. Earth pressure around the pipe pile

downward frictional force along pile side, which counteracted partial lateral pressure. Then, the earth pressure gradually reduced to a steady value.

3.2 Dynamic Response Analysis

Figure 6 shows the radial peak ground accelerations throughout the pile penetration process, which were recorded at five different depths (i.e., H = 125 mm, 250 mm, 375 mm, 500 mm, and 550 mm) and three different lateral distances (i.e., X = 1D, 3D, and 5D) from pile. With the increase of penetration depth, the absolute value of the peak acceleration was increasing. This could be attributed to decreasing of penetration depth at the final stage of each driving, during which the main blow energy was transferred into vibration wave rather than kinetic energy for pile penetration.

The incremental filling ratio (IFR) is considered as an effective tool to quantify soil plugging effect, which can be calculated by Eq. (1) as below:

$$\text{IFR} = \frac{\Delta L_P}{\Delta L} \tag{1}$$

in which ΔL_P = unit increment of soil plug length; ΔL = unit increment of pile penetration depth.

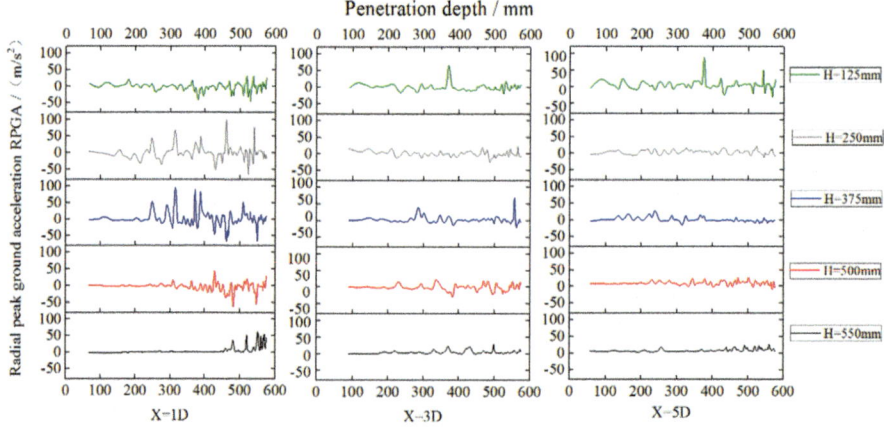

Fig. 6. Radial peak ground acceleration around the model pile

Figure 7 shows the IFR values versus the pile penetration depth for the four tests. The data of four sub tests were almost consistent with each other in tendency, which in turn verified the reproducibility and reliability of the model test. Generally, the development of IFR can be classified into two stages. In the initial stage (penetration depth less than 300 mm), the IFR values decreased substantially with obvious fluctuations. Soil plug inside the open-ended pile was mainly subjected to the upward extruding force of the sandy soil beneath the pile tip, and the friction force on the inner side of the pile was subordinate. In the second stage (penetration depth over 300 mm), with more sand entering into the pile, the shaft resistance on the inner pile surface increased, which resulted in low growth rate of soil plugging; thereby, the IFR value tended to be stable and had a slight decrease tendency. Because of the relatively loose sand and length limit of model pile, the pipe pile did not reach a fully plugged state and the IFR was not reduced to zero.

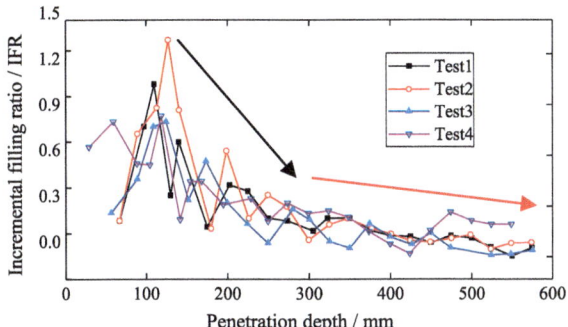

Fig. 7. Variation of IFR values with pile penetration depth

4   Testing Results of Fixed Soil Plug Lengths

4.1   Soil Squeezing Effect

Figure 8 illustrates the lateral earth pressure at a distance of 1D from pile during driving closed- and open-ended piles with different fixed soil plug lengths. A series of dash lines represented the peak values of earth pressures with different fixed soil plug lengths. Variations of earth pressures in the case of fixed soil plug lengths were consistent with each other. It indicated that the fixed soil plug lengths hardly affected the lateral earth pressure around the pile. It can be inferred that with the certain dimension of the model pile with fixed soil plug lengths, the peak values of earth pressures were mainly affected in the stage of soil plug formation. At this stage, each hammer blow produced relative displacement of soil plug and pipe pile. When soil plug was fully formed, no matter how much the actual length of soil plug was, no relative displacement happened after driving. Therefore, no obvious variation of peak earth pressure appeared after soil plug reached the limited height.

4.2   Dynamic Response Analysis

Figure 9 shows the variation of peak acceleration versus lateral distance when pile tip reached a depth of 200–300 mm in the case of the fixed soil plug length equivalent to 300 mm. The maximum value appeared at 2D rather than the nearest distance of 1D from the pile. The peak accelerations at 2D and 4D were obviously greater than those at 1D, 3D and 5D, causing a wave variation on attenuation curves. It showed that the

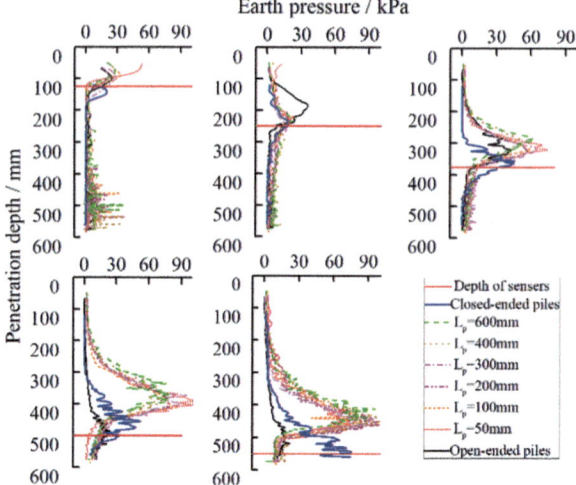

Fig. 8. Variation of earth pressure with penetration depth for different pile types

vibration responses of soil adjacent to the pile were not exponentially attenuated. Acceleration fluctuation was affected by wave crests and wave nodes, which was consistent with the theoretical results of Zhang et al. (1996) but inconsistent with the theory proposed by Wiss (1974); Wiss (1974) indicated that vibration waves exhibited exponential attenuation as the distance from the pile increased.

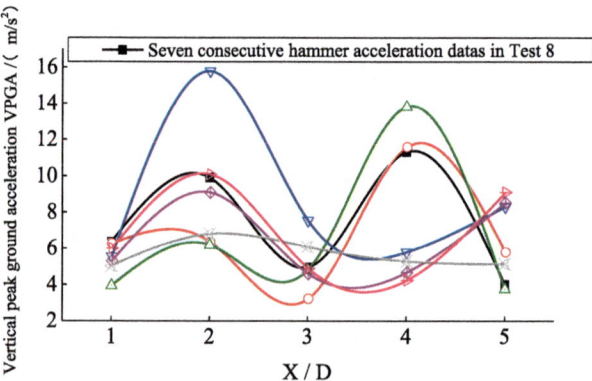

Fig. 9. Variation of peak accelerations with radial distances

5 Conclusion

Via physical model tests, characteristics of lateral earth pressure and ground-borne wave acceleration during driving open-ended pile into sand were investigated. Generally, the following major conclusions can be drawn:

(1) Variation of peak acceleration was closely related to the variation of IFR during piling. In the first stage of IFR curve, the main energy of pile driving contributed to pile penetration and formation of soil plug; when the IFR value stabilized, the driving energy was almost transformed into vibration wave.
(2) In the first stage of the model testing, the soil plug length had limited influence on lateral earth pressure when the soil plug has been formed. In the second stage, both lateral earth pressure and acceleration show no significant change with different fixed soil plug lengths.
(3) During pile driving, the peak accelerations at a distance of 1D, 3D and 5D from pile were greater than those at 2D and 4D. It indicates that the surface fluctuation did not conform to a simple exponential decay trend within 5D distance from pile. This was distinct from those empirical formula proposed in literature and deserves further study.

References

Qi, C.M., Xu, B.B., Xie, W.P.: Evaluation for surrounding vibration due to pilling. Soil Eng. Found **20**(5), 61–63 (2006)

Paik, K., Salgado, R., Lee, J., et al.: Behavior of open- and closed-ended piles driven into sands. J. Geotech. Env. Eng. **129**(4), 296–306 (2003)

Miller, G.A., Lutenegger, A.J.: Influence of pile plugging on skin friction in overconsolidated clay. J. Geotech. Geoenv. Eng. **123**(6), 525–533 (1997)

Tan, Y., Lan, H.: Vibration effects attributable to driving of PHC pipe piles. J. Perform. Constr. Facil. **26**(5), 679–690 (2012)

Holeyman, A., Bertin, R., Whenham, V.: Impedance of pile shafts under axial vibratory loads. Soil Dyn. Earthq. Eng. **44**(1), 115–126 (2013)

Zheng, J.J., Nie, C.J., Lu, Y.E.: Analytical solutions of cylindrical cavity expansion problems considering plugging effects. Chin. J. Rock Mech. Eng. **25**(S2), 4004–4008 (2006)

Zhou, J., Chen, X.L., Zhou, K.M.: model test and numerical simulation of driving process of open-ended jacked pipe piles. Chin. J. Rock Mech. Eng. **29**(s2), 3839–3846 (2010)

Guo, L.Q., Wang, L., Chen, F.Q.: Soil plugging effect of sleeves driven by high frequency hammers. J. Southwest Jiaotong Univ. **01**, 47–54 (2013)

Tao, G.L., Cui, J.H.: Improved wave equation calculation method considering effect of soil plug and its application. Water Resour. Power **07**, 96–99 (2013)

Wiss, J.F.: Vibration during construction operations. J. Constr. Div. **100**, 239–246 (1974)

Wang, Y.X., Sun, J.: Influence of pile driving on properties of soils around pile and pore water pressure. Chin. J. Rock Mech. Eng. **23**(1), 153–158 (2004)

Zhang, J.X., Su, H., Wu, G.Q.: Analysis of ground vibration caused by pile driving. Mech. Eng. **04**, 36–38 (1996)

Bi-objective Optimization of Site Investigation Program for Liquefaction Hazard Mapping

Mengfen Shen[1], Qiushi Chen[1(✉)], C. Hsein Juang[1], Wenping Gong[2],
and Xiaohui Tan[3]

[1] Clemson University, Clemson, SC 29634, USA
qiushi@clemson.edu
[2] China University of Geosciences, Wuhan 430074, Hubei, China
[3] Hefei University of Technology, Hefei 230009, Anhui, China

Abstract. The development of a liquefaction hazard map generally requires field data from site investigations. In this study, a bi-objective optimization framework is proposed for selecting an optimal site investigation program considering both the accuracy of the liquefaction hazard map and the site investigation efforts. To validate the proposed framework, a three-dimensional synthetic soil field with extremely detailed soil properties is generated and the corresponding liquefaction hazard map is used as the benchmark. Both regular and random sampling-based site investigation programs with varying site investigation efforts are considered and are used to infer input parameters of the subsequent random field-based liquefaction hazard mapping. It is found that the random field-based liquefaction hazard maps generally overestimate the hazard when validated against the benchmark liquefaction hazard map. When site investigation efforts (quantified by the number of sounding sites) are the same, regularly spaced site investigation programs yield more accurate hazard maps than those by random sampling-based investigation programs. An optimal site investigation program is recommended for the study site and the proposed framework can be applied to optimize site investigation of other sites.

Keywords: Optimization · Site investigation · Liquefaction hazard mapping
Random field

1 Introduction

Soil liquefaction is one of the major geohazards triggered by an earthquake. It can cause lateral spreading, ground settlement and sand boiling, which in turn may damage infrastructures and induce economic and life losses. An accurate liquefaction hazard map is of great importance for planning and prevention of such a geohazard.

To map liquefaction hazards over a region, geostatistical tools and random field model-based methods have been developed [1–5]. Random field model-based methods map liquefaction hazards as well as the associated uncertainties, which can be particularly useful for probabilistic analysis of liquefaction hazard. With the random field model-based methods, the sampling data obtained by a site investigation program serves as important inputs that provide statistical and spatial correlation information of

© Springer Nature Singapore Pte Ltd. 2018
T. Qiu et al. (Eds.): GSIC 2018, *Proceedings of GeoShanghai 2018 International Conference: Advances in Soil Dynamics and Foundation Engineering*, pp. 86–93, 2018.
https://doi.org/10.1007/978-981-13-0131-5_10

the mapped liquefaction hazards over the region of interest. A site investigation program with an insufficient amount of sampling data could lead to an inaccurate statistical and spatial characterization of the data, and thus may yield an inaccurate liquefaction hazard map, whereas, a site investigation program with too many sampling data may be cost-inefficient. Determining the optimal site investigation program thus requires the simultaneous consideration of two conflicting objectives, i.e., maximum the accuracy of the liquefaction hazard map while minimizing the cost of the investigation efforts.

The objective of this study is to develop a bi-objective optimization framework for selecting an optimal site investigation program through which both the accuracy of the liquefaction hazard mapping and the site investigation effort are simultaneously accounted for and optimized. Methodologies and numerical results are presented in detail in the following sections.

2 Methodology

2.1 Liquefaction Potential Index (LPI)

The liquefaction potential index (LPI) developed by Iwasaki [6] is adopted in this study to quantify liquefaction hazard at a location.

$$\text{LPI} = \int_0^{20} \omega(z) F_L = \int_0^{20} (10 - 0.5z) F_L \tag{1}$$

$$F_L = \begin{cases} 0 & \text{FS} \geq 1.2 \\ 1 - \text{FS} & \text{FS} \leq 0.95 \\ 2 \times 10^6 e^{-18.427\text{FS}} & 0.95 < \text{FS} < 1.2 \end{cases} \tag{2}$$

where z is the depth in meters and it is commonly evaluated for the top 20 m of a soil profile; FS is the factor of safety against liquefaction. In this study, a CPT-based liquefaction model by Robertson and Wride [7, 8] is used to calculate FS.

2.2 Random Field Model

The random field model is based on the recent work by Chen and co-workers [4, 5, 9–12], where the spatial variabilities of soil properties are characterized and simulated across scales. Here, we only consider single-scale spatial variability. The conditional random field model preserves the known data precisely and allows all subsequently generated data to be conditioned upon the known information. The distribution of the unknown data points to be simulated is given as:

$$(Z_n | Z_p = z) \sim N \left(\Sigma_{np} \cdot \Sigma_{pp}^{-1} \cdot z, \sigma_n^2 - \Sigma_{np} \cdot \Sigma_{pp}^{-1} \cdot \Sigma_{pn} \right) \tag{3}$$

where $Z_p = z$ is a vector of previously defined or simulated data; Z_n is the next point to be simulated; $N(\cdot)$ denotes the vector of random variables following a normal distribution; σ_n^2 is the prior variance of the next simulated point; Σ_{np}, Σ_{pn} and Σ_{pp} are the covariance matrices and the subscripts 'n' means next point to be simulated, and 'p' represents previously simulated points.

In this work, the spatial structure is described using a form of covariance known as the semivariogram $\gamma(\boldsymbol{h})$, which is equal to half the variance of two variables separated by a vector distance $\boldsymbol{h}$

$$\gamma(\boldsymbol{h}) = \mathrm{Var}[Z(\boldsymbol{u}) - Z(\boldsymbol{u} + \boldsymbol{h})] \tag{4}$$

where $Z(\boldsymbol{u})$ and $Z(\boldsymbol{u} + \boldsymbol{h})$ are the values of the variable under consideration at locations $\boldsymbol{u}$ and $\boldsymbol{u} + \boldsymbol{h}$, respectively.

In practice, various analytical semivariogram models and their linear combinations are typically fitted to the empirical semivariogram that is calculated from field data. Among which, the exponential model is widely used and thus adopted in this study:

$$\gamma(h) = 1 - \exp(-h/a) \tag{5}$$

where h is a scalar form of the vector distance that can account for both separation distance and geometric anisotropy [14]; a is the range parameter and $3a$ is the practical range, i.e., the distance at which the exponential semivariogram levels off [13].

2.3 Procedure for Optimizing Site Investigation Programs

The procedure for optimizing site investigation programs is summarized as follows.

Step 1: Obtain the benchmark LPI field. To evaluate the accuracy of the liquefaction hazard mapping, a spatially correlated clean sand equivalent tip resistance $(q_{c1N})_{cs}$ field is first generated as shown in Fig. 1(a). The dimension of this field is 1000 m × 1000 m × 20 m, which has 4,010,000 soil elements. Given this synthetic $(q_{c1N})_{cs}$ field, LPI values for a given earthquake shaking scenario (e.g., peak ground acceleration $a_{\max} = 0.3$ g, moment magnitude $M_w = 7.0$) can be obtained at every location, which is denoted as the "true LPI field" in the following analysis. This true LPI field will serve as the benchmark to validate the accuracy of the random field-based LPI fields based on different investigation programs.

Step 2: Design site investigation programs. Two types of site investigation programs, regular and random sampling, are studied in this study. For the regular sampling, the number of sampling data in x and y directions (N_x and N_y) are the same and the data is equally spaced in both directions. For the random sampling, the sampling data is randomly taken from the synthetic digital field. In this study, the site investigation effort (SIE) is defined as the total number of sampling data.

Step 3: Map of LPI fields using random field models. With the site investigation program, sampling data can be "taken" from the synthetic digital $(q_{c1N})_{cs}$ field. Then, the corresponding LPI values can be calculated. These LPI values will be treated as known data to infer random field model parameters and the subsequent random field realizations can be performed. LPI is assumed to follow a lognormal distribution and is transformed to normal distribution during the conditional sequential random field simulation process [14]. The transformation has previously been shown to preserve the prescribed spatial structure for lognormally distributed

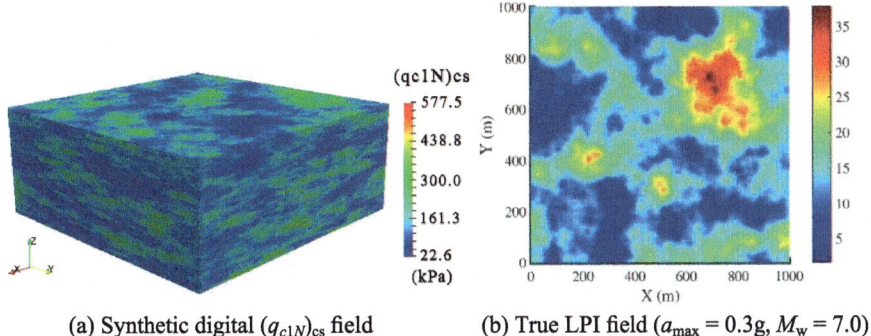

(a) Synthetic digital $(q_{c1N})_{cs}$ field   (b) True LPI field ($a_{max} = 0.3g$, $M_w = 7.0$)

Fig. 1. Synthetic digital soil field & benchmark liquefaction potential field [5, 10].

variables [14]. A total of 1,000 Monte Carlo simulations (MCSs) are performed for each site investigation program and the average of the simulated LPI random fields will be validated against the "true LPI field".

Step 4: Evaluate the accuracy of the simulated LPI random field. Three frequently used evaluation criteria are adopted in this study: the bias factor criterion, the mean absolute percentage error (MAPE) criterion, and the root mean square deviation (RMSD) criterion.

$$\text{bias factor} = \frac{1}{n}\sum_{i=1}^{n}\frac{(\text{LPI}_{\text{sim}})_i}{(\text{LPI}_{\text{true}})_i} \tag{6}$$

$$\text{MAPE} = \frac{1}{n}\sum_{i=1}^{n}\left|\frac{(\text{LPI}_{\text{true}})_i - (\text{LPI}_{\text{sim}})_i}{(\text{LPI}_{\text{true}})_i}\right| \tag{7}$$

$$\text{RMSD} = \sqrt{\frac{1}{n}\sum_{i=1}^{n}\left[(\text{LPI}_{\text{true}})_i - (\text{LPI}_{\text{sim}})_i\right]^2} \tag{8}$$

where n is the number of data, LPI_{true} is the LPI value of true LPI field, and LPI_{sim} is the simulated or predicted LPI value by the random field-based method. The simulation or prediction overestimates LPI if the bias factor is greater than 1, and the smaller MAPE and RMSD is preferred, which means a more accurate prediction.

Step 5: Conduct bi-objective optimization. As the objectives with respect to the accuracy of the LPI random field map and the site investigation effort conflict with each other, a single best solution can be found on a Pareto front, showing a trade-off between the conflicting objectives. Then the knee point, which represents the best compromise solution among all non-dominated designs, can be found on the Pareto front [15–18]. The non-dominated sorting genetic algorithm version II (NSGA-II) developed by Deb et al. (2002) [15] is used in this study to obtain the Pareto front, and the minimum distance approach is adopted for the knee point [16].

3 Numerical Results

3.1 Investigation Programs for Liquefaction Hazard Mapping

In the following analysis, the regular sampling is designed as $N_x = N_y \in \{6, 7, 8... 19, 20\}$, and the corresponding site investigation efforts are $\text{SIE}_{\text{regular}} = N_x = N_y = \{36, 49, 64... 361, 400\}$. A total of 15 site investigation programs are considered. For random sampling, the site investigation effort $\text{SIE}_{\text{random}}$ is distributed in the range of 36 to 400 for a total of 120 site investigation programs.

To illustrate the regular and random sampling, one particular site investigation program with a total of 36 samples is shown in Fig. 2(a) and (b) for regular and random sampling, respectively. It is noted that while the 36 locations are randomly generated, only one realization is utilized. The random effect in generating sampling location is not considered in the current study.

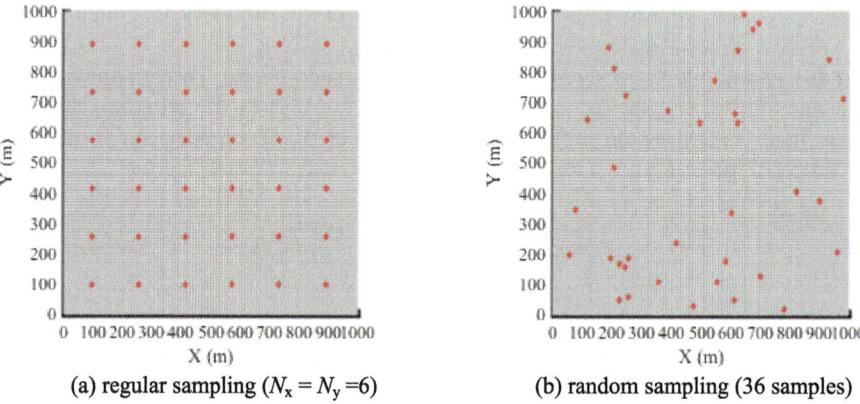

(a) regular sampling ($N_x = N_y = 6$)   (b) random sampling (36 samples)

Fig. 2. Illustrative site investigation programs of 36 samples (red dots are sampling data).

3.2 Accuracy Evaluation of Simulated LPI Random Field

By comparing the simulated LPI random fields to the true LPI field, the accuracy of the liquefaction hazard mapping is evaluated and the results are shown in Fig. 3. Figure 3(a) shows the range parameter a vs. the site investigation effort SIE. The range parameter of the true LPI field is $a_x = a_y = 114.63$ m. It is observed that under the same site investigation effort, the range parameter of simulated LPI random field by regular sampling is closer to the true range parameter, and has less variability than those obtained by random sampling. Figure 3(b) shows that the bias factors of all the simulated LPI random fields are greater than 1, which means they overestimate the LPI field. Generally, the simulated LPI random fields by regular sampling are more accurate and have less variability in accuracy than the ones by random sampling.

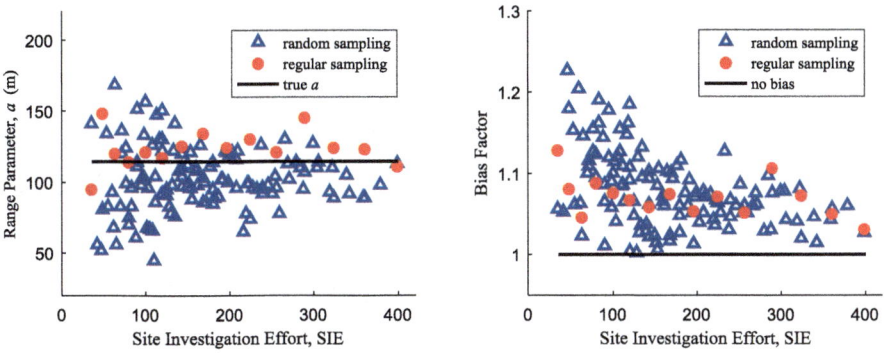

(a) range parameter vs. site investigation efforts   (b) bias factor vs. site investigation efforts

Fig. 3. Accuracy evaluations of simulated LPI random fields.

3.3 Optimal Site Investigation Program

The objective of this study is to seek an optimal site investigation program with respect to the accuracy of liquefaction hazard mapping (i.e., MAPE and RMSD) and the site investigation effort (SIE). As shown in Fig. 4, under the same site investigation effort, the simulated LPI random fields by regular sampling investigation program are more accurate than those by random sampling. To achieve the same simulation accuracy, the regular sampling requires less site investigation effort than the random sampling. The optimal site investigation program indicated by knee point is the regular sampling investigation program with $N_x = N_y = 11$. It is noted that the selected optimal investigation program is specific for the study site and further study will be needed for sites with different dimensions. The optimal site investigation program and corresponding simulated LPI random field (averaged from 1,000 MCS) are shown in Fig. 5(a) and (b), respectively. Based upon a comparison between Figs. 5(b) and 1(b), it is observed that

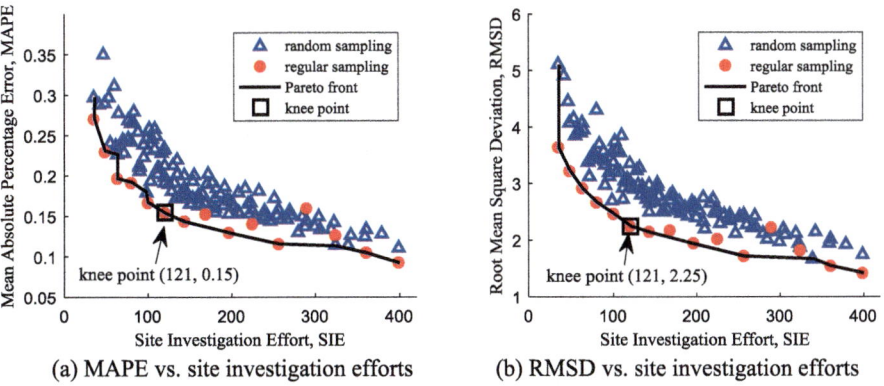

(a) MAPE vs. site investigation efforts   (b) RMSD vs. site investigation efforts

Fig. 4. Bi-objective optimization of site investigation programs

the simulated LPI random field can generally capture the liquefaction hazard "pattern" of the field, and the MAPE and RMSD are 0.15 and 2.25, respectively.

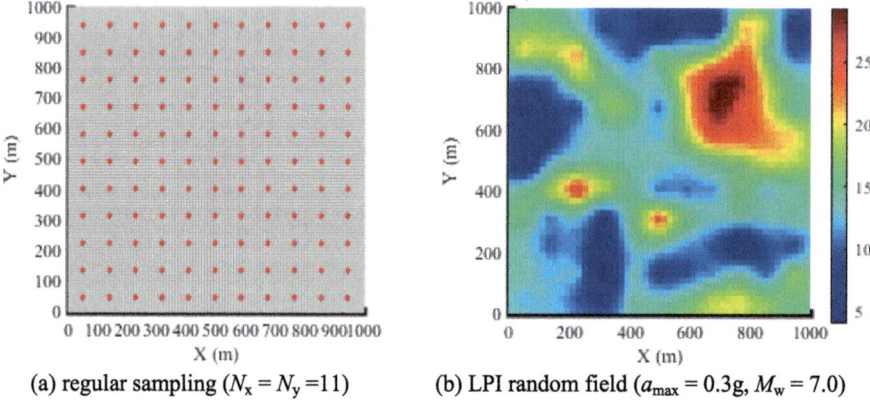

(a) regular sampling ($N_x = N_y = 11$)   (b) LPI random field ($a_{max} = 0.3g$, $M_w = 7.0$)

Fig. 5. Optimal site investigation program and corresponding LPI random field.

4   Conclusions

This paper proposes a framework for selecting an optimal site investigation program that considers both the accuracy of the liquefaction hazard mapping and the site investigation effort. It is found that the random field-based liquefaction hazard maps generally overestimate the hazard when validated against the benchmark liquefaction hazard map. When site investigation efforts (quantified by the number of sounding sites) are the same, regularly spaced site investigation programs yield more accurate hazard maps than those by random sampling-based investigation programs. For the particular site (1000 m × 1000 m) considered in this study, the optimal investigation program is a regular spaced site investigation program with 121 sampling locations. The proposed framework can be applied to other sites for selecting the optimal site investigation program.

Acknowledgements. The authors would like to acknowledge the financial support provided by the U.S. Geological Survey (Grant No. G17AP00044). Clemson University is acknowledged for the generous allotment of computer time in the Palmetto high-performance computing facility.

References

1. Baise, L.G., Higgins, R.B., Brankman, C.M.: Liquefaction hazard mapping—statistical and spatial characterization of susceptible units. J. Geotech. Geoenv. Eng. **132**(6), 705–715 (2006)
2. Lenz, J.A., Baise, L.G.: Spatial variability of liquefaction potential in regional mapping using CPT and SPT data. Soil Dyn. Earthq. Eng. **27**(7), 690–702 (2007)

3. Thompson, E.M., Baise, L.G., Kayen, R.E.: Spatial correlation of shear-wave velocity in the San Francisco Bay Area sediments. Soil Dyn. Earthq. Eng. **27**(2), 144–152 (2007)
4. Chen, Q., Wang, C., Juang, C.H.: CPT-based evaluation of liquefaction potential accounting for soil spatial variability at multiple scales. J. Geotech. Geoenv. Eng. **142**(2), 04015077 (2015)
5. Juang, C.H., Shen, M., Wang, C., Chen, Q.: Random field-based regional liquefaction hazard mapping—data inference and model verification using a synthetic digital soil field. Bull. Eng. Geol. Env., 1–14 (2017)
6. Iwasaki, T.: A practical method for assessing soil liquefaction potential based on case studies at various sites in Japan. In: Proceedings of Second International Conference on Microzonation Safer Construction Research Application, vol. 2, pp. 885–896 (1978)
7. Robertson, P.K., Wride, C.E.: Evaluating cyclic liquefaction potential using the cone penetration test. Can. Geotech. J. **35**(3), 442–459 (1998)
8. Robertson, P.K.: Performance-based earthquake design using the CPT. In: Proceedings of IS-Tokyo, pp. 3–20 (2009)
9. Chen, Q., Wang, C., Juang, C.H.: Probabilistic and spatial assessment of liquefaction-induced settlements through multiscale random field models. Eng. Geol. **211**, 135–149 (2016)
10. Chen, Q., Shen, M., Wang, C., Juang, C.H.: Verification of random field-based liquefaction mapping using a synthetic digital soil field. In: Geotechnical Frontiers 2017, pp. 236–245 (2017)
11. Wang, C., Chen, Q., Shen, M., Juang, C.H.: On the spatial variability of CPT-based geotechnical parameters for regional liquefaction evaluation. Soil Dyn. Earthq. Eng. **95**, 153–166 (2017)
12. Wang, C., Chen, Q.: A hybrid geotechnical and geological data-based framework for multiscale regional liquefaction hazard mapping. Géotechnique, 1–12 (2017)
13. Goovaerts, P.: Geostatistics for Natural Resources Evaluation. Oxford University Press, New York (1997)
14. Chen, Q., Seifried, A., Andrade, J.E., Baker, J.W.: Characterization of random fields and their impact on the mechanics of geosystems at multiple scales. Int. J. Numer. Anal. Meth. Geomech. **36**(2), 140–165 (2012)
15. Deb, K., Pratap, A., Agarwal, S., Meyarivan, T.: A fast and elitist multiobjective genetic algorithm: NSGA-II. IEEE Trans. Evol. Comput. **6**(2), 182–197 (2002)
16. Khoshnevisan, S., Gong, W., Wang, L., Juang, C.H.: Robust design in geotechnical engineering–an update. Georisk Assess. Manag. Risk Eng. Syst. Geohazards **8**(4), 217–234 (2014)
17. Gong, W., Luo, Z., Juang, C.H., Huang, H., Zhang, J., Wang, L.: Optimization of site exploration program for improved prediction of tunneling-induced ground settlement in clays. Comput. Geotech. **56**, 69–79 (2014)
18. Gong, W., Tien, Y.M., Juang, C.H., Martin, J.R., Zhang, J.: Calibration of empirical models considering model fidelity and model robustness—focusing on predictions of liquefaction-induced settlements. Eng. Geol. **203**, 168–177 (2016)

Longitudinal Nonlinear Seismic Response of Shield Tunnel Based on Generalized Response Deformation Method

Miao Yu$^{(\boxtimes)}$ and Ruan Bin

School of Civil Engineering and Mechanics,
Huazhong University of Science and Technology, Wuhan 430074, China
miaoyu@hust.edu.cn

Abstract. The Sanyang Yangtze River Shield Tunnel in Wuhan is selected for the research. The precise beam spring model of the shield tunnel is established, and considering the nonlinearity of soil, the longitudinal seismic response analysis of the shield tunnel is carried out based on the generalized response displacement method. The maximum opening width at ring intersegment and the internal force distribution of the pipe ring along the longitudinal direction are presented with different intensities and frequency spectra of earthquake motions. The results show that the uneven distribution of the terrain and the soil layer has an important influence on the opening value and the internal force of the pipe rings. Due to that the bedrock with larger stiffness is set and the constraint of silty-fine sand with relatively small stiffness to tunnel is smaller, a sudden change of the opening value is obvious at the intersection of bedrock and silty-fine sand.

Keywords: Shield tunnel · Generalized response deformation method
Longitudinal seismic response · Opening value · Nonlinear of soil

1 Introduction

The scale of construction of tunnels has developed rapidly, and many of these tunnels are located in strong earthquake areas in China. In the event of strong earthquakes, the safe operation of tunnels will be seriously threatened and the damage of tunnels will bring huge economic losses and casualties. Therefore, it is necessary to study the seismic response characteristics of underground tunnels under strong earthquakes. This research is based on the project of crossing-river shield tunnel built in Sanyang road in Wuhan, China. The Sanyang tunnel is the main river crossing route connecting Hankou district to Wuchang district.

The tunnel longitudinal analysis methods consist of the response displacement method based on the beam-spring model and three-dimensional finite element method by solid element (or shell structure). The former can describe the macroscopic characteristics of the tunnel as a whole, but the uncertainty of the foundation spring stiffness is large; the latter can accurately represent the meso-reaction of the structural components, but the modeling is complex and computationally large. In the past earthquake

© Springer Nature Singapore Pte Ltd. 2018
T. Qiu et al. (Eds.): GSIC 2018, *Proceedings of GeoShanghai 2018 International Conference: Advances in Soil Dynamics and Foundation Engineering*, pp. 94–102, 2018.
https://doi.org/10.1007/978-981-13-0131-5_11

history, it is generally believed that the underground structure is constrained by the surrounding soil and the vibration amplitude is smaller than the ground structure, so it is more difficult to be affected by the earthquake disaster. The seismic design analysis of the tunnel structure is much less concerned than the ground structure. The above subjective understanding can be controversial due to lack of physical evidence and difficulty in investigation.

About 25 tunnels have been destroyed in the event of the 1923 Kanto [1]. Four tunnels on the South Pacific were destroyed in the 1952 Kern County earthquake [2]. In the event of a major earthquake that occurred in recent two decades, there are some underground structures that are severely damaged(e.g., the 1995 Kobe earthquake; the 1999 Kocaeli earthquake; the 1999 Chi-Chi earthquake; the 2008 Wenchuan earthquake). In the 2004 Niigata earthquake in Japan, dozens of tunnels were severely damaged by stratigraphic crusts, some of which were closer to the epicenter, and the damage was more severe and needed to be repaired after the earthquake [3].

Through the study of the seismic response of the immersed tunnel, it is found that the damage of the tunnel depends more on the domination of the surrounding soil [4]. Further, the quasi-static method is used to analyze the seismic response of the underground structure. Among them, there are many methods: seismic coefficient method, reaction displacement method, reaction acceleration method and pushover method [5–7]. The response displacement method has been incorporated into the Chinese national norms [8]. More recently, with the increase of computer computing ability, more scholars have studied the seismic response characteristics of tunnels through the establishment of three-dimensional model [9–11].

Based on ABAQUS software secondary development platform, a numerical simulation model of large-scale two-dimensional free-field of Sanyang Road river crossing shield tunnel with unit partition, viscoelastic artificial boundary setting, different ground motion input methods and modified soil dynamic constitutive model is constructed. Then, a fine beam - spring structure model for the longitudinal seismic response analysis of shield tunnel based on generalized response displacement method is established. This paper emphasizes the influence of different intensity and different hypocentral distance on the maximum opening width at ring intersegment.

2 Selection of Bedrock Ground Motions

A total of more than 600,000 sets of three-component ground motion observations were accumulated between June 1996 and August 2016 in the KiK-net network by all 697 sites. This research emphasizes the seismic effect of tunnel structure under strong earthquake. Therefore, from these observation records, ground motions selection was conducted in accordance with following three conditions: 1. from the drilling depth of 100 m or more; 2. ground acceleration 100 gal or more; 3. magnitude $6.5 \sim 7.5$ M. A total of 59 horizontal earthquake records correspond to the requirements are grouped according to the hypocentral distance. There are 21 records possessing the hypocentral distance from 0 to 60 km, 31 records from 60 to 120 km, and 7 records from 120 to 180 km. The Fourier amplitudes were averaged for each group of seismic records, and the record with the smallest variance is found as a typical ground motion.

Figure 1 shows the Fourier spectra and time-history curves corresponding to the selected seismic records. The gray lines denote the Fourier spectra of all records for each group and black solid line denotes the average in Fig. 1(a). The red line is the record according to the hypocentral distance. Figure 1(b) shows the horizontal and vertical acceleration time-histories. The selected three records are named E1, E2 and E3 from near-field earthquake to far-field earthquake. The bedrock surface takes two-way input, and the horizontal amplitude of the horizontal seismic velocity is 1: 0.65. The peak value of the actual ground record is adjusted to 0.2 g, 0.3 g and 0.4 g correspond to the small earthquake (SE), the middle earthquake (ME) and the high earthquake (HE), respectively.

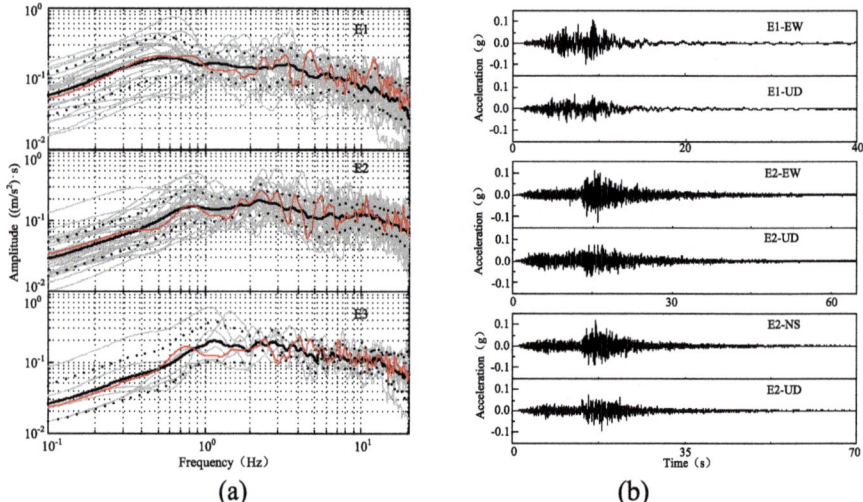

(a)   (b)

Fig. 1. Fourier spectrum and acceleration time-histories

3 Numerical Model of Free-Field Site

The free-field f the Sanyang shield tunnel with two-dimensional (2D) cross-section along the tunnel longitudinal is shown in Fig. 2. The deposits can be roughly divided into three layers: Banket, Silty clay and Silty-fine sand, and the total thickness of the deposits is about 90 m. The deposits is mixed with silty soil and lens-like soil and the red solid line is the location of the tunnel at the site.

A large number of experimental studies have shown that the three-parameter Davidenkov constitutive model can describe the nonlinear dynamic characteristics of various soils ideally. Thus, the modified Davidenkov viscoelastic dynamic constitutive model is used to describe the mechanical behavior of soil under cyclic loading based on ABAQUS [12]. The normalized shear modulus and the damping ratio curves obtained from the resonant column test for typical undisturbed soil samples of the Sanyang shield tunnel site are shown in Fig. 3.

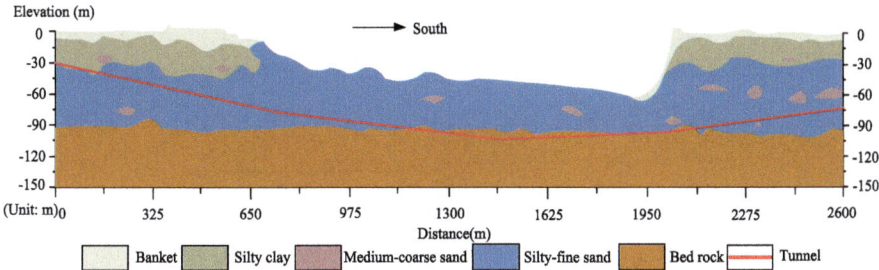

Fig. 2. The cross-section along the longitudinal axis of the Sanyang Shield Tunnel

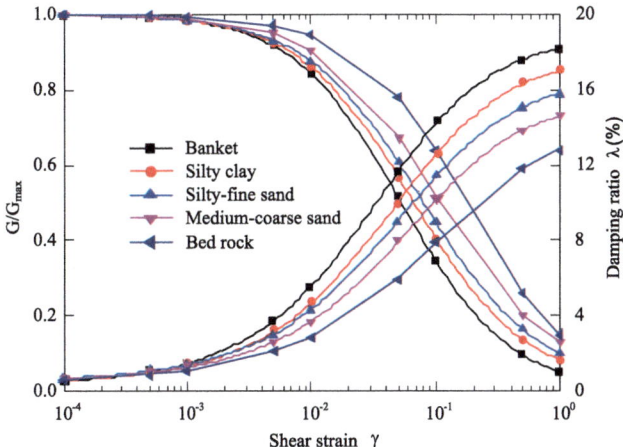

Fig. 3. G/Gmax $\sim \gamma$ and $\lambda \sim \gamma$ curves of the soil

The model is 2.6 km long horizontally and about 150 m high, the number of soil elements is large Therefore, the four-node bilinear interpolation plane strain reduction integral unit (CPE4R) is used to proceed simulation. In the grid division, the change of the depth of the wavelength along the soil depth is taken into account and the maximum size of the mesh (h_{max}) is 1/8 to 1/10 of the corresponding wavelength [13]. The mesh is dominated by quadrilateral elements, and partial positions are triangular elements for transition. The boundary is processed by setting a series of simple physical elements composed of linear springs and dampers along the boundary to absorb the scattering energy and reflected waves from the artificial boundary, so as to achieve the transmission process of the simulated wave to exit the artificial boundary [14]. Schematic diagram of working boundary setting is shown in Fig. 4.

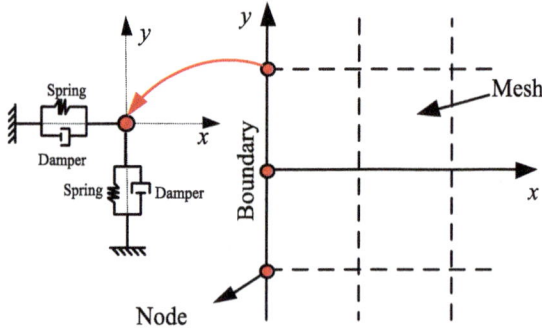

Fig. 4. Schematic diagram of working boundary setting

4   Generalized Response Displacement Method for Shield Tunnel

Based on the results of the nonlinear seismic response analysis of the 2D free-field in the preceding section, the displacement time-history response of the intermediate point of the segments is extracted at the tunnel position and a series of displacement time-history responses are then input to the foundation spring fixed end [15]. The combination of two horizontal components with the same peak along the longitudinal axis of the tunnel and in the transverse direction gives the calculation results conservatively [16]. Therefore, the vertical and horizontal displacement time-history responses of the input are consistent. Each segment of the beam-spring model has a ring with width of 2 m, and soil springs are set in the middle of the segment. The model takes one unit per 1 m. The segment rings are linear elastic Timoshenko's beams that allow for transverse shear deformation, which are modeled by selecting of hollow three-dimensional (3D) linear B31 beam element type in the code ABAQUS. The horizontal shear strain of the cross-section is constant, that is, the cross section after the deformation is still flat and not distorted. The outer diameter of the cross section of the segment ring is 15.2 m and the inner diameter is 13.9 m. The whole model is 2600 m in length. Considering the influence of the longitudinal joints of the tunnel gap, the beam elements are connected by a tension and compression spring and a torsion spring (a total of 2598 joint springs including tension and compression springs and torsion springs). Because of the difference in the properties of the soil around the tunnel, the stiffness of the soil in the four directions around the tunnel is considered for each section of the segment ring and the model has 5200 soil springs [17].

The main structural parameters of shield tunnel are shown in Table 1. When the segment ring is pulled, the tensile force of the joint is provided by the bolt. The axial tensile stiffness of the joint is regarded as the sum of the stiffness of each joint bolt. When the segment ring is pressed, only the pipe section is pressed and the joint bolt is no longer forced. The axial compressive stiffness of the joint can be regarded as the compression stiffness of the pipe section. When the joints are bent, the compressive stress is borne by the connecting bolts in the tension zone, and the compressive stress is borne by the pipe section concrete alone in the compression zone. The stress of the concrete is always in the elastic state. The section deformation conforms to the flat section and the small deformation assumption. Figure 5 shows the constitutive schematic of the tension and compression spring and torsion spring.

Table 1. Structure parameters of shield tunnel

Elastic modulus of C60 concrete E_c(MPa)	Elastic modulus of C60 concrete E_c(MPa)	Diameters of bolt d_0(mm)	Length of bolt l(mm)	Bolt tensile yield stress σ_y(MPa)	Bolt tensile ultimate stress σ_m(MPa)	Joint bolt M42 numbers N
2.06E5	3.6E4	36	750	640	800	42

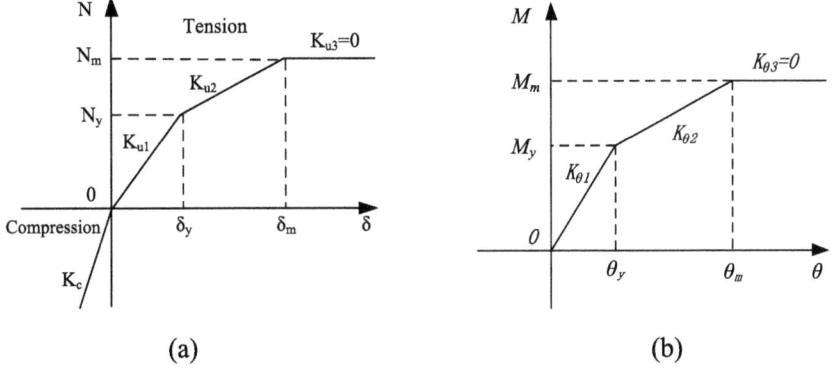

(a)   (b)

Fig. 5. The constitutive schematic of joint springs: (a) tension and compression spring (b) torsion spring

5 Longitudinal Seismic Response Characteristics of Shield Tunnel

Segment rings longitudinal gapping is one of the most common diseases during shield tunnel operation, which is the main cause of leakage. Therefore, it is very important to study the change of the opening amount of the crack under the response of the earthquake, which has fundamental significance for the operation and maintenance of the shield tunnel. This section gives maximum opening width at ring intersegment along the longitudinal direction of the tunnel, that is, the maximum opening value of the two ends of the adjacent two segments.

Figure 6, 7 and 8 show the maximum opening width at ring intersegment under different ground motions (E1, E2 and E3). It can be seen from the figures that the trends of the maximum opening width along the longitudinal direction of the tunnel are similar. The opening width at the joint of the tunnel into the silt and bedrock has twice sudden change, respectively, at 1400 and at 2000 m. It is noteworthy that the difference of opening width with a variety of frequency characteristics is larger, and the maximum opening width under E3 wave having medium-high frequency content is obviously smaller than that under E1 wave that possesses low-medium frequency content under

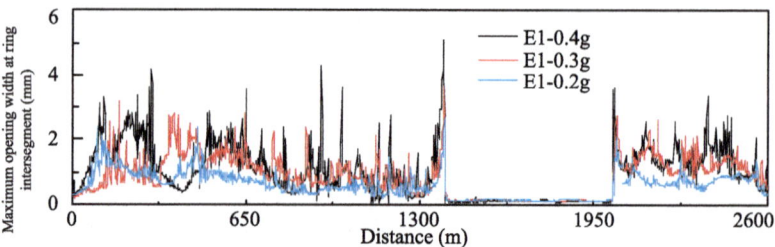

Fig. 6. Maximum opening width at ring intersegment under E1

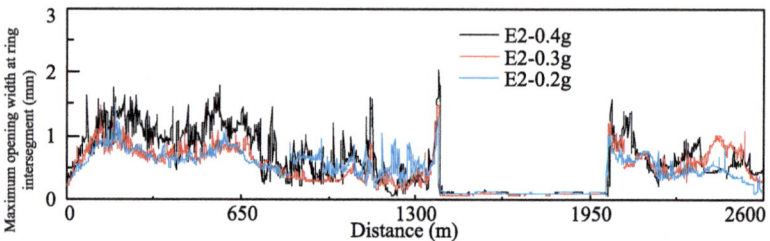

Fig. 7. Maximum opening width at ring intersegment under E2

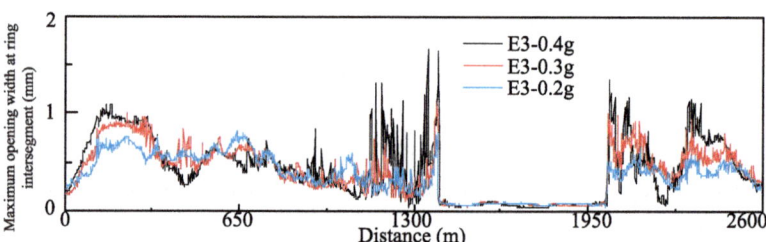

Fig. 8. Maximum opening width at ring intersegment under E3

the same ground motion intensity. It is also obvious that with the increase of ground motion intensity, the opening width value also increases, but not with the linear growth of ground motion intensity, which also shows the influence of site space heterogeneity on the effect of the field amplification. In the northern section of the tunnel, in some positions are also large, which may be caused by the large response of the soft soil to the long-term earthquakes.

6 Conclusions

This study takes the river crossing shield tunnel built in Sanyang road, Wuhan city as the engineering background, and the integral longitudinal seismic response of river tunnel is analyzed based on generalized response displacement method. The opening width at the ring intersegment is adopted as an representing index and the following conclusions were reached:

1. The effect of the nonlinear of soil on the response of the tunnel has a very large coupling with the increase of the ground motion intensity. The maximum value of the opening width varies with different intensities and frequency characteristics of earthquake. The result under input motions with significant low-frequency content is more conservative.
2. Due to limited number of earthquake records, although the maximum opening width at ring intersegment does not reach the allowable limit of opening amount of the general waterproof measures under all earthquake records, engineering measures should be taken at the positions where large opening width occurs.
3. The remarkable large opening width regions were located at the northern section of the tunnel or the transition zone from soft soil to bed rock layer and the maximum value of opening width located in the transition position.

References

1. Kawashima, K.: Seismic design of underground structures in soft ground: a review. In: Proceedings of the International Symposium on Tunneling in Difficult Ground Conditions, vol. 60, Tokyo, Japan, pp. 3–21 (1999)
2. Stein, R.S., Thatcher, W.: Seismic and aseismic deformation associated with the 1952 Kern County, California, earthquake and relationship to the Quaternary history of the White Wolf fault. J. Geophys. Res. Solid Earth 86(B6), 4913–4928 (1981)
3. Konagai, K., Takatsu, S., Kanai, T., et al.: Kizawa tunnel cracked on 23 October 2004 Mid-Niigata earthquake: an example of earthquake-induced damage to tunnels in active-folding zones. Soil Dyn. Earthq. Eng. 29(2), 394–403 (2009)
4. Okamoto, S., Tamura, C., Kato, K., et al.: Behaviors of submerged tunnels during earthquakes. In: Proceedings of the Fifth World Conference on Earthquake Engineering, vol. 1, Rome, Italy, pp. 544–553 (1973)
5. Wang, J.N.: Seismic Design of Tunnels: A Simple State-of-the-Art Design Approach. Parsons Brinckerhoff, New York (1993)
6. Hashash, Y.M.A., Hook, J.J., Schmidt, B., et al.: Seismic design and analysis of underground structures. Tunn. Undergr. Space Technol. 16(4), 247–293 (2001)
7. Liu, J.B., Liu, X.Q., Li, B.: A pushover method for seismic response analysis and design of underground structures. China Civ. Eng. J. 41(4), 73–80 (2008). (in Chinese)
8. The National Standards Compilation Group of People's Republic of China.: Code for seismic design of buildings (CNS GB50011-2010). China Architecture and Building Press, Beijing, China (2010). (in Chinese)
9. Do, N.A., Dias, D., Oreste, P., et al.: Three-dimensional numerical simulation of a mechanized twin tunnels in soft ground. Tunn. Undergr. Space Technol. 42, 40–51 (2014)
10. Peng, C., Wu, W., Zhang, B.: Three-dimensional simulations of tensile cracks in geomaterials by coupling meshless and finite element method. Int. J. Numer. Anal. Meth. Geomech. 39(2), 135–154 (2015)
11. Zhang, Z.X., Liu, C., Huang, X., et al.: Three-dimensional finite-element analysis on ground responses during twin-tunnel construction using the URUP method. Tunn. Undergr. Space Technol. 58, 133–146 (2016)

12. Zhao, D.F., Ruan, B., Chen, G.X., et al.: Validation of the modified irregular loading-reloading rules based on Davidenkov skeleton curve and its equivalent strain algorithm implemented in ABAQUS. Chin. J. Geotech. Eng. (online, 2017). (in Chinese)
13. Liu, J.B., Liao, Z.P.: Elastic wave motion in discrete grids (II): Comparison of common finite element models. Earthq. Eng. Eng. Vib. **9**(2), 1–11 (1989). (in Chinese)
14. Zang, X.L., Li, X.J., Chen, G.X., et al.: An improved method of the calculation of equivalent nodal forces in viscous-elastic artificial boundary. Chin. J. Theor. Appl. Mech. **48**(5), 1126–1135 (2016). (in Chinese)
15. Geng, P., He, C., Yan, Q.X.: The current situation and prospect of seismic analysis methods for tunnel structure. China Civ. Eng. J. **46**(S1), 262–268 (2013). (in Chinese)
16. Anastasopoulos, I., Gerolymos, N., Drosos, V., et al.: Nonlinear response of deep immersed tunnel to strong seismic shaking. J. Geotech. Geoenviron. Eng., ASCE **133**(9), 1067–1090 (2007)
17. The National Standards Compilation Group of People's Republic of China: Code for seismic design of urban rail transit structures (CNS GB50909-2014). China Architecture and Building Press, Beijing, China (2014) (in Chinese)

Seismic Responses of the Large-Scale Deep Shaft in Shanghai Soft Soils

Zhiyi Chen[1,2] and Bu Zhang[1(✉)]

[1] Department of Geotechnical Engineering, Tongji University,
Shanghai 20092, China
{zhiyichen, zhangbu}@tongji.edu.cn
[2] Key Laboratory of Geotechnical and Underground Engineering
of Ministry of Education, Shanghai 200092, China

Abstract. In recent years many researchers focus on the seismic responses of underground structures mainly about the lined tunnels and the subway stations as well as the other framed structures. However, few efforts were put on those of the vertical shafts. In this paper, the main purpose is to study the seismic response of the large-scale deep vertical shaft. Three-dimensional dynamic time history analysis was carried out to explore seismic responses of the vertical shaft in shanghai soft soils. Shaft deformation (such as circumferential deformation and lateral deformation) and inner forces (hoop force and vertical force) were then studied. Results reveal that the shaft behaves like a rigid body in the earthquake event since the lateral deformation is inclined. The shaft hoop force and deformation are influenced remarkably by the shaft floor which makes the hoop force decrease and the hoop bending moment increase from a third of the depth distance to the shaft floor.

Keywords: Shaft · Underground structure · Seismic response
Inner force

1 Introduction

In recent years, underground structures suffer different degrees of destruction during the earthquake events. Especially after the collapse of Daikai subway station in the 1995 Kobe earthquake, more and more researchers show their interests on the seismic performance of underground structures. However, most of the researches are mainly focus on the lined tunnels and the framed subway stations, and there are few studies about the vertical type underground structures such as the vertical shafts. In the construction of modern sponge cities, vertical shafts play an important role in collecting combined sewage and storm flow into the deep drainage tunnel. Therefore, it is essential to study the seismic responses and dynamic characteristics of the vertical shaft, which are consequently the main purpose of the presented paper.

At present, some scholars have studied the seismic resistance of shafts. Ando et al. [1] and Sawada et al. [2] studied the large-circular shaft with 20 m diameter, 40 m depth and 0.3 m thickness built in sand and soft soil by shaking table test and numerical simulation. The distribution of vertical displacement and vertical and circumferential

© Springer Nature Singapore Pte Ltd. 2018
T. Qiu et al. (Eds.): GSIC 2018, *Proceedings of GeoShanghai 2018 International Conference: Advances in Soil Dynamics and Foundation Engineering*, pp. 103–111, 2018.
https://doi.org/10.1007/978-981-13-0131-5_12

stress distribution is analyzed. Shiba [3] not only proposed a modified simplified seismic calculation method but also compared it with numerical simulation results, and the results show that the modified simplified seismic calculation method works well. Mayoral et al. [4] conducted a series of dynamic time-historical analysis about shafts with different size which are situated in the very soft clay which has high plasticity in Mexico, and the variation of vertical displacement and the inner force along the depth is obtained.

Considering the large diameter and depth of the shafts as well as the important role in construction of the Shanghai sponge city, this paper aims to study the dynamic response characteristics of a large depth vertical underground structure in shanghai soft soil by implementing the three-dimensional dynamic time history analysis. This paper mainly studied the seismic performance of the shaft, including vertical and circumferential deformation as well as vertical and hoop force characteristics.

2   Three Dimensional Dynamic Time History Analysis

2.1   Modelling and Material Parameters

The three dimensional dynamic time history analysis of the shaft is carried out by ABAQUS [5] Standard with implicit Newmark-β method, the widely used finite element software. The horizontal dimension of the model is 400 m $\times$ 400 m, which is more than five times than shaft diameter [6]. The bottom of the model is placed at 120 m away from the surface, which is corresponded to the top of the rigid bedrock according to the follow-up soil information. The diameter of the shaft is 30 m, and the depth is 60 m. The thickness of the shaft wall is 3 m. The thickness of the shaft baseboard is 4 m. Figure 1 shows the discretization model used for shaft dynamic numerical analyses.

In this model, the Shanghai soft soil deposits is divided into four layers for simplicity based on the previous research [7]. For the shear wave velocity at 120 m depth is more than 300 m/s, the depth of the bedrock is assumed at 120 m. The soil seismic response is implemented by an elastic model. The shaft is assumed to exhibit an elastic behavior throughout the entire analysis. Table 1 summarizes the soil deposits and shaft parameters.

The soil material damping is introduced by Rayleigh damping with 5% damping ratio, which is defined in Eq. (1):

$$\zeta_i = \frac{\alpha_R}{2w_i} + \frac{\beta_R w_i}{2} \tag{1}$$

where ζ_i is the damping ratio($\zeta_i = 5\%$), w_i is the circular frequency of the model i, α_R and β_R are Rayleigh damping coefficients proportional to the mass and stiffness. For this purpose, the first and second natural frequency of the free field are computed by implementing the modal analysis, then the mass damping coefficient $\alpha_R = 0.264743$ and stiff damping coefficient $\beta_R = 0.007795$ are obtained according the above equation.

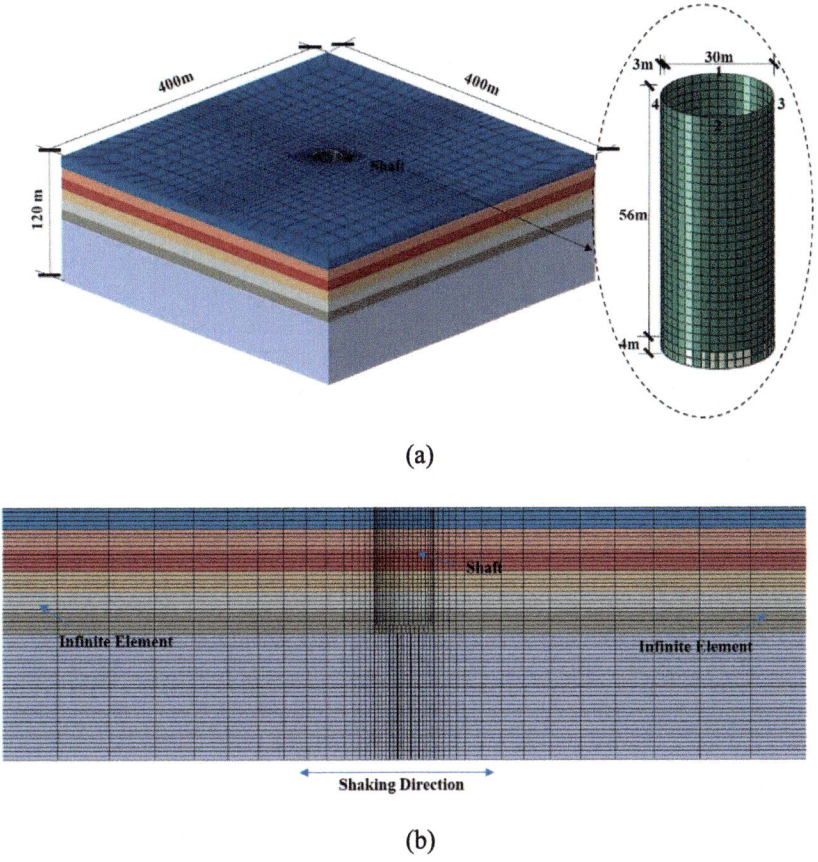

(a)

(b)

Fig. 1. The discretization model used for shaft dynamic numerical analyses: (a) 3D view of the whole model and the shaft; (b) Side view from the profile (Transverse cross-section).

Table 1. Soil deposits and shaft concrete parameters.

Soil layer	Type	Depth (m)	Density (kg/m³)	Shear wave velocity (m/s)	Poisson's ratio	Shear modulus G(Pa)	Elasticity modulus E(Pa)
1	Clay	20	1750	150	0.3	3.9E + 07	1.01E + 08
2	Sand	20	1800	200	0.3	7.2E + 07	1.87E + 08
3	Sand	20	1850	250	0.3	1.16E + 08	3.02E + 08
4	Sand	60	1900	300	0.3	1.71E + 08	4.45E + 08
–	Concrete	–	2500	–	0.2	–	3.45E + 10

Eight-node continuum solid element C3D8R is selected for the soils and shaft baseboard. The shaft lining is defined as four-node shell element S4R. All these elements are provided and validated by ABAUQS element library [5]. Considering the grid and time discretization is important for the accuracy of the dynamic finite element simulations, the maximum element size in the direction of wave propagation should not exceed 1/10 of the wave length, so a maximum grid length 2 m is used. To avoid the wave front reaching two consecutive element in the same computational moment, the maximum time increment is limited in no more than 0.01 s [8].

2.2 Boundary Conditions

In the dynamic time history analysis, the ground motions imposed at the bottom of the discretization model in horizontal direction, and the displacement of vertical direction is constrained. In order to weaken the effect of the boundary reflection in dynamic analysis, the boundary conditions on the model sides are taken as infinite boundaries [9, 10]. The boundary at the top of the analytical model is free [11]. In most cases, the interface condition between underground structures and soils is between the full-slip and no-slip. In this model, the interfaces between the shaft and the soils are in no-slip conditions by setting "tie" constraint, which is a simplification of reality and may overestimate the responses of the shaft.

2.3 Ground Motion

Considering the uncertainty and variability of ground motion, the main focus of this paper is on the seismic response of the large-scale deep shaft. In this model, the Shanghai artificial seismic motion is introduced at the model base as horizontal acceleration, and the maximum horizontal acceleration is 0.1 g. Figure 2 shows the information about the Shanghai artificial seismic motion.

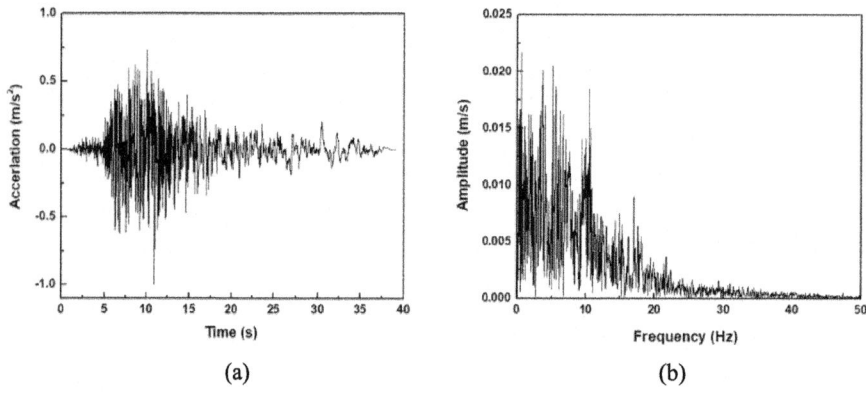

(a)   (b)

Fig. 2. (a) Input motion time history; (b) Amplification of base acceleration with different frequencies of Shanghai artificial seismic motion.

2.4 Model Validation

In order to validate the three-dimensional numerical model, this model was tested by comparing the previous results by Sawada et al. [2]. They have done a small shaking table test and a series of numerical simulations about the shaft joints. The model plane size is 70 cm long and 70 cm wide. The depth of this model is 45 cm. This model contains two soil layers. The thickness of the upper soil is 25 cm. The upper soil is modeled as an elastic material with unit weight of 9.7kN/m³, Young's modulus of 291kN/m², Poisson's ratio of 0.4 and the damping ratio of 4%. The thickness of the lower soil is 20 cm. The lower soil is also modeled as an elastic material with unit weight of 9.8kN/m³, Young's modulus of 541kN/m², Poisson's ratio of 0.4 and the damping ratio of 4%. The diameter of the shaft is 10 cm, and the depth is 40 cm. The thickness of the floor is 5 mm, and the thickness of the wall is 3 mm. The shaft is modeled as an elastic material with unit weight of 11kN/m³, Young's modulus of 3070kN/m², Poisson's ratio of 0.49 and the damping ratio of 4%. The input ground motion in this model is the sine wave with frequency of 6 Hz and peak acceleration of 0.5 m/s². The comparison results of the maximum vertical axis stress and circumferential stress along the shaft depth are shown in Fig. 3. The results of this model were in good agreement with the results of Sawada, and revealed the validity of this model.

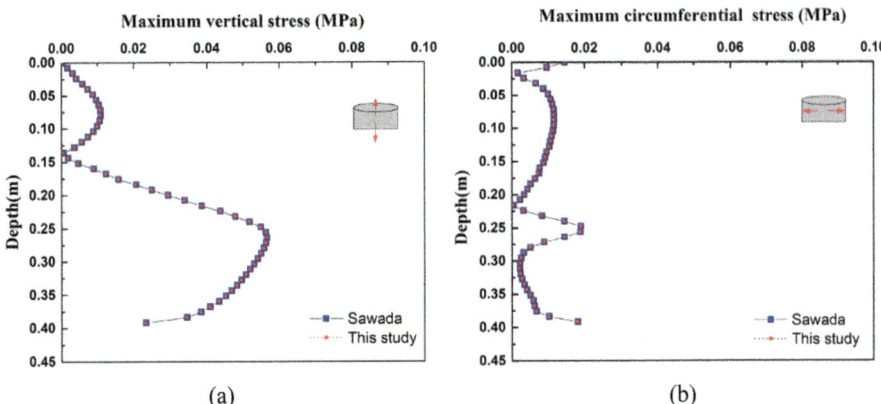

Fig. 3. (a) Vertical axis stress; (b) Circumferential stress along the shaft depth

3   Seismic Responses: Shaft Deformation

3.1   Lateral Deformation

Figure 4 shows the corresponding lateral displacement of the shaft along the depth when the shaft top maximum displacement occurs. As the shaft stiffness is much more than the adjacent soils, it can be noticed that the lateral displacement along the depth is a sloping line, which indicates that the shaft moves as a rigid body in the earthquake. The maximum displacement at the top of the shaft is 2.17 cm, meanwhile the displacement of the shaft foot is 1.36 cm.

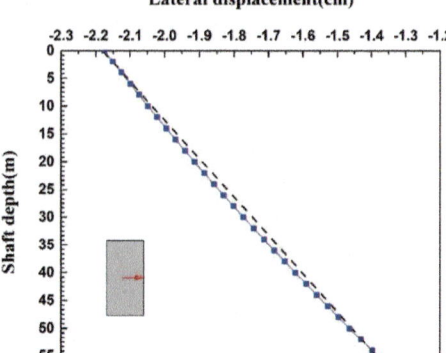

Fig. 4. Lateral displacement of the shaft along the depth

3.2 Radial Deformation

In order to investigate the radial deformation of the shaft, there are four rows of reference points selected at the shaft lining transverse sections, as Fig. 1 shows: two rows of reference points in the plane parallel to the direction of vibration, while the other two rows of reference points in the plane perpendicular to the direction of vibration. Then Fig. 5 shows the corresponding radial displacement of the shaft along the depth at the moment when the shaft top maximum displacement occurs. The transverse section of the shaft is mainly compressed along the circumference and the magnitude of the radical displacement is very small within 1.2 mm. The compressed radial displacement of the shaft increases with the depth increased from 0 to 40 m. Inversely, because of the slight floor heave of the shaft bottom, the compressed radical displacement of the shaft decreases along the depth from 40 to 56 m.

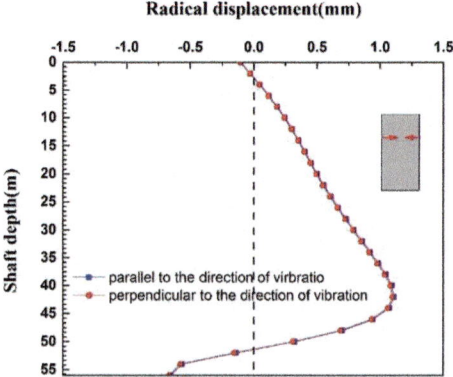

Fig. 5. Radical displacement of the shaft along the depth

4   Seismic Responses: Shaft Inner Forces

4.1   Vertical Inner Forces

In order to investigate the inner forces of the shaft, three inner forces in shaft transverse sections are analyzed when the shaft top maximum displacement occurs. Figure 6(a) shows the corresponding axial force, which represents the gravity load of the shaft in vertical direction along the shaft depth. The axial force increases along the depth increase. Figure 6(b) shows the corresponding shear force along the shaft depth. The general trend of shear force nearly increases along the depth with slightly increase at the top and the bottom. Figure 6(c) shows the corresponding bending moment along the shaft depth. The bending moment increases from the depth 0 to 25 m, while it decreases to 0 from the depth 25 m to 40 m. After then, it increase reversely from the depth 40 m to the bottom. The maximum bending moment occurs at the shaft bottom, which should be noticed in design.

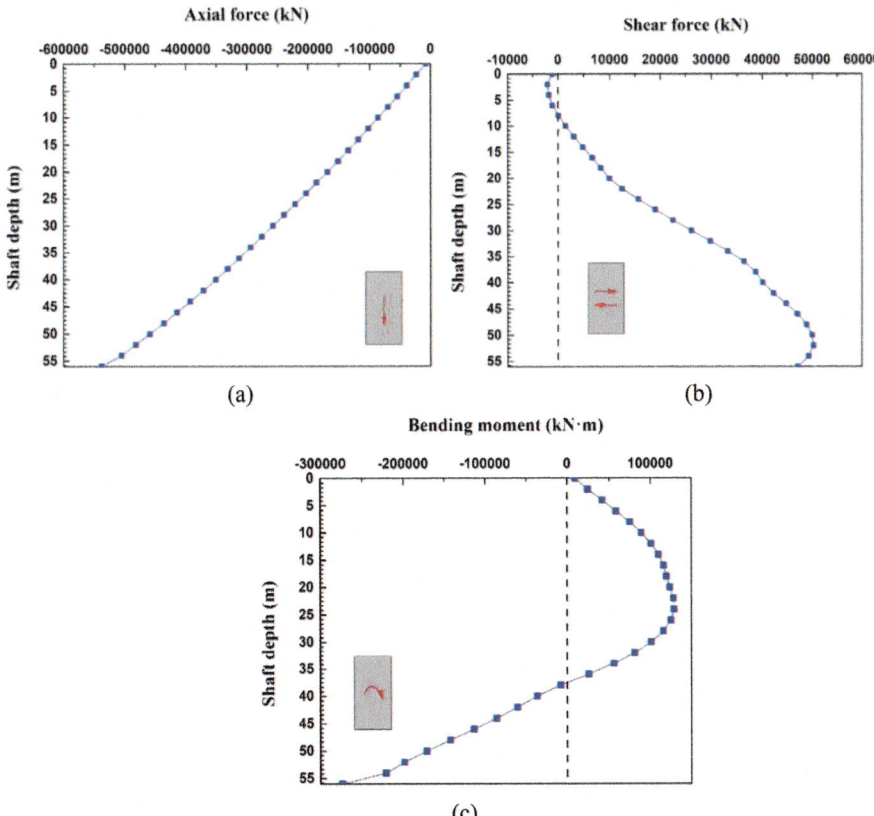

Fig. 6.   (a) Axial force; (b) Shear force; (c) Bending moment along the shaft depth

4.2   Radial Inner Forces

Figure 7 shows the axial force and bending moment in circumferential direction of the shaft along the depth at the moment when the shaft top maximum displacement occurs. Figure 7(a) gives the circumferential axial forces of the shaft along the depth. The circumferential axial force increases with the depth increase from the 0 to 40 m. By contrast, the circumferential axial force decreases with the depth increase from 40 m to the bottom of the shaft. It is affected by the upward displacement of the shaft bottom. Figure 7(b) shows the circumferential bending moment of the shaft along the depth. The circumferential bending moment basically remains unchanged along the depth. However, there is a great change in the section from the 40 m depth to the bottom mainly because of the slightly floor heave of the shaft bottom.

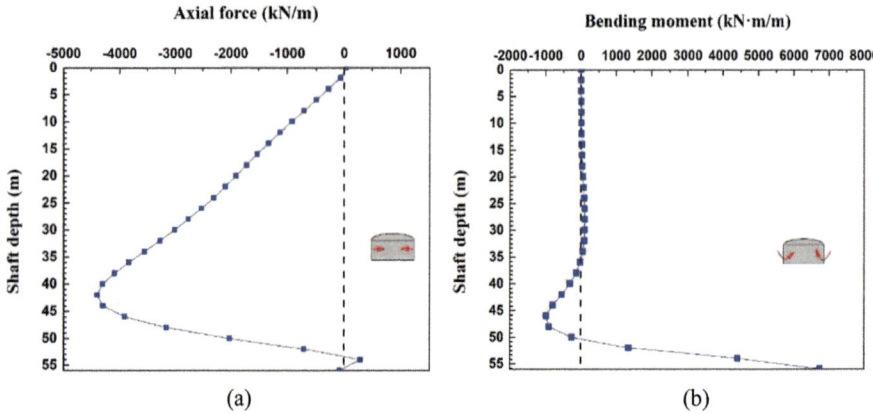

Fig. 7.   (a) Axial force; (b) Bending moment in circumferential direction along the depth

5   Conclusions

Seismic behavior of the shaft in Shanghai soft soil is investigated by using dynamic time history finite element analysis. The deformation and the inner force of the shaft are analyzed. Following conclusions can be drawn from the seismic responses of the shaft:

(1) The Lateral Displacement of the Shaft Along the Depth Is a Sloping Line in the Dynamic Analysis and the Seismic Response of the Shaft Behaves like a Rigid Body at This Input Seismic Motion

(2) The radical deformation of the shaft is very small because of the great stiffness of the shaft.

(3) The shaft maximum bending moment in vertical direction appears at the bottom, while the maximum transverse axial force appears at the depth of 40 m (about 2/3 depth of the shaft).

Acknowledgement. This research was supported by the National Natural Science Foundation of China (Grant No. 41472246, 51778464), Key laboratory of Transportation Tunnel Engineering (TTE2014-01), and "Shuguang Program" supported by Shanghai Education Development Foundation and Shanghai Municipal Education Commission. All supports are gratefully acknowledged.

References

1. Ando, K., Fujiwara, Y., Shiba, Y., Akito, H., Matsumoto, M., Koizum, A.: Study on the seismic behavior of a deep shielded shaft and the effect of flexible joints by shaking table tests. In: The 66th Annual Academic Lecture Meeting of JSCE (2011). (in Japanese)
2. Sawada, M., Shiba, Y., Akito, H., Tokumaru, D., Koizum, A.: Study on the seismic behavior of a deep shielded shaft and the effect of flexible joints by numercil simulation of the vibration tests. In: The 66th Annual Academic Lecture Meeting of JSCE (2011). (in Japanese)
3. Shiba, Y.: A theoretical study on the analytical model used in the seismic design for underground vertical shaft structures. JSCE J. Earthq. Eng. **69**(4), 55–72 (2012)
4. Mayoral, J.M., Sarmiento, N., Ernesto, C.: Seismic response of tunnel shafts. In: Second European Conference on Earthquake Engineering and Seismology, Istanbul (2014)
5. ABAQUS, Users Manual V. 6.10-1. Dassault Systemes Simulia Corp., Providence, RI (2010)
6. Zafeirakos, A., Gerolymos, N.: On the seismic response of under-designed caisson foundations. Bull. Earthq. Eng. **11**(5), 1337–1372 (2013)
7. Huang, Y., Ye, W., Chen, Z.: Seismic response analysis of the deep saturated soil deposits in shanghai. Environ. Geol. **56**(6), 1163–1169 (2009)
8. Hatzigeorgiou, G.D., Beskos, D.E.: Soil-structure interaction effects on seismic inelastic analysis of 3-D tunnels. Soil Dyn. Earthq. Eng. **30**, 851–861 (2010)
9. Chen, Z.Y., Chen, W., Bian, G.Q.: Seismic performance upgrading for underground structures by introducing shear panel dampers. Adv. Struct. Eng. **17**(9), 1343–1358 (2014)
10. Yang, Y.B., Hung, H.H.: A 2.5d finite/infinite element approach for modelling visco-elastic bodies subjected to moving loads. Int. J. Numer. Meth. Eng. **51**(11), 1317–1336 (2001)
11. Huo, H., Bobet, A., Fenmandez, G., Ramirez, J.: Load transfer mechanisms between underground structure and surrounding ground-evaluation of the failure of the Daikai station. J. Geotech. Geoenviron. Eng. **131** (2005)

In-situ Measurement of Metro-Induced Building Vibrations over a Metro Station

Shi-Jin Feng$^{(\boxtimes)}$, Guo-Wei Dong, and Xiao-Lei Zhang

Department of Geotechnical Engineering, Tongji University, Shanghai, China
fsjgly@tongji.edu.cn

Abstract. In recent years, many Chinese cities have begun to develop over-track buildings above metro stations for business and residence. The frequently passing trains can induce excessive vibrations to over-track buildings and adversely affect the living quality of the staff. Considering the current need of reliable experimental data for the optimization of the superstructures over metro stations, in-situ measurements of vibrations at different floors inside an over-track building were conducted at Wuxi, China. The time-domain, frequency-domain, amplitude and vibration level of accelerations are quantified and compared. It is found that the vertical accelerations are significantly greater than the horizontal accelerations. The dominant frequencies of vibrations transmitting to the over-track buildings are mainly concentrated between 20 and 60 Hz. The results can also be used for the validation of numerical prediction models for metro-induced building vibrations.

Keywords: In-situ measurement · Metro-induced vibrations
Over-track buildings · Metro station

1 Introduction

More and more over-track buildings have been built above metro stations for people's life and work. But the frequently moving trains can generate excessive vibrations that transmit into the over-track buildings, causing harm to the nearby buildings, breakdown of sensitive equipment and discomfort to people.

The rapid development of metro has attracted lots of researchers focusing on the prediction models for metro-induced building vibrations. Lopes et al. (2014) proposed a sub-structuring numerical approach which includes 2.5D FEM-PML and 3D FEM, for the prediction of building vibrations due to underground railways. Xu et al. (2015) developed and verified 2D and new types of 3D prediction models of environmental vibrations induced by with a direct fixation track and steel spring floating slab track. Ma et al. (2016) studied the long-term vibration effects caused by metro trains on historic buildings with a metro train–track–tunnel–soil 3D dynamic FE model.

Many experimental studies of metro train-induced vibrations on adjacent buildings have been conducted as well. Wei et al. (2011) measured the subway induced vibration of a pair of tunnels near a subway station as well as a building nearby in Shanghai. Sanayei et al. (2013) studied ground-borne vibration induced by trains and subways at different sites in the Boston area. Zou et al. (2016) conducted field measurements of

© Springer Nature Singapore Pte Ltd. 2018
T. Qiu et al. (Eds.): GSIC 2018, *Proceedings of GeoShanghai 2018 International Conference: Advances in Soil Dynamics and Foundation Engineering*, pp. 112–119, 2018.
https://doi.org/10.1007/978-981-13-0131-5_13

vibration during subway operations in five measurement setups in a metro depot at Guangzhou. However, the experimental investigations of the metro-induced vibrations inside over-track buildings above metro stations are very limited.

This paper intends to present detailed experimental investigations of metro-induced vibrations in a superstructure above a metro station simultaneously in three directions, with time-domain, frequency-domain, amplitudes and vibration level analysis. The in-situ measurement was firstly conducted in an over-track building above a metro station in Wuxi, China. Then, the measurement results are analyzed.

2 In-situ Measurement Process

2.1 Experimental Site

The in-situ measurement was carried out in Wuxi, a famous city in eastern China. The over-track building is right above and built together with a metro station in Wuxi Metro Line 2. The location of the building and the metro station is showed in Fig. 1a.

The test building is a reinforced concrete frame structure with three floors on the ground (F1–F3) and one basement below the ground (B1). The main structural columns have sectional dimensions of 0.9 m by 0.9 m mostly and are located on a grid of size 6.0 m by 6.0 m in general. The cast-in-place reinforced concrete floors are 0.5 m thick below the ground while 0.12 m thick above the ground. The concrete beams with different dimensions hold the floors. Figure 1b shows the interior space of the third floor. The metro station is founded on concrete bored piles with 0.7 m side walls and a 0.9 m baseplate. The station is right below the building as basement two (B2) and basement three (B3). The height of every floor in the joint-built construction is 5.5 m. Moreover, Fig. 3a presents the cross section of the joint-built construction.

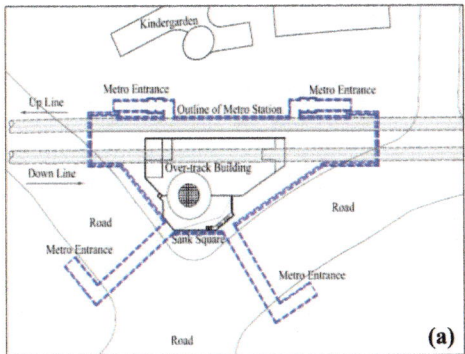 

Fig. 1. The test building and metro station: (a) plan view, (b) the third floor.

2.2   Underlying Soil Properties

According to geological exploration, the underlying soil of the test site is divided into 6 layers, as shown in Fig. 3a. Detailed parameters are listed in Table 1.

Table 1. Underlying soil properties.

Soil layer	Soil type	Thickness (m)	Young's modulus (MPa)	Poisson ratio	Density (kg/m^3)
(1)	Backfill soil	2.5	–	–	–
(2)	Clay	3.5	356.1	0.30	1980
(3)	Silt	2.4	204.7	0.30	1920
(4)	Silty sand	5.6	203.0	0.28	1910
(5)	Clay	10.0	450.4	0.32	2010
(6)	Silty clay	37.6	405.7	0.32	1970

2.3   Metro Train and Track

The daily operation trains of Wuxi Metro are B car-type. The train contains 6 carriages. Figure 2 shows the geometry of a typical carriage. The length of a typical carriage is 19 m, and the spacing of wheels and bogies are 2.3 m and 12.6 m, respectively. As listed in Table 2, l_i denotes the characteristic dimensions of the B car-type train.

Wuxi Metro adopts the standard CHN60 rail, and DTIII-2 fastener in ordinary track. The spacing of sleepers is 0.6 m. Generally speaking, the speed of the train will be no higher than 60 km/h.

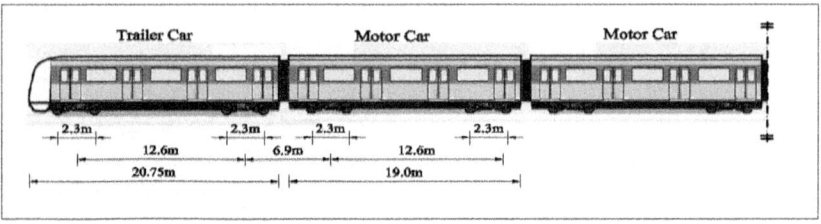

Fig. 2. The geometry of typical carriages of B car-type train.

Table 2. Characteristic dimensions of the B car-type train.

Characteristic	l_1	l_2	l_3	l_4
Dimensions (m)	2.3	6.9	12.6	19

2.4   Measurement Arrangement

To obtain propagation characteristics of vibration in three directions at the same time, 10 measuring points were arranged at the center of the floor slabs and the foot of the columns from B2 to F3. As shown in Fig. 3a, every floor has two test points.

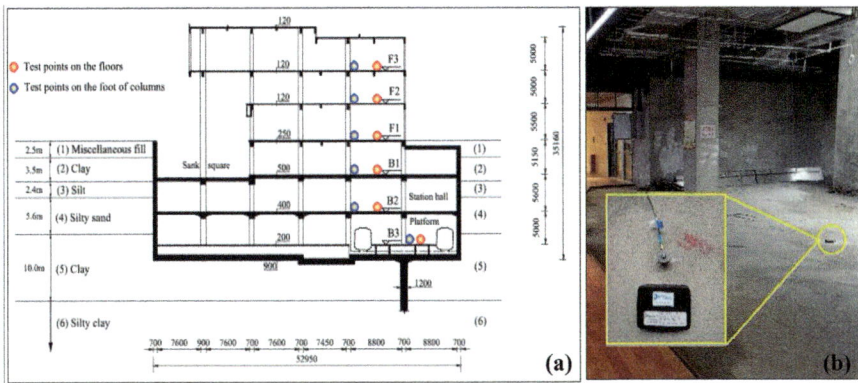

Fig. 3. Test points arrangement and accelerometers: (a) test point arrangement, (b) three-component accelerometer.

All the accelerometers in this experiment can be used to test three-component vibrations, as shown in Fig. 3b. The measurement range is 0–5 g and its effective frequency range is 0.3–6000 Hz. The adopted vibration signal acquisition instruments are NI cDAQ-9184 which has 16 channels and NI cDAQ-9191 which has 4 channels. Five instruments are used in parallel in this experiment, thus providing 32 channels to collect vibration data. The sampling frequency was set to be 1652 Hz.

The vibrations at all test points would be collected simultaneously for 30 s when the train started to leave the metro station. Moreover, in order to obtain accurate vibration data, background vibration is collected before a train passed by.

3   Characteristics of Metro-Induced Building Vibrations

3.1   Time Domain Characteristics of Vibration Acceleration

For the imitation of space, only the accelerations of floor slabs are presented here. A set of typical vibration data of test Point 1–5 is selected for analysis. For simplicity, only the Point-F1's time history of vertical, transversal and longitudinal vibration accelerations is shown in Fig. 4a. It can be summarized that the transversal and longitudinal vibration intensity is far less than that of vertical vibration. It indicates that metro-induced building vibrations should primarily focus on vertical vibrations.

The Fig. 4a–e shows the measuring points' time history of vertical vibration accelerations from B2 to F3, respectively. The metro-induced vibrations last for about 11 s. A succession of vibration peak values, induced by periodic wheel-rail interaction, is clear. Due to the superposition effects of the adjacent bogies, the peak values in the middle part are clearly larger than those of the beginning and final part. By comparing the accelerations of different floors, it could be found that the vibration rises dramatically from B1 to F1, and reaches the largest amplitude at the first floor. The reason could be that the underground structure has a larger rigidity than the over-ground structure for the lateral confinement of the surrounding soil.

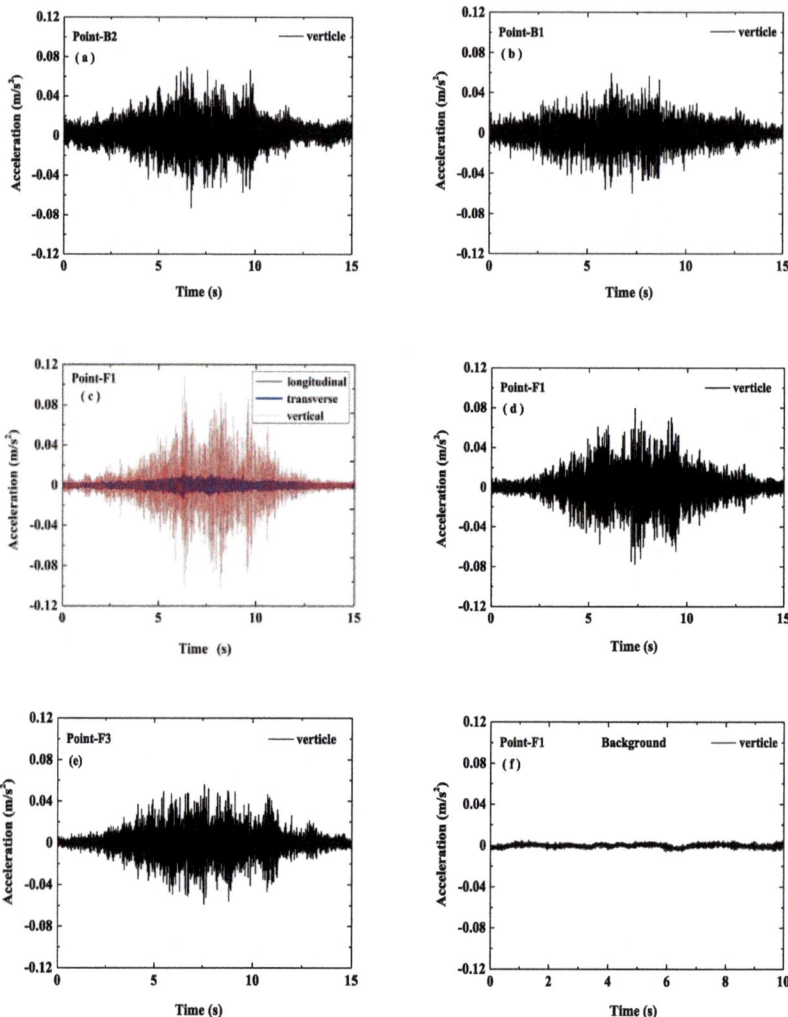

Fig. 4. Time histories of vibration accelerations of the measuring points at: (a) B2, (b) B1, (c) F1, (d) F2, (e) F3; (f) background vibration at F1.

Additionally, the background vibration at F1's test point is shown in Fig. 4f. The amplitude of the background vibration only reaches 0.007 m/s². Compared with the metro-induced vibration, the background vibration inside the building is quite little.

3.2   Frequency Domain Characteristics of Vibration Acceleration

By the Fourier Transformation, the acceleration frequency spectra are shown in Fig. 5a–e, corresponding to the time history of vertical vibration adopted in previous section (five measuring points from B2 to F3). The dominant frequency components of the vibrations transmitted to the building vary from 20 to 60 Hz. In general, the peak

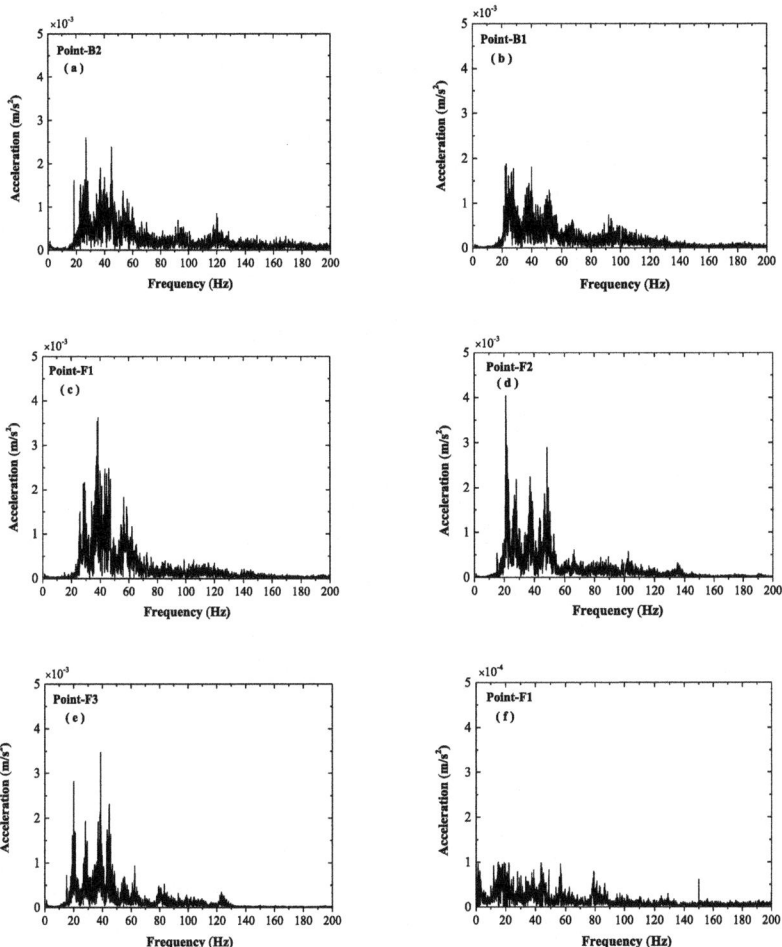

Fig. 5. Frequency spectra of vibration accelerations of the measuring points at: (a) B2, (b) B1, (c) F1, (d) F2, (e) F3; (f) background vibration at F1.

frequencies are around 23, 28, 40, 44 and 50 Hz. The peak frequencies between 35 and 45 Hz have a clear enhance at first floor. However, at second floor, the peak frequencies around 40 Hz decrease slightly while the peak frequencies around 23 Hz and 50 Hz increase obviously. Meanwhile, the vibration components below 20 Hz and above 60 Hz at all floors are quite little.

The frequency spectrum of background vibration of Point-F1 is shown in Fig. 5f. The amplitude is far less than those of metro-induced vibrations. Thus, the experimental results have high accuracy in the absence of other vibration sources.

3.3   Analysis of Amplitude and Vibration Level

Figure 6a shows the average vibration amplitudes (red points) and ten samples (black points). The average amplitude keep rising from B2 to F1 (0.052 m/s², 0.055 m/s² and 0.095 m/s², respectively), and attains the maximum value at F1. From F1 to F3, the average vibration amplitude decreases gradually with the increase of the floor (0.095 m/s², 0.076 m/s² and 0.068 m/s², respectively).

In order to evaluate the influence of the metro-induced vibration in the over-track building, the vibration levels in vertical direction at all floors are calculated, according to ISO2631-2 and GB10070. Figure 6b shows the average vibration levels (red points) and ten samples (black points). The average vibration level falls down from B2 to B1 and reaches the minimum value (69.01 dB). But it jumps to the maximum value (74.66 dB) at F1 with a sharp rise of 5.65 dB. From F1 to F3, the average vibration level decreases gradually (75.13 dB, 74.82 dB and 73.18 dB, respectively), which indicates that the vibration attenuates as the height increases.

Moreover, the average vibration levels from F1 to F3 exceed the Chinese criterion limit for the business district in the night (72 dB) and almost reach the criterion limit for the daytime (75 dB). Therefore, some effective measures should be taken to mitigate the metro-induced vibrations in the over-track building.

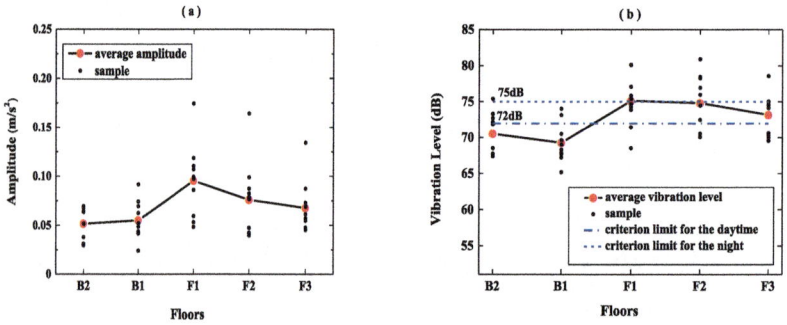

Fig. 6. Amplitude and vibration level from B2 to F3: (a) amplitude, (b) vibration level.

4   Conclusions

The in-situ measurement was carried out to investigate the propagation law of metro-induced vibrations of over-track building above a metro station. The experiment results would contribute to the study of building vibration induced by the metro and provide reliable vibration data for designers to develop safe and effective measures to mitigate the vibration in the over-track buildings. Conclusions are as follows:

(1) The vertical vibration inside the over-track building is far greater than the longitudinal and the transversal vibration.

(2) The dominant frequency components of the metro-induced vibrations transmitted to the building vary from 20 to 60 Hz. The peak frequencies are around 23, 28, 40, 44 and 50 Hz.

(3) The vibration rises dramatically from basement one to first floor, and reaches the largest amplitude at first floor. The peak frequencies between 35 and 45 Hz have a clear enhance at first floor.

(4) The average amplitude keeps rising from basement two to first floor and attains the maximum value at first floor. However, it decreases gradually from first floor to third floor.

(5) The average vibration level decreases from basement two to basement one and reaches the minimum value at basement one. But it has a rapid rise and attains the maximum value at first floor. From first floor to third floor, the average vibration level decreases gradually, which indicates that the vibration attenuates as the height increases.

(6) The average vibration levels from first floor to third floor exceed the Chinese criterion limit for the business district in the night. Thus the metro-induced vibrations of the over-track building should be paid more attention.

Acknowledgement. This research was supported by Wuxi Metro Group Company Ltd. The authors would like to thank Jian-Ping Li, Hong-Yu Gao, Zhang-Long Chen, Ming-Qing Peng, Yong Zhao and Yi-Cheng Li for their assistance in the in-situ measurement.

References

Wei, D., Shi, W., Han, R., et al.: Measurement and research on subway induced vibration in tunnels and building nearby in Shanghai. In: International Conference on Multimedia Technology, pp. 1602–1605 (2011)

Sanayei, M., Maurya, P., Moore, J.A.: Measurement of building foundation and ground-borne vibrations due to surface trains and subways. Eng. Struct. **53**(10), 102–111 (2013)

Zou, C., Wang, Y.M., Moore, J.A., et al.: Train-induced field vibration measurements of ground and over-track buildings. Sci. Total Environ. **575**, 13–39 (2016)

Lopes, P., Costa, P.A., Calçada, R., et al.: Numerical modeling of vibrations induced by railway traffic in tunnels: from the source to the nearby buildings. Soil Dyn. Earthq. Eng. **61–62**(3), 269–285 (2014)

Xu, Q., Xiao, Z., Liu, T., et al.: Comparison of 2D and 3D prediction models for environmental vibration induced by underground railway with two types of tracks. Comput. Geotech. **68**, 169–183 (2015)

Ma, M., Liu, W., Qian, C., et al.: Study of the train-induced vibration impact on a historic Bell Tower above two spatially overlapping metro lines. Soil Dyn. Earthq. Eng. **81**, 58–74 (2016)

Characteristics of Surrounding Soils Influenced by High Frequency Vibratory Pile Pulling

Zhaohui Qin, Longzhu Chen[(✉)], Chunyu Song, and Zhaowei Ding

Shanghai Jiao Tong University, Shanghai, China
lzchen@sjtu.edu.cn

Abstract. A field test was performed on steel sheet piles penetrated and extracted in soils by an excavator vibrator. The maximum vibration frequency of the vibrator was 50 Hz and the study conducted near a railway construction site. The changes of ground displacement, horizontal deformation of deep soils and pore water pressure of soils due to pile sinking and pulling were measured and data were recorded until a steady state was reached. The ground settlement generated during the pile pulling showed that the value decreases with increasing horizontal distance, and increases with time until a constant is reached after three days. After pile extraction, the surrounding soils move closer to the pile, the horizontal deformation of deep soils decreases with increasing distance in the horizontal direction and depth in vertical direction. In addition, the deformation also resulted in significant rebound with time. Vibratory pile pulling caused excess pore water pressure of the surrounding soils, similar to the change rule of the horizontal deformation, pore water pressure decreases with increasing depth in the vertical direction and distance in the horizontal direction. The excess pore water is shown to dissipate completely within three hours after extraction.

Keywords: Field test · High frequency vibratory pile pulling
Ground displacement · Horizontal deformation · Pore water pressure

1 Introduction

Vibratory pile driving and pulling is an alternative pile installation and removal technique where piles attached to a vibrator are inserted into or pulled out of the ground by vertical vibrations. This technique is often used in temporary earth retaining structures, particularly in steel sheet pile supports.

The technique of vibratory pile driving has been widely investigated over the last century and current research mainly focuses on aspects including development of the vibratory driver [1–3], pile driving models [4–8] and improving pile driving [9–11]. In recent years, research efforts have switched to environment problems caused by pile driving such as ground settlement or heave, and excess pore water pressure [12, 13]. Previous studies have used laboratory models, field records, in situ testing and mathematical models to investigate different aspects of vibratory pile driving.

© Springer Nature Singapore Pte Ltd. 2018
T. Qiu et al. (Eds.): GSIC 2018, *Proceedings of GeoShanghai 2018 International Conference: Advances in Soil Dynamics and Foundation Engineering*, pp. 120–129, 2018.
https://doi.org/10.1007/978-981-13-0131-5_14

Through laboratory testing, Bement and Selby [14] investigated the compaction settlement of granular soils during vibratory pile driving. Vogelsang et al. [15, 16] studied the change rule of pore water pressure using a half-model test. Using a field case and in situ test, Clough and Chameau [17] found that settlements are correlated with the density of sands and the level of acceleration. Glatt et al. [18] observed that settlements occurred within a limited range and these settlements were related to ground acceleration. Svinkin [19] studied the relationship between pore water pressure and the heave of ground surface. Mijers and Tol [20] also showed that the settlements at depth may be larger than on the surface. Mahutka et al. [21], Osinov et al. [22] and Chrisopoulos et al. [23] investigated the settlement and displacement of soils caused by vibratory pile driving using finite elements.

The technique of vibratory pile pulling is equally important in the construction of temporary earth retaining structures but has not been comprehensively studied. This study reports on a field test of steel sheet piles penetrated and extracted in soils by an excavator vibrator in which the maximum vibration frequency reached 50 Hz. The change of ground displacement, horizontal deformation of deep soils and the pore water pressure of soils due to the pile sinking and pulling, were measured and data were recorded until a steady state was reached. The characteristics of the surrounding soils influenced by vibratory pile pulling were determined in this paper.

2 Description of the Test

2.1 Testing Site and Soil Conditions

The field test was carried out at a high speed rail way construction site, in Qingdao, in Shandong province, China. The soil conditions encountered in the test site generally consist of backfill, fine sand, muddy silty clay with varying amounts of marine shell fragments, silty clay and coarse sand. The thicknesses of each of the soil layers and N-values with depths in the test site were measured, as shown in Fig. 1. The water table was two meters below the ground surface.

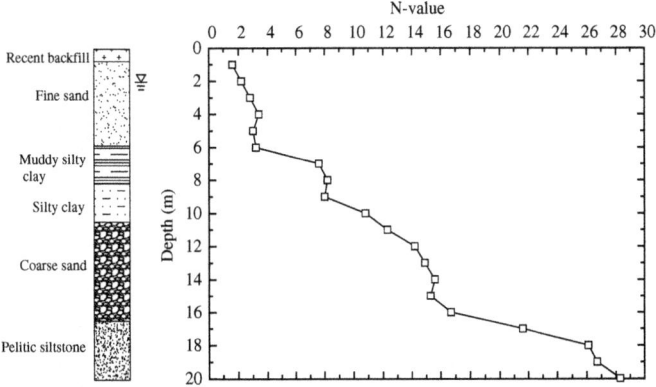

Fig. 1. Soil layers and N-value versus depth.

The fine sand had a mean particle size (D50) of 0.17 mm, a uniformity coefficient of 3.67 and a coefficient of curvature of 0.91. Comparatively, the coarse sand had a mean particle size (D50) of 0.92 mm, a uniformity coefficient of 9.63 and a coefficient of curvature of 1.48. Particle size distributions are shown in Fig. 2.

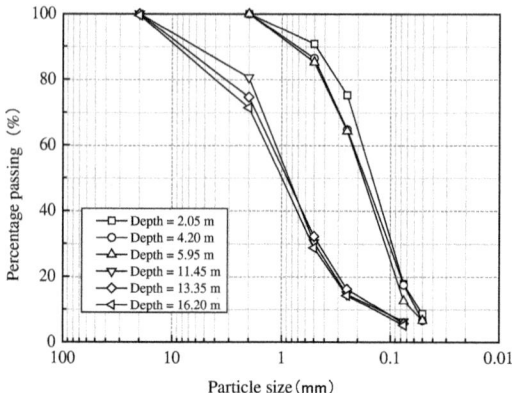

Fig. 2. Particle size distribution of the sands.

The physical and mechanical properties of the muddy silty clay and silty clay were tested in the laboratory and the results are described in Table 1.

Table 1. Physical and mechanical properties of the muddy silty clay and silty clay

Depth (m)	Soli type	Water content	Density	Void ratio	Saturation	Plasticity index	Liquidity index	Coefficient of compressibility	Friction	Cohesion
6.20	Muddy silty clay	31.4	1.85	0.93	92	12.6	1.02	0.47	9.3	32
7.35	Muddy silty clay	27.8	1.84	0.89	85	12.2	0.78	0.41	8.5	32
8.40	Silty clay	22.9	2.00	0.68	92	14.4	0.21	0.18	11.6	45
9.20	Silty clay	22.7	2.04	0.64	96	13.8	0.26	0.21	12.5	34
10.35	Silty clay	22.1	2.03	0.64	94	13.9	0.19	0.18	12.5	33

2.2 Vibratory Equipment and the Pile

The excavator-modified vibratory system was composed of a modified excavator in which the arm length was lengthened and balance weight extended. A hydraulic vibratory hammer was used instead of the bucket. The pile was a U-profile larssen IV steel sheet pile which was 12 m in length. Both the profile and length of the pile are representative of the most common piles in China. The profiles of the vibratory equipment and pile are shown in Fig. 3.

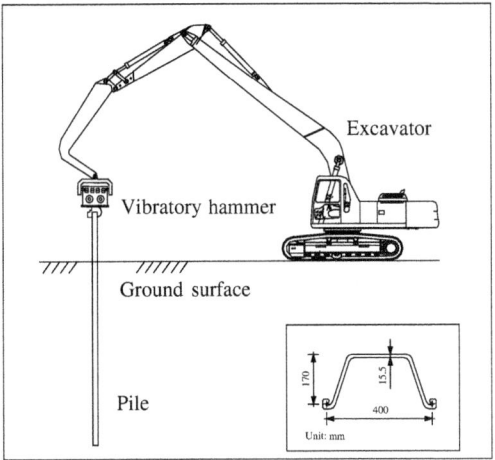

Fig. 3. The profiles of the excavator driving system and U-type Larssen steel sheet pile.

2.3   Instrumentation and Monitoring

Five high precision dial indicators at the ground surface were evenly spaced away from the pile at distances from 0.5 to 2.5 m to observe the ground displacements. Three inclinometer pipes with nineteen meters in length were buried at distances of 1.0, 1.5 and 5.0 m from the pile to measure the horizontal deformations of deep soils. Pore water piezometers of monitoring pore water pressure were embedded at six test points at distances of 1.5, 2.5, 5.0, 7.5, 10.0 and 15.0 m from the pile respectively. At each point, five sensors were buried from top to bottom at depths of 2.5, 5.0, 8.0, 11.5, and 14.0 m. A cross sectional view of the test site is presented in Fig. 4.

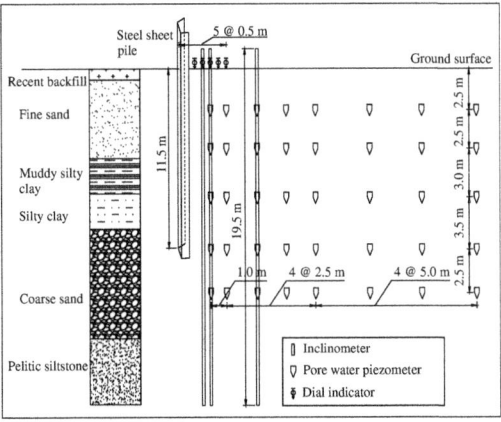

Fig. 4. Cross sectional view of the test site.

After a pile was pulled out the ground, the data of ground displacement, the horizontal deformation of the deep soils, and the pore water pressure of soils were collected and were recorded until a stable state was reached.

3   Results and Analyses

3.1   Ground Displacements

After the pile had been completely pulled out from the earth, the ground displacements at different observing points were recorded and the results at different recording time are shown in Fig. 5. A small amount of settlement occurred after pile extraction, the maximum value was 2.45 mm at a distance 0.5 m away from the pile. With increasing distance, the settlement values gradually decreased to almost zero at the position 2.5 m away from the pile.

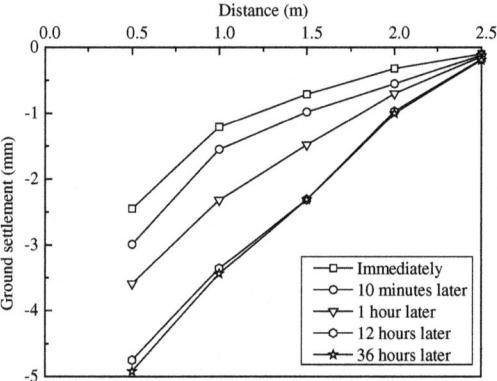

Fig. 5. Ground settlements of different observing points at unequal times.

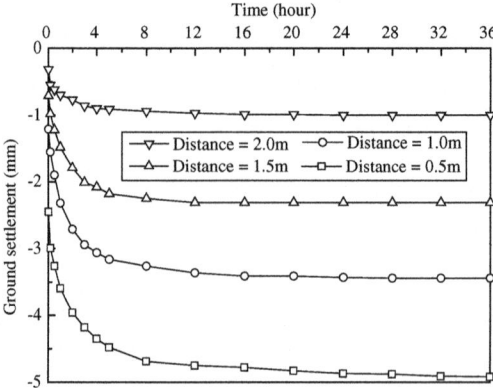

Fig. 6. Ground settlements changes with time at different observation points.

From the observed data it can be seen with time that the ground settlements have a growth tendency. Figure 6 shows the time history curves of ground settlements at four observation points. At the beginning, all settling rates of the four series were fast and the rates decreased with the time. At about twelve hours later, the settlements gradually reached to constant level.

3.2   Horizontal Deformation of Deep Soils

The change of horizontal deformation of three inclinometer pipes with depth before and after pile extraction is shown in Fig. 7. The dashed lines represent the soil deformations before the pile extraction, and due to the previous pile penetration, these lines deviate from the center axis of pile. The solid lines with different colors represent the soil deformations after the pile extraction. It was observed that pile extraction caused the surrounding soils to move close to the pile although the overall deviations were small and with time, the deformations arose with considerable rebound.

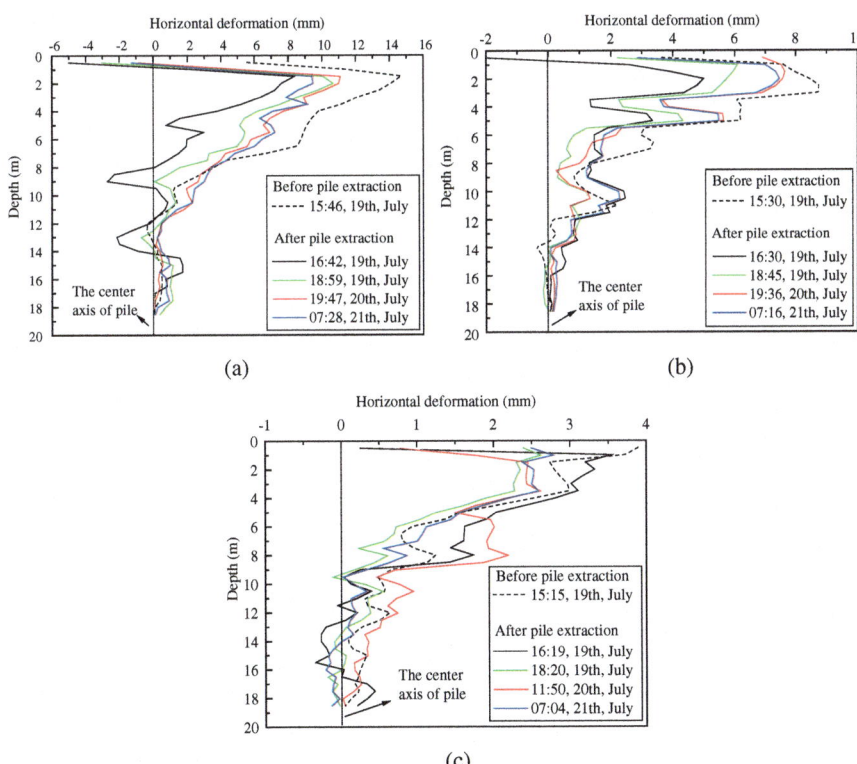

Fig. 7. Horizontal deformations at different measuring points changing with depth at unequal times: (a) distance from the pile was 1.0 m (b) distance from the pile was 1.5 m, and (c) distance from the pile was 5.0 m.

Figure 8 shows the net deformation of three inclinometer pipes changing with depth at the moment of finishing pile extraction. In the horizontal direction, the net deformation decreased with increasing distance from the pile, and there are almost no deformations in the measuring point five meters away from the pile. In the vertical direction, the net deformation decreased with increasing soil depth until about two meters beneath the pile tip.

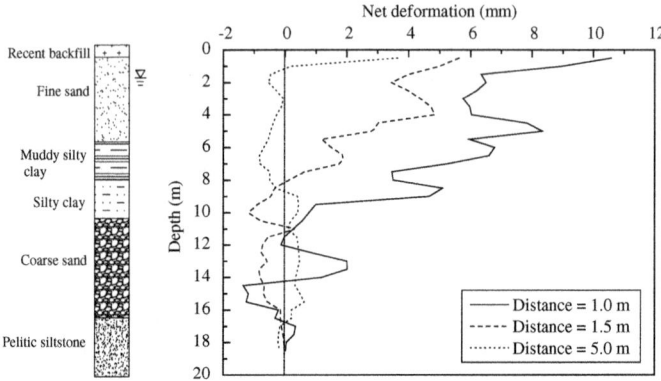

Fig. 8. The net horizontal deformations of different measuring points with changing depth.

3.3 Pore Water Pressure

Figure 9 shows the pore water pressure distribution with depth at the first monitoring point 1.5 m away from the pile both before and after pile extraction. Pile extraction caused the pore water pressure to increase and the maximum increment which was 8.0 kPa at a depth of 2.5 m.

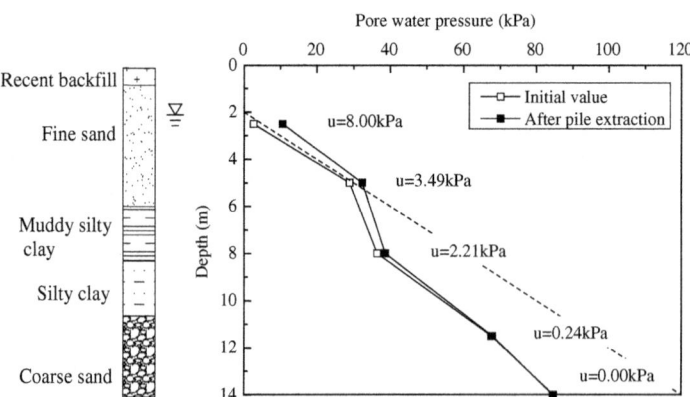

Fig. 9. Pore water pressure distribution with varying depths before and after pile extraction.

Curves showing the increment in pore water pressure with distance at different depths are shown in Fig. 10. The increment of pore water pressure at equal depth decreases with distance in the horizontal direction except at a depth of 5 m. The maximum disturbed radius was observed to be fifteen meters. At a depth of 5 m, the increment of the pore water pressure at the distance of 1.5 m was slightly lower than at the distance of 2.5 m which may be caused by malfunction of the pore water piezometer at the distance of 1.5 m.

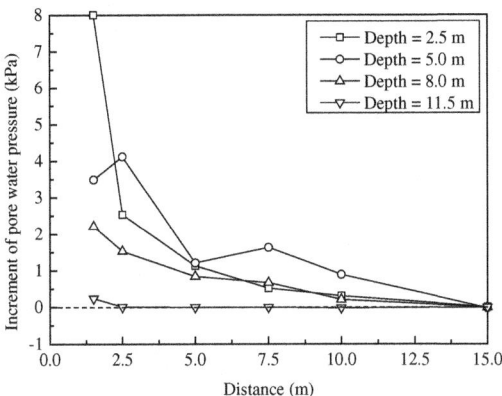

Fig. 10. Increment of pore water pressure at different depths and distances.

Figure 11 shows the increment of pore water pressure with different depths at the first monitoring point. It was found that excess pore water pressure dissipates with time until zero, and the rate of dissipation also decreases. The complete dissipation time at different depths is unequal. The time at depths of 2.5 and 5.0 m is two and one half hours respectively. The time at depths of 8.0 and 11.5 m is more than one hour and less than half an hour, respectively.

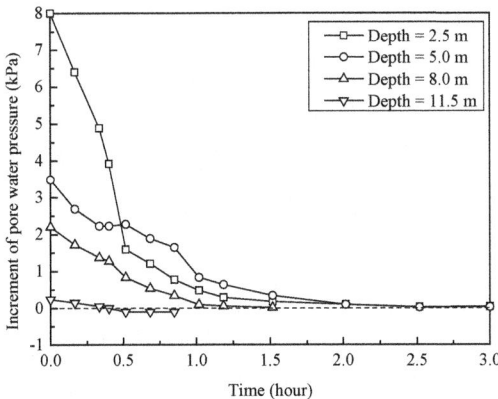

Fig. 11. Increment of pore water pressure at different depths dissipating with the time.

From the observed data it can be seen with time that the ground settlements have a growth tendency. Figure 6 shows the time history curves of ground settlements at four observation points. At the beginning, all settling rates of the four series were fast and the rates decreased with the time. At about twelve hours later, the settlements gradually reached to constant level.

4   Conclusion

A field test of high frequency vibratory pile pulling was conducted and several experimental parameters including variations of ground displacement, deep soil horizontal deformation and pore water pressure induced by pile extraction were studied. Based on the experimental observations, the following conclusions can be made.

(1) The ground close to the pile will generate settlement after pile extraction. Subsidence decreases with the distance and the maximum disturbed radius is 2.5 m away from the pile. The subsidence can be sustained until 12 h later.

(2) After pile pulling, surrounding soils move close to the pile and the deviations arise with a certain amount of rebound with time. The horizontal deformations decrease with increasing horizontal distance and depth in vertical direction.

(3) In surround soils, excess pore water pressures are generated due to the forced oscillation. The maximum increment of pore water pressure was 8.0 kPa with the depth of 2.5 m at the monitoring point 1.5 m from the pile. The value decreases with increasing distance in the horizontal direction and depth in vertical direction. The excess pore water pressure dissipates to zero in a short period of time.

References

1. Jonker, G.: Vibratory pile driving hammers for pile installations and soil improvement projects. In: 19th Annual Offfshore Technology Conference, pp. 549–560 (1987)
2. Holeyman, A.: Vibratory Driving analysis. In: 6th International Conference on the Application of Stress-wave Theory to Piles, pp. 479–494 (2000)
3. Viking, K.: Vibro-driveability-A field study driven sheet piles in non-cohesive soils. PhD Thesis, Royal Institute of Technology, Stockholm, Sweden
4. Wong, D., O'Neill, M.W., Vipulanandan, C.: Modeling of vibratory pile driving in sand. Int. J. Numer. Anal. Meth. Geomech. **16**(4), 189–210 (1990)
5. Holeyman, A., Legrand, C., Van Rompaey, D.: A method to predict the driveability of vibratory driven piles. In: 3rd International Conference on the Application of Stress-wave Theory to Piles, pp. 1101–1112 (1996)
6. Van Baars, S.: Design of sheet pile installation by vibration. Geotech. Geol. Eng. **22**, 391–400 (2004)
7. Mens, A.J.M., Korff, M., Van Tol, A.F.: Validating and improving models for vibratory installation of steel sheet piles with field observations. Geotech. Geol. Eng. **30**, 1085–1095 (2012)
8. Holeyman, A., Whenham, V.: Critical review of the Hypervib1 model to assess pile vibro-drivability. Geotech. Geol. Eng. **3**, 1–19 (2017)

9. Rodger, A.A., Littlejohn, G.S.: A study of vibratory driving in granular soils. Geotechnique **30**(3), 269–293 (1980)

10. O'Neill, M.W., Vipulanandan, C., Wong, D.: Laboratory modeling of vibro-driven piles. J. Geotech. Eng. **116**(8), 1190–1209 (1990)

11. Lee, S.H., Kim, B.-I., Han, J.T.: Prediction of penetration rate of sheet pile installed in sand by vibratory pile driver. KSCE J. Civil Eng. **16**(3), 316–324 (2012)

12. Drabkin, S., Lacy, H., Kim, D.S.: Estimating settlement of sand caused by construction vibration. J. Geotech. Eng. **122**(11), 920–928 (1996)

13. Massarsch, K.R.: Static and dynamic soil displacement caused by pile driving. In: 4th International Conference on the Application of Stress-wave Theory to Piles, pp. 15–24 (1992)

14. Bement, R.A.P., Selby, A.R.: Compaction of granular soils by uniform vibration equivalent to vibrodriving of piles. Geotech. Geol. Eng. **15**(2), 121–143 (1997)

15. Vogelsang, J., Hube, G., Triantafyllidis, T.: Requirements, concepts, and selected results for model tests on pile penetration. In: Holistic Simulation of Geotechnical Installation Processes - Benchmarks and Simulations, pp. 1–30. Springer, Cham (2015)

16. Vogelsang, J., Huber, G., Triantafyllidis, T.: Experimental investigation of vibratory pile driving in saturated sand. In: Holistic Simulation of Geotechnical Installation Processes - Theoretical Results and Applications, pp. 101–123. Springer, Cham (2017)

17. Clough, G.W., Chameau, J.L.: Measured effects of vibratory sheetpile driving. J. Geotech. Eng. Division **106**(10), 1081–1099 (1980)

18. Glatt, J., Roboski, J., Finno, R.: Sheet pile-induced vibration at the Lurie excavation project. Geotech. Eng. Transp. Projects **2004**, 2130–2138 (2004)

19. Svinkin, M.R.: Mitigation of soil movements from pile driven. Pract. Periodical Struct. Design Constr. **11**(2), 80–85 (2006)

20. Meijers, P. Van Tol, A.F.: The Raamsdonksveer sheet pile test: observed settlements due to installation of vibratory driven sheet piles. In: International conference on vibratory pile driving and deep soil compaction, pp. 353–362 (2006)

21. Mahutka, K.P., Grabe, J.: Numerical prediction of settlements and vibrations due to vibratory pile driving using a continuum model. In: International conference on vibratory pile driving and deep soil compaction, pp. 243–252 (2006)

22. Osinov, V., Chrisopoulos, A.S., Triantafyllidis, T.: Numerical study of the deformation of saturated in the vicinity of a vibrating pile. Acta Geotech. **8**, 439–446 (2013)

23. Chrisopoulos, S., Vogelsang, J., Triantafyllidis, T.: FE Simulation of model tests on vibratory pile driving in saturated sand. In: Holistic Simulation of Geotechnical Installation Processes - Theoretical Results and Applications, pp. 124–149, Springer, Cham (2017)

Observations on Model Uncertainty of Robertson-Wride Model for Liquefaction Potential Evaluation

Yixun Ge and Jie Zhang[✉]

Department of Geotechnical Engineering, Tongji University,
Shanghai 200092, China
cezhangjie@gmail.com

Abstract. The Robertson-Wride (RW) model is widely used for liquefaction potential assessment. This paper studies the model uncertainty of the Robertson-Wride (RW) model for liquefaction potential assessment. It is found that the accuracy of the RW model may vary from one region to another. The RW model is more accurate in predicting case histories before 2010 than in predicting case histories after 2010. There is a substantial amount of uncertainty in how to interpret past case histories, and such uncertainty has important effects on model uncertainty characterization.

Keywords: Soil liquefaction · Model uncertainty · Robertson-Wride model

1 Introduction

In 2010 and 2011, a series of earthquakes hit New Zealand, which caused extensive damages. During these earthquakes, about half of the damages were caused by soil liquefaction [1, 2]. After the earthquakes, several liquefaction potential assessment models were assessed by the case histories collected from the New Zealand earthquakes. However, none of them can predict the liquefaction phenomena accurately [3]. The above observations may imply that such liquefaction potential models may be less accurate in predicting the liquefaction potential of soils in the aforementioned New Zealand earthquake zones. The accuracy of a liquefaction potential assessment model can be quantitatively represented by a model bias factor, z. It might be interesting to assess the factors affecting the accuracy of a liquefaction potential assessment model through model uncertainty characterization, which is the purpose of this paper. As an example, the model uncertainty of the 2009 RW model [4–6] will be assessed. This paper is organized as follows. First, the method of model calibration is introduced. Then, the calibration database is briefly described. Finally, the factors affecting the model uncertainty of the RW model are examined.

© Springer Nature Singapore Pte Ltd. 2018
T. Qiu et al. (Eds.): GSIC 2018, *Proceedings of GeoShanghai 2018 International Conference: Advances in Soil Dynamics and Foundation Engineering*, pp. 130–137, 2018.
https://doi.org/10.1007/978-981-13-0131-5_15

2 Method of Model Calibration

Let F_{sa} denote the actual factor of safety (FOS) against soil liquefaction, and let F_s denote the computed FOS from a liquefaction potential assessment model. To consider the model uncertainty, the actual FOS can be connected to the computed FOS via a model bias factor z as follows [7].

$$F_{sa} = \frac{F_s}{z} \tag{1}$$

Let μ and δ denote the mean and the coefficient of variation (COV) of the model bias factor, respectively. The values of μ and δ characterize the model uncertainty. For example, if μ is larger than 1, it means the model on average overestimates the actual FOS, and vice versa. The uncertainty associated with the model error increases with δ. Let $\mathbf{D}$ denote a calibration database. The values of μ and δ can be estimated using the weighted maximum likelihood method as follows [8].

$$\ln[L(\mu, \delta | \mathbf{D})] = w_L \sum_{i=1}^{n_L} \ln \left\{ 1 - \Phi \left[\frac{\left[\ln(F_{si}) - \ln\left(\frac{\mu}{\sqrt{1+\delta^2}}\right) \right]}{\sqrt{\ln(1+\delta^2)}} \right] \right\}$$
$$+ w_{NL} \sum_{j=1}^{n_{NL}} \ln \Phi \left[\frac{\left[\ln(F_{sj}) - \ln\left(\frac{\mu}{\sqrt{1+\delta^2}}\right) \right]}{\sqrt{\ln(1+\delta^2)}} \right] \tag{2}$$

In the above equation, n_L and n_{NL} are the numbers of liquefied and non-liquefied cases, respectively. w_L and w_{NL} are the weighting factors to consider the effect of the sampling bias on model calibration, which can be calculated as follows [7, 9].

$$w_L = \frac{Q_p}{Q_s} \tag{3}$$

$$w_{NL} = \frac{1 - Q_p}{1 - Q_s} \tag{4}$$

where Q_s is the proportion of liquefied cases in the database, and Q_p is the proportion of liquefied cases in the real world. In this study, $Q_p = 0.456$ is adopted [7].

3 Calibration Database

Boulanger and Idriss [10] complied a comprehensive CPT-based liquefaction case histories database incorporating recent earthquakes in 2016. In this study, this database will be used for model calibration. There are 253 cases from 20 earthquake events around the world with moment magnitude in the range of 5.90 to 9.00, as summarized in Table 1. The number of case histories in each earthquake event ranges from 2 to 76.

Table 1. List of case histories in the database of Boulanger and Idriss [10].

No.	Earthquake	M_w	Overall	Effective	FP	FN
1	1964 Niigata, Japan	7.60	3	2	0	0
2	1968 Inaguaha, New Zealand	7.20	2	2	0	0
3	1975 Haicheng, China	7.00	2	2	0	1
4	1976 Tangshan, China	7.60	19	16	1	2
5	1979 Imperial Valley, California	6.53	4	4	0	0
6	1980 Mexicali, Mexico	6.33	4	4	4	0
7	1981 Westmorland, California	5.90	5	5	3	0
8	1983 Nibonkai-Chubu, Japan	7.70	3	3	0	0
9	1983 Borah Peak, Idaho	6.88	4	2	2	0
10	1987 Edgecumbe, New Zealand	6.60	17	17	1	3
11	1987 Superstition Hills 01, California	6.22	2	2	0	0
12	1987 Superstition Hills 02, California	6.54	4	4	1	1
13	1989 Loma Prieta, California	6.93	76	70	5	5
14	1994 Northridge, California	6.69	2	2	0	0
15	1995 Hyogoken-Nanbu, Japan	6.90	23	18	0	1
16	1999 Kocaeli, Turkey	7.51	15	14	1	0
17	1999 Chi-Chi, Taiwan	7.62	11	11	0	0
18	2010 Darfield, New Zealand	7.00	25	25	4	2
19	2011 Christchurch, New Zealand	6.20	25	25	9	1
20	2011 Tohoku, Japan	9.00	7	7	0	2
	Overall		253	235	31	18

FP: False positive cases, i.e., cases liquefied while $F_S > 1$; FN: False negative cases, i.e., case didn't liquefy while $F_S < 1$

4   Accuracy of the RM Model in Different Regions

To have a rough idea about the accuracy of the RW model in different regions, we first calculate the FOS of each case history in the calibration database using the RW model. Note the 2009 RW model is only applicable when the clean-sand equivalence of normalized cone tip resistance, $q_{t1N,cs}$, is less than 160 kPa [4, 5]. Out of the Boulanger and Idriss database, 18 are beyond such a range, and hence are excluded from consideration. The rest cases can be further divided into four categories. If $F_s > 1$ and the soil did not liquefy, it is a correct prediction. Similarly, if $F_s < 1$ and the soil liquefied, it is also a correct prediction. In addition, there are also two types of wrong predictions, i.e., $F_s > 1$ but the soil liquefied, and $F_s < 1$ but the soil did not liquefy, which are denoted as false positive (FP) cases and false negative (FN) cases in this study, respectively. The numbers of FP cases and FN cases during each earthquake are also summarized in Table 1. For ease of understanding, Fig. 1 plots the numbers and proportions of FP and FN in different earthquakes. As can be seen from this figure, the proportions of FP and FN in different regions are quite different. For example, in the 1989 Loma Prieta earthquake, the numbers of FN cases and FP cases are the same. In the 2011 Christchurch earthquake, there is only one FN case but nine FP cases. The ratios of false predictions also vary from one region to another. For example, in the 1995 Hyogoken-Nanbu earthquake, the proportion of false predictions is only 5.6%. For comparison, the proportion of false predictions in the 2011 Christchurch earthquake is 40%. These data show that the accuracy of the RW model may vary from one region to another.

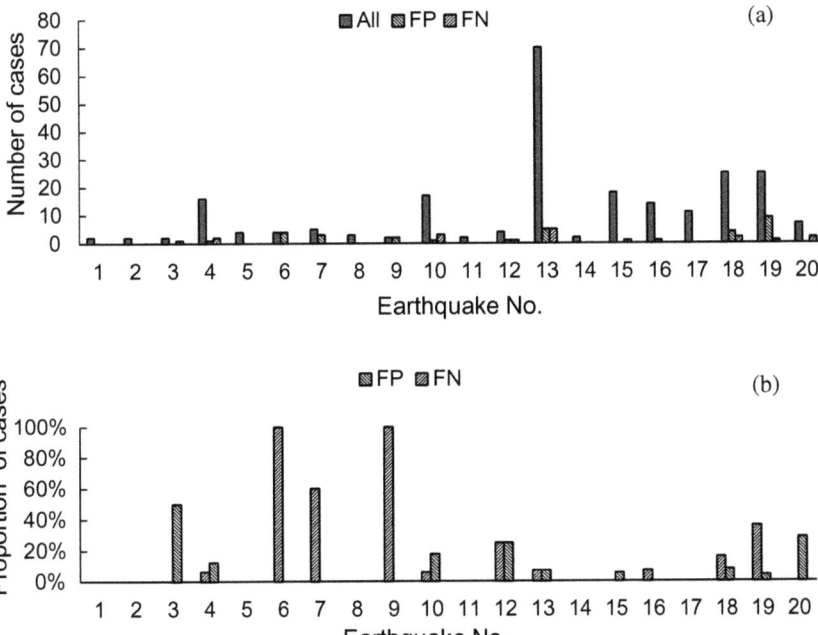

Fig. 1. (a) Numbers of different cases in different earthquakes; (b) Proportions of different cases in different earthquakes.

5 Model Uncertainty of the RW Model

The model uncertainty of the RW model is first assessed using the effective cases (i.e., cases with $q_{c1N,cs} < 160$ kPa) in Table 1 based on Eq. (2). The values of μ and δ in such a case are 0.988 and 0.710, respectively, which are summarized in Table 2. The mean of the model bias factor is close to 1.00, indicating the model is almost unbiased. The COV of the model bias factor is 0.710, indicating there is a substantial amount of model uncertainty associated with the RW model.

Table 2. Calibrated means and COVs of two categories.

Boulanger and Idriss database	μ	δ
Before 2010 (178 cases)	0.932	0.647
After 2010 (57 cases)	1.203	0.975
All cases (235 cases)	0.988	0.710

5.1   Impact of Earthquakes After 2010

The RW model used in this study was mainly developed based on case histories before 2010. To assess the effect of recent case histories, the data in Table 1 are separated into two categories, i.e., cases before 2010 and cases after 2010. We then calibrate the model uncertainty of the RW model using these two categories of data, and the results are also summarized in Table 2. For case histories before 2010, the obtained values of μ and δ are 0.932 and 0.647, respectively, indicating that the RW model on average underestimates the FOS of these case histories. For case histories after 2010, the values of μ and δ are 1.203 and 0.975, respectively, indicating that the RW model on average overestimates the FOS of these case histories. The COV of the model uncertainty associated with the case histories after 2010 is greater than that of case histories before 2010, indicating that the RW model has a greater model error when for predicting the case histories after 2010.

To understand the above phenomena, Figs. 2(a) and (b) plot the case histories before 2010 and after 2010, respectively, in which *CSR* denotes the cyclic stress ratio. For ease of comparison, the boundary curve defined by the RW model is also plotted in these figures. The solid points denote liquefied cases, and hollow points denote non-liquefied cases. The solid points below the boundary curve are FP cases, and the hollow points above the boundary curve are FN cases. In Fig. 2(a), the false predictions distribute evenly around the boundary curve. In Fig. 2(b), there are a larger proportion of FP cases, which indicates RW model tends to overestimate these cases. Comparing these two figures, there are a greater proportion of false predictions in Fig. 2(b), and these points are further away from the boundary curve, which indicates greater model uncertainty.

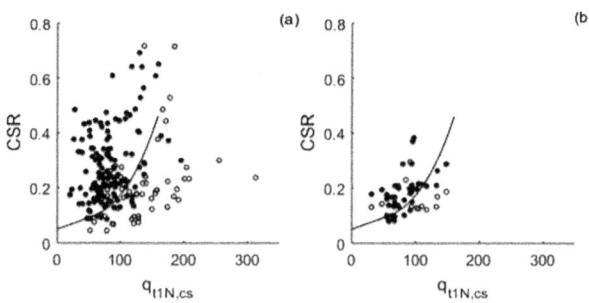

Fig. 2. (a) case histories before 2010; and (b) case histories after 2010

5.2   Impact of Method for Interpreting the Liquefaction Case Histories

When compiling the liquefaction case histories, subjective judgement cannot be fully avoided. In 2006, a liquefaction case history database was compiled by Moss [11]. Comparing these two databases, except cases after 2010, the case histories whose $q_{cIN, cs} < 160$ kPa in the two databases can be divided into three categories, i.e., (1) cases from the same sites and the same soil layers; (2) cases from the same sites but different soil layers; and (3) cases from different sites. In this study, if difference between the depths of the two case histories from the same site registered in two databases is less than 0.3 m, the soil layers of the two case histories are considered the same. The three categories of case histories in the Boulanger and Idriss database [10] are shown in Figs. 3(a), (b), and (c), respectively. The three categories of case histories in the Moss database [11] are shown in Figs. 4(a), (b), and (c), respectively. The numbers of different categories of case histories in the two databases are summarized in Table 3.

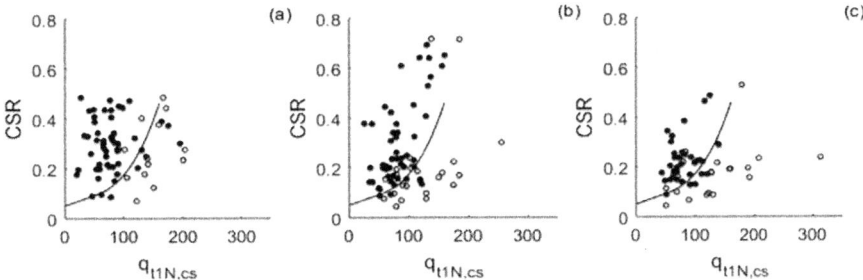

Fig. 3. Case histories in Boulanger and Idriss database [10]: (a) cases from the same sites and the same soil layers; (b) cases from the same sites but different soil layers; (c) cases from different sites and Robertson-Wride model prediction boundary curve.

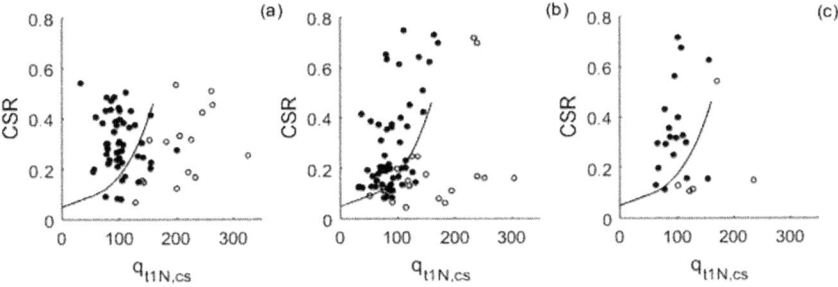

Fig. 4. Case histories in the Moss database [12]: (a) cases from the same sites and the same soil layers; (b) cases from the same sites but different soil layers; (c) cases from different sites and Robertson-Wride model prediction boundary curve.

Table 3. Three categories of Boulanger and Idriss and Moss databases

Database	Category	Number[a]	μ	δ
Boulanger and Idriss [10]	Cases from the same sites and same soil layers	53	0.747	0.669
	Cases from the same sites but different soil layers	63	0.942	0.572
	Cases from different sites	51	1.029	0.593
Moss [11]	Cases from the same sites and same soil layers	53	1.371	0.387
	Cases from the same sites but different soil layers	62	1.066	0.456
	Cases from different sites	24	1.279	0.422

[a]That one case history may correspond to one or two case histories results in the small difference in the numbers of the same category.

In the first category, although two cases registered in the two databases are from the same sites and the same soil layers, significant differences are observed among vertical total overburden stress, vertical effective overburden stress, normalized tip resistance and sleeve friction. In addition, the moment magnitude, the peak ground surface acceleration and ground water table depth are also not exactly the same. The above differences indicate significant uncertainty exists in interpreting the past case histories. When the first category case histories from the Boulanger and Idriss database are used, the calibrated mean and the COV of the model uncertainty are 0.747 and 0.669, respectively. When the first category case histories from the Moss database are used, the calibrated mean and the COV of the model uncertainty are 1.371 and 0.387, respectively. The significant differences in the model calibration results reflect the significant effects of the method on how to interpreting the case histories on model uncertainty characterization.

6 Summary and Conclusions

This paper studies the model uncertainty of the Robertson and Wride (RW) model for liquefaction potential assessment. It is found that the accuracy of the RW model may vary from one region to another. The RW model is more accurate in predicting case histories before 2010 than in predicting case histories after 2010. There is a substantial amount of uncertainty in how to interpret past case histories, and such uncertainty has important effects on model uncertainty characterization.

Acknowledgement. This research was substantially supported by the Natural Science Foundation of China (41672276, 51538009), National 973 Basic Research Program of China (2014CB049100), and the Shanghai Rising-star Program (15QA1403800).

References

1. Cubrinovski, M., Green, R.A., Allen, J., Ashford, S., Bowman, E., Bradley, B.: Geotechnical reconnaissance of the 2010 Darfield (Canterbury) earthquake. Bull. N. Z. Soc. Earthq. Eng. **43**(4), 243–320 (2010)
2. Cubrinovski, M., Bradley, B., Wotherspoon, L., Green, R., Bray, J., Wood, C.: Geotechnical aspects of the 22 February 2011 Christchurch earthquake. Bull. N. Z. Soc. Earthq. Eng. **44** (4), 205–226 (2011)
3. Wotherspoon, Liam M., Orense, Rolando P., Green, Russell A., Bradley, Brendon A., Cox, Brady R., Wood, Clinton M.: Assessment of liquefaction evaluation procedures and severity index frameworks at Christchurch strong motion stations. Soil Dyn. Earthq. Eng. **79**, 335–346 (2015)
4. Robertson, P.K., Wride, C.E.: Evaluating cyclic liquefaction potential using the cone penetration test. Can. Geotech. J. **35**, 442–459 (1998)
5. Robertson, P.K.: Performance based earthquake design using the CPT. In: Performance-Based Design in Earthquake Geotechnical Engineering (2009)
6. Youd, T.L., Idriss, I.M.: Liquefaction resistence of soil: summary report from the 1996 NCEER and 1998 NCEER/NSF workshops on evaluation of liquefaction resistence of soils. J. Geotech. Geoenviron. Eng. **127**(4), 297–313 (2001)
7. Juang, C.H.., Khoshnevisan, S., Zhang, J.: Maximum likelihood principle and its application in soil liquefaction assessment. In: Risk and Reliability in Geotechnical Engineering
8. Ku, C.-S., Juang, C.H., Chang, C.-W., Ching, J.: Probabilistic version of the Robertson and Wride method for liquefaction evaluation: development and application. Can. Geotech. J. **49** (1), 27–44 (2012)
9. Cetin, K.O., Der Kiureghian, A., Seed, R.B.: Probabilistic models for the initiation of seismic soil liquefaction. Struct. Saf. **24**(9), 67–82 (2002)
10. Boulanger, R.W., Idriss, I.M.: CPT-based liquefaction triggering procedure. J. Geotech. Geoenviron. Eng. **142**(2), 04015065 (2016)
11. Moss, R.E.S., Seed, R.B., Kayen, R.E., Stewart, J.P.: CPT-Based Probabilistic and Deterministic Assessment of In Situ Seismic Soil Liquefaction Potential (2006)

Shaking Table Investigation of Seismic Performance of Micropiles

Hadis Jalilian[1(✉)], Jian Hua Yin[1], and Ali Komak Panah[2]

[1] The Hong Kong Polytechnic University, Hong Kong, Hong Kong
hadissjalilianmashhoud@polyu.edu.hk
[2] Tarbiat Modares Uinversity of Iran, Tehran, Iran

Abstract. Micropile foundations are progressively recognized worldwide because of their significant advantages such as flexibility, ductility, capacity to withstand extension forces and installability in different ground conditions causing minimal noise and disturbance. Owing to this increasing application of micropile systems, particularly in seismic prone areas, studying their response to seismic excitation is of great importance. In this study, shaking table experiments were carried out on a small-scale physical model of micropiles. The physical model consisted of a 4×4 group of micropiles installed in loose sand. The scaled horizontal acceleration time history of 1995 Kobe earthquake was applied to the base of soil container and response of soil and micropile system was monitored in terms of horizontal acceleration and variation of bending moment induced along two model micropiles. Analysis of the results of shaking table tests showed that acceleration on both soil surface and micropiles cap were amplified with respect to the input excitation acceleration. The horizontal acceleration response on micropiles cap was smaller than response of soil surface. Maximum bending moment was observed at the mid-length of both model micropiles and corner micropile experienced higher bending moment compared to center micropile.

Keywords: Micropile · Sand · Shaking table test · Acceleration response
Bending moment

1 Introduction

Micropiles are defined as piles of small diameter (i.e. less than 300 mm) that can be installed at different inclination angles, in low overhead clearance and almost all types of ground conditions with causing minimal noise and disturbance in the surrounding environment [1]. Owing to their ability to resist axial and lateral loads [2–6] and also slenderness and ductile steel core, micropiles show high flexibility and follow the shock-induced ground displacements while remaining integrated with soil [7, 8]. This is confirmed by field observations after earthquake events such as the 1995 Kobe Earthquake in which friction small diameter piles showed a good performance under seismic loading [9]. As a result, application of micropiles in seismic prone areas as foundation support of new structures and also underpinning and strengthening existing foundations has significantly increased in recent decades [10–18].

© Springer Nature Singapore Pte Ltd. 2018
T. Qiu et al. (Eds.): GSIC 2018, *Proceedings of GeoShanghai 2018 International Conference: Advances in Soil Dynamics and Foundation Engineering*, pp. 138–147, 2018.
https://doi.org/10.1007/978-981-13-0131-5_16

Seismic behaviour of micropiles and the parameters influencing it has been investigated in some studies. According to some of these studies, application of micropile systems, particularly inclined ones, in seismic retrofitting projects as well as remediation purposes was effective [7, 10, 19–26]. However, numerical methods were used in most of the previous studies for probing the response of micropile systems to dynamic excitements [9, 11–14, 17, 24–27]. There were promising findings including positive group effect of micropile groups during dynamic loading [15] and also a delay in development of pore water pressure in liquefiable soil during an earthquake caused by installation of micropiles [23]. Also, peak amplitude of earthquake, configuration of micropile systems (number and inclination angle), slenderness ratio, predominant frequency and mass of superstructure and connection conditions of micropiles were found to be parameters of the highest influence on seismic performance of micropiles [27]. Results of the study conducted by Shahrour et al. [15] also indicated that presence of the superstructure induces significant inertial effect and a large increase in the bending moments in the upper parts of the micropiles. In addition, connection conditions at the head and tip of micropiles [10] and elastoplastic behavior of soil surrounding micropiles [26] was also found to play a major role in their response to seismic loading. Favorable performance of slender micropiles under seismic loading was also confirmed by the results of another numerical investigation [25].

There are a few experimental studies reported in the literature that investigated seismic performance of micropile systems. Benslimane et al. [28] conducted centrifugal model tests and observed a positive group effect increasing with number of micropiles and the inclination angle for the frequency of excitation considered in their experimental study. Moreover, results of a series of shaking table micropile model tests performed in the University of Canterbury in New Zeland by Yang et al. [29], pointed out that micropiles followed the motion of the soil at weak base shakings (<0.25 g). Neglection of soil-pile interaction and pile inertia effect at this level of shaking gave a reasonable approximation in estimating peak bending moment distributions of the micropile. But in strong base shaking (≥ 0.25 g), it became difficult to make good predictions of bending moments using frequency domain methods because of non-linear soil behavior. Further experiments seem to be essential in order to elucidate the seismic performance of micropile systems owing to limited number of experimental studies on this subject. Hence, this study aims at investigating the response of a small-scale physical model of micropiles to a seismic wave through shaking table experiments.

2   Shaking Table Experiments

The uniaxial earthquake shaking table facility in The Hong Kong Polytechnic University (PolyU) was used to perform the 1 g shaking table tests for this study. Table 1 presents the characteristics of this shaking table facility. To reproduce the free field shaking response with minimal boundary effects, a laminar soil container was used that was fabricated using 15 rectangular steel frames. These frames were made by welding hollow sections of 50 mm × 50 mm × 3 mm. The internal dimensions and height of the laminar soil container were 1.4 m, 0.9 m and 0.75 m, respectively

(Fig. 1). After welding the first rectangular frame to a bottom steel plate, the remaining frames were mounted on it. To reduce the friction and allow two adjacent frames to easily slide on each other, 1-mm-thick Teflon layers were glued to the contact surface of the frames. Moreover, a three-dimensional steel frame with cross bracing was set up outside the laminar container for preventing the excessive translational movements of the container. The beams/column and braces of this frame were made of 50 mm × 50 mm × 3 mm hollow square and 40 mm × 40 mm × 3 mm L-shaped sections, respectively.

Table 1. Characteristics of the shaking table in the Hong Kong Polytechnic University

Property	Value
Plan dimensions	3 m × 3 m
Maximum allowable load at 1 g horizontal acceleration	10 ton
Maximum overturning moment	10 ton-m
Working frequency range	1–50 Hz

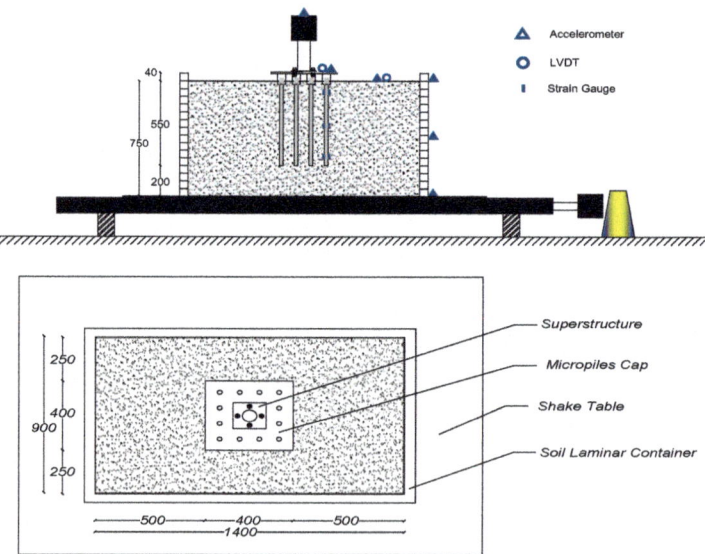

Fig. 1. Schematic view of physical model used for shaking table tests (dimensions in mm): (a) Side view; (b) Plan view.

The soil deposit used for shaking table tests was prepared by pouring poorly graded (SP) river sand imported from China in the laminar soil container(Table 2). Since the average density and moisture content of the sand fill prepared was 1.50 g/cm^3 and 5%, respectively, it can be considered dry cohesion-less soil with low relative density [30].

Table 2. Properties of the soil used in shaking table tests

Property	G_s	C_u	C_c	e_{max}	e_{min}	D_r	c	φ
Value	2.625	3.12	0.93	0.83	0.59	32.48%	3.32 kPa	32°

Full-scale vertical load tests carried out on isolated micropiles in the CEBTP site of Saint Rémy-lès-Chevreuse CEBTP in the French FOREVER national project were considered as the prototype for this study [3]. A scaling factor equal to n = 8.5 was selected according to the scaling relations, specifications of prototype and laminar box. Thus, using the scaling laws proposed for 1 g shaking table tests, which are displayed in Table 3, the physical model for this study was designed and manufactured [31]. Micropile system consisted of a 4 × 4 group of PVC pipes of 23 mm diameter, 3 mm thickness and 550 mm length. The micropile group were connected to a 400 mm 400 mm square steel cap of 8 mm thickness at a spacing of 90 mm. There was a rotational degree of freedom in the direction of shaking for the head of micropiles. To avoid interactions between micropiles cap and soil surface, the cap was free of contact with soil during the tests. In addition, the friction between micropile and surrounding soil was mobilized by gluing sand particles along the entire micropile length.

Table 3. 1 g shaking table scaling relations used to design the small-scale physical model

Quantity	Model
Geometrical dimensions	$1/n$
Horizontal displacement	$1/n^{1.5}$
Mass density	1
Frequency of input motion	$1/n^{0.75}$
Acceleration amplitude	1
EI of pile/width	$1/n^{3.5}$

The instrumentation system used to monitor response of the physical model in this study included 8 accelerometers (Model Bruel & Kjaer 4382) and 6 pairs of half bridge circuited strain gauges(BX120-15AA). As depicted in Fig. 1(a), these sensors were installed at different locations in order to record horizontal acceleration time history and variation of bending strain along two model micropiles during the shaking event. It is noteworthy that the bending moment generated at each level was calculated by multiplying the average bending strain measured at each level by EI/c of the model micropile, which was previously determined in a point loading test. E, I and c denote the Young's modulus, second moment of area and the maximum distance to the neutral axis(i.e. radius) of model micropile. A Lab View 8.2 analogue to digital dynamic data logger was set to scan at 1000 times per second per channel and collected tests data results.

The scaled acceleration time history of 1995 Kobe was considered as the input wave to the shaking table. Time intervals of the original earthquake record were reduced by the factor of 0.201 based on the shaking table tests scaling laws. So, the input wave used in the experiment was of higher frequency and shorter duration. In addition, the magnitude of the earthquake wave was decreased by a reduction scaling factor equal to 0.25. Figure 2 depicts the acceleration time history and Fourier spectrum of the shaking table input wave. However, there was an inevitable slight difference between the wave input to the shaking table and the wave recorded at the base of soil container. The acceleration time history and Fourier spectrum of the wave recorded at the base of soil container are displayed in Fig. 3. It can be seen in Figs. 2 and 3 that there was a 7.65-percent difference between the amplitude of the shaking table input and the wave recorded at the base of laminar soil container. It should b mentioned that the wave recorded at the base of soil container is considered as the base excitation in this study. In addition, the Fourier amplitude of the waves under consideration increased from 0.025 in shaking table input wave to 0.029 in the wave recorded at the base of soil container. However, there was no significant variation in the dominant frequency of the waves under consideration; Dominant frequency of the former wave was 7.20 Hz and slightly increased to 7.26 Hz in the latter.

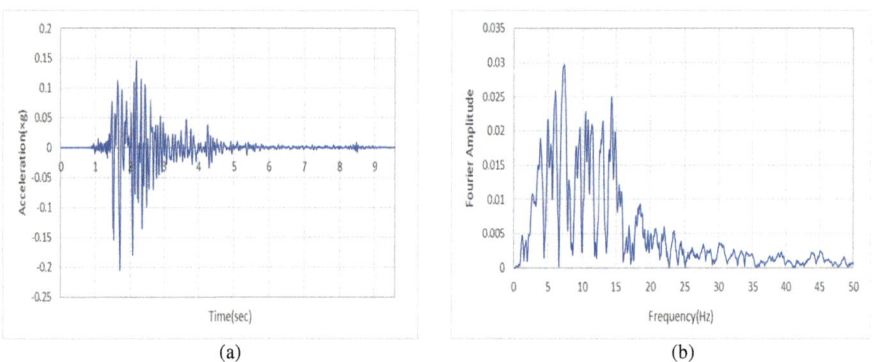

(a)   (b)

Fig. 2. Shaking table input wave used in this study: (a) Acceleration time history, (b) Fourier spectrum

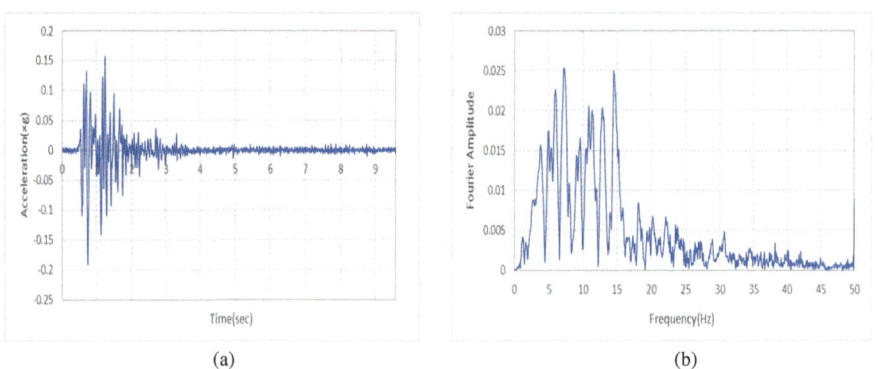

(a)   (b)

Fig. 3. The excitation recorded at the base of soil container used in this study: (a) Acceleration time history, (b) Fourier spectrum

3   Results

Response of soil and micropiles cap to the input excitation was studied in terms of horizontal acceleration. Time history of horizontal acceleration recorded on soil surface in the direction of shaking and its attributed response spectrum are was displayed in Fig. 4.

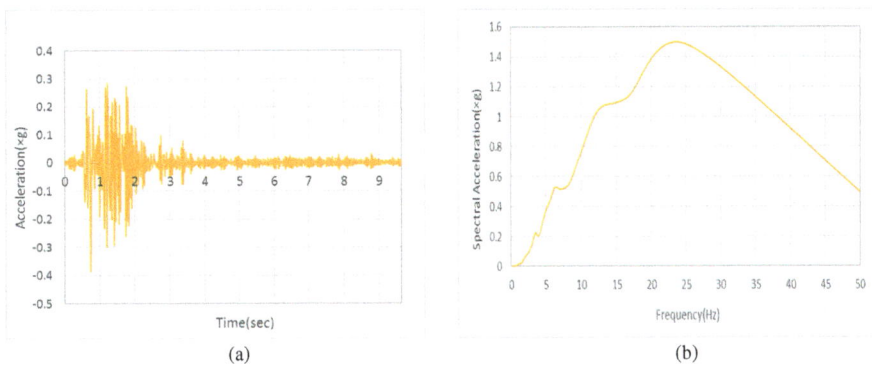

(a)   (b)

Fig. 4. Response recorded on soil surface: (a) Horizontal acceleration time history, (b) Response spectrum.

It can be seen in Figs. 3 and 4 that the absolute maximum acceleration of the input excitation and soil surface (i.e. PGA) were 0.19 g and 0.39 g, respectively. Hence, it is apparent that the magnitude of acceleration on soil surface was amplified with respect to the input excitation acceleration. Low relative density of soil used in the experiments is the reason for the amplification of acceleration in the soil body. In addition, response spectrum calculated for the response wave on soil surface peaks at 1.48 g at a

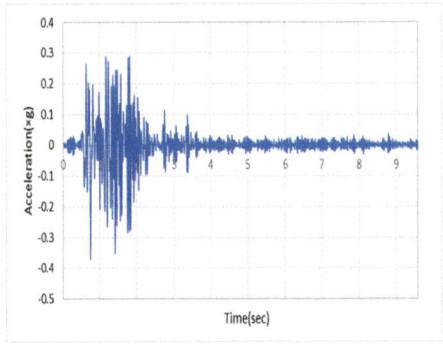

Fig. 5. Horizontal acceleration time history recorded on micropiles cap

frequency equal to 25 Hz. Response of micropile cap to the seismic waves was also monitored in terms of horizontal acceleration and illustrated in Fig. 5.

According to this figure, the horizontal acceleration on micropiles cap was also amplified with respect to the input excitation acceleration. It can be seen in Fig. 5 that the scaled Kobe earthquake wave led to a maximum horizontal acceleration of 0.37 g on the micropiles cap. Moreover, a comparison of the horizontal acceleration response on soil surface and micropile cap indicates that the acceleration response of micropile cap to the input base motion was smaller than the soil surface.

Subsequently, natural frequency of the micropile cap was determined. After computing Fourier spectra of recorded acceleration waves at the base of soil laminar container and micropiles cap, Fourier amplitude of response wave recorded on micropiles cap was normalized by the amplitudes computed for the input excitation. The frequency at which Fourier amplitude ratio peaked was considered as the natural frequency of vibration of micropiles system. Response spectra of horizontal acceleration time history recorded on micropiles cap was also computed. Figures 6 and 7 illustrate normalized Fourier spectrum and response spectrum of horizontal acceleration on micropiles cap, respectively. It is clear in Fig. 6 that the natural frequency of vibration micropiles cap equal 32.41 Hz. Also, according to Fig. 7, the recorded

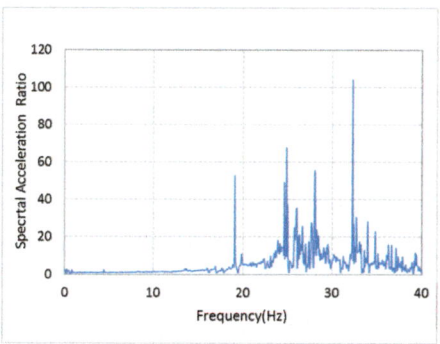

Fig. 6. Normalized Fourier spectrum of acceleration on micropile cap

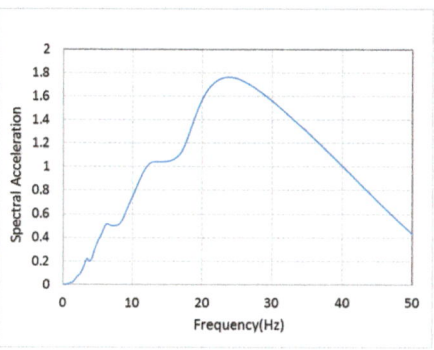

Fig. 7. Response spectrum of acceleration on micropile cap

acceleration time history on micropiles cap reached a maximum response spectral acceleration of 1.75 g.

To examine the seismic response of the instrumented micropiles to the input wave

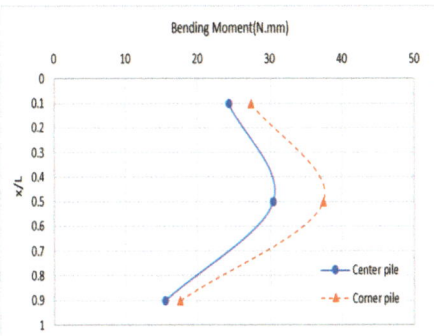

Fig. 8. Bending moment envelope in the instrumented micropiles

in terms of bending moment, distribution of this parameter along the length of these micropiles was determined as displayed in Fig. 8.

It is clear that bending moment in corner micropile rises from 27.37 N.mm close to the head of micropile to 37.39 N.mm in its mid-length and then drops to 17.54 N.mm in vicinity of its tip. It is also clear that bending moment induced in center micropile increased from to 24.35 N.mm near the head to 30.45 in the mid part and then decreased to 15.58 N.mm near the tip. As a result, center micropile experienced an overall lower bending moment along entire its length compared to corner micropile. This observation can be attributed to the edge and shadowing effect in the micropile group.

4   Conclusion

This paper presents the results of shaking table test carried out on a small-scale physical model of micropiles. The physical model consisted of a 4 × 4 vertical micropile group installed in sand of low relative density and was subjected to the scaled 1995 Kobe earthquake acceleration time history. Response of physical model to the shaking event was monitored in terms of horizontal acceleration. According to the test results and their analysis, magnitude of acceleration on both soil surface and micropiles cap was amplified with respect to the input excitation acceleration. The response of micropiles cap was slightly lower than soil surface response. Variation of bending moment along the instrumented micropiles showed that maximum bending moment was induced in the mid-length of both model micropiles. Also, bending moment induced in the corner micropile exceeded center micropile due to edge and shadowing effect.

References

1. FHWA: Micropile design and construction guidelines-Reference manual, Publication No. FHWA NHI-05-039. Federal Highway Administration, U.S. Department of Transportation (2005)
2. Bruce, D.A., Dimillio, A.F., Juran, I.: Micropiles: the state of practice. Ground Improv. **1**(1), 25–35 (1997)
3. Juran, I., Weinstein, G., Sourisseau, L.: Synthesis of the results and recommendations of the French National project on micropiles (FOREVER), pp. 45–51 (2007)
4. Taylor, G.E., Gularte, F.B., Gularte, G.G.: Seismic retrofit of fourth street & riverside viaducts with micropiles. In: Maher, A., Yang, D.S. (eds.) Soil Improvement for Big Digs, ASCE, vol. 81, pp. 313–325. Geotechnical Special Publication, (1998)
5. Zelenko, B.H., Bruce, D.A., Schoenwolf, D.A., Traylor, R.P.: Micropile applications for seismic retrofit preserves historic structure in old San Juan, Puerto Rico. In: Johnsen, L., Berry, D. (eds.) Grouts and Grouting, ASCE, vol. 80, pp. 43–62. Geotechnical Special Publication (1998)
6. Misra, A., Oberoi, R., Kleiber, A.: Micropiles for seismic retrofitting of highway interchange foundation. In: Proceedings of 2nd International Workshop on Micropiles, Yamaguchi University, Ube City, Japan, pp. 215–223 (1999)
7. Lizzi, F.: The pali radice (root pile): a state of the art report. In: Proceedings of the Symposium on Recent Development in Ground Improvement Techniques, Bangkok, pp. 417–432 (1999)
8. Thorburn, S., Littlejohn, G.S.: Underpinning and retention. Springer Science Business Media (1993)
9. Wong, J.C.: Seismic behavior of micropiles. M.Sc thesis, Washigton State University (2004)
10. Sadek, M., Shahrour, I.: Influence of the head and tip connection on the seismic performance of micropiles. Soil Dyn. Earthq. Eng. **26**, 461–468 (2006)
11. Turan, A., El Naggar, H.: Lateral behavior of micropile groups under static and dynamic loads. In: Proceedings of the 4th Canadian Conference on Geohazards (2008)
12. Alsaleh, H., Shahrour, I.: Influence of plasticity on the seismic soil–micropiles–structure interaction. Soil Dyn. Earthq. Eng. **29**, 574–578 (2009)
13. Noorzad, R., Saghaee, G.R.: Seismic analysis of inclined micropiles using numerical method. In: Proceedings of International Foundation Congress and Equipment Expo (2009)
14. Fattah, M.Y., Al-Shakarchi, Y.J., Kadhim, Y.M.: Investigation on the use of micropiles for substitution of defected piles by the finite element method. J. Eng. **16**(3), 5300–5314 (2010)
15. Shahrour, I., Sadek, M., Ousta, R.: Seismic behavior of micropiles: used as foundation support elements: three-dimensional finite element analysis. J. Transp. Res. Board **1772**, 84–90 (2001)
16. Sadek, M., Shahrour, I., Mroueh, H.: Influence of micropile inclination on the performance of a micropile network. Ground Improv. **10**(4), 165–172 (2006)
17. Kawamura, M., Jiang, J.Q.: Dynamic stiffness properties of micropile foundation (1999)
18. Itani, M., Kawamura, T., Onodera, S., Oshita, T.: Centrifugal model test on pile group effect of pile foundation reinforced by micropile. PWRI (2000)
19. Nishitani, M., Umebar Fukui, J.: Development of seismic retrofitting technologies for existing foundations (2001)
20. Takahiro, K., Etsuro, S., Masao, S., Jiro, F., Takeshi, O.: Dynamic behavior of an existing foundation reinforced with micropiles. In: Proceedings of JSCE (2003)
21. Kishishita, T., Saito, E., Miura, F.: Dynamic-response characteristics of structures with micropile foundation system. In: 12th World Conference on Earthquake Engineering (2000)

22. Wang, Z., Mei, G.: Numerical analysis of seismic performance of embankments supported by micropiles. J. Mar. Georesour. Geotechnol. **30**(1), 52–62 (2012)
23. Ousta, R., Shahrour, I.: Three-dimensional analysis of the seismic behaviour of micropiles used in the reinforcement of saturated soil. Int. J. Numer. Anal. Meth. Geomech. **25**, 183–196 (2001)
24. Sadek, M., Isam, S.: Three-dimensional finite element analysis of the seismic behavior of inclined micropiles. Soil Dyn. Earthq. Eng. **24**, 473–485 (2004)
25. Saghaee, G.R., Noorzad, R.: Effects of slenderness ratio on seismic behavior of vertical micropiles. In: Proceedings of Geo-Risk 2011, pp. 352–360 (2011)
26. Shahrour, I., Alsaleh, H., Souli, M.: 3D elastoplastic analysis of the seismic performance of inclined micropiles. Comput. Geotech. **39**, 1–7 (2012)
27. Ghorbani, A., Hasanzadehshooiili, H., Ghamari, E.: Comprehensive three-dimensional finite element analysis, parametric study and sensitivity analysis on the seismic performance of soil–micropile–superstructure interaction. Soil Dyn. Earthq. Eng. **58**, 21–36 (2014)
28. Benslimane, A., Juran, A., Hanna, A., Drabkin, S., Perlo, S., Frank, R.: Seismic retrofitting using micropile systems: centrifugal model studies. In: Proceedings of 4th International Conference on Case Histories in Geotechnical Engineering, Missouri, pp. 1260–1268 (1998)
29. Yang, J.X., McManus, K.J, Berrill, J.B.: Kinematic soil-micropile interaction. In: 12th World Conference on Earthquake Engineering, Auckland, New Zealand, pp. 1–8 (2000)
30. Chau, K.T., Shen, C.Y., Guo, X.: Nonlinear seismic soil–pile–structure interactions: shaking table tests and FEM analyses. Soil Dyn. Earthq. Eng. **29**(2), 300–310 (2000)
31. Iai S.: Similitude for shaking table tests on soil-structure-fluid model in 1 g gravitational field. Report of the Port and Harbor Research Institute 27(3) (2002)

Effects of Near-Fault Pulse-Like Ground Motion on Site Liquefaction and Settlement

Wenkui Dong[1,2,3], Guangyun Gao[1,2(✉)], Juan Chen[1,2], and Song Jian[4]

[1] Department of Geotechnical Engineering, Tongji University,
Shanghai 200092, China
[2] Key Laboratory of Geotechnical and Underground Engineering
of Ministry of Education, Tongji University, Shanghai 200092, China
gaoguangyun@263.net
[3] Center for Built Infrastructure Research (CBIR),
School of Civil and Environmental Engineering,
University of Technology Sydney, Ultimo, NSW 2007, Australia
[4] College of Civil and Transportation Engineering,
Hohai University, Nanjing 210098, China

Abstract. A soil column of 20 m is established to simulate the seismic responses of free site by using the open source finite element software Open-Sees. The characteristics of site liquefaction are investigated under near-fault ground motions, especially under the ground motion component with directivity effect. Based on the index of unit settlement of site, the influence of different types of near-fault ground motions on the susceptibility of site liquefaction and the seismic settlement is analyzed. The results show that the near-fault pulse-like ground motion is more likely to induce the occurrence of site liquefaction and cause a larger permanent vertical deformation, especially the double-sided directivity-pulse motions. The liquefaction susceptibility and the settlement of the site under one-side pulse-like motions are lower than that under double-sided directivity-pulse motions because of its pseudo-static characteristic.

Keywords: Directivity effect · Pulse-like ground motion
One-side pulse-like motion · Site liquefaction
Liquefaction-induced deformation

1 Introduction

The near-fault ground motion is wholly different with that away from the source area. It is significantly influenced by rapture mechanism, direction of propagation and the point position, etc., and thus has the effects of rapture directivity, hanging wall and fling-step.

Several studies have examined the characters of near-fault ground motions. Feng et al. (2004) analyzed the acceleration peak ratios, equivalent velocity pulse, response spectrum and the attenuation law of near-fault ground motions. The basic properties of directivity effects, hanging wall effects and fling-step effects were captured using FLAC technique to simulate the near-fault peak velocity field by Pan (2006). The effects of near-fault ground motions on buildings were investigated as well. For instance,

© Springer Nature Singapore Pte Ltd. 2018
T. Qiu et al. (Eds.): GSIC 2018, *Proceedings of GeoShanghai 2018 International Conference: Advances in Soil Dynamics and Foundation Engineering*, pp. 148–157, 2018.
https://doi.org/10.1007/978-981-13-0131-5_17

Hall et al. (1995) studied the characteristic of near-fault ground motion including the frequency components, duration time, incremental velocity and displacement; and indicated that these properties would have dramatic influences on structural nonlinear response. In the aspect of slope sliding, the influences of near-fault effects on the slope dynamic response were analyzed through building dynamic analysis models of soil slope by FLAC technique by Song (2013). Meanwhile, the characters of near-fault ground motion and parameters of the slope were taken into account to calculate the slope sliding displacement (Song et al. 2014).

However, relatively little attention has been given to effects of near-fault ground motion on site liquefaction and large deformation. New Zealand suffered the $M7.1$ Darfield earthquake and the $M6.3$ Christchurch earthquake in the year 2010 and 2011. Contrary to the common sense, earthquake with smaller magnitude caused more severe damage. On the basis of these engineering cases, Green (2008) assessed the influence of rupture directivity on the inducement of liquefaction in sand, holding the view that the strike normal components tended to induce larger cyclic stresses in the soil than the strike parallel components; and the former had fewer numbers of equivalent cycles as compared to the latter. These trends are somewhat compensating in their influence on the inducement of liquefaction and the rupture directivity pulses had a slightly larger potential to induce liquefaction. Carter (2014) analyzed the numbers of equivalent cycles n and magnitude scaling factors (MSFs) under the separate and joint work of strike normal and parallel ground motion component; and concluded that directivity effects resulted in less severe liquefaction during the Darfield earthquake than the Christchurch earthquake. These two papers primarily indicate the effects of near-fault ground motion on liquefaction. Pitifully, their conclusions are contradictive.

In this paper, a soil column is established to simulate the seismic responses of free site by the open source finite element software OpenSees; In order to take into account the liquefaction properties and the accumulation of shear deformation under cyclic loading, the soil constitutive model is established based on the boundary surface plasticity model by Wang et al. (2014). Results of this paper may provide reference for further study of near-fault field liquefaction.

2 Modeling and the Selection of Ground Motions

2.1 Modeling

Considering the view that the vibration response of the free site could be stimulated through shear beams by Phillips (2012), this paper reference to numerical modeling by Wang (2014) creating the soil column model of 20 m length, as shown in Fig. 1. Furthermore, the model consists of twenty cubic elements BrickUP which is three-dimensional and fully coupling (OpenSees command language 2006). To simulate the real seismic process more closely, the nodes in each same level are tied together in both degrees of freedom, only the bottom nodes are fixed. In addition, the top surface is drained thus the pore water pressure is always zero, while the undersurface and other lateral the profiles of the model are water-proof.

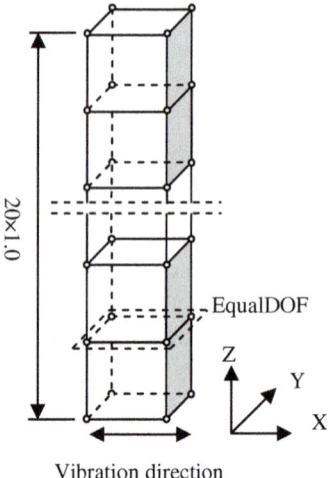

Fig. 1. The model of the site

2.2 Selection of Ground Motions

The model is subject to unilateral excitation. The unilateral excitation exerted on the model is selected from the Darfield and Christchurch earthquake events in New Zealand. Figure 2 shows the liquefaction area in Christchurch during the Darfield and Christchurch earthquake.

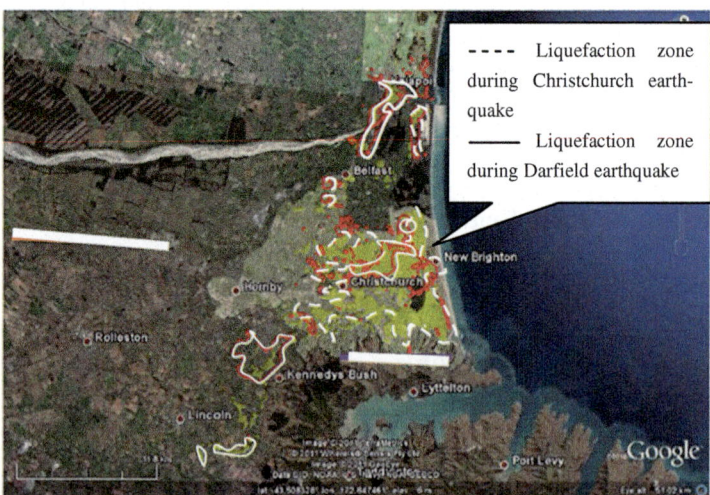

Fig. 2. The liquefaction caused in Christchurch during the Darfield and Christchurch earthquake

The heavy line in the west of the city and distance 37 km is the fault zone of the Darfield earthquake, and the other one belongs to the Christchurch earthquake which is 9 km away in the southern. It is obvious that the Christchurch earthquake with much lower magnitude resulted in severe liquefaction. So 40 near-fault (epicentral distance is less than 25 km) ground motions of these events are selected from the Next Generation Attenuation (NGA) database as shown in Table 1. It is noteworthy that the peak ground velocity (PGV) of each ground motion is adjusted to the same value for variable controlling.

Table 1. Near-fault motions

Fault-like motions			Non-pulse motions		
NGA sequence NO.	Station	Distance* (km)	NGA sequence NO.	Station	Distance* (km)
6887	CBG	18.1	6886	CAC	14.5
6897	DSLC	8.5	6888	CCC	19.9
6906	GDLC	1.2	6889	CH	18.4
6911	HORC	7.3	6890	CCHS	17.6
6927	LINC	5.1	6893	DFHS	11.9
6959	CR	19.5	6915	HVPS	24.5
6960	RHS	13.6	6930	LRSC	12.5
6962	ROLC	1.5	6952	PHS	18.7
6966	SL	22.3	6953	PRPS	24.6
6969	SMTS	20.9	6961	RKAC	16.5
6975	TPLC	6.1	6965	SBRC	24.3
8064	CCC	3.3	8062	CAC	14.4
8119	PRPS	2.0	8063	CBG	5.6
8123	CR	5.1	8066	CH	4.9
			8067	CCHS	4.5
			8090	HDPS	4.4
			8099	KNS	17.9
			8102	LINC	18.5
			8110	MQZ	16.1
			8118	PHS	9.1
			8124	RHS	9.4
			8126	ROLC	24.3
			8130	SL	5.6
			8134	SMTS	11.3
			8142	TPLC	16.6
			8157	HVPS	3.4

*Closest distance to the ruptured area on the fault

3   The Influence of Pluse-like Ground Motion on Site Liquefaction and Settlement

3.1   Influence of Pulse-like and Non-pulse Ground Motions on Site Liquefaction and Settlement

Near-fault pulse-like ground motion is mainly induced by the forward directivity effect with a short ground motion duration and high ground peak velocity (PGV). A series of site response analysis were performed and the permanent displacements of the soil column were shown in Fig. 3. The mean values of the displacement by different types of ground motion in these seismic events are listed in Table 2.

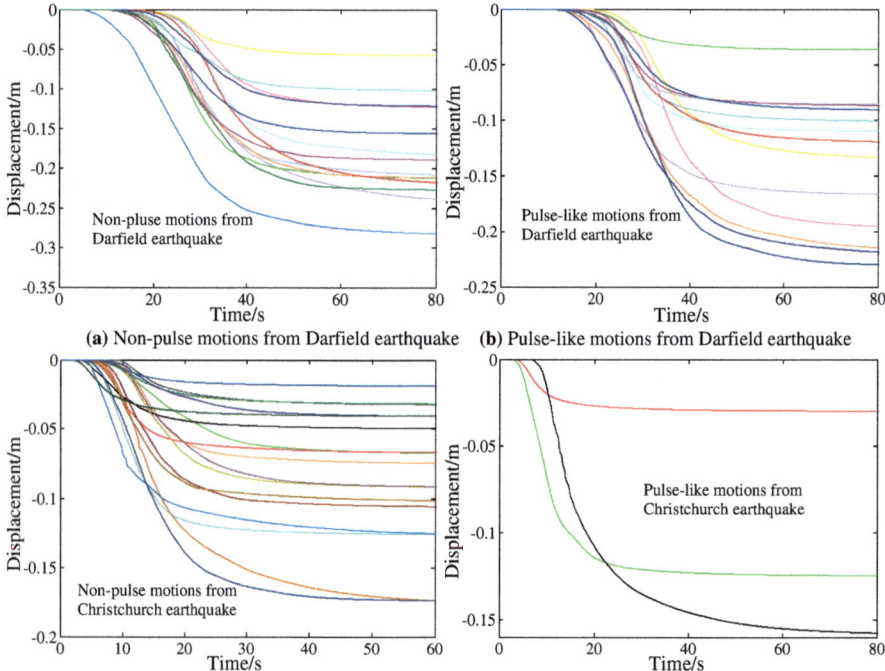

Fig. 3. The settlement of the soil column under pulse-like and non-pulse ground motions during the Darfield and Christchurch earthquake

Table 2. Mean values of permanent displacement for near-fault pulse-like and non-pulse ground motions

Types	Mean values of settlements/cm	Mean durations/s
Non-pulse of Christchurch earthquake	8.25	37.28
Pulse-like of Christchurch earthquake	10.38	25.52
Non-pulse of Darfield earthquake	13.72	40.37
Pulse-like of Darfield earthquake	15.08	34.32

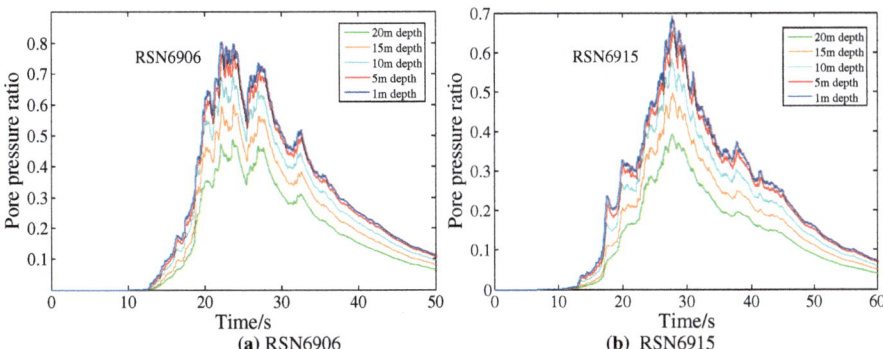

Fig. 4. The pore pressure ratio in the soil column of the ground motions RSN6906 and RSN6915

It can be observed that the pulse-like ground motion tends to yield a greater permanent displacement than non-pulse ground motion, while its duration time of displacement development is shorter than that of non-pulse ground motion. This explains why Christchurch earthquake causes more severe damage than Darfield earthquake.

To further assess the influence of near-fault motions on site liquefaction triggering and the permanent deformation. The pulse-like ground motion RSN6906 and non-pulse ground motion RSN6915 were selected as the input loading. The pore water pressure in the soil column are measured and shown in Fig. 4.

Figure 4 shows that, the pore water pressure induced by pulse-like ground motion fluctuated just like a pulse while non-pulse ground motion does not share the same character. Besides, we can see the soil column under pulse-like ground motion suffers higher pore water pressure and shorter duration time on the same condition, which indicates that liquefaction can be triggered more easily by pulse-like ground motion, and the permanent deformation would then be induced as well.

3.2   Influence of Fault-Parallel and Fault-Normal Component of Pulse-like Motion on Site Liquefaction and Settlement

Only the fault-normal component of pulse-like motion tends to produce large velocity pulses due to the directivity effect in seismology. To figure out how far the components work in triggering site liquefaction and the displacement, all the 14 pulse-like ground motions from these two events were selected. According to the NGA database, the September 2010 Darfield earthquake strikes 85°, almost east-west trend, and next year's Christchurch earthquake strikes 59°. It was found that when the fault-normal and fault-parallel motions were treated individually, the former yielded a higher displacement than the latter in 13 records, as shown in Table 3. Only one record had an opposite result.

Figure 5 shows the time history of the pore pressure ratio induced by RSN6969 in the column with the depth of 20 m and 1 m, and the settlement-time curve are shown in

Table 3. The influence of fault-parallel and fault-normal component of pulse-like motions on site settlement

RSN	6887	6897	6906	6911	6927	6928	6959
Fault-normal component	0.20	0.17	0.09	0.11	0.10	0.07	0.24
Fault-parallel component	0.19	0.17	0.08	0.09	0.10	0.04	0.22
Ratio	1.06	1.04	1.13	1.17	1.01	1.81	1.06
RSN	6962	6969	6966	6975	8064	8119	8123
Fault-normal component	0.12	0.23	0.20	0.12	0.12	0.04	0.16
Fault-parallel component	0.09	0.16	0.22	0.09	0.11	0.03	0.08
Ratio	1.36	1.41	0.90	1.34	1.14	1.17	2.06

the Fig. 6. It is clear that the pore water pressure induced by fault-normal component is larger in both depth, as well as the permanent displacement, which is completely contrast to the conclusion put forward by Carter who believe fault-normal component are less demanding on soil and resulting less severe liquefaction. Most notably, the developing speed of fault-normal component is faster as well when compared to fault-parallel component.

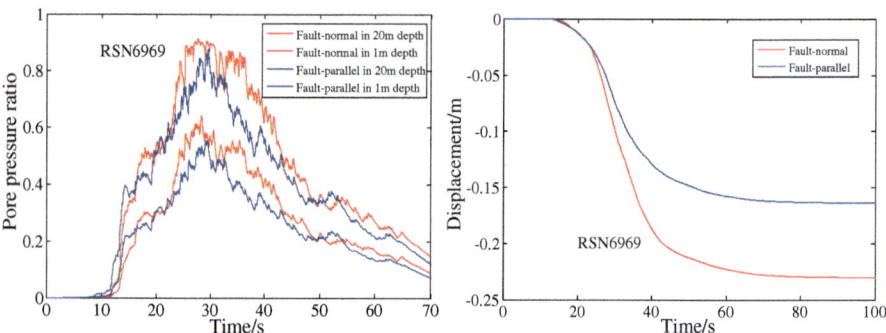

Fig. 5. The pore pressure ratio with the depth of 20 m and 1 m

Fig. 6. Displacement curve

3.3 Influence of One-Side Pulse-like Ground Motion on Site Liquefaction and Settlement

Pulse-like ground motions are more likely to be created by the forward directivity effect which always performed with double-sided pulse as is mentioned above. While a small part of the pulse-like ground motions manifested as one-side pulse-like which mainly induced by the fling-step effect have much higher PGV, as shown in Fig. 7.

The influence on site liquefaction and settlement under one-side pulse-like ground motion is totally different with two-side pulse-like ground motion. Due to the limited space, this paper selects all four one-side pulse-like ground motions and some

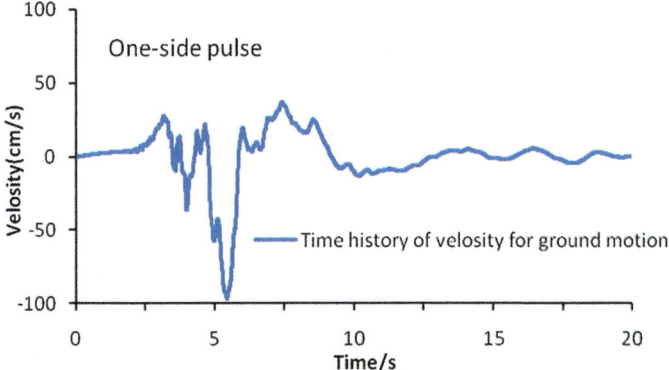

Fig. 7. Time history of velocity of one-side pulse ground motion

pulse-like motions with double-sided pulse from these two earthquake events. However, because of the huge gap of lasting time of these ground motions, the concept of unit settlement h, that is the value of settlement in per unit time, is involved to measure the influence. The ground motion duration is defined as the time interval when the displacement developing from 5% to 95% of the final settlement.

$$h = H/s \tag{1}$$

Where, H is the final settlement of the soil column in cm. s is the duration of the motions in sec.

Table 4 shows the maximum pore water pressure ratio in the 1 m depth, the final settlement, the duration and the unit settlement of the soil column under the motions with one-side pulse-like and double-sided pulse. Obviously, the pore water pressure ratio induced by one-side pulse-like motions is much less than double-sided pulse motions, as well as the unit settlement, the former causes the mean value of 0.23, and twice the value 0.46 produced by the latter counterparts.

Table 4. The maximum pore water pressure ratio and unit settlement of the soil column under one-side and double sided pulse

Types	One-sided pulse				Double-sided pulse				
RSN	6906	6928	6942	8119	6887	6897	6962	6969	8064
e_{max}*	0.79	0.60	0.78	0.65	0.91	0.82	0.85	0.92	0.95
Duration s (sec)	28.68	27.56	50.70	22.92	46.00	40.53	26.95	38.81	18.29
Final settlement H (cm)	7.60	6.51	13.94	3.50	20.29	17.11	11.67	16.38	10.94
unit settlement h	0.27	0.24	0.27	0.15	0.44	0.42	0.43	0.42	0.60

*e_{max} means the maximum pore water pressure ratio of the soil column.

4   Conclusions

The soil column of 20 m is established in this paper to simulate the seismic responses of free site by the open source finite element software OpenSees. The soil constitutive model is based on boundary surface plasticity model. Influences on site liquefaction and settlement with the pulse-like and non-pulse ground motions are analyzed, and the effects of components with or without the directivity effect are surveyed as well. In addition, the liquefaction effect under one-side pulse-like ground motions is primarily investigated. Mainly results are as follows:

(1)   The pulse-like ground motion tends to trigger site liquefaction more easily and cause larger permanent settlement, which might due to the concentration of seismic energy and the high PGV.

(2)   The fault-normal component of pulse-like motion which contains direction effects also has more potential to induce site liquefaction when compared to the fault-parallel counterparts. Notably, the result is greatly different from the conclusion of Carter's.

(3)   In the liquefiable site, contrary to the past cognition,the damage caused by one-side pulse-like ground motion is less severe than that of double-sided ground motion. This is because that the soil liquefaction could hardly be triggered without cyclic loading can hardly, even one-side pulse-like ground motion with high PGV.

(4)   The reason why the damage caused by Christchurch earthquake was much severe than Darfield earthquake is not only because of the shorter epicentral distance or the secondary damage, but also the types of the ground motions.

Aknowledgements. The research of this project is partly supported by the natural science foundation of china (no. 41372271). However, any opinions, findings, and conclusions expressed in this paper are those of the authors and do not necessarily reflect the views of the national science foundation.

References

Carter, L., Green, R., Bradley, B., et al.: The influence of near-fault motions on liquefaction triggering during the Canterbury earthquake sequence. In: Soil Liquefaction During Recent Large-Scale Earthquakes (2014)

Feng, Q.M., Shao, G.B.: Research on attenuation of near fault peak strong ground motion velocity and displacement. Earthq. Eng. Eng. Vibr. **24**(4), 13–19 (2004). (In Chinese)

Green, R.A., Lee, J., White, T.M.: The significance of near-fault effects on liquefaction, Beijing (2008)

Hall, J.F., Heaton, T.H., Halling, M.W., et al.: Near-source ground motion and its effects on flexible buildings. Earthq. Spectra **11**(4), 569–605 (1995)

Pan, B., Xu, J.D., et al.: The simulation of near fault ground motions in the region of Beijing. Seismolog. Geol. **28**(4), 623–634 (2006). (In Chinese)

Mazzoni, S., Mckenna, F., Scott, M.H.: The Open System for Earthquake Engineering Simulation (OpenSEES) User Command-Language Manual. Pacific Earthquake Engineering Research Center (PEER), University of California (2006)

Phillips, C., Hashash, Y.M.A., Olson, S.M.: Significance of small strain damping and dilation parameters in numerical modeling of free-field lateral spreading centrifuge tests. Soil Dyn. Earthq. Eng. **42**, 161–176 (2012)

Song, J.: Dynamic response of soil slope under near-fault ground motions. China Earthq. Eng. J. **35**(1), 62–68 (2013)

Song, J.: Predictive models for permanent displacement of slopes induced by near-fault pulse-like ground motions. Rock Soil Mech. **35**(5), 1340–1347 (2014)

Wang, R., Zhang, J., Wang, G.: A unified plasticity model for large post-liquefaction shear deformation of sand. Comput. Geotech. **59**(3), 54–66 (2014)

Influence of Consolidation Pressure on Cyclic and Post-cyclic Response of Fine-Grained Soils with Varying Mineralogical Compositions and Plasticity Characteristics

Beena Ajmera[1](✉), Binod Tiwari[2], Brian Yamashiro[3], and Quoc-Hung Phan[3]

[1] California State University, Fullerton, 800 N. State College Blvd., E-318, Fullerton, CA 92831, USA
bajmera@fullerton.edu
[2] California State University, Fullerton, 800 N. State College Blvd., E-419, Fullerton, CA 92381, USA
btiwari@fullerton.edu
[3] California State University, Fullerton, 800 N. State College Blvd., Fullerton, CA 92831, USA
byamashiro@gmail.com, qphan001@csu.fullerton.edu

Abstract. In this study, nine different soil samples with varying mineralogical compositions with a wide range of plasticity characteristics were consolidated to five different normal stresses prior to testing in the cyclic simple shear apparatus, where cyclic loading in the form of a sinusoidal wave form with a frequency of loading of 0.5 Hz and varying amplitudes was applied. The shear strength of the samples immediately after cyclic loading were also determined and compared to the static shear strength to determine the reduction in shear strength resulting from the cyclic loading. The results were used to determine the cyclic strength curves for each sample at 2.5%, 5%, and 10% double amplitude shear strain. These curves demonstrated that an increase in the consolidation pressure corresponded to an increase in the cyclic resistance. The post-cyclic undrained shear strength measurements suggested that an increase in the consolidation pressure would cause a lower reduction in undrained shear strength as a result of cyclic loading.

Keywords: Cyclic response · Degradation ratio · Cyclic mobility
Post-cyclic behavior · Cyclic simple shear

1 Introduction

There are several important problems in geotechnical engineering that require information about the peak shear strength of soils after cyclic loading. Some examples of these problems include the design of structural foundations after earthquake loading, the variation in the ultimate bearing capacity in driven piles as well as the evaluation of stability of slopes subjected to wave action [1]. Previous work has examined how

© Springer Nature Singapore Pte Ltd. 2018
T. Qiu et al. (Eds.): GSIC 2018, *Proceedings of GeoShanghai 2018 International Conference: Advances in Soil Dynamics and Foundation Engineering*, pp. 158–167, 2018.
https://doi.org/10.1007/978-981-13-0131-5_18

various factors will influence the cyclic and post-cyclic behavior of soils. The factors that have been examined may be related to either the characteristics of the soil and its stress history, characteristics of the cyclic loading function applied, and/or other parameters for the testing conducted. Some of the characteristics of the soil and the stress history that have been studied include the plasticity index [2–9], the over consolidation ratio [10–13], the initial static shear stress [6, 12, 14], the consolidation pressure, [2, 8, 14, 15] degree of saturation [16], percent of non-plastic or plastic fines [17–25], and mineralogy [9, 26–35]. The impact of varying the loading frequency [14, 36–38], the number of cycles of loading [6, 39–41], and allowing drainage between cycles [42–44] on the cyclic and post-cyclic behavior have also been examined. However, the conclusions drawn from these studies are often contradictory suggesting that there is still substantial work that needs to be undertaken in order to thoroughly understand the behavior of clayey soils during cyclic loading and immediately after cyclic events. For example, consider the impact of confining pressure on the cyclic resistance. Bray and Sancio [2] and Bray et al. [8] showed that an increase in the confining pressure would result in an increase in the cyclic resistance in soils with the same plasticity index. Ding et al. [15] also concluded that the cyclic resistance of marine clays from Bohai Bay, China will increase with an increase in the confining pressure. However, Thammathiwat and Chim-oye [14] found that an increase in the confining pressure caused a reduction in the cyclic resistance in normally consolidated soils subjected to the same frequency of loading.

The contradictory conclusions may be a result of the limited number of soils tested. Another source of the inconsistencies in the conclusions is the differences in the choice of various parameters and the testing procedures. In this study, the results of a systematic evaluation of the impact of the confining pressure on the cyclic and post-cyclic resistance of natural soils with varying mineralogical compositions, plasticity characteristics and depositional characteristics is presented.

2 Background on Locations of Natural Samples

2.1 Cyclic Failure Site at Lokanthali, Nepal

Four of the natural samples considered in this study were collected from a cyclic failure site at Lokanthali, Nepal, which is located at the southeast border of Kathmandu. The site is located along the Araniko Highway, which serves as the main overland connection to China and Tibet from Nepal. The cyclic failure was evidenced during the M_w 7.8 2015 Gorkha (Nepal) earthquake that occurred on 15 April 2015. The main shock was followed by hundreds of aftershocks several of which were with M_w greater than 6.0 [45–47]. The earthquake and its aftershocks caused substantial damage to many buildings and hydropower plants, obstructed transportation networks, partially blocked the natural flow of rivers, and triggered thousands of landslides in the affected region [48]. One particular landslide occurred in Lokanthali causing the collapse of a 200 m stretch of the Araniko Highway and resulting in the partial or full damage of over 20 buildings in the surrounding area [49]. Soil samples were collected from this cyclic failure site and transported to the geotechnical engineering laboratory at California State University, Fullerton under an USDA permit for laboratory testing.

2.2   Montmorillonite Seam at Portuguese Bend Landslide, CA, USA

The remaining natural samples that were tested as part of this study were collected from the montmorillonite seam of the Portuguese Bend Landslide in the Rancho Palos Verdes Peninsula, CA, USA. Encompassing a total area of approximately 270 acres, the Portuguese Bend Landslide is the cause of substantial damage to roads, homes, utilities, and other infrastructure within and around the slide area [50]. The landslide is underlain by montmorillonitic clays that were formed from volcanic ash deposits. The natural samples obtained in this study were collected from this layer and were tested in the geotechnical engineering laboratory of CSU, Fullerton.

3   Soil Properties and Testing Procedures

3.1   Soil Properties

The procedures outlined in ASTM D4318-10 [51] were adapted in order to determine the Atterberg limits of each of these soil samples. Table 1 summarizes the Atterberg limits and the USCS classification for each of the soils tested in this study. The mineralogical composition of the samples were found using X-ray diffraction tests. The results indicated that the kaolinite was the predominant clay mineral in the soils collected from Lokanthali, while montmorillonite was the predominant clay mineral in the soils collected from the Portuguese Bend Landslide.

Table 1. Atterberg limits and USCS classification for the soils tested in this study.

Soil ID	Location	LL	PI	USCS classification
SN1	Lokanthali, Nepal	32	12	CL
SN2	Lokanthali, Nepal	35	12	CL
SN3	Lokanthali, Nepal	46	17	ML
SN4	Lokanthali, Nepal	44	13	ML
PB1	Portuguese Bend Landslide, USA	58	22	MH
PB2	Portuguese Bend Landslide, USA	73	33	MH
PB3	Portuguese Bend Landslide, USA	55	25	MH

3.2   Simple Shear Testing Procedures

Static and cyclic simple shear testing was conducted on the portion of the soil sample passing through the U.S. Sieve No. 40 (particles smaller than 0.042 mm). Sufficient de-aired distilled water was added to the soil sample in order to obtain an initial liquidity index of unity. The resulting soil slurry was placed in an airtight container to allow the samples to hydrate for a period of at least 24 h after which the samples were used to conduct the static and cyclic simple shear testing. The procedures for the static simple shear testing are described in ASTM D6528-07 [52], while cyclic simple shear testing procedures previously detailed by the authors have been followed as no ASTM standard for cyclic simple shear testing is available [26–31].

Norwegian Geotechnical Institute (NGI)-type static and cyclic simple shear devices were utilized in this study. The portion of the slurried soil was placed into a rubber membrane and confined by a stack of Teflon®-rings to form a sample with a diameter of 63.5 mm at a height of 25.4 mm. The sample was then subjected to incremental consolidation stresses until the desired consolidation pressure was achieved. In this study, the samples were consolidated to pressure of 100 kPa. The consolidation process was monitored through real-time logarithm of time versus vertical deformation curves, which were used to determine the end of the primary consolidation before the vertical stress was increased. Upon the completion of the primary consolidation at the final consolidation stress, the samples in the static simple shear apparatus were sheared to failure under undrained, constant volume conditions at a strain rate of 5% per hour. On the other hand, samples in the cyclic simple shear apparatus were subjected to cyclic loading under undrained, constant volume, stress-controlled conditions. A sinusoidal loading function with a frequency of 0.5 Hz and an amplitude determined from a designated cyclic stress ratio (CSR) was applied. The cyclic stress ratio is defined as the ratio of the amplitude of the cyclic loading to the consolidation pressure. In this study, several samples from the same batch mixture were tested under different cyclic stress ratios for each consolidation pressure. The cyclic loading was applied until the sample experienced 10% double amplitude shear strain. If the sample did not experience this double amplitude shear stain in 500 cycles, the cyclic loading phase was terminated at 500 cycles. Immediately following the cyclic loading phase, the samples were subjected to static undrained, constant volume shearing at a rate of 5% per hour. This phase is referred to as the post-cyclic shearing phase. Both static test and the post-cyclic shearing phase of the cyclic test were terminated when the peak strength was measured or after a maximum shear strain of 25%. Soil SN1 from Lokanthali, Nepal was also tested at consolidation pressures of 50 kPa, 200 kPa, 400 kPa and 800 kPa.

4   Results and Discussion

Shown in Fig. 1 is the typical results obtained from the cyclic simple shear tests. The results in Fig. 1 are for SN3 from Lokanthali, Nepal under a cyclic stress ratio of 0.20. The figure contains the applied sinusoidal cyclic loading function. The application of this cyclic loading causes a decrease in the effective normal stress and a corresponding increase in the pore pressure. The resulting variations for both the effective normal stress and the pore pressure are shown in Fig. 1. The resulting hysteresis loops for this sample is also shown in Fig. 1.

The results of cyclic simple shear test were used to develop cyclic strength curves, or plots of the cyclic stress ratio versus the number of cycles to particular failure criteria. In this study, cyclic strength curves were developed using the number of cycles required to cause 2.5%, 5% and 10% double amplitude shear strains in the soil samples.

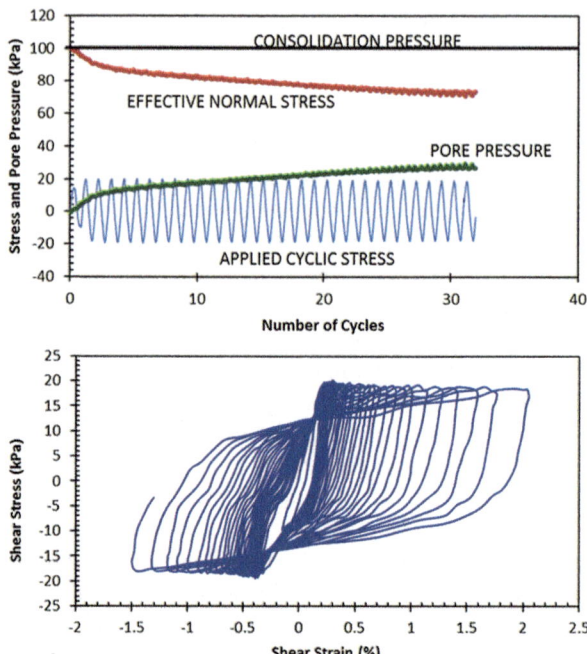

Fig. 1. Applied cyclic loading function, resulting effective normal stress and pore pressure as well as hysteresis loops for SN3 from Lokanthali, Nepal under a consolidation pressure of 100 kPa at a cyclic stress ratio of 0.20.

Shown in Fig. 2 are the cyclic strength curves for all of the samples under a consolidation pressure of 100 kPa. These cyclic strength curves are for 5% double amplitude shear strain. Figure 2 shows that as the plasticity index increases, the cyclic resistance offered by the soil mass also increases. Moreover, soils having montmorillonites exhibited higher cyclic resistance compared to the soil containing kaolinite or illite. The variation in the cyclic resistance with confining pressure is demonstrated in Fig. 3, which contains the cyclic strength curves at 5% double amplitude shear strain. Figure 3 shows that the cyclic resistance was highest in the soils subjected to a consolidation pressure of 50 kPa. In the soils with consolidation pressures greater than 50 kPa, the cyclic resistance did not vary significantly, as shown by the upper and lower bound cyclic strength curves in Fig. 3. This may be attributed to the curvature in the undrained failure envelope. Specifically, Tiwari and Ajmera [53] demonstrated that the undrained failure envelope was curved when the vertical stress was less than 100 kPa, but was linear when the vertical stresses were greater than 100 kPa. Thus, it would be expected to see greater variations in the cyclic resistances at vertical stresses less than 100 kPa, but little differences in the cyclic resistance in terms of cyclic stress ratio when the vertical stress was greater than 100 kPa.

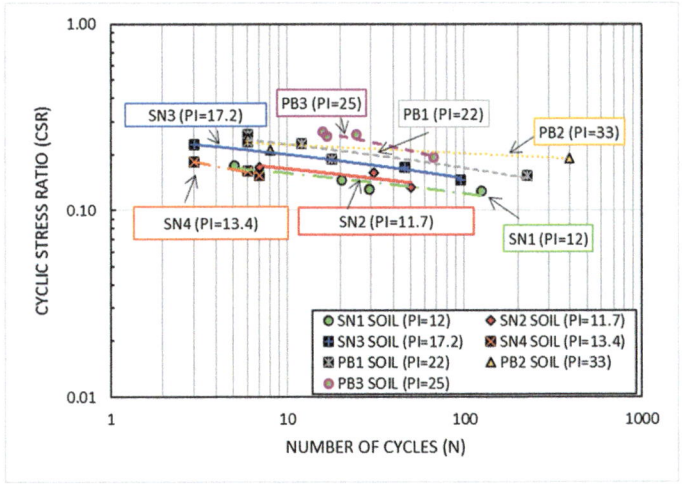

Fig. 2. Cyclic strength curves at 5% double amplitude shear strain for the samples tested in this study at the consolidation pressure of 100 kPa.

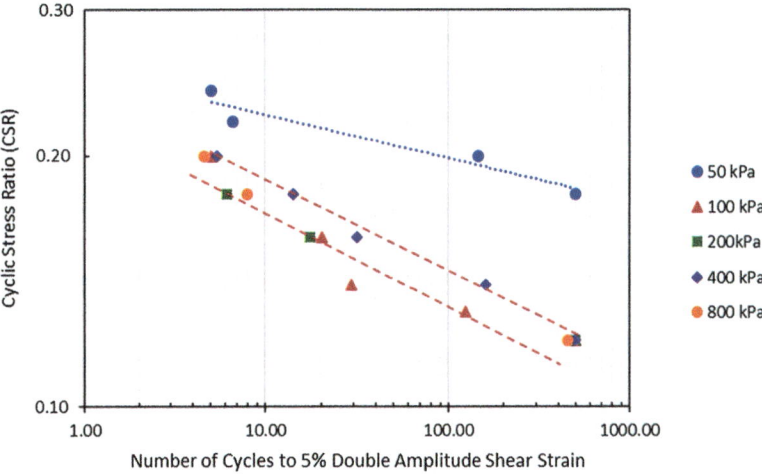

Fig. 3. Cyclic strength curves at 5% double amplitude shear strain for SN1 from Lokanthali, Nepal at different consolidation pressures.

The results from the post-cyclic shearing phase were used to determine the reduction in the shear strength due to cyclic loading. Figure 4 contains the variation in the ratio of the post-cyclic undrained shear strength to the static undrained shear strength with respect to the consolidation pressure. The figure shows that an increase in the consolidation pressure results in lower reductions in the post-cyclic shear strength. Furthermore, the maximum reduction in shear strength due to cyclic loading appears to be approximately 50%.

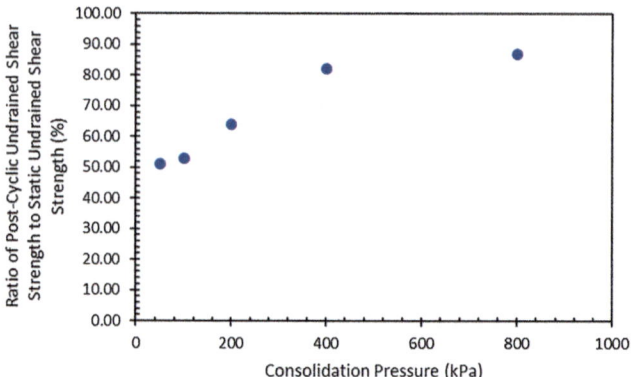

Fig. 4. Variation in the ratio of the post-cyclic undrained shear strength to static undrained shear strength with consolidation pressure for SN 1 from Lokanthali, Nepal.

5   Conclusions

The results of cyclic simple shear testing on nine natural samples collected from a cyclic failure site in Lokanthali, Nepal and from the Portuguese Bend Landslide, USA were presented in this study. The following conclusions were drawn from the results obtained for the experimental studies:

- An increase in the plasticity index caused an increase in the cyclic resistance in the soil samples tested.
- The cyclic resistance in terms of cyclic stress ratio was found to be greatest when the vertical stress was 50 kPa, but the cyclic resistance did not vary significantly when the vertical stress was greater than 100 kPa.
- The ratio of the post-cyclic undrained shear strength to the static undrained shear strength was found to increase with an increase in the consolidation pressure.

References

1. Sangrey, D.A., France, J.W.: Peak strength of clay soils after a repeated loading history. In: Proceedings of the Ninth International Symposium on Soils Under Cyclic and Transient Loading, vol. 1, pp. 421–430 (1980)
2. Bray, J.D., Sancio, R.B.: Assessment of the liquefaction susceptibility of fine-grained soil during the Loma Prieta earthquake. J. Geotech. Geoenviron. Eng. **132**(9), 1165–1177 (2006)
3. Guo, T., Prakash, S.: Liquefaction of silt and silt-clay mixtures. J. Geotech. Geoenviron. Eng. **125**(8), 706–710 (1999)
4. El Hosri, M.S., Biarez, H., Hicher, P.Y.: Liquefaction characteristics of silty clay. In: Proceeding of Eighth World Conference on Earthquake Engineering, vol. 3, pp. 277–284 (1984)

5. Hyodo, M., Ito, S., Yamamoto, Y., Fujii, T.: Cyclic shear behaviour of marine clays. In: Proceedings of the Tenth International Offshore and Polar Engineering Conference, vol. 2, pp. 606–611 (2000)
6. Ishihara, K., Yasuda, S.: Cyclic strengths of undisturbed cohesive soils of Western Tokyo. In: Proceedings of the International Symposium on Soils under Cyclic and Transient Loading, pp. 57–66 (1980)
7. Hyodo, M., Yamamoto, Y., Fujii T.: Cyclic shear failure and strength of undisturbed marine clays. In: Proceedings of the Eighth International Offshore and Polar Engineering Conference, vol. 1, pp. 567–563 (1998)
8. Bray, J.D., Sancio, R.B., Riemer, M., Durgunohlu, H.T.: Liquefaction susceptibility of fine-grained soils. In: Proceedings of the International Conferences on Earthquake Geotechnical Engineering, vol. 1, pp. 655–662 (2004)
9. Gratchev, I.B., Sassa, K., Fukuoka, H.: How reliable is the plasticity index for estimating the liquefaction potential of clayey sands? J. Geotech. Geoenviron. Eng. 132(1), 124–127 (2006)
10. Andersen, K.H., Brown, S.F., Foss, I., Pool, J.H., Rosenbrand, W.F.: Effect of cyclic loading on clay behaviour. Norw. Geotech. Inst. Publ. 113, 1–6 (1976)
11. Boulanger, R.W., Idriss, I.M.: Evaluation of cyclic softening in silts and clays. J. Geotech. Geoenviron. Eng. 133(6), 641–652 (2007)
12. Andersen, K.H.: Bearing capacity under cyclic loading-offshore, along the coast, and on land. The 21st Bjerrum Lecture presented in Oslo, 23 November 2007. Can. Geotech. J. 46, 513–535 (2009)
13. Yasuhara, K., Murakami, S., Song, B., Seiji, Y., Hyde, A.F.L.: Postcyclic degradation of strength and stiffness for low plasticity silt. J. Geotech. Geoenviron. Eng. 129(8), 485–488 (2003)
14. Thammathiwat, A., Chim-oye, W.: Behavior of strength and pore pressure of soft Bangkok clay under cyclic loading. J. Sci. Technol. 9(4), 21–28 (2004)
15. Ding, J., Liu, H., Hu, L.: Response of marine clay to cyclic loading. In: Proceedings of the 17th International Offshore and Polar Engineering Conference, pp. 1188–1192 (2007)
16. Orense, R.P., Altun, S., Ansal, A.: Cyclic shear behavior and seismic response of partially saturated slopes. Soil Dyn. Earthq. Eng. 42, 71–79 (2012)
17. Nabeshima, Y., Matsui, T.: Role of plastic and non-plastic fines on cyclic shear behavior of saturated sands. In: Proceedings of the Thirteenth International Offshore and Polar Engineering Conference, pp. 440–444 (2003)
18. Chang, N.Y., Yeh, S.T., Kaufman, L.P.: Liquefaction potential of clay and silty sands. In: Proceedings of the Third International Conference on Soil Mechanics and Foundations Engineering, vol. 5, pp. 80–133 (1982)
19. Dezfulian, H.: Effects of silt content on dynamic properties of sandy soil. In: Proceedings of the Eighth World Conference on Earthquake Engineering, pp. 63–70 (1982)
20. Amini, F., Qi, G.Z.: Liquefaction testing of stratified silty sands. J. Geotech. Geoenviron. Eng. 126(3), 208–217 (2000)
21. Shen, C.K., Vrymoed, J.L., Uyeno, C.K.: The effects of fines on liquefaction of sands. In: Proceedings of the Ninth International Conference on Soil Mechanics and Foundations Engineering, vol. 2, 99. 381–385 (1997)
22. Troncoso, J.H., Verdugo, R.: Silt content and dynamic behavior of tailings sand. In: Proceedings of the Twelfth International Conference on Soil Mechanics and Foundations Engineering, pp. 1311–1314 (1985)
23. Troncoso, J.H.: Failure risks of abandoned tailings dams. In: Proceedings of International Symposium on Safety and Rehabilitation of Tailings Dams, pp. 82–89 (1990)

24. Finn, W.D.L., Ledbetter, R.H., Wu, G.: Liquefaction in silty soils: design and analysis, vol. 44, pp. 17–33. Geotechnical Special Publication (1994)
25. Vaid, V.P.: Liquefaction of silty sands, vol. 44, pp. 1–16. Geotechnical Special Publication (1994)
26. Ajmera, B.: Factors influencing the post-earthquake shear strength. Ph.D. dissertation, Virginia Polytechnic Institute and State University (2015)
27. Ajmera, B., Brandon, T., Tiwari, B.: Effect of mineralogy on the post-earthquake shear strength of clay-like materials. Association of State Dam Safety Officials Dam Safety, vol. 1 (2014)
28. Ajmera, B., Tiwari, B., Brandon, T.: Cyclic and post-cyclic behavior of clay-like materials. In: Proceedings of the Twelfth International Conference on Geo-Disaster Reduction, vol. 1, pp. 5–10 (2014)
29. Ajmera, B., Brandon, T., Tiwari, B.: Cyclic strength of clay-like materials. In: Proceedings of Sixth International Conference on Earthquake Geotechnical Engineering (2015)
30. Ajmera, B., Tiwari, B., Brandon, T.: Influence of mineralogy and plasticity on the cyclic and post-cyclic behavior of normally consolidated soils. In: Proceedings of Geotechnical and Structural Engineering Congress, pp. 1522–1531 (2016)
31. Ajmera, B., Tiwari, B.: Damping and shear moduli of laboratory prepared mineral mixtures. In: Proceedings of Geotechnical Frontiers Geotechnical Special Publication, vol. 281, pp. 10–18 (2017)
32. Ajmera, B., Tiwari, B., Pandey, P.: Use of pore pressure response to determine shear strength degradation from cyclic loading. In: Proceedings of the Geotechnical Frontiers Geotechnical Special Publication, vol. 281, pp. 19–26 (2017)
33. Ajmera, B., Brandon, T., Tiwari, B.: Influence of index properties on shape of cyclic strength curve for clay-silt mixtures. Soil Dyn. Earthq. Eng. 102, 46–55 (2017)
34. Kuwano, J., Nukano, H., Sugihara, K., Yabe, H.: Factors affecting undrained cyclic strength of sand containing fines. In: Proceedings of the 31st Annual Meeting of the Japanese Geotechnical Society, pp. 989–990 (1996)
35. Beroya, M.A.A., Aydin, A., Katzenbach, R.: Insight into the effects of clay mineralogy on the cyclic behavior of silt-clay mixtures. Eng. Geol. 106, 154–162 (2009)
36. Silver, M.L.: Laboratory triaxial testing procedures to determine the cyclic strength of soils. Report for the U.S. Nuclear Regulatory Commission (1977)
37. Silver, M.L., Chan, C.K., Ladd, R.S., Lee, K.L., Tiedemann, D.A., Townsend, F.C., Valera, J.E., Wilson, J.H.: Cyclic triaxial strength of standard test sand. J. Geotech. Eng. Div. 102(5), 511–523 (1976)
38. Ozaydin, K., Erguvanli, A.: The generation of pore pressures in clayey soils during earthquakes. In: Proceedings of the Seventh World Conference on Earthquake Engineering, vol. 3, pp. 326–330 (1980)
39. Lefebvre, G., LeBoeuf, D.: Rate effects and cyclic loading of sensitive clays. J. Geotech. Eng. 113(5), 476–489 (1987)
40. Koutsoftas, D.C.: Effect of cyclic loads on undrained strength of two marine clays. J. Geotech. Eng. Div. 104(5), 609–620 (1978)
41. Soroush, A., Soltani-Jigheh, H.: Pre- and post-cyclic behavior of mixed clayey soils. Can. Geotech. J. 46, 115–128 (2009)
42. Yasuhara, K., Andersen, K.H.: Effect of cyclic loading on recompression of overconsolidated clay. In: Proceedings of the Twelfth International Conference on Soil Mechanics and Foundation Engineering, vol. 1, pp. 485–488 (1989)
43. Hannam, A.D., Javed, K.: Design of foundations on sensitive champlain clay subjected to cyclic loading. J. Geotech. Geoenviron. Eng. 134(7), 929–937 (2008)

44. Brown, S.F., Lashine, A.K.F., Hyde, A.F.L.: Repeated load triaxial testing of a silty clay. Géotechnique **25**(1), 95–114 (2007)
45. Hashash, Y.M.A., Tiwari, B., Moss, R.E.S., Asimaki, D., Clahan, K., Kieffer, D.S., Dreger, D.S., Macdonald, A., Madugo, C.M., Mason, H.B., Pehlivan, M., Rayamajhi, D., Acharya, I., Adhikari, B.: Geotechnical field reconnaissance: Gorkha (Nepal) earthquake of April 25 2015 and related shaking sequence. Geotechnical extreme event reconnaissance. GEER Association Report No. GEER-40, ver. 1.1, 2, p. 152 (2015)
46. Tiwari, B., Ajmera, B., Yamashiro, B.: Causes of cyclic shear failure at Lokanthali of Araniko highway after M_w 7.8 2015 Gorkha earthquake. In: Proceedings of Fifth International Conference on Forensic Geotechnical Engineering, pp. 78–91 (2016)
47. Tiwari, B., Pradel, D., Ajmera, B., Yamashiro, B., Diwakar, K.: Case study: numerical analysis of landslide movement at Lokanthali during the $M_w = 7.8$ Gorkha (Nepal) earthquake. J. Geotech. Geoenviron. Eng. (in Press)
48. Tiwari, B., Ajmera, B., Dhital, S.: Characteristics of moderate to large scale landslides triggered by the M_w 7.8 Gorkha earthquake and its aftershocks. Landslides **14**(4), 1297–1318 (2017)
49. Tiwari, B., Pradel, D.: Ground deformation at Lokanthali, Kathmandu due to Mw 7.8 2015 Gorkha earthquake, vol. 278, pp. 333–342. Geotechnical Special Publication (2017)
50. Ozsvath, D.L.: The slippery slope of litigating geologic hazards: California's Portuguese Bend. National Center for Case Study Teaching in Science (1999)
51. ASTM D4318.: Standard test methods for liquid limit, plastic limit and plasticity index of soils. ASTM International (2010)
52. ASTM D6528.: Standard test method for consolidated undrained direct simple shear testing of cohesive soils. ASTM International (2007)
53. Tiwari, B., Ajmera, B.: Curvature of failure envelopes for normally consolidated clays. In: Proceedings of the Third Landslide Forum – Landslide Science for a Safer Geo-environment, vol. 2, pp. 117–122 (2014)

Dynamic Response of Lightweight Cellular Concrete MSE Walls

Binod Tiwari[1(✉)], Beena Ajmera[2], and Diego Villegas[3]

[1] California State University, Fullerton, 800 N. State College Blvd., E-419,
Fullerton, CA 92831, USA
btiwari@fullerton.edu
[2] California State University, Fullerton, 800 N. State College Blvd., E-318,
Fullerton, CA 92831, USA
bajmera@fullerton.edu
[3] Cell Crete Corporation, 135 E. Railroad Ave., Monrovia, CA 91016, USA
dvillegas@cell-crete.com

Abstract. Due to the numerous benefits that lightweight cellular concrete offers, its use has become increasingly popular in several construction applications. One specific application is as the backfill of retaining walls or MSE walls in order to reduce earth pressure, construction cost and time. Previous studies examined the static and dynamic properties of lightweight cellular concrete with varying test and dry unit weights through laboratory testing. Furthermore, numerical modelling evaluated the deformation characteristics of geo-grid reinforced lightweight cellular concrete backfills. However, results from physical model tests to validate these results are unavailable in the literature. Given the increasing use of these geo-grid reinforced lightweight cellular concrete backfills in seismically active regions, it is critical to observe the seismic performance of these structures. As such, shake table tests were performed on a 1.2 m tall lightweight cellular concrete (material with a test unit weight of 4 kN/m^3) MSE wall reinforced with a geo-grid layer at approximately 0.6 m height. The models were instrumented with a series of accelerometers and strain gauges. Additionally, in order to simulate self-weight of the MSE walls of different heights, vertical stresses of approximately 5 kPa and 8 kPa were applied to the geo-gird reinforced lightweight cellular concrete before the structure was subjected to a series of sinusoidal ground motions with varying amplitudes and frequencies. Additionally, the geo-grid reinforced lightweight cellular concrete was also subjected to the ground motions recorded from the 1994 Northridge earthquake. The entire geo-gird reinforced lightweight cellular concrete backfill was found to displace as a single monolithic unit with no significant relative displacements between the geo-grid reinforcement layer and the lightweight cellular concrete. Furthermore, no fractures were observed during the application of the ground motions.

Keywords: Lightweight cellular concrete · MSE wall · Dynamic response
Deformation · Shake table

© Springer Nature Singapore Pte Ltd. 2018
T. Qiu et al. (Eds.): GSIC 2018, *Proceedings of GeoShanghai 2018 International Conference: Advances in Soil Dynamics and Foundation Engineering*, pp. 168–175, 2018.
https://doi.org/10.1007/978-981-13-0131-5_19

1   Background

Lightweight cellular concrete (LCC) has become increasingly popular in several construction applications due to the numerous benefits it has to offer. Among these benefits is that LCC in addition to having low density is also highly durable, has high freeze-thaw resistance and has low permeability. Several of the applications of LCC include landslide repair works, earthquake shock absorbent near tunnels and pipelines, pavement material and engineered fills in approaches to bridges [1–4].

LCC is produced by introducing air voids into the traditional components of concrete – water, aggregates, and cement. Protein-based or synthetic-based foaming agents are used to create the air voids. The agents create the air voids by trapping the air when reacting with the other components [2, 4–7]. The percent of air voids created within the traditional mixture of concrete will be directly related to the amount of foaming agent present in the mixture with most LCC materials containing between 10% and 70% air voids [7]. The unit weight of the material will be directly dependent on the percent of air voids present in the LCC and can be as low as 3.1 kN/m^3 [1].

Results are available in the literature for several static properties of LCC including the hydraulic conductivity [1, 8, 9], the drying shrinkage [1, 10], the thermal expansion [1], and unconfined compressive strengths [2, 10, 11]. One of the most comprehensive studies on the static properties of LCC is that by Tiwari et al. [12]. They presented the shear strength parameters for undrained and drained conditions, coefficients of permeability, and at-rest earth pressures for LCC materials having four different unit weights. In addition, the results were used to develop relationships to estimate the unconfined compressive strengths as well as the total friction angle and cohesion intercept with the unit weight of the LCC material. Tiwari et al. [12] further concluded that the effective friction angle and cohesion intercept were independent of the unit weight of the LCC material over the range of stresses that they tested. The dynamic properties of LCC were examined in detail by Tiwari et al. [13], which presented the results of cyclic simple shear tests on LCC materials having four different unit weights. The cyclic simple shear tests were conducted on LCC materials subjected to a series of fifteen strain-controlled undrained sinusoidal cyclic motions with varying amplitudes at four different consolidation pressures. Tiwari et al. [13] determined that the maximum shear modulus of LCC will increase with an increase in the dry unit weight as well as with an increase in the consolidation pressure. It was further found that the damping ratio would decrease with an increase in the shear strain until a threshold shear strain was achieved, beyond which increase in the shear strain corresponded to an increase in the damping ratio.

Although results regarding the static and dynamic properties of LCC are available in the literature, the behavior of LCC in construction applications has not been studied in sufficient detail. Furthermore, with an increase in the use of LCC in several geotechnical engineering applications, it is necessary to examine how the LCC materials will behave at seismic loading. The focus of this paper will be the on the behavior of LCC material as the backfill of mechanically stabilized earth (MSE) retaining walls at seismic loading. This was examined by Pradel and Tiwari [14] using numerical analyses with FLAC, but experimental testing and validation of the results they obtained is currently not available. As such, shake table experiments were conducted in this study on two small scale LCC retaining walls reinforced with a geo-gird at the mid-height. The cyclic loading applied in the shake table experiments had varying frequencies and amplitudes in order to obtain an understanding of the performance of the LCC retaining wall under seismic loading.

2 Model and Testing Details

2.1 Casting of LCC Retaining Wall

LCC is casted using two concurrent processes, the first of which will create the foaming agent while the second will produce the neat cement slurry. To create the foaming agent, 40 parts of water are combined with one part of Elastizell Foam Concentrate, which is a protein based biodegradable surfactant and a by-product of the food industry. The mixture is mechanically agitated through small nozzle while subjected to high-pressure compressed air action. The production of the neat cement slurry is achieved by mixing cement and water, per the design specifications, together in a progressive cavity pump coupled with a customized concrete mixer. The air-filled cellular concrete is then produced by adding the pre-formed foam to the neat cement slurry in a Proprietary Blending System. This material was, then, used to cast the LCC retaining wall.

In this study, the LCC material had a unit weight of approximately 4 kN/m³. Tiwari et al. [12, 13] reported the static and dynamic properties, respectively, of this material. To cast the retaining wall, the air-filled cellular concrete slurry obtained from the blending system was poured into a rectangular wooden mold having length of 1.8 m and width of 1.2 m. The height of the wooden mold was 1.2 m. A geo-grid layer was installed at mid-height to reinforce the LCC retaining wall. The LCC retaining wall was cured for a period of 25 days within the wooden mold, after which period the mold was removed and the LCC retaining wall was air-dried for at least 3 days before being subjected to any testing.

2.2 Shake Table Testing

Following the curing process of the LCC retaining wall, the model was placed on a shake table. As a safety measure, the LCC retaining wall was surrounded by a metal cage, as shown in Fig. 1. This was done to ensure that the model would not slip off the shake table during the shaking and cause safety issue. It is noted that the metal cage would not prevent the model from deforming freely and thus, the results are not impacted by this safety measure. The LCC retaining wall was then instrumented with three accelerometers located at the center of the model at different depths corresponding to (1) the top of the model, (2) mid-way between the top of the model and the location of the geo-grid, and (3) on top of the geo-grid. In addition, four strain gauges were also mounted to the metal cage to record the deformation at (1) the top of the model, (2) mid-way between the top of the model and location of the geo-grid, (3) at the geo-grid, and (4) mid-way between the geo-grid and the bottom of the model.

Unfortunately, the strain gauge mid-way between the top of the model and the geo-grid did not function properly. As such, the results from this strain gauge will not be presented in this paper. Figure 2 illustrates the locations of the instruments.

Fig. 1. Photograph of LCC retaining wall on the shake table with surrounding metal cage safety measure. LCC retaining wall pictured is subjected to vertical stress of 5 kPa.

The LCC retaining wall was then subjected to a surcharge stress of 5 kPa following which a series of sinusoidal cyclic loads were applied to the model. These cyclic loads had amplitudes of 0.1 g or 0.2 g and frequencies of 2 Hz or 3 Hz. In addition, a ground motion recorded during the 1994 Northridge Earthquake was also applied to the LCC retaining wall. Higher amplitudes and lower frequencies were not applied to avoid catastrophic failure due to safety issue. The sequence of cyclic loads were repeated three times for the model. Following the completion of the shake table testing, the LCC retaining was trimmed in order to remove approximately 15.2 cm from each side of the model. The resulting LCC retaining wall was 0.91 m wide by 1.5 m long with a height of 1.2 m. This model was instrumented in a similar way as in the larger model described previously and subjected to the same triplicates of the cyclic loads.

3   Results and Discussion

Displacement time histories were calculated for each applied ground motion using the acceleration time histories recorded at each of the three accelerometers. It is noted that the raw data from the accelerometers was first filtered and baseline corrected prior to use in any calculation. Furthermore, the calculations were preformed not only for the

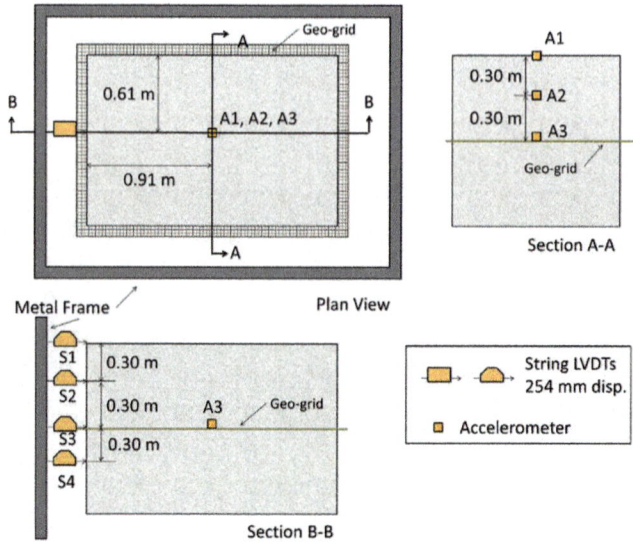

Fig. 2. Instrumentation diagram showing the locations of the accelerometers, denoted with A, and strain gauges, denoted with S.

direction of shaking, but also for the remaining two directions. However, it was determined that the results in the horizontal direction perpendicular to the shaking direction as well as the results in the vertical direction were. As such, only the results from the direction of shaking are presented in this paper.

A typical displacement time history obtained from the model subjected to a surcharge stress of 5 kPa is presented in Fig. 3. The displacement time histories presented in Fig. 3 correspond to the sinusoidal cyclic load with an amplitude of 0.1 g at a frequency of 3 Hz. Results for other loading frequencies and surcharge stress also demonstrated similar result, although is not included in this paper. The results in Fig. 3 indicate the total displacement experienced by the accelerometer at the specified location. In-depth examination of the figure indicates that the relative displacement between the geo-grid (location A3) and the top of the LCC retaining wall (location A1) is insignificant. This suggests that the entire LCC retaining wall moved monolithically.

Figure 4 presents the typical displacement time histories recorded in the strain gauges. The results were obtained from the model subjected to surcharge stress of 5 kPa and sinusoidal cyclic loads with an amplitude of 0.2 g with a frequency of 3 Hz. Results for other loading frequencies and surcharge stress also demonstrated similar result, although is not included in this paper. Unlike the accelerometers, the strain gauges measure the relative displacement of the LCC retaining wall and the shake table. The results in Fig. 4 indicate that the top of the model experiences greater displacements than those experienced by geo-grid. However, the magnitude of the differences is quite small and can be considered negligible in relation to the height of the model. This is in agreement with results obtained from the displacement time histories determined from the accelerometer data.

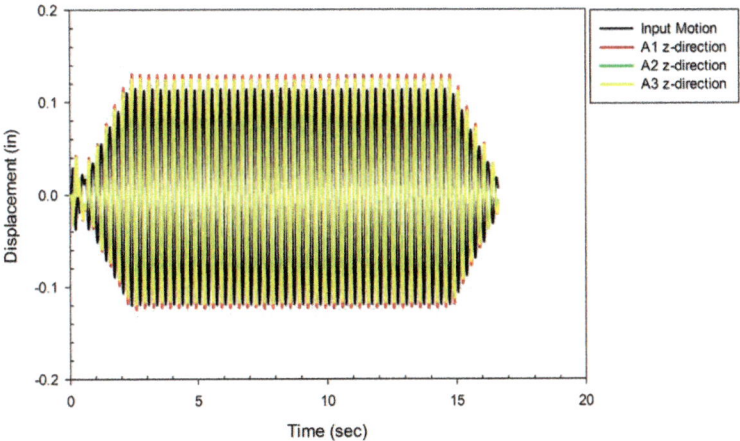

Fig. 3. Displacement time histories obtained from the double integration of accelerometer data collected during the application of sinusoidal cyclic load with amplitude of 0.1 g and frequency of 3 Hz in LCC retaining wall model with a surcharge stress of 5 kPa.

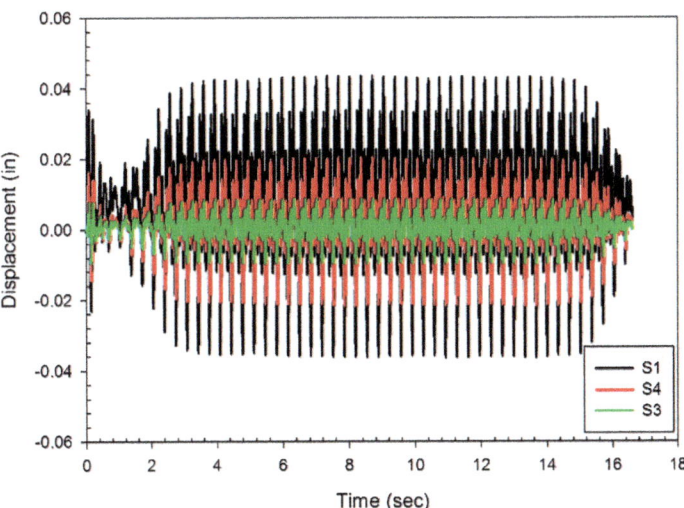

Fig. 4. Displacement time histories recorded in strain gauges during the application of sinusoidal cyclic load with amplitude of 0.1 g and frequency of 3 Hz in LCC retaining wall model with a surcharge stress of 5 kPa.

The negligible relative displacements between the top of the LCC retaining wall model and the geo-grid appears to suggest that the function of the geo-grid in the model is simply to reduce crack propagation. While the geo-grid may also assist to secure the facing material, it does not appear to contribute to reinforcing the LCC material.

The results obtained from Pradel and Tiwari [14] agree with those presented in this study. However, their results were based on FLAC analyses of a hypothetical LCC retaining wall structures founded on soft clays.

4 Conclusions

Shake table tests were conducted on two LCC retaining wall models. These models were constructed using the LCC materials having a unit weight of approximately 4 kN/m^3 with a geo-grid reinforcement placed at mid-height. The models were subjected to surcharge stresses corresponding to 5 kPa and 8 kPa respectively. They were subjected to a series of sinusoidal cyclic loads as well as a motion recorded during the 1994 Northridge earthquake. The displacement response was determined by double integrating the acceleration time histories. In addition, the displacements were also recorded using strain gauges. Results from both methods suggested that displacement of the top of the LCC model was nearly the same as the displacement of the geo-grid at mid-height. Therefore, it was determined that the geo-grid was not providing reinforcement in the LCC, but instead was only aiding to reduce crack propagation and to secure the facing material.

References

1. Aberdeen Group: Cellular Concrete. Publication No. C630005. Aberdeen Group, Boston (1963)
2. LaVallee, S.: Cellular Concrete to the Rescue. Aberdeen Group, Boston (1999). Publication No. C99A039
3. Tikalsky, P.J., Popisil, J., MacDonald, W.: A method for assessment of the freeze-thaw resistance of preformed foam cellular concrete. Cem. Concr. Res. **34**(5), 889–893 (2004)
4. Maruyama, R.C., Camarini, G.: Properties of cellular concrete for filters. Int. J. Eng. Technol. **7**(3), 223–228 (2015)
5. Tian, Y.: Experimental study on aerated concrete produced by iron tailings. Adv. Mater. Res. **250**, 853–856 (2011)
6. Albayrak, M., Yorukoglu, A., Karahan, S., Atlihan, S., Aruntas, H.Y., Girgin, T.: Influence of zeolite additive on properties of autoclaved aerated concrete. Build. Env. **42**, 3161–3165 (2007)
7. Panesar, D.K.: Cellular concrete properties and the effect of synthetic and protein foaming agents. Constr. Build. Mater. **44**, 575–584 (2013)
8. Loudon, A.G.: The thermal properties of lightweight cellular concretes. Int. J. Lightweight Concr. **11**(2), 71–85 (1979)
9. Neville, A.M.: Properties of Concrete, 4th edn. Pearson Educational Limited, Essex (2002)
10. Narayanan, N., Ramamurthy, K.: Structure and properties of aerated concrete: a review. Cem. Concr. Compos. **22**, 3210329 (2000)
11. Zaidi, A.A.M., Rahman, A.I., Zaidi, N.H.A.: Behavior of fiber reinforced foamed concrete: indentation test analysis. In: Proceedings of the Seminar on Geotechnical Engineering, pp. 92–101 (2008)

12. Tiwari, B., Ajmera, B., Maw, R., Cole, R., Villegas, D., Palmerson, P.: Mechanical properties of lightweight cellular concrete for geotechnical applications. J. Mater. Civil Eng. **29**(7), 06017007 (2017)
13. Tiwari, B., Ajmera, B., Villegas, D.: Dynamic properties of lightweight cellular concrete for geotechnical applications. J. Mater. Civil Eng. **30**(2), 04017271 (2017)
14. Pradel, D., Tiwari, B.: The use of MSE walls backfilled with lightweight cellular concrete in soft ground seismic areas. In: Proceedings of the Third International Conference of Deep Foundations, vol. 1, pp. 107–114 (2015)

Experimental Evaluation of Encased Stone Column Technique for Liquefaction Mitigation

Piyush Punetha[1(✉)] and Ganesh Kumar Shanmugam[2]

[1] AcSIR, CSIR-Central Building Research Institute, Roorkee 247667, India
punetha.piyush@yahoo.in
[2] GE Group, CSIR-Central Building Research Institute, Roorkee 247667, India
ganeshkumar@cbri.res.in

Abstract. The present paper investigates the effectiveness of using geotextile encased stone column for mitigating the liquefaction phenomenon in saturated sandy deposits. Reduced scale 1-g model tests were conducted using a uniaxial shake table to study the behavior of loose saturated sand reinforced with encased stone columns when subjected to harmonic (sinusoidal) loading. Additionally, the response of saturated sand reinforced with stone column, with and without geotextile encasement is also studied and compared. The test results show that the installation of stone column in the saturated loose sand increases the liquefaction resistance of sand. The presence of geotextile allows quicker dissipation of pore water which results in an improved liquefaction resistance of sand. Moreover, the acceleration amplitude influences the response of both the unreinforced and reinforced sand. The increase in acceleration amplitude increases the magnitude of excess pore water pressure ratio. Furthermore, the presence of stone column also reduces the settlement of the shallow foundation.

Keywords: Liquefaction · Encased stone column · Shaking table tests
Geotextile

1 Introduction

The occurrence of liquefaction in saturated loose sand deposits has caused substantial damage to buildings, bridges and other structures during the past earthquakes [1–3]. Numerous techniques have been developed till date to prevent this phenomenon, which include in situ densification, replacement of liquefaction susceptible soil, construction of gravel or rock drains, chemical stabilization, grouting etc. The primary aim of these techniques is to either reduce the generation of excess pore water pressure, to improve the shear deformability of soil or both. Despite widespread use of these techniques, the studies pertaining to their effectiveness in mitigating liquefaction are still very limited.

Numerous case studies, theoretical analysis and model tests have shown that the stone columns can be effectively used for liquefaction mitigation [4–11]. However, there are shortcomings in the use of stone columns such as contamination due to the migration of fines, inadequate confinement from surrounding soil etc. These shortcomings can be addressed by providing geosynthetic encasement around the stone

© Springer Nature Singapore Pte Ltd. 2018
T. Qiu et al. (Eds.): GSIC 2018, *Proceedings of GeoShanghai 2018 International Conference:*
Advances in Soil Dynamics and Foundation Engineering, pp. 176–184, 2018.
https://doi.org/10.1007/978-981-13-0131-5_20

column [12]. However, the studies pertaining to the use of geosynthetic encased stone columns for liquefaction mitigation are limited.

Therefore, the present paper investigates the effectiveness of using geotextile encased stone column for mitigating the liquefaction phenomenon in saturated sandy deposits. Reduced scale 1-g model tests were conducted using a uniaxial shake table to study the behavior of loose saturated sand reinforced with encased stone column when subjected to harmonic (sinusoidal) loading. Additionally, the response of saturated sand reinforced with stone column, with and without geotextile encasement is also studied and compared.

2 Materials

2.1 Soil

A uniformly graded river sand has been used in the present study, which has been collected from nearby the CSIR-CBRI campus. Table 1 shows the physical properties of the sand. The mean particle size, uniformity coefficient and coefficient of curvature of sand are 0.21 mm, 2.67 and 1.2 respectively. The sand is classified as poorly-graded sand (SP) as per Indian standards [13]. Figure 1 shows the particle size distribution curve of the sand.

Table 1. Physical properties of river sand.

Property	Value
D_{50} (mm)	0.21
Coefficient of uniformity (C_u)	2.67
Coefficient of curvature (C_c)	1.20
Specific gravity (G)	2.67
Maximum unit weight (γ_{max}) kN/m^3	16.00
Minimum unit weight (γ_{min}) kN/m^3	14.20
Maximum void ratio (e_{max})	0.88
Minimum void ratio (e_{min})	0.67

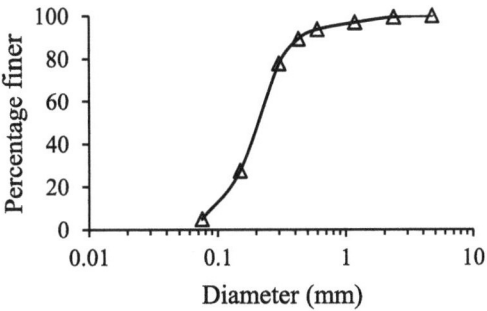

Fig. 1. Grain size distribution curve of sand.

2.2 Aggregates

Uniformly graded angular granite chips of size range 2 mm to 10 mm have been used to construct the stone columns. The stone aggregates in all the tests were compacted to a dry density of 16 ± 0.2 kN/m^3 corresponding to a relative density of 73% (representing field situation). The peak angle of frictional resistance of stone aggregates obtained using the large direct shear box tests is 41.5°.

2.3 Geotextile

A 1.5 mm thick nonwoven needle punched polypropylene geotextile has been used to encase the stone columns. Table 2 shows the physical properties of the nonwoven geotextile. The geotextile serves a dual function of confining the stone columns and at the same time providing drainage path for water.

Table 2. Physical properties of nonwoven geotextile.

Property	Value
Thickness (mm)	1.5
Mass/unit area (g/m^2)	200
Wide width tensile strength MD (kN/m)	14
Wide width tensile strength CD (kN/m)	12
Elongation (%)	55
Opening size (mm)	0.085
Permeability (m/s)	0.0036
Permittivity (s^{-1})	1.34

3 Experimental Procedure

The tests were performed in a rigid plexiglass tank with dimensions 1.4 m $\times$ 1 m $\times$ 1 m, mounted over a uniaxial shake table. Figure 2(a) shows the schematic diagram of the test assembly. The rigidity of the container could generate significant boundary effects in the soil sample [14, 15]. When the container (filled with soil) is shaken horizontally (one-dimensional shaking), a shear (SH) wave gets generated at the base which propagates up through the soil. However, the motion of the soil near the boundary is restricted by the rigid walls of the container. Consequently, the soil near the boundary undergoes compression and extension and generates P waves [16, 17]. Therefore, a 50 mm thick PU foam was placed near the boundary of the tank at right angles to the shaking direction, to reduce the boundary effects [18]. The base of the tank was initially made rough by gluing sand to ensure the transfer of shear stress (or input motion) from the base of the tank to the sand sample. The saturated sand sample was then prepared inside the plexiglass tank using water sedimentation technique to achieve a relative density of 25% [19]. Initially, the height of sample inside the tank was fixed at 650 mm and corresponding dry weight of sand required to occupy the desired volume (at required relative density) inside the tank was calculated. The amount of water required to form saturated sand sample was evaluated and filled in the tank. Subsequently, dry sand was poured into the tank from a constant height from the water surface in three stages. The height of filling was fixed after several trials. The sample was then left for 24 h for complete saturation.

For the tests involving stone column, the saturated sample was prepared initially using a similar procedure. Thereafter, stone columns were installed in the saturated sand using a replacement technique. A hollow PVC pipe of outer diameter 111 mm was driven inside the prepared sand bed and subsequently, the sand inside the PVC pipe was removed. The bore hole (thus formed) was then filled with stone aggregates in three layers and compacted to achieve a density of 16 ± 0.2 kN/m^3. Four stone columns of diameter (D) 111 mm were installed at a center-to-center spacing of 222 mm (2D). Figure 2(b) shows the plan of the test assembly used for the tests involving saturated sand reinforced with stone columns. The entire assembly was then left for 24 h.

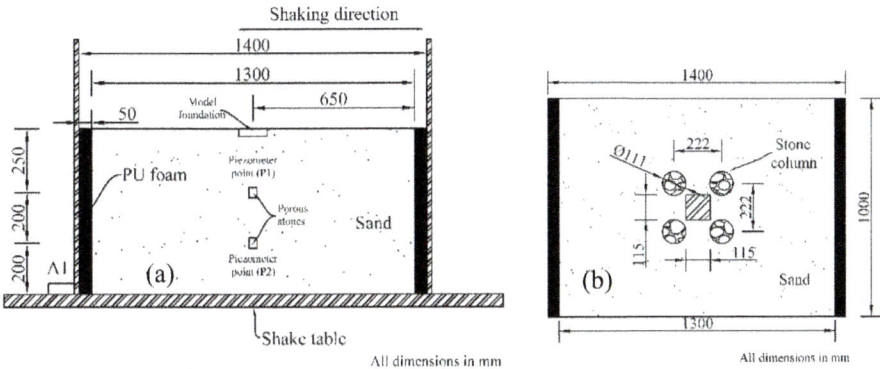

Fig. 2. (a) Schematic diagram of the test assembly; (b) plan of the test assembly for test involving stone column.

The geotextile encased stone columns were constructed using a similar procedure used for installing stone columns (without encasement). Encasement was carried out by wrapping the geotextile around the PVC pipe. The edges of the geotextile were stitched to have a maximum seam strength and were closed at the bottom. Then the pipe with geotextile was carefully driven inside the saturated sand bed till it reached the bottom of the tank. The aggregates were then filled inside the geotextile tube and compacted simultaneously with the withdrawal of the PVC pipe at regular intervals. Similar to ordinary stone column test series, the number, diameter and spacing of encased stone columns were 4, 111 mm and 222 mm respectively and the entire assembly was left for 24 h. Figures 3(a) and (b) show the entire assembly for the tests involving saturated sand reinforced with ordinary (without encasement) and encased stone columns respectively.

The test assembly was then subjected to a harmonic (sinusoidal) loading with an acceleration amplitude ranging between 0.1–0.2 g and 5 Hz input frequency. The duration of shaking was fixed at 40 s (i.e. 200 cycles) for each of the tests. Two tests

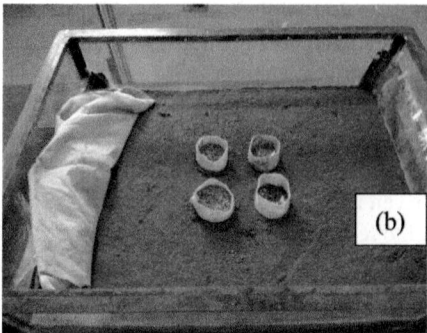

Fig. 3. Test assembly for the shake table test on saturated sand reinforced with (a) ordinary stone columns (b) encased stone columns

were conducted for each sample i.e. unreinforced saturated sand (S), saturated sand reinforced with stone column (SC), saturated sand reinforced with encased stone column (ESC). A glass tube piezometer was used to measure the pore water pressure inside the saturated sand sample at a height of 0.2 m and 0.4 m from the base of the tank. The piezometer was connected to the tank using rubber tubes. The mouth of the tubes was connected to filter paper wrapped porous stones inside the tank (Fig. 2(a)). The porous stones were placed at the center of the tank to minimize the influence of rigid wall in the pore water pressure measurements [19]. A model foundation (shallow) of size 115 mm × 115 mm × 30 mm was then placed at the top of the sand sample such that its top portion is positioned at the same level as that of the surface of sand sample (as shown in Fig. 2a). The settlement of the model footing was measured after the tests to investigate the effectiveness of stone columns in reducing the settlement. A MEMS accelerometer (A1) was attached to the base of the tank to record the acceleration transmitted to the tank (Fig. 2a).

4   Results and Discussion

The results of the shake table tests are presented in terms of variation of excess pore water pressure ratio with time (after the start of shaking) obtained from piezometer at 0.2 m and 0.4 m from the base of the tank. The excess pore water pressure ratio (r_u) is defined as the ratio of excess pore water pressure (U) to the effective overburden pressure (σ'_{vo}) (Eq. 1).

$$r_u = (U/\sigma'_{vo}) \tag{1}$$

Figure 4 compares the variation of excess pore water pressure ratio with time obtained from shaking table tests for unreinforced saturated sand (S), sand reinforced with stone column (SC) and sand reinforced with geotextile encased stone column (ESC) at 0.2 m from the base of the tank at an input acceleration amplitude of 0.1 g. It can be observed that the excess pore water pressure ratio reduces significantly on

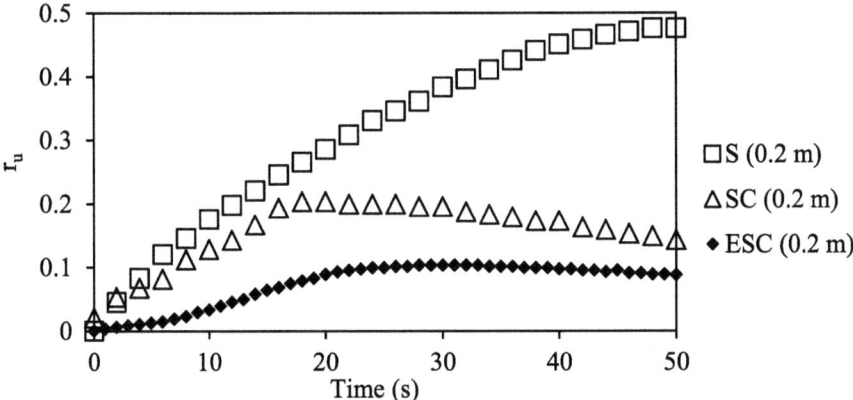

Fig. 4. Variation of excess pore water pressure ratio (r_u) with time at 0.2 m from base of the tank at an input acceleration amplitude and frequency of 0.1 g and 5 Hz respectively.

installation of stone columns. The maximum value of r_u for saturated sand sample is 0.48, however, on installation of stone column, the value reduces to 0.2 and on encasing with geotextile, the value further reduces to 0.1. This indicates that the stone column and geotextile encased stone column increases the liquefaction resistance of sand by 58% and 79% respectively at 0.1 g.

The reduction in maximum value of r_u may be attributed to the drainage path provided by the stone column inside the saturated sand sample (or drainage of water through the stone column). The permeability of aggregates (gravels) is greater than the sand, therefore, the excess pore water pressure developed during the shaking gets dissipated immediately and consequently, the maximum pore water pressure ratio decreases. The drainage increases further in case of the encased stone column due to geotextile confinement which confines the stone column and prevents the entry of finer materials from the surrounding soil into the stone column for additional input acceleration conditions. From Fig. 4, it can also be observed that the time required for reaching maximum r_u increases for geotextile encased stone columns. In other words, the encasement of stone column by geotextile increases the number of cycles to reach the peak pore water pressure and also prevents column contamination additionally. This can be verified from the comparative test results of same unreinforced stone column under 0.2 g conditions.

Moreover, the settlement of the model foundation after the test is 40 mm, 20 mm and 22.25 mm for S, SC and ESC respectively. This indicates that the stone column significantly reduces the settlement of the model foundation. This may be due to the reduction in the generation of excess pore water pressure after installation of stone column as observed previously. Another reason may be the densification of the sand due to the construction of stone columns. The difference in settlement for encased stone column and ordinary stone column may be due to the confining effect of geotextile material which will be verified with the future experimental test results.

Figure 5 compares the variation of excess pore water pressure ratio with time (after the start of shaking) for SC and ESC at 0.1 g and 0.2 g measured at 0.2 m and 0.4 m

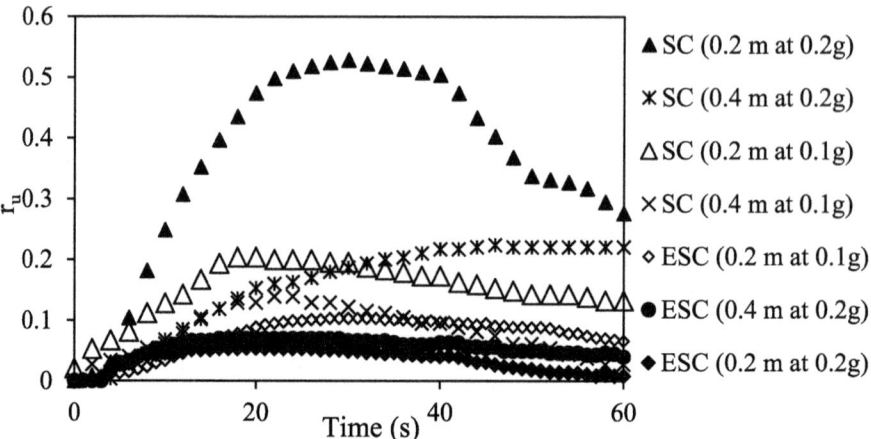

Fig. 5. Variation of excess pore water pressure ratio (r_u) with time for stone column and geotextile encased stone column at acceleration amplitudes and frequency of 0.1 g and 0.2 g, and 5 Hz respectively at an acceleration amplitude of 0.1 g and 5 Hz.

from the base of the tank. It can be observed that the maximum excess pore water pressure ratio increases with an increase in input acceleration amplitude for SC. However, the maximum excess pore water pressure ratio remains almost identical at 0.1 g and 0.2 g input acceleration amplitude for ESC. As predicted, the maximum value of r_u for SC is higher at 0.2 m from the base of the tank than at 0.4 m from bottom, however, it is nearly identical at 0.2 m and 0.4 m from the base for ESC. It must be noted that at 0.2 m from the tank base, the excess pore water pressure ratio increases with time and reaches a peak value and subsequently decreases at a slower rate than the pre-peak zone (dissipation rate). However, at 0.4 m from the tank base, the pore water pressure ratio increases up to the peak and then reduces at a slower rate as compared to dissipation rate for 0.2 m from the tank base. In case of ESC, r_u shows almost identical curves at 0.2 m and 0.4 m depth from the bottom of the tank which indicates that the inclusion of geotextile allows quicker dissipation and delays the development of peak pore water pressure.

5   Conclusions

The following conclusions may be drawn from the present study:

- The construction of stone columns significantly increases the liquefaction resistance of saturated sand deposit. Moreover, the encasement of stone columns with geo-textile further increases the liquefaction resistance. In the present study, the stone columns and geotextile encased stone columns increased the liquefaction resistance of loose saturated sand by 58% and 79% at 0.1 g input acceleration amplitude.

- The input acceleration amplitude significantly influences the behavior of stone columns. With an increase in acceleration amplitude, the excess pore water pressure ratio increases.
- The test results indicate that the presence of stone columns reduces the settlement of the shallow foundation. However, no considerable improvement in the mitigation of foundation settlement is observed in case of geotextile encased stone column as compared to the ordinary stone column. Nevertheless, the long-term performance of the encased stone columns will be much better than the ordinary stone columns which might fail due to column contamination, inadequate confinement from surrounding soil.

Acknowledgement. The authors would like to thank the Director, CSIR-Central Building Research Institute, Roorkee for giving permission to publish this research work. The authors would also like to thank the Head, Geotechnical Engineering Division, CSIR-CBRI for his continuous support during this research work. The authors also wish to thank the anonymous reviewers for their valuable time and suggestions.

References

1. Seed, R.B., et al.: Preliminary report on the principal geotechnical aspects of the October 17, 1989 Loma Prieta earthquake. Earthquake Engineering Research Center, University of California (1990)
2. Ishihara, K., Haeri, S.M., Moinfar, A.A., Towhata, I., Tsujino, S.: Geotechnical aspects of the June 20, 1990 Manjil earthquake in Iran. Soils Found. **32**(3), 61–78 (1992)
3. Kramer, S.L.: Geotechnical Earthquake Engineering. Prentice Hall, New York (1996)
4. Seed, H.B., Booker, J.R.: Stabilization of potentially liquefiable sand deposits using gravel drains. J. Geotech. Geoenv. Eng. **103**(ASCE 13050) (1977)
5. Sasaki, Y., Taniguchi, E.: Shaking table tests on gravel drains to prevent liquefaction of sand deposits. Soils Found. **22**(3), 1–14 (1982)
6. Iai, S., Koizumi, K., Noda, S., Tsuchida, H.: Large scale model tests and analyses of gravel drains. In: Proceedings of Ninth World Conference on Earthquake Engineering, Tokyo, Japan (1988)
7. Mitchell, J.K., Wentz, F.J.: Performance of improved ground during the Loma Prieta Earthquake, vol. 91, no. 12. Earthquake Engineering Research Center, University of California (1991)
8. Mitchell, J.K., Baxter, C.D., Munson, T.C.: Performance of improved ground during earthquakes. In: Soil Improvement for Earthquake Hazard Mitigation, pp. 1–36. ASCE (1995)
9. Ashford, S., Rollins, K., Bradford, V.S., Weaver, T., Baez, J.: Liquefaction mitigation using stone columns around deep foundations: full-scale test results. Transp. Res. Rec. J. Transp. Res. Board **1736**, 110–118 (2000)
10. Adalier, K., Elgamal, A., Meneses, J., Baez, J.I.: Stone columns as liquefaction countermeasure in non-plastic silty soils. Soil Dyn. Earthq. Eng. **23**(7), 571–584 (2003)
11. Adalier, K., Elgamal, A.: Mitigation of liquefaction and associated ground deformations by stone columns. Eng. Geol. **72**(3), 275–291 (2004)
12. Murugesan, S., Rajagopal, K.: Studies on the behavior of single and group of geosynthetic encased stone columns. J. Geotech. Geoenv. Eng. **136**(1), 129–139 (2009)

13. IS 1498: Classification and Identification of Soils for General Engineering Purposes. Bureau of Indian Standards, New Delhi (2007)
14. Whitman, R.V., Lambe, P.C.: Effect of boundary conditions upon centrifuge experiments using ground motion simulation. Geotech. Test. J. **9**, 61–71 (1986)
15. Lee, C.J., Wei, Y.C., Kuo, Y.C.: Boundary effects of a laminar container in centrifuge shaking table tests. Soil Dyn. Earthq. Eng. **34**(1), 37–51 (2012)
16. Zeng, X., Schofield, A.N.: Design and performance of an equivalent-shear-beam container for earthquake centrifuge modelling. Geotechnique **46**, 83–102 (1996)
17. Lombardi, D., Bhattacharya, S.: Shaking table tests on rigid soil container with absorbing boundaries. In: Proceedings of 15th World Conference on Earthquake Engineering, Lisbon, Portugal (2012)
18. Lombardi, D., Bhattacharya, S., Scarpa, F., Bianchi, M.: Dynamic response of a geotechnical rigid model container with absorbing boundaries. Soil Dyn. Earthq. Eng. **69**, 46–56 (2015)
19. Maheshwari, B.K., Singh, H.P., Saran, S.: Effects of reinforcement on liquefaction resistance of Solani sand. J. Geotech. Geoenv. Eng. **138**(7), 831–840 (2012)

Numerical Evaluation of Cyclic Response of Shallow Foundation Resting on Liquefiable Soil

Sunita Kumari[✉], Amrendra Kumar, and Sanjeev Kumar Suman

Department of Civil Engineering, National Institute of Technology Patna,
Patna 800005, Bihar, India
{sunitafce,sksuman}@nitp.ac.in,
amrendraroy2k8@gmail.com

Abstract. As the earthquake events are unpredictable, the accurate estimation of settlement and excess pore pressure is challenging for the soil foundation system against earthquake loading. Therefore, it is essential to study the effect of an earthquake before the construction of any structure on a particular site. In the present paper, response of a shallow foundation resting on liquefiable soil is numerically modeled to evaluate its response under cyclic loading. Two different conditions have been numerically simulated and the effect of spacing in between two foundations has been studied along with light and heavy foundation. The porous media theory based on coupled approach has been considered for analyzing the system. A computer code for finite element analysis has been developed in FORTRAN 90. The inelastic behavior which includes dilatancy and hardening behavior of the soil is modeled using Pastor–Zienkiewicz Mark III model. The settlement and effective pore pressure obtained are validated with centrifuge test results. It is observed that settlements during the cyclic loading time remained almost the same, regardless of the foundations spacing. The post-shaking settlements, however, were different for different spacing. Foundations settled often less when located close to each other.

Keywords: Liquefaction · Shallow foundation · Shear modulus
Permeability

1 Introduction

Major damage of superstructure situated on shallow foundation is one of the most dangerous phenomena due to liquefaction other than catastrophic failure reported in past. Building of pore pressure is associated with reduction in shear strength during earthquake causes decline of bearing capacity and induced seismic settlement of shallow foundation. Generally, structures were located close to each other in urban area. Considerable damages have been reported for such foundations, may be due to improper design of shallow footing on liquefiable soils. Accordingly, numerous studies have been carried out to assess these complicated mechanisms. Several of reinforced concrete buildings were damaged by excessive settlement and tilting as a result of liquefaction, during the Niigata 1964 earthquake reported by Yoshimi and Tokimatsu [1], Seed and

© Springer Nature Singapore Pte Ltd. 2018
T. Qiu et al. (Eds.): GSIC 2018, *Proceedings of GeoShanghai 2018 International Conference: Advances in Soil Dynamics and Foundation Engineering*, pp. 185–195, 2018.
https://doi.org/10.1007/978-981-13-0131-5_21

Idriss [2] and Nagase and Ishihara [3]. Adachi et al. [4] and Acacio et al. [5] reported the occurrence of subsoil liquefaction in the city of Dagupan during the 1990 Luzon Philippines earthquake resulting excessive settlement in many buildings. Tokimatsu et al. [6] found that the excessive settlements occurred in corner building as compare to the building situated between two buildings. Also, these settlements are dependent on the separation between the adjacent buildings. Tokimatsu et al. [7] reported the stability of structures were depend on the dimensions, confining pressure, and the shear stress imposed by the buildings and their adjacent structures. Yoshida et al. [8] studied the Adapazari, Turkey (1999) earthquake, the damaged to structures by the liquefaction due to shallow and relatively thin layers of saturated sand.

In addition to the field studies, numerous experimental efforts including shaking table and centrifuge experiments were conducted for a better understanding of this problem. Yoshimi and Tokimatsu [1] suggested different factors like pore pressure development and structure's width, height and contact pressure, controlling the settlement of the structure. Liu and Dobry [9] conducted several centrifuge tests to study the mechanism of pore pressure buildup and foundation settlement of soil. Adalier et al. [10] discussed the mechanisms involved in shallow foundation settlement and the influence of stone column as liquefaction countermeasure. Dashti et al. [11] performed centrifuge test and confirmed the combined role of the shear strains imparted by the superstructure and the post-liquefaction volumetric strains as dominant mechanisms of total settlement. Coelho [12] mentioned that the initial static shear stress imparted by rigid foundation influences excess pore pressure history. Mason et al. [13] examined seismic soil– foundation–structure interaction of framed structures through centrifuge experiments. It was found that structure-soil-structure interaction (SSSI) can be beneficial or detrimental, depending on the earthquake motion and the structural system. Tsukamoto et al. [14] conducted two series of seismic 1 g shaking table tests on rigid circular foundations to examine the effects of shaking duration and the group effects of foundations. Settlements during the shaking time remained almost the same, regardless of the foundations spacing. The post-shaking settlements, however, were different for different spacing. Foundations settled often less when located close to each other. Hayden et al. [15] conducted centrifuge experiments to observe the performance of adjacent structures affected by liquefaction. It was concluded that adjacent structures tended to tilt away from one another and settled less than isolated structures. Mehrzad et al. [16] investigated the effect of soil permeability and contact pressure on foundation response through the centrifuge and numerical models and reported that settlement of foundations increased with the increase of soil permeability. All of the reviewed studies deal with liquefaction-induced settlement of isolated foundations, without adjacent structures. However, in urban areas, structures are located in close proximity which may affect the seismic response of foundations. There are limited studies which address the influence of adjacent structures on settlement mechanism of shallow foundations.

In the present study, the behavior of shallow foundation (light and heavy) has been simulated using coupled algorithm. The governing equilibrium and continuity equations have been used to develop the finite element equation. Newmark-beta time integration scheme is used to solve these equations. Two different conditions have been numerically simulated and the effect of spacing in between two foundations has been studied along with light and heavy foundation. The Pastor–Zienkiewicz Mark III model has been used to explain the inelastic behavior of soils under isotropic cyclic loadings. Dashpot system is attached to transmitting boundary to absorb the wave energy generated due to boundary condition.

2 General Formulation

In the present study, the Finite element method based discretized u-P formulation Zienkiewicz and Taylor [17] has been used for fully coupled dynamic analysis of liquefaction-induced settlement of foundation resting on saturated porous media subjected to earthquake loading.

Initially, static analysis is performed using consolidation form of the equation:

$$P(u) - QP = f_u \tag{1}$$

$$Q^T \dot{U} + S\dot{P} + HP = f_p \tag{2}$$

After static analysis, dynamic analysis is performed using the following equation:

$$M\ddot{U} + P(u) - Qp = f_u \tag{3}$$

$$G\ddot{U} + Q^T\dot{U} + S\dot{P} + HP = f_p \tag{4}$$

In which, M is the mass matrix, G is the dynamic coupling matrix, S is the compressibility Matrix, H is the permeability Matrix, Q is the coupling Matrix, f_u is the force matrix, f_p is the force matrix for fluid phase of an element, U and P are the vectors for the nodal value of u and p.

$P(u)$ is the nonlinear internal force vector given by:

$$P(u) = \int_w B^T \sigma' d\Omega \tag{5}$$

In which, B is the displacement-strain transformation matrix for the finite element method, σ' is a vector of all effective stress components at the integration point, Ω is the domain concerned. The inelastic behavior which includes dilatancy and hardening behavior of the soil is modeled using Pastor–Zienkiewicz Mark III model (Kumari and Sawant [18]). After dynamic analysis again consolidation is done using static analysis for dissipation of pore pressure. Viscous damping has been incorporated into the dynamic equation also called Rayleigh damping which prevents back propagation of wave into the soil domain.

$$L_m = \alpha_1 M + \alpha_2 K \tag{6}$$

The coefficients α_1 and α_2 can be obtained by selecting a damping ratio ξ_r and a certain frequency ω_i such that

$$\xi_r = \frac{\alpha_1}{2\omega_i} + \frac{\alpha_2 \omega_i}{2} \tag{7}$$

In which, ξ_r is damping ratio and ωi is frequency corresponding to i^{th} mode of system.

Park and Hashash [19] explained that if the damping ratio is constant throughout the soil domain, scalar values of α_1 and α_2 can be evaluated using two significant natural modes. In the present study, the values of ω_1 and ω_2 are considered as frequency corresponding to first and third peak of the frequency amplitude response. For $\xi_{r1} = \xi_{r2}$ and $\omega_2 > \omega_1$, α_1 and α_2 are given by,

$$\alpha_1 = \frac{2\xi_r \omega_1 \omega_2 (\omega_2 - \omega_1)}{(\omega_2^2 - \omega_1^2)}, \quad \alpha_2 = \frac{2\xi_r (\omega_2 - \omega_1)}{(\omega_2^2 - \omega_1^2)} \tag{8}$$

If the value of α_1 is set to zero then damping matrix becomes proportional to stiffness whereas for α_2 equal to zero it becomes mass proportional. In the present analysis, the typical variation of damping ratio with frequency for mass proportional, stiffness proportional and Rayleigh damping is evaluated for $\omega_1 = 14$ rad/s and $\omega_2 = 24$ rad/s.

3 Verification of Numerical Code

The proposed FEM based solution algorithm has been validated by comparing the numerical results of settlement and excess pore water pressure (EPP) with the centrifuge model test results conducted by Elgamal et al. [20]. It is considered best for modeling and observing soil liquefaction phenomena which creates stress conditions in the model which closely simulate those in the full-scale prototype. A 10 m saturated medium Nevada sand having relative density (D_r) of 40% has been considered to check the accuracy and correctness of the proposed model. Herein, the load (40 kPa, or about 2 m of an equivalent soil overburden) was simply applied at ground level in the form of a distributed surficial vertical stress over a 2 m × 2 m area. A 10 cycle, 1.5 Hz uniform sinusoid with average acceleration amplitude of 0.2 g is used as input parameter. The results of both centrifuge and numerical modeling show the good agreement and hence verify the model. However, the relative density and dynamic response of soil changes during shaking upon large settlements and densification. This numerical model does not update soil properties due to shaking-induced densification in a single time-domain analysis. Hence, capturing the dilation cycles that lead to sharp acceleration spikes and drops in excess pore pressures became difficult after the initial cycles that caused substantial settlement and densification. This is one of many possible explanations for the observed differences between numerical predictions and experimental results (Figs. 1 and 2).

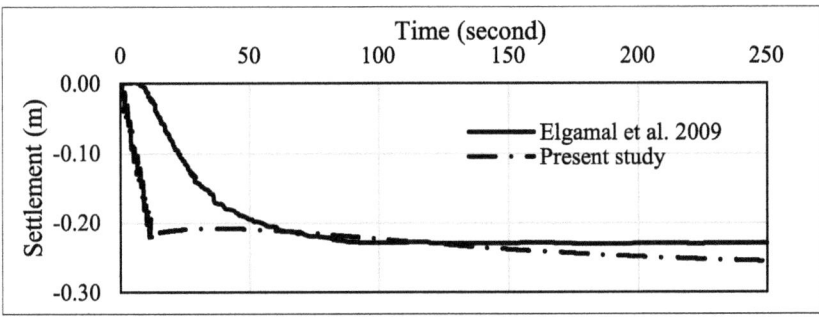

Fig. 1. Settlement of foundation at the center of footing

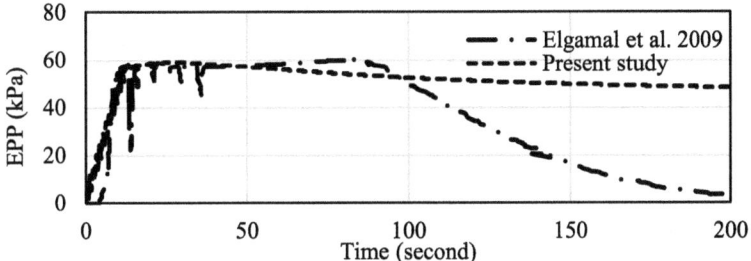

Fig. 2. Excess pore pressure verses time at 4 m depth

4   Results and Discussion

The saturated loose soil strata of size 48 m × 12 m has been considered for the analysis of shallow foundation. Plane strain condition is assumed to reduce the computational efforts. The soil domain in XZ plane is discretized into 144 elements of uniform finite element mesh. The 8–4 node mixed element having eight displacement nodes and four pore pressure nodes were used for finite element analysis. As a result, displacements are continuous biquadratic and pore pressures are continuous bilinear in the element. Two different conditions have been numerically simulated and the effect of spacing in between two foundations has been studied along with light and heavy foundation. Case 1 demonstrates, no spacing in between light and heavy foundation (Fig. 3), whereas Case 2 is the situation when the spacing between these foundations is 8 m (Fig. 4).

4.1   Case 1 (Response of Foundations Without Spacing)

This condition satisfies when the two foundations have no spacing in between them. The analysis is performed at 1 Hz frequency and 0.1 g acceleration. The sinusoidal input ground motions has been imparted to the model with the peak base shaking amplitude of 0.1 g and frequency 1 Hz. Figures 5 and 6 shows the excess pore pressure

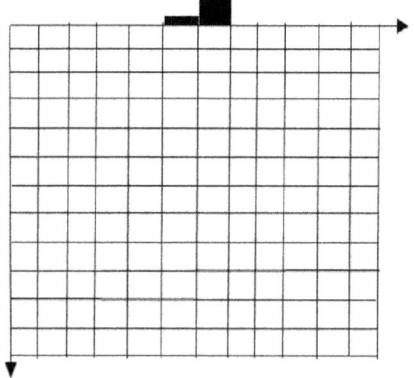

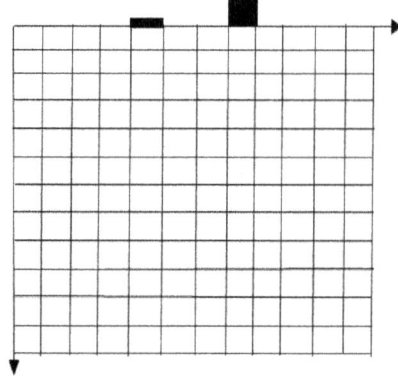

Fig. 3. Case 1 Light and heavy foundation are placed in contact

Fig. 4. Case 2 Light and heavy foundation are placed at 8 m

generated with respect to time in case of heavy and light footing respectively at different depth i.e. at 1.5 m, 3.5 m, 7.5 m and 9.5 m.

The time history of EPP shows that liquefaction occurs at lower depth only in case of heavy foundation may be due to amplification of frequency. Immediately after excitation ceased, vertical and horizontal hydraulic gradient was developed and surrounding water flowed towards the foundations. After a relatively long time, large positive EPP was generated under the foundations due to seepage. Once the water pressure equalized in each level and horizontal hydraulic gradients vanished, the soil began to reconsolidate starting from the base of soil strata towards the soil surface.

Settlement in the form of non-dimensional (Z/H) quantity at different depth of light and heavy foundations is shown in Figs. 7 and 8. Foundations settlement commenced immediately after the first cycle of shaking and continued with time. Foundations continued to settle, even after excitation ceased, until the end of reconsolidation in the upper layers. The rate of settlement accumulation decreased after shaking ceased. Negligible settlement of foundation occurred during soil reconsolidation. Such behavior was also observed and reported in the recent centrifuge experiments (Dashti et al. [21]). Partial drainage and inertial force imposed by foundation surcharge seem to be two dominant mechanisms of foundation settlement during shaking. The vertical settlement is always greater than horizontal settlement and settlement of heavy foundation is more than that of light foundation.

4.2   Case 2 (Response of Foundations with Spacing of 8 m)

Case-2 demonstrates the condition when spacing between light and heavy foundation is 8 m. The results are taken different frequency and different acceleration. The results of peak value of EPP and Z/H ratio shown below in different graphs.

Figure 9 shows the EPP of heavy and light foundation at 0.5 Hz frequency and 0.1 g acceleration for different depth. The EPP is taken maximum value from the curve of EPP verse time history. Results show the both light and heavy foundation do not

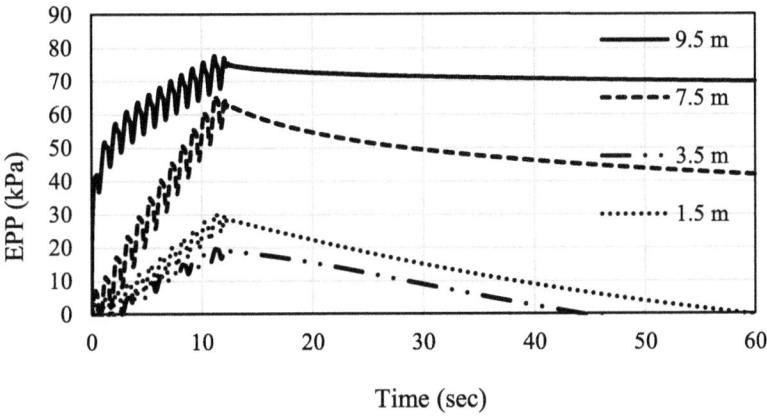

Fig. 5. EPP verse time below the center of heavy footing at different depth

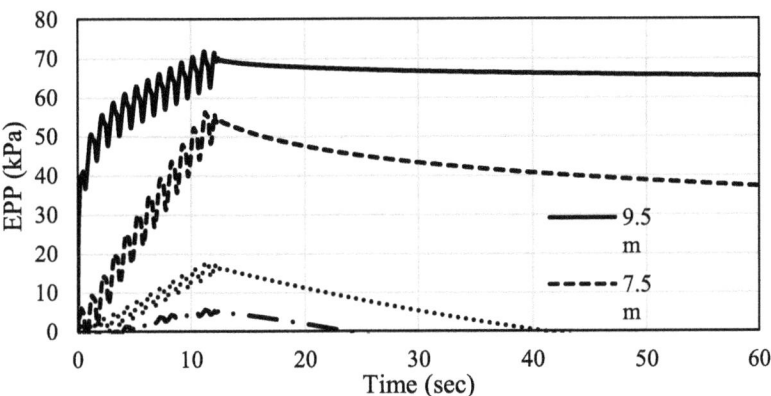

Fig. 6. EPP verse time below the center of light footing at different depth

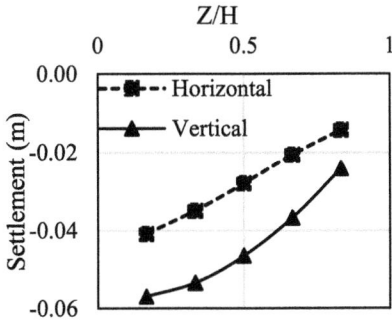

Fig. 7. Settlements of light foundation at 1 Hz and 0.1 g

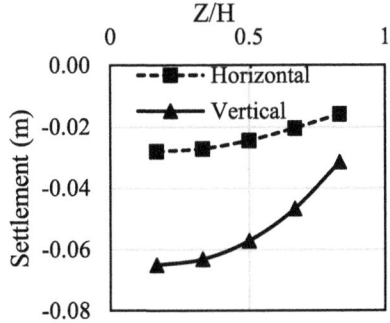

Fig. 8. Settlements of heavy foundation at 1 Hz and 0.1 g

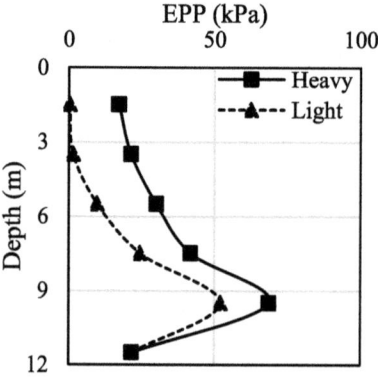

Fig. 9. EPP of foundation at 0.5 Hz and 0.1 g

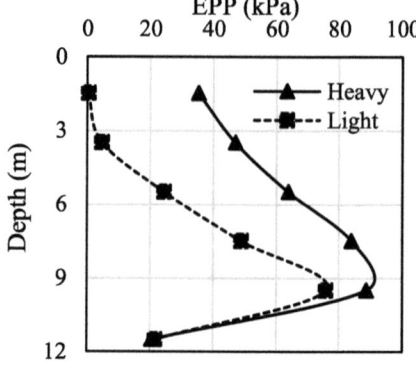

Fig. 10. EPP of foundation at 0.5 Hz and 0.2 g

liquefy at that frequency and acceleration. But liquefaction will occur at all the point for higher value of acceleration at same frequency shown in Fig. 10 for heavy foundation (Fig. 12).

Figure 11 shows the EPP of heavy and light foundation at 1 Hz frequency and 0.1 g acceleration for different depth. The EPP is taken maximum value from the curve of EPP verse time history. Results show the both light and heavy foundation do not liquefy at that frequency and acceleration. But liquefaction will occurs all the point at higher value of acceleration at same frequency shown in Fig. 10 for heavy foundation.

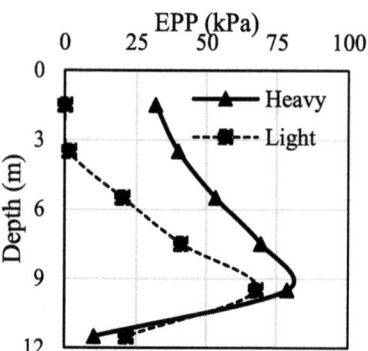

Fig. 11. EPP of foundation at 1 Hz and 0.1 g

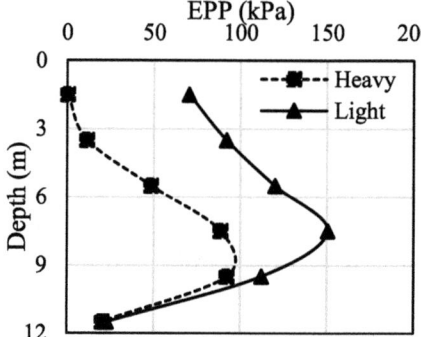

Fig. 12. EPP of foundation at 1 Hz and 0.2 g

Thereafter, EPP started to dissipate and the solidification front moved from the bottom of the container towards the ground surface. The upward flow of fluid caused seepage-induced liquefaction in the sand; thus, there was a delay in EPWP dissipation

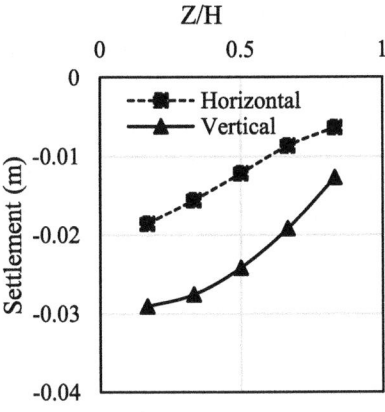

Fig. 13. Settlements of light foundation at 0.5 Hz and 0.1 g

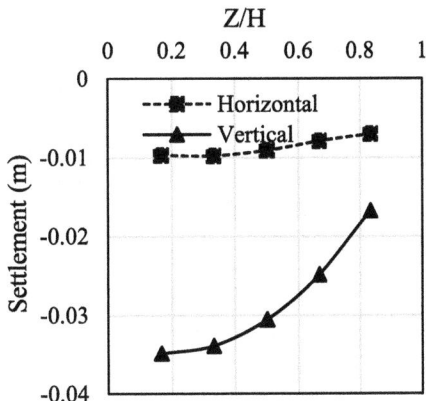

Fig. 14. Settlements of heavy foundation at 0.5 Hz and 0.1 g

in the upper layers. In fact, the bottom layers reconsolidated (or solidified) and settled first and the upper layers settled after a while. Total settlement of the ground surface was dependent to shaking intensity whereas more settlement occurred by the stronger shaking events. In the two strongest events in which large amounts of sand were liquefied, about 70% of total settlement occurred during the reconsolidation phase. It should be noticed that sand's relative density changed in each stage of loading. Ground surface and foundations experienced permanent settlements in loading stages. Therefore, the relative density of sand in each stage was higher than that in the previous stage. The subsequent increase of relative density per loading stage may change soil behavior and liquefaction-induced settlements (Figs. 13 and 14).

Hausler [22], Liu and Dobry [9] stated that deep densification may cause the shallow foundation settles more. In this situation, the acceleration that is transferred through the dense layer enhances the dynamic effect on the structure resulting in higher shear stress beneath the structure. Therefore, if the decrement effect of the dense layer on the settlement does not outweigh the effect of the dense layer on earthquake exaggeration, more settlement of the shallow foundation is expected. Coelho (2004) who concluded that using a dense layer depth more than a critical value will result in more settlement.

5   Conclusion

In this study dynamic fully coupled analyses have been carried out using FORTRON 90 in order to investigate the liquefaction- induced settlement of shallow foundations with different surcharge resting on liquefiable soil under seismic loadings. A bounding surface critical state constitutive model has been employed to capture the undrained behavior of Nevada saturated sands. Based on the results of the numerical investigations, the following conclusions can be drawn which can shed light on the issue of foundation design based on their performance. After verification of the numerical model, a comprehensive parametric study has been conducted by which the influence of different parameters such as spacing of surcharge, acceleration and frequency are investigated. Although EPP value beneath the shallow footing is lower for low value of surcharge. Vertical settlement is more than horizontal settlement for all cases. The generation of excess pore water pressure is increase up to 8 m depth for each value of surcharge. The excess pore water pressure decreases at higher depth due to more permeability. Effect of the surcharge placement also play important role.

References

1. Yoshimi, Y., Tokimatsu, K.: Settlement of buildings on saturated sand during earthquakes. Soils Found. **17**(1), 23–38 (1977)
2. Seed, H.B., Idriss, I.M.: Analysis of soil liquefaction: Niigata earthquake. J. Soil Mech. Found. Div. ASCE **93**(3), 83–108 (1967)
3. Nagase, H., Ishihara, K.: Liquefaction-induced compaction and settlement of sand during earthquakes. Soils Found. **28**(1), 65–76 (1988)
4. Adachi, T., Iwai, S., Yasui, M., Sato, Y.: Settlement and inclination of reinforced-concrete buildings in Dagupan City due to liquefaction during the 1990 Philippine earthquake. In: Proceedings of 10th World Conference on Earthquake Engineering, Madrid, vol. 1, pp. 147–152 (1992)
5. Acacio, A., Kobayashi, Y., Towhata, I., Bautista, R.T., Ishihara, K.: Subsidence of building foundation resting upon liquefied subsoil: case studies and assessment. Soils Found. **41**(6), 111–128 (2001)
6. Tokimatsu, K., Kuwayama, S., Tamura, S., Miyadera, Y.: Vs Determination from steady state Rayleigh wave method. Soil Found. **31**(2), 153–163 (1991)
7. Tokimatsu, K., Kojima, J., Kuwayama, A.A., Midorikawa, S.: Liquefaction-induced damage to buildings I 1990 Luzon Earthquake. J. Geotech. Engrg. **120**(2), 290–307 (1994)
8. Yoshida, N., Tokimatsu, K., Yasuda, S., Kokusho, T., Okimura, T.: Geotechnical aspects of damage in Adapazari city during 1999 Kocaeli, Turkey earthquake. Soils Found. **41**(4), 25–45 (2001)
9. Liu, L., Dobry, R.: Seismic response of shallow foundation on liquefiable sand. J. Geo-tech. Geoenviron. Eng. ASCE **123**(6), 557–567 (1997)
10. Adalier, K., Elgamal, A., Meneses, J., Baez, J.I.: Stone columns as liquefaction countermeasure in non-plastic silty soils. Soil Dyn. Earthq. Eng. **23**(7), 571–584 (2003)
11. Dashti, S., Bray, J.D., Pestana, J.M., Riemer, M.R., Wilson, D.: Mechanisms of seismically-induced settlement of buildings with shallow foundations on liquefiable soil. J. Geotech. Geoenviron. Eng. **136**(1), 151–164 (2010)
12. Coelho, P., Haigh, S.K., Madabhushi, S.P.: Centrifuge modeling of liquefaction of saturated sand under cyclic loading. In: Proceedings of the International Conference on CBS 2004, Bochum, Germany (2004)
13. Mason, H.B., Trombetta, N.W., Chen, Z., Bray, J.D., Hutchinson, B.L., Kutter, B.L.: Seismic soil–foundation–structure interaction observed in geotechnical centrifuge experiments. Soil Dyn. Earthq. Eng. **48**, 162–174 (2013)
14. Tsukamoto, Y., Ishihara, K., Sawada, S., Fujiwara, S.: Settlement of rigid circular foundations during seismic shaking in shaking table tests. Int. J. Geomech. **12**(4), 462–470 (2012)

15. Hayden, C., Zupan, J., Allmond, J., Kutter, B.: Centrifue tests of adjacent mat-supported buildings affected by liquefaction. J. Geotech. Geoenviron. Eng. **141**(3), 04014118 (2015)
16. Mehrzad, B., Haddad, A., Jafarian, Y.: Centrifuge and numerical models to investigate liquefaction-induced response of shallow foundations with different contact pressures. Int. J. Civ. Eng. (2016). https://doi.org/10.1007/s40999-016-0014-5
17. Zienkiewicz, O.C., Taylor, R.L.: Basic Formulation and Linear Problems, 4th edn. McGraw-Hill, Maidenhead (1989)
18. Kumari, S., Sawant, V.A.: Simulation of liquefaction phenomenon in semi-infinite domain under harmonic loading. Int. J. Geotech. Eng. **9**(3), 251–264 (2015)
19. Park, D., Hashash, Y.M.A.: Soil damping formulation in nonlinear time domain site response analysis. J. Earthq. Eng. **8**(2), 249–274 (2004)
20. Elgamal, A., Yang, Z., Parra, E., Ragheb, A.: Liquefaction-induced settlement of shallow foundations and remediation: 3D numerical simulation. J. Earthq. Eng. **9**(1), 17–45 (2005). Online (2009)
21. Dashti, S., Bray, J.D.: Numerical simulation of building response on liquefiable sand. J. Geotech. Geoenviron. Eng. **139**(8), 1235–1249 (2013)
22. Hausler, E.A.: Influence of Ground Improvement on Settlement and Liquefaction: A Study Based on Field Case History Evidence and Dynamic Geotechnical Centrifuge Tests. University of California, Berkeley (2002)

Three-Dimensional Analysis of a Row of Holes as Active Wave Barrier in Saturated Soil

Gang Shi[✉] and Yonghui Li

School of Civil Engineering, Zhengzhou University, Zhengzhou, China
13526584056@163.com

Abstract. Traffic, machine operations and other human activities can generate ground vibrations, which can cause distress to adjacent structures, and disturb the operation of precision instruments. Generally, these adverse effects of vibrations can be eliminated or prevented by installation of various types of wave barriers, such as a row of holes. In the paper, the investigation is focused on the effects of using a row of holes for the reduction of nearby vibration response generated by the motion of a machine foundation on saturated soil. A 3D semi-analytical BEM model is established, where the TLM (thin layered method) basic solutions of saturated soil are employed as poroelastodynamic fundamental solutions. Then the effects of the model parameters on effectiveness of vibration isolation are investigated and discussed in detail. The results show that a row of holes can isolate the ground vibrations successfully. Increasing the radius and the depth, decreasing the net spacing between two successive holes can all result in an increase in the screening effectiveness. The distance between the machine foundation and wave barriers can affect the screening effectiveness, and the larger the distance is, the poor the screening effectiveness is.

Keywords: A row of holes · Semi-analytical BEM · Saturated soil
Active wave barriers · Screening efficiency

1 First Section

1.1 A Subsection Sample

Machine operations, traffic, blasting and other human activities can generate ground vibrations, which can adversely affect nearby structures and sensitive machinery. Generally, these adverse effects can be eliminated or prevented by installation of various types of wave barriers. Even though a trench is a more effective wave barrier than a row of holes or piles, the use of trenches is restricted because of construction depth and stability. Then holes are studied for vibration barriers in the paper.

Woods et al. [1] were the first to study experimentally the screening effectiveness of a row of hollow cylindrical piles as barrier. Liao and Sangrey [2] investigated the model piles as passive barriers by studying the propagation of sound waves in fluid media. Haupt [3] performed model tests by using open and in-filled trenches, hollow piles in a row to reduce the ground vibrations. Aviles and Sanchez-Sesma [4, 5] gave an analytical investigation on the screening efficiency of 8 solid piles on P-wave, SH-wave and SV-wave. Baroomand and Kaynia [6] presented semi-analytical solutions

© Springer Nature Singapore Pte Ltd. 2018
T. Qiu et al. (Eds.): GSIC 2018, *Proceedings of GeoShanghai 2018 International Conference: Advances in Soil Dynamics and Foundation Engineering*, pp. 196–205, 2018.
https://doi.org/10.1007/978-981-13-0131-5_22

of the problem involving Rayleigh waves. Kattis et al. [7, 8] used 3D BEM to study the problem of vibration isolation by a row of solid piles and tubular piles, which are only treated as long cylindrical cavities. Gao et al. [9] used integral equation to study the isolation effect of a row of piles on incident Rayleigh wave. Tsai et al. [10] used 3D BEM to investigate the screening effectiveness of circular piles in a row as active wave barriers. Lu et al. [11] studied the isolation of the vibration due to moving loads using pile rows embedded in a homogenous and layered poroelastic half-space by using Muki and Sternberg's method.

This paper presents a study on the problem of the ground vibration isolation efficiency by a row of holes as active barriers in saturated soil. The investigation is accomplished with the aid of the semi-analytical BEM in cooperation with the basic solution of the TLM. Effects of model parameters on effectiveness of vibration isolation by a row of holes are investigated and discussed in detail.

2 Active Vibration Isolation of Foundations by a Row of Holes

This section deals with the active vibration isolation problem associated with a row of holes case in a homogenous poroelastic half-space, schematically described by Fig. 1.

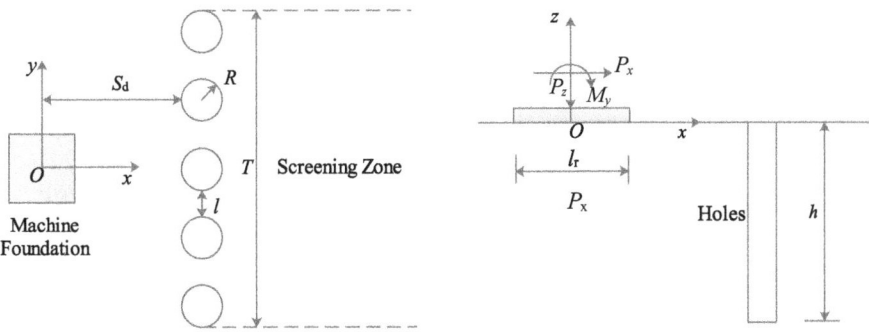

Fig. 1. A row of holes vibration isolation system as active barrier

The source of disturbance is a rigid, massive, surface rectangular machine foundation in perfect bonding with the soil which is subjected to the system of forces P_x, P_y, P_z and moments M_x, M_y given by

$$P_x = \tilde{P}_x e^{-i\omega t} \ P_y = \tilde{P}_y e^{-i\omega t} \ P_z = \tilde{P}_z e^{-i\omega t} \ M_x = \tilde{M}_x e^{-i\omega t} \ M_y = \tilde{M}_y e^{-i\omega t} \quad (1)$$

The motion of the machine foundation can generate ground vibrations, and these vibrations can be reduced by a row of holes installed close to the foundation. In order to determine the screening effectiveness of the wave barrier, the whole dynamic system of Fig. 1 can be divided into two parts, the saturated soil half-space with a row of holes and the rigid foundation, which are related to each other through the compatibility and equilibrium equations at their interface.

Herein, a three-dimensional poroelastic semi-analytical BEM is put forward, where the fundamental solutions by using TLM is employed, to investigate the isolation efficiency of a row of holes in homogenous poroelastic half-space. Unlike other conventional poroelastic BEM procedures, this semi-analytical BEM does not require the discretization at the ground surface, which can decrease the computational complexity. For detail description, one can result Shi et al. [12]. Moreover, a SPMD parallel algorithm is developed to accelerate the semi-analytical BEM computation.

2.1 Boundary Element Equations of Machine Foundation

When utilizing 3D semi-analytical BEM to study the ground vibrations generated by the foundation, the interface between the foundation and soil half-space is divided into boundary elements. The displacements of i-th boundary node of the contact area are given by

$$\tilde{u}_{rxi} = \tilde{\Delta}_x \quad \tilde{u}_{ryi} = \tilde{\Delta}_y \quad \tilde{u}_{rzi} = \tilde{\Delta}_z - \tilde{\alpha}_x y_i - \tilde{\alpha}_y x_i \tag{2}$$

where the subscript r refers to quantities at the soil-foundation interface, $\tilde{\Delta}_x$, $\tilde{\Delta}_y$ and $\tilde{\Delta}_z$ correspond to the displacements at the center of foundation, while $\tilde{\alpha}_x$ and $\tilde{\alpha}_y$ are the small amplitude of foundation rotation about the axis x and the axis y, respectively, x_i and y_i are the coordinates of the i-th boundary node. Further, equilibrium of forces acting on the foundation takes the form

$$\tilde{P}_x = -m\omega^2\tilde{\Delta}_x + \sum_{k=1}^{E_1} A_k \tilde{t}_x^k \quad \tilde{P}_y = -m\omega^2\tilde{\Delta}_y + \sum_{k=1}^{E_1} A_k \tilde{t}_y^k \quad \tilde{P}_z = -m\omega^2\tilde{\Delta}_z + \sum_{k=1}^{E_1} A_k \tilde{t}_z^k \tag{3}$$

$$\tilde{M}_x = -J_x\omega^2\tilde{\alpha}_x + \sum_{k=1}^{E_1} y^k A_k \tilde{t}_z^k \quad \tilde{M}_y = -J_y\omega^2\tilde{\alpha}_y + \sum_{k=1}^{E_1} x^k A_k \tilde{t}_z^k \tag{4}$$

where m is the mass of foundation, J_x and J_y are the mass moments of inertia about x axis and y axis, respectively, A_k represents the area of the k-th boundary element of foundation, $\tilde{t}_x$, $\tilde{t}_y$ and $\tilde{t}_z$ are the interface tractions of the k-th boundary element of foundation. Equations (2)–(4) can be written in a compact matrix form as

$$\begin{cases} \tilde{U}_r = C\tilde{\Delta} \\ \tilde{P} = D\tilde{\Delta} + H^r \tilde{T}_r \end{cases} \tag{5}$$

The boundary element equation for soil medium of the contact area takes the form

$$H_{f_1}\begin{Bmatrix} \tilde{U}_{f_1} \\ \tilde{P}_{f_1} \end{Bmatrix} = G_{f_1}\begin{Bmatrix} \tilde{T}_{f_1} \\ \tilde{Q}_{f_1} \end{Bmatrix} \tag{6}$$

where the subscript f_1 refers to quantities at the soil-foundation interface, $\tilde{U}_{f_1}$, $\tilde{P}_{f_1}$, $\tilde{T}_{f_1}$ and $\tilde{Q}_{f_1}$ are the displacement, pore pressure, traction and flux vectors field, respectively.

The compatibility and equilibrium equations at the soil-foundation interface read

$$\tilde{U}_r = \tilde{U}_{f_1} \quad \tilde{Q}_{f_1} = 0 \quad \tilde{T}_{f_1} + n_{f_1}\tilde{P}_{f_1} + \tilde{T}_r = 0 \tag{7}$$

in which n_{f_1} is the unit normal to the rigid foundation.

Expanding Eqs. (5) and (6) and taking into account the Eq. (7), one can obtain

$$\left\{ \begin{array}{c} \tilde{P} \\ 0 \end{array} \right\} = \begin{bmatrix} D & -H^r n \\ 0 & 0 \end{bmatrix} \left\{ \begin{array}{c} \tilde{\Delta} \\ \tilde{P}_{f_1} \end{array} \right\} + \begin{bmatrix} -H^r & 0 \\ 0 & I \end{bmatrix} G_{f_1}^{-1} H_{f_1} \begin{bmatrix} C & 0 \\ 0 & I \end{bmatrix} \left\{ \begin{array}{c} \tilde{\Delta} \\ \tilde{P}_{f_1} \end{array} \right\} \tag{8}$$

where I is the identity matrix.

Solutions of Eq. (8) can provide the displacements of rigid foundation and the pore pressures at the soil-foundation interface, while the displacement at the soil-foundation interface can be obtained with the aid of Eqs. (5) and (6).

2.2   Boundary Element Equations of a Row of Holes

Considering the two parts of the whole dynamic system of Fig. 1, boundary element equation for the soil medium takes the form

$$H_{f_a} \left\{ \tilde{U}_{f_1}^T \quad \tilde{U}_{f_2}^T \quad \tilde{P}_{f_1}^T \quad \tilde{P}_{f_2}^T \right\}^T = G_{f_a} \left\{ \tilde{T}_{f_1}^T \quad \tilde{T}_{f_2}^T \quad \tilde{Q}_{f_1}^T \quad \tilde{Q}_{f_2}^T \right\}^T \tag{9}$$

where the subscript f_2 refers to quantities of the wave barrier boundary. Because the holes' surfaces are free of tractions and impermeable, one has

$$\tilde{Q}_{f_2} = 0 \quad n_{f_2}\tilde{P}_{f_2} + \tilde{T}_{f_2} = 0 \tag{10}$$

in which n_{f_2} is the unit normal to the holes surfaces.

Considering the boundary conditions of the soil-foundation interface and the holes surfaces, elimination $(\tilde{U}_{f_1}, \tilde{U}_{f_2})$ and $(\tilde{T}_{f_1}, \tilde{T}_{f_2})$ between Eqs. (5, 8) and (9) results in the equation

$$\left\{ \begin{array}{c} \tilde{P} \\ 0 \\ 0 \\ 0 \end{array} \right\} = \begin{bmatrix} D & 0 & -H^r n_{f_1} & 0 \\ 0 & 0 & 0 & 0 \\ 0 & 0 & 0 & 0 \\ 0 & 0 & 0 & n \end{bmatrix} \left\{ \begin{array}{c} \tilde{\Delta} \\ \tilde{U}_{f_2} \\ \tilde{P}_{f_1} \\ \tilde{P}_{f_2} \end{array} \right\} + \begin{bmatrix} -H^r & 0 & 0 & 0 \\ 0 & 0 & I & 0 \\ 0 & 0 & 0 & I \\ 0 & I & 0 & 0 \end{bmatrix} G_{f_a}^{-1} H_{f_a} \begin{bmatrix} C & 0 & 0 & 0 \\ 0 & I & 0 & 0 \\ 0 & 0 & I & 0 \\ 0 & 0 & 0 & I \end{bmatrix} \left\{ \begin{array}{c} \tilde{\Delta} \\ \tilde{U}_{f_2} \\ \tilde{P}_{f_1} \\ \tilde{P}_{f_2} \end{array} \right\} \tag{11}$$

2.3   The Screening Effectiveness of a Row of Holes

In order to investigate the screening effectiveness of a row of holes, the amplitude attenuation ratio A_{RF} is defined as the ratio of the vertical displacement component of the soil surface in the presence of wave barrier to that in the absence of wave barrier. Further, the average surface amplitude attenuation ratio behind barrier is defined as

$$A_R = \frac{1}{A} \int_A A_{RF}(x, y) dx dy \tag{12}$$

where A is the area behind wave barrier, and in the paper, A is a rectangular with a length $5\lambda_R$ and the same width with the wave barrier. Herein, λ_R is the Rayleigh wave length of soil medium.

The soil is assumed to be poroelastic and the material data are given in Table 1.

Table 1. Material data of soil medium

μ (MPa)	κ (m^4/Ns)	ρ_s (kg/m^3)	v	v_u	B	ρ_a (kg/m^3)	ρ_f (kg/m^3)	ϕ
19.4	1.0×10^{-10}	2700	0.38	0.49	0.94	150	1000	0.40

A row of seven holes with circular cross-section as active barrier is examined. Each hole has a radius r and a depth h, while the net spacing between two successive holes is l. Further, the machine foundation is assumed to be massless and has a length l_r and width t_r, and the distance between foundation and a row of holes is S_d. All the geometric parameters is normalized by Rayleigh wavelength λ_R as $r^* = r/\lambda_R = 0.15$, $h^* = h/\lambda_R = 1.0$, $l^* = l/\lambda_R = 0.1$, $S_d^* = S_d/\lambda_R = 2.0$, $l^* = l/\lambda_R = 0.2$ and $t_r^* = t_r/\lambda_R = 0.2$.

Figure 2 depicts the contour of A_{RF} and the corresponding variation of A_{RF} along x-axis for different C^*. Herein, $C^* = 2y/T$ where T is the width of a row of holes. Thus, $C^* = 0$ indicates the center of a row of holes, while $C^* = 1$ refers to the edge of the wave barrier.

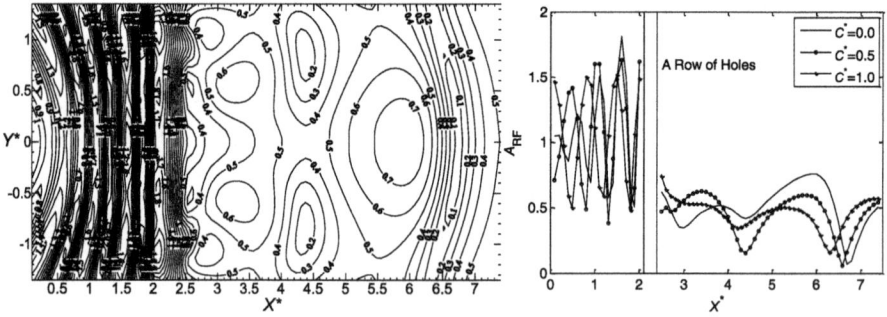

Fig. 2. A row of holes vibration isolation system as active barrier: (a) contour of A_{RF} and (b) graphs of A_{RF} for different C^* along x-axis

From these figures, it can be observed that $A_{RF} < 1$ behind the wave barrier, while $A_{RF} > 1$ in front of wave barrier, indicating that the incident waves have been scattered by a row of holes and the barrier can successfully isolate the ground vibrations generated by machine foundation in poroelastic soil. Further, there is a fluctuation phenomenon of A_{RF} behind the barrier, and the center of the wave barrier doesn't has the best screening effectiveness, which should be noticed when designing the wave barrier of a row of holes.

3   Active Vibration Isolation of Foundation by a Row of Holes

In this section, the influence of various key parameters on active vibration isolation by a row of holes has been studied. In what follows, unless otherwise stated, the normalized geometrical parameters of the row of holes are assumed to be constant and take the values mentioned above.

3.1   Effects of the Normalized Distance S_d^*

In this section, the active vibration isolation problem by a row of holes with different normalized distance S_d^* is examined in detail. In what follows, S_d^* takes the values of 1.0–5.0, respectively, and the results are depicted in Fig. 3.

From Fig. 3, it can be found an increase of the normalized distance S_d^* results in a decrease of the achieved reduction isolation efficiency behind barrier. Thus, it is suggested that the wave barrier should be installed as close to the source of disturbance as possible.

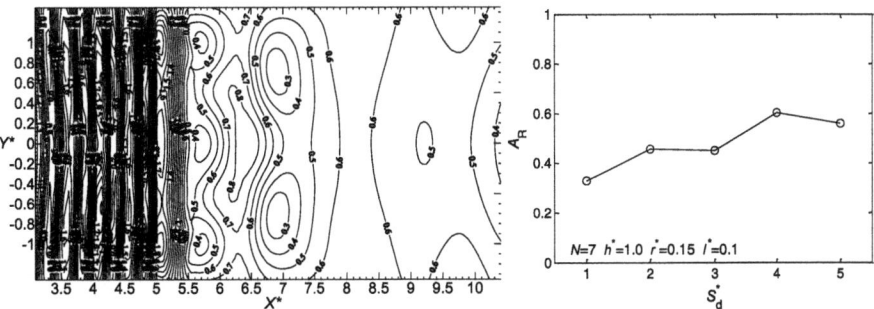

Fig. 3. The screening effectiveness of a row of holes for different S_d^*: (a) contour of A_{RF} and (b) effect of S_d^* on average isolation effectiveness A_R

3.2   Effects of Geometrical Parameters of a Row of Holes

In what follows, the active vibration isolation problems by a row of holes with different geometrical parameters are examined in detail. The geometric variables of the barrier are summarized in Table 2. It should be notice that when one parameter changes, other parameters should remain unchanged.

Table 2. Geometric parameters of a row of holes

Parameter	Calculating values	Parameter	Calculating values
The number N	2, 3, 4, 5, 6, 7, 8	The radius r^*	0.05, 0.1, 0.15, 0.2
The nest spacing l^*	0.05, 0.1, 0.15, 0.2, 0.25, 0.3	The depth h^*	0.4, 0.7, 1.0, 1.3

Effect of the Number of Holes N. The number of holes in a row is an important variable of practical concern for the vibration isolation by a row of holes. For the purpose of this investigation, it is assumed that the number n varied from 2 to 8, and the results are shown in Fig. 4.

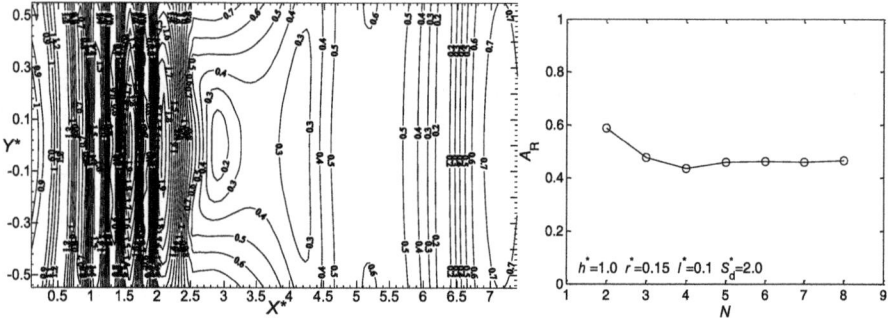

Fig. 4. The screening effectiveness of a row of holes for different N: (a) contour of A_{RF} and (b) effect of N on average isolation effectiveness A_R

From these figures, one can observe that when the number of holes is small, increasing the number of holes can increase the screening efficiency of the barrier. While when $N > 3$, the increasing of the number of holes has less effect on the vibration isolation effectiveness. However, increasing the number of holes can achieve a larger screened zone behind wave barrier, which should be noticed in the design.

Effect of the Net Spacing l^*. The influence of the normalized net spacing l^* on the screening effectiveness of a row of holes is shown in Fig. 5. We can see that a row of holes behaves like a unit system to instead of a set of independent holes when the net spacing l^* is small, while when l^* is large, the row of holes behaves like a set of independent holes, and the screening efficiency appears to be very poor. Moreover, the larger the normalized net spacing l^* is, the poorer the screening efficiency of the barrier is. Hence, it is suggested that the net spacing may be as small as possible to ensure a good screening efficiency.

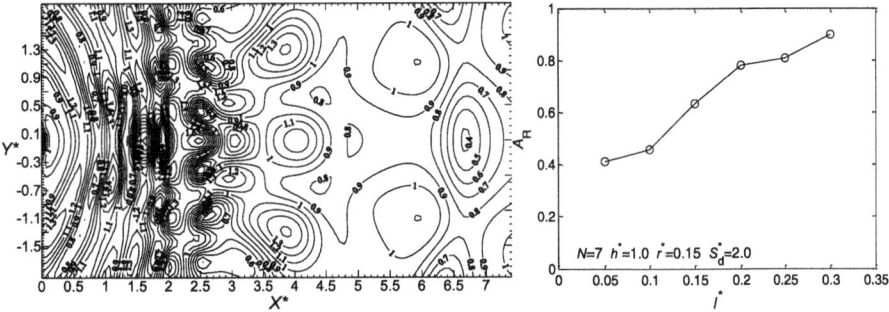

Fig. 5. The screening effectiveness of a row of holes for different l^*: (a) contour of A_{RF} and (b) effect of l^* on average isolation effectiveness A_R

Effect of the Depth of Holes h^*. The depth of holes is also an important variable in practice, which is concerned with costs and difficulties of construction. Figure 6 gives the results of the effect of the holes depth h^* on vibration isolation effectiveness.

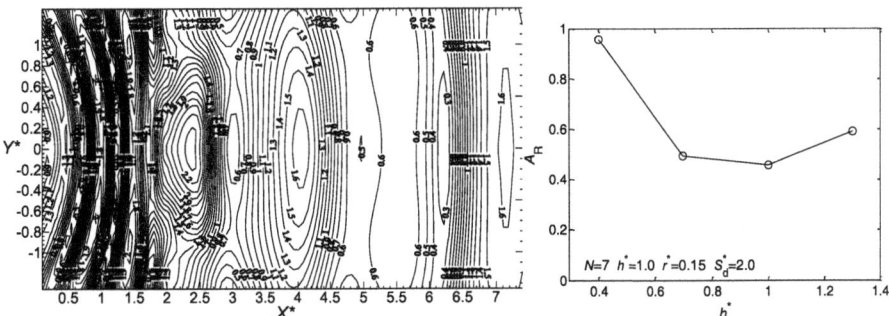

Fig. 6. The screening effectiveness of a row of holes for different h^*: (a) contour of A_{RF} and (b) effect of h^* on average isolation effectiveness A_R

When hole depth h^* is small, for example $h^* = 0.4$ in this case, A_{RF} is larger than 1.0 in some area behind the wave barrier, meaning that the row of holes is failure to isolate the ground vibrations. Moreover, the increase of the hole depth results in an obvious increase of screening efficiency behind barrier when $h^* \leq 1.0$, while it produces a slightly decrease of screening efficiency when $h^* > 1.0$. Therefore, it is suggested that the holes depth h^* should take the value of a Rayleigh wavelength in order to ensure a good screening effectiveness and excessive depth is not requisite.

Effect of the Radius of Holes r^*. The influence of the radius of holes r^* on the screening effectiveness of a row of holes is shown in Fig. 7. From these figures, one can see that when the holes radius r^* is very small, i.e. $r^* = 0.05$ in this case, A_{RF} behind the barrier tends to 1.0, meaning that the barrier fails to isolate the ground vibrations due to the machine foundation. When increasing the holes radius r^*, one can

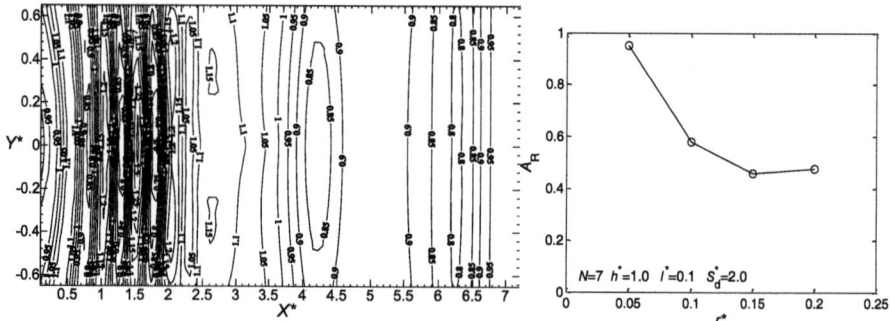

Fig. 7. The screening effectiveness of a row of holes for different r^*: (a) contour of A_{RF} and (b) effect of r^* on average isolation effectiveness A_R

obtain a better screening efficiency as well as a larger screening zone. However, excessive holes radius may increase the costs and construction difficulties dramatically.

Effect of the Arrangement of a Row of Holes for Same Barrier Width. In practice, the barrier width T may be confirmed firstly. Under this circumstance, how to design a row of holes to ensure a good screening efficiency becomes a critical problem. Herein, four cases are studied in this section to reveal the effect of the arrangement of a row of holes on the screening effectiveness, which are given as follows

Case 1: $l^* = 0.30$, $r^* = 0.15$, $N = 5$ Case 2: $l^* = 0.18$, $r^* = 0.15$, $N = 6$
Case 3: $l^* = 0.10$, $r^* = 0.15$, $N = 7$ Case 4: $l^* = 0.157$, $r^* = 0.1$, $N = 8$

The results are listed in Table 3. One can observe that the holes radius and the net spacing between two successive holes are the major factors in the design of a row of holes, which can affect the screening effectiveness significantly.

Table 3. The average surface amplitude attenuation ratio

Case	Case 1	Case 2	Case 3	Case 4
A_R	0.889	0.750	0.457	1.105

4 Conclusions

A 3D semi-analytical BEM model is established, to investigate the effects of using a row of holes as active barrier to reduce the nearby ground vibration generated by the disturbance of a rigid foundation in saturated soil. Based on the obtained results, the following conclusions can be made.

(1) A row of holes can isolate the ground vibrations successfully in saturated soil.
(2) Increasing the radius, the depth of the holes, decreasing the net spacing between two successive holes can all result in an increase in the screening effectiveness.

(3) The distance between the machine foundation and wave barriers can affect the screening effectiveness obviously, and the larger the distance is, the poor the screening effectiveness is.

(4) In practical engineering, we can increase the number of holes to achieve a larger screened zone and the holes depth should take the value of a Rayleigh wavelength.

Acknowledgments. This work is supported by the National Natural Science Foundation of China with Grant no. 51308506.

References

1. Woods, R.D., Barnet, N.E., Sangesser, R.H.: A new tool for soil dynamics. J. Geotech. Eng. Div. ASCE **100**(11), 1234–1247 (1974)
2. Liao, S., Sangrey, D.A.: Use of piles as isolation barriers. J. Geotech. Eng. Div. ASCE **104**, 1139–1152 (1978)
3. Haupt, W.A.: Model tests on screening of surface waves. In: Balkema, A.A. (ed.) Proceedings of the 10th International Conference in Soil Mechanics and Foundation Engineering, vol. 3, pp. 215–222. Publications Committee of X.ICSMFE, Rotterdam, Stockholm (1981)
4. Aviles, J., Sanchez-Sesma, F.J.: Piles as barriers for elastic waves. J. Geotech. Eng. **109**(9), 1134–1146 (1983)
5. Avilles, J., Sanchez-Sesma, F.J.: Foundation isolation from vibration using piles as barriers. J. Eng. Mech. **114**(11), 1854–1870 (1988)
6. Baroomand, B., Kaynia, A.M.: Vibration isolation by an array of piles. In: Soil Dynamics and Earthquake Engineering, pp. 683–691. Computational Mechanics Publications, Southampton (1991)
7. Kattis, S.E., Polyzos, D., Beskos, D.E.: Vibration isolation by a row of piles using a 3-D frequency domain BEM. Int. J. Numer. Methods Eng. **46**(5), 713–728 (1999)
8. Kattis, S.E., Polyzos, S., Beskos, D.E.: Modelling of pile wave barriers by effective trenches and their screening effectiveness. Soil Dyn. Earthq. Eng. **18**, 1–10 (1999)
9. Gao, G., Li, Z., Qiu, C., et al.: Three-dimensional analysis of rows of piles as passive barriers for ground vibration isolation. Soil Dyn. Earthq. Eng. **26**(11), 1015–1027 (2006)
10. Tsai, P.-H., Feng, Z., Tinlon, J.: Three-dimensional analysis of the screening effectiveness of hollow pile barriers for foundation-induced vertical vibration. Comput. Geotech. **35**, 489–499 (2008)
11. Lu, J.F., Xu, B., Wang, J.H.: A numerical model for the isolation of moving-load induced vibrations by pile rows embedded in layered porous media. Int. J. Solids Struct. **46**, 3771–3781 (2009)
12. Shi, G., Gao, G.Y.: Two-dimensional analysis of open trench as passive barrier in saturated soil. Environmental vibrations: Prediction, monitoring, mitigation and evaluation, 265–270 (2009)

Topographic Effects on the Seismic Coefficient and Earthquake-Induced Permanent Displacement of Earth Slopes

Jian Song[(✉)] and Yufeng Gao

Key Laboratory of Ministry of Education for Geomechanics and Embankment Engineering, Hohai University, Nanjing 210098, China
jiansonghh@163.com

Abstract. Topographic irregularities have found to considerably affect the amplitude and frequency content of ground motions, but this effect is not included in current pseudo-static and displacement-based methods of seismic slope stability analysis. In this study, two-dimensional (2D) seismic response of step-like slopes under vertically propagating in-plane shear waves (SV waves) is assessed to investigate the topographic effects on seismic coefficient and permanent displacement of earth slopes. The analyses are performed for slopes with different heights, inclinations, soil types subject to artificial input motions with various frequencies. The decoupled method is adopted to separately calculate the seismic response of slopes and the permanent displacements. The slip surface of slopes and the characteristics of sliding mass are firstly derived through static stability analysis, and the seismic coefficient of sliding mass is then evaluated by assuming no sliding scenario occurs. The permanent seismic displacements are finally computed by the Newmark method. The one-dimensional (1D) seismic analysis is also performed for the sliding mass, and the results are compared with those from the 2D analysis to provide some insights into the topographic effects on the assessment of the seismic stability of earth slopes. It is found that the conventionally used ratio of slope height to the wavelength may not consistently perform well to describe the topographic effects for slopes over a rigid bedrock, where the soil layer amplification is dominant. The ratio of the fundamental site period to the predominant frequency of the input motion can be used to characterize the topographic effects on the seismic coefficient and earthquake-induced permanent displacement of earth slopes. In addition, the 1D analysis does not consistently provide conservative results compared with 2D analysis for the full cover sliding cases.

Keywords: Earth slopes · Earthquake-Induced landslides · Topographic effects
Seismic coefficient · Newmark displacement

1 Introduction

Newmark's [1] original methodology is generally referred to as rigid-block analysis. For deformable sliding mass, the dynamic response of the sliding mass should be taken into account. To this end, the decoupled method is widely used in which the dynamic

© Springer Nature Singapore Pte Ltd. 2018
T. Qiu et al. (Eds.): GSIC 2018, *Proceedings of GeoShanghai 2018 International Conference: Advances in Soil Dynamics and Foundation Engineering*, pp. 206–214, 2018.
https://doi.org/10.1007/978-981-13-0131-5_23

response analysis is firstly performed to compute the horizontal seismic coefficient time history by assuming no failure surface [2], and then the sliding displacement is computed by the rigid-block analysis with the horizontal seismic coefficient time history (multiplies g) as the input motion. Both two-dimensional (2D) finite element analysis [3] and one-dimensional (1D) soil column [4] can be used to model this dynamic response, while the 1D analysis is more commonly used in previous studies. Topographic effects on the amplitude and frequency content of earthquake ground motions have been well observed in various studies and destructive earthquakes [5, 6]. However, most previous studies on the topographic effects mainly focused on estimating the modification in ground motions along the slope surface and few attempts to the slope stability. It is therefore important to understand the topographic effects on the seismic slope stability and to investigate the adequacy of 1D analysis to accurately predict the horizontal seismic coefficient of a 2D slope.

In this study, we investigate the topographic effects on the seismic coefficient and permanent displacement of earth slopes by comparing both 1D and 2D results. Several slope models with various characteristics are analyzed to examine the interaction between topographic amplification and soil layer amplification and its influence on the evaluation of seismic slope stability.

2 Analysis Methodology

2.1 Critical Slip Surface and Horizontal Yield Coefficient

As an initial step, the critical slip surface and the horizontal yield coefficient (k_y) of slopes were determined through pseudo-static analyses using FLAC [7]. The k_y was obtained to result in a factor of safety equal to 1.0 by incrementally increasing the horizontal component of gravitational acceleration of the numerical model. The sliding surface can be characterized by searching the positions where have a large shear strain increment. The Mohr–Coulomb model was used in the pseudo-static analysis. The physical and mechanical properties used for the soil materials are presented in Table 1, and the slope models analyzed in this study are listed in Table 2. Figure 1 shows an example of the determined critical slip surface for different types of soil. The sandy slope (cohesionless soil) are generally associated with a shallower slip surface while a deeper sliding surface occurs for the clayey slope (cohesive soil). The influence of the difference in sliding mass geometry on the seismic coefficient and earthquake-induced permanent displacement need to be clarified.

2.2 Dynamic Response and Horizontal Seismic Coefficient of Sliding Mass

The dynamic response of the sliding mass is represented by the average horizontal acceleration within the sliding mass, i.e., the seismic coefficient time history (k). This is illustrated in Fig. 2(a), and can be given by:

Table 1. Physical and mechanical parameters of soil.

Parameters	Stiff clay	Dense sand	Relative stiff clay	Relative dense sand
Dry density (kg/m³)	2000	2000	1800	1800
Poisson's ratio	1/3	1/3	1/3	1/3
Shear wave velocity V_s (m/s)	400	400	200	200
Cohesion (kPa)	50	2	26	0.5
Friction angle (°)	25	42	18	37
Dilation angle (°)	0	0	0	0

Table 2. Slope model parameters.

Soil type	Slope height, H (m)	Slope inclination, i (°)	Soil layer thickness, Z (m)	Factor of safety	Yield coefficient (k_y)	Sliding mass depth (m)	Soil column depth (m)
Stiff clay	40	15	100	2.677	0.37	16.5	84.4
Stiff clay	40	30	100	1.549	0.22	11.7	85.1
Stiff clay	40	45	100, 150, 200, 250	1.119	0.07	10.7	86.3, 134.3, 186.3, 234.3
Stiff clay	40	45	Half-space	1.119	0.07	10.7	Half-space
Stiff clay	20	30	80	2.141	0.41	7.5	73.6
Relative stiff clay	20	30	80	1.369	0.16	7.1	72.5
Dense sand	40	30	100	1.549	0.15	3.0	82.2
Dense sand	20	30	80	1.701	0.20	1.7	71.2
Relative dense sand	20	30	80	1.365	0.11	0.8	71.3

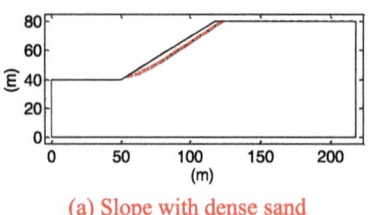

(a) Slope with dense sand

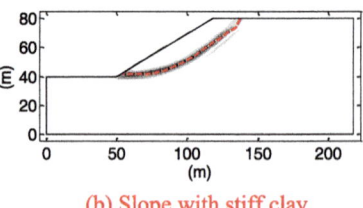

(b) Slope with stiff clay

Fig. 1. Critical slip surface of sandy and clayey slopes (H = 40 m, i = 30°, V_S = 400 m/s).

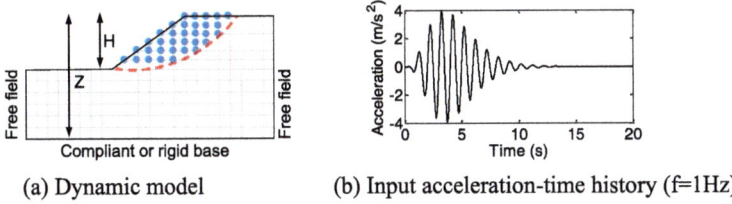

(a) Dynamic model   (b) Input acceleration-time history (f=1Hz)

Fig. 2. Dynamic analysis model of slopes and the input ground motions used in this study.

$$k(t) = a_{avg}(t) = \sum_{slidingmass} m_i a_i(t) / \sum_{slidingmass} m_i g \qquad (1)$$

where m_i and a_i are the mass and acceleration of nodal points within the sliding mass. The length of the dynamic model is 1000 m for H = 40 m, and is 500 m for H = 20 m. The base of the mesh is assumed to be a rigid boundary to consider the interaction between the soil layer and topographic effects. A compliant base where a quiet absorbing boundary is used at the base of model for one case of the homogeneous half-space. The free-field boundaries are applied to both the lateral boundaries to minimize reflections from the lateral boundaries (Fig. 2a). We use Chang's time histories with an amplitude of 0.4 g and dominant frequencies range from 0.2 to 10 Hz as inputs (Fig. 2b). The soil is modelled as a linear viscoelastic material with a target damping ratio of 5% using Rayleigh damping. The element size is taken as 4 m for V_s = 400 m/s and 2 m for V_s = 200 m/s (a tenth of the minimum wavelength of input motion) to assure the accurate representation of wave transmission through the model.

3   Topographic Effects on the Seismic Coefficient

3.1   Verification of Numerical Accuracy

The numerical accuracy of the model is verified with that from [5]. The slope model studied in [5] is in a homogenous half-space. The comparison result is shown in Fig. 3, where the ratio of acceleration amplitude in the slope crest from 2D analysis and 1D analysis versus various normalized frequencies (i.e., ratio of slope height to the wavelength, H/λ) is shown. The results agree well with each other at all H/λ values.

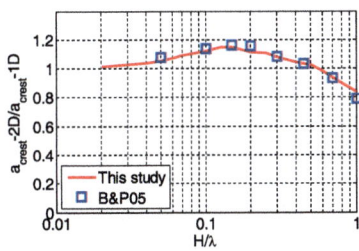

Fig. 3. A comparison of the result from this study with [5] (i = 45°, homogeneous half-space).

3.2   Parametric Analysis of Slope Geometry

Slope Inclination. The amplitude ratio of acceleration in slope crest (a_{crest}) and seismic coefficient (multiplies g) from 2D analysis to the peak acceleration of the input motion versus normalized frequencies H/λ for various slope inclinations is shown in Fig. 4. Also shown is the amplitude ratio of seismic coefficient from fully 2D analysis (k_{max}-2D) and 1D analysis (k_{max}-1D) using the averaged depth of the sliding mass and the soil column of the sliding mass to the bottom boundary. We can see the amplification is larger for crest acceleration than k. This is expected and consistent with [8] because the amplification is reduced from the crest to bottom of slope surface, and k is the averaging result of the acceleration of sliding mass. Note that the difference between the crest acceleration and k is dependent on the normalized frequencies H/λ. The maximum amplification occurs at H/λ equal to 0.1. This corresponds to the first natural period of the 1D soil column behind the crest (equal to $4Z/V_s = 1.0$ s), which means that soil layer amplification is dominant. In addition, the maximum amplification in 2D k_{max} increases slightly as increasing the slope inclination, with a maximum amplification near 500%. Also can be seen from the figure is that the 1D result is not consistently conservative. There are some cases (H/λ between about 0.15 and 0.25) that the 2D k_{max} is larger than 1D k_{max}, and the maximum ratio also increases slightly with increasing the slope inclination (with a maximum ratio of 1.4 for i = 45°).

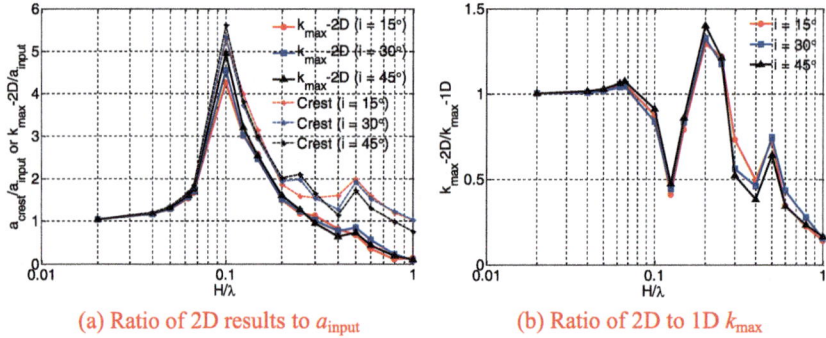

(a) Ratio of 2D results to a_{input} (b) Ratio of 2D to 1D k_{max}

Fig. 4. Amplification of a_{crest} and k_{max} of the sliding mass from 1D and 2D analysis for various slope inclinations (stiff clay, H = 40 m, V_s = 400 m/s and Z = 100 m).

Slope Height. Figure 5 shows amplification of the acceleration amplitude in slope crest and the 2D and 1D k_{max} for different slope heights. We can see the normalized frequency at the maximum amplification is different for the two slope heights. The amplification ratios shift to larger H/λ values for H = 40 compared to H = 20 m. Therefore, for slopes in a rigid bedrock, the normalized frequency H/λ may not be consistently good for describing the topographic effects. In this case, the soil layer and topographic effects are coupled together. The maximum ratio of 2D to 1D k_{max} is similar for different slope heights.

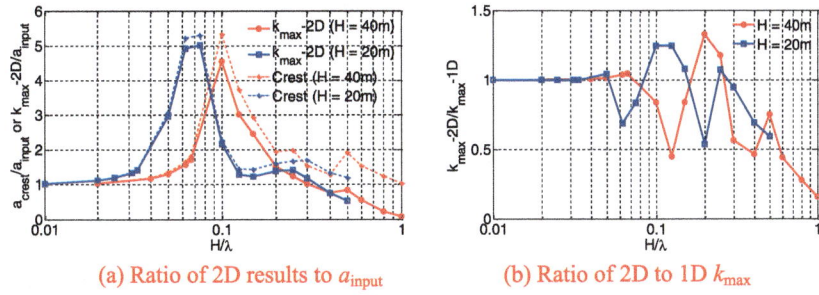

(a) Ratio of 2D results to a_{input} (b) Ratio of 2D to 1D k_{max}

Fig. 5. Amplitude amplification of a_{crest} and k_{max} of the sliding mass from 1D and 2D analysis for various slope heights (stiff clay, i = 30°, V_s = 400 m/s, Z = 100 and 80 m).

3.3 Parametric Analysis of Soil Layer

Soil Type. As an example shown in Fig. 1, the sandy slopes are generally associated with a shallower slip surface while a deeper sliding surface occurs for the clayey slope. Therefore, the influence of this difference in the sliding mass geometry on the seismic coefficient is investigated, as shown in Fig. 6. Observe that the amplification in crest acceleration and 2D k_{max} is consistent for two soil types. This is related to the distribution of the coupled amplification (topographic and soil layer) within slope. The amplification reduces significantly from the crest to the bottom of the slope surface, but it is similar from the slope surface to some deeper bodies along the horizontal direction. The slight difference observed in the ratio of 2D to 1D k_{max} is likely the result of the different sliding mass depth in the 1D analysis (Table 2).

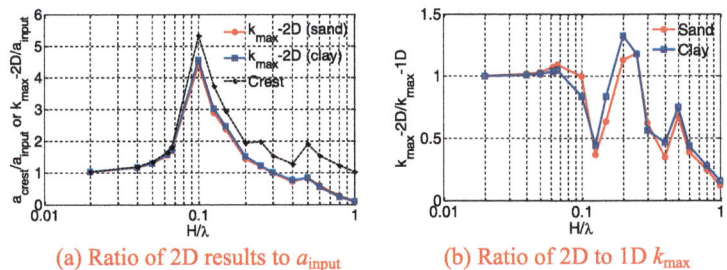

(a) Ratio of 2D results to a_{input} (b) Ratio of 2D to 1D k_{max}

Fig. 6. Amplification of a_{crest} and k_{max} of the sliding mass from 1D and 2D analysis for different soil types (H = 40 m, i = 30°, V_s = 400 m/s, Z = 100 m).

Shear Wave Velocity of Soil. Figure 7 shows the amplification of the acceleration amplitude in slope crest and k_{max} of the sliding mass from 1D and 2D analysis for different clay soils (different shear strengths and shear wave velocities). The amplification in crest acceleration and k_{max} is similar for the two clay soils, although there is a slight difference observed in the ratio of 2D to 1D k_{max}. Therefore, when there is a rigid bedrock beneath the earth slope, the stiffness and strength of the soil have little

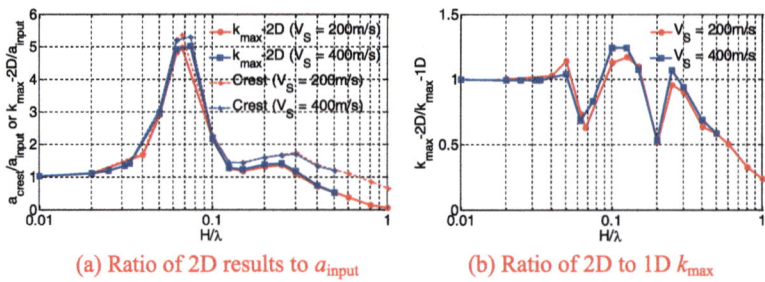

Fig. 7. Amplification of a_{crest} and k_{max} of the sliding mass from 1D and 2D analysis for soils with various shear wave velocities (clay, H = 20 m, i = 30°, Z = 80 m).

influence on the k estimation. Note that this may be different for an elastic bedrock, in which the shear wave velocity ratio between soil and bedrock is important.

Thickness of Soil Layer. Figure 8 shows the amplification of the acceleration amplitude in slope crest and k_{max} of the sliding mass from 1D and 2D analysis for different soil layer thicknesses, along with the result of the homogenous half-space. Similar to the observation in different slope heights, the maximum amplification ratio is not at a constant H/λ value, and there is a shift to larger H/λ values for a thinner soil layer. In addition, the topographic effects is more prominent for a thinner soil layer from the ratio of 2D to 1D k_{max}. The amplification in a homogenous half-space is much lower than the case of a rigid bedrock. Although the assumption of a rigid bedrock could have aggravated the interaction between topographic and soil layer effects, it is recommended that a more realistic elastic bedrock base is used to examine these coupled effects on the assessment of seismic landslide hazard.

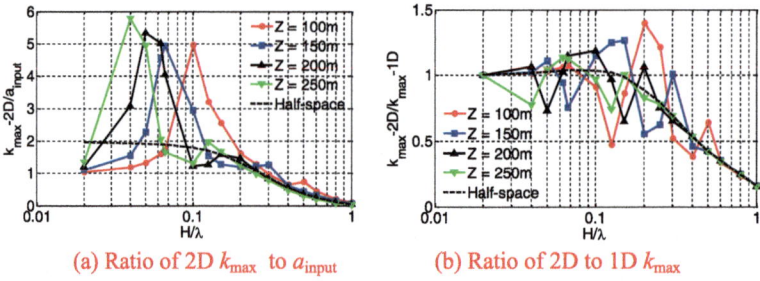

Fig. 8. Amplification of k_{max} of the sliding mass from 1D and 2D analysis for various soil layer thicknesses (stiff clay, H = 40 m, i = 45°, V_s = 400 m/s).

4 Topographic Effects on the Permanent Displacement

One important observation above is that H/λ may not be consistently good to describe the topographic effects for slope in a rigid bedrock. Therefore, we use the period ratio, defined as the ratio of the natural period of the soil column in 1D analysis to the predominant frequency of input motion to characterize the topographic effects, as shown in Fig. 9. We can see that the topographic effects can be well determined by the period ratio. A slight amplification is observed at a period ratio range of approximate 0.4–0.7, and another significant amplification occurs at a period ratio range of 1.2–2.2. The maximum mean ratio is 1.2 for k_{max} and 1.6 for permanent displacement. At other period ratios, topographic de-amplification relative to 1D analysis occurs. Also interesting to note that there are two de-amplification valleys at the first natural period and 1/3 of first natural period (i.e., the second natural period) of the soil layer and one de-amplification peak at 1/4 of first natural period. It is noted that all the identified critical slip surfaces for the homogeneous earth slopes in this study are full cover sliding through the slope surface, where the 1D analysis was considered to provide a significant conservative estimate of seismic coefficient and permanent displacement [8].

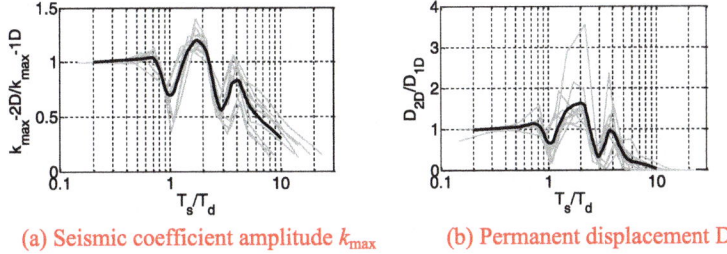

(a) Seismic coefficient amplitude k_{max} (b) Permanent displacement D

Fig. 9. k_{max} and D from 1D and 2D analysis versus period ratio (gray thin lines are the results for different slope models and the black thick line represents the mean value of the ratio).

5   Conclusions

This paper presents some decoupled analysis results to investigate the topographic effects on the estimation of seismic coefficient and earthquake-induced permanent displacement of earth slopes. A rigid bedrock base is used to consider the interaction between the soil layer and topographic effects. We found that the conventionally used normalized frequency H/λ may not consistently perform well to describe the topographic effects for slope in a rigid bedrock, where the soil layer amplification is dominant. The period ratio of the natural period of the soil column with an averaged thickness from the sliding mass to the bottom boundary to the predominant frequency of the input motion can be used to characterize the topographic effects on the seismic coefficient and earthquake-induced permanent displacement of earth slopes. In addition, the 1D analysis does not consistently provide conservative results compared with 2D analysis for the full cover sliding cases.

Acknowledgements. This research has been supported by the National Natural Science Foundation of China (Grant No. 41602280 and 41630638), National Key Basic Research Program of China (Grant No. 2015CB057901), National Key Research and Development Program of China (Grant No. 2016YFC0800205), China Postdoctoral Science Foundation (Grant No. 2016M601708), the Fundamental Research Funds for the Central Universities in China

(Grant No. 2016B01114), a Project Funded by the Priority Academic Program Development of Jiangsu Higher Education Institutions (PAPD).

References

1. Newmark, N.M.: Effects of earthquakes on dams and embankments. Geotechnique **15**, 139–159 (1965)
2. Chopra, A.K.: Earthquake response of earth dams. J. Soil Mech. Found. Div. ASCE **93**(2), 65–81 (1967)
3. Makdisi, F.I., Seed, H.B. Simplified procedure for estimating dam and embankment earthquake induced deformations. J. Geotech. Eng. Div. ASCE, 849–867 (1978)
4. Rathje, E.M., Antonakos, G.: A unified model for predicting earthquake-induced sliding displacements of rigid and flexible slopes. Eng. Geol. **122**(1), 51–60 (2011)
5. Bouckovalas, G.D., Papadimitriou, A.G.: Numerical evaluation of slope topography effects on seismic ground motion. Soil Dyn. Earthq. Eng. **25**(7), 547–558 (2005)
6. Tripe, R., Kontoe, S., Wong, T.K.C.: Slope topography effects on ground motion in the presence of deep soil layers. Soil Dyn. Earthq. Eng. **50**(7), 72–84 (2013)
7. Itasca Consulting Group. FLAC: fast Lagrangian analysis of continua (2005)
8. Rathje, E.M., Bray, J.D.: One- and two-dimensional seismic analysis of solid-waste landfills. Can. Geotech. J. **38**(4), 850–862 (2001)

Experimental Investigation on Liquefaction Resistance of Soil Mixtures Incorporating Intergrain Contact State

Qi Wu, Guo-xing Chen[✉], and Kai Zhao

Institute of Geotechnical Engineering,
Nanjing Tech University, Nanjing 210009, China
gxc6307@163.com

Abstract. In order to investigate the liquefaction resistance CRR of fine-coarse mixtures, a series of undrained cyclic triaxial tests are performed on the mixtures with wide-range fines content FC and relative density D_r. Considering intergrain contact state, the mixtures are categorized as coarse-like soils, in-transition soils and fines-like soils. Skeleton void ratio e_{sk} is used as an alternative index to describe the force structure of the mixtures, and the parameters b and m are introduced to characterize the contribution of the fines and coarse grains fraction to the transition between coarse-like and fines-like behavior in the mixtures. The test results show that as FC increasing, the CRR of the mixtures with the same D_r decreases first, thereafter, it remains constant; in addition, the CRR of the mixtures with different FC and D_r decrease with the increase of e_{sk}. Moreover, by analyzing the test data obtained in this study, as well as the published results of liquefaction resistance for different kinds of soil mixtures, it is found that e_{sk} can be used as an index to uniquely evaluate the CRR of the mixtures, and a power relationship between CRR and e_{sk} is then obtained.

Keywords: Fines-coarse mixtures · Liquefaction resistance
Intergrain contact state · Fines content · Skeleton void ratio

1 Introduction

Recent earthquake case histories indicate that natural soils and man-made sandy deposits that contain a significant amount of finer-grains (silty sands or clayed sands) and/or gravel do liquefy and cause lateral spreads [1–3]. Experience grained from past literatures on clean sands does not always directly translate to such broadly graded soils. Recognition of this has lead to a lot of laboratory and field studies to evaluate the effect of increasing fines content on: (a) liquefaction resistance, (b) collapse potential, (c) steady state strength, (d) shear wave velocity, etc. Results from laboratory studies on clean sands mixed with fines content show that, there is no uniformly correlation between liquefaction resistance and fines content. Polito and Martin [4] found that the liquefaction resistance of fines-coarse mixtures with the same void ratio first decreases with increasing fines content, as the fines content continues to increase, the liquefaction resistance begins to increase; but for the mixtures with the same relative density, the liquefaction resistance keeps almost constant with the increase of fines content, and

© Springer Nature Singapore Pte Ltd. 2018
T. Qiu et al. (Eds.): GSIC 2018, *Proceedings of GeoShanghai 2018 International Conference: Advances in Soil Dynamics and Foundation Engineering*, pp. 215–223, 2018.
https://doi.org/10.1007/978-981-13-0131-5_24

then decreases as long as fines content more than 30%. Hsiao et al. [5] assumed that the liquefaction resistance of fines-coarse mixtures with the same relative density decreases with the fines content increasing. Therefore, it is necessary to investigate the effect of different particle compositions and density state on the liquefaction resistance of fines-coarse mixtures systemically.

This paper presents the results of an investigation into the effects of fines content on the liquefaction resistance of fine-coarse mixtures. Nearly 123 cyclic triaxial tests were performed on mixtures with different fines content ranging from 0 to 100%. The liquefaction resistance was evaluated using the fines content and skeleton void ratio of the specimen.

2   Theory of Intergrain Contact State

In order to describe the effect of fines content FC on the mechanical properties and responses of fines-coarse mixtures, the theory of intergrain contact state was proposed by Thevanayagam [6]. With the increase of FC, the intergrain contact state of the mixtures can be constituted by infinite different ways, a few extreme limiting contact state of microstructure and the relevant roles of coarser and finer grains are as follows:

Case 1: the contact state is when the fine grains are fully confined within the void spaces between the coarse grains with no contribution whatsoever in supporting the coarse grain skeleton. They may largely play the role of "filler" of intergranular voids. The mechanical behavior is affected primarily by the coarse grain contacts.

Case 2: a part of fines grains act as a load transfer vehicle between "some" of the coarse grain particles in the soil matrix while the remainder of the fines play the role of "filler" of voids.

Case 3: fines grains play an active role of "separator" between an significant number of coarse grains contacts and therefore begin to dominate the strength characteristics, while the coarse grains may act as reinforcing elements embedded within the fines grain matrix.

Case 4: the fines grains carry the contact and shear forces while the coarse grains are fully dispersed.

In order to distinguish this four different intergrain contact states, minimum in-transition fines content (FC_{in-min}), threshold fines content (FC_{th}) and maximum in-transition fines content (FC_{in-max}) were employed, FC_{in-min}, FC_{th} and FC_{in-max} can be calculated as follows [6]:

$$FC_{in-min} = (e_{c-max} - e)/(e_{c-max} + 1); \quad FC_{th} = e/e_{f-max}; \quad FC_{in-max}$$
$$= 1 - \pi(1+e)/6s^3 \tag{1}$$

Where e is the global void ratio of the mixtures, e_{c-max} and e_{f-max} are the maximum void ratio of fines graines and coarse graines, and s = 1 + a/R_d, a = 10, R_d = D_{50}/d_{50}, D_{50} and d_{50} are mean grain size of pure fines grain and coarse grain.

Skeleton void ratio (e_{sk}) was used to describe the idealized packing conditions of the dominant grains fraction [6]. For the mixtures with distinct coarse and fines grains, e_{sk} is defined as the volumetric ratio between the voids formed by the soil skeleton and the volume of grains that make up the skeleton. When the contact state of the mixtures is case 1, e_{sk} can be described as:

$$e_{sk} = (e + FC)/(1 - FC) \tag{4}$$

when the contact state of mixtures is case 2, e_{sk} can be calculated as follows:

$$e_{sk} = [e + (1 - b) \cdot FC]/[1 - (1 - b) \cdot FC] \tag{5}$$

when the contact state of mixtures is case 3, e_{sk} can be shown as:

$$e_{sk} = e/[FC + (1 - FC)/R_d^m] \tag{6}$$

when the contact state of mixtures is case 4, e_{sk} can be described as:

$$e_{sk} = e/FC \tag{7}$$

where $0 < b$, $m < 1$, b and m denote the portion of the fines grains and coarse grains that contributes to the active intergrain contact.

3   Undrained Cyclic Triaxial Tests

The fines-coarse mixtures used in this study with FC of 0%, 10%, 20%, 25%, 30%, 35%, 40% 50%, 60%, 70%, 85% and 100% was man-made, silt grains with a sub-angular shape and a mean particle size of 0.051 mm was used as the fines fraction. The gravelly soils with a sub-angular shape and a particle size of 10 - 0.075 mm were used as the coarse fraction. The physical properties of the testing materials used in this study are given in Table 1. The specimens were prepared by adding fines in various percentages to the gravelly soil without fines grains. The grain size distribution of fines-coarse mixtures with different FC are presented in Fig. 1.

Table 1. Index properties of fines grain and coarse grain of the mixtures

	Coarse grain	Fines grain
Soil type	Gravel soil	silty
Mean grain size D_{50}/mm	1.849	0.056
Effective grain size d_{10}/mm	0.567	0.014
Uniformity coefficient C_u	4.306	4.130
Specific gravity G	2.680	2.720
Maximum index density ρ_{max}/(g·cm^{-3})	1.740	1.679
Minimum index density ρ_{min}/(g·cm^{-3})	1.437	1.271
Maximum void ratio e_{max}	0.865	1.140
Minimum void ratio e_{min}	0.540	0.620

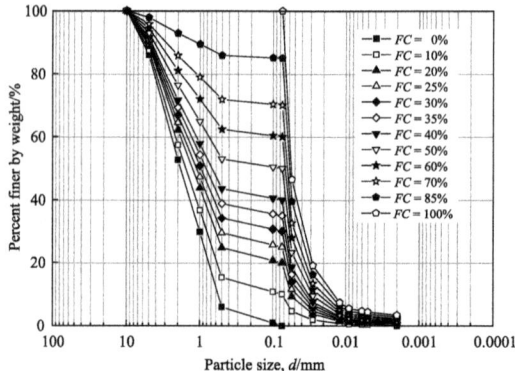

Fig. 1. Grain size distribution of the fines-coarse mixtures

Specimens were 100 mm in diameter and 200 mm in height. The specimens were formed using waist tamping method. Once the preparation of the specimen was completed, initial saturation of the specimen was done by passing carbon dioxide through the specimen for 1 h. After that the confining pressure and the back pressure maintain a difference of 20 kPa for 10 min, then, The specimen is then saturated by increasing the cell pressure and back pressure by 50 kPa every 10 min until cell pressure up to 400 kPa. At the end of this process the cyclic machine was switched on. The machine is capable of applying sufficient back pressure till it was ensured that Skempton's B parameter is higher than 0.95. The specimen were then isotropically consolidated to 100 kPa initial effective confining stress (σ'_{3c}).

The testing apparatus used in the current study is GDS hollow cylinder torsional shear apparatus (HCA). All tests were conducted at a cyclic loading frequency (f) of

Table 2. Cases of undrained cyclic triaxial tests for fine-coarse mixtures

ID	$FC/\%$	$D_r/\%$	Contact state	ID	$FC/\%$	$D_r/\%$	Contact state	ID	$FC/\%$	$D_r/\%$	Contact state
S1	0	30	Case 1	S15	30	15	Case 2	S29	50	70	Case 3
S2	0	50	Case 1	S16	30	22	Case 2	S30	60	30	Case 3
S3	0	70	Case 1	S17	30	30	Case 2	S31	60	50	Case 3
S4	10	30	Case 1	S18	30	50	Case 3	S32	60	70	Case 3
S5	10	50	Case 1	S19	30	70	Case 3	S33	70	30	Case 4
S6	10	70	Case 1	S20	35	15	Case 2	S34	70	50	Case 4
S7	20	30	Case 1	S21	35	30	Case 3	S35	70	70	Case 4
S8	20	50	Case 1	S22	35	50	Case 3	S36	85	30	Case 4
S9	20	70	Case 1	S23	35	70	Case 3	S37	85	50	Case 4
S10	25	15	Case 2	S24	40	30	Case 3	S38	85	70	Case 4
S11	25	22	Case 2	S25	40	50	Case 3	S39	100	30	Case 4
S12	25	30	Case 1	S26	40	70	Case 3	S40	100	50	Case 4
S13	25	50	Case 1	S27	50	30	Case 3	S41	100	70	Case 4
S14	25	70	Case 1	S28	50	50	Case 3				

1 Hz. The specimens were then loaded with a sinusoidal axial stress at the appropriate cyclic stress ratio $\left(CSR = \sigma_d/2\sigma'_{3c}\right)$ until they liquefied, σ_d is axial cyclic stress amplitude. Experimental program are shown in Table 2.

4 Test Results and Discussion

The typical undrained cyclic triaxial testing results for fines-coarse mixtures with $FC = 10\%$, 30%, 50%, and 85% are shown in Fig. 2, respectively. As shown in Fig. 2 (a), when the intergrain contact state is case 1, the mixtures is coarse-like soils, excess pore pressure ratio R_u developed fast at first and then became stable, showing a "fast-steady" development model, the experimental results show that the double-axial strain ε_{DA} of the specimen is more than 2.5% when $R_u = 1$, which is basically consistent with the experimental results of gravelly soils [9]. As shown in Fig. 2(b) and (c), when the intergrain contact state is case 2 or 3, the mixtures is in-transition soils, R_u developed linearly as number of cycles (N) increasing, and in the early stage, ε_{DA} remained almost unchanged, when $R_u > 0.7$, ε_{DA} increased gradually, and when $R_u = 1.0$, ε_{DA} could reach 5%. As shown in Fig. 2(d), when the intergrain contact state

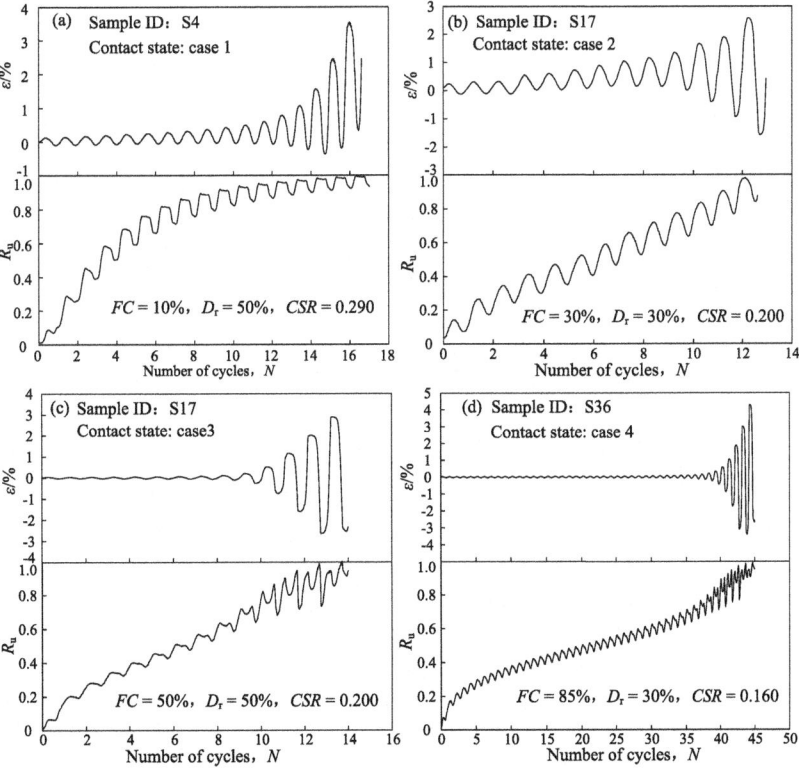

Fig. 2. Undrained cyclic triaxial test results of fines-coarse mixtures with different intergrain contact state

is case 4, the mixtures is fines-like soils, as N increasing, R_u grown fast at the beginning and then grown at a steady rate, and finally developed sharply, showing a "fast-steady-rapid" development model, when $R_u < 0.8$, ε_{DA} remained constant and close to 0, and when $R_u > 0.8$, ε_{DA} increased rapidly with few N, when $R_u = 1$, ε_{DA} approached 7.5%, this results were basically the same with silty [10]. In conclusion, with the action of cyclic axial loading, the initial liquefaction ($R_u = 1$) can be reached, the skeleton of mixtures shown a flow failure, and the mixtures no longer have the shear strength. Therefore, "$R_u = 1$" was used as the liquefaction criterion for fines-coarse mixtures.

Liquefaction resistance ratio (CRR) of fines-coarse mixtures was determined as CSR at which initial liquefaction occurred ($R_u = 1$) for 15 cycles of loading. Figure 3 shows the influence of FC on the CRR of the mixtures with the same relative density D_r. It can be seen from the figure that the CRR of mixtures deceases with the increase of FC, and then remains constant regarding whether the dense state at loose ($D_r = 30\%$), medium ($D_r = 50\%$) or dense ($D_r = 70\%$), this is consistent with the effect of FC on the CRR of the mixtures from the experimental results of Karim and Alam [7]. It is noteworthy that, when $D_r = 30\%$, as long as FC is greater than 35%, the CRR of mixtures keep basically unchanged; and when $D_r = 50\%$ or 70%, once FC more than 30%, the CRR of mixtures will remain constant. This difference may be due to the fact that, as D_r difference, the transition process of intergrain contact state of fines-coarse mixtures is different as FC increasing.

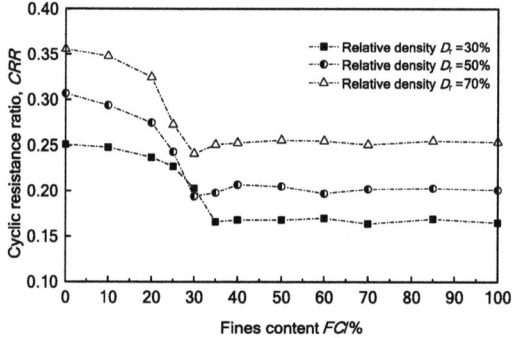

Fig. 3. The cyclic resistance of fine-coarse mixtures with different fines content

According to the theory of intergrain contact state, the contact state of fines-coarse mixtures with the same void ratio e varies with the change of FC, thus leading to the change of mechanical properties and responses of the mixtures, which indicate that e can't reasonably describe the CRR of the mixtures. Polito and Martin II [4] found that there is no clear correlation between e and CRR for Monterey fines-coarse mixtures, at the same time, within each FC, a decrease in CRR with an increase in e was observed. The test results of Hsiao et al. [5] demonstrate that the CRR of fines-coarse mixtures is not correlated with e, regardless of whether the mixtures have the same D_r or undrained shear strength. This indicates that e does not describe the effect of FC on the CRR of the mixtures well.

Figure 4 shows the relationship between liquefaction resistance CRR and skeleton void ratio e_{sk} of fines-coarse mixtures. From the figure we can see, when $b = 0.15$ and $m = 0.80$, the CRR of fines-coarse mixtures with different e_{sk} are distributed in a narrow band, regardless of FC and D_r, and the CRR of mixtures decreases rapidly with increasing e_{sk}, which is consistent with the relationship between CRR and e_{sk} presented by Polito and Martin II [4] or Papadopoulou and Tika [8]. However, the test results of Polito and Martin II [4] or Papadopoulou and Tika [8] showed that the distribution of CRR of fines-coarse mixtures for $FC > 50\%$ was no consistent with the distribution of CRR for $FC < 50\%$. A unified relationship between CRR and e_{sk} of fines-coarse mixtures can't be established, this is because Polito and Martin II [4] or Papadopoulou and Tika [8] do not take into account the effect of fines grain or coarse grain on e_{sk} when the grain contact state is case 2 or case 3, thus overestimating the e_{sk} of in-translation soils. While the e_{sk} based on the theory of intergrain contact state of the mixtures can describe the CRR with different intergrain contact state uniformly, and CRR can be reasonably expressed as a power function form of e_{sk}:

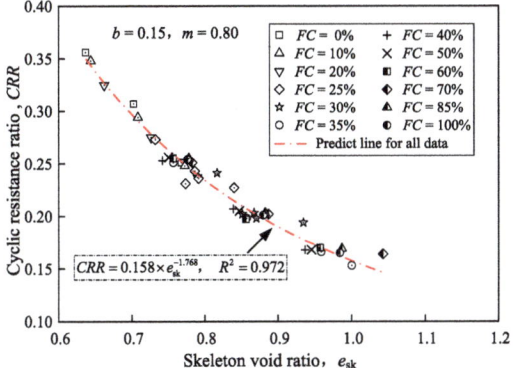

Fig. 4. CRR versus e_{sk} of fine-coarse mixtures

$$CRR = 0.158 \times e_{sk}^{-1.768}, R^2 = 0.972 \tag{8}$$

To further validate the CRR prediction method using e_{sk}, the CRR of four types of soil mixtures were re-evaluated from the published undrained cyclic triaxial test results [4, 5, 8]. Figure 5 shows the relationship between CRR and e_{sk} for different types of soil mixtures. As shown in the figure, although the physical properties and liquefaction criteria of the mixtures are different, the CRR decreases rapidly with the increase of e_{sk}, and the CRR can be reasonably expressed as the power function of e_{sk}:

$$CRR = A \times e_{sk}^{-B} \tag{9}$$

In summary, e_{sk}, synthesizing the nature of grain-size distribution, density state, and intergrain contact, is an apprizate index of physical properties for evaluating the CRR of different types of soil mixtures.

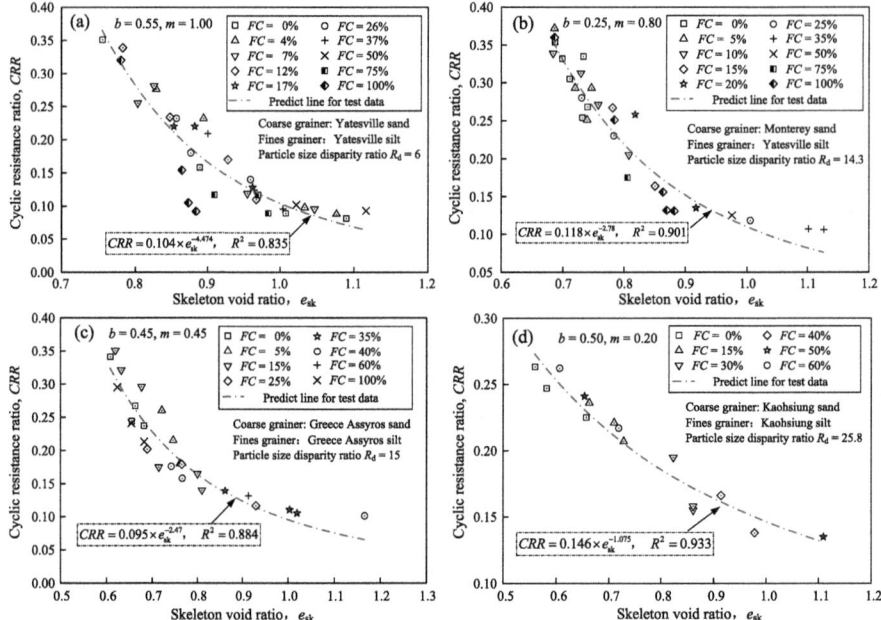

Fig. 5. *CRR* versus e_k for different types of fines-coarse mixtures (a. data from Polito and Martin II [4]; b. data from Polito and Martin II [4]; c. data from Papadopoulou and Tika [8]; d. data from Hsiao et al. [5])

5 Summary

A series of undrained cyclic triaxial tests were performed on the fines-coarse mixtures with 12 different *FC* in the range of 0% to 100%. The testing results and findings are summarized as following:

(1) As *FC* increasing, intergrain contact state of the mixtures can be divided into 4 cases, and according to the theory of intergrain contact state, the mixtures can be classified as fines-like soils, in-translation soils and coarse-like soils.
(2) Liquefaction resistance *CRR* of fine-coarse mixtures with the same D_r decreased first, thereafter, it remained constant, as *FC* increasing.
(3) *CRR* decreases with the increase of e_{sk}, *CRR* and e_{sk} show a good power relationship, which means e_{sk} based on the theory of intergrain contact state can characterize *CRR* of different types of fines-coarse mixtures reasonably.

References

1. Ishihara, K.: Stability of natural deposits during Earthquakes. In: Proceedings of the Eleventh International Conference on Soil Mechanics and Foundation Engineering, San Francisco (1985)
2. Munenori, H., Akihiko, U., Junryo, O.: Liquefaction characteristics of a gravelly fill liquefied during the 1995 hyogo-ken nanbu earthquake. J. Jpn. Geotech. Soc. Soils Found. **37**(3), 107–115 (1997)
3. Chu, B.L., Hsu, S.C., Lai, S.E., et al.: Soil liquefaction potential assessment of the Wufeng area after the 921 Chi-Chi earthquake (in Chinese). Report of National Science Council (2000)
4. Polito, C.P., Martin Ii, J.R.: Effects of nonplastic fines on the liquefaction resistance of sands. J. Geotech. Geoenviron. Eng. **127**(5), 408–415 (2001)
5. Hsiao, D.H., Phan, V.T.A., Hsieh, Y.T., et al.: Engineering behavior and correlated parameters from obtained results of sand–silt mixtures. Soil Dyn. Earthq. Eng. **77**, 137–151 (2015)
6. Thevanayagam, S.: Intergrain contact density indices for granular mixes—II: Liquefaction resistance. Earthq. Eng. Eng. Vibr. **6**(2), 135–146 (2007)
7. Karim, M.E., Alam, M.J.: Effect of non-plastic silt content on the liquefaction behavior of sand–silt mixture. Soil Dyn. Earthq. Eng. **65**, 142–150 (2014)
8. Papadopoulou, A., Tika, T.: The effect of fines on critical state and liquefaction resistance characteristics of non-plastic silty sands. Soils Found. **48**(5), 713–725 (2008)
9. Evans, M.D., Zhou, S.P.: Liquefaction behavior of sand-gravel composites. J. Geotech. Eng. **121**(3), 287–298 (1995)
10. Singh, S.: Liquefaction characteristics of silts. Geotech. Geol. Eng. **14**(1), 1–19 (1996)

The Influence of Bedrock Surface Depth on Seismic Site Response in Deep Sediment Layers

Jiao Zhu, Guoxing Chen$^{(\boxtimes)}$, and Dingfeng Zhao

Institute of Geotechnical Engineering, Nanjing Tech University,
Nanjing 210009, China
gxchen@njtech.edu.cn

Abstract. A reasonable determination of seismic bedrock surface has a significant impact on the evaluation of design ground motion parameters. In this study, based on the borehole profiles in the downtown area of Suzhou (China), nine interfaces of soil layers, whose shear wave velocities range from 400 m/s to 800 m/s, are selected as the seismic bedrock surfaces. The equivalent linear viscoelastic model is employed to approximate nonlinear soil behavior, and the effects of bedrock surface depth on site surface motion are analyzed. The results show that: the surface peak ground acceleration (PGA) increases as the shear wave velocity of bedrock surface increases. However, the magnitude of increase reduces as the PGA of bedrock motion increases. Moreover, the surface acceleration response spectra (Sa) become larger as bedrock surface depth increases, and the values become smaller as the PGA of bedrock motion increases. In addition, under near-field medium bedrock motion, the depth of bedrock surface has significant influence on the values for the period less than 1.0 s. However, under far-field strong bedrock motion, the depth of bedrock surface has significant influence on the values for the period less than 4.0 s. The values under far-field ground motion are obviously higher than those under near-field ground motion. Therefore, it is appropriate to choose the interface of soil with shear velocity not less than 700 m/s as the seismic bedrock surface.

Keywords: Seismic bedrock interface · Deep sediment layers
Seismic site response · Ground motion parameters

1 Introduction

Local site conditions have a profound influence on surface ground motions and seismic damage distribution, which was verified in a series of constructive earthquakes. This is mainly manifested in two aspects: the amplification of ground motion in soft soil is much larger than that in dense and hard soil [1]; the local variation of topography has a great influence on the propagation of seismic wave, resulting in significant difference in the spatial distribution of ground motion [2]. Therefore, reliable predictions of seismic site response and design ground motion parameters play an important role in predicting seismic damage and taking further effective anti-seismic measures.

© Springer Nature Singapore Pte Ltd. 2018
T. Qiu et al. (Eds.): GSIC 2018, *Proceedings of GeoShanghai 2018 International Conference: Advances in Soil Dynamics and Foundation Engineering*, pp. 224–232, 2018.
https://doi.org/10.1007/978-981-13-0131-5_25

Currently, the study of seismic site response mainly includes theoretical and empirical analysis. The empirical analysis is usually based on measured ground motion records, such as SSR and HVSR [3]. The empirical analysis results are reliable, however, they depend on the ground motion records. Considering the diversity of sites, it is difficult to give the analytical solution to site with complex terrain and geomorphology or different properties of soils. Numerical simulation has become a mainstream method for analyzing the seismic characteristics of complex heterogeneous site [4]. In that the size of the borehole data, which could be adopted for the existing site characterizations, is usually small because of the limited budget for site investigation. For deep soft site, it's not feasible to drill to obtain the depth of bedrock surface. Hence, a reasonable determination of bedrock surface has a significant impact on the evaluation of design ground motion parameters.

The interface of elastic homogeneous bedrock and nonhomogeneous soil is taken as the seismic bedrock surface. In different provisions, although the determination of bedrock surface all depends on the shear wave velocity of soil, the specific value of velocity is significantly different. According to the China Code for Evaluation of seismic safety for engineering sites GB17741-2005, the soil layer with velocity not less than 500 m/s is selected. However, according to the China Code for seismic design of nuclear power plants GB50267-97, the soil layer with velocity not less than 700 m/s is selected. Moreover, according to the relevant US nuclear codes, such as ASCE4-98, ASCE43-05 and NUREG-0800, the soil layers with velocity not less than 1100 m/s, 2400 m/s and 2400 m/s are selected, respectively [5–7]. The thickness of site overburden layer is determined by the depth of bedrock surface, which is also the reason why the characteristics of ground motion (amplitude and frequency content) are affected. Therefore, in this study, based on two typical deep borehole profiles in Suzhou region, interfaces of soils with different shear wave velocities are selected as seismic bedrock interface, in order to gain insight into the effects of bedrock surface depth on the resulting surface ground motion.

2   Site Characterization

In deep site, soft soil is always exist, and the natural period of site is larger. Suzhou is located at the southern margin of the Yangtze River delta in china, and is covered by extensive sediment of marine and alluvial lacustrine facies. Deep overburden layers exist in this region, and the general strata are characterized as multi-sedimentary rhythms. Two typical borehole profiles (profile No. 1 and 2), with actual drilling depth of 250 m and 180 m, are selected to perform seismic site responses. Profile No. 1 has revealed 16.3 m thick completely weathered mudstone and 15 m thick intensely weathered mudstone, below which 30 m thick sandstone is constructed. Profile No. 2 is close to profile No. 1, and the drilled soils in the two borehole profiles are similar. Thus, according to the stratigraphic characteristics of profile No. 1, 43 m thick silty clay, 16.3 m thick completely weathered mudstone and 15 m intensely weathered mudstone are constructed in profile No. 2. Eventually, the depths of both profiles are extended to 280 m.

The natural frequency of deposit is closely related to the shear wave velocity of soil [8]. Thus, it is necessary to use the measured shear wave velocities. Due to the limitation of test condition, the depths of measured shear wave velocity for profile No. 1 and No. 2 are only 200 m and 167 m, respectively. The shear wave velocities of soil at deeper depths are obtained by gradual extrapolation method [9]. Figure 1 shows the soil distribution characteristics and corresponding shear wave velocities for the two borehole profiles.

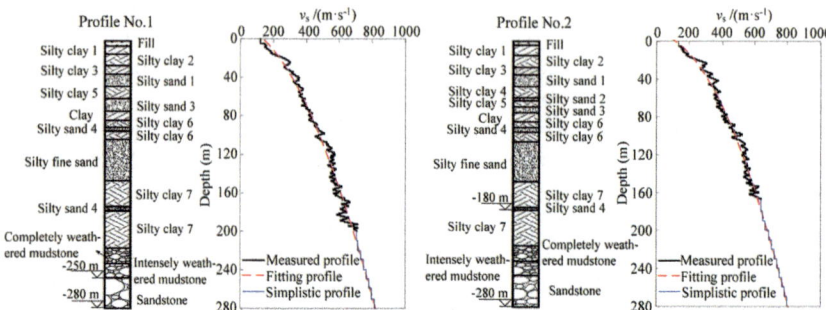

Fig. 1. Soil profiles and corresponding shear wave velocity profiles used in the analysis.

According to the borehole data, as well as the resonant column test results for typical soil samples in Suzhou, Fig. 2 shows the normalized shear modulus degradation and damping ratio curves as functions of the shear strain. Due to the lack of soil sample, the corresponding curves of sandstone are achieved by reference to the published literature [10].

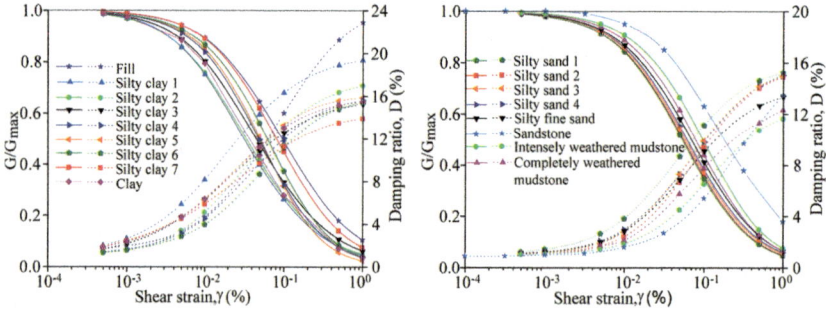

Fig. 2. G/Gmax and damping ratio D curves versus shear strain γ for soils in Suzhou region

3   Selection of Seismic Bedrock Interface

In order to quantitatively study the influence of bedrock surface depth on seismic site response in deep sediment layers, nine interfaces of soil layers are selected as the seismic bedrock surfaces. Specifically, in profile No. 1, the shear wave velocities of selected bedrock surfaces are 405–800 m/s, and the corresponding depths are 72–260 m. In profile No. 2, the shear wave velocities are 412–800 m/s, and the corresponding depths are 81–270 m, as shown in Table 1.

Table 1. Depths and corresponding shear-wave velocities of the seismic bedrock surfaces.

Borehole	Bedrock surface	1	2	3	4	5	6	7	8	9
Profile No. 1	Depth (m)	72	88	107	118	165	191	200	230	260[a]
	υ_s (m/s)	405	452	505	556	606	654	709[a]	750[a]	800[a]
Profile No. 2	Depth (m)	81	93	105	145	164	170	200[a]	240[a]	270[a]
	υ_s (m/s)	412	459	508	546	602	650[a]	700[a]	750[a]	800[a]

Note: [a] represents the extrapolation value.

4   Seismic Bedrock Motion

A large number of earthquake damage investigations indicated that the seismic site responses under far-field and near-field earthquakes exhibit significant differences.

Therefore, in this paper, a suite of four horizontal acceleration records were selected as the bedrock motions, including the ground motion records on Tarzana site during the 1994 Ms 6.6 Northridge earthquake, Tekirdag site during the 1999 Ms 7.4 Turkey earthquake, Jingning site during the 2008 Ms 8.0 Wenchuan earthquake, and one artificial wave, in order to reflect the influence of bedrock motion characteristics on seismic site response.

According to the China Code for Evaluation of seismic safety for engineering sites GB17741-2005, the peak accelerations of bedrock motions with 63%, 10% and 2% probability of exceedance in 50 years in Suzhou region are approximately 0.020 g, 0.070 g and 0.135 g, respectively, which are defined as a low-level earthquake (LLE), a moderate-level earthquake (MLE) and a high-level earthquake (HLE), respectively. Thus, the PGAs of the selected four bedrock motions are adjusted to be equal to 0.020 g, 0.070 g, and 0.135 g, corresponding to LLE, MLE, and HLE, respectively. Figure 3 shows the acceleration time histories and Fourier spectra of bedrock motions. For near-field earthquake, the mid- and high-frequency components of the Tarzana record are abundant. The frequency content of artificial wave is similar to that of the Tarzana record, the mid- and high-frequency components are also abundant. However, for far-field earthquake, the low-frequency component is extremely abundant.

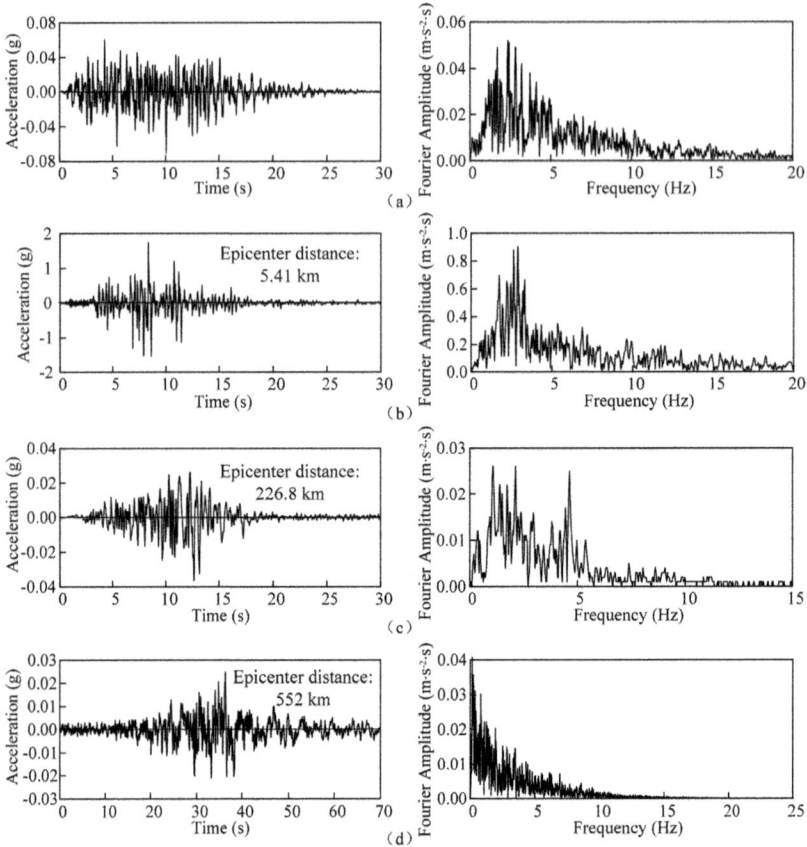

Fig. 3. Acceleration time-histories and Fourier spectra of bedrock motions: (a) Artificial wave; (b) Tarzana record; (c) Tekirdag record and (d) Jingning record.

5   Results and Analysis

In what follows, we perform seismic site response analyses using Proshake, in which the seismic site response is simplified as the one-dimensional wave problem of vertical incident S wave in horizontally layered soil. The equivalent linear viscoelastic model is employed to approximate nonlinear soil behavior.

5.1   Surface PGAs

The surface PGAs of borehole profiles No. 1 and 2 when the interfaces of soil layers with different shear wave velocities are selected as the seismic bedrock surfaces are shown in Fig. 4. With the gradual increase of shear wave velocity of bedrock surface, the corresponding bedrock surface depth increases gradually, and the surface PGAs increase monotonically. However, this monotonicity dissipates with the increase of bedrock PGA. This indicates that the thicker the overburden layer in deep site, the more significant the amplification effect of ground motion. When choosing the interface of soil with shear velocity of 800 m/s as the seismic bedrock surface, the surface PGAs increases dramatically. It may be due to the significant differences between the dynamic shear modulus and damping ratio of constructed sandstone at this depth and those of the overburden layers. Thus when the upper interface of sandstone is selected as bedrock surface, the corresponding surface PGAs increase remarkably. If not taking the abrupt change of constructed sandstone formed by depth extension into consideration, the surface PGAs will remain constant when the shear wave velocity of bedrock surface is approximately 700 m/s. The further increase of shear wave velocity has little impact on the surface PGAs. Accordingly, from the perspective of surface PGA, it is more appropriate to choose the interface of soil with shear velocity not less than 700 m/s as the seismic bedrock surface.

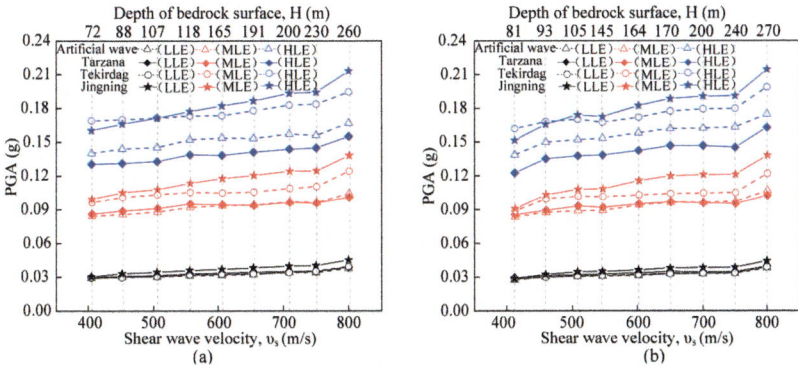

Fig. 4. PGAs of surface ground under bedrock motions from different soil interfaces for: (a) borehole profile No. 1 and (b) borehole profile No. 2.

In addition, surface PGAs are closely related to the spectral characteristics of input ground motion. For the deep site, the seismic site response under far-field earthquake is obviously higher than that under near-field earthquake. Specifically, the surface PGAs are similar when inputting the artificial wave and the Tarzana record with similar frequency content or inputting the two far-field ground motions (Tekirdag and Jingning record). Moreover, the difference of surface PGAs induced by different frequency content of input ground motion increases as the PGA of bedrock motion and the bedrock surface depth increase.

5.2 Surface Spectral Acceleration

Figure 5 shows the surface Sa spectra of borehole profile No. 1 when the interfaces of soil layers with different shear wave velocities are selected as seismic bedrock surfaces. With the gradual increase of shear wave velocity of bedrock surface, the surface Sa spectra also gradually increase, which indicates that the amplification effect of ground motion in deep site increases with the thickness of overburden layer. Without considering the constructed sandstone, the influence of shear wave velocity of bedrock surface on Sa spectra will gradually become stable. Compared with the results when the shear wave velocity of bedrock surface is about 700 m/s, if the shear wave velocity is less than 700 m/s, the magnitude of shear wave velocity has great influences on the Sa spectra in the period range of T < 1.0 s under near-field ground motions or in the period range of T < 4.0 s under far-field ground motions. However, when the interface of soil with shear wave velocity of 750 m/s is chosen as the bedrock surface (interface 8), the increase of Sa spectra is very small, and the values at short periods are consistent with those of choosing interface 7 as the bedrock surface. Thus, the changes of Sa spectra also show that it is appropriate to choose the interface of soil with shear velocity not less than 700 m/s as the seismic bedrock surface.

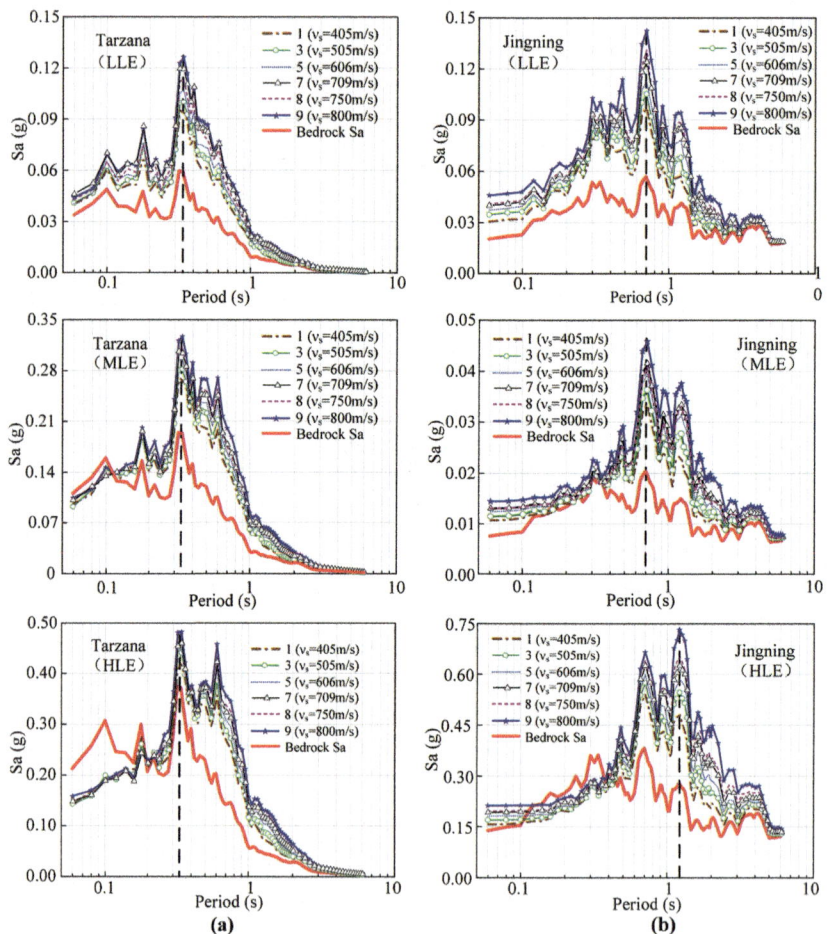

Fig. 5. Surface acceleration response spectra of borehole profile No. 1 under: (a) Tarzana record and (b) Jingning record.

The low-frequency components of surface ground motions become richer as the PGAs of bedrock motions increase, and the high-frequency components are filtered out. However, this variation trend is closely related to the bedrock motion characteristics. Thus, under the near-field Tarzana record, the Sa spectra change from single peak to double peaks. Corresponding to LLE, MLE, and HLE, the predominant periods of surface ground motions are all 0.34 s. However, under the far-field ground motions, the peaks of Sa spectra at short period reduce significantly, and the predominant periods move to long periods. Corresponding to LLE, MLE, and HLE, the predominant periods are 0.70 s, 0.70 s, 1.20 s for Jingning record. Compared with the bedrock Sa spectra, the amplification effect of surface Sa spectra under far-field ground motion is greater than that under near-field ground motion, and the amplification effect of surface Sa under LLE is more remarkable than that under MLE.

6 Conclusions

Based on typical borehole profiles in Suzhou region, the influence of bedrock surface depth on seismic site response in deep sediment layers are performed. The conclusions are as follows:

The surface PGAs and Sa spectra increase with the depth of bedrock surface. Under the near-field Tarzana record, the bedrock surface depth has an impact on the Sa spectra in the period range of $T < 1.0$ s for MLE and HLE cases. However, under the far-field Jingning record, the bedrock surface depth has a great influence on the Sa spectra in the period range of $T < 4.0$ s.

Compared with the bedrock Sa spectra, the surface Sa spectra under far-field ground motion are obviously higher than those under near-field ground motion. Moreover, the surface spectra for LLE cases are amplified in the whole period range, and those for MLE and HLE cases are smaller than the bedrock Sa spectra in the short period range of $T \leq 0.4$ s.

In summary, it is appropriate to choose the interface of soil with shear velocity not less than 700 m/s as the seismic bedrock surface.

References

1. Aki, K.: Local site effects on weak and strong ground motion. Tectonophysics **218**(1), 93–111 (1993)
2. Lanzo, G., Silvestri, F., Costanzo, A., D'nofrio, A., Martelli, L., Pagliaroli, A., Sica, S., Simonelli, A.: Site response studies and seismic microzoning in the Middle Aterno valley (L'Aquila, Central Italy). Bull. Earthq. Eng. **9**(5), 1417–1442 (2011)
3. Mittal, H., Kumar, A., Singh, S.K.: Estimation of site effects in Delhi using standard spectral ratio. Soil Dyn. Earthq. Eng. **50**, 53–61 (2013)
4. Guoxing, C., Dandan, J., Jiao, Z., Jian, S., Xiaojun, L.: Nonlinear analysis on seismic site response of Fuzhou basin, China. Bull. Seismol. Soc. Am. **105**, 928–949 (2015)
5. American Society of Civil Engineers: Seismic Analysis of Safety-Related Nuclear Structures and Commentary, ASCE 4-98 (1998)

6. American Society of Civil Engineers: Seismic Design Criteria for Structures, Systems, and Components in Nuclear Facilities, ASCE/SEI 43-05 (2005)
7. U.S. Nuclear Regulatory Commission: Standard Review Plan for the Review of Safety Analysis Reports for Nuclear Power Plants, NUREG-0800, Revision 3 (2007)
8. Hashash, Y.M.A., Park, D.: Non-linear one-dimensional seismic ground motion propagation in the Mississippi embayment. Eng. Geol. **62**, 185–206 (2001)
9. Jiao, Z., Guoxing, C., Hangang, X., Xuening, L.: Spatial variation characteristics of shear wave velocity structure and its application for quaternary deep sediment layers in Suzhou region. Chin. J. Geotech. Eng. (2017). http://kns.cnki.net/kcms/detail/32.1124.TU. 20170525.1614.008.html. (in Chinese)
10. Schneider, J.F.: Guidelines for Determining Design Basis Ground Motions. Electric Power Research Center (1993)

Shaking Table Experimental Study on Time-Frequency Characteristics of Subway Stations Constructed in Different Materials

Dingfeng Zhao, Ji Shen, Guoxing Chen$^{(\boxtimes)}$, and Jiao Zhu

Nanjing Tech University, Nanjing 210009, China
gxc6307@163.com

Abstract. Seismic behaviors of the subway stations modeled by micro-concrete and plaster in liquefied soils under strong ground motion are compared by conducting shaking table tests. More severe damage has been caused to the plaster structure than that to the micro-concrete one. From a pure time-domain point of view, model soil-structure interaction system stated in these collected signals may be ignored, damages to the structures and ground failure are generally more time-frequency dependent than time-dependent. Based on the optimal wavelet packet transform, accelerations of the two subway stations are decomposed into time-frequency domains. The results show that the liquefied soils make the fundamental frequencies of the model soil-structure systems decrease in the feature band of $0.1 \sim 3.125$ Hz, and damages of the model structures are responded first to feature band of $12.5 \sim 25$ Hz that includes their own fundamental frequencies. Combing time histories of pore pressure ratio around the model structure and time-frequency-amplitude of collected accelerations on the sidewall, it can be inferred that the plaster structure may damage much earlier than the micro-concrete structure. The following greater amplitudes make the plaster structure suffer more damages.

Keywords: Shaking table test · Time-frequency
Optimal wavelet packet transformation · Damage to the model structure
Different materials

1 Introduction

Underground facilities are crucial parts of transportation networks. Their seismic performance is generally satisfactory because of the lower levels of shaking at depth. However, several cases have been reported that underground structures have experienced considerable damages in recent large earthquakes [1, 2]. Model testing is one of the most efficient methods to provide insight into the finer points that may significantly affect the stability of underground openings. In these modelling tests, different materials were used to model the prototype concrete, such as aluminum, micro-concrete, plaster and so on [3–5]. Because of the high strength and stiffness of the aluminum and micro-concrete, damages of these model structures were rarely reported. Interestingly, in a shaking table test carried out by Chen et al. [4], a plaster subway station in liquefied ground suffered serious damages. Increasing attention has been given to

© Springer Nature Singapore Pte Ltd. 2018
T. Qiu et al. (Eds.): GSIC 2018, *Proceedings of GeoShanghai 2018 International Conference: Advances in Soil Dynamics and Foundation Engineering*, pp. 233–241, 2018.
https://doi.org/10.1007/978-981-13-0131-5_26

investigate the seismic performance of underground structures with different physical properties. On the other hand, most of the collected time-varying signals in modelling tests were analyzed either in a pure time-domain or in a frequency domain, recession in model soil-structure system indicated in these collected signals may be ignored. Furthermore, development of damages to the structures and ground failure will always be with the frequencies changed over time. Seismic behaviors of the model soil-structure system may be seen more clearly in time-frequency domains. Wavelet packet transform has time-frequency analysis capabilities by both changing the time window and frequency window. It provides an effective tool in describing analyzing transient response, because it possesses high time and frequency resolution, noise robust and high signal reconstruction [6, 7]. For this purpose, wavelet-based signal processing technique is used to decompose these measured time-varying accelerations into time-frequency domains, damage characteristics of the two material subway stations in liquefied soils under a strong ground motion are compared and discussed.

2 Optimal Wavelet Packet Transformation

Wavelets can be constructed with different smoothness, symmetry and support properties. Daubechies N are compactly supported wavelets that offer more flexibility in blasting, earthquake and other non-stationary signal problems [8]. In this paper, Daubechies wavelets are adopted to transform the measured accelerations from pure time-domains to time-frequency domains. When a wavelet packet is selected, decomposition level becomes a crucial problem in signature extraction. Excessive decomposition layers will increase correlation with frequency band, and these overlapping bands will reduce the accurate analysis of the signal. Entropy cost function $C(f_{j,n})$ provides a criterion to define the best basis [9]. This criterion has predominantly been used in application due to its good discrimination properties with high energy concentration. $C(f_{j,n})$ is defined as:

$$C(f_{j,n}) = -\sum_{k=0}^{t} |f_{j,n,k}|^2 \ln |f_{j,n,k}|^2 \tag{1}$$

Where $f_{j,n,k}$ is the wavelet packet coefficient in the $f_{j,n}$ of depth j. The algorithm for finding the best basis is supposed that the data vector is of length 2^j, and the depth of the wavelet packet is $\leq j$. $C(f_{j,n})$ is the entropy cost function of $f_{j,n}$, and $f_{j+1,2n}$ and $f_{j+1,2n+1}$ are the children nodes of $f_{j,n}$. In search for the best basis, comparisons are made between two adjacent generations from the bottom depth to the top depth in the library tree. If $C(f_{j,n}) \leq C(f_{j+1,2n}) + C(f_{j+1,2n+1})$ is satisfied, then discard the children nodes and reserve $f_{j,n}$.

3   Design of the Two Shaking Table Tests

To evaluate seismic behaviors of structures with different materials in liquefiable ground during a strong earthquake, two shaking table tests were constructed and tested utilizing the independently developed laminar soil box, frequencies of the input motion coverage for 0.1 ~ 50 Hz [4, 5]. Layout of sensors is shown in Fig. 1, real time signals recorded on three accelerometers (A1-1, A1-3 and A1-4) on the model structure and two pore pressure transducers (W10 and W6) near the model structure were used in this paper.

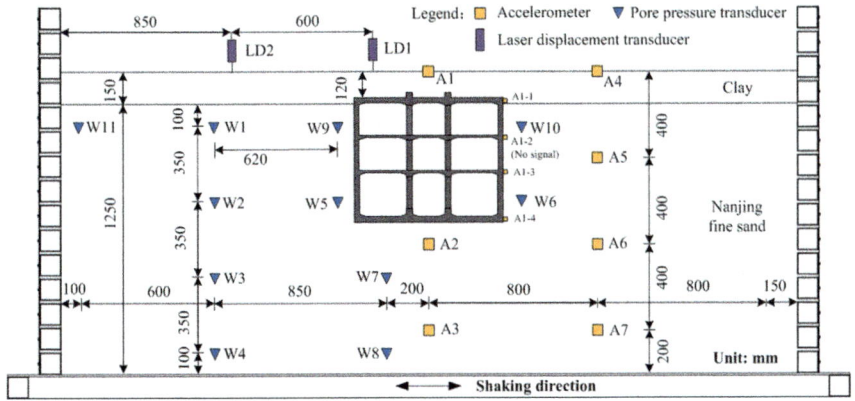

Fig. 1. Sensors embedded in the observation section of the two shaking table tests

Table 1. Mechanical properties of model structures used in the two shaking table tests.

Material	Mixture ratio	Compressive strength (Mpa)	Young's modulus (GPa)
Micro-concrete	water: cement: lime: coarse sand = 0.5:1:0.58:5	7.5	7.5
Plaster	water: plaster = 1:0.85	2.9	2.0

Table 2. Mechanical properties of model soils used in the two shaking table tests.

Test	Material	Unit weight γ (kN/m^3)	Friction angle (°)	Shear modulus (Mpa)	Relative density
MCON	Clay	16.6	8.0	10.1	–
	Fine sand	17.5	20.0	10.6	41%
PL	Clay	16.5	8.1	10.9	–
	Fine sand	17.6	20.3	11.6	43%

For the two model structures, geometries were designed to be the same, steel rebars were modelled by the same galvanized steel wires, but prototype concretes were modeled by micro-concrete and plaster, respectively. Material properties of micro-concrete and plaster are listed in Table 1, the elastic modulus and the compressive strength of the micro-concrete are 3.75 and 2.59 times higher than those of the plaster. Furthermore, taking inertia forces in accordance with the prototype structure and loading capacities of the two model structures into consideration, additional mass of 480 kg and 60 kg were distributed on the micro-concrete and the plaster structure, respectively. Model soils in both tests consisted of clay with 150 mm, and saturated Nanjing fine sand with 1250 mm. In both tests, the model soils were saturated by the water sedimentation method, mechanical properties are listed in Table 2. Shaking table tests for the micro-concrete and the plaster subway station structures are called the MCON structure test and the PL structure test in short. Taking a white noise of 0.02 g to get the fundamental frequency of the soil-structure system. After that, the Shifang ground motion was selected to simulate the main shock, acceleration time-history and Fourier spectrum are shown in Fig. 2.

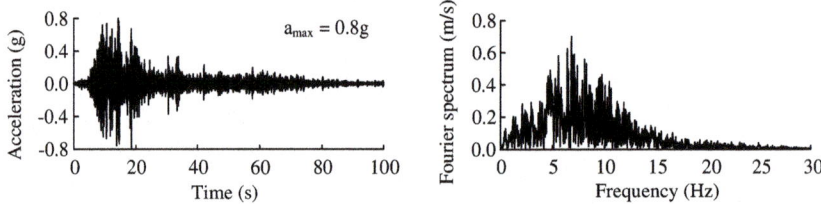

Fig. 2. Acceleration time-history and Fourier spectrum of the input motion

Fig. 3. Comparison on the macroscopic phenomena after shaking between the two tests: (a) the MCON structure test and (b) the PL structure test

Figure 3 presents the macroscopic phenomena after the two shaking table tests. Water boils and waterspouts in model soils were observed. Moreover, differences in seismic responses of the two model structures were achieved. In the MCON structure test, structural sidewall was undamaged, some diagonal cracks were observed on the columns. In the PL structure test, cracks in the side wall and plaster spalling of the columns were observed.

4 Time-Frequency Characteristics of Accelerations for the Two Soil-Structure Systems

Fundamental frequency of the model soil-structure system can be obtained from the white noise excitations, sampling frequency was 100 Hz. The fundamental frequency of the soil-MCON structure system was 21.38 Hz and 19.47 Hz of the soil-PL structure system, respectively. Figure 4 shows acceleration time-histories collected on sidewalls of the two model structures under the Shifang motion. From a pure time-domain of view, peak acceleration collected from points A1-1, A1-3 and A1-4 of the MCON structure were about 1.47, 1.58 and 1.60 times higher than those of the PL structure. Influence of material physical properties on their seismic behaviors was revealed but limited.

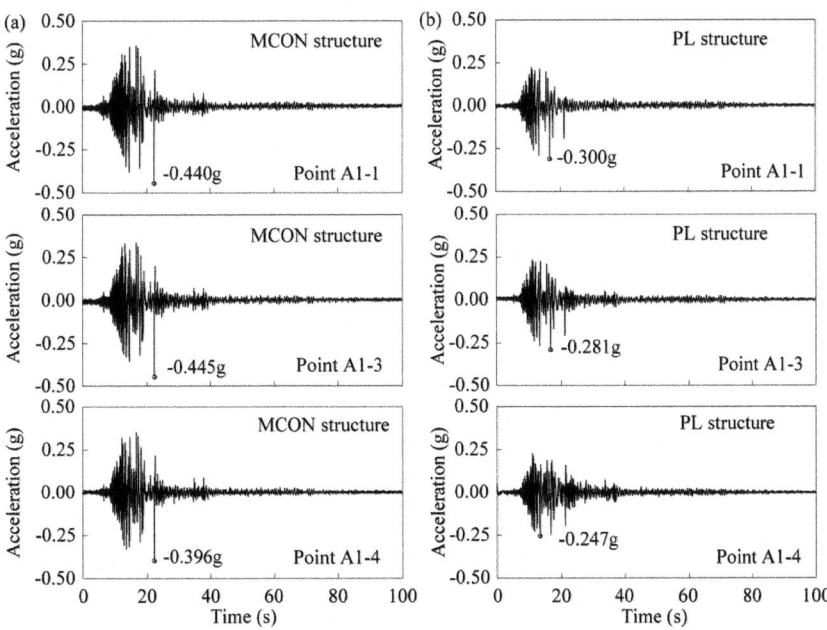

Fig. 4. Acceleration time-histories collected on sidewalls of the two model structures modeled by different materials

Based on the best basis algorithm, considering accurate resolutions both in frequency and time domain, acceleration responses shown in Fig. 4 were transformed using Daubechies 25 wavelet functions of deep 4 (for the MCON structure) and Daubechies 10 wavelet functions of deep 4 (for the PL structure). Reconstruct errors of these decomposed acceleration responses were all $<10^{-5}$. In consideration of the operating frequencies (0.1 $\sim$ 50 Hz), they were divided into 0.1 $\sim$ 3.125 Hz, 3.125 $\sim$ 6.25 Hz, 9.375 $\sim$ 12.5 Hz, 12.5 $\sim$ 25.0 Hz and 25.0 $\sim$ 50.0 Hz, respectively. In which, the sum of frequency energies was mainly distributed in 0.1 $\sim$ 25 Hz, about 99.4% for the MCON structure and about 100.0% for the PL structure relative to the total energies. Frequency bands in 0.1 $\sim$ 25 Hz were known as the feature bands. Figure 5 compares the feature band energy densities of the measured accelerations collected from the two model structures with different materials. Feature band energy density $P_{j,n}$ is represented as:

$$P_{j,n} = \frac{E_{j,n}}{\sum_m E_{j,n}} = \frac{\sum_{k=0}^{t} |f_{j,n,k}|^2 \times \Delta t}{E_{\text{total}}} \tag{2}$$

Where $E_{j,n}$ is feature band energy at node (j,n); m is the number of decomposed nodes; Δt is the time interval of an acceleration response.

By using optimal wavelet packet transformation, influence of structural material properties on seismic behaviors of the model soil-structure interaction system was achieved. For the two model structures, feature band energy densities of measured accelerations were both concentrated in feature band of 0.1 $\sim$ 3.125 Hz. It was much higher in the PL structure test than that in the MCON structure. Interestingly, feature band energy densities of 12.5 $\sim$ 25 Hz, including their own fundamental frequencies, were much higher in the stiffer MCON structure acceleration than those in the PL structure acceleration.

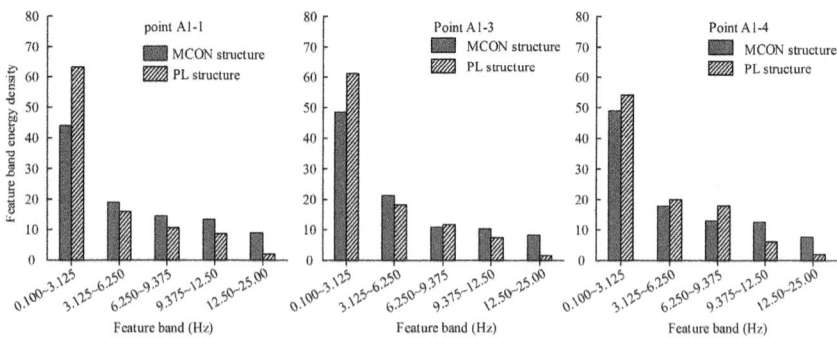

Fig. 5. Comparison on feature band energy densities of measured accelerations collected from two model structures with different racking stiffness

For further study, the relationship between the time-frequency-amplitude (point A1-1, A1-2, and A1-3 and A1-4) and excess pore pressure ratios at point W10 and W6 are illustrated in Fig. 6. Pore pressure ratio of the model soil around the MCON structure increased much later than that around the PL structure. As the movement of the adjacent model soil was restricted by the stiff model structure, the starting of shear modulus degradation in the model soil was limited by the relative stiffness between the model structure and soil. Moreover, as model soils were all liquefied in the two tests, this contributed to the phenomenon that most of the energies were moved to the lower frequencies of 0.1 ~ 3.125 Hz. As seen in Fig. 6, it could be also noted that feature band of 12.5 ~ 25 Hz including the fundamental frequencies were responded first. When pore pressure ratios increased dramatically, frequencies were filtered from the higher to the lower component. In the MCON structure test, when pore pressure ratio reached about 0.6 at Point A1-1 and about 1.0 at Pint A1-3 and Point A1-4, feature band of 12.5 ~ 25 Hz were filtered. In the PL structure test, feature band of 12.5 ~ 25 Hz filtered when pore pressure ratio reached only about 0.2 at Point A1-1 and about 0.4 at

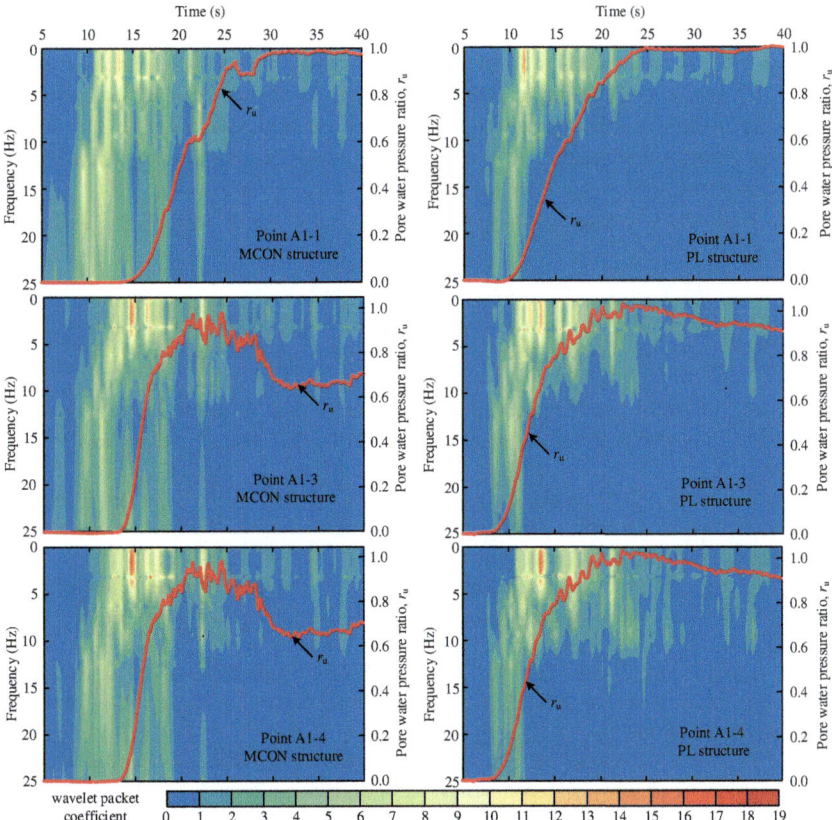

Fig. 6. Comparison on time-frequency-amplitude of the measured accelerations and time histories of the pore pressure ratios around the model structures

Point A1-3 and Point A1-4. Frequency band of 12.5 ~ 25 Hz of accelerations in the MCON structure test lasted longer than that in the other one. The modulus of model soil was highly degraded due to the increasing excess pore pressure, but amplitudes in 12.5 ~ 25 Hz of the MCON structure accelerations still exist. This may indicate damages occurred much earlier (at about 12 s) in the PL structure as the fundamental frequency was filtered but the soil still possessed higher modulus. The following greater amplitudes may make the PL structure suffered more damages. This may explain the test phenomena that the PL structure is damaged much more seriously than the MCON structure.

5 Conclusion

Based on the optimal wavelet packet transform, accelerations of the model structures were decomposed into time-frequency domains. Difference in seismic behaviors of two material subway stations was achieved, and the following conclusions were drawn. For the two tests, high frequencies of 12.5 ~ 25 Hz including the fundamental frequencies were responded first. Soil-structure interaction made the pore pressure ratios raised around the plaster model structure more quickly. Due to the high recession in the model soils, densities of measured accelerations were both moved to the feature band of 0.1 ~ 3.125 Hz. Combing time histories of pore pressure ratio around the model structure and time-frequency-amplitude of collected accelerations on the sidewalls, damages occurred much earlier in the PL structure, as the feature band including its fundamental frequency was filtered quickly, but the model soils still had higher modulus. Following greater amplitudes would make the PL structure suffer more damages.

References

1. Pitilakis K., Tsinidis G.: Performance and seismic design of underground structures. In: Earthquake Geotechnical Engineering Design, pp. 279–340. Springer International Publishing (2014)
2. Uenishi, K., Sakurai, S.: Characteristic of the vertical seismic waves associated with the 1995 Hyogo-ken Nanbu (Kobe), Japan earthquake estimated from the failure of the Daikai underground station. Earthq. Eng. Struct. Dyn. 29(6), 813–822 (2000)
3. Zhou, J., Jiang, J.H., Chen, X.L.: Micro- and macro-observations of liquefaction of saturated sand around buried structures in centrifuge shaking table tests. Soil Dyn. Earthq. Eng. 72, 1–11 (2015)
4. Chen, G.X., Wang, Z.H., Zuo, X., et al.: Shaking table test on the seismic failure characteristics of a subway station structure on liquefiable ground. Earthq. Eng. Struct. Dyn. 42, 1489–1507 (2013)
5. Chen, G.X., Zuo, X., Wang, Z.H., et al.: Large scale shaking table test study of the dynamic damage behavior of subway station structures in liquefiable foundation under near-fault and far-field ground motions. China Civil Eng. J. 43(12), 120–126 (2010). (in Chinese)
6. Iyama, J.: Estimate of input energy for elasto-plastic SDOF systems during earthquakes based on discrete wavelet coefficients. Earthq. Eng. Struct. Dyn. 34(15), 1799–1815 (2005)

7. Ding, Y.L., Li, A.Q., Deng, Y.: Parameters for identification of wavelet packet energy spectrum for structural damage alarming. J. Southeast Univ. Natural Sci. Ed. **41**(4), 824–828 (2011). (in Chinese)

8. Zhong, G.S., Ao, L.P., Zhao, K.: Influence of explosion parameters on wavelet packet frequency band energy distribution of blast vibration. J. Cent. South Univ. **19**(9), 2674–2680 (2012)

9. Coifmann, R.R., Wickerhauser, M.V.: Entropy-based algorithms for best basis selection. IEEE Trans. Inf. Theory **38**(2), 713–718 (1992)

Dynamic Responses of Saturated Transversely Isotropic Ground Subjected to High-Speed Train Load

Chenxiao Xu[1,2], Guangyun Gao[1,2(✉)], and Qingsheng Chen[3]

[1] Department of Geotechnical Engineering, Tongji University,
Shanghai 200092, China
gaoguangyun@263.net
[2] Key Laboratory of Geotechnical and Underground Engineering of Ministry
of Education, Tongji University, Shanghai 200092, China
[3] Department of Civil and Environmental Engineering,
National University of Singapore, Singapore 119077, Singapore

Abstract. A *u-p* format 2.5D finite element method (FEM) is proposed to investigate the ground vibration and distribution of pore water pressure in transversely isotropic saturated medium subjected to high-speed train load. The governing equations are derived based on Biot's theory in frequency domain by applying Fourier transform with respect to time. The Galerkin method is then employed to develop the 2.5D FE model. Verification of the model is carried out by comparing the predictions with published data. Effects of soil parameters on the dynamic responses of ground are investigated in detail. Results indicate that the increase in vertical elastic modulus contributes much more to the decrease in displacement amplitude of ground and maximum amplitude of pore water pressure in soil, compared to that of horizontal elastic modulus. Both horizontal and vertical Poisson ratios have insignificant effects on ground vibration, while the change of Poisson ratio lead to a great change in the maximum amplitude of pore water pressure. Effects of shear modulus on both ground vibration and maximum amplitude of pore water pressure are negligible.

Keywords: Transversely isotropic saturated medium · Ground vibration
Pore water pressure · 2.5D FEM

1 Introduction

High-speed train is a convenient and comfortable trip mode, but it also causes environmental vibration and affects the living and working environment. Finite element method (FEM) appears to be one of the most suitable approaches with versatility, efficiency and availability, without compromising on accuracy [1]. While the 2.5D FEM has been well recognized as a useful and effective approach in investigating the dynamic ground responses induced by the moving load, when the structure itself and applied load are invariant along the train moving direction [2–6]. However, those studies simplified the soil as a single-phase elastic medium, while transversely isotropic ground is widely distributed in practice.

© Springer Nature Singapore Pte Ltd. 2018
T. Qiu et al. (Eds.): GSIC 2018, *Proceedings of GeoShanghai 2018 International Conference:
Advances in Soil Dynamics and Foundation Engineering*, pp. 242–250, 2018.
https://doi.org/10.1007/978-981-13-0131-5_27

   Anisotropy is now recognized as an important feature when describing the behavior
of many types of soils. Kuwano [7] and Nishimura [8] adopted experimental methods
to study the properties of transversely isotropic soil. Analytical solutions are proposed
by Zymnis [9] and Ai [10], to solve wave propagation in transversely isotropic ground.
A coupling method of finite and hierarchical infinite elements was presented to solve a
non-homogeneous cross-anisotropic half-space subjected to a non-uniform circular
loading [11]. However, to date, a limited attention has been paid to the effects of soil
parameters on dynamic responses of saturated transversely isotropic soil induced by
high-speed train.
   This paper develops a 2.5D FEM to investigate the dynamic responses of transversely isotropic saturated soil induced by high-speed train load. The flow viscoelastic
boundary is employed. Applying Fourier transform with respect to time and train
moving direction, the u-p format 2.5D finite element equations in frequency domain are
derived based on Biot theory and Galerkin method. The track structure is simplified as
an Euler beam resting on the half-space, and the ground is modeled by 2.5D quadrilateral elements. Using this method, effects of soil parameters on the ground vibration
and pore water pressure in soil induced by high-speed train are investigated in detail.

2 Mathematical Formulations

Track and ground are assumed to be continuous and uniform along train moving
direction in 2.5D finite elements, which are simplified as Euler-Bernoulli beam and
transversely isotropic saturated porous medium, respectively. The finite element model
in this paper is the same as that employed in Reference [5], where x is the train moving
direction, y is the direction perpendicular to track, and z is the vertical direction, the
track center is origin of coordinates. High-speed train moves along the track with a
velocity c, the expression of train loads can be seen in Reference [5]. The underground
water level is assumed to be at the ground surface, and the height of embankment is set
to be 1.0 m.
   Based on the Hooke's law, stress-stain relationship and effective stress principle of
soil, the relationship between stresses and displacements of soil are given as:

$$
\begin{cases}
\sigma_x = C_{11}\dfrac{\partial u}{\partial x} + C_{12}\dfrac{\partial v}{\partial y} + C_{13}\dfrac{\partial w}{\partial z} - p, \ \tau_{yz} = C_{44}\left(\dfrac{\partial v}{\partial z} + \dfrac{\partial w}{\partial y}\right) \\[2mm]
\sigma_y = C_{12}\dfrac{\partial u}{\partial x} + C_{11}\dfrac{\partial v}{\partial y} + C_{13}\dfrac{\partial w}{\partial z} - p, \ \tau_{zx} = C_{44}\left(\dfrac{\partial w}{\partial x} + \dfrac{\partial u}{\partial z}\right) \\[2mm]
\sigma_z = C_{13}\dfrac{\partial u}{\partial x} + C_{13}\dfrac{\partial v}{\partial y} + C_{33}\dfrac{\partial w}{\partial z} - p, \ \tau_{xy} = C_{66}\left(\dfrac{\partial u}{\partial y} + \dfrac{\partial v}{\partial x}\right)
\end{cases}
\tag{1}
$$

where u, v and w are displacements of soil skeleton in x, y and z directions, respectively; σ_{ij} is the stress of porous medium; p is the excess pore water pressure;
$C_{ij}(i, j = 1, 2, 3, 4, 6)$ is the mechanical parameter of transversely isotropic soil (e.g.,
moduli and Poisson ratios). Elastic modulus in complex form is introduced to account
for the material damping.

According to wave propagation in fluid-saturated porous medium, the dynamic motion equations can be expressed as follows [5]:

$$\sigma_{ij,j} + F_i = \rho u_i'' + \rho_f W_i'', \ \rho = \rho_s(1-n) + n\rho_f, \ W_i = n(w_i - u_i) \tag{2}$$

$$-\frac{n}{K_f}p' = W_{i,i}' + u_{i,i}' \tag{3}$$

$$-p_i = p_f u'' + \frac{\rho_f}{n}W_i'' + \frac{\rho_f g}{K_d}W_i' \tag{4}$$

in which F_i is the body force; ρ and ρ_f denote the bulk density of the porous medium and the density of the pore fluid, respectively; ρ_s is the density of the solid skeleton and n is the porosity; W_i is the average displacement of the pore fluid relative to the solid skeleton; w_i and u_i denote the infiltration displacements of pore fluid and the average displacement of solid skeleton, respectively; g is the acceleration of gravity; K_f and K_d are the bulk modulus of pore fluid and the permeability, respectively; $(')$ indicates differentiation with respect to time t.

Fourier transformation with respect to time is performed on Eq. (4) to eliminate time derivatives, and then Eq. (4) is transformed into the frequency domain. By using the derivative nature of Fourier transform, the following equation can be obtained:

$$W_i = F(\omega^2 \rho_f \bar{u}_i - \bar{p}_{,i}), \ F = nK_d/(i\omega\rho_f gn - \omega^2 K_d \rho_f) \tag{5}$$

in which, ω represents circular frequency; variables with a bar above indicate the components in frequency domain.

Performing Fourier transformation with respect to time on Eqs. (1) and (2), then combining with Eq. (5), the balance equations of saturated transversely isotropic soil in frequency domain are given by:

$$\begin{cases} (C_{11}\dfrac{\partial^2}{\partial x^2} + C_{66}\dfrac{\partial^2}{\partial y^2} + C_{44}\dfrac{\partial^2}{\partial z^2})\bar{u} + (C_{12} + C_{66})\dfrac{\partial^2 \bar{v}}{\partial x \partial y} + \\[2mm] (C_{13} + C_{44})\dfrac{\partial^2 \bar{w}}{\partial x \partial z} - \bar{P}_{,x} + \omega^2 \rho \bar{u} + \omega^2 \rho_f F(\omega^2 \rho_f \bar{u} - \bar{P}_{,x}) = 0 \\[2mm] (C_{12} + C_{66})\dfrac{\partial^2 \bar{u}}{\partial x \partial y} + (C_{66}\dfrac{\partial^2}{\partial x^2} + C_{11}\dfrac{\partial^2}{\partial y^2} + C_{44}\dfrac{\partial^2}{\partial z^2})\bar{v} + \\[2mm] (C_{13} + C_{44})\dfrac{\partial^2 \bar{w}}{\partial y \partial z} - \bar{P}_{,y} + \omega^2 \rho \bar{v} + \omega^2 \rho_f F(\omega^2 \rho_f \bar{v} - \bar{P}_{,y}) = 0 \\[2mm] (C_{13} + C_{44})\dfrac{\partial^2}{\partial y \partial z}\bar{v} + (C_{44}\dfrac{\partial^2}{\partial x^2} + C_{44}\dfrac{\partial^2}{\partial y^2} + C_{33}\dfrac{\partial^2}{\partial z^2})\bar{w} + \\[2mm] C_{44} + C_{13})\dfrac{\partial^2 \bar{u}}{\partial x \partial z} - \bar{P}_{,z} + \omega^2 \rho \bar{w} + \omega^2 \rho_f F(\omega^2 \rho_f \bar{w} - \bar{P}_{,z}) = 0 \end{cases} \tag{6}$$

Similarly, Fourier transformation with respect to time is performed on Eq. (3). Then by substituting the obtained results into Eq. (5), the balance equation of fluid in frequency domain is expressed as:

$$(F\omega^2\rho_f K_d + K_d)(\frac{\partial \bar{u}}{\partial x} + \frac{\partial \bar{v}}{\partial y} + \frac{\partial \bar{w}}{\partial z}) + n\bar{P} - FK_d(\frac{\partial^2}{\partial x^2} + \frac{\partial^2}{\partial y^2} + \frac{\partial^2}{\partial z^2})\bar{P} = 0 \quad (7)$$

Stress boundary condition and flow boundary condition in frequency domain of the FEM model are given as:

$$\sigma_x l + \tau_{xy} m + \tau_{xz} n = f_x, \tau_{yx} l + \sigma_y m + \tau_{yz} n = f_y, \tau_{xz} l + \tau_{zy} n + \sigma_z m = f_z \quad (8)$$

$$\bar{q} = \rho_f g \bar{v}_n = -K_d(\frac{\partial}{\partial x} l + \frac{\partial}{\partial y} m + \frac{\partial}{\partial z} n)\bar{P} \quad (9)$$

where $f_i(i = x, y, z)$ are components of external forces; l, m, n are corresponding directions cosine; $\bar{q}$ and $\bar{v}_n$ are flow and velocity of pore water, respectively.

Combining the constitutive equations and applying the Galerkin method to Eqs. (6)–(9), and then incorporating the developed shape function and performing wave-number expansion on the resulting equation in x-direction, the 2.5D FEM governing equations in wave-number domain and frequency domain can be derived by conventional finite element method, which are given by:

$$\left(K_{up} - M_{up}\right)\tilde{\bar{u}} + \left(Q'_{up} - Q_{up}\right)\tilde{\bar{P}} = \tilde{\bar{f}}^s_{up}, \quad \left(H_{up} + S_{up}\right)\tilde{\bar{P}} + Q''_{up} = \tilde{\bar{f}}^q_{up} \quad (10)$$

where K_{up} is stiffness matrix; M_{up} is mass matrix; Q''_{up}, Q'_{up} and Q_{up} are solid and fluid coupling matrixes, respectively; H_{up} and S_{up} are Jacobian matrixes; $\tilde{\bar{f}}^s_{up}$ and f^s_{up} are equivalent node load vectors; $\bar{u}$ is node displacement matrix; variables with a wave above indicate the component in wave-number domain.

Boundary condition has a non-negligible influence on calculation accuracy. Referring to Gao et al. [5], this paper adopted a 2.5D viscoelastic dynamic artificial boundary to model the wave propagation in transversely isotropic saturated ground.

3 Numerical Results and Discussion

Numerical results obtained from this paper are degenerated to compare with Gao's results [5], by assuming that $E_{hh} = E_{hv}, \mu_{hh} = \mu_{hv}, G_{hh} = G_{hv} = E_{hh}/2(1 + v_{hh})$. High-speed train moves on the track at a speed of $0.5V_s$ (V_s is the shear wave velocity of soil). The dynamic responses at 1.0 m beneath the ground surface are investigated. Figure 1 shows the normalized vertical displacement $w^* = wG/F_{max}$ and normalized pore water pressure $P = p/F_{max}$ predicted by current model and that of Gao et al. [5], where F_{max} and G are the train wheel load and soil shear modulus, respectively. It is observed that these two results agree dramatically well with each other for both normalized vertical displacement and pore pressure against the increase of time.

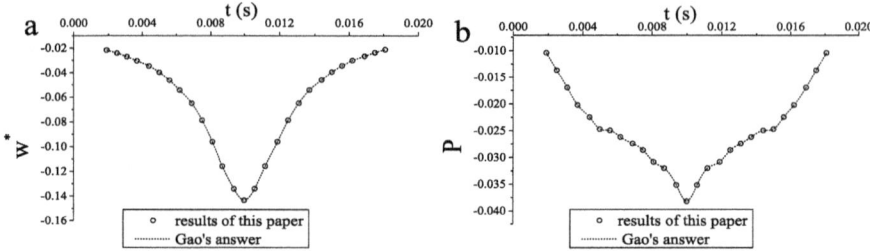

Fig. 1. Verification of the 2.5D FEM: (a) Normalized displacement time history curve; (b) Normalized pore pressure time history curve

3.1 Mathematical Description of Calculation Model

The track is normally used in Beijing-Shanghai high-speed railway. Its bending rigidity is EI = 38.0496 kN•m, and the comprehensive quality is 401 kg/m. CRH5 is the adopted model of high-speed train. In one unit, there are 8 carriages, including 2 motor cars and 6 trailers. The length of train is 205.2 m, including 16 wheel sets in total. Length of trailer and motor car are 25 m and 27.6 m, respectively. The distance of bogie center and the fixed wheelbase are 17.5 m and 2.7 m, respectively. The average axle load is 17 t. The soil parameters are shown in Table 1. The scenario of the high-speed train moving at a speed of 300 km/h is investigated herewith.

Table 1. Calculation parameters of transversely isotropic saturated soil

Parameters	Values	Parameters	Values
Density ρ(kg/m^3)	1800	Porosity n	0.54
Horizontal elastic modulus E_{hh} (MPa)	5.46	Vertical elastic modulus E_{hv} (MPa)	7.86
Horizontal Poisson ratio μ_{hh}	0.35	Vertical Poisson ratio μ_{vh}	0.34
Material damping β	0.05	Bulk elastic modulus of fluid K_f (Pa)	2×10^9
Dynamic permeability coefficient K_d (m/s)			2×10^{-7}

3.2 Effects of Horizontal and Vertical Elastic Moduli

By fixing vertical elastic modulus as 7.86 MPa, the horizontal elastic moduli with 9.06 MPa, 7.86 MPa, 6.66 MPa and 5.46 MPa are used in the model respectively. Figure 2 shows that variation of horizontal elastic moduli leads to limited effects on the response of vertical displacements and pore water pressure. It is shown in Fig. 2(a) that the difference among attenuation curves of displacement within 7 m away from track center is almost negligible, and it only increases marginally as the increase of horizontal elastic modulus in further distance. In Fig. 2(b), the maximum amplitude of pore pressure is slightly higher at the depth of around 2 m beneath the track center with relatively higher horizontal elastic moduli, in contrast to that with lower ones.

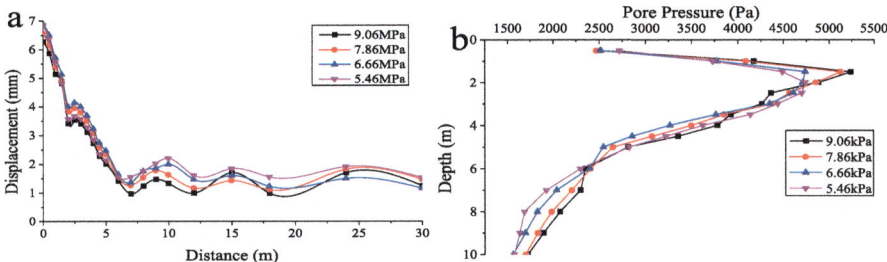

Fig. 2. Effects of horizontal elastic modulus: (a) Attenuation of displacement on ground surface from the track centre; (b) Distribution of pore water pressure along depth

By fixing horizontal elastic modulus as 5.46 MPa, vertical elastic moduli setting as 7.86 MPa, 6.66 MPa, 5.46 MPa and 4.26 MPa are used in the model respectively. Figure 3 indicates that vertical elastic modulus has significant effects on the dynamic responses of ground. With increase of vertical elastic modulus, it is observed that displacement decreases greatly, and it is also shown that pore water pressure decreases enormously above the depth of around 2 m. By comparing Figs. 2 and 3, vertical elastic modulus shows much greater influence on ground vibration and development of pore water pressure compared to horizontal elastic modulus. The reason may be that, in contrast to horizontal elastic modulus, the vertical one presents a larger restriction on the movements of soil particles. Movement of soil particles alters greatly with the change of vertical elastic modulus, leading to huge alteration of displacement.

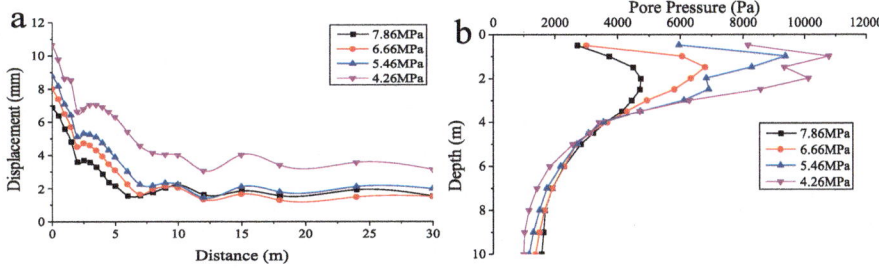

Fig. 3. Effects of vertical elastic modulus: (a) Attenuation of displacement on ground surface from the track centre; (b) Distribution of pore water pressure along depth

3.3 Effects of Horizontal and Vertical Poisson Ratio

By fixing vertical Poisson ratio as 0.32, horizontal Poisson ratios with 0.30, 0.35, 0.40 and 0.45 are used in this model respectively. Figure 4(a) indicates that obvious distinction of displacement only exists within 2 m away from track center, and the variation of horizontal Poisson ratio results in insignificant influence on the displacement response with the further away (>2 m) from the track centre on the ground surface. Vibration attenuates greatly within the distance of 5 m and fluctuates thereafter as the distance increases. In contrast, Fig. 4(b) presents that the increase of horizontal Poisson ratio leads to a great increase in pore water pressure, especially within the depth of around 5.2 m, while the effect of the variation in horizontal Poisson ratio on pore pressure is almost negligible below the depth of 5.2 m.

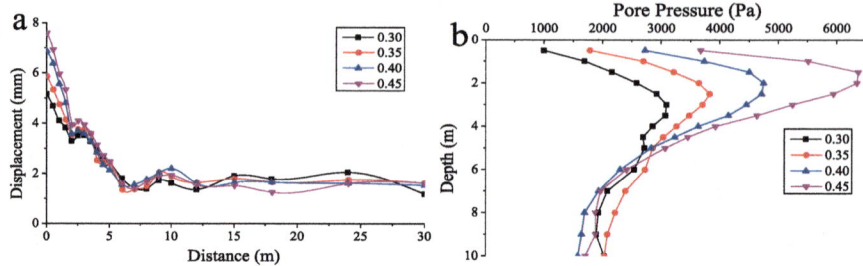

Fig. 4. Effects of horizontal Poisson ratio: (a) Attenuation of displacement on ground surface from the track centre; (b) Distribution of pore water pressure along depth

By fixing horizontal Poisson ratio as 0.32, vertical Poisson ratios with 0.30, 0.35, 0.40 and 0.45 respectively, are used in the model. Figure 5(a) indicates that the increase in vertical Poisson ratio results in a slight increase in displacement within the distance of around 17 m away from the track center, beyond which its effects are negligible. Figure 5(b) shows that influence of vertical Poisson ratio on distribution of pore pressure is much less significant compared to that of horizontal Poisson ratio.

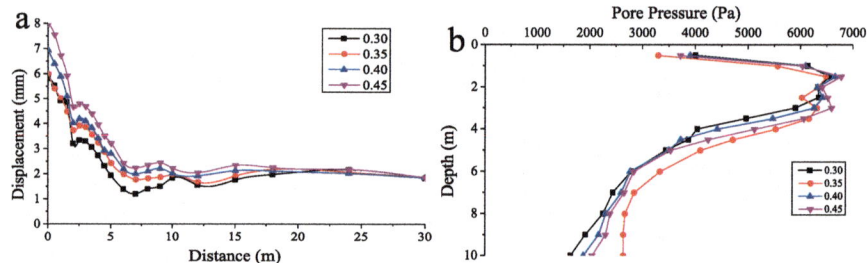

Fig. 5. Effects of vertical Poisson ratio: (a) Attenuation of displacement on ground surface from the track centre; (b) Distribution of pore water pressure along depth

3.4   Effects of Shear Modulus

Vertical shear modulus of transversely isotropic soil is an independent variable, while the horizontal shear modulus is as same as that of isotropic soil. Figure 6(a) shows the attenuation manner of the displacement on ground surface with various shear moduli, i.e., 2.0 MPa, 2.45 MPa, 2.90 MPa, 3.35 MPa. It is observed that the increase of shear modulus significantly contributes to the decrease in displacement within the distance of 2.5 m away from track center, beyond which such effects are insignificant. Ground displacements attenuate greatly as the distance increases within 5 m, and subsequently they present a slight fluctuation. Figure 6(b) shows that the pore water pressure at shallow depth (<4 m) decreases with the increasing shear moduli, while the effects of shear modulus on development of pore pressure beyond depth of 4 m are negligible.

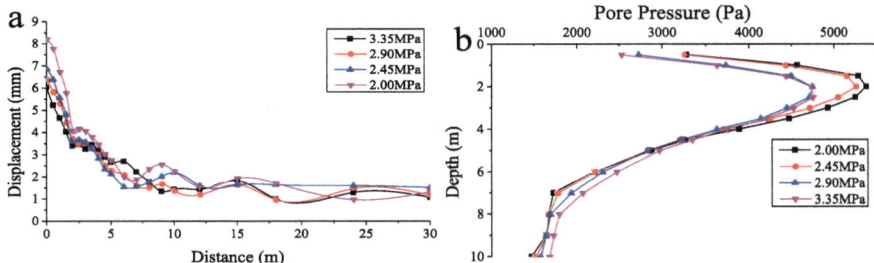

Fig. 6. Effects of shear modulus: (a) Attenuation of displacement on ground surface from the track centre; (b) Distribution of pore water pressure along depth

4   Conclusions

The *u-p* format 2.5D FEM and dynamic viscoelastic boundary in transversely isotropic saturated medium are presented in this paper. Ground vibration and distribution of pore water pressure induced by high-speed train are discussed. Results indicate that the increase of vertical elastic modulus contributes much more to the decrease of displacement amplitude of ground surface and maximum amplitude of pore water pressure, compared to the horizontal one. Both horizontal and vertical Poisson ratios have insignificant effects on ground vibration, while the change of horizontal Poisson ratio lead to a great change in the maximum amplitude of pore water pressure. Effects of shear modulus on both ground vibration and maximum amplitude of pore water pressure are negligible.

Acknowledgements. The research was supported by National Natural Science Foundation of China (no. 41772288).

References

1. Beskou, N.D., Theodorakopoulos, D.D.: Dynamic effects of moving loads on road pavements: a review. Soil Dyn. Earthq. Eng. **31**(4), 47–67 (2011)
2. Yang, Y.B., Hung, H.H.: A 2.5D finite/infinite element approach for modeling visco-elastic bodies subjected to moving loads. Int. J. Numer. Methods Eng. **51**(11), 17–36 (2001)
3. Takemiya, H., Bian, X.C.: Substructure simulation of inhomogeneous track and layered ground dynamic interaction under train passage. J. Eng. Mech. Div. **131**(7), 699–711 (2005)
4. Bian, X.C., Jiang, H., Chang, C., et al.: Track and ground vibrations generated by high-speed train running on ballastless railway with excitation of vertical track irregularities. Soil Dyn. Earthq. Eng. **76**, 29–43 (2015)

5. Gao, G.Y., Chen, Q.S., He, J.F., et al.: Investigation of ground vibration due to trains moving on saturated multi-layered ground by 2.5D finite element method. Soil Dyn. Earthq. Eng. **40**(3), 87–98 (2012)

6. Romero, A., Tadeu, A., Galvín, P., et al.: 2.5D coupled BEM–FEM used to model fluid and solid scattering wave. Int. J. Numer. Methods Eng. **101**(2), 148–164 (2015)

7. Kuwano, R., Wicaksono, R.I., Mulmi, S.: Small strain stiffness of coarse granular materials measured by wave propagation. In: Proceedings of 4th International Symposium on Deformation Characteristics of Geomaterials, pp. 749–756. IS-Atlanta (2008)

8. Nishimura, S.: Cross-anisotropic deformation characteristics of natural sedimentary clays. Géotechnique **64**(12), 981–996 (2014)

9. Zymnis, D.M., Whittle, A.J., Chatzigiannelis, I.: Effect of anisotropy in ground movements caused by tunnelling. Geotechnique **63**(13), 83–102 (2013)

10. Ai, Z.Y., Ren, G.P.: Dynamic analysis of a transversely isotropic multilayered half-plane. Soil Dyn. Earthq. Eng. **83**, 162–166 (2016)

11. Abedrrahim, H.: Coupling of finite and hierarchical infinite elements: application to a non-homogeneous cross-anisotropic half-space subjected to a non-uniform circular loading. Intern. J. Numer. Anal. Methods Geomech. **37**(15), 52–73 (2012)

Numerical Study on Settlement of Building with Shallow Foundation Under Earthquake Loading

Hao Wang[1], Guangyun Gao[1(✉)], and Jian Song[2]

[1] Key Laboratory of Geotechnical and Underground Engineering of Ministry of Education, Tongji University, Shanghai 200092, China
gaoguangyun@263.net
[2] Key Laboratory of Ministry of Education for Geomechanics and Embankment Engineering, College of Civil and Transportation Engineering, Hohai University, Nanjing 210098, China

Abstract. Earthquake-induced ground shaking can cause settlement in free-field sites and subsidence of ground structures. In recent years, numerous cases of liquefaction-induced settlement of buildings with shallow foundations have been observed on liquefiable soils. To investigate the soil liquefaction and building settlement, a three-dimensional (3D) soil-foundation-structure numerical model is built to simulate the development of pore water pressure and soil deformation in liquefiable ground during earthquakes. The bounding surface constitutive model and variable permeability function are adopt in numerical simulation with the finite element program OpenSees. The model is validated by the results of a centrifuge shaking table test under earthquake loading from a literature, and the simulation accuracy of pore water pressure and ground settlement in free-field are evaluated. On the basis of the 3D numerical model, the seismic responses of shallow foundation are simulated to certify the effectiveness of numerical simulation of foundation settlement and pore water pressure underneath the center of foundation. The soil-foundation-structure-interaction (SFSI) can aggravate the foundation settlement, foundation inclination and the instability of development of pore water pressure at sites adjacent to the foundation.

Keywords: Liquefaction · Foundation settlement · OpenSees
Earthquake loading · Bounding surface constitutive model
Variable permeability function

1   Introduction

Liquefaction, which could induce massive building destruction (e.g., settlement and tilt of building) is frequently caused by earthquake motions in the liquefiable ground. To investigate the mechanism of liquefaction-induced building settlement, many shaking table and centrifuge experiments, reproducing the behaviors of soils and structures in the actual sites have been conducted (e.g., Pyke et al. 1975; Liu and Dobry 1997; Hausler 2002; Ueng et al. 2010; Dashti et al. (2009a, b)). In these tests, many factors

© Springer Nature Singapore Pte Ltd. 2018
T. Qiu et al. (Eds.): GSIC 2018, *Proceedings of GeoShanghai 2018 International Conference: Advances in Soil Dynamics and Foundation Engineering*, pp. 251–260, 2018.
https://doi.org/10.1007/978-981-13-0131-5_28

influencing the buildings with shallow foundation have been analyzed, including ground motion characteristics, property of soil masses and structure features.

Apart from these well-documented experimental researches, numerical method is also an effective and practical tool to analyze the impact of key factors on soil lique-faction triggering, strength loss and deformation. With the development of soil con-stitutive model, simulation of soil's large post-liquefaction shear deformation has been realized in numerical studies (Wang et al. 2014). Many numerical models were set up and explored the effects of some factors, including the relative density of sand, the thickness of liquefied layer, the ground motion intensity parameters, structural stiffness etc. (Shahir and Pak 2010; Dashti and Bray 2012; Karamitros et al. 2012; Karimi and Dashti (2015, 2016)). These models takes soil-structure interaction into account and can represent some main aspects of the corresponding physical models in laboratory.

In this paper, to investigate liquefaction-induced settlement of building with shal-low foundation in liquefiable ground, 3D numerical model employing the unified plasticity constitutive model (Wang et al. 2014) and variable permeability function (Shahir et al. 2014) is established. The model is validated through data from centrifuge tests (Dashti et al. 2009a, b), and is further exploited to evaluate pore water pressure and foundation settlement in soils, and to study the building response.

2   Models and Parameters

2.1   Centrifuge Testing Overview

Dashti et al. (2009a, b) conducted a series of centrifuge tests to investigate the mechanism of seismically induced settlement of buildings with shallow foundations. In this paper, the case T3-30 of centrifuge experiment is utilized as the physical contrast experiment to build the 3D numerical model. The centrifuge acceleration is 55 g, and both experimental data and model sizes are prototype scale in this paper. As shown in Fig. 1, structure A represented the building with shallow foundation and was a single-degree-of-freedom (SDOF) structural model.

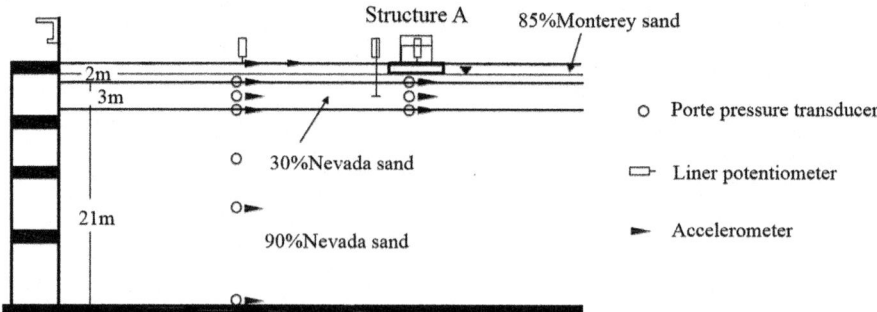

Fig. 1. Schematic diagram of centrifuge experiment model

There are three different layers of sand in total, the dense Nevada sand ($D_r \approx 90\%$, $e_{min} \approx 0.51$, $e_{max} \approx 0.78$) with prototype thickness of 21 m in the lower deposit, the loose Nevada sand ($D_r \approx 30\%$) with prototype thickness of 3 m in the intermediate deposit and the dense Monterey sand ($D_r \approx 85\%$, $e_{min} \approx 0.54$, $e_{max} \approx 0.84$) with prototype thickness of 2 m in the upper deposit. The data recorded at 15 m away from the structure A was regard as the free-field, since the influence from SFSI was negligible at this place. The solution used in the centrifuge experiment was equivalent to water.

The dimensions of the SDOF model were 6 m × 9 m × 5 m. The foundation base of SDOF model was founded at the depth of 1 m, with a contact pressure of 80 kPa. In the cases of T3-30M and T3-30L, two sets of earthquake motions were imported from the bottom of centrifuge model, as shown in Fig. 2. More details can be found in Dashti et al. (2009a, b).

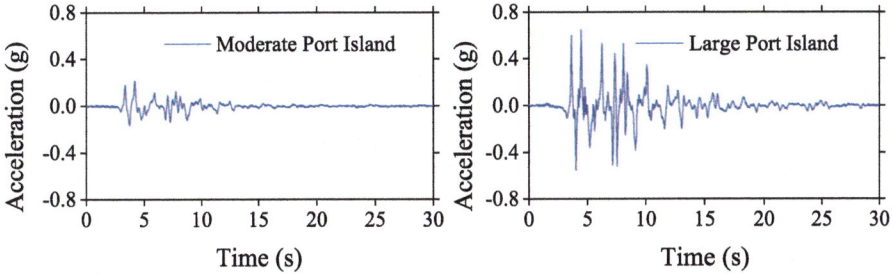

Fig. 2. Input earthquake motions

2.2 Numerical Model

Based on the open source finite element framework OpenSees, a 3D numerical model was built to capture the dynamic behaviors of soils and structure under earthquake loading. According to the centrifuge model, the 3D SFSI model was shown in Fig. 3. Dimension of this numerical model was 36 m × 27 m × 26 m, and the upper structure was 9 m × 6 m × 4.7 m. There are totally 4240 nodes and 3780 elements used in the 3D model.

The constitutive model CycliqCPSP at the OpenSees platform (Wang et al. 2014) was selected to simulate the seismic response of saturated sands in the 3D numerical model. The constitutive model can properly capture the large shear deformation of saturated sand under the zero effective stress state. The model parameters of Nevada sand are listed in Table 1. Monterey sand with a layer thickness of 2 m never liquefied during the earthquake loading in the test. Liquefaction and settlement of sand mainly occur in the intermediate layer. In addition, due to a lack of experimental data of Monterey sand, the precise CycliqCPSP parameters of Monterey sand is difficult to obtain but physical properties of Nevada sand and Monterey sand are similar. Therefore, the model parameters of Monterey sand were utilized as that of Nevada sand, which may have little influence on the computed results. The 8-noded BrickUP elements, used to model the soil elements in the simulation, were solid-fluid fully coupled

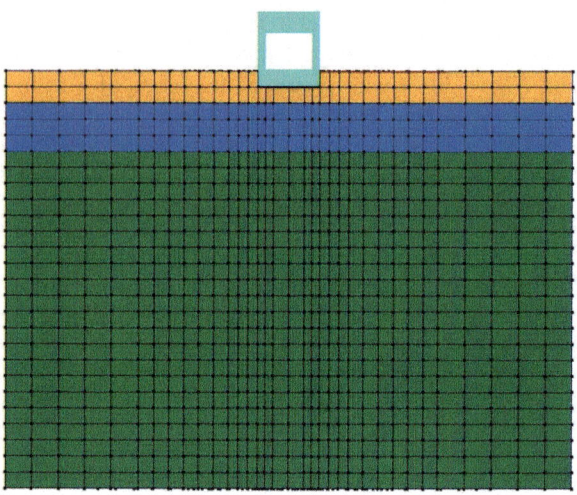

Fig. 3. Schematic diagram of numerical model

up elements. The SDOF structure was simplified to a linear elastic frame, and constitutive model of ElasticIsotropic material and stdBrick element were adopted to emulate the structure model. The equivalent material parameters are shown in Table 2.

Table 1. Parameters of CycliqCPSP constitutive model (Wang 2016)

G_0	κ	h	M	$d_{re,1}$	$d_{re,2}$	d_{ir}	α	$\gamma_{d,r}$	n^b	n^d	λc	e_0	ζ
225	0.004	1.7	1.35	0.8	30	0.6	10	0.05	1.1	8.0	0.029	0.843	0.7

The values of sand permeability coefficient were selected as that in Karimi and Dashti (2015, 2016). In the saturated sand deposit, cyclic loading can result in the growth of pore water pressure, with the decreasing in the void ratio. Then the permeability coefficient would vary significantly, which would in return affect the buildup of pore water pressure and drainage. Shahir and Pak (2010) and Shahir et al. (2014) proposed a variable permeability function, as shown in Eq. (1), which described the relationship between permeability coefficient and excess pore pressure ratio.

$$\frac{k}{k_0} = \begin{cases} 1 + (\alpha - 1)r_u^{\beta_1}, & r_u < 1 \text{ in build-up phase} \\ \alpha, & r_u = 1 \text{ in liquefied state} \\ 1 + (\alpha - 1)r_u^{\beta_2}, & r_u < 1 \text{ in dissipation phase} \end{cases} \qquad (1)$$

$$r_u = \Delta u / \sigma'_{v0} \qquad (2)$$

Where k is the permeability coefficient of sand, r_u is the excess pore pressure ratio, Δu is the excess pore water pressure, σ'_{v0} is the initial effective stress and α, β_1 and β_2

Table 2. Equivalent parameters of ElasticIsotropic material

	Young's modulus E(Pa)	Poisson's ratio υ	Density ρ(kg/m^3)
Column	6.8×10^7	0.35	2.7
Beam and foundation	2.72×10^4	0.35	1.3

are three positive constants related to materials. $\alpha = 10$, $\beta_1 = 2$ and $\beta_2 = 10$ were used in this paper, the same as that in Shahir et al. (2014).

The equalDOF command was used to connect soil nodes and structure nodes and to reproduce the boundary condition as the laminar shear model box. The steps of analysis included: (1) gravity applied on soil model without structure; (2) structure modeled on the soil model; (3) gravity applied on 3D model; (4) seismic loading; (5) the process of post-earthquake pore pressure dissipation.

3   Results and Discussions

3.1   Settlement and Pore Pressure in Free-Field (FF)

Figure 4 shows the comparison of ground surface settlement between numerical calculation and test data in FF. The simulation of settlement in this paper is more approximate to the experiment result in Dashti et al. (2009a) than that in Karimi and Dashti (2015). The reasons are that application of the unified plasticity constitutive model can represent the relationship between large shear strain and volumetric strain of post-liquefaction sand better, and the variable permeability function employed in FEM agrees more with actual situation than constant permeability coefficient.

As shown in Fig. 5, the development tendency of surface settlement and Arias intensity (I_a) are in very good consistency. I_a can be used as an intensity index of ground motion to be correlated with settlement.

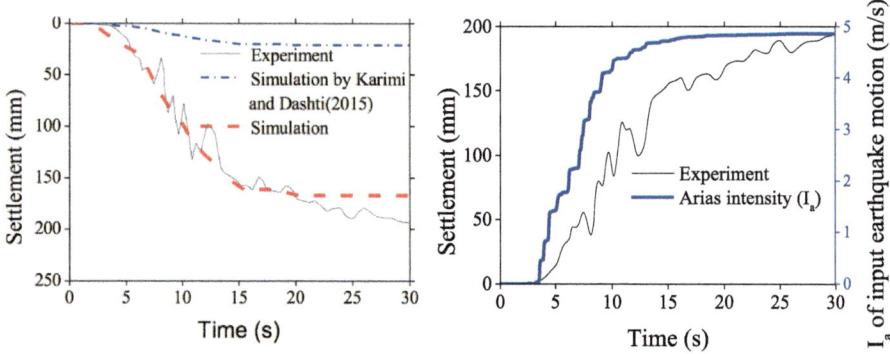

Fig. 4. Ground surface settlement (FF) (T3-30L)

Fig. 5. Comparison between settlement and Arias intensity (T3-30L)

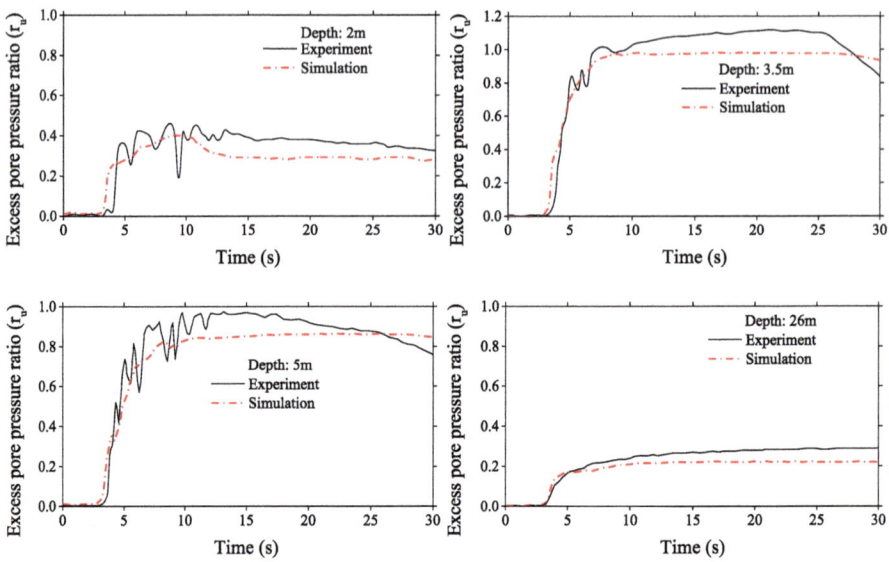

Fig. 6. Excess pore water pressure at different depths (T3-30M)

In Fig. 6, the simulation of excess pore pressure time histories at different depths was very close to that in experiment. The depth of 2 m is located at the interface between the upper layer and the intermediate layer, near the ground surface. Hence, the excess pore pressure rose relatively slowly, and the maximum of r_u only reached approximate 0.4. Depths of 3.5 m and 5 m are in the intermediate deposit, thus r_u reached 1.0 eventually and liquefaction occurred. The r_u at depth of 26 m was always relatively small, because a larger depth leads to greater effective stress, and liquefaction is more hardly to occur.

3.2   Response of Structure and Soil Adjacent to Foundation

Figure 7 presents the comparison between numerical simulation and centrifuge testing in the cases of T3-30M and T3-30L. Compared to surface settlement in Fig. 4, foundation settlement is much larger in T3-30L. One of the reasons is that SFSI can push soil under foundation into surrounding areas during earthquake loading, further aggravating foundation settlement. The gap between experiment and simulation in T3-30M is much narrower than that in T3-30L. The results indicate that the 3D numerical model performs better in moderate earthquake than in strong earthquake.

In Fig. 8, the pore water pressure did not accumulate and dissipate steadily. The closer to the foundation bottom the position is, the more significant the shocking of pore pressure curve is. And, even negative pore pressures appeared at depth of 2 m. From the results shown in Figs. 6 and 8, the SFSI had a significant influence on pore pressures underneath foundation. This can be explained by that SFSI induced larger soil deviatoric deformation and water absorption in shear, thus negative pore pressure happened. On the whole, the accuracy of simulation results of pore pressure is

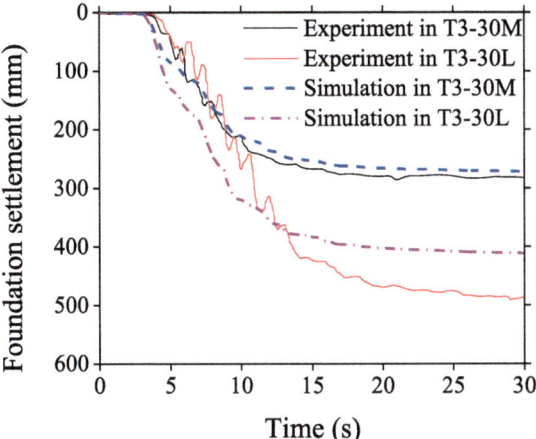

Fig. 7. Foundation settlement

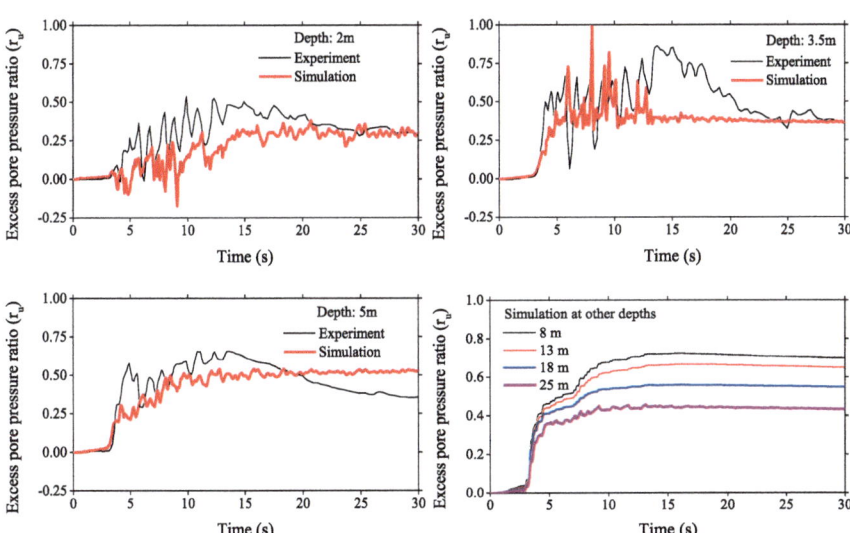

Fig. 8. Excess pore water pressure ratio at different depth underneath foundation (T3-30L)

acceptable. The simulation of pore pressure at other depths reveals that r_u decreased and the occurrence of liquefaction appeared more difficult with increasing depth, which is expected.

Figure 9 shows the acceleration time histories at different positions of structure A. From the bottom of 3D model to the top of structure A, the acceleration response was amplified gradually. PGA of top and foundation of structure A were 0.38 g and 0.30 g, corresponding to an amplification of 1.75 and 1.39 respectively.

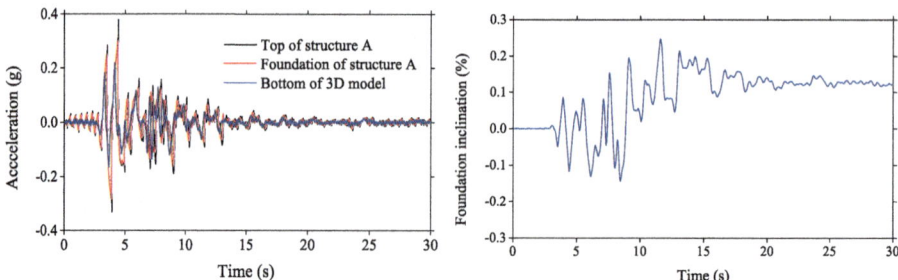

Fig. 9. Acceleration time histories of structure A (T3-30M)

Fig. 10. Foundation inclination of structure A (T3-30M)

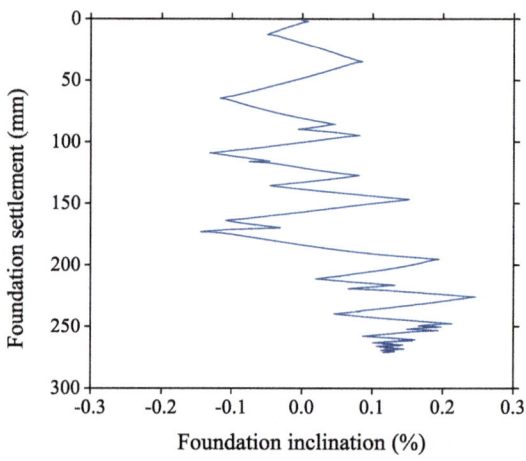

Fig. 11. Foundation inclination versus foundation settlement (T3-30M)

The variation of foundation inclination with time is shown in Fig. 10. The seismic inertial force acting on structure generated overturning moment, making the foundation swaying all the time. This rotation effect was closely related to SFSI, further resulting in additional soil deformation underneath foundation. This viewpoint can also be concluded from Fig. 11. The period of rapid subsidence of foundation was consistent with that when foundation rotated strongly and SFIS was manifested further in the process.

4   Conclusions

In this paper, a 3D FEM model employing the unified plasticity model and variable permeability function was established using the OpenSees finite element framework, aiming at capturing the seismic response of building with shallow foundation in liquefiable ground, especially the foundation settlement. Following conclusions can be drawn:

(a) Based on the comparison of settlement and pore water pressure between numerical simulation and centrifuge testing, the effectiveness and feasibility of 3D model is proved. Arias intensity may be used as a predictive parameter of site settlement in future, however, more data are still needed to verify this view.
(b) The validity of foundation settlements of 3D model is better in moderate earthquake than that in strong earthquake. The acceleration response of building is amplified obviously. There are Negative pore pressures observed underneath the center of foundation.
(c) The SFSI can aggravate soil deviatoric deformation, inducing larger foundation settlement and foundation inclination. Meanwhile, the instability of development of pore water pressure adjacent to the foundation is influenced intensely by SFSI.

Acknowledgments. This paper is supported by the National Natural Science Foundation of China (No. 41372271).

References

Dashti, S., Bray, J.D., Pestana, J.M., Riemer, M., Wilson, D.: Mechanisms of seismically induced settlement of buildings with shallow foundations on liquefiable soil. J. Geotech. Geoenviron. Eng. **136**(1), 151–164 (2009a)

Dashti, S., Bray, J.D., Pestana, J.M., Riemer, M., Wilson, D.: Centrifuge testing to evaluate and mitigate liquefaction-induced building settlement mechanisms. J. Geotech. Geoenviron. Eng. **136**(7), 918–929 (2009b)

Dashti, S., Bray, J.D.: Numerical simulation of building response on liquefiable sand. J. Geotech. Geoenviron. Eng. **139**(8), 1235–1249 (2012)

Hausler, E.A.: Influence of ground improvement on settlement and liquefaction: a study based on field case history evidence and dynamic geotechnical centrifuge tests. Ph.D. thesis, Univ. of California, Berkeley (2002)

Karamitros, D.K., Bouckovalas, G.D., Chaloulos, Y.K.: Insight into the seismic liquefaction performance of shallow foundations. J. Geotech. Geoenviron. Eng. **139**(4), 599–607 (2012)

Karimi, Z., Dashti, S.: Numerical and centrifuge modeling of seismic soil–foundation–structure interaction on liquefiable ground. J. Geotech. Geoenviron. Eng. **142**(1), 04015061 (2015)

Karimi, Z., Dashti, S.: Seismic performance of shallow founded structures on liquefiable ground: Validation of numerical simulations using centrifuge experiments. J. Geotech. Geoenviron. Eng. **142**(6), 04016011 (2016)

Liu, L., Dobry, R.: Seismic response of shallow foundation on liquefiable sand. J. Geotech. Geoenviron. Eng. **123**(6), 557–567 (1997)

Pyke, R., Seed, H.B., Chan, C.K.: Settlement of sands under multidirectional shaking. J. Geotech. Geoenviron. Eng. 101(ASCE# 11251 Proceeding) (1975)

Shahir, H., Pak, A.: Estimating liquefaction-induced settlement of shallow foundations by numerical approach. Comput. Geotech. **37**(3), 267–279 (2010)

Shahir, H., Mohammadi-Haji, B., Ghassemi, A.: Employing a variable permeability model in numerical simulation of saturated sand behavior under earthquake loading. Comput. Geotech. **55**, 211–223 (2014)

Wang, R., Zhang, J.M., Wang, G.: A unified plasticity model for large post- liquefaction shear deformation of sand. Comput. Geotech. **59**, 54–66 (2014)

Wang, R.: A unified plasticity model for large post-liquefaction shear deformation of sand and its numerical implementation. In: Single Piles in Liquefiable Ground. Springer, Heidelberg (2016)

Ueng, T.S., Wu, C.W., Cheng, H.W., Chen, C.H.: Settlements of saturated clean sand deposits in shaking table tests. Soil Dynam. Earthq. Eng. **30**(1), 50–60 (2010)

Seismic Response of Segmental Lining Tunnel by Using Shaking Table Test and Numerical Simulation

Juntao Chen[1(✉)], Weiguo He[2], Chaoye Song[2], Haitao Yu[1], and Yong Yuan[1]

[1] Tongji University, Shanghai 200092, China
chen1609@foxmail.com
[2] China Railway Tunnel Survey and Design Institute Co., Ltd., Tianjing 300133, China

Abstract. Inner lining is generally proposed to improve the earthquake reliability of segmental lining tunnel. This paper focuses on the seismic response of segmental lining in rock ground and the influence of inner lining on the dynamic performance of tunnel structure. A series of large-scale (1:10) shaking table tests and numerical simulations are carried out to simulate the seismic response of the segmented tunnel with and without inner lining in rock mass. Key technical details of the test setup are introduced with particular focuses on: shaking table system, similitude design and relations between real tunnel and scaled model, modeling method, fabrication of model, sensor layout and input motions. In addition, a series of cases are conducted to study the influence of earthquake intensity, input motion and input direction on the seismic performance of tunnel section. Besides, numerical simulations corresponding to the test cases are conducted to validate the test result and to further study the seismic performance of segmental lining tunnel. Based on the result, peak acceleration, Fourier spectrum and joint extension are analyzed. Close attentions are paid to the effect of inner lining on segmental linings. Results indicate that inner lining leads to higher peak acceleration at transverse direction due to the increase of stiffness. In addition, the inner lining reduces joint extension significantly.

Keywords: Segmental lining · Inner lining · Seismic performance
Shaking table test · Numerical simulation

1 Introduction

Assemblage of prefabricated segments forms lining ring of a tunnel, which provides safe and efficient construction. It is considered as flexible lining system with respect to cast-in-place lining of tunnel, to withstand earth pressure from surrounding strata. Since the dynamic performance of tunnel under earthquake is different from that of tunnel under normal loads, a series of efforts had been put on the investigating dynamic behavior of tunnel.

A typical example of these researches was the seismic behavior of shield tunnel [1]. In which shaking table test was conducted to investigate the transverse response of the

© Springer Nature Singapore Pte Ltd. 2018
T. Qiu et al. (Eds.): GSIC 2018, *Proceedings of GeoShanghai 2018 International Conference: Advances in Soil Dynamics and Foundation Engineering*, pp. 261–269, 2018.
https://doi.org/10.1007/978-981-13-0131-5_29

model tunnel. To address the issue multi-scale method was proposed to simulate the seismic performance of assemblages of lining rings [2]. With the help of supercomputer this multi-scale method in the 3-D numerical simulation was successfully applied on studying the seismic response of water convey tunnel. It is revealed that the extension of ring joint and 'ovaling' deformation in transverse direction under strong earthquake should be considered. Besides, a series of experiments was carried out to investigate the dynamic behavior between shield tunnel and working shaft, by making use of large-scale E-Defense shake table at Hyogo Earthquake Engineering Research Center [3]. Besides, numerical simulations were conducted to simulate the shaking table test [4]. Numerical simulations were also conducted to validate the longitudinal rigidity of model tunnel [5].

To date, only few researches have been carried out to study the seismic response of segmental lining tunnel in rock mass and inner lining. Therefore, in this paper, a series of the shaking-table tests and numerical simulations are carried out to study the seismic performance of segmental lined tunnel with inner lining in rock mass. The physical model is composed of segmental rings and surrounding rock. Segmental lining is supposed to be installed in rock mass; meanwhile, comparison portion of the tube with inner lining is also located in the same rock mass. In addition, a series of study cases with different earthquake motions and intensities are fully considered.

2 Prototype

2.1 Engineering Geology

Tunnel might cross strata in soft ground or in rock mass. For instance, a typical longitudinal profile of the surrounding ground is shown in Fig. 1(a), where a subway tunnel is located in rock mass. Figure 1(b) gives the shearing velocity of each stratum taken from geological survey. Generally, the soil stratum is around 10.0 m in thickness, and its shearing velocity is relatively small, about 260.0 m/s. Below the soil is the stratum of 5.0 m thick weathered rock. The shearing velocity within this depth increases to 600 m/s. At last, is much stronger stratum of rock mass and the shearing velocity of it reaches to 1200.0 m/s.

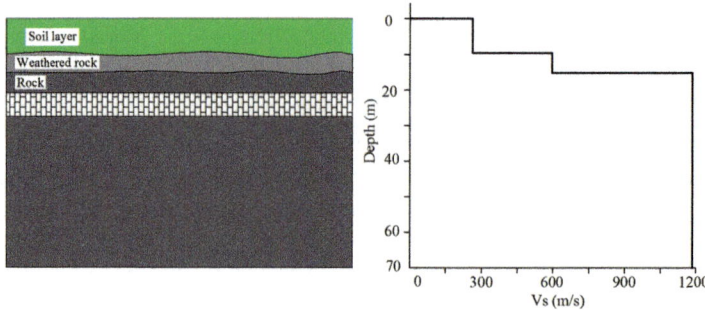

Fig. 1. Typical geological condition of a segmental lining tunnel: (a) longitudinal profile and (b) shearing velocity of ground stratum.

The tunnel structure is assembled by a number of segments both at longitudinal and transverse direction. Therefore, compared to uniform tubes with continuous lining, the segmented tunnel has relatively lower capability to earthquake loadings. As a result, to improve the static and dynamic performance of tunnel structure, a concrete inner lining is proposed to install at the inside of segments. The segmented ring and inner lining have the thickness of 350 mm and 250 mm respectively.

3 Seismic Study

3.1 Input of Earthquakes

Intensity of Earthquake. In this study, input scenarios can be grouped according to the intensity of ground motion. Two design levels of earthquake intensities were considered, including DE (Design Earthquake) and SE (Strong Earthquake). It is assumed that Peak Ground Acceleration (PGA) of DE is 0.15 g at the level of 10% probability of exceedance in 50 years. For the SE, the PGA is assumed to be 0.31 g at the level of 2% of probability of exceedance in 50 years. Therefore, in this study, all the peak acceleration of input motions was amplified to 0.15 g and 0.31 g.

Characteristics of Earthquake. In this study, the in-site synthetic motion is one of the most important earthquake motions, which was given from Evaluation of Seismic Safety for the engineering site. According to the in-site geological condition and the regional seismic activity, the synthetic motion was generated based on the results of a probabilistic seismic hazard analysis and numerical simulation method according to conditions and seismic motion characteristics at the tunnel site. Besides, the natural earthquake motion from rock ground, Wenchuan Wolong motion, is also considered in this study.

Direction of Excitation. Former researches have proved that the most demanding situation is applying input motions both at the horizontal longitudinal and transverse direction. Therefore, both transverse and longitudinal input motions were firstly considered in this study. Besides, single input cases are also conducted to study the influence of transverse and longitudinal motions on the seismic performance of discontinuous segmented tunnel. The vertical earthquake motion has small influence on the seismic performance of tunnel structure that can be neglected. Therefore, three kinds of input scenarios were considered in this study, both longitudinal and transverse motions, single longitudinal motion and single transverse motion. The test cases and details are listed in Table 1.

3.2 Shaking-Table Tests

Law of Similarity and Model Material. The law of similarity between prototype and test model can be derived based on the Buckingham-π theorem and related regulations on model experiment in geotechnical engineering, which has been widely used for designing the shaking table test by other researchers. At the beginning, the basic

Table 1. Test cases of the test.

Test case	Intensity	Input motion	PGA	Main frequency	Input direction
1	DE	Synthetic	0.15 g	0.316–17.4	X
2	DE	Synthetic	0.15 g	0.316–17.4	Z
3	DE	Synthetic	0.15 g	0.316–17.4	XZ
4	DE	Wenchuan	0.15 g	2.5–7.5	XZ
5	SE	Synthetic	0.31 g	0.316–17.4	X
6	SE	Synthetic	0.31 g	0.316–17.4	Z
7	SE	Synthetic	0.31 g	0.316–17.4	XZ
8	SE	Wenchuan	0.31 g	2.5–7.5	XZ

similarity ratio including geometry S_l, Young's modulus S_E and density S_ρ should be determined first, and the others can be deduced by the similarity theory. In this study, the large-scale test with the geometry ratio of 1:10 is proposed to better simulate the complex components of segmented tunnel. The similitude relations and ratios used in the test are shown in Table 2. To achieve the requirement of the law of similarity model materials are proposed to simulate the structure. For instance, foam concrete is used to simulate the real concrete. The elastic ratio of the two materials is 1/50. In addition, the dynamic parameters including amplitude, frequency, and velocity should also be scaled to meet the law of similarity.

Table 2. Similitude relations and ratios developed for the shaking table test

Items	Symbol and relational expression	Similarity ratio
Geometry	S_l	1/10
Strain	S_ε	1
Young's modulus	S_E	1/50
Unit weight	S_ρ	1/5
Mass	$S_m = S_\rho\, S_l^3$	0.0002
Velocity	$S_v = S_E^{1/2}\, S_\rho^{-1/2}$	0.316
Time	$S_t = S_l\, S_E^{-1/2}\, S_\rho^{1/2}$	0.316
Frequency	$S_w = S_E^{1/2}\, S_l^{-1}\, S_\rho^{-1/2}$	3.16
Stress	$S_\sigma = S_E$	0.02
Acceleration	$S_a = S_E\, S_l^{-1}\, S_\rho^{-1}$	1

Layout of Monitoring Sensors. The dynamic information, i.e. acceleration, strain, contact pressure between surrounding rock and structure, deformation of the segment joints, etc. was collected by a set of sensors during shaking table tests. Figure 2 shows the layout of the monitoring sections for different kinds of sensors. Note that there are three kinds of vertical lines arranged along the tunnel axis. The vertical lines show that one, two and three kinds of sensors are installed in the section. In addition, the short lines with 'D1' or 'D2' at the bottom of shield tunnel represent the sections for recording the joints' deformation. The terminology used to describe the sensors is:

A, S, P and D are for the acceleration, strain, pressure and deformation section respectively. In addition, the number '1' means section one, such as P2 indicating the pressure Sect. 2.

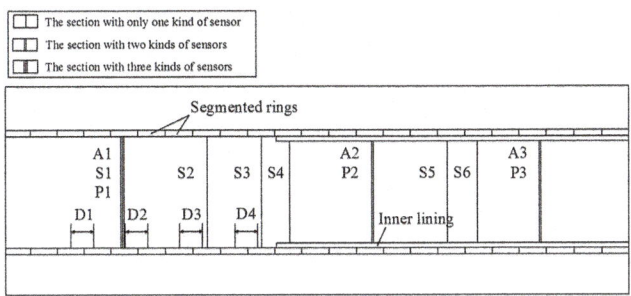

Fig. 2. Layout for the sections with different sensors

3.3 Numerical Simulation

Apart from the shaking table test, the numerical simulation is another most important research approach to study the seismic response of underground structure, because more cases and different excitation scenarios can be completed by using numerical simulation. Therefore, in this study, a series of full three-dimensional analyses were also carried out in an attempt to investigate seismic response of segmented tunnel and validate the shaking table test result. Figure 3 shows the whole model, which includes the surrounding ground, tunnel segments and longitudinal bolts. All the components are discretized with eight-node hexahedral solid elements. The finite element model is built according to numerical platform ABAQUS. The C3D8R solid element is used to model the tunnel and surrounding rock. The whole mesh includes 17,3192 elements. As can be seen in the figure, the details of ring segments and bolts are also simulated. The interaction between two adjacent ring segments' surfaces is simulated by 'hard contact', which indicates that sliding and limited separation is allowed but no penetration.

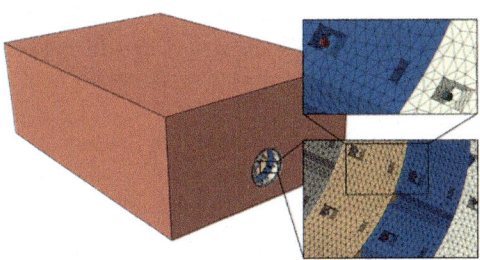

Fig. 3. Numerical model of segmented tunnel and surrounding rock

Note that the material parameters used in the numerical simulation are the same to the shaking table test, as shown in Table 3. In addition, the tunnel structure is located in relatively good surrounding ground, only elastic behavior of the materials is considered. Thus, elastic constitutive models are employed to simulate the rock and concrete.

Table 3. Material parameters used for numerical simulation

Items	Young's modulus (MPa)	Possion's ratio	Unit weight (kN/m³)
Rock model	0.11×10^3	0.32	460
Shield tunnel segment	0.71×10^3	0.2	950
Inner lining	0.67×10^3	0.2	930

4 Results

A number of test results are obtained from the experiment and numerical simulation. To provide a global understanding on the seismic performance of the tunnel and yet limit the amount of data included in the paper, only the acceleration response, spectrum and joint extension are discussed. In addition, in this paper, close attentions are mainly paid to making comparisons between the tunnel sections with and without inner lining.

4.1 Acceleration

Acceleration response is firstly proposed to study the influence of inner lining on shield tunnel structure. To limit the amount of data, only peak acceleration and Fourier spectrum of section A1 (without inner lining) and A2 (with inner lining) are discussed in this section. In addition, only the key locations including arch crown at the top of tunnel cross-section and invert at the bottom of tunnel cross-section are studied. Figure 4(a) and (b) show the peak accelerations at horizontal X and Z direction respectively. The plots indicate a good agreement between shaking table test and numerical simulations, even the peak value of numerical simulation is little bit larger than that of experiment. Moreover, for both invert and arch crown, the peak acceleration of the section with inner lining is relatively larger than the peak value without inner lining, particularly at the horizontal X direction. However, the inner lining plays little influence on the acceleration response at Z direction. The inner lining increases the transverse stiffness significantly, which also raises the basic vibrational frequency of tunnel structure. Therefore, the stiffer tunnel section with inner lining leads to higher peak acceleration.

Figure 5(a) and (b) gives the Fourier spectrum of the two tunnel sections at X and Z direction respectively. Figure 5(a) shows that, for the tunnel section with and without inner lining, the major amplitude is mainly distributed at the frequency range from 0 Hz to 25 Hz, and the main frequency is the same, 3.3 Hz. However, compared to the segmented tunnel without inner lining, the amplitude of the tunnel section with inner lining is relatively larger, especially from 1 Hz to 8 Hz and 12 Hz to 27 Hz. By contrast, for the Fourier spectrum at Z direction, the two plots show little differences

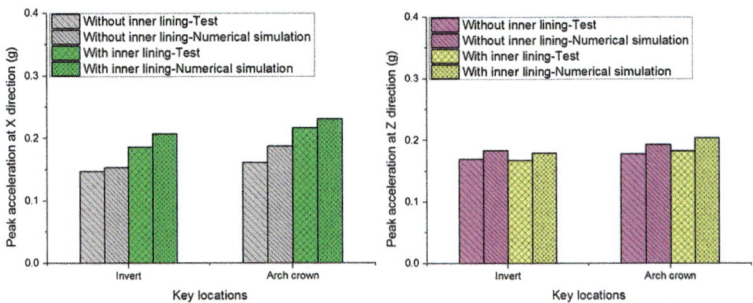

Fig. 4. Peak acceleration: (a) In X direction, (b) In Z direction.

between each other. The result indicates that the inner lining generally enlarges the dynamic response at transverse direction both on peak acceleration and Fourier amplitude; while, the inner lining has little effect on the response at longitudinal direction.

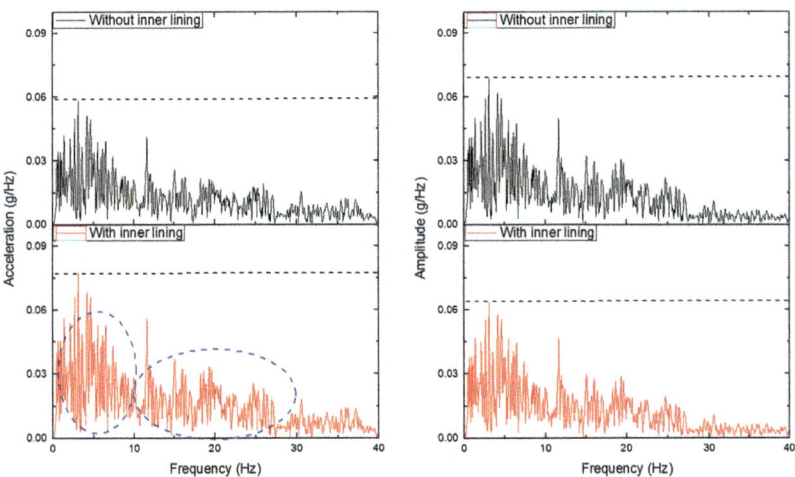

Fig. 5. Fourier spectrum: (a) In X direction, (b) In Z direction.

4.2   Extension of Joint

Joint extension of shield tunnel is one of the most important factors to evaluate the safety and stability of tunnel structure. Note that it is hard to monitor the joint deformation of the segmental lining with inner lining, which is between surrounding ground and inner lining. Therefore, the joint extension of tunnel with inner lining is obtained from numerical simulation. Figure 6(a) and (b) show the peak extension of key locations under different earthquake intensities, DE and SE, namely case 3 and case 7 in Table 1. Obviously, compared to the tunnel section with inner lining, the peak extension of the tunnel without inner lining is much larger. The peak value is about 3 times to that with inner lining. Besides, for the two sections, the distributions of peak extension are different. The large extension generally occurs at the spandrel, abutment and sidewall basement for the section without inner lining; while, larger extensions mainly appear at invert and abutment for the other portion. Therefore, the result indicates that the inner lining decreases the joint extension significantly.

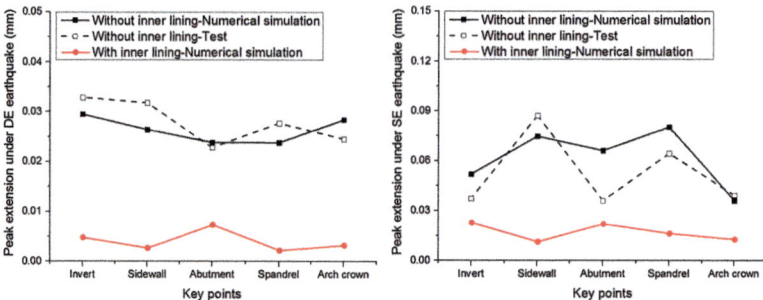

Fig. 6. Peak extension of segment joints: (a) DE, (b) SE.

5   Summary

A series of shaking table tests and numerical simulations are conducted to study the seismic response of segmental ling tunnel in rock mass and the influence of inner lining on the seismic performance of segmental tunnel. The large-scale shaking table test with scale ratio of 1:10 is able to better model complex structures, such as segmented rings and joints. Detailed information on experimental setup, similitude design, fabrication of model system, and simulation of different earthquake excitations has been fully addressed and would provide the references for further similar investigations, especially the longitudinal and transverse stiffness equivalent methods for design segment models. Meaningful data are obtained and discussed from tests and numerical simulations, i.e. acceleration responses, spectrum and joint extension.

References

1. He, C., Koizumi, A.: Dynamic behavior in transverse direction of shield tunnel with considering effect of segment joints. In: Proceedings of 12th World Conference on Earthquake Engineering, Auckland, New Zeland (2000)
2. Yu, H.T.: Multiscale method for long tunnels subjected to seismic loading. Int. J. Numer. Anal. Meth. Geomech. **37**(4), 374–398 (2013)
3. Kawamata, Y.: Dynamic behaviors of underground structures in E-Defense shaking experiments. Soil Dyn. Earthq. Eng. **82**, 24–39 (2016)

4. Kandeda, K., Shigeno, Y., Suzuki, T., et al.: Numerical simulation of underground structure behaviors using E-Defense large shaking table. In: Proceedings of the 16th World Conference on Earthquake Engineering, 16WCEE 2017, Santiago, Chile (2017)
5. Bao, Z.: Longitudinal rigidity of shield tunnels based on numerical investigation. In: SEE Tunnel: Promoting Tunneling in SEE Region, ITA WTC 2015 Congress and 41st General Assembly, Croatia (2015)

Study on Pounding Effect Between Long Span Suspension Bridge and Approach Bridge Under Earthquake

Kun Yan[1,2,3(✉)], Lixin Wang[1,2,3], and Jiajian Zhu[1,2,3]

[1] Guangdong Earthquake Agency, Guangzhou 510070, China
kevin_yan12@hotmail.com
[2] Key Laboratory of Guangdong Province, Earthquake Early Warning and Safety Diagnosis of Major Projects, Guangzhou 510070, China
[3] Key Laboratory of Earthquake Monitoring and Disaster Mitigation Technology, Guangzhou 510070, China

Abstract. Expansion joints are important connecting members between long-span suspension bridge and approach bridge. Since the longitudinal vibration is coupled between the suspension bridge and approach bridge under a strong earthquake, accidences of pounding and falling may happen between the girders on both sides of the expansion joints. In this study the influence of pounding effect on seismic response of bridge is analyzed by nonlinear time history seismic method based on a real long-span suspension bridge. The influences of pounding stiffness and initial width of expansion joint are both considered in this paper. The results show that pounding stiffness has a great impact on the bending moment at the bottom section of piers and the longitudinal displacement of approach bridge. Meanwhile the analysis tells that different anti-push stiffening between two approach bridges will lead to a variety of pounding effect. Under the designed earthquake intensity, long-period seismic earthquake creates a very large pounding force between stiffening girder of suspension bridge and box girder of approach bridge, which may cause local damage of the girders.

Keywords: Suspension bridge · Approach bridge · Pounding effect
Expansion joint

1 Introduction

Strong earthquake has always been one of the major natural disasters that seriously threaten human society. Due to the randomness and suddenness of earthquakes, the consequences are often catastrophic and often result in huge casualties and economic losses. As a key part of the lifeline project, the anti-seismic performance of long-span suspension bridge is an important foundation for ensuring the structural safety and the unobstructed transportation channel in the earthquake relief [1]. The seismic responses of long-span suspension bridges and its approach bridges is interdependent because of the coupling effect of expansion joints. The longitudinal seismic response of suspension bridge and approach bridge may cause a pounding between stiffening girder of

© Springer Nature Singapore Pte Ltd. 2018
T. Qiu et al. (Eds.): GSIC 2018, *Proceedings of GeoShanghai 2018 International Conference: Advances in Soil Dynamics and Foundation Engineering*, pp. 270–278, 2018.
https://doi.org/10.1007/978-981-13-0131-5_30

suspension bridge and box girder of approach bridge, even damages and collapses. Pounding damage of bridges and buildings have appeared widely during the moderate or strong earthquakes in the past several decades. Such as many bridge decks and abutments were damaged by collisions in 1971 San Femando earthquake in the United States. In the 1994 Northridge earthquake in the United States, the bridge pier and expansion joint of the highway bridge, 14 km from the epicenter, had severely collided and damaged.

A large number of scholars have studied the pounding effect of continuous beam bridge under earthquake action [2–5]. The results show that the gap of expansion joint, mass ratio of the pounding bodies, and natural period ratio of adjacent bridge have a certain influence on seismic response of the continuous beam bridge. However, there are relatively few studies on the pounding between long-span bridges and approach bridges at expansion joints. In order to reveal the influence on seismic response of pounding between the stiffening girder of suspension bridge and the adjacent approach bridge at the extension joint, taking a long-span suspension bridge as engineering background, the nonlinear finite element model considering pounding effect is established. And the effects of pounding stiffness, initial gap of expansion joint and other factors on the pounding effect are studied, and the influence of each parameter on the pounding effect is discussed.

2   Finite Element Model of Long - Span Suspension Bridge

2.1   Finite Element Model

A single span suspension which middle span is 1108 m and its adjacent approach bridges are considered here as the research object. And the left and right approach bridges are both six-span continuous rigid frame bridge with 62.5 m span. Finite element model of the object is shown in Fig. 1. In this model the bridge pylons, stiffening girders, box girders and bridge piers are simulated by beam elements, while the main cables and hangers are simulated by tension-only bar element whose stress stiffening effect is also considered. The supporting system of the steel box girder of suspension bridge is arranged as follows: a pair of vertical bearings, a pair of transverse wind support and a pair of longitudinal limit displacement devices are arranged at the ends of stiffening girder. The expansion joint is arranged between stiffening girder of suspension bridge and box girder of approach bridge.

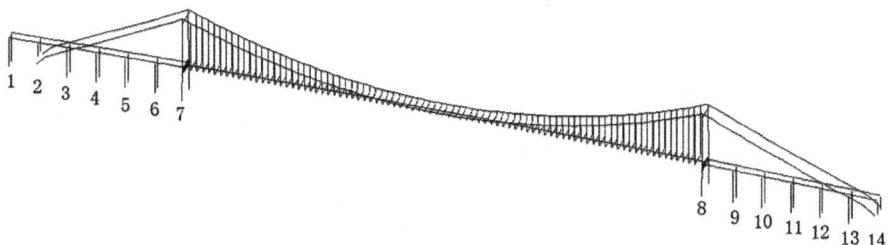

Fig. 1. Finite element model of suspension bridge and approach bridge

2.2   Dynamic Characteristics and Anti-push Stiffening

The dynamic characteristics and seismic response analysis of the suspension bridge are carried out on the basis of static geometric nonlinear initial equilibrium state. The natural frequencies of the bridge are presented in Table 1.

Table 1. Natural frequency of the bridge

Vibration parts	Order	Frequency/HZ	Mode	Vibration parts	Order	Frequency/HZ	Mode
Stiffening girder	1	0.069	1st L − S	Stiffening girder	5	0.200	1st V − S
	2	0.095	1st V − A + F		6	0.210	1st L − A
	3	0.136	1st V − A + F		14	0.297	1st T − S
	4	0.148	1st V − S		23	0.352	1st T − A
Right approach bridge	12	0.286	F	Left approach bridge	27	0.374	L − S
	18	0.319	L − S		40	0.511	F
	47	0.656	L − A		69	0.852	L − A

L–Lateral bending; V–Vertical bending; F–Vertical floating; T–Twist; S–Symmetrical; A–Antisymmetrical

According to the finite element model, longitudinal anti-push stiffness of stiffening girder of suspension bridge can be obtained as K_M = 6400 kN/m, and the total longitudinal anti-push stiffness of box girders of left and right approach bridges are K_L = 3.7 × 10^5 kN/m and K_R = 1.19 × 10^5 kN/m respectively. It can be found that although these two approach bridges have the same span, difference between their longitudinal anti-push stiffening is significant. And it is probably caused by the inconsistency of arrangement of pier and bearing. The height of the piers of left approach bridge placed on the anchors is short and consolidated with box girder. As the longitudinal anti-push stiffening is controlled by the shortest pier, anti-push stiffening of left approach bridge is significantly greater than that of right approach bridge.

2.3   Simulation of Pounding Element of Expansion Joint

In order to simulate the pounding effect of stiffening girder of suspension bridge and box girder of approach bridge, contact pounding elements are added at the expansion joints, as shown in Fig. 2. Kelvin-Voigt collision model is used to simulate the pounding between the adjacent rigid bodies, which is composed of a linear spring k_k and a damper c_k in parallel. The contact force during the collision is given by Anagnostopoulos [6] as follows,

$$\begin{cases} F_c = k_k(g_d - g_p) + c_k \dot{g}_d & g_d - g_p \geq 0 \\ F_c = 0 & g_d - g_p = 0 \end{cases} \tag{1}$$

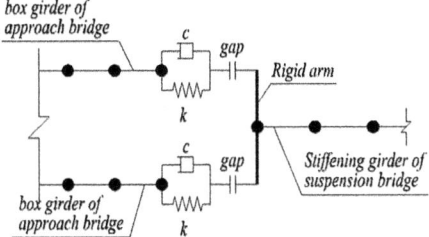

Fig. 2. Pounding element model

where g_p is the initial gap of expansion joint, g_d is the relative displacement of the adjacent girder at expansion joint under the earthquake action and k_k is contact stiffness which is the axial stiffness of the shorter girder.

According to the law of conservation of energy, the relationship between damping coefficient c_k and the coefficient of restitution e can be established as follows,

$$c_k = 2\xi\sqrt{k_k\frac{m_1 m_2}{m_1 + m_2}} \text{ and } \xi = \frac{-\ln e}{\sqrt{\pi^2 + (\ln e)^2}} \tag{2}$$

where m_1 and m_2 are the masses of two colliding bodies respectively.

As shown in Table 2 ten typical seismic waves are selected as the longitudinal seismic input here. It is assumed that the bridge is located in the 8 degree seismic intensity zone, considering the rare earthquake, so the peak value of each seismic wave is adjusted to 0.4 g. And it should be noted that each seismic response presented in the follow analysis is the average value of the calculated results obtained from these 10 seismic waves. Since the predominant periods of the selected seismic waves range from 0.24 s to 1.42 s, the frequency band of the input is wide enough, and the analysis results will be universal and representative.

Table 2. Typical seismic waves

No.	Earthquake	Station	Mw	PGA/g	Tg/s
1	1940 Imperial Valley	EL Centro	6.95	0.281	0.46
2	1952 Kern County	Taft Lincoln School	7.36	0.159	0.36
3	1989 Loma Prieta	Coyote Lake Dam	6.93	0.485	0.64
4	1994 Northridge	Hollywood	6.69	0.251	0.88
5	1994 Northridge	Old Ridge Route	6.69	0.514	0.54
6	1995 Kobe	Nishi-Akashi	6.9	0.483	0.46
7	1995 Kobe	Shin-Osaka	6.9	0.225	0.66
8	1999 Chi-Chi	TCU052	7.62	0.447	1.14
9	1999 Chi-Chi	TCU068	7.62	0.371	1.42
10	1999 Chi-Chi	TCU129	7.62	1.005	0.24

3 Influence of Pounding Stiffness on Pounding Effect

Pounding stiffness is a very important parameter in the analysis of pounding effect [4]. Many literatures suggest that axial stiffness of the pounding body is chosen as pounding stiffness [7]. However, the length of box girder of approach bridge and stiffening girder of suspension bridge reach 370 meters and 1100 meters respectively. Axial stiffness of box girder and stiffening girder is small, and the maximum is 3.7×10^5 kN/m, which is not suitable as the pounding stiffness. In order to further investigate the influence of pounding stiffness on pounding effect, pounding stiffness of expansion joint is chosen within a certain range (between 1×10^5 kN/m and 2×10^7 kN/m). The initial gap of expansion joints taken as 0.3 m.

The minimum relative displacements of the girders on both sides of expansion joints are shown in Fig. 3, where the pounding stiffness of 0 means that the pounding between the girders is not taken into account. It can be seen that the relative displacement is very sensitive to pounding stiffness of expansion joints. The relative displacement will decrease as the increase of pounding stiffness, and at last tends to the initial gap. If the pounding stiffness is closed to the axial stiffness of girder, the relative displacement value will exceed the maximum value of the initial gap, which indicating that the expansion joint is elastically compressed.

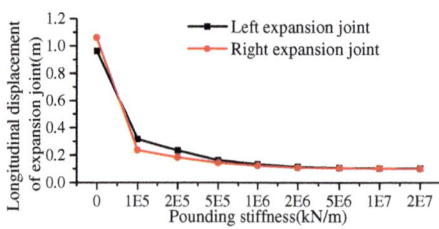

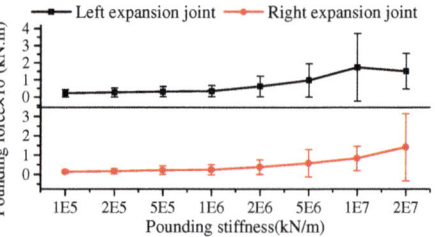

Fig. 3. The minimum value of relative displacement of expansion joint

Fig. 4. Pounding force of expansion joint

The pounding force and its standard variance with different pounding stiffness are shown in Fig. 4. It can be seen that pounding force increases with pounding stiffness increases. And when the pounding stiffness is small, standard variation of pounding force is small. When the pounding stiffness k_k is greater than 1×10^6 kN/m, the pounding force increases abruptly with the increase of pounding stiffness, and the discretization of the results calculated by each seismic wave is also larger. The maximum pounding force of left expansion joint under the TCU068 seismic wave is 6.07×10^5 kN, and average stress of girder section under the pounding force is calculated in the following analysis. The cross-sectional area of girder stiffening of suspension bridge is 1.56 m^2. It can be calculated that the average stress of stiffening girder section is 389.2 MPa under the action of the maximum pounding force, which has surpassed the yield strength of the stiffening girder steel Q345C. The cross-sectional area of box girder of approach bridge is 13.4 m^2, and the average stress of box girder section is 45.4 MPa under the maximum pounding force, which is close to the compressive strength of concrete C50. Because TCU068 belongs to long-period seismic wave, the seismic response of suspension bridge is very obvious and the pounding force is also very large. A large pounding force may cause local damages to box girders and stiffening girder at both ends of expansion joints, which requires sufficient attention.

Bending moment at the bottom section of pier 2# and 12# of approach bridge and bending moment at the bottom section of pylon 7# and 8# of suspension bridge are shown in Figs. 5 and 6 respectively. It can be seen that bending moment at the bottom of pier and pylon with pounding is slightly different from that without pounding. Considering the pounding effect, bending moment at the bottom section of pier 2# of left approach bridge is reduced, while bending moment of pier 12# of right approach bridge is increased. It can be seen from Fig. 5 that pounding stiffness of expansion joints has little effect on bending moment of the pylon. The main reason is that the stiffening girder of suspension bridge is a floating system, and only a small part of the pounding force is transmitted to the bridge pylon. Therefore the pounding has no significant effect on the internal force of the bridge pylon.

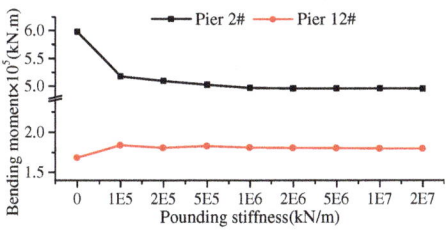

Fig. 5. Bending moment of piers

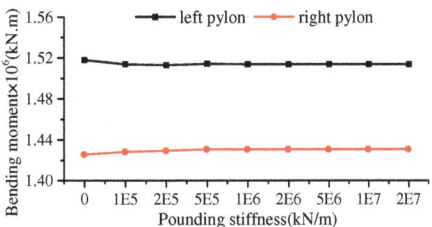

Fig. 6. Bending moment of pylons

4 Influence of Initial Width of Expansion Joint on Pounding Effect

The impact of the initial width of expansion joint on the pounding effect and relative displacement of adjacent girder under the action of earthquake is worth studying because it relates to the design of the expansion joint. So seismic response with pounding is analyzed by changing the initial width of the expansion joints in this section. It is assumed that initial width of expansion joint ranges from 0.05 m to 0.3 m [8]. The maximum longitudinal displacements of the approach bridge and the top pylon of suspension bridge are shown in Figs. 7 and 8 respectively. It can be seen that the effect of initial gap of expansion joint on longitudinal displacement of approach bridge is greater than that at the top of pylon. However the influence of the initial gap of expansion joints on the longitudinal displacement of the approach bridge is also very

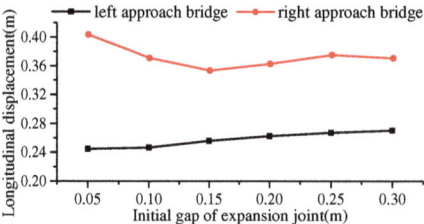
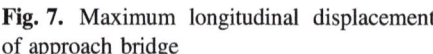
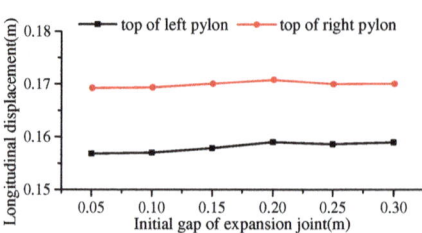

Fig. 7. Maximum longitudinal displacement of approach bridge

Fig. 8. Maximum longitudinal displacement at the top of pylon

weak. Therefore the changes of longitudinal displacement of the top of pylon are small whether considering the pounding and initial width of the expansion joints or not.

Bending moment at the bottom section of approach bridge pier with different initial width of expansion joints is shown in Fig. 9. It can be seen that the bending moment of pier 2# of left approach bridge increases with the initial width of the expansion joints increases, while the bending moment of pier 12# of right approach bridge presents a contrary tendency. This is mainly due to the significant differences in the longitudinal anti-pushing stiffness of the approach bridges on both sides. Therefore not all seismic waves can cause the pounding between the stiffening girder and the approach bridge when the expansion joint width increases. And the number of pounding under each seismic wave is also different at each expansion joint. For example, 6 of the 10 seismic waves make the stiffening girder pounding with the left approach bridge when the width of expansion joint is 0.3 m, but only 5 waves make the pounding on the right.

Pounding force and its standard variance with different initial width is shown in Fig. 10, where the pounding force is not involved in calculation when pounding force is 0 with no pounding. It can be seen that with the increase of initial width, the pounding force of left expansion joint does not change obviously, while the dispersion of it increases. The pounding force on right side has almost no change with different initial width, except pounding force is the largest when initial width is 0.05 m. And the dispersion of pounding force is reduced when the initial width increases.

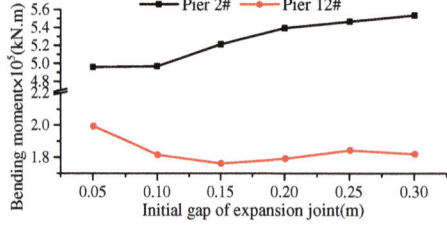
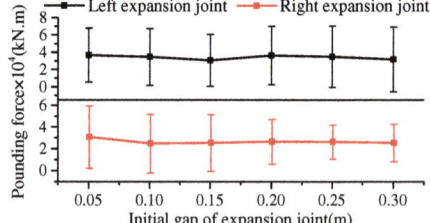

Fig. 9. Bending moment of approach bridge piers

Fig. 10. Pounding force of expansion joint

5   Conclusions

In this paper, a typical long-span suspension bridge and approach model are used to analyze the longitudinal pounding effect of stiffening girder of suspension bridge and box girder of approach bridge based on ten selected earthquake waves. The main influencing factors are also discussed above. The conclusions are as follows:

(1) As pounding stiffness of expansion joints increases, the relative displacement of girder ends on both sides of expansion joints decreases, while the maximum pounding force increases. The effect of initial width of expansion joint on pounding effect is not obvious, and the pounding does not occur under some of seismic waves with the increase of initial width.

(2) Different anti-push stiffness of the approach bridges on both sides of the suspension bridge leads to significant differences in the pounding effect between the stiffening girder and the approach bridge under the earthquake. The greater the anti-push stiffness of approach bridge is, the greater pounding force and bending moment of the piers, and the smaller longitudinal displacement of the box girder will be.

(3) Seismic responses of suspension bridge caused by long-period seismic waves are more obvious than that caused by other waves, and very large pounding forces will be generated between the stiffening girder and the approach bridge under such excitations. Under the designed earthquake intensity, the average stress of the girder may exceed its allowable compressive stress. The pounding effect analysis of long period ground motion on long-span suspension bridge is worthy of further study.

Acknowledgements. This work was financially supported by Science and Technology Planning Project of Guangdong Province, China (2015A020217007), Science for Earthquake Resilience (XH16031), and Technology Foundation for Selected Overseas Chinese Scholar, Ministry of Personnel of China (2013-277), and Independent Project Fund of China Earthquake Administration (CEA) Key Laboratory of Earthquake Monitoring and Disaster Mitigation Technology. Such financial aids are gratefully acknowledged.

References

1. Fan, L.: Bridge seismic. Tongji University Publication, Shanghai (1997)
2. Malhotra, P.K.: Dynamics of seismic pounding at expansion joints of concrete bridges. J. Eng. Mech. **124**(7), 794–802 (1998)
3. Chouw, N., Hao, H.: Study of SSI and non-uniform ground motion effect on pounding between bridge girders. Soil Dyn. Earthq. Eng. **25**(7–10), 717–728 (2005)
4. Jankowski, R., Wilde, K., Fujino, Y.: Reduction of pounding effects in elevated bridges during earthquakes. Earthq. Eng. Struct. Dyn. **29**(2), 195–212 (2015)
5. Ruangrassamee, A., Kawashima, K.: Relative displacement response spectra with pounding effect. Earthq. Eng. Struct. Dyn. **30**(10), 1511–1538 (2001)
6. Anagnostopoulos, S.A., Spiliopoulos, K.V.: An investigation of earthquake induced pounding between adjacent buildings. Earthq. Eng. Struct. Dyn. **21**(4), 289–302 (1992)

7. Li, J., Fan, L.: Longitudinal seismic response and pounding effects of girder bridges with unconventional configurations. China Civ. Eng. J. **38**(1), 84–90 (2005)
8. Deng, Y., Peng, T., Li, J.: Parametric study on longitudinal seismic pounding response for a long-span multi-tower suspension bridge. J. Vib. Shock **30**(4), 205–210 (2011)

Effects of Pile Group Configuration on the Seismic Response of Buildings Considering Soil-Pile-Structure Interaction

Ruoshi Xu$^{(\boxtimes)}$ and Behzad Fatahi

University of Technology Sydney (UTS), PO Box 123, Ultimo 2007, Australia
Ruoshi.Xu@uts.edu.au

Abstract. Mid-rise buildings supported by different configurations of end-bearing pile foundations can fulfil the design requirements addressed in the modern building codes and selecting the most efficient configuration is a challenging task. In this study, the influence of group configuration (but keep the area replacement ratio constant) on the seismic response of mid-rise buildings resting on end-bearing pile foundations considering seismic soil-pile-structure interaction (SSPSI) is investigated. A soil-pile-structure system is simulated in FLAC^{3D} to carry out the fully nonlinear seismic analysis in the time domain. The elastic-perfectly plastic structural behaviour is considered while the variation of soil shear modulus due to cyclic shear strain and the corresponding damping ratio is simulated adopting hysteretic damping algorithm, and the soil plastic behaviour is modelled using Mohr-Coulomb criterion. The results of the seismic pile responses, namely the envelopes of shear forces and bending moments along the piles and the lateral pile displacements, and the building responses including the base shear, and the lateral building displacements are reported and discussed. It is observed that for the cases analysed, the pile group configuration influences the seismic response of the system. Provided that the same volume of concrete (i.e. constant area replacement ratio) is used for all cases, by increasing the number of piles but with smaller diameters, more seismic loads may be attracted to the system due to the kinematic interaction between the piles and the surrounding soil and consequently causing the increase in the base shear and lateral displacements of the building.

Keywords: End-bearing pile foundation · Pile group configuration
Seismic soil-pile-structure interaction · Kinematic interaction

1 Introduction

Deep foundations, such as end-bearing pile foundations, are commonly employed in soft soil conditions to transmit superstructure loads to deeper strata with a higher bearing capacity and stiffness. It is a complex process to determine the seismic response of a soil-pile-structure system due to the inertial and kinematic interactions and the non-linearity of the soil. Many research studies have highlighted the importance of considering the seismic soil-pile-structure interaction (SSPSI) in the seismic analysis to ensure the design is safe and cost-effective [1–4]. Although some work has been done on

© Springer Nature Singapore Pte Ltd. 2018
T. Qiu et al. (Eds.): GSIC 2018, *Proceedings of GeoShanghai 2018 International Conference: Advances in Soil Dynamics and Foundation Engineering*, pp. 279–287, 2018.
https://doi.org/10.1007/978-981-13-0131-5_31

the influence of the foundation types on the seismic response of superstructures [5–7], investigations on the impact of the pile group configuration of end-bearing pile foundation on the seismic response of soil-pile-structure systems are rare. Gazetas et al. [8] conducted research work about the influence of the group configuration on the seismic response of the soil-pile-structure system using Beam-on-Dynamic-Winkler-Foundation method, however, linear behaviour of the adopted system was assumed.

Considerable attention is received by numerical simulation in the study of SSPSI as it can model the complex conditions of a site by adopting nonlinear soil behaviour and heterogeneous material conditions. Moreover, a fully coupled response of the soil-pile-structure system can be captured using the direct method, since the entire system can be analysed in one attempt. In contrast to the sub-structuring technique, the direct method of analysis does not require superposition, thus more accurate and true nonlinear analysis becomes possible [9]. To achieve this target, a numerical simulation is performed using FLAC^{3D} software [10] which is a three dimensional finite difference program. The influence of pile group configuration on the seismic responses of a fifteen storey building and its end-bearing pile foundation is investigated. The results can shed light on the seismic behaviour of soil-pile-structure system, typically end-bearing pile foundation and can help geotechnical and structural engineers to optimise their designs.

2   Adopted Numerical Soil-Pile-Structure Model

A fifteen storey moment resisting building is designed representing the conventional mid-rise buildings according to relevant Australian building codes [11–13]. Moreover, cracked section factors are considered by modifying the stiffness of the structural members according to ACI318 [14]. Tables 1 and 2 show designed structural members and material properties of the building, respectively. The weight and natural period of the building are 36 MN and 1.28 s, respectively.

Table 1. Designed structural members for the adopted moment resisting building

Levels	Cross sections of structural members		
	Columns	Beams	Slabs
11–15	0.7 × 0.7 m	0.55 × 0.55 m	0.25 m (Thickness)
6–10	0.75 × 0.75 m	0.6 × 0.6 m	
1–5	0.8 × 0.8 m	0.65 × 0.65 m	

Table 2. Adopted material properties of the building

Materials	Properties	Value
Concrete	Mass density (kg/m$^{3)}$	2400
	Specified compressive strength (MPa)	40
Steel reinforcement	Mass density (kg/m^3)	7800
	Yield strength (MPa)	500

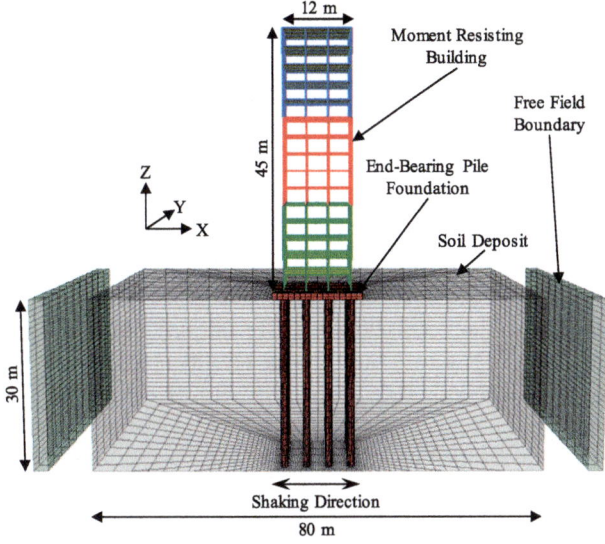

Fig. 1. The general layout of the numerical soil-pile-structure model

As presented in Fig. 1, the columns and beams are modelled by beam structural elements while the floor slabs are modelled by shell elements. The elastic-perfectly plastic behaviour is assigned to structural members and 5% structural damping ratio is assigned to the structural model.

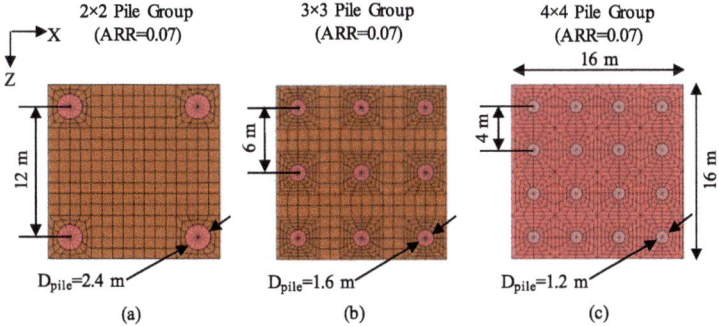

Fig. 2. Adopted foundations: (a) 2 × 2 pile group, (b) 3 × 3 pile group and (c) 4 × 4 pile group

Figure 2 shows three end-bearing pile foundations made of reinforced concrete, which are designed according to AS2159 [15] while keeping the size of foundation raft and the area replacement ratio (ARR) consistent. ARR is determined using Eq. (1).

$$ARR = \frac{A_{pile}}{A_{raft}} \tag{1}$$

where A_{pile} is the total area of the piles and A_{raft} is the area of the foundation raft. Moreover, each corner pile aligns with each corner column of the building as per practical norms. Solid elements are adopted, and the elastic-perfectly plastic behaviour with yielding criteria is considered to control any inelastic behaviour in the foundations. Note that all foundation rafts are located on the ground surface whereas the pile toes are socketed in the bedrock.

The superstructure and foundation sit in a soft clayey soil with the unit weight of 14.4 kN/m3, the shear wave velocity of 150 m/s and the effective cohesion and friction angle of 3 kPa and 29°, respectively. A 30 m deep soil deposit is modelled using solid elements assuming the bedrock level at the bottom of the deposit as the majority of soil amplification and attenuation occurs in this range. Mohr-Coulomb criterion is used to simulate plastic flow in the deposit while hysteretic damping algorithm is applied to capture the cyclic nonlinear behaviour of the soil following the actual stress-strain path during cyclic loading as suggested by Vucetic and Dobry [16] and Masing rule assumption for loading/unloading. To capture possible pile-soil gapping and to incorporate the different mechanical behaviours of the piles and soil during the seismic analysis, interface elements are applied at the contacting surfaces between the foundation and soil. A set of normal and shear spring-slider systems with Mohr-Coulomb criterion is adopted with interface parameters estimated following the procedure recommended by Itasca [10].

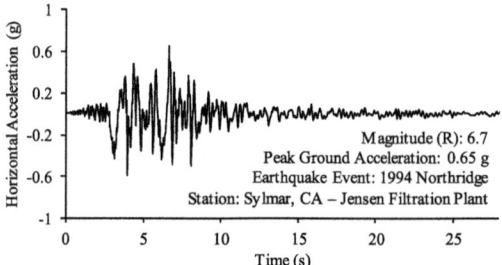

Fig. 3. Adopted acceleration record of the 1994 Northridge Earthquake

To conduct seismic analysis, free field boundaries are applied to the four sides of the deposit as shown in Fig. 1, so that wave reflection due to artificial boundaries can be alleviated. Also, a rigid boundary is assigned to the deposit bottom to simulate the strong bedrock, while the acceleration record of the 1994 Northridge Earthquake (see Fig. 3) is applied to the bedrock level to perform seismic analysis in the time domain.

3   Results and Discussion

Figures 4 and 5 show the dimensionless normalised shear force and bending moment envelopes along the pile depth of corner piles (one sample only) under the influence of the applied earthquake, respectively. The normalized shear forces $(\hat{Q}(z))$ and bending moments $(\hat{M}(z))$ are calculated using Eqs. (2) and (3), respectively [8].

$$\hat{Q}(z) = \frac{Q_{max}(z)}{\rho_p d^3 a} \tag{2}$$

$$\hat{M}(z) = \frac{M_{max}(z)}{\rho_p d^4 a} \tag{3}$$

where $Q_{max}(z)$ and $M_{max}(z)$ are the maximum shear forces and bending moments imposed on the pile at the depth z during the earthquake, ρ_p and d are the mass density and diameter of the pile, respectively, and a is the peak ground acceleration (PGA) of the applied earthquake.

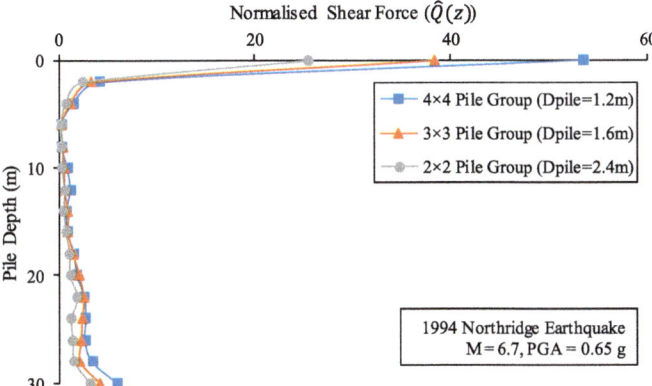

Fig. 4. Normalised shear force envelope on the piles under the influence of the 1994 Northridge Earthquake considering three different group configurations

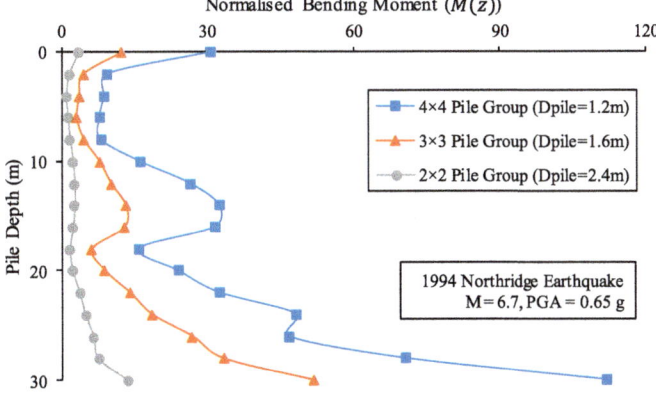

Fig. 5. Normalised bending moment on the piles under the applied earthquake excitation in conjunction with three different group configurations

As shown in Figs. 4 and 5, the pile head and toe, which were mainly influenced by inertial and kinematic interaction, respectively, are the critical sections of the piles due to significant shear forces and bending moments. The inertial interaction stemmed from the motion of the superstructure generates notable shear forces and bending moments in the top part of the pile near the pile head, whereas, the kinematic interaction induces considerable bending moments and shear forces due to the different movement between the piles and the soil during the seismic wave propagation in the bottom section of the pile near the pile toe. Referring to Figs. 4 and 5, the normalised kinematic loads near the pile toes are increased with the number of the piles employed while keeping the same concrete volume for piles. In this study, the increased kinematic loads stem from two contributors: (i) the flexure strength of the pile resisting the soil movement and (ii) the soil reaction applied to the pile shaft. Indeed, due to the load bearing mechanism of the adopted foundations, the piles experience their maximum bending moment at the

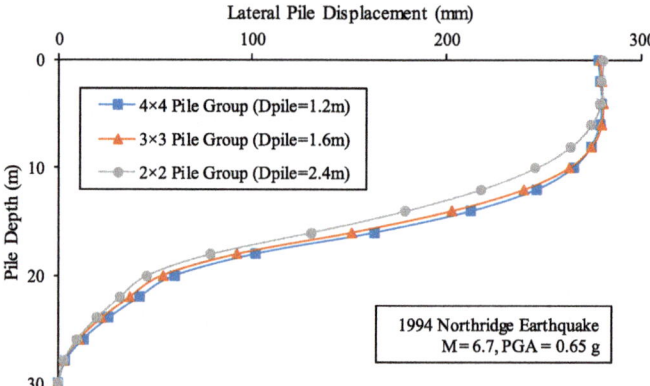

Fig. 6. The maximum lateral pile displacement induced by the applied earthquake with three different pile group configurations

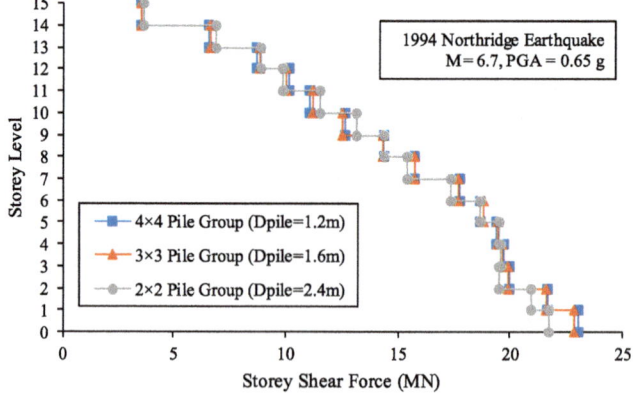

Fig. 7. Shear force generated along the building under the influence of the applied earthquake with three different pile group configurations applied

toes, where they are connected to the strong bedrock. Figure 6 shows that the average lateral pile displacement increases slightly with the number of piles (but same area replacement ratio) applied since more shear forces and bending moments are developed along the piles referring to Figs. 4 and 5.

Figure 7 shows the storey shear forces developed along the superstructure considering three different pile group configurations under the influence of the 1994 Northridge Earthquake. Referring to Fig. 7, the base shear increases slightly with the number of piles used. For example, the base shear is 21.7 MN when 4 large piles (i.e. 2 × 2 pile group) are used while the corresponding value is 23.1 MN (i.e. 6.2% increase) when the 16 smaller piles (i.e. 4 × 4 pile group) is employed. This observation can be attributed to the fact that more kinematic loads are generated due to larger contacting surfaces between 4 × 4 pile group in comparison to 2 × 2 pile group leading to larger transmitted forces to the superstructure. Obviously, bending moments and shear forces acting on the pile head would increase with base shear.

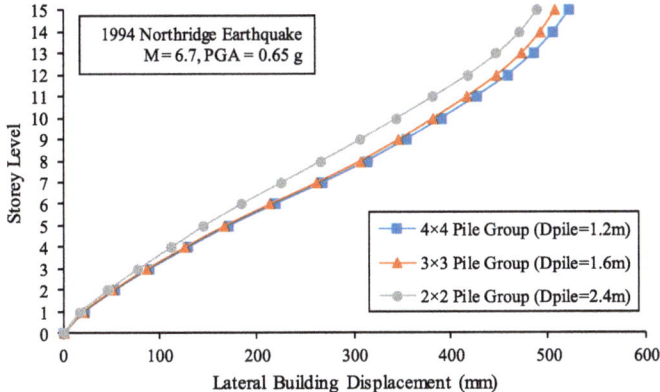

Fig. 8. Lateral building displacement while the roof level reaches the maximum displacement under the influence of the 1994 Northridge Earthquake considering three numbers of piles used

Figure 8 depicts the building lateral displacements recorded when the roof level reached the absolute maximum lateral displacement. Building displacement stems from structural distortion and foundation rotation. Since in this study, end-bearing pile foundations are adopted, the lateral displacement is influenced significantly by the structural distortion which is mainly due to the shear forces developed in the structure. Therefore, the variations of the lateral building displacements (Fig. 8) follow a similar trend to the base shear variations (Fig. 7). For example, when 2 × 2 pile group, is adopted the maximum building displacement is 488 mm while the corresponding value is 522 mm for the case of 4 × 4 pile group (i.e. 7% increase).

4 Conclusions

In this study, the influence of pile group configuration on the seismic response of the soil-pile-structure system is investigated. The soil-pile-structure system is simulated using FLAC3D and the coupled fully nonlinear analysis is conducted in the time domain considering seismic soil-pile-structure interaction (SSPSI).

The results indicate that the pile group configurations can influence the seismic response of the foundation notably. Specifically, the shear forces and bending moments attracted by the piles increase with the number of the piles and this increase is due to kinematic interaction between piles and surrounding soil, stemmed from the rigidity of the pile and soil reaction on the pile shaft. In terms of the impact on the building, as observed from results the increase in base shear is 6.2% when the number of the piles adopted increased from 4 (larger piles) to 16 (smaller piles) and therefore the building displacement increases accordingly. Thus, the pile group configurations can impact the seismic response of the soil-pile-structure system. Moreover, the less number of piles but larger diameter can result in reduced shear forces and bending moments in piles, which is recommended to be adopted by practising engineers. However, sufficient thickness of the foundation raft is needed to be ensured to tolerate the additional bending moments generated in the raft due to the increase in the spacing between piles.

References

1. Hokmabadi, A.S., Fatahi, B.: Influence of foundation type on seismic performance of buildings considering soil–structure interaction. Int. J. Struct. Stab. Dyn. **16**(08), 1550043 (2016)
2. Kampitsis, A., Sapountzakis, E., Giannakos, S., Gerolymos, N.: Seismic soil–pile–structure kinematic and inertial interaction—A new beam approach. Soil Dyn. Earthq. Eng. **55**, 211–224 (2013)
3. Hokmabadi, A.S., Fatahi, B., Samali, B.: Assessment of soil–pile–structure interaction influencing seismic response of mid-rise buildings sitting on floating pile foundations. Comput. Geotech. **55**, 172–186 (2014)
4. Hokmabadi, A.S., Fatahi, B., Samali, B.: Physical modeling of seismic soil-pile-structure interaction for buildings on soft soils. Int. J. Geomech. **15**(2), 04014046 (2014)
5. Chu, D., Truman, K.Z.: Effects of pile foundation configurations in seismic soil-pile-structure interaction. In: 13th World Conference on Earthquake Engineering, Canadian Association for Earthquake Engineering, p. 1551 (2004)
6. Nguyen, Q.V., Fatahi, B., Hokmabadi, A.S.: Influence of size and load-bearing mechanism of piles on seismic performance of buildings considering soil–pile–structure interaction. Int. J. Geomech. **17**(7), 04017007 (2017)
7. Nguyen, Q.V., Fatahi, B., Hokmabadi, A.S.: The effects of foundation size on the seismic performance of buildings considering the soil-foundation-structure interaction. Struct. Eng. Mech. **58**(6), 1045–1075 (2016)
8. Gazetas, G., Fan, K., Kaynia, A.: Dynamic response of pile groups with different configurations. Soil Dyn. Earthq. Eng. **12**(4), 239–257 (1993)
9. Borja, R.I., Wu, W.-H., Amies, A.P., Smith, H.A.: Nonlinear lateral, rocking, and torsional vibration of rigid foundations. J. Geotech. Eng. **120**(3), 491–513 (1994)

10. Itasca, F.D.: V5. 0, Fast Lagrangian Analysis of Continua in 3 Dimensions, User's Guide. Itasca Consulting Group, Minneapolis, Minnesota (2011)
11. AS1170.1: Structural Design Actions Part 1: Permanent, Imposed and Other Actions. Standards Australia, Sydney (2002)
12. AS1170.4: Structural Design Actions Part 4: Earthquake Actions in Australia. Standards Australia, Sydney (2007)
13. AS3600: Concrete Structures. Standards Australia, Sydney (2009)
14. ACI318: Building Code Requirements for Structural Concrete and Commentary (ACI 318-08). American Concrete Institute, Washington DC (2008)
15. AS2159: Piling Design and Installation. Standards Australia, Sydney (2009)
16. Vucetic, M., Dobry, R.: Effect of soil plasticity on cyclic response. J. Geotech. Eng. ASCE **117**(1), 89–107 (1991)

Deterministic Seismic Hazard Analysis
of Central Gujarat Region

Payal Mehta[✉], Tejas P. Thaker, and H. B. Raghvendra

School of Technology, Pandit Deendayal Petroleum University,
Raisan, Gandhinagar 382007, Gujarat, India
payalmehta2910@gmail.com

Abstract. The microzonation study of a particular region can be carried out by Deterministic Seismic Hazard Analysis (DSHA) approach. Gujarat is situated in a stable continental region of India which has experienced two severe destructive earthquakes in Kutch 1819 (M_w 7.8) and Bhuj 2001 (M_w 7.7) and seven earthquakes of magnitude $M_w \geq 6.0$ during past two centuries. In the present paper, earthquake catalogue and seismotectonic model for the Central Gujarat region have been developed considering 350 km radius around Vadodara city as a centre covering the latitude 18° N to 26° N and longitude 68° E to 77° E. The seismic data has been compiled for past approximately 350 years covering a span of 1668 to 2017 from the various agencies and published literatures. Homogenization of earthquake catalogue has been carried out considering appropriate conversion relationships, which is further validated through the recorded earthquake data. All magnitudes have been converted into equal magnitude i.e. moment magnitude M_w. Declustering of the catalogue has been carried out using various methods like Gardner and Knopoff (1974), Uhrhammer (1986) to remove the foreshocks and aftershocks events. The tectonic details have been collected and compiled from various published literatures and other sources on common platform using Geographical Information System (GIS). Seismotectonic model for Central Gujarat has been developed from the final catalogue. Further, peak ground acceleration values at rock level have been determined considering DSHA approach for Central Gujarat region which varies from 0.167 g to 0.355 g and compared with standard code of practice.

Keywords: Earthquake catalogue · Central Gujarat · Seismotectonic model
Deterministic Seismic Hazard Analysis

1 Introduction

Earthquake is the one of the most destructive natural hazards which can neither be predicted nor can be controlled. India has experienced many devastating earthquake events from the historic era. Gujarat is situated in a stable continental region (SCR) of India which has also witnessed many severe damaging earthquakes in the past. Seismic hazard analysis is a crucial base for earthquake resistant design of any structure and microzonation study of the particular region. In the present paper, Vadodara city is considered as a centre of the study region covering 350 km radius around the centre with latitude 18° N to 26° N and longitude 68° E to 77° E. The earthquake catalogue

© Springer Nature Singapore Pte Ltd. 2018
T. Qiu et al. (Eds.): GSIC 2018, *Proceedings of GeoShanghai 2018 International Conference: Advances in Soil Dynamics and Foundation Engineering*, pp. 288–299, 2018.
https://doi.org/10.1007/978-981-13-0131-5_32

has been compiled for the past 350 years from the published literatures and various agencies. Homogenized earthquake catalogue has been developed considering appropriate conversion relationships, which is further validated through the recorded earthquake data. This catalogue will be useful for carrying out numerous seismic studies in the region. Further, the catalogue is used to carry out the Deterministic Seismic Hazard Analysis (DSHA) of the Central Gujarat region and to evaluate peak ground acceleration (PGA) at the rock level of the study region. Seismotectonic model has been developed for the Central Gujarat region compiling the tectonic details in common platform using Geographic Information System (GIS). A rock level PGA model has been developed by dividing the study area in a grid of size 1 km × 1 km. This PGA model will be used to design the earthquake resistant structures and mitigation of seismic hazards. The PGA model generated which will be useful as input for non linear ground response modelling of the region.

2   Seismicity and Geology of the Study Region

According to IS:1893-2016, Gujarat is distinct under the four seismic zones i.e. zone II, III, IV, V. Major part of the Kutch region falls in Zone V, remaining part of Kutch and narrow fringe of northern Kathiawar falls in Zone IV, while rest of the part of the study region falls in Zone III. Based on the seismotectonics and geography of the region, Gujarat state can be divided into three major seismogenic regions namely, Saurashtra, Mainland Gujarat and Kutch. The Saurashtra region has experienced random seismicity at different places such as Junagadh, Jamnagar, Dwarka, Rajkot, Ghogha and Bhavnagar. Mainland Gujarat, which covers two rift zones namely Narmada rift zone and Cambay rift zone, has experienced moderate seismicity. Kutch is a seismically most active intracontinental region with high intensity. The geology of Gujarat mainly consists of Precambrian basement without any rocks of Paleozoic era (Merh 1995; Biswas 1987; Biswas 2005). The younger rocks of Jurassic, Cretaceous, Tertiary and Quaternary are deposited over the Precambrian basement (Biswas 1987, 2005). Saurashtra, some portion in Kutch and major portion of South Gujarat is covered by Deccan basalt with intervening Cretaceous and Tertiary rocks at many places. Stratigraphically, Mainland Gujarat comprises of Precambrian crystalline, sedimentary rocks of Cretaceous, Tertiary and Quaternary periods and the Deccan Trap. Quaternary deposits of around 100 m and 400 m thickness covers major area of the central part of the mainland (Merh 1995). Proterozoic rocks cover the NE part of the mainland and southern part is mostly covered by Deccan basalt (Sinha-Roy et al. 1995). In the Saurashtra region, major part is covered by basalt and sedimentary sequence which is as old as Upper Jurassic era. Kutch shows good development of Mesozoic and Tertiary sequence (Merh 1995).

The Kutch region is associated with the largest intraplate Bhuj 2001 earthquake. In the Saurashtra region of Gujarat, 1938 Paliyad sequence of maximum magnitude 5.7 was a major seismic activity with more than 190 shocks felt in one and a half month (Bapat et al. 1989). A number of moderate sized earthquakes have been produced due to the Cambay fault which is situated on the eastern boundary of Saurashtra. Most of the seismic activities have been noticed along the Narmada fault in the southern portion of Mainland. The area also experienced significant earthquakes like Surat (1856, M_w 5.7; 1871, M_w 5.0 and 1935, M_w 5.7), Ahmedabad (1864, M_w 5.0), Bharuch (1970, M_w 5.4). Recently few events of $M_w < 4$ have been observed along Cambay West fault and Tapti North fault. Thus the seismicity of Gujarat has increased multifold in last few decades.

3   Earthquake Catalogue Compilation

The earthquake catalogue has been compiled from the various sources such as, Institute of Seismological Research (ISR), Gujarat; International Seismological Centre (ISC), U.K.; Geological Survey of India (GSI); India Meteorological Department (IMD), New Delhi; United States Geological Survey (USGS), National Earthquake Information Centre; Oldham (1883); Chandra (1992); Malik et al. (1999) for the time interval of 1668 to 2017.

3.1   Homogenization of Earthquake Magnitude

Heterogeneous data having different magnitude scale have been reported in the present catalogue. The catalogue consists of three magnitude scale i.e., Body wave magnitude in m_b (USGS) & M_b (ISC), Surface wave magnitude M_s, which were converted into equal magnitude scale i.e. moment magnitude M_w. Hardly any attempt has been made to check out the reliability of the conversion equations developed for the Central Gujarat region. Various relationships for the conversion of various magnitudes into moment magnitude have been identified and analyzed. The appropriate relationships have been selected after validating them with the actual data. Table 1 depicts six relationships which are identified from literature to convert body wave magnitude m_b (USGS) into moment magnitude M_w. Among these relationships, Johnston (1996) and Das et al. (2013) relationships have been selected in the present catalogue as they give -10% to $+10\%$ and -8% to $+8\%$ variations between observed and calculated M_w respectively and the higher value of M_w has been considered.

Table 1. Conversion of body wave magnitude m_b (USGS) into moment magnitude M_w

Sr. No.	Details of relationship	Study area	Relationship	% Variation between observed and calculated M_w
1.	Scordilis (2006)	Global data	$M_w = 0.85(\pm0.04)$ $m_b + 1.03(\pm0.23)$, $3.5 \leq m_b \leq 6.2$	-20% to $+20\%$ (see Fig. 1)
2.	Johnston (1996)	Stable continental regions of the world	$\log(M_0) = 18.28 + 0.679$ $(m_b) + 0.077(m_b)^2$ $M_w = 2/3 \log M_0 - 10.7$	-10% to $+10\%$ (see Fig. 2)

(continued)

Table 1. (*continued*)

Sr. No.	Details of relationship	Study area	Relationship	% Variation between observed and calculated M_w
3.	Das et al. (2011)	Global data	$m_{b,NEIC} = 0.61$ (±0.005) $M_{w,HRVD} + 1.94$ (±0.02); $3.8 \leq m_{b,NEIC} \leq 6.5$	-30% to $+30\%$ (see Fig. 3)
4.	Das et al. (2012)	Northeast, India	$M_{w,HRVD} = 1.37$ (±0.006)$m_{b,NEIC} - 1.77$ (±0.157); $4.6 \leq m_{b,NEIC} \leq 6.8$	Valid for $m_b \geq 4.6$, Covers only 50% of data, so it doesn't give the actual representation as compare to other relationships (see Fig. 4)
5.	Das et al. (2013)	Peninsular India	$M_w = 1.155 m_{b,obs} - 0.67$	-8% to $+8\%$ (see Fig. 5)
6.	Yadav et al. (2009)	Northeast, India	$M_{w,HRVD} = 0.80(\pm0.06)$ $m_{b,NEIC} + 1.03(\pm0.31)$	-15% to $+20\%$ (see Fig. 6)

Table 2 shows relationships which are used for the conversion of body wave magnitude M_b (ISC) into moment magnitude M_w. The lower bound of Thingbaijam et al. (2008); Yadav et al. (2009) relationships have been selected to convert M_b (ISC) into M_w for the present catalogue and the higher value of M_w has been considered.

Table 2. Conversion of body wave magnitude M_b (ISC) into moment magnitude M_w

Sr. No.	Details of relationship	Study area	Relationship	% Variation between observed and calculated M_w
1.	Das et al. (2011)	Global data	$m_{b,ISC} = 0.65(\pm0.003)$ $M_{w,HRVD} + 1.65(\pm0.02)$; $2.9 \leq m_{b,ISC} \leq 6.5$	-25% to $+5\%$ (see Fig. 7)
2.	Thingbaijam et al. (2008)	Northeast, India	$M_{W,GCMT} = 1.3691(\pm0.211)$ $m_{b,ISC} - 1.7742(\pm1.139)$; for $M_{W,GCMT} > 4.4$; $m_{b,ISC} < 6.7$	Upper bound: -25% to $+1\%$ Lower bound: 60% data: linear relation, 40% data: 0 to -10% (see Fig. 8)
3.	Yadav et al. (2009)	Northeast, India	$M_{w,HRVD} = 1.08(\pm0.05)$ $m_{b,ISC} - 0.24(\pm0.29)$	Upper bound: 0 to -10% Lower bound: 80% data: linear relation, 20% data: -5% to -5% (see Fig. 9)

Table 3 shows seven relationships to convert body wave magnitude M_b (ISC) into moment magnitude M_w. Ambraseys and Douglas (2004) and Thingbaijam et al. (2008) lower bound relationship have been identified to convert the surface wave magnitude (M_s) into moment magnitude (M_w) for the present catalogue and higher value of M_w is considered in the catalogue.

Table 3. Conversion of surface wave magnitude M_s into moment magnitude M_w

Sr. No.	Details of relationship	Study area	Relationship	% Variation between observed and calculated M_w
1.	Scordilis (2006)	Global data	$M_W = 0.67(\pm0.005)M_S + 2.07(\pm0.03), 3.0 \leq M_S \leq 6.1$	−10% to +10% (see Fig. 10)
2.	Johnston (1996)	Stable continental regions of the world	$\log(M_0) = 24.66\text{-}1.083(M_S) + 0.192(M_S)^2 (M_S \geq 3.6);$ $M_W = 2/3 \log M_0 - 10.7$	0 to +8% (see Fig. 11)
3.	Das et al. (2011)	Global data	$M_{w,HRVD} = 0.67(\pm0.00005) M_S + 2.12 (\pm0.0001),$ $3.0 \leq M_S \leq 6.1$	0 to +12% (see Fig. 12)
4.	Das et al. (2012)	Northeast, India	$M_{w,HRVD} = 0.6389 (\pm0.0006) M_{s,ISC} + 2.20 (\pm0.016), 4.1 \leq M_{s,ISC} \leq 6.9$	+3% to +%5 (valid for $M_s > 4.1$)
5.	Ambraseys and Douglas (2004)	North India	$\log M_0 = 19.38 + 0.93M_S,$ for $M_S \leq 5.94$ $M_W = 2/3 \log M_0 - 10.7$	80% data: Linear relation 20% data: 0 to +8% (see Fig. 13)
6.	Thingbaijam et al. (2008)	Northeast, India	$M_{w,GCMT} = 0.7042(\pm0.0356) M_{S,ISC} + 1.8197(\pm0.1896),$ for $M_{S,ISC} < 7.5$	Upper bound: 0 to +12% Lower bound: 0 to −4% (see Fig. 14)
7.	Yadav et al. (2009)	Northeast, India	$M_{w,HRVD} = 0.62(\pm0.03) M_{S,ISC} + 2.28(\pm0.13),$ for $M_S < 6.2$	−10% to +12%

3.2 Declustering of Earthquake Catalogue

Declustering of the catalogue is to separate the main and associated events. Three different declustering methods i.e. Gardner and Knopoff (1974), Uhrhammer (1986) and Grunthal have been applied to the present catalogue using temporal and spatial windowing approaches. The epicentres falling within these windows are removed. Uhrhammer (1986) algorithm shows less percentage of events to be declustered so conservatively it is taken into consideration for preparation of final catalogue. The final catalogue consists of 322 earthquake events after removal of foreshocks and aftershocks.

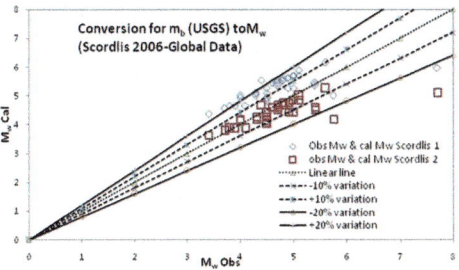

Fig. 1. Observed m_b (USGS) and predicted M_w (Scordilis)

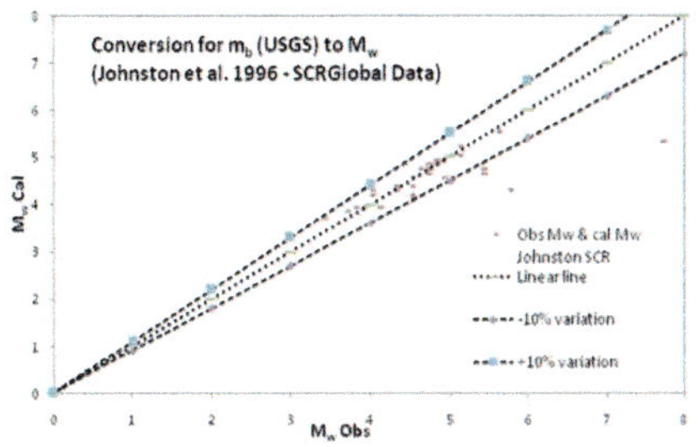

Fig. 2. Observed m_b (USGS) and predicted M_w (Johnston)

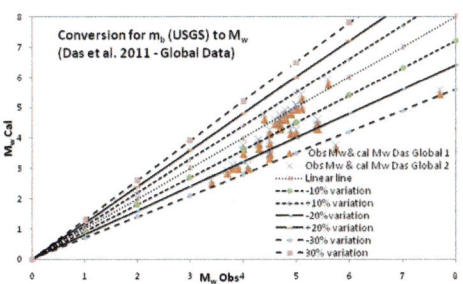

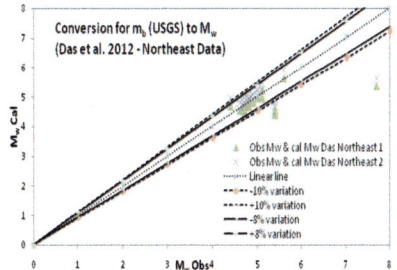

Fig. 3. Observed m_b (USGS) and predicted M_w (Das)

Fig. 4. Observed m_b (USGS) and predicted M_w (Das)

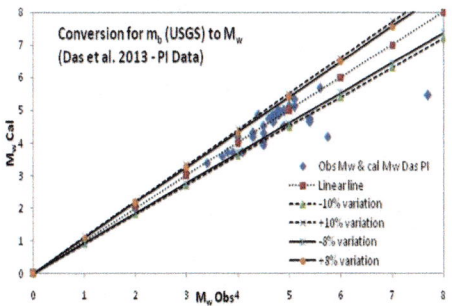

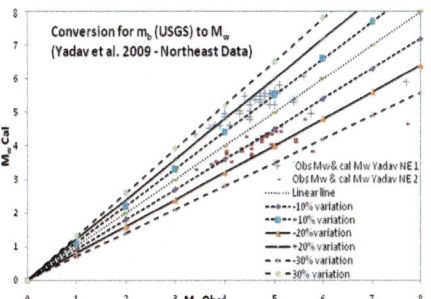

Fig. 5. Observed m_b (USGS) and predicted M_w (Das)

Fig. 6. Observed m_b (USGS) and predicted M_w (Yadav)

4 Seismotectonic Model of the Study Region

Major faults and lineaments have been identified from the Seismotectonic Atlas of India and its environs (2000) and from other published literatures for the study region. Seventeen major faults, ten lineaments and seventeen minor tectonic features have been

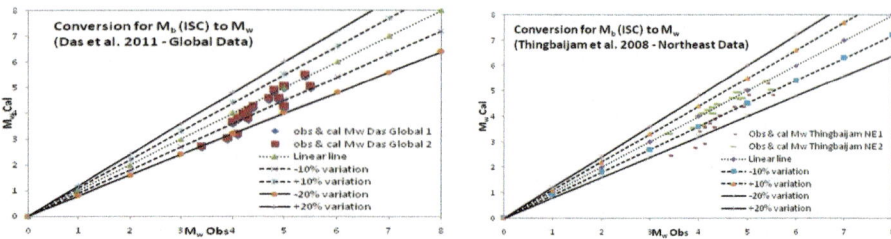

Fig. 7. Observed M$_b$ (ISC) and predicted M$_w$ (Das)

Fig. 8. Observed M$_b$ (ISC) and predicted M$_w$ (Thingbaijam)

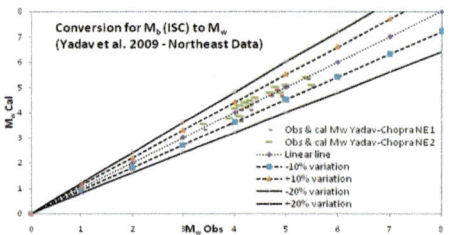

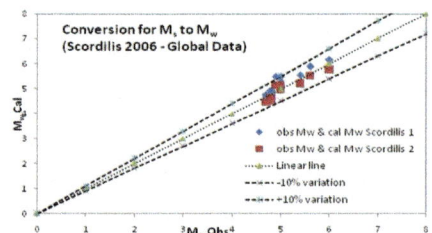

Fig. 9. Observed M$_b$ (ISC) and predicted M$_w$ (Yadav)

Fig. 10. Observed M$_s$ and predicted M$_w$ (Scordilis)

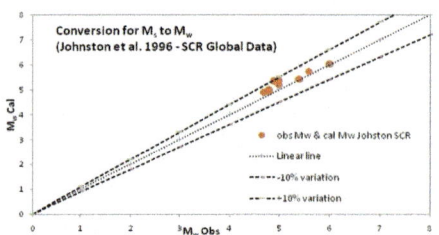

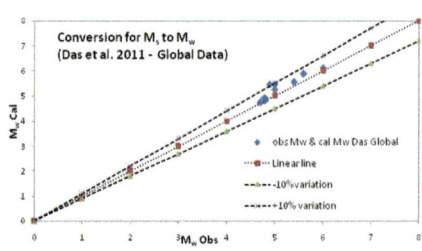

Fig. 11. Observed M$_s$ and predicted M$_w$ (Johnston)

Fig. 12. Observed M$_s$ and predicted M$_w$ (Das)

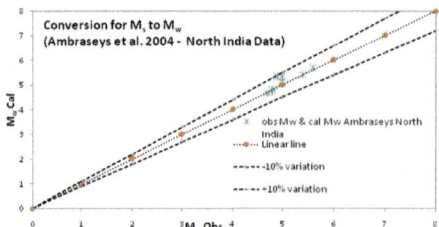

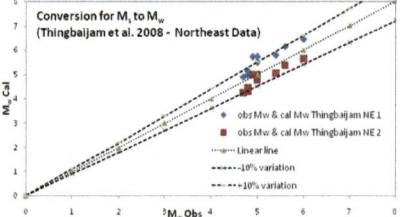

Fig. 13. Observed M$_s$ and predicted M$_w$ (Ambraseys)

Fig. 14. Observed M$_s$ and predicted M$_w$ (Thingbaijam)

compiled on the common platform of GIS with the earthquake events of magnitude $M_w \geq 1.5$. Seismotectonic model of the study region has been developed (see Fig. 15). Marginal fault, Son Narmada fault, Cambay West fault and Tapti North fault are the major faults of the Central Gujarat region having past earthquake events of magnitude $M_w \geq 5$. Marginal fault, Son Narmada fault, Cambay West fault and Tapti North fault are the near field sources having epicentral distance of 13.83 km, 61.75 km, 89.14 km and 99.15 km respectively from the centre of Vadodara. Far field sources like Kutch Mainland fault, Island Belt fault and Allah Bund Fault are major tectonic features on the west side of the Central Gujarat region. Around thirteen earthquakes have been reported which is having moment magnitude $M_w \geq 6$ in the study area. The epicenters of past 350 years earthquakes for the time interval of 1668 to 2017 have been observed in the seismotectonic map for the Central Gujarat region.

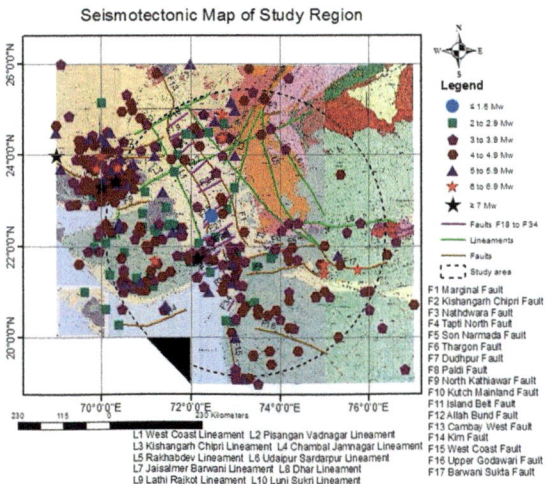

Fig. 15. Seismotectonic model of Central Gujarat region

5   Deterministic Seismic Hazard Analysis

The Deterministic Seismic Hazard Analysis (DSHA) is one of the approaches to determine maximum possible ground motion at a site considering the past seismicity and seismotectonic setup of any region. Historic earthquake data and attenuation relationship play important role in the Deterministic Seismic Hazard Analysis. In the present study, each seismic source has assigned maximum earthquake magnitude based on maximum reported earthquake in the past. Vadodara region is divided into grid of size 1 km × 1 km. The shortest distance from the centre of each grid cell to each fault has been calculated. Six attenuation relationships have been considered in the present study. Two region specific relationships for peninsular India (PI) and other four based on Eastern North America (ENA) are considered in the study. Due to the regional tectonics similarity between the intraplate regions, the relationships developed for the ENA are applicable to the intraplate regions of India (Bodin et al. 2004). ENA and PI are having similar features in terms of seismogenic activities and known seismotec-tonics (Schweig et al. 2003). Cramer and Kumar (2003) compared the attenuation relations of ENA with ground motion of the 2001 Bhuj earthquake and suggested that the ground motion attenuation in ENA and PI is comparable. Raghukanth and Iyenger (2007); NDMA (2010) two region specific ground motion attenuation relationships and Hwang and Huo (1997); Toro et al. (2002); Campbell (2003); Pezeshk et al. (2011) ENA based ground motion prediction equations (GMPEs) have been considered in the present study. Equal weightage has been assigned to each relationship. Table 4 depicted the earthquake model with the major faults and the maximum magnitude assigned with them. The maximum PGA value is from Marginal Fault i.e. 0.247 g for the hypocentral distance of 17.07 km and maximum earthquake of 5.7 moment mag-nitude. Other major vulnerable faults are Son Narmada Fault and Cambay West Fault with PGA of 0.088 g and 0.084 g respectively. The contour map of the study region in a grid size of 1 km × 1 km for PGA (g) at rock level using DSHA approach has been developed (see Fig. 16).

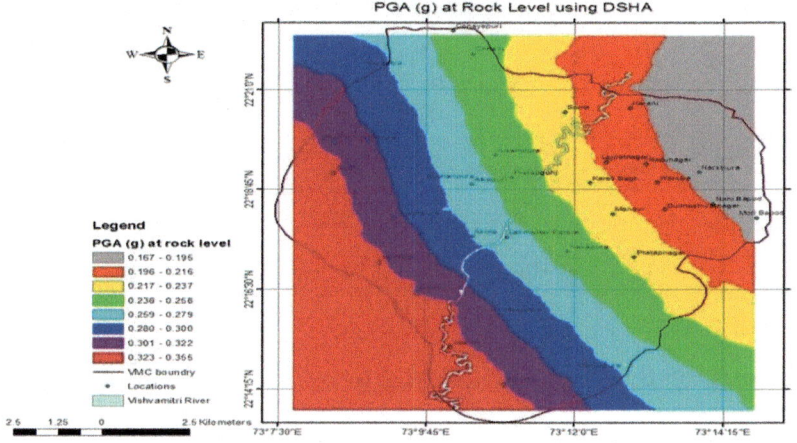

Fig. 16. PGA model at rock level of Central Gujarat region using DSHA

Table 4. Earthquake model for deterministic approach

Major faults with notations	Epicentral distance (R_{epi}) in km	Hypocentral distance (R_{hyp}) in km	Maximum earthquake (M_{wmax})	PGA (g) at rock level
Marginal fault (F1)	13.83	17.07	5.7	0.247
KishangarhChipri fault (F2)	303.3	303.67	5.4	0.003
Nathdwara fault (F3)	289	289.17	6.1	0.007
Tapti North fault (F4)	99.15	99.65	5.4	0.017
Son Narmada fault (F5)	61.75	62.55	6.5	0.088
Paldi fault (F8)	66.66	67.41	5	0.019
North Kathiawar fault (F9)	232.2	232.42	5.8	0.007
Kutch Mainland fault (F10)	266.5	266.69	7.7	0.035
Island Belt fault (F11)	264.7	265.12	6.3	0.010
Allah Bund fault (F12)	300.9	301.07	7.8	0.032
Cambay West fault (F13)	89.14	89.7	7	0.084
Kim fault (F14)	233.1	233.31	5.7	0.007
West Coast fault (F15)	222	222.23	5.7	0.007
Upper Godawari fault (F16)	233.9	234.11	4.8	0.003
BarwaniSukta fault (F17)	163.2	170.69	6.2	0.018

6   Discussions and Conclusions

The Deterministic Seismic Hazard Analysis (DSHA) for the Central Gujarat region has been carried out considering Vadodara as a centre covering 350 km radius around it. All available seismotectonic features and 350 years past earthquake data have been taken into account for the analysis. Six attenuation relationships have been considered in the study. Marginal Fault, Son Narmada Fault, Cambay West Fault and Tapti North Fault are the near field sources of Central Gujarat region associated with the earthquake events of moment magnitude $M_w \geq 5$ and located at an epicentral distance within 100 km from the centre of Vadodara region. Other major vulnerable far field sources within 300 km epicentral distance are Kutch Mainland Fault and Allah Bund Fault, which are associated with the earthquake events of moment magnitude $M_w \geq 7$.

The PGA model at rock level has been generated by dividing the study region in 1 km × 1 km grid. Nine major faults namely Marginal Fault, Son Narmada Fault, Cambay West Fault, Kutch Mainland Fault, Allah Bund Fault, Paldi Fault, BarwaniSukta Fault, Tapti North Fault and Island Belt Fault are having PGA more than 0.01 g. The PGA at rock level estimated for Central Gujarat region varies from 0.167 g to 0.355 g using deterministic approach, which is approximately 55% higher as compare to Indian codal provision i.e. 0.160 g (IS 1893:2016). This PGA model will be further useful for the city planning, risk reduction, urban rehabilitation, seismic mitigation planning and detailed microzonation study of the region. Further studies are needed to evaluate PGA at surface level considering local site conditions. Different studies have been carried out using DSHA approach for the peninsular India.

References

Ambraseys, N.N., Douglas, J.: Magnitude calibration of north Indian earthquakes. Geophys. J. Int. **159**, 165–206 (2004)

Bapat, A., Deshpande, N.V., Das, P.B., Bhavnarayana, V.: Pre-impoundment seismicity studies around Sardar Sarovar Site. CBIP 55 R & D session (1989)

Biswas, S.K.: Regional tectonic framework, structure and evolution of the western marginal basins of India. Tectonophysics **135**, 307–327 (1987)

Biswas, S.K.: A review of structure and tectonics of Kutch basin, western India, with special reference to earthquake. Curr. Sci. **88**, 1592–1600 (2005)

Bodin, P., Malagnini, L., Akinci, A.: Ground motion scaling in the Kutch Basin, India, deduced from aftershocks of the 2001 Mw 7.6 Bhuj earthquake. Bull. Seismol. Soc. Am. **94**, 1658–1669 (2004)

Campbell, K.W.: Prediction of strong ground motion using the hybrid empirical method and its use in the development of ground-motion (attenuation) relations in eastern North America. Bull. Seismol. Soc. Am. **93**, 1012–1033 (2003)

Chandra, U.: Seismotectonics of Himalaya. Curr. Sci. **62**(1–2), 40–71 (1992)

Cramer, C.H., Kumar, A.: 2001 Bhuj India, earthquake engineering seismoscope recordings and eastern North America ground motion attenuation relations. Bull. Seismol. Soc. Am. **93**, 1390–1394 (2003)

Das, R., Wason, H.R., Sharma, M.L.: Global regression relations for conversion of surface wave and body wave magnitudes to moment magnitude. Nat. Hazards **59**, 801–810 (2011)

Das, R., Wason, H.R., Sharma, M.L.: Temporal and spatial variations in the magnitude of completeness for homogenized moment magnitude catalogue for northeast India. J. Earth Syst. Sci. **121**(1), 19–28 (2012)

Das, R., Wason, H.R., Sharma, M.L.: General orthogonal regression relations between body-wave and moment magnitudes. Seismol. Res. Lett. **84**(2), 219–224 (2013)

Gardner, J.K., Knopoff, L.: Is the sequence of earthquakes in southern California, with aftershocks removed poissonian. Bull. Seismol. Soc. Am. **64**(5), 1363–1367 (1974)

GSI: Seismotectonic Atlas of India and its Environs. Geological Survey of India (2000)

Hwang, H., Huo, J.R.: Attenuation relations of ground motion for rock and soilsites in eastern United States. Soil Dyn. Earthq. Eng. **16**, 363–372 (1997)

IS 1893 (Part 1): Indian Standard Criteria for Earthquake Resistant Design of Structures. Part 1 General Provisions for Buildings. Bureau of Indian Standards, New Delhi (2001)

Johnston, A.C.: Seismic moment assessment of earthquakes in stable continental regions-I, instrumental seismicity. Geophys. J. Int. **1996**(124), 381–414 (1996)

Malik, J.N., Sohoni, S., Karanth, V., Merh, S.: Modern and historic seismicity of Kutch Peninsula, western India. J. Geol. Soc. India **54**, 545–550 (1999)

Merh, S.S.: Geology of Gujarat, p. 222. GeolSocInd, Banglore (1995)

National Disaster Management Authority: Development of probabilistic seismic hazard map of India, Technical report Published by Government of India, Working committee of experts (WCE), NDMA (2010)

Oldham, T.: A catalogue of Indian earthquakes from the earliest times to the end of 1869 A.D. Mem. Geol. Surv. India **XIX**(3), 163–215 (1883)

Pezeshk, S., Zandieh, A., Tavakoli, B.: Hybrid empirical ground-motion prediction equations for eastern north America using NGA models and updated seismological parameters. Bull. Seismol. Soc. Am. **101**(4), 1859–1870 (2011)

Raghukanth, S.T.G., Iyengar, R.N.: Estimation of seismic spectral acceleration in Peninsular India. J. Earth Syst. Sci. **116**(3), 199–214 (2007)

Schweig, E., Gomberg, J., Petersen, M., Ellis, M., Bodin, P., Mayrose, L., Rastogi, B.K.: The Mw 7.7 Bhuj earthquake: global lessons for earthquake hazard in intra-plate regions. J. Geol. Soc. India **61**(3), 277–282 (2003)

Scordilis, E.M.: Empirical global relations converting Ms and mb to moment magnitude. J. Seismol. **10**, 225–236 (2006)

Sinha-Roy, S., Malhotra, G., Guha, D.S.: A Transect across Rajasthan precambrian terrain in relation to geology, tectonics and crustal evolution of south central Rajasthan, In: Sinha-Roy, S., Gupta, K. R. (eds.) Continental Crust of Northwestern and Central India, Geological Society of India, vol. 31, pp. 63–90 (1995)

Thingbaijam, K., Nath, S., Yadav, A., Raj, A., Walling, M., Mohanty, W.: Recent seismicity in northeast India and its adjoining region. J. Seismol. **12**, 107–123 (2008)

Toro, G. R.: Modification of the Toro et al. (1997) attenuation equations for large magnitudes and short distances, Risk Engineering, Boulder (2002)

Uhrhammer, R.: Characteristics of Northern and Southern California Seismicity, Earthq. Notes **57** (1986). 21 pages

Yadav, R., Bormann, P., Rastogi, K., Das, C., Chopra, S.: A homogeneous and complete earthquake catalog for northeast India and the adjoining region. Seismol. Res. Lett. **80**(4), 609–627 (2009)

Investigation on the Reliquefaction Behaviors of Sand Using Shaking Table Tests

Bin Ye[✉] and Hailong Hu

Department of Geotechnical Engineering, College of Civil Engineering,
Tongji University, Shanghai, China
yebin@tongji.edu.cn

Abstract. The reliquefaction behaviors of sand have been investigated by many scholars and the research results of different scholars are inconsistent. In order to test the validity of sand's reliquefaction behaviors, electromagnetic shaking table tests were carried out with different shaking durations. In each test, prepared sandy ground was shaken several times in succession till it wouldn't reliquefy. Before and after each shaking event, the density of sand was calculated by measured sandy ground's height so that the influence of density on sand reliquefaction behaviors could be explored. The test results demonstrate that density plays a crucial part in sand reliquefaction behaviors. Sand reliquefied zero to three times according to various shaking durations. Sand even experienced the transformation from liquefied to not liquefied state during the first shaking event if the shaking duration lasted long enough. During each reliquefaction shaking event, circles required to trigger reliquefaction were fewer than that in first liquefaction, inferring that sand reliquefaction decreased despite of an increase in density.

Keywords: Reliquefaction · Shaking table

1 Introduction

Liquefaction of sandy ground is a common geo-disaster when earthquake happens. It is one of the main factors causing the subsidence of ground and destruction to structures. Traditional liquefaction research mainly focus on the first liquefaction of sand. However, some laboratory tests reveal that sand can liquefy more than once when cyclic loadings were imposed in sand successively. What's more, the resistance of sand's second liquefaction (or called "reliquefaction") may be obviously lower than its first liquefaction even if the sand has been densified after the first liquefaction due to the dissipation of excess pore water pressure. The researchers who revealed this phenomenon firstly were Finn et al. (1970). They conducted reliquefaction tests on sand by simple and triaxial shear tests, and the results showed that sand reliquefies more easily than its first liquefaction despite the increase in density.

After that, the phenomenon that sand reliquefies more easily was proved by (Ishihara and Okada 1978) and Oda et al. (2001) with triaxial tests. They also further investigated the mechanism of sand reliquefaction. The same reliquefaction behavior was also discovered by Wang et al. (2013). They conducted triaxial tests on silt and showed that silt reliquefaction resistance decrease too. However, Yamada et al. (2010)

T. Qiu et al. (Eds.): GSIC 2018, *Proceedings of GeoShanghai 2018 International Conference: Advances in Soil Dynamics and Foundation Engineering*, pp. 300–307, 2018.
https://doi.org/10.1007/978-981-13-0131-5_33

carried several reliquefaction tests successively on the same sand specimen with triaxial tests. They found that the reliquefaction resistance of sand was both able to increase and decrease. Whether the reliquefaction resistance increased or not depended on the level of anisotropy developed during the last liquefaction. Besides the triaxial tests, some researchers investigated the reliquefaction behaviors of sand ground through model tests. Ha et al. (2011) used a shaking table to vibrate a sand ground five times successively with the same input motion. It was found that sand reliquefaction resistance in the second vibration was obviously lower than its liquefaction resistance in the first vibration. However, from the third vibration on, the reliquefaction resistance began to increase, and the sand ground couldn't reliquefy during the fourth and fifth vibration. Ye et al. (2007) also conducted similar shaking table tests on sand ground. They vibrated the sand ground three times successively with the same input motion but found that sand ground's reliquefaction resistance didn't decrease.

In general, researchers have made great efforts to investigate sand reliquefaction behaviors, but the current research results are in consistent. In this paper, we tried to investigate the sand reliquefaction behaviors by shaking table tests. Firstly, a series of shaking table tests were carried out on a sandy specimen. During each test, sandy ground was shaken several times successively till it wouldn't reliquefy. Before and after each shaking event, sandy specimen's density was calculated to ascertain density's effect on its reliquefaction behavior. As a result, sand reliquefaction behaviors, including whether sand being able to reliquefy or not, sand reliquefaction resistance and reliquefaction times, were investigated.

2 Materials and Methods

2.1 Tested Sand

The tested sand in this study was Fujian standard sand. This sand is widely used in geotechnical experiments in China. To produce a sand specimen more prone to liquefy, grains larger than 355 μm in size were sieved out from the tested sand. Figure 1 shows the grain size distribution curves of the tested sand, and Table 1 presents its physical properties.

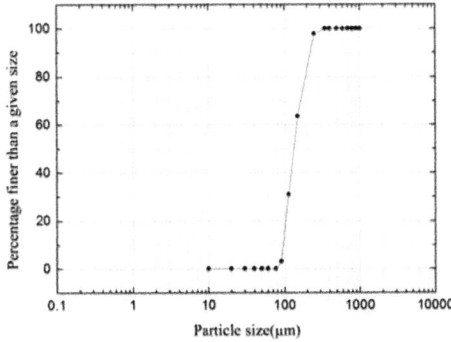

Fig. 1. Particle size distribution curve of tested sand.

Table 1. Physical properties of tested sand

Specific gravity of soil particles, G_s (g/cm^3)	2.64
Maximum void ratio, e_{max}	0.958
Minimum void ratio, e_{min}	0.611

2.2 Experimental Apparatus

The shaking-table device used in this study was a small electromagnetic shaking table. A model box with an interior dimension of 50 cm (in length) × 20 cm (in width) 30 cm (in height) was fixed on the shaking table. Two 4 cm thick sponges were attached to the shorter edges of the model box respectively to reduce the boundary effect. A pore pressure transducer was buried in the center of the ground to measure the generated excess pore water pressure (EPWP). Another accelerometer was fixed on shaking table to measure the input motion. Three transparent plastic rules were attached to the outside of the model box to record the height of the model ground before and after each shaking event. The average height of the three recorded heights was used to calculate the density of the sandy ground in each shaking event. The positions of rules, pore pressure transducer are showed in Fig. 2.

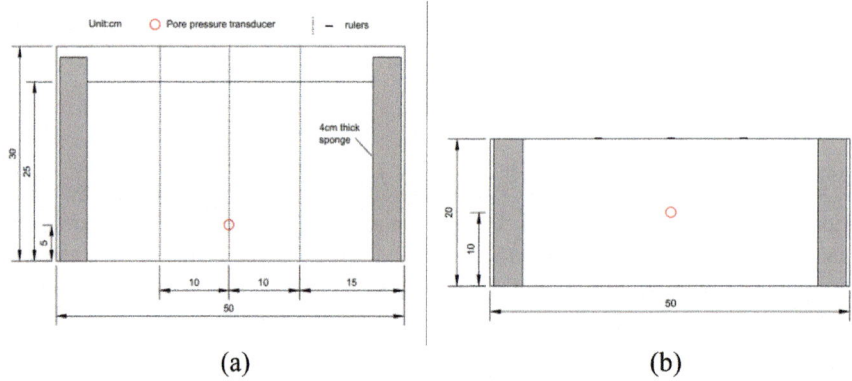

Fig. 2. Schematic model and instrumentation layout: (a) front view; (b) top view.

2.3 Preparation of the Model Ground

Specimens with a height of 25 cm were prepared by water sedimentation method in the model box for each test. Specimens preparation mainly involved following steps: (1) The measurement sensors were settled in planed positions. (2) Water was added into the model box to a depth of 25 cm. (3) A 425-μm wire sieve was fixed in the water surface. (4) The saturated sand was slowly transferred by a scoop into the model box;

the sand moved through the wire sieve and was then deposited by gravity in the water. According to the mass and volume of the sand in the model box, the calculated relative density of the specimens was approximately 42%.

2.4   Procedure of Shaking Table Test

A total of 11 groups of tests were carried out to investigate liquefaction behaviors of sandy ground. The only difference between these 11 groups of tests was the shaking duration, which is listed in Table 2. In each test, the prepared ground was shaken several times successively until it no longer reliquefied. There was an interval of ten minutes between two contiguous shaking events to let the excess pore water pressure generated in the ground dissipate completely. The ground subsidence was measured before and after each shaking event.

Table 2. Shaking duration of each test

Test number	Shaking duration	Test number	Shaking duration
1	1.5 s	7	7.5 s
2	2.5 s	8	8.5 s
3	3.5 s	9	9.5 s
4	4.5 s	10	10.5 s
5	5.5 s	11	42 s
6	6.5 s		

3   Test Results and Discussions

3.1   Input Acceleration, Pore Water Pressure and Settlement Response

Figure 3 presents the input motion measured by the accelerometer fixed on the shaking table in Test 2. The input acceleration was approximate to sine wave with a frequency of 4 Hz. As shown in the figure, the input acceleration reached its full amplitude after the first two circles and maintained for 2.5 s. When the input motion stopped, the wave would decay gradually as a result of inertia. The input motions for other tests were similar to Fig. 6 and only differentiated in duration.

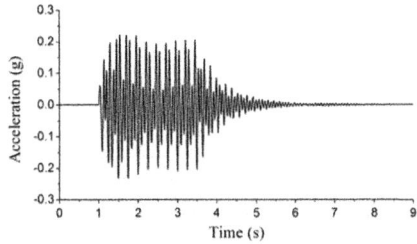

Fig. 3. Time history of the input motion measured in test 2

Sand liquefaction happens when EPWP equals the initial vertical effective stress firstly in a shaking event. Figure 4 shows EPWP time histories for each shaking event for Test 2, Test 10 and Test 11. During the first shaking of Test 2, EPWP increased slowly in the first two circles and rapidly reached at r_u line in the third circle. While during the second and third shaking event, circles needed for EPWP reaching at r_u line were 1 and 2, both fewer than which in first shaking event. EPWP begun to decay before it reached r_u line, meaning that sand couldn't liquefy during the fourth shaking event. Another interesting phenomenon was that, during the first shaking event in Test 2, EPWP maintained liquefied value for a period before it begun to dissipate. While during the second shaking event, this period become shorter and finally disappeared during the third shaking event.

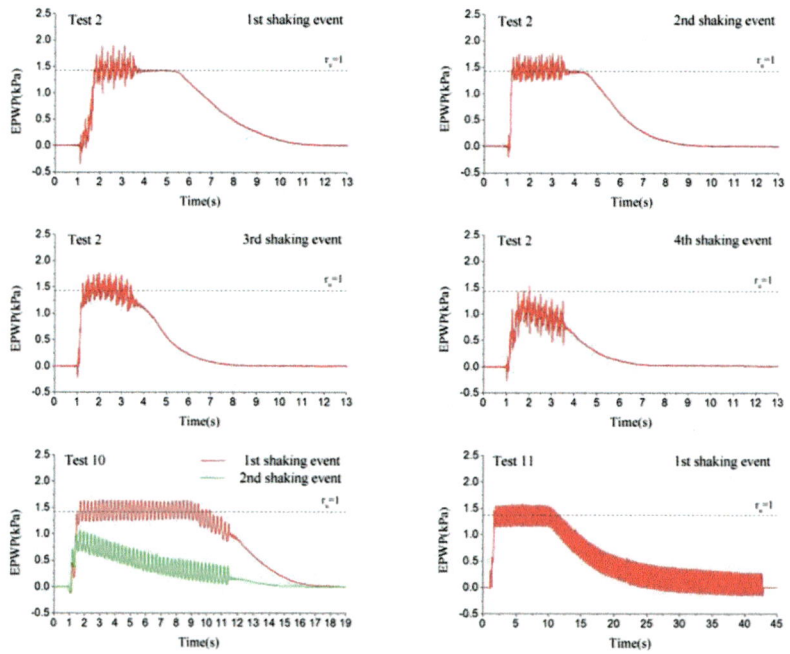

Fig. 4. Excess pore water pressure generation measured in Test 2, Test 10 and Test 11.

This phenomenon can be explained as follows. At the moment that input shaking motion stopped in a liquefied shaking event, sand particles are still in suspension and begin to sink. Sand particles which are near the bottom of model box complete their sinking firstly and the upper particles fall on it. At the same time, new soil structure forms from the bottom to the top in model box. Before the new structure forms to the plane that pore pressure transducer lied in, all particles upper to the horizontal plane of pore pressure transducer are still in suspension; therefore EPWP maintains the same liquefied value. When new structure upper to the horizontal plane of pore pressure transducer forms, sand particles' gravity is undertaken by the new structure gradually, so EPWP begin to dissipate. As shaking event increases, sand ground becomes denser at the moment that input shaking motion stops so that the distance for suspended sand particles to fall shortened. As a result, the new structure forms faster and the time for EPWP to maintain liquefied value become shorter.

Compared with Test 2 in which sand reliquefied twice, sand could only experience its first liquefaction in Test 10. As shown in Test 10 in Fig. 4, the recession of EPWP appeared in the later period of shaking during the first shaking event and EPWP was already unable to reach r_u line during the second shaking event. In test 11, EPWP even decayed to 0 kPa during the first shaking event as the shaking duration lasted long enough. It also could be concluded from these tests that the decay of EPWP from r_u line in a liquefied shaking event implied that sand couldn't reliquefy in the next shaking event. In addition, the EPWP of last shaking event still increased faster than it of first shaking event during the first circle.

Figure 5 provides the total settlement and corresponding density of sand ground after each shaking event in some tests. The filled and hollow points denote that the ground liquefied and did not liquefy during the shaking event, respectively. As illustrated in Fig. 5, it is clear that greater increases in density can be observed in the shaking events with longer shaking durations. As sand became denser and denser, the settlement after each shaking event become smaller and smaller in a certain test. There also existed a specific total settlement in sand marked by dotted line in Fig. 5 that sand couldn't reliquefy once total settlement had surpassed this value. As total settlement has one-to-one relationship with density, this also suggested that there is a specific density called liquefaction boundary density that sand can't reliquefy once current density has exceeded it.

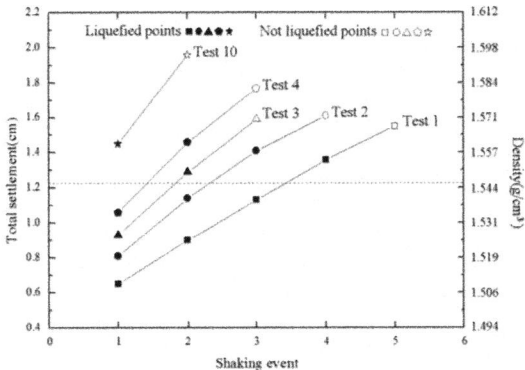

Fig. 5. The total settlement of five listed tests after each shaking event.

3.2 Characteristics of Reliquefaction Times and Reliquefaction Resistance

Figure 6 shows the relation between reliquefaction times and shaking durations of Test 1 to Test 10. As illustrated in Fig. 6, reliquefaction times decreased as shaking duration extended on the whole. The maximum reliquefaction times were 3 in Test 1 and reliquefaction times decreased linearly from 3 to 1 during the first three tests and then maintain 1 until Test 7. Starting with Test 8, sand ground wouldn't reliquefy anymore. Considering what has discussed in Fig. 5, conclusions can be drawn that whether sand can reliquefy depends on its current density. If the current density is smaller than liquefaction boundary density after several shaking events, sand will reliquefy during next shaking event. However, sand won't reliquefy if current density has reached liquefaction boundary density even if sand has just experienced its first shaking event.

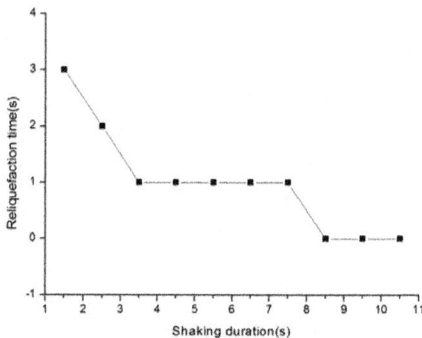

Fig. 6. Variation of reliquefaction times with different shaking durations.

Table 2 reports the number of circles required to trigger liquefaction of the first four shaking events of all 11 tests. During the first shaking event, the number of cycles required to trigger sand liquefaction was three and it all was reduced to one during second shaking event and third shaking event of Test 1. However, during the third and fourth shaking event in Test 1 and Test 2 respectively, sand experienced its last liquefaction and the number increased to two. To sum up, the number of circles required to trigger liquefaction during the first shaking event was larger than that required to trigger liquefaction during second, third and fourth shaking events in all tests if sand could reliquefy. These results suggest that if can be reliquefied, the reliquefaction resistance of sand is lower than its first liquefaction no matter in which shaking event in spite of an increase in density (Table 3).

Table 3. Number of cycles required to trigger liquefaction during each of four shaking events

Test number	1st shaking event	2nd shaking event	3rd shaking event	4th shaking event
1	3	1	1	2
2	3	1	2	Not liquefied
3	3	1	Not liquefied	Not liquefied
4–7	3	1	Not liquefied	Not liquefied
8–11	3	Not liquefied	Not liquefied	Not liquefied

4   Conclusions

In this study, electromagnetic shaking table tests were carried out with different shaking durations to test the validity of sand ground's liquefaction behavior. The main conclusions are drawn as follows:

1. Weather sand ground reliquefied or not depended on the shaking duration. In short shaking duration tests, sand ground could reliquefy several times. With shaking duration increasing, reliquefaction times decreased gradually until sand ground couldn't reliquefy.
2. There exists a liquefaction boundary density that can be used to tell whether a sandy ground can reliquefy or not. If the sand density after liquefaction is lower than the liquefaction boundary density, it can reliquefy, while if the sand density exceeds the liquefaction boundary density, the sand will not reliquefy again.
3. If sandy ground can reliquefy, the EPWP ascended faster faster than that during the first liquefaction, implying that sand reliquefaction resistance is lower than its first liquefaction resistance despite of an increase in density.

References

Finn, W.D.L., Bransby, P.L., Pickering, D.J.: Effects of strain history on liquefaction of sands. Soil Mech. Found. Div. **96**(6), 1917–1934 (1970). Proceeding of ASCE

Ishihara, K., Okada, S.: Effects of stress history on cyclic behavior of sand. Soils Found. **18**(4), 31–45 (1978)

Oda, M., Kawamoto, K., Suzuki, K., Fujimori, H., Sato, M.: Microstructural interpretation on reliquefaction of saturated granular soils under cyclic loading. J. Geotech. Geoenviron. Eng. **127**(5), 416–423 (2001)

Ha, I.S., Olson, S.M., Seo, M.W., Kim, M.M.: Evaluation of reliquefaction resistance using shaking table tests. Soil Dyn. Earthq. Eng. **31**, 682–691 (2011)

Wang, S., Yang, J., Onyejekwe, S.: Effect of previous cyclic shearing on liquefaction resistance of Mississippi River Valley silt. J. Mater. Civ. Eng. **25**(10), 1415–1423 (2013)

Ye, B., Ye, G.L., Zhang, F., Yashima, A.: Experiment and numerical simulation of repeated liquefaction-consolidation of sand. Soils Found. **47**(3), 547–558 (2007)

Yamada, S., Takamori, T., Sato, K.: Effects on reliquefaction resistance produced by changes in anisotropy during liquefaction. Soils Found. **50**(1), 9–25 (2010)

Dynamic Properties of Mixtures of Waste Materials

Yujie Qi[1], Buddhima Indraratna[2(✉)], and Jayan S. Vinod[1]

[1] University of Wollongong, Wollongong, Australia
[2] Centre for Geomechanics and Railway Engineering,
University of Wollongong, Wollongong, Australia
indra@uow.edu.au

Abstract. The stockpiling of waste mining by-products, i.e. steel furnace slag (SFS) and coal wash (CW) has brought significant environmental hazard and attracted research attention to reuse them in a more innovative way. In recent years, SFS+CW mixtures have been successfully applied in geotechnical projects, while the inclusion of rubber crumb (RC, from waste tyres) will extend them into dynamic projects. Thus the investigation of the geotechnical properties of SFS+CW+RC mixtures under dynamic loading is in urgent need. In this paper, the dynamic properties (i.e. shear modulus and damping ratio) have been explored based on extensive drained cyclic triaxial tests. The influences of number of loading cycles, RC contents, shear strain level, and the effective confining pressure have been presented. The dynamic properties of SFS+CW +RC mixtures presented in this paper will be essential for the application of the mixtures in the seismic isolation projects or railway foundation.

Keywords: Waste materials · Dynamic loading · Shear modulus
Damping ratio

1 Introduction

Steel furnace slag (SFS) and coal wash (CW) are by-products from steel making and coal mining industries, respectively. They are very common waste materials in Australia, and in the Wollongong region (Australia) alone, the production of SFS and CW could be several million tons per year [1]. Rubber crumbs (RC) are granulated materials from waste tires. The stockpiles of waste tires can lead to serious environmental hazards and have caused great public concern to reuse them. One of the best ways to deal with this problem is to reuse these waste materials into civil engineering projects.

As the detrimental properties of these waste materials (i.e. the swelling potential of SFS, the particle breakage of CW, and the low shear strength and high deformation of RC), they are usually blended with other materials when used in civil engineering. For instance, SFS are usually mixed with fly ash or cement to be served as landfill or used in unbound pavements [2, 3], and the SFS+CW mixtures have been successfully used in practical engineering applications such as port reclamation [1, 4] and landfill projects [5, 6]. With the high damping property, RC is usually blended with sand or other soils applied in seismic conditions or as an integral part of vibration damping systems for

© Springer Nature Singapore Pte Ltd. 2018
T. Qiu et al. (Eds.): GSIC 2018, *Proceedings of GeoShanghai 2018 International Conference: Advances in Soil Dynamics and Foundation Engineering*, pp. 308–317, 2018.
https://doi.org/10.1007/978-981-13-0131-5_34

machine foundations and railroads [7, 8]. Moreover, Indraratna et al. [9] developed an energy absorbing layer for subballast by adding RC into SFS+CW mixtures.

Since subballast is subjected to cyclic loading, it is of great importance to understand the cyclic loading behavior of SFS+CW+RC mixtures, especially the shear modulus and damping ratio. The aim of this paper is to investigate the influence of RC contents $R_b(\%)$, the loading cycles, the shear strain level, and the effective confining pressure on the shear modulus and damping ratio of SFS+CW+RC mixtures based on drained cyclic loading triaxial tests.

2   Laboratory Investigations

2.1   Materials

The SFS and CW used in this study were provided by Illawarra Coal and Australia Steel Milling Services, respectively. RC was from waste tires and three different sizes (0–2.3 mm, 0.3–3 mm, and 1–7 mm) were used. The particle size distribution (PSD) of SFS, CW, and RC are shown in Fig. 1. According to the unified soil classification system, SFS and CW can be classified as well-graded gravel with silty-sand (GW-GM), and well-graded sand with gravel (SW), respectively, while RC can be referred to as granulated rubber.

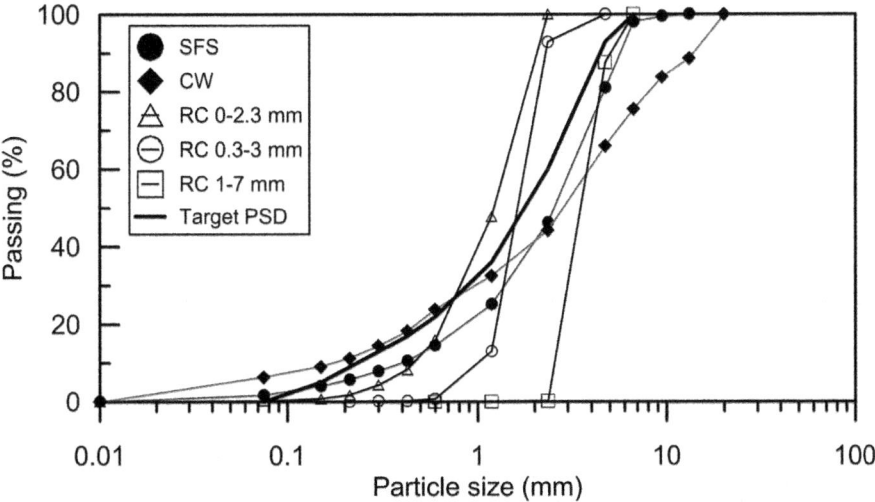

Fig. 1. PSD of SFS, CW, RC, and the target PSD for SFS+CW+RC mixtures

2.2   Specimen Preparation and Testing Program

To exclude the influence of gradation, all the mixtures tested in this study were mixed to the same gradation (the target PSD), also shown in Fig. 1. Please note that the waste materials (i.e. SFS, CW, and RC) were blended by weight, and the content rate of SFS and CW was set to be SFS:CW = 7:3 as with this rate the waste mixtures can maintain higher shear strength and less particle breakage [9], then 0%, 10%, 20%, 30%, and 40% RC were added to the SFS+CW mixtures. All the specimens were prepared at the optimum moisture content and compacted to achieve an initial dry unit weight equivalent to 95% of their maximum dry density to simulate typical field conditions of subballast.

A series of stress-controlled drained cyclic triaxial tests were carried out for the SFS +CW+RC mixtures following the procedure suggested by ASTM D5311/D5311M [10]. The specimens were compacted in three layers and had 50 mm in diameter by 100 mm high. In this study, an appropriate range of effective confining pressure (i.e. $\sigma'_3 = 10, 40,$ and 70 kPa) was used to simulate the field conditions of railway subballast depending on the typical axle loads (heavy haul) and heights of track embankments in the state of NSW [11, 12]. Moreover, to simulate the good drainage condition and the long term permanent settlement response of the subballast layer, the cyclic loading tests were conducted under drained condition, which was in agreement with Suiker et al. [15].

Cyclic loading tests were conducted following three stages, i.e. saturation, con-solidation, and cyclic loading. During the saturation stage, the specimens were flooded with de-aired water and then the back pressure was applied at an increasing rate of 1 kPa/min until 500 kPa was reached. This stage was completed when the Skempton's B-value exceeded 0.98, and then isotropic consolidation was carried out under the desired effective confining pressure of 10, 40 or 70 kPa. After consolidation, the cyclic loading stage was conducted at CSR = 0.8 (cyclic stress ratio, Eq. 1), using a loading frequency of $f = 5$ Hz. The deviator stress used in this study is governed by σ'_3 and cyclic stress ratio, CSR. For CSR = 0.8, the confining pressures of $\sigma'_3 = 10, 40,$ and 70 kPa correspond to deviator stresses of 16, 64, and 112 kPa, respectively. These values are in line with the observed stress conditions generated in typical freight tracks [12]. All the cyclic loading tests were continued for 50000 cycles.

$$CSR = \frac{\sigma_a}{2\sigma'_3} \tag{1}$$

Where, CSR is the cyclic stress ratio; σ_a is the peak cyclic axial stress; and σ'_3 is the effective confining pressure.

3 Test Results

The shear modulus G and the damping ratio D are the two key parameters needed to estimate the stiffness and energy absorbing capacity of soil. Damping is the loss of energy within a vibrating or a cyclically loaded system which is usually dissipated in the form of heat or breakage for granular materials; it is commonly used to measure the damping capacity for energy dissipation during dynamic or cyclic loading. The defi-nition of the shear modulus and damping ratio is presented in Fig. 2; where the area of the hysteretic loop A_2 in the shear stress-shear strain plain represents the energy dis-sipated during a loading cycle, while four times the area of the triangle A_1 is the maximum elastic energy absorbed during the cycle [13].

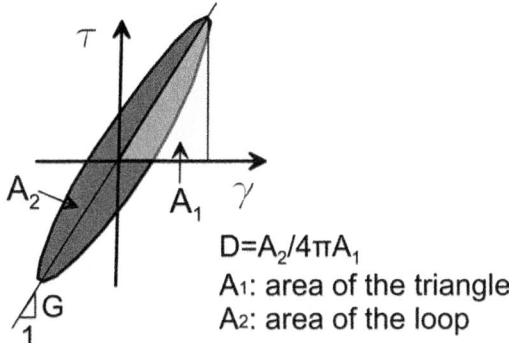

$$D=A_2/4\pi A_1$$
A_1: area of the triangle
A_2: area of the loop

Fig. 2. Definition of shear modulus and damping ratio (after Kokusho [13])

3.1   The Effects of $R_b(\%)$ and the Effective Confining Pressure

Figure 3 shows the shear modulus and damping ratio of SFS+CW+RC mixtures with different $R_b(\%)$ versus loading cycles in logarithm. It can be noted that the addition of RC has a significant influence on the shear modulus and damping ratio of SFS+CW +RC mixtures. As with previous studies of rubber-sand mixtures (e.g. [7, 8, 14]), the shear modulus decreases with increasing $R_b(\%)$ because of the low stiffness of rubber materials. Unlike shear modulus, the damping ratio of SFS+CW+RC mixtures increases with $R_b(\%)$ indicating the high damping properties of rubber materials. However, the SFS+CW+RC mixtures with $R_b \geq 10\%$ tend to achieve a similar damping ratio after 10000 cycles (Fig. 3a). This is because the inclusion of RC increases the area of the hysteretic loop, but as R_b increase the hysteretic loop becomes more inclined, which then causes a rapid increase in the area of the triangle A_1, and this also suggests that the damping capacity of the waste mixtures with $R_b \geq 10\%$ is similar at high loading cycles, while for the waste mixtures without rubber the value of the damping ratio is stable albeit a little fluctuation after 10 cycles.

The shear modulus calculated from the test result of traditional subballast (well-graded sand with gravel) tested by Suiker et al. [15] is also shown in Fig. 3(b). The test conditions were the same with this study except the deviator stresses applied were different. Here only the result of $q = 175$ *and* 91 kPa are presented. It can be seen that the shear modulus increases as the deviator stress increases, and thus it can be estimated that when $q = 112$ kPa at $\sigma'_3 = 70$ kPa (same with this study), the value of the shear modulus of subballast would be similar with SFS+CW+RC mixtures having 0% RC. Therefore, only the waste mixtures with $R_b \leq 10\%$ have acceptable stiffness comparing with traditional subballast.

The influence of effective confining pressure σ'_3 on the shear modulus and damping ratio of SFS+CW+RC mixtures is presented in Fig. 4(a) and (b), respectively. It is clear that with the same $R_b(\%)$, as σ'_3 increases the shear modulus of the waste mixtures increases while the damping ratio decreases.

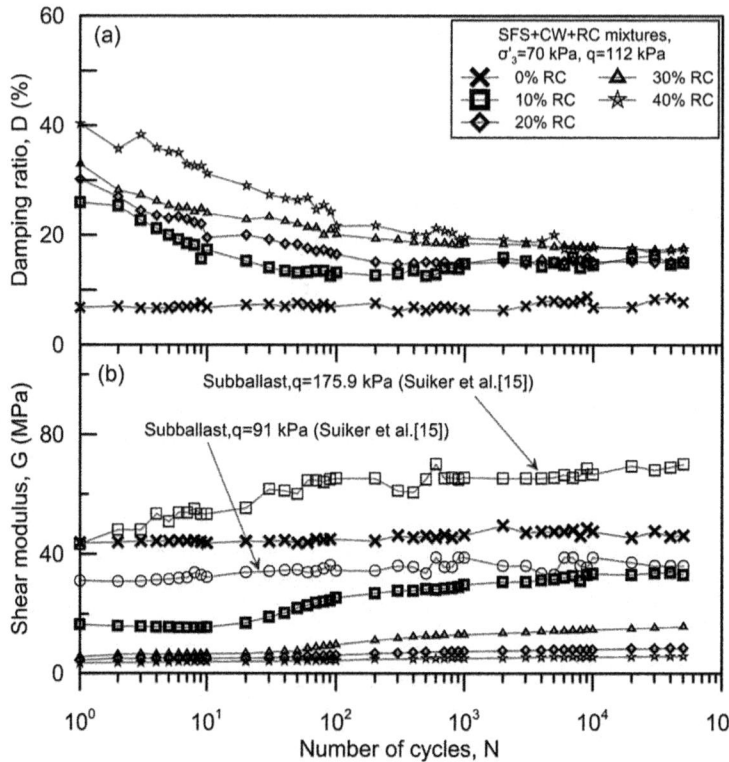

Fig. 3. Cyclic loading results of traditional subballast and SFS+CW+RC mixtures with different $R_b(\%)$: (a) damping ratio, and (b) shear modulus

Figure 5 shows the evolution of the shear modulus and damping ratio at 1000 and 10000 cycles varying with $R_b(\%)$ and σ'_3. It is evident from Fig. 5 that of the variation of shear modulus and damping ratio with R_b at 1000 cycles is similar to that at 10000 cycles. Note that the effect of confining pressures on shear modulus weakened as R_b increases, which in line with past studies such as Nakhaee and Marandi [16]. This is because as more RC included, the waste mixes tend to behave more elastic, and the influence of the confining pressure become insignificant [7]. Obviously, the behaviour of shear modulus and the damping ratio is governed mainly by R_b. When $R_b < 20\%$, the shear modulus decreases and the damping ratio increases as R_b increase. However, when $R_b > 20\%$ both the shear modulus and the damping ratio only change a little indicating that the rubber crumbs has formed the skeleton of the specimen and the specimen behaves rubber-like. It is worthy to note that when R_b increases in the range of $10\% \leq R_b \leq 20\%$, only a minor increase happens to the damping ratio, suggesting that 10% RC is sufficient for the purpose of energy absorbing.

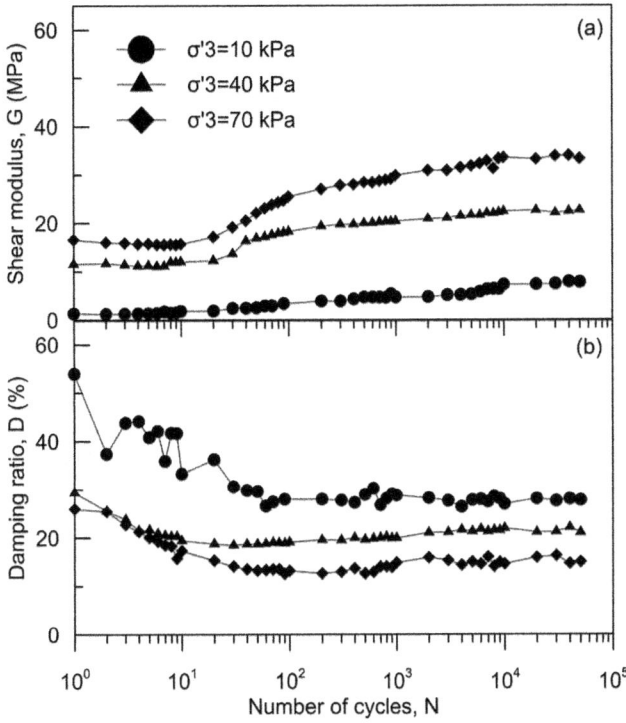

Fig. 4. Cyclic loading results of SFS+CW+RC mixtures with $R_b = 10\%$ under different confining pressures: (a) shear modulus, and (b) damping ratio

3.2 The Effects of Cyclic Loading Cycles

The effect of cyclic loading cycles on the shear modulus and damping ratio of SFS +CW+RC mixtures can be observed in Figs. 3 and 4. It can be seen that at $\sigma'_3 = 70$ kPa the shear modulus of the waste mixtures with $R_b \geq 10\%$ stays stable during the first 10 cycles and increases at a reducing rate after 10 cycles suggesting that the stiffness of these waste mixtures increases with the contraction of the specimen. For the waste mixtures with $R_b = 0\%$ the shear modulus fluctuates marginally after 10 cycles indicating a stable stiffness of the waste mixtures with no rubber. In Figs. 3(a) and 4(b), note that the damping ratio of SFS+CW+RC mixtures with $R_b \geq 10\%$ decreases as the loading cycles increase albeit at a reducing rate, while the damping ratio of the waste mixture without rubber keep stable as the cyclic test continuing. It is worthy to note that both the shear modulus and the damping ratio of SFS+CW+RC mixtures stabilized after 10000 cycles.

3.3 The Effects of Shear Strain Level

Figure 6(a and b) shows the effect of shear strain level on the shear modulus and damping ratio of the waste mixtures as well as traditional subballast (after Suiker et al. [15])

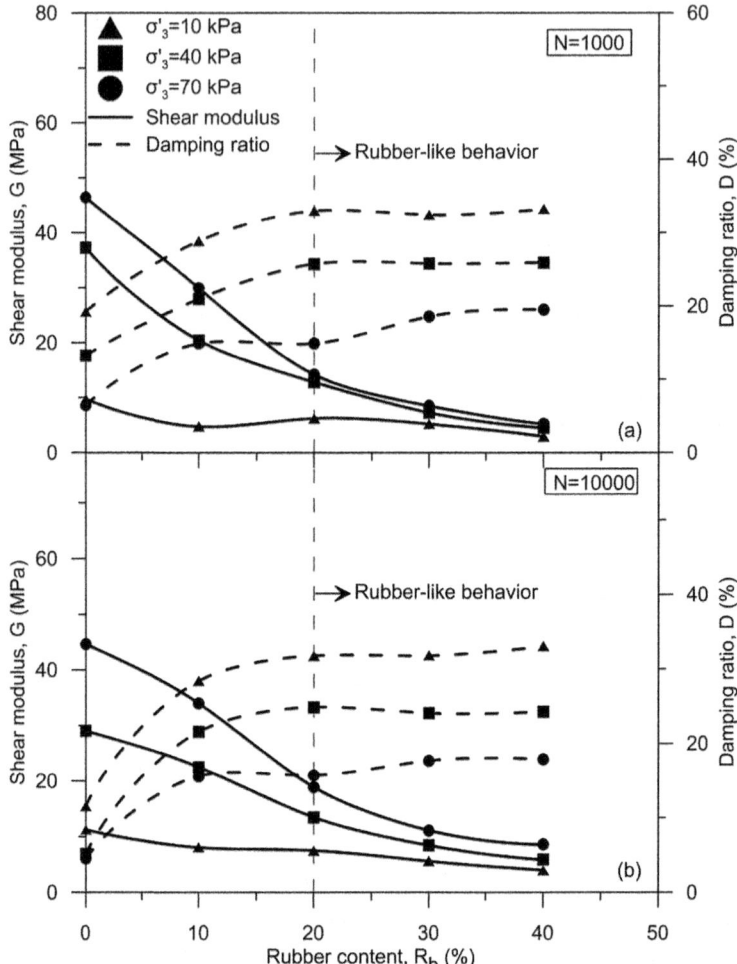

Fig. 5. Shear modulus and damping ratio of waste mixtures (SFS:CW = 7:3) changing with R_b at (a) N = 1000 cycles, and (b) N = 10000 cycles

and sand-RC mixtures (after Li et al. [8]). It is evident from Fig. 6(a) that G decreases with an increase in the shear strain regardless of the magnitude of RC. The variation of G with shear strain for SFS+CW+RC compares well with the subballast material reported by Suiker et al. [15]. The value of shear modulus for sand-RC mixtures decreases as the RC contents increase and the confining pressure decreases, therefore it can be argued that with $RC \leq 10\%$ and at $\sigma'_3 = 70$ kPa, the shear modulus of the sand-RC mixtures is much lower than SFS+CW+RC mixtures. Figure 6(b) shows the damping ratio of SFS+CW mixtures increases with shear strain level and a sharp increase occurs when the shear strain reach a certain level. In addition, the damping ratio at this level of RC (10%) is comparable to that obtained with sand-RC mixtures (Fig. 6).

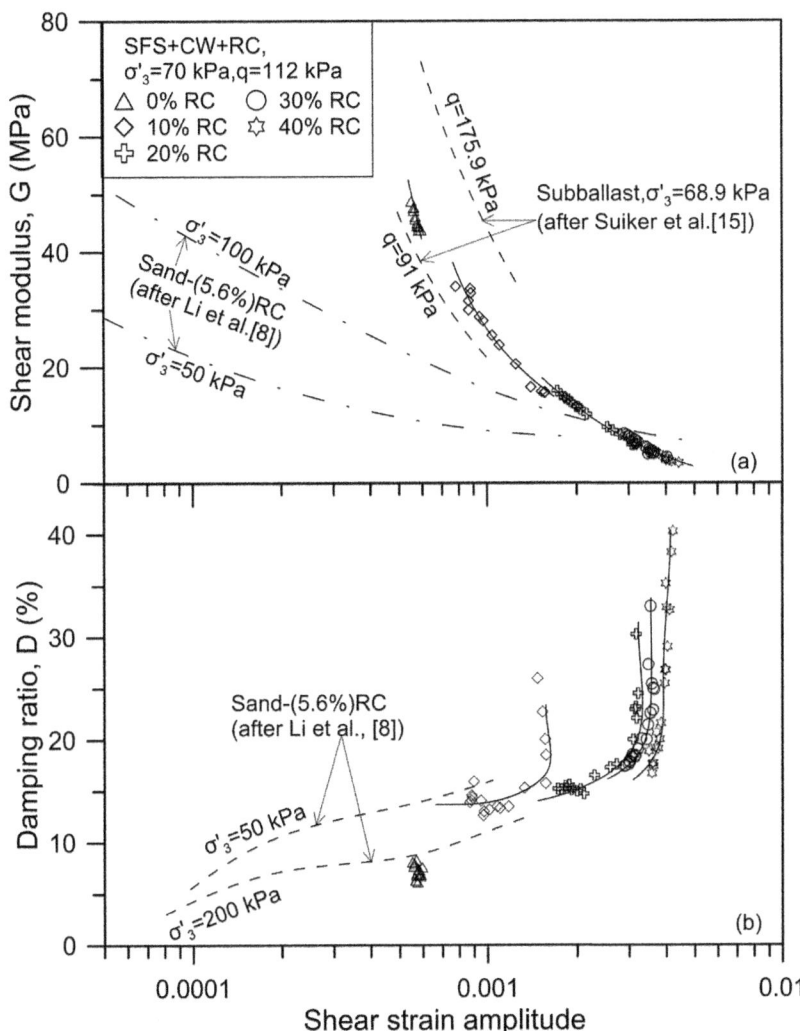

Fig. 6. Shear modulus of SFS+CW+RC mixtures changing with shear strain amplitude and comparison with traditional subballast [15] and sand-RC mixtures [8]; (b) Damping ratio of SFS +CW+RC mixtures changing with shear strain amplitude.

4   Conclusions

This paper investigates the influence of R_b, σ'_3, and the cyclic loading cycles on the shear modulus and damping ratio of SFS+CW+RC mixtures (SFS:CW = 7:3) based on drained cyclic loading triaxial tests. The test result reveals that the addition of RC caused the shear modulus to decrease and the damping ratio to increase indicating that the stiffness of the waste mixtures decreased, while the absorbed energy dissipated to heat or breakage became more efficient. It was also found that the behavior of shear modulus and damping ratio was controlled by the percentage of the waste mixtures inside the mixtures. The particles that form the skeleton of the specimens changed from rigid particles (SFS and CW) to RC gradually as R_b increased, and the transition point was around $R_b = 20\%$. By comparing the shear modulus with traditional subballast, the SFS+CW+RC mixture having 10% RC is a promising structural fill to be used as a subballast layer. Moreover, increasing the confining pressure will cause shear modulus to increase and damping ratio to decrease. The shear modulus decreases with the shear strain level, while the damping ratio increases as the shear strain increases. The shear modulus and damping ratio of all the specimens stabilized after 10000 cycles.

Acknowledgements. The first author would like to acknowledge the financial assistance provided by the China Scholarship Council. The assistance provided by industry (ASMS and South 32) in relation to the procurement of material used in this study is gratefully acknowledged. The assistance in the laboratory from Mr. Richard Berndt and the occasional technical feedback from Dr. Ana Heitor are appreciated.

References

1. Chiaro, G., Indraratna, B., Tasalloti, S.M.A, Rujikiatkamjorn, C.: Optimisation of coal wash-slag blend as a structural fill. In: Ground Improvement, vol. 168, no. GI1, pp. 33–44 (2013)
2. Yildirim, I.Z., Prezzi, M.: Geotechnical properties of fresh and aged basic oxygen furnace steel slag. J. Mater. Civ. Eng. **27**(12), 04015046-1–04015046-11 (2015)
3. Juan, L.M., Claisse, P., Ganjian, E.: Effect of steel slag and Portland cement in the rate of hydration and strength of blast furnace slag pastes. J. Mater. Civ. Eng. **23**(2), 153–160 (2011)
4. Tasalloti, S.M.A., Indraratna, B., Rujikiatkamjorn, C., Heitor, A., Chiaro, G.: A laboratory study on the shear behavior of mixtures of coal wash and steel furnace slag as potential structural fill. Geotech. Test. J. **38**(4), 361–372 (2015)
5. Indraratna, B., Gasson, I., Chowdhury, R.N.: Utilization of compacted coal tailings as a structural fill. Can. Geotech. J. **31**(5), 614–623 (1994)
6. Heitor, A., Indraratna, B., Kaliboullah, C.I., Rujikiatkamjorn, C., McIntosh, G.: Drained and undrained shearing behavior of compacted coal wash. J. Geotech. Geoenviron. Eng. **142**(5), 04016006-1–04016006-10 (2016)
7. Zheng, Y.F., Kevin, G.S.: Dynamic properties of granulated rubber/sand mixtures. Geotech. Test. J. **23**(3), 338–344 (2000)
8. Li, B., Huang, M.S., Zeng, X.W.: Dynamic behavior and liquefaction analysis of recycled-rubber sand mixtures. J. Mater. Civ. Eng. **28**(11), 04016122 (2016)
9. Indraratna, B., Qi, Y.J., Heitor, A.: Evaluating the properties of mixtures of steel furnace slag, coal wash, and rubber crumbs used as subballast. J. Mater. Civ. Eng. (2017, accepted). https://doi.org/10.1061/(asce)mt.1943-5533.0002108
10. ASTM D5311/D5311M: American Society for Tests and Materials 2008, Standard Test Method for Load Controlled Cyclic Triaxial Strength of Soil. ASTM International, West Conshohocken, PA, USA (R2012)
11. Indraratna, B., Raut, A.K., Khabbaz, H.: Constriction-based retention criterion for granular filter design. J. Geotech. Geoenviron. Eng. **133**(3), 266–276 (2007)

12. Indraratna, B., Salim, W., Rujihiatkamjorn, C.: Advanced Rail Geotechnology-Ballasted Track. CRC Press/Balkema, Leiden (2011)
13. Kokusho, T.: Cyclic triaxial test of dynamic soil properties for wide strain range. Soil Found. **20**(2), 45–60 (1980)
14. Nakhaei, A., Marandi, S.M., Kermani, S.S., Bagheripour, M.H.: Dynamic properties of granular soils mixed with granulated rubber. Soil Dyn. Earthq. Eng. **43**, 124–132 (2012)
15. Suiker, A.S.J., Selig, E.T., Frenkel, R.: Static and cyclic triaxial testing of ballast and subballast. J. Geotech. Geoenviron. Eng. **131**(6), 771–782 (2005)
16. Nakhaee, A., Marandi, S.M.: Reduce the forces caused by earthquake on retaining walls using granulated rubber-soil mixture. IJE Trans. B: Appl. **24**(4), 337–350 (2011)

Dynamic Flow Characteristics of Liquefied Sand Under the Extrema Large Deformation in the Cyclic Torsional Shear Tests

Haiyang Zhuang[✉], Qifei Liu, Xuchao Xue, and Guoxing Chen

Institute of Geotechnical Engineering, Nanjing Tech University,
Nanjing 210009, China
zhuang7802@163.com

Abstract. The dynamic flow deformation of liquefied saturated sand may cause serious damages to the ground and underground structures. However, to investigate the static fluid characteristics of the post-liquefied saturated sand, most studies on this problem have applied the cyclic loading to the saturated sand samples first, and then used the monotonic loading. The above test loading process is different from the actual stress state of soil in site. To investigate the dynamic flow characteristics of the liquefied saturated Nanjing sand under the cyclic loading, a series of undrained cyclic torsional shear tests are performed by using the hollow column torsional shear apparatus, with the largest shear strain up to 100%. At the same time, different effective confining pressure, initial shear stress and cyclic loading amplitude are loaded on the soil samples respectively. It is found that the soil sample has been in shear dilation state at the end of "zero effective stress" stage which is only determined by the excess pore pressure ratio. In other words, the response of excess pore pressure should be hysteretic to the shear dilation of the samples. It also proves that the initial shear stress should have the greatest influence on the relationship curves of the apparent viscosity-strain rate and the shear strain rate in the "zero effective stress" stage. Meanwhile, the dynamic apparent viscosity of sand in the "zero effective stress" stage under the cyclic loading is larger than the static apparent viscosity of post-liquefied sand when the shear strain rate reaches a relatively small value. Based on the test results, the dynamic apparent viscosity of liquefied sand should not be predicted by the empirical equation derived from the static fluid characteristics of the post-liquefied saturated sand.

Keywords: Liquefaction flow deformation · Initial shear stress
Effective confining pressure · Loading amplitude · Torsional shear test

1 Introduction

Under the cyclic loadings, the alternating stress will lead to the increase of the pore water pressure, making the stress state in the saturated sand gradually change to the "zero effective stress" stage. When the effective stress of saturated sand reaches to zero for the first time, it is considered that the saturated sand reaches the state of initial liquefaction.After the initial liquefaction of the inclined site, the joint action of the

© Springer Nature Singapore Pte Ltd. 2018
T. Qiu et al. (Eds.): GSIC 2018, *Proceedings of GeoShanghai 2018 International Conference: Advances in Soil Dynamics and Foundation Engineering*, pp. 318–330, 2018.
https://doi.org/10.1007/978-981-13-0131-5_35

earthquake loading and the driving static shear stress will lead to large flow deformation of the ground. For instance, the Niigata earthquake in 1964 caused the inclined liquefiable site sliding a several meters (Hamada et al. 1986). During the Nihonkai-Chubu earthquake in 1983, the driving static shear stress made the surface oblique and the site adjacent to the bank of river sliding with a generate enormous horizontal permanent deformation, which caused severe damages on the ground and underground structures (Takayasu et al. 1986).

According to the stress state of liquefaction, Seed et al. (1983) defined the initial liquefaction as the state of the effective stress in the sample reaches to zero for the first time in undrained cyclic torsional shear tests. Thus, the process of sand liquefaction can be divided into the process of pre-liquefaction and post-liquefaction. Most previous studies focus mainly on the generation conditions of initial liquefaction, the influence factors, the evaluation criteria, and the stress-strain response of the pre-liquefaction stage. However, fewer scholars study on the stress-train response of the post-liquefaction stage. In the recent decades, some scholars have carried out a series of researches on the way to predict the large deformation of post-liquefied saturated sand. For example, some complicated mechanics methods have also been developed to study the constitutive relation between the stress and strain of post-liquefaction saturated sand (Yang et al. 2003; Andrianopoulos et al. 2010; Zhuang et al. 2012; Wang et al. 2014; Chen et al. 2015). To better understand and predict liquefaction-induced lateral permanent deformation, the large liquefaction-induced deformation properties of saturated sands after the initial liquefaction have also be described by dynamic fluid mechanics theories (Chen et al. 2009) and at the micro level (Xu et al. 2012, Shamy et al. 2010).

By analyzing the experimental results, Sasaki et al. (1992) considered that the post-liquefied saturated sand has the characteristics of the fluid. The post-liquefied saturated sand is very similar to the fluid, so the deformation direction of the liquefied layer can be controlled by the total hydraulic gradient. This result was further confirmed by Miyajima (1995). Chen et al. (2004) carried out the preliminary flow test by using MTS dynamic triaxial apparatus. Chen et al. (2004) firstly introduced the definitions of apparent viscosity and shear strain rate in fluid mechanics to analyze the dynamic flow characteristics of the post-liquefaction saturated sand. Wang et al. (2013) conducted a series of experiments with dynamic cylindrical torsional shear apparatus. They obtained the stress-strain relationship curves and defined the average flow coefficient and liquidity level which can describe the fluidity of saturated sand.

However, in the past shear tests, the maximum shear strain amplitude is limited to 20%, which is much smaller than the shear strain of the inclined site according to the existing seismic damage data. In view of this, a series of cyclic torsional shear tests on the saturated Nanjing sand have been conducted with the double amplitude of cyclic shear strain up to about 100%, which is loaded by the dynamic hollow cylinder apparatus produced by GDS company in the United Kingdom. On the basis of these tests, the dynamic flow characteristics of saturated Nanjing sand after the initial liquefaction are analyzed in the view of dynamic fluid mechanics theories.

2 Test Apparatus and Test Procedure

In this paper, we performed the experiments using the dynamic hollow cylinder apparatus produced by GDS company in the United Kingdom. As shown in Fig. 1, the apparatus consists of a pressure chamber, servo host system, hydraulic servo control loading system, analog and digital signal-control system, and computer-controlled system. It can be applied to load the specimen in four separate directions, that is, the axial force W, torque force MT, inside horizontal confining pressure Pi, and peripheral pressure Po.

Fig. 1. The GDS torsional shear apparatus and the loading modes

In the experiments, the soil samples were taken from Nanjing area, which belong to the typical floodplain facies sedimentary soil. It mainly comprises the detrital quartz, a small amount of chlorite, muscovite and other clay minerals (Wu 2014). Moreover, its particles are flaky. The grain-size distribution curve of Nanjing sand is shown in Fig. 2, and its material properties are shown in Table 1.

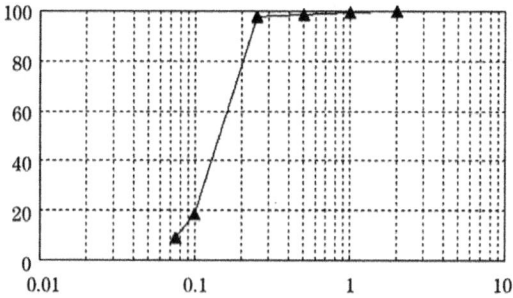

Fig. 2. Grain-size distribution curve of Nanjing sand

Table 1. Physical parameters of Nanjing sand compared with those of Toyoura sand.

Material	Specific gravity, G_S	Mean diameter D_{50} (mm)	Relative density, D_r	Fines content, (%)	Max void ratio, e_{max}	Min void ratio, e_{min}
Nanjing sand	2.70	0.17	42–46%	0.9	1.14	0.62

The hollow cylindrical specimens, prepared from air-dried Nanjing sand, were 100 mm in outer diameter, 60 mm in inner diameter and 200 mm in height. To achieve the desired relative density, the specimens were divided equally into eight sublayers along their height. Then, the dried sands were placed into a specimen preparation mold through a funnel. Each sublayer was compacted to the controlled height, and the top surfaces of each sublayer were coarsely scratched before adding the next sublayer of soil to make a good contact between two sublayers. The specimens were flushed with carbon dioxide gas followed by de-aired water until the Skempton's pore pressure coefficient was greater than 0.95 with a back-pressure of 200 kPa. Then, the specimens were isotropically consolidated by increasing the effective stress up to the expected values given in Table 2. After isotropic consolidation, the stress state was changed by applying a drained monotonic torsional shear stress up to a specified value. Finally, undrained cyclic torsional loading was applied at a constant double amplitude of the shear stress until the double amplitude shear strain reached to 100%, with the stress being controlled. The preparation process of specimen is shown in Fig. 3.

Table 2. Loadings planned for the tests.

Test	D_r	p'_0 (kPa)	$CSR = \tau_{cyclic}/p'_0$	$SSR = \tau_{static}/p'_0$	τ_{max} (kPa)	τ_{min} (kPa)	Loading pattern
1	44.3	80	0.3125	0	+25	−25	Reversal
2	45.2	100	0.25	0	+25	−25	Reversal
3	45.4	120	0.208	0	+25	−25	Reversal
4	45.8	150	0.167	0	+25	−25	Reversal
5	42.8	100	0.14	0	+14	−14	Reversal
6	43.6	100	0.18	0	+18	−18	Reversal
7	44.1	100	0.30	0	+30	−30	Reversal
8	44.3	100	0.18	0.3	+21	−15	Reversal
9	43.1	100	0.18	0.6	+24	−12	Reversal
10	44.7	100	0.18	0.10	+28	−8	Reversal
11	44.2	100	0.18	0.14	+32	−4	Reversal
12	45.6	100	0.18	0.18	+36	+0	Intermediate
13	45.1	100	0.18	0.22	+40	+4	Non-reversal
14	45.4	100	0.18	0.27	+45	+9	Non-reversal

Fig. 3. The specimen preparation process before the test

Two types of cyclic loading patterns were employed in this study, that is, stress reversal and non-reversal loadings. Under reversal loading, the combined shear stress value of the initial static shear stress (τ_s) and the cyclic shear stress amplitude (τ_d) was reversed from positive ($\tau_{max} = \tau_s + \tau_d > 0$) to negative ($\tau_{min} = \tau_s - \tau_d < 0$), or vice versa. The type of loading in which the combined shear stress always remained positive ($\tau_{min} = \tau_s - \tau_d \geq 0$) was called non-reversal loading. Based on the above-mentioned loading methods, the test conditions were planned and are listed in Table 2.

According to the test results, in the absence of initial static shear stress, the relationships between excess pore pressure ratio, apparent viscosity, shear stress and shear strain are shown in Fig. 4. In order to draw the relationship curve of apparent viscosity and shear strain in the logarithmic coordinate system, the absolute value of the apparent dynamic viscosity value was employed in the graphs. In order to facilitate the collation and analysis of the data, the post-liquefied cyclic torsional shear curve of each cycle is divided into two "half cycle". In Fig. 4, the starting point and ending point of each "half cycle" are marked. During the process of cyclic loading of post-liquefaction sample, each cycle consists of two loading procedures and two unloading procedures. In the loading stage, when the excess pore pressure ratio of the specimen is close to 1 (1–2, 4–5), the shear strain of the specimen develops rapidly under a very low shear stress. Meanwhile, the sample is in the stage of low strength with the shear strain rate increasing gradually and the value of apparent viscosity being small. With a decrease in the excess pore pressure ratio of the specimen (2–3, 5–6), the relationship curve of shear stress and shear strain rate has an inflection point and the shear strain rate decreases to zero. Moreover, the apparent dynamic viscosity of the sample increases rapidly, and the shear strain of the specimen increases slowly at this stage. In the reversal unloading stage (3–4, 6–7), the sample enters the stage of reversal loading with the excess pore pressure ratio increasing gradually and the apparent viscosity decreasing gradually.

In the Fig. 4, two "half cycle" in which the single shear strain value reaches to 10% are named "$\gamma = 10\%$ the first half cycle" and "$\gamma = 10\%$ the second half cycle" respectively, and the two "half cycle" in which the single shear strain value reaches to 20% are named "$\gamma = 20\%$ the first half cycle" and "$\gamma = 20\%$ the second half cycle" respectively, and so on.

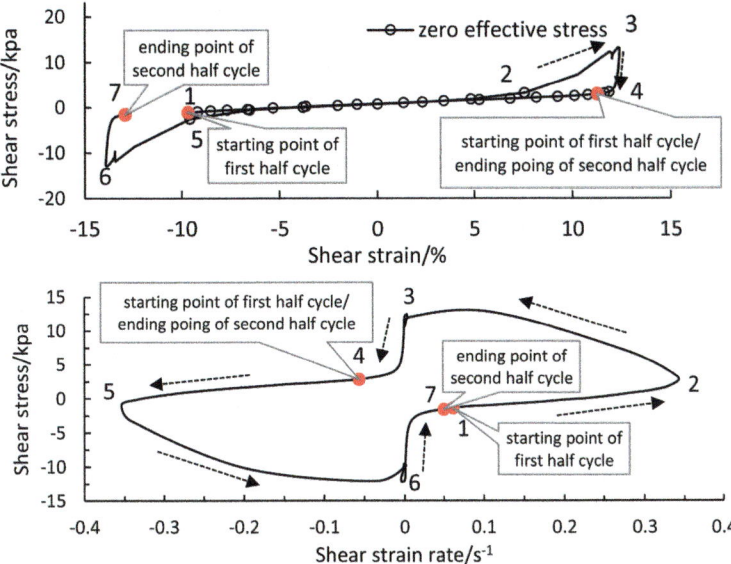

Fig. 4. The relationship curves between apparent viscosity, pore pressure ratio, stress and shear strain rate

3 Analysis of the Test Results

3.1 Effect of the Effective Confining Pressure

In fluid mechanics, the shear strain rate can be used to analyze the dynamic flow characteristics of the post-liquefied saturated sand. The international unit of apparent dynamic viscosity is Pa.s. Since it was found that the apparent dynamic viscosity of saturated sands is relatively larger, the unit kPa.s is taken in this paper.

Figure 5 shows the relationship curves of the post-liquefaction apparent viscosity and the shear strain rate under different initial effective confining pressure. At the end of the "zero effective stress" stage (4–5 loading stage), the apparent viscosity decreases sharply and then increases with the increase in shear strain rate, which has also been found in all other tests. The above phenomenon should be caused by insufficient loading capacity of instrument when the state of the sample converted from the liquefied state into non-liquefied state. The apparent viscosity in this stage mainly changes between 1 kPa.s and 20 kPa.s. To the effect of the shear strain amplitude changes, the apparent viscosity decreases more and more slowly with the increase in shear strain rate, especially when the effective confining pressure is as small as 80 kPa or 100 kPa. Figure 6 shows the Apparent viscosity-strain rate relationship of saturated Nanjing sand when $\gamma = 20\%$, which shows that the effect of the effective confining pressure is so little that its effect law is not obvious.

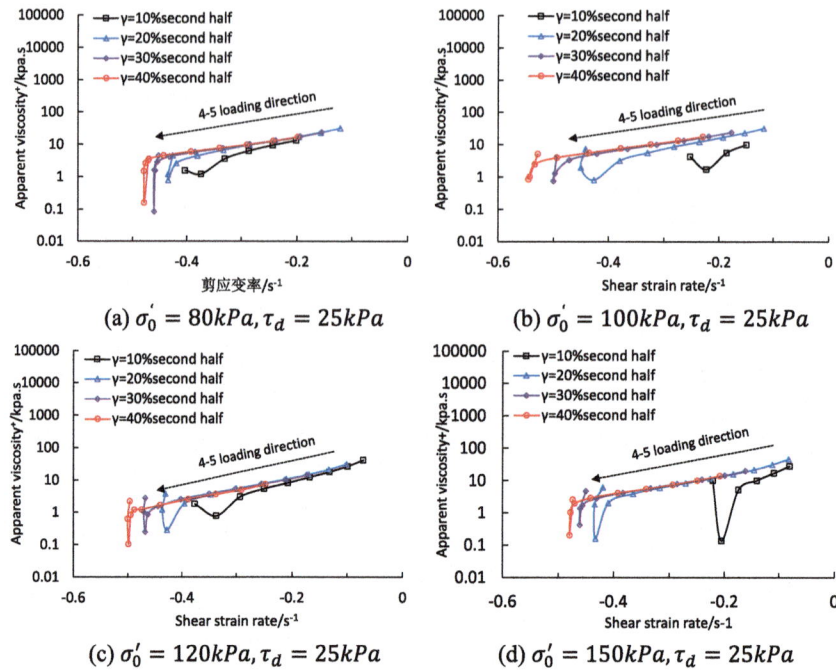

Fig. 5. Apparent viscosity-strain rate relationship curves of saturated Nanjing sand under different effective confining pressure

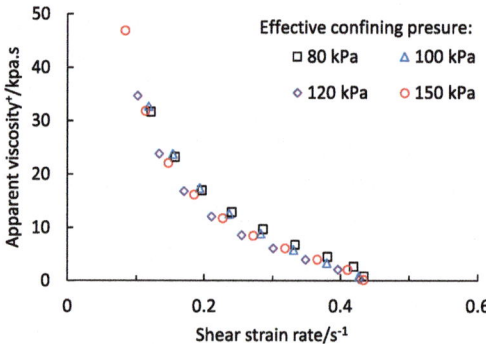

Fig. 6. Apparent viscosity-strain rate relationship of saturated Nanjing sand when $\gamma = 20\%$

3.2 Effect of the Cyclic Loading Amplitude

Figure 7 shows the relationship curves of the post-liquefaction apparent viscosity and the shear strain rate under the different cyclic loading amplitudes. In general, the relationship curves of the apparent viscosity and shear strain rate in the "zero effective stress" stage is similar to those under the different effective confining pressures.

However, when the shear strain amplitude changes from 20% to 40%, the relationship curves of apparent viscosity and shear strain rate moves down, when the cyclic loading amplitude is as small as 14 kPa or 18 kPa, this change trend is contrary to those under the cyclic loading amplitude as large as 25 kPa or 30 kPa. How to explain the above different change law under the different cyclic loading amplitude should be studied in the following research work.

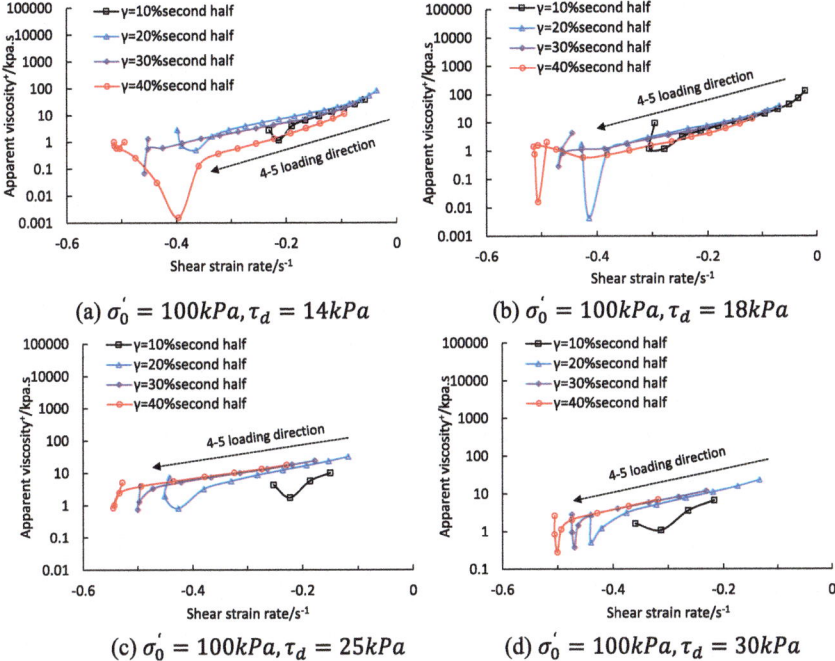

Fig. 7. Apparent viscosity-strain rate relationship curves of saturated Nanjing sand under different cyclic loading amplitude

Figure 8 shows the apparent viscosity-strain rate relationship of saturated Nanjing sand under the different cyclic loading amplitudes when the shear strain amplitude is 20%. When the cyclic loading amplitude increases from 18 kPa to 30 kPa, the apparent viscosity-strain rate relationship curves move upward, which proves that the cyclic loading amplitude has a great influence on the relationship curves. The equations for the fitting curves are shown in Fig. 8.

In the previous experimental researches, most of the researchers applied cyclic loading to the saturated sand samples, and then used the monotonic load to test the static fluid characteristics of the post-liquefied saturated sand. As a result, an equation has been given by Zhou et al. (2015), which is well fit the test results given by Kawakami et al. (1994) and Hamada et al. (1994), as

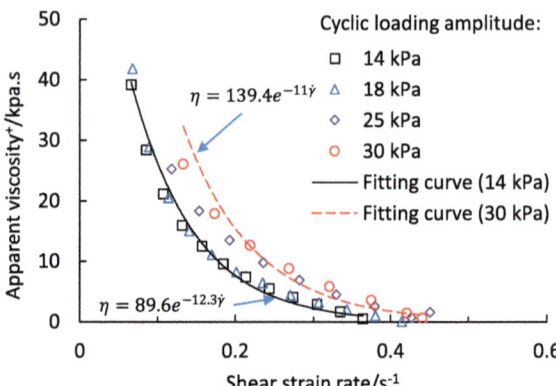

Fig. 8. Apparent viscosity-strain rate relationship curves under the different cyclic loading amplitude when $\gamma = 20\%$

$$\eta = \left(1 - \frac{u}{\sigma_0'}\right) \cdot 2.5 \cdot \sigma_0' \cdot \dot{\gamma}^{-0.94} + 0.05 \cdot 2.5 \cdot \sigma_0' \cdot \dot{\gamma}^{-0.94} \tag{1}$$

where u/σ_0' is extra pore pressure ratio, which is set equal to 1.0 in this paper where the fluid characteristics of the post-liquefied saturated sand in the "zero effective stress" stage is investigated under the cyclic loading process. Accordingly, Fig. 9 shows the fitting curves for the apparent viscosity-strain rate relationship when the effective confining pressure is 100 kPa, which is compared with the curve predicted by the Eq. (1). The apparent viscosity predicted by Eq. (1) is obviously smaller than that measured by the tests in this paper when the shear strain rate is smaller than 0.27. Figure 9 also shows that the apparent viscosity decreases quickly when the sample is in the "zero effective stress" stage under the cyclic loading process. This suggests that the

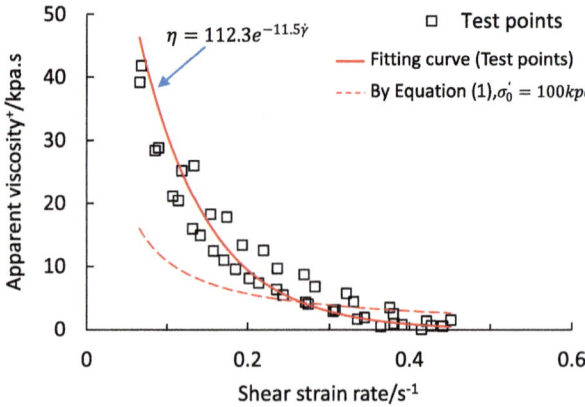

Fig. 9. Fitting curves for the apparent viscosity-strain rate relationship when $\sigma_0' = 100\,\text{kPa}$

apparent viscosity should not be predicted by the empirical equation achieved by the static fluid characteristics of the post-liquefied saturated sand.

3.3   Effect of the Initial Shear Stress

When the initial static shear stress is applied to the specimen, the stress-strain curve of saturated Nanjing sand will generate unidirectional excursion. The stress-strain relationship curve of the first half cycle is more representative of the dynamic flow characteristics of the post-liquefied saturated Nanjing sand. Therefore, Fig. 10 shows

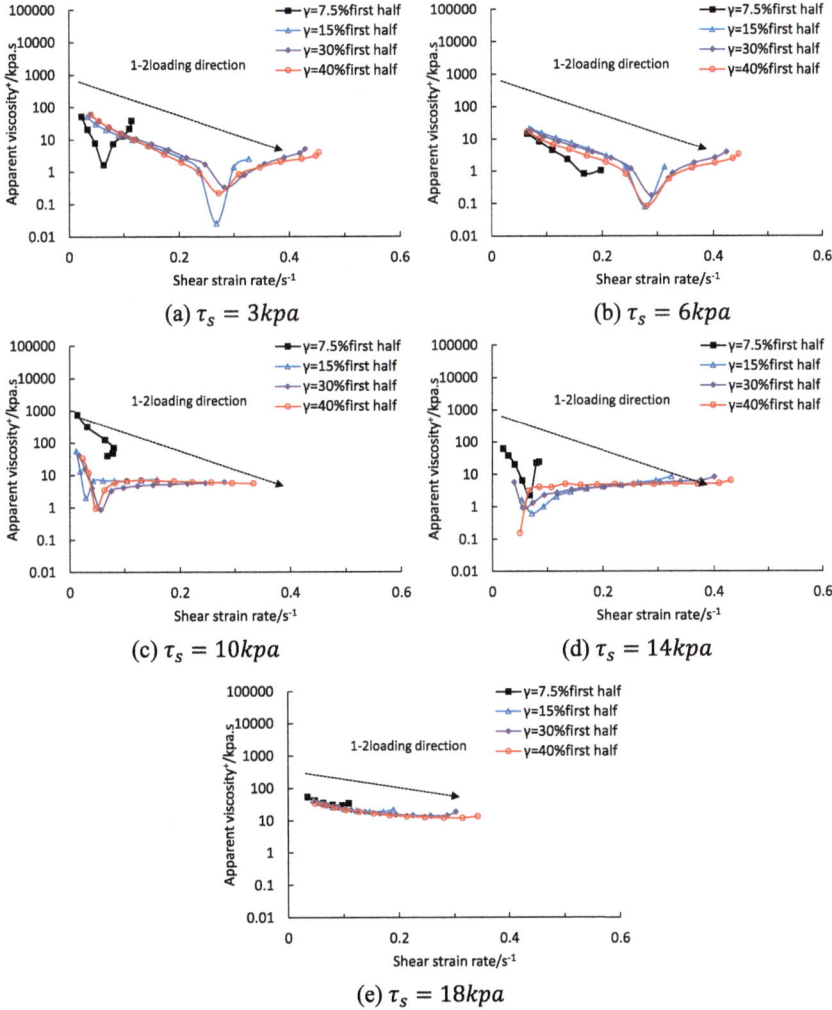

Fig. 10. Apparent viscosity-strain rate relationship curves of saturated Nanjing sand under different initial static shear stress

the relationship curves of the apparent viscosity and the shear strain rate under different initial static shear stress in the first half cycle.

Compared with the effective confining pressure and the load amplitude, the initial static shear stress has the greatest effect on the relationship curves of the apparent viscosity and shear strain rate in "zero effective stress" stage. On the whole, when the initial shear stress varies from 3 kPa to 6 kPa, the relationship curves of the apparent viscosity and shear strain rate has a decreasing trend at the beginning of the "zero effective stress" stage and then an increasing at the end of this stage. However, when the initial shear stress changes from 10 kPa to 14 kPa, the relation curves decrease quickly at first and then the apparent viscosity is constant with the change of shear strain rate. On the whole, in the dilatancy stage and the reversal unloading stage, the initial static shear stress has little effect on the relationship curves of apparent viscosity and shear strain rate. When the initial shear stress increases to 18 kPa, with the loading pattern changes from reversal cyclic loading to the intermediate cyclic loading, the apparent viscosity of post-liquefied Nanjing sand decreases between 100 kPa.s and 10 kPa.s in "zero effective stress" stage. The apparent viscosity in this stage is larger than that of another corresponding test results (generally between 20 kPa.s and 1 kPa.s).

Figure 11 shows the apparent viscosity-strain rate relationship of saturated Nanjing sand under different initial shear stress when the shear strain amplitude is 15%. Compared with the effect of the confining pressure and the cyclic loading amplitude, the initial shear stress has the greatest influence on the apparent viscosity-strain rate relationship of saturated Nanjing sand, especially when the initial shear stress increases from 6 kPa to 10 kPa. When the initial shear stress is no more than 6 kPa, the apparent viscosity decreases gently with the increasing of shear strain rate and the initial shear stress has very little influence on the relationship curves. However, when the initial shear stress increases to 10 kPa and 14 kPa, the apparent viscosity decreases quickly first and then maintains in a very small value without large change with the increase of shear strain rate. Especially, when the initial shear stress increases to 18 kPa continuously, the apparent viscosity-strain rate relationship curves move upward and the apparent viscosity decreases most slowly with an increasing of the shear strain rate. As

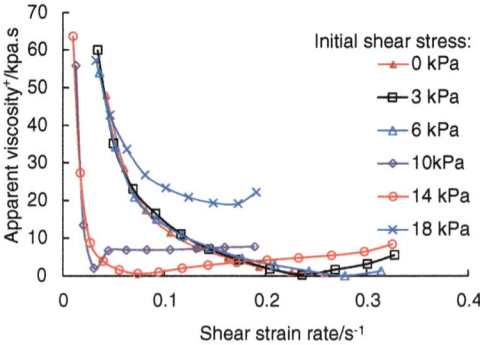

Fig. 11. Apparent viscosity-strain rate relationship curves under the different initial shear stress when $\gamma = 15\%$

a result, it has been found that the sample isn't liquefied completely when the initial shear stress is 18 kPa (Zhuang et al. 2016), which should be the main reason for the above finding.

4 Conclusions

In this paper, the effect of the initial shear stress, the cyclic loading amplitude and the effective confining pressure are considered in the cyclic torsional shear tests to investigate the dynamic flow characteristics of the liquefied saturated Nanjing sand, with the maximal shear strain amplitude up to about 100%. Some new findings and main conclusions are proposed as follows:

It is obvious that the relationship curves of shear stress and shear strain rate of the liquefied saturated Nanjing sand are nonlinear in the "zero effective stress" stage. Particularly, an inflection point appears at the end of the relationship curves of the specimen, which means that the shear stress increases rapidly with the shear strain rate remaining unchanged. This proves that the sample has actually enter the dilation state at the end of the "zero effective stress" stage which is only determined by the excess pore pressure ratio. In other words, at the end of the "zero effective stress" stage, the excess pore pressure response should be hysteretic to the stress response.

These results indicate that the initial shear stress should have the greatest influence on the relationship curves of the apparent viscosity-strain rate and the shear strain rate in the "zero effective stress" stage, and the effect of the cyclic loading amplitude should be taken in the second place. However, the effect of the effective confining pressure is so little that it can be neglected in this paper. In general, when the cyclic shear strain amplitude increases with the increase of the number of loading cycles, the apparent viscosity and shear strain rate relationship curves are on an ascending trend. However, the effect of the initial shear stress is so complicated that the further research should be done to explain the finding in this paper.

Compared with the test results of the sample under the monotonic load after liquefaction under the cyclic loading, the dynamic apparent viscosity in the "zero effective stress" stage is larger when the shear strain rate is smaller than some value. It proves that the dynamic viscosity of cyclic liquefied sand should not be predicted by the empirical equation achieved by the static fluid characteristics of the post-liquefied saturated sand.

Acknowledgments. The study on which the paper is based was supported by the Natural Science Foundation of China (51778290, 51278246). The authors wish to gratefully acknowledge these supports.

References

Andrianopoulos, K.I., Papadimitriou, A.G., Bouckovalas, G.D.: Explicit integration of bounding surface model for the analysis of earthquake soil liquefaction. Int. J. Numer. Anal. Methods Geomech. **34**(15), 1586–1614 (2010)

Chen, J., O-tani, H., Hori, M.: Stability analysis of soil liquefaction using a finite element method based on particle discretization scheme. Comput. Geotech. **67**, 64–72 (2015)

Chen, W.H.: Flow-slip deformations induced by seismic liquefaction and preliminary test results. J. Nat. Disasters **13**(3), 75–80 (2004)

Chen, Y.M., Liu, H.L., Shan, G.J., et al.: Liquefaction and post-liquefaction flow behavior of sand. Chin. J. Geotech. Eng. **31**(9), 1408–1413 (2009)

Hamada, M., Yasuda, S., Isoyama, R., et al.: Observation of permanent displacements induced by soil liquefaction. Proc. JSCE **3**(6), 211–220 (1986)

Hamada, M., Sato, H., Kawakami, T., et al.: A consideration of the mechanism for liquefaction-related large ground displacement. In: Proceedings of the 5th US–Japan Workshop on Earthquake Resistant Design of Lifeline Facilities and Countermeasures Against Soil Liquefaction, Salt Lake City, pp. 217–232 (1994)

Kawakami, T., Suemasa, N., Hamada, M., et al.: Experimental study on mechanical properties of liquefied sand. In: Proceedings of the 5th US–Japan Workshop on Earthquake Resistant Design of Lifeline Facilities and Countermeasures Against Soil Liquefaction, Salt Lake City, pp. 285–299 (1994)

Miyajima, M., Kitaura, M., Koike, T., et al.: Experimental study on characteristics of liquefied ground flow. In: The First International Conference on Earthquake Geotechnical Engineering, Balkema, pp. 969–974 (1995)

Seed, H.B., Idriss, I.M., Arango, I.: Evaluation of liquefaction potential using field performance data. J. Geotech. Eng. **3**, 90–98 (1983)

Shamy, U.E., Denissen, C.: Microscale characterization of energy dissipation mechanisms in liquefiable granular soils. Comput. Geotech. **37**(7), 846–857 (2010)

Sasaki, Y., Towhata, I., Tokida, K.I., Yamada, K., et al.: Mechanism of permanent displacement of ground caused by liquefaction. Soils Found. **32**(3), 79–96 (1992)

Takayasu, T., Aita, Y., Fukudome, T., et al.: Ground failure caused by the Nihonkai Chubu earthquake, 1983 in relation to subsurface geology. Memo. Geol. Soc. Jpn., 237–256 (1986)

Wang, G., Xie, Y.N.: Modified bounding surface hyproplasticity model for sands under cyclic loading. J. Eng. Mech. **140**(1), 91–101 (2014)

Wang, Z.H., Zhou, E.Q., Lv, C., et al.: Liquefaction mechanism of saturated gravelly soils based on flowing property. Chin. J. Geotech. Eng. **35**(10), 1816–1822 (2013)

Wu, H.R.: Study of Nanjing Sand Particle Shape and Static Liquefaction Characteristics. Nanjing University, Nanjing (2014)

Xu, X.M.: The Study of Microscopic Mechanism of Saturated Sand Liquefaction Evaluation. Zhejiang University, Zhejiang (2012)

Yang, Z.H., Elgamal, A., Parra, E.: Computational model for cyclic mobility and associated shear deformation. J. Geotech. Geoenviron. Eng. **129**(12), 1119–1127 (2003)

Zhou, E.Q., Wang, Z.H., Chen, G.X., et al.: Constitutive model for fluid of post-liquefied sand. Chin. J. Geotech. Eng. **37**(1), 112–118 (2015)

Zhuang, H.Y., Huang, C.X., Zuo, Y.F.: Sensitivity analysis of model parameters for predicting liquefied large deformation of sand. Chin. J. Rock Soil Mech. **33**(1), 280–286 (2012)

Zhuang, H.Y., Hu, Z.H., Wang, R., et al.: Cyclic torsional shear loading tests on extremely large post-liquefaction flow deformation of saturated Nanjing sand. Chin. J. Geotech. Eng. **38**(12), 2164–2174 (2016)

Seismic Responses Analysis of Base-Isolated LNG Storage Tank

Yu Zhou[1(✉)], Xiongyan Li[3,4], and Zhiyi Chen[1,2]

[1] Department of Geotechnical Engineering, Tongji University,
Shanghai 200092, China
mmmmonsterzy@163.com
[2] Key Laboratory of Geotechnical and Underground Engineering
of Ministry of Education, Shanghai 200092, China
[3] The Key Laboratory of Urban Security and Disaster Engineering,
Ministry of Education, Beijing University of Technology, Beijing 100124, China
[4] Beijing Key Laboratory of Earthquake Engineering and Structural Retrofit,
Beijing University of Technology, Beijing 100124, China

Abstract. In order to study seismic responses of the base-isolation system, a three-dimensional 160 thousand cubic meters finite element model of LNG storage tank is established by the general finite element code ABAQUS. Then, the nonlinear dynamic analyses are conducted on the base-isolation LNG storage tank. The liquid filling ratio and the earthquake motion periods have been taken into accounts for further investigating the efficiency of the isolation system. To accurately simulate the interaction between the inner steel tank and the outer concrete tank, a thermal insulation layer is also simulated in details. Meanwhile, the impulsive and convective hydrodynamic forces towards outer tank wall, which are induced by a horizontal earthquake motion, are considered through the additional mass method. The connection elements "cartesian" and "align" are combined to simulate isolators to achieve a reasonable and accurate analysis for the LNG storage tank. The numerical results show that the lead rubber bearings have excellent effectiveness in decreasing most seismic responses. This is more remarkable in the cases of larger volume of liquid and stronger earthquake motion. The efficiency of the isolation system is better in the earthquake ground motion with shorter predominant periods. Whereas, the impulsive effect of liquid mass is strengthened in earthquakes with long predominant periods and then to magnify the displacement of tank walls.

Keywords: LNG storage tank · Base-isolated system · Seismic analysis

1 Introduction

LNG (Liquified Natural Gas, LNG for short) is a clean, safe and high-quality energy, of strategic significance to change our country's structural system of energy industry and improve ecological environment. However, there are still many difficulties and crucial problems of independent design and construction techniques for LNG tank needed to be further studied owing to relative late start point. Therefore, insuring its seismic safety are important steps to design LNG equipment.

© Springer Nature Singapore Pte Ltd. 2018
T. Qiu et al. (Eds.): GSIC 2018, *Proceedings of GeoShanghai 2018 International Conference: Advances in Soil Dynamics and Foundation Engineering*, pp. 331–339, 2018.
https://doi.org/10.1007/978-981-13-0131-5_36

At present, the studies on seismic mitigation and isolation of storage tanks are mainly focused on the vertical oil storage steel tanks, through numerical simulations. For example, Tsai et al. [1] and Luo et al. [2] carried out the numerical analysis of base-isolated LNG storage tanks considering the influences of earthquakes with long predominant periods; Shrimali and Jangid [3, 4] performed the finite element analysis taking the damping properties and bearing types into accounts to study the different effectiveness of the isolation system; Panchal and Jangid [5] conducted parametric analysis to investigate the sensitivity of tanks isolated by variable friction pendulum system (VFPS) to near field ground motion. Zhang et al. [6] proposed an isolation system composed of an annular damper reaction wall, lead rubber bearings and viscous dampers to decrease seismic responses of LNG tanks. The above-mentioned researches developed simplified analysis models of outer concrete tank and inner steel tank separately to study the seismic mitigation and isolation. Actually, the circumferential thermal insulation layer between outer tank wall and inner tank wall can transmit normal forces and constraint deformation, exerting a non-negligible effect on the numerical simulation results. Given these conditions, the springs are utilized to simulate the thermal insulation layer and the additional mass method is applied to model the liquid part for conducting seismic analysis in this paper. Furthermore, the performance of the isolation system is analyzed considering the parameters including the liquid mass and earthquake motion periods.

2   The Finite Element Model Establishment of LNG Tank

2.1   Dimensions of the LNG Tank

The research object in this paper is based on a 160 thousand extra-large LNG full-containment storage tank in certain area of Jiangsu province in China. The whole simplified model and detailed dimensions is presented in Fig. 1.

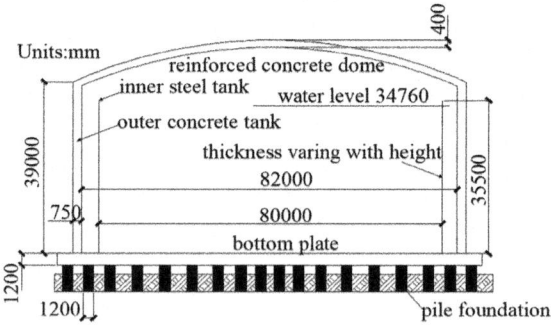

Fig. 1. Simplified model of LNG tank

Typical extra-large LNG full-containment storage tank is composed of outer pre-stressed reinforced concrete tank, reinforced concrete bearing platform, inner steel tank, reinforced concrete dome and circumferential thermal insulation layer between two tank walls. The influence of prestress is not considered so as to simplify analysis process in this paper.

2.2 Modeling

The LNG tank with outer reinforced concrete tank and inner steel tank is of large scale and has various structural details. In view of the complicity of seismic analysis, the reinforced concrete is developed by the embedded model, which utilizes the rebar layer elements to disperse the steel into the concrete to model reinforced concrete. 9622 quadrilateral shell elements S4R and 18 triangular shell elements S3 are used to simulate the outer tank, with the circumferential sizes of grids are about 2.02 m and the vertical sizes vary from 1.5 m to 1.2 m from the bottom to the top. 10400 quadrilateral shell elements S4R are used to simulate the inner tanks, with the circumferential sizes of grids are about 1.18 m and the vertical sizes are about 1.21 m along the height. 5103 linear beam elements B31 with length of 1.2 m are used to simulate the pile foundation.

With low temperature liquid stored in inner steel tank, the structure needs to be equipped with strict thermal insulation system to control the inner tank's temperature. In this case, the elastic blanket and perlite layer are laid in the annular space between the two tank walls to control the temperature. These materials used in the insulation layer is of small mass and stiffness, having insignificant dynamic impacts on the structure. Therefore, not establishing special geometric elements in the model but utilizing spring elements to simulate the non-negligible function of transmitting the normal force for insulation layer. In the practical implement, the spring2 elements with only one degree of freedom along the connecting direction achieve the function of transferring normal force. In view of mechanics of materials, the axial deformation is the only factor to define the stiffness as the following form:

$$\varepsilon = \frac{F}{EAl} \tag{1}$$

where ε is the axial strain along the connecting direction for spring elements. E is the elastic modulus of insulation layer, which is assumed to be the elastic modulus of perlite. A is the unit area of circumferential space covered equally by all spring

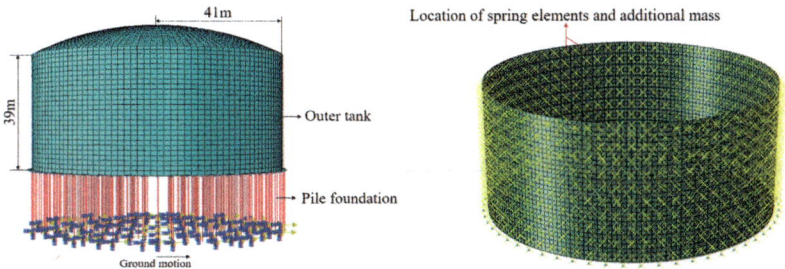

Fig. 2. Numerical models of LNG storage tank

elements, 1 is the length of every spring, F is the normal force transmitted by the insulation layer. EA equals to 3.222×10^{10} N/m.

As shown in the Fig. 2, the layers of the spring elements are divided into 10 parts, with total amount of 72 per 5° along the circle and 720 spring elements totally arranged between the inner and outer tank walls.

2.3   Material Parameters

The outer tank wall and bearing platform use the concrete damaged plasticity model as the constitutive model. While for pile foundations without pile-soil interaction, which are not the key object of research, the concrete is reckoned as elastic material for reducing the computational cost. The material parameters are defined according to the practical engineering parameters. Outer tank concrete's density is 2500 kg/m³, with elastic modulus of 3.504×10^{10}Pa and Poisson's ratio of 0.2. The density of rebars in the outer tank's wall is 7800 kg/m³, and the Poisson's ratio is 0.3. For inner steel tank, the density, elastic modulus and Poisson's ratio is 7850 kg/m³, 2.1×10^{11} Pa and 0.3 respectively. The LNG filling the tank has a density of 450 kg/m³.

2.4   Boundary Conditions and Load

The tank's bottom plate is fixed with pile foundation before isolation. Meanwhile, the acceleration of gravity is applied in the vertical direction, loosening the constraint in the corresponding direction when subjected to ground motions. The Housner's [7] classical liquid-solid coupling theory gives the hydrodynamic pressure towards the tank wall under earthquake. In this paper, the hydrodynamic pressure is equivalent to the inertia force of point mass to simulate the dynamic fluid effects when subjected to unidirectional or bi-directional seismic motions, namely additional mass method [7] (Fig. 3). And then the equivalent additional mass can be deduced based on the relation between the inertia force and mass. The point mass elements are added to the inner tank wall, sharing the same location with the spring elements of insulation layer shown in the Fig. 2.

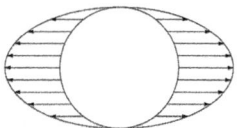

Fig. 3. Circumferential distribution of additional mass under unidirectional earthquake motion

2.5   Base Isolation Layer

The lead rubber bearings are assumed to isolate ground shaking. Referring to the design procedure for LRBs in buildings and bridges of Kelly et al. [8] and the design code of China (GB 20688.3-2006), 319 round isolators with diameter of 0.8 m and height of 0.275 m arranged between each pile and superstructure. The first and second shape factor is 33.33 and 5.56 respectively.

The translational connector "Cartesian" and rotational connector "Align" are combined to constraint the rotation of lead rubber bearings and keep the relative motions available in two horizontal directions and vertical direction. Defining the identical initial stiffness as 7515.60 kN/m in two horizontal directions and the vertical compression stiffness as 3074.61 kN/m represent the elastic characteristics of the combined connector. The equivalent damping ratio is 0.156 horizontally for the combined connector. The restoring force model is assumed to be bilinear elastic model to describe the actual nonlinear force-deformation behavior and simplify the analysis. Based on the bilinear model, the kinematic hardening model is specified in the plastic part of the connector. The post-yield stiffness is 894.45 kN/m, which determine the shape of restoring force model of the lead rubber bearings. And the equivalent horizontal stiffness is 1203.52 kN/m.

3 Modal Analysis of LNG Tank

Compared with the results of modal analysis between the tanks with isolators and without isolators, the comparison of first three orders are listed in the Table 1. It is observed that the natural periods of vibration are all prolonged despite the volume of liquid, while with the increasing of the liquid level, the prolonged rate is much larger.

Table 1. Periods of first three orders at different liquid filling ratio.

Mode	Empty tank(s)			Half tank(s)			Full tank(s)		
	Before isolation	After isolation	$\frac{T_2}{T_1}$	Before isolation	After isolation	$\frac{T_2}{T_1}$	Before isolation	After isolation	$\frac{T_2}{T_1}$
1	1.1581	1.4826	0.280	1.17281	1.5041	0.283	1.2140	1.5599	0.285
2	1.0495	1.3412	0.278	1.0599	1.3572	0.281	1.0894	1.3971	0.283
3	1.0491	1.3410	0.278	1.0594	1.3568	0.281	1.0887	1.3966	0.283

4 Time History Analysis

In order to analyze the nonlinear seismic responses of LNG storage tank, time history analysis is conducted based on Newmark-β time integration procedure by ABAQUS [9]. According to seismic design criterion for LNG tank, time history analysis needs to be done at two different earthquake fortification level, namely operating basis earthquake (OBE) and safety shutdown earthquake (SSE). The peak acceleration of two fortification are 145 Gal for OBE and 230 Gal for SSE. Two natural waves and one artificial wave are selected by seismic design spectrum for each fortification level in statistical sense. The waves are input at the bottom of pile foundation in the form of acceleration. Two of six simulated waves after normalization are plotted in Fig. 4.

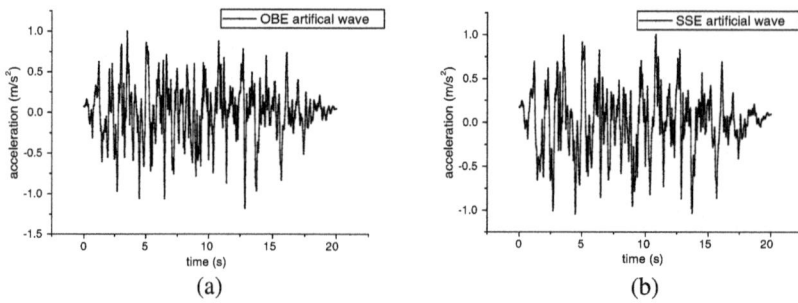

Fig. 4. Artificial time history curves after normalization.

Define the effectiveness of isolation system is expressed as index η:

$$\eta = \frac{\Delta - \mu}{\Delta} \times 100\% \tag{2}$$

where Δ and μ is the seismic responses including acceleration, velocity and displacement without isolation and with isolation respectively. $\eta < 0$ means that the seismic response is magnified, while $0 < \eta < 100$ means the shrinkage of the seismic response and the larger η is, the higher the isolator's efficiency is.

4.1   Seismic Response Analysis of Different Liquid Levels

Same artificial wave is input in one horizontal direction for different liquid level of empty tank, half tank and full tank, with the duration time of 20 s. Compare the seismic responses and efficiency of three different liquid levels of the outer tank as following:

Acceleration of Tank Wall
Results of acceleration response are demonstrated in Table 2, extracted from same point A on outer tank wall at the height of 15.975 meters. The peak acceleration presents obvious relation as: full tank > half tank > empty tank, while after isolation it all shrinks to 0.75 m/s². It means that the effectiveness of isolation: full tank > half tank > empty tank.

Table 2. Acceleration of outer tank wall at different liquid levels.

Liquid levels	Full tank	Half tank	Empty tank
Peak acceleration before isolation (m/s^2)	3.75583	3.3476	2.6827
Peak acceleration after isolation (m/s^2)	0.7367	0.7504	0.7550
Decreasing rate $\eta(\%)$	80.39	77.58	71.86

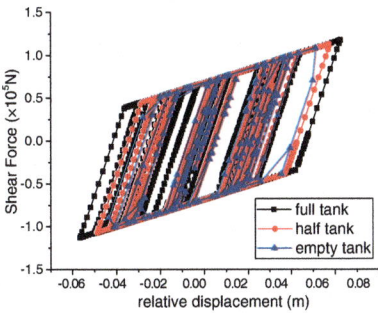

Fig. 5. Hysteretic curves of isolators at different liquid levels

Hysteretic Curve Analysis

The impulsive action of hydrodynamic force and intensity of seismic response are different at three filling ratios, consequently, isolators' effectiveness are different. The hysteretic curves are plotted in the Fig. 5. Combining the results of acceleration, hysteretic curves and modal analysis, following conclusions can be drawn: with the liquid storage level increasing, the effectiveness of isolation system appears to be better.

Combine the results of acceleration, hysteretic curves and modal analysis, following conclusions can be drawn: with the liquid storage level increasing, the effectiveness of isolation system appears to be better.

4.2 Seismic Analysis of Seismic Wave with Different Predominant Periods

In this chapter, the analyses for seismic waves with different predominant periods are performed to study the effects of predominant periods on the earthquake responses. Results below are all extracted at 100% liquid filling ratio. The duration time of natural waves RSN13 and RSN90 are both 30 s, with predominant periods 0.66 s and 0.62 s individually. The duration time of artificial wave is 20 s and predominant period is 0.42 s.

Acceleration of Tank Wall

Table 3 demonstrates that the isolators are effective in controlling acceleration regardless types of seismic waves, and the efficiency for natural wave RSN13 is the worst of three waves. The effectiveness appear to be: artificial wave > natural wave RSN90 > natural wave RSN13.

Table 3. Acceleration of outer tank wall.

Seismic waves	Peak acceleration	Peak acceleration	Decreasing rate
	Before isolation (m/s²)	After isolation (m/s²)	$\eta(\%)$
Artificial wave	3.7558	0.7367	80.39
Natural wave RSN13	3.4276	1.2660	63.06
Natural wave RSN90	2.5607	0.7207	71.85

Displacement of Tank Wall

Table 4 shows that the results of displacement change to a small degree and are even magnified for natural wave RSN13 and RSN90, while the decreasing rate of displacement for artificial wave is only 7.2%

Table 4. Displacement of outer tank wall.

Seismic waves	Peak displacement Before isolation (mm)	Peak displacement After isolation (mm)	Decreasing rate $\eta(\%)$
Artificial wave	0.1910	0.1772	7.20
Natural wave RSN13	0.2710	0.2939	−8.44
Natural wave RSN90	0.9011	0.9120	−1.21

From analyses in chapter 4.2, it can be concluded that the accelerations are controlled effectively, and as the predominant period increases, the effectiveness will be worse. This is because long-period ground shaking has similar predominant period to the natural period of base-isolated tank, inducing the effectiveness to decrease. Secondly, the insignificant or even magnified mitigation effect of displacement mean larger sloshing and impulsive action on tank walls. The phenomenon is because the convective mass which results in sloshing effects has longer period than solid wall and is more sensitive to long-period ground shaking.

5   Conclusions

A three-dimensional numerical model of 160 thousand cubic meter LNG storage tank is developed to study the seismic responses with various parameters such as different liquid filling ratio and seismic waves with different predominant periods, and comes into following conclusions:

(1) Lead rubber bearing is effective in decreasing response of tank wall acceleration, velocity and base shear force, however, it is not obvious in controlling displacement of tank wall and even amplify the displacement in some cases of long-period seismic wave. The reason is attributed to that the convective part of liquid mass has long periods and tends to result in larger sloshing and impulsive effects towards wall, inducing the amplification of tank wall' displacement.

(2) As the liquid filling ratio increases, the decreasing rate of response becomes higher. It can be speculated that the liquid imposes a sloshing and impulsive action on tank wall when subjected to the earthquake motion, equivalent to dampers dissipating energy and absorbing vibration, similar to the mechanism of Tuned Liquid Damper in passive structural control theory.

Acknowledgement. This research was supported by the National Natural Science Foundation of China (Grant No. 41472246,51778464), Key laboratory of Transportation Tunnel Engineering (TTE2014-01), and "Shuguang Program" supported by Shanghai Education Development Foundation and Shanghai Municipal Education Commission. All supports are gratefully acknowledged.

References

1. Tsai, C.S., Chiang, T.C., Chen, B.J.: Seismic behavior of MFPS isolated structure under near-fault sources and strong ground motions with long predominant periods. Seism. Eng. ASME **466**, 73–79 (2003)
2. Luo, D.Y., Sun, J.G., Hao, J.F., et al.: The effect analysis of the LNG storage tanks with pile foundation and base isolation under the long period earthquake. Earthq. Eng. Earthq. Dyn. **35**(6), 170–176 (2015)
3. Shrimali, M.K., Jangid, R.S.: Non-linear seismic response of base-isolated liquid storage tanks to bi-directional excitation. Nuclear Eng. Design **217**(1–2), 1–20 (2002)
4. Shrimali, M.K., Jangid, R.S.: A comparative study of performance of various isolation systems for liquid storage tanks. Int. J. Struct. Stab. Dyn. **2**(4), 573–591 (2002)
5. Panchal, V.R., Jangid, R.S.: Variable friction pendulum system for seismic isolation of liquid storage tanks. Nucl. Eng. Design **238**(6), 1304–1315 (2008)
6. Zhang, R.F., Weng, D.G., Ni, W.B.: Application of annular damper reaction wall in seismic isolated LNG tanks. In: Proceedings of the ASME 2010 Pressure Vessels & Piping Division/K–PVP Conference, Bellevue (2010)
7. Housner, G.W.: Dynamic pressure on accelerated fluid containers. Bull. Seismol. Soc. Am. **47**(1), 15–35 (1957)
8. Kelly, T.E., Robinson, W.H., Skinner, R.I.: Seismic Isolation for Designers and Structural Engineers. Robinson Seismic Ltd., Wellington (2006)
9. ABAQUS: Theory and analysis user's manual, version 6.14. Providence: Dassault Systèmes SIMULIA (2014)

Seismic Response and Safety Evaluation of the Underground Slab-Wall Structure with Large Length

Zhiyi Chen[1,2] and Peng Jia[1(✉)]

[1] Department of Geotechnical Engineering, Tongji University,
Shanghai 200092, China
919526724@qq.com
[2] Key Laboratory of Geotechnical and Underground Engineering
of Ministry of Education, Shanghai 200092, China

Abstract. The shaft of Shanghai Zhongshan Park is 30 m deep with the large length and an irregular slab-wall structure. So the form of destruction and weak parks of the structure can't be analysed by the existing simplified method due to its irregular structure. In this paper, the three-dimensional finite element model of surrounding soils and the shaft structure is established. By using the time history analysis method, this paper analyzes the dynamic response of the structure to seismic wave with different spectrum characteristics and acceleration amplitude. Then evaluate the safety of the structure by the interlayer displacement angle and the force of section during the earthquake. The result shows that the dynamic response of the structure is related to the acceleration amplitude and the spectrum characteristics of the seismic wave and the wave which has more long-period components causes larger reaction. Overall, the structure is safe in the medium earthquake, but there is shear damage in the end of mid-board walls. In addition, there is plastic deformation in the discontinuous parts of the structure in the strong earthquake. It shows that discontinuous parts of the structure are more sensitive in the earthquake.

Keywords: Underground structure · Dynamic time history analysis
Seismic response

1 Introduction

With the rapid development of economy and the acceleration of urbanization process, the number of underground structures is increasing. At the same time, with the improvement of construction level and the design level, the underground structures become deeper and more complex. The shaft of Shanghai Zhongshan Park is 30 m deep with the large length and an irregular slab-wall structure. It may suffer overall destruction or local damage in earthquake and the weak parts are also difficult to be confirmed [1–3]. In order to promote the development of urban underground space, reduce the economic loss and casualty caused by the earthquake, the study of the seismic response of the underground structure is very necessary. However, the problem

© Springer Nature Singapore Pte Ltd. 2018
T. Qiu et al. (Eds.): GSIC 2018, *Proceedings of GeoShanghai 2018 International Conference: Advances in Soil Dynamics and Foundation Engineering*, pp. 340–348, 2018.
https://doi.org/10.1007/978-981-13-0131-5_37

is that the seismic response can not be analysed accurately through the existing simplified methods [4, 5].

So in this paper, the dynamic time history analysis will be used to analyze the dynamic response of the shaft in Shanghai Zhongshan Park and evaluate its safety.

2 Project Overview

The shaft of Shanghai Zhongshan Park is about 75 m in length, 40 m in width and 30 m in depth. It is also a part of the multi-layer intersecting tunnel in normal service stages. The service life of this structure is designed for 100 years. The structure safety degree is grade one and its seismic fortification intensity is 7°. The fortification classification is key fortification class, in addition, the site soils belong to class IV, and the seismic fortification grade is three.

There is an engineering drawing of the structure (see Fig. 1).

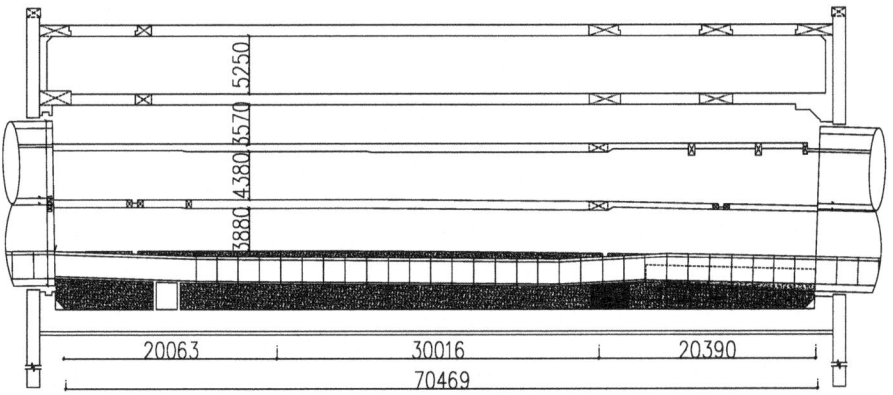

Fig. 1. The engineering deawing of the longitudinal section.

3 Numerical Simulation

In order to study the dynamic response of the shaft structure during the earthquake, the finite element model of the structure and soils is established by ABAQUS. The whole model measures 500 m long and 250 m wide, which is 50 m under the bottom of the structure, approximately 80 m beneath ground level. In addition, to absorb the reflected waves, it is surrounded by infinite elements.

3.1 Structure Model

According to the design drawings,the shaft structure is divided into five layers. The figure below shows the finite element model of the structure without side walls (see Fig. 2) and the horizontal section (see Fig. 3), under the premise of ignoring the smaller holes on each floor.

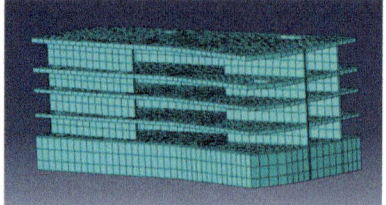

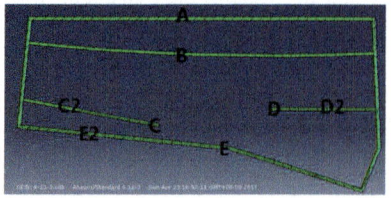

Fig. 2. The finite element model.   **Fig. 3.** The horizontal section.

The shaft is a reinforced concrete structure. This paper mainly studies the holistic dynamic response under earthquake, in which, reinforced concrete is calculated as one material, in consideration of the properties of both steel and concrete. According to the principle of equivalent stress, the tensile strength of reinforced concrete is calculated as follows [6]:

$$R_t = R_c(1 + \frac{R_s}{R_c} \cdot \frac{V_s}{V}) \tag{1}$$

where R_t is the tensile strength of reinforced concrete, R_t is the tensile strength of concrete, R_s is the tensile strength of steel, V_s is the sectional area of steel and V is the sectional area of reinforced concrete.

The tensile strength of reinforced concrete is approximately 6 MPa as calculated, and other parameters of material are similar to the concrete. Under strong earthquakes, the structure is supposed to be in plastic state, so dynamic plastic-damage constitutive model is used to simulate the reinforced concrete. The reinforced concrete's density is 2500 kg/m³, its elastic modulus is 31.5 Gpa, its Poisson's ratio is 0.2, its tensile strength is 6 Mpa, its compressive strength is 20 MPa, its dilation angle is 30°, its eccentricity is 0.1, its compressive strength ratio is 1.16, its stress invariant ratio is 0.667 and its coefficient of viscosity is 0.0005.

3.2 Soils Model

According to the report of engineering geological exploration, the soil is divided into 11 layers, and the thickness of each layer is determined by the report, too. This paper uses the equivalent linear model with damping to simulate the soil layers. The soils' dynamic elastic modulus is calculated as follows:

$$G = \rho V^2 \tag{2}$$

$$E = 2 \cdot (1 + \sigma) \cdot G \tag{3}$$

where ρ is the density of soils, V is shear wave's velocity, G is dynamic shear modulus, σ is Poisson's ratio and E is dynamic elastic modulus.

This paper selects Rayleigh damping as the form of soils' damping. According to the modal analysis function in ABAQUS, the first-order natural frequency of the soils is 0.635 s^{-1} and the second-order natural frequency is 1.69 s^{-1}. The Rayleigh damping is calculated as follows:

$$\xi = \frac{\alpha}{2} \cdot \frac{1}{\omega_1} + \frac{\beta}{2} \cdot \omega_2 \qquad (4)$$

where ω_1 and ω_2 are main order natural frequencies and ξ is damping ratio, which is 0.05 [7]. The following Table 1 gives material parameters of the soil layers.

Table 1. Material parameters of the soils.

Soil number	Thickness (m)	Density (kg/m³)	Poisson's ratio	Shear wave velocity (m/s)	Elastic modulus (MPa)	α	β
①₁	1.6	1850	0.3	140	75.4208	0.29	0.0068
②₁	1.4	1820	0.3	140	74.19776	0.29	0.0068
③	4.2	1720	0.3	140	70.12096	0.29	0.0068
④	9.4	1680	0.3	180	113.2186	0.29	0.0068
⑤₁	11.2	1780	0.3	180	119.9578	0.29	0.0068
⑥	3.12	1960	0.3	180	132.0883	0.29	0.0068
⑦₁	7.3	1890	0.3	180	127.3709	0.29	0.0068
⑦₂	15	1870	0.3	230	205.7598	0.29	0.0068
⑧₁₋₁	11.7	1800	0.3	230	198.0576	0.29	0.0068
⑧₁₋₂	6.5	1840	0.3	230	202.4589	0.29	0.0068
⑧₂	9.5	1850	0.3	280	301.6832	0.29	0.0068

3.3   Selection of Seismic Waves

The "Code for seismic design of urban rail transit structure" stipulates that no less than three sets of seismic waves should be used when structural analysis is carried out using dynamic time history analysis. So El Centro wave, Kobe wave and Shanghai artificial wave are selected in this paper. There are three waves' acceleration-time curve and Fourier spectrum curve (see Figs. 4, 5 and 6).

In the paper, the dynamic response of the structure is analyzed in the cases of medium earthquake and strong earthquake. In the cases of medium earthquake, the acceleration amplitude of the seismic waves should be adjusted to 0.1 g. For example, the ground peak acceleration of the El Centro wave is 0.281 g, so the adjustment factor is 0.356. Similarly, the Kobe wave's adjustment factor is 0.287, and the Shanghai artificial wave's adjustment factor is 0.813. In the case of strong earthquake, the acceleration amplitude of the seismic waves should be adjusted to 0.2 g. The Shanghai artificial wave is chosen to analyze the dynamic response under the strong earthquake, and its adjustment factor is 1.626.

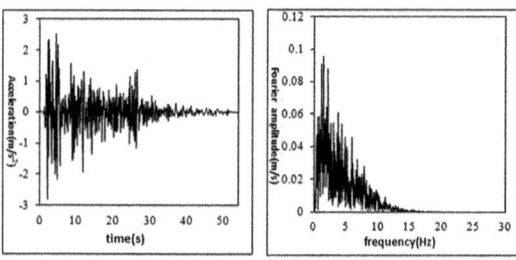

Fig. 4. El Centro wave's acceleration-time curve and Fourier spectrum curve.

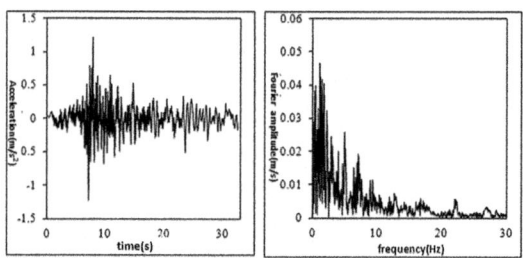

Fig. 5. Kobe wave's acceleration-time curve and Fourier spectrum curve.

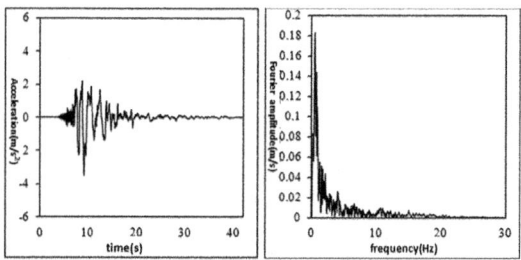

Fig. 6. Shanghai artificial wave's acceleration-time curve and Fourier spectrum curve.

4   Dynamic Response Under the Medium Earthquake

Previous studies have shown that the seismic waves inputted transversely are more dangerous than those inputted longitudinally. So the three seismic waves are inputted transversely in the bottom of the model.

4.1   Displacement Response

The dynamic response of the structure under the medium earthquake is analyzed by interlayer displacement and interlayer displacement angle. The following Table 2 gives the displacement response of the structure under three seismic waves.

Table 2. Displacement response of the structure.

Number of layers	Kobe wave		Shanghai artificial wave		El Centro wave	
	Interlayer displacement (mm)	Interlayer displacement angle	Interlayer displacement (mm)	Interlayer displacement angle	Interlayer displacement (mm)	Interlayer displacement angle
−1	4.35	1/1437	2.94	1/2126	1.46	1/4280
−2	3.36	1/1301	2.05	1/2132	1.05	1/4162
−3	4.49	1/1154	2.76	1/1877	1.35	1/3837
−4	4.25	1/984	2.51	1/1665	1.25	1/3344
−5	1.71	1/4058	0.74	1/9378	0.45	1/16667
Sum	18.1		11		5.54	

Based on the above data, the curves of the maximum value of interlayer displacement and interlayer displacement angle in each layer are drawn (see Figs. 7 and 8).

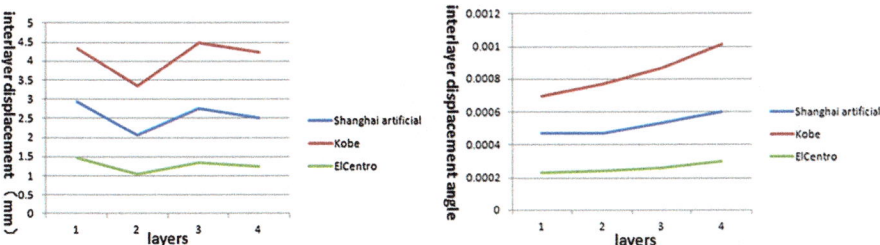

Fig. 7. Maximum interlayer displacement.   **Fig. 8.** Maximum value of interlayer displacement angle.

The maximum interlayer displacement angle of the structure increases from top to bottom. It is also found that the dynamic response of the structure is different under three seismic waves, although the three seismic waves have the same acceleration amplitude. So the dynamic response is related to the spectral characteristics of seismic waves and the wave which has more long-period components causes larger reaction.

4.2 Safety Evaluation

The maximum interlayer displacement angle of each layer is less than 1/550, which is the limit value provided by the "Code for seismic design of urban rail transit structure". So the structure is safe overall in the medium earthquake. The largest dynamic response is caused by the Kobe wave, so this paper selects it to check the structure's safety. The following Table 3 gives the internal force of each section.

Table 3. Internal force of each section.

Location	Axial force (kN/m)		Moment (kN · m/m)		Shear force (kN/m)	
	Static	Dynamic	Static	Dynamic	Static	Dynamic
A	2346	2687.5	25.7	29.875	724	737.1
B	3137.5	3343.5	148.1	150.98	799	807.35
C	4525.5	5252	292.8	432.3	546	757.5
C2	2810.5	2979	92.65	115.85	132	150.15
D	5055	5888	37.67	156.67	142.3	292.8
D2	3699	3766	46.66	162.66	114.6	214
E	4130.5	5862.5	76.9	202.9	127.1	584.95
E2	2246	2743	98.65	182.15	130.9	359.4

It is found that the bending capacity and the vertical compressive bearing capacity of each section is enough, but the shear capacity of the section C is insufficient. It means that the structure may be partially destroyed under medium earthquake. It is also found that the growth of internal force in discontinuous parts is larger than that in continuous parts.

5   Dynamic Response Under the Strong Earthquake

5.1   Displacement Response

The displacement response of the structure under the strong earthquake is analyzed by the interlayer displacement and the interlayer displacement angle, too. The following Table 4 gives the displacement response of the structure under the seismic wave.

Table 4. Displacement response of the structure.

Number of layers	Interlayer displacement (mm)	Interlayer displacement angle
−1	5.61	1/1114
−2	4.13	1/1058
−3	6.08	1/852
−4	5.35	1/781
−5	1.67	1/4156
Sum	22.58	

The displacement response under the strong earthquake is similar to that under the medium earthquake. The maximum interlayer displacement angle under the strong earthquake is 1/781, which is about twice as much as that under the medium earthquake (see Figs. 9 and 10). It is less than 1/250, which is the limit value provided by the "Code for seismic design of urban rail transit structure". So the structure is safe overall under the strong earthquake.

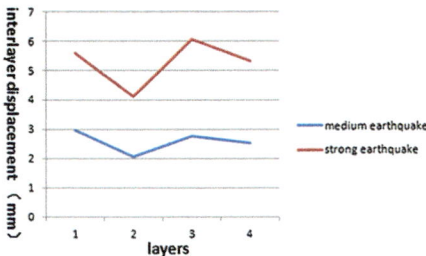

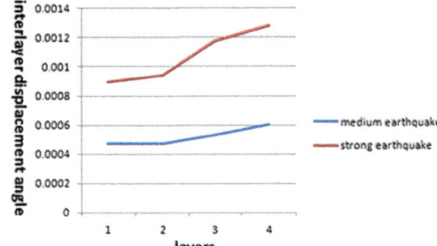

Fig. 9. Maximum interlayer displacement.

Fig. 10. Maximum value of interlayer displacement angle.

5.2 Plasity Developing

There is plastic deformation in the structure under the strong earthquake (see Figs. 11 and 12). Firstly the plastic deformation appers in the end of mid-board walls. Then it appers in the junction of mid-board walls and side walls.

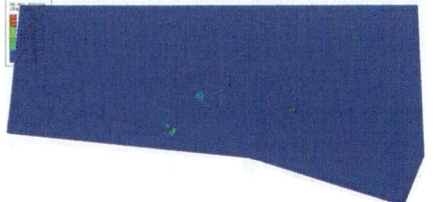

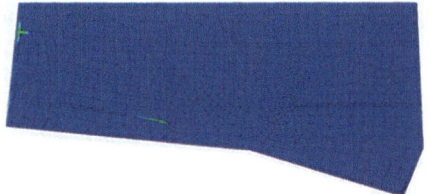

Fig. 11. Appearance of the plastic deformationin.

Fig. 12. Devolopment of the plastic deformation.

6 Conclusions

In this paper, the three-dimensional finite element model of surrounding soils and the shaft structure is established. By using the time history analysis method, this paper analyzes the dynamic response of the structure to seismic wave with different spectrum characteristics and acceleration amplitude. Based on the analytical results, the following conclusions can be drawn:

(1) The dynamic response is related to the acceleration amplitude and the spectral characteristics of the seismic wave. The dynamic response of the structure increases with increasing acceleration amplitude. And the seismic wave which has more long-period components causes larger reaction.

(2) Overall, the structure is safe, but there are partial failures. The maximum interlayer displacement angle of the structure fulfil the requirements provided by the

"Code for seismic design of urban rail transit structure". But there is shear damage in the end of mid-board walls.

(3) The discontinuous parts of the structure are more sensitive in earthquake. Due to the lack of constraints, the growth of internal force in discontinuous parts is larger than that in continuous parts. And there is plastic deformation in discontinuous parts under the strong earthquake.

Acknowledgments. This research was supported by the National Natural Science Foundation of China (Grant No. 41472246, 51778464), Key laboratory of Transportation Tunnel Engineering (TTE2014-01), and "Shuguang Program" supported by Shanghai Education Development Foundation and Shanghai Municipal Education Commission. All supports are gratefully acknowledged.

References

1. Hashash, Y.M., Hook, J.J., Schmidt, B., John, I., Yao, C.: Seismic design and analysis of underground structures. Tunn. Undergr. Space Technol. **16**, 247–293 (2010)
2. Huang, F., Li, M., Ma, Y., Han, Y., Tian, L., Yan, W., Li, X.: Studies on earthquake precursors in China: a review for recent 50 years. Geod. Geodyn. **8**, 1–12 (2017)
3. Wang, W.L., Wang, T.T., Su, J.J., Lin, C.H., Seng, C.R., Huang, T.H.: Assessment of damage in mountain tunnels due to the Taiwan Chi-Chi earthquake. Tunn. Undergr. Space Technol. **16**, 133–150 (2001)
4. Li, S., Zuo, Z., Zhai, C., Xie, L.: Comparison of static pushover and dynamic analyses using RC building shaking table experiment. Eng. Struct. **136**, 430–440 (2017)
5. Kuesel, T.R.: Earthquake design criteria for subways. J. Struct. Div. **95**, 1213–1231 (1969)
6. Gao, D., Zhang, L., Nokken, M.: Compressive behavior of steel fiber reinforced recycled coarse aggregate concrete designed with equivalent cubic compressive strength. Constr. Build. Mater. **141**, 235–244 (2017)
7. Clough, R., Penzien, J.: Dynamics of Structures, 2nd edn. Higher Education Press, Beijing (2006)

Evaluation of SPT-Based Liquefaction Assessment Methods in China

Yi-Fan Zhang[✉], Rui Wang, Jian-Min Zhang, and Jian-Hong Zhang

Tsinghua University, Beijing 100084, China
yf-zhang15@mails.tsinghua.edu.cn

Abstract. Various SPT-based empirical methods for liquefaction assessment have been developed over the past decades. However, the effectiveness of these methods for liquefaction assessment in China needs further investigation. This paper evaluates the efficacy of currently commonly used SPT-based methods for liquefaction assessment in China based on existing earthquake liquefaction data. Through calculating and analyzing the prediction success rate (*PSR*) of in-site liquefied and non-liquefied sites, the applicability of several major SPT-based methods for liquefaction assessment in China is discussed. The analysis results show that the JRA method is in general over-conservative for application in China and is more applicable to liquefaction sites with large sand depth and deep groundwater table; meanwhile the NCEER, I-B and T-Y methods tend to underestimate liquefaction potential in China, and are relatively effective for sites with great acceleration or large magnitude; furthermore, the CSDB methods have a relatively high success rate of liquefaction judgement based on data from historic earthquake data before the 1990's, but a lower success rate for the two most recent earthquakes, especially in the 2003 Bachu earthquake. Based on these findings, further expansion of China's existing liquefied database for liquefaction assessment is suggested to improve the CSDB methods, and the JRA, NCEER, I-B and T-Y methods also require further improvement based on China's liquefaction data when applied to China.

Keywords: Liquefaction assessment · SPT · China · Earthquake

1 Introduction

The evaluation of liquefaction potential is an important issue for seismic design. Standard penetration test (SPT) has been widely used in development of liquefaction assessment methods. In 1971, a relatively complete SPT-based liquefaction assessment method was first proposed by Seed [1–4], in which *CSR* (cyclic stress ratio) was determined by ground motion conditions and the *CRR* (cyclic resistance ratio) boundary curves of soil were established based on the laboratory tests and SPT data. Later on, the United States, Japan, and China have developed various empirical or semi-empirical methods for liquefaction assessment using SPT data [5–11], which are different from each other in evaluation procedure, parameters selection, factors consideration and the database used.

© Springer Nature Singapore Pte Ltd. 2018
T. Qiu et al. (Eds.): GSIC 2018, *Proceedings of GeoShanghai 2018 International Conference: Advances in Soil Dynamics and Foundation Engineering*, pp. 349–357, 2018.
https://doi.org/10.1007/978-981-13-0131-5_38

China has widely distributed liquefiable sediment in Bohai coast, Yangtze delta and Pearl delta areas [12], with frequent seismic activity, which have led to a large number of recorded liquefaction cases. Due to the empirical nature of SPT-based liquefaction assessment methods, the effectiveness of these various methods for application in China needs to be evaluated based on existing liquefaction database. Such evaluation will be able to provide a basis for the choice of liquefaction assessment method in seismic design and improvement of existing methods.

In this paper, China's historic earthquake liquefaction data is collected and adopted in evaluating the effectiveness of several currently existing SPT-based liquefaction assessment methods.

2   Analysis of Liquefaction Data in China

This paper totally collects 314 liquefaction sites, including 201 liquefied sites and 113 non-liquefied sites, and 1571 SPT liquefaction data from 10 Chinese historic earthquakes, which has been the most systematic and complete collation and statistics for China's earthquake liquefaction data in recent thirty years. The liquefaction data in China considered in this study includes: (1) China's original liquefaction database, in which liquefaction data are from the 1966 Xingtai earthquake, the 1970 Tonghai earthquake, and four other earthquakes from 1962 to 1970 [13]; liquefaction data in (2) the 1975 Haicheng Earthquake [13]; (3) the 1976 Tangshan Earthquake [13]; (4) the 1999 Chi-chi Earthquake [14]; and (5) the 2003 Bachu Earthquake [15]. Gravelly soils that were reported to have liquefied in the 2008 Wenchuan earthquake are not considered here.

The data mentioned above provides a basis for examining existing SPT-based methods, with the following basic characteristics:

(1) Earthquake magnitude ranges from 6.3 to 7.8, and the largest magnitude occurred in the Tangshan and Tonghai earthquakes, while the Chi-chi earthquake recorded the highest peak acceleration of 0.8 g.
(2) Liquefaction can happen in regions where seismic intensity is 6.
(3) The maximum liquefaction depth reaches 25.2 m, meanwhile the maximum groundwater depth is 5 m, both occurred in the Chi-chi earthquake.
(4) The maximum SPT-N value triggering liquefaction is 22.2, occurring in the Tangshan earthquake.
(5) Liquefied soils mainly consist of silt and fine sand, with high fines content in the Chi-chi earthquake due to sand-mud interbedding.

More information about liquefaction data characteristics of each historic earthquake in China is shown in Table 1.

Table 1. Liquefaction data characteristics of each historic earthquake in China

Earthquake	Magnitude	a_{max}	Average groundwater depth (m)		Average sand depth (m)		Average SPT-N value		Number of liquefied sites	Number of non-liquefied sites	Characteristics
			Liq	N-Liq	Liq	N-Liq	Liq	N-Liq			
Original database	6.3–7.8	0.1–0.5	0.99	0.97	1.87	2.48	8.46	17.64	38	21	Shallow groundwater table and sand depth
1975 Haicheng	7.3	0.1–0.2	1.71	1.40	7.49	10.2	7.50	14.90	7	5	Small acceleration of ground motion and deep sand layer
1976 Tangshan	7.8	0.1–0.7	1.47	2.71	4.42	7.22	7.93	22.13	55	37	Large magnitude and great acceleration
1999 Chi-chi	7.6	0.2–0.8	1.51	2.78	7.4	10.9	8.80	16.30	80	24	Great acceleration, deep sand layer and high fine content
2003 Bachu	6.8	0.1–0.4	2.32	2.78	4.35	8.63	13.52	29.04	21	26	Small magnitude, deep groundwater table and great SPT-N value

3 SPT-Based Liquefaction Assessment Methods

Currently, commonly used methods for liquefaction assessment include: (1) the Tokimatsu-Yoshimi (T-Y) method (1983) [5]; (2) the Japan Rail Association (JRA) method (1996) [6, 7]; (3) the NCEER method (2001) [8]; (4) the Idriss-Boulanger (I-B) method (2008) [9]; and (5) Chinese Code for Seismic Design of Buildings method (CSDB, 2001&2010) [10, 11]. Some of these can be used in combination with laboratory test results, while others rely solely on in-situ test data.

The T-Y, NCEER and I-B methods are all developed based on Seed's liquefaction assessment method, evaluating the cyclic stress ratio (*CSR*) and the cyclic resistance ratio (*CRR*) respectively before calculating a safety factor against liquefaction (*FL*). The differences between these three methods are mainly reflected in the factors considered and the database used. The JRA method is established based on seismic intensity and designed seismic grouping of Japan, mainly using liquefaction data in the 1995 Kobe earthquake. The CSDB (2001&2010) methods are developed based on empirical correlation between underground water depth, sand depth and SPT-N value in the Chinese updated liquefaction database, which is expanded with the liquefaction data in the 1975 Haicheng and 1976 Tangshan earthquakes based on Chinese original liquefaction database. In the CSDB 2010 method, horizontal peak ground acceleration a_{max} and earthquake magnitude *M* are adopted to replace the seismic intensity and design seismic grouping, respectively. A comparison of these SPT-based methods is made in Table 2.

Table 2. Comparison of different SPT-based liquefaction assessment methods.

Method	Earthquake intensity	Ground motion duration	Ground water depth	Sand depth	Unit weight	Particle size	Fine content	Influence of overburden pressure	Influence of SPT procedure	Database used
T-Y	a_{max}^a	M^c	h^f	z^h	γ^j	NA	FC^l	C_N	NA	Niigata^p
JRA	I^b	EQ type^d	h	z	γ	D_{50}^k	FC	C_N	NA	Kobe^q
NCEER	a_{max}	M	h	z	γ	NA	FC	C_N	$C_E\ C_B\ C_R\ C_S^o$	Seed
I-B	a_{max}	M	h	z	γ	NA	FC	$C_N\ \&\ K_\sigma^n$	$C_E\ C_B\ C_R\ C_S$	Cetin^r
CSDB2001	I	EQ group^e	d_u^g	d_s^i	NA	NA	ρ_c^m	$d_u\ \&\ d_s$	NA	China^s
CSDB2010	a_{max}	M	d_u	d_s	NA	NA	ρ_c	$d_u\ \&\ d_s$	NA	China

Note: NA = not applicable.
[a] Horizontal peak ground acceleration a_{max}; [b] Seismic intensity I; [c] Earthquake magnitude M; [d] Earthquake type defined in the Design Code and Explanations for Roadway Bridges in Japan [6, 7]; [e] Design earthquake grouping defined in the Code for Seismic Design of Buildings in China [10, 11]; [f] Ground water depth h; [g] Ground water depth d_u; [h] Sand depth z; [i] Sand depth d_s; [j] Unit weight of soil γ; [k] Average particle size D_{50}; [l] Fine congtent FC; [m] Clay congtent ρ_c; [n] Overburden correction factor $C_N\&K_\sigma$; [o] Energy ratio correction factor C_E, Borehole diameter correction factor C_B, Short rod correction factor C_R, Sampling method correction factor C_S; [p] The T-Y method is based on Seed's research results and the CRR curves are established by the laboratory test on the frozen samples from Niigata [5]; [q] The liquefaction data in the 1995 Kobe earthquake in Japan, which is the main liquefaction data used in the JRA method [7]; [r] Liquefaction database updated by Cetin based on Seed's database [9, 16]; .[s] Updated Chinese liquefaction database including the liquefaction data in the 1975 Haicheng and 1976 Tangshan earthquakes [6, 7, 13].

4   Evaluation Results and Analysis

In this paper, prediction success rate (*PSR*) is adopted to evaluate the effectiveness of several currently existing SPT-based liquefaction assessment methods. Prediction success rate is the probability of success in judging a certain event that happened in the past, such as liquefaction or non-liquefaction [17]. If there are K in-site liquefaction events in total, the number of in-site liquefied sites which are predicted to be liquefied is K_{11} and non-liquefied is K_{12}; the same for in-site non-liquefied sites, which are predicted to be liquefied is K_{21} and non-liquefied is K_{22}, the prediction success rate of liquefied sites (*PSR$_L$*), non-liquefied sites (*PSR$_{NL}$*) and the combination of liquefied and non-liquefied sites (*PSR$_C$*) can be calculated as shown in the following Table 3.

Table 3. The calculation method of prediction success rate (*PSR*).

		Field investigation		Liquefaction prediction	PSR$_L$	PSR$_{NL}$	PSR$_C$
		liq	non-liq	combined			
Liquefaction prediction	liq	K_{11}	K_{21}	$K_{11} + K_{21}$	$\dfrac{K_{11}}{K_{11}+K_{12}}$	$\dfrac{K_{22}}{K_{21}+K_{22}}$	$\dfrac{K_{11}+K_{22}}{K}$
	non-liq	K_{12}	K_{22}	$K_{12} + K_{22}$			
Field investigation	combined	$K_{11} + K_{12}$	$K_{21} + K_{22}$	K			

Note: liq = liquefied sites; non-liq = non-liquefied sites; combined = the combination of liquefied and non-liquefied sites; and the same below.

In the evaluation results of China's original liquefaction database shown in Fig. 1, the CSDB 2010 method performs best, while the four foreign methods perform not as well, mainly because that different databases are used in the process of establishing different methods [5–11]. The results by the T-Y, NCEER and I-B methods are similar, and underestimate the liquefaction potential in China's original liquefaction database, which is caused by that all these three methods are established based on Seed's procedures and similar database are used [5, 8, 9]. The two CSDB methods (2001&2010) perform similarly for both liquefied sites and non-liquefied sites. However, the JRA method performs conservative, where non-liquefied sites are often assessed as liquefied sites.

In the Haicheng earthquake, the predictions of liquefaction sites by the CSDB methods (2001&2010) are 100% correct, while the success rates of liquefaction assessment by the other four methods are much lower (Fig. 2), which is also due to the different databases used in different methods. Similar to the results from China's original liquefaction database, the T-Y, NCEER and I-B methods underestimate liquefaction potential, and the JRA method performs conservatively. Due to the small acceleration of ground motion and the deep sand layer, the underestimation or overestimation of liquefaction potential by different methods is more significant in the Haicheng earthquake compared with the original database.

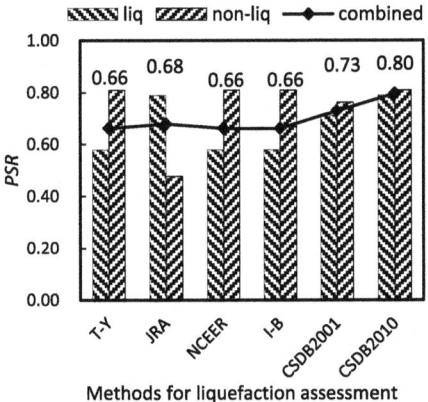 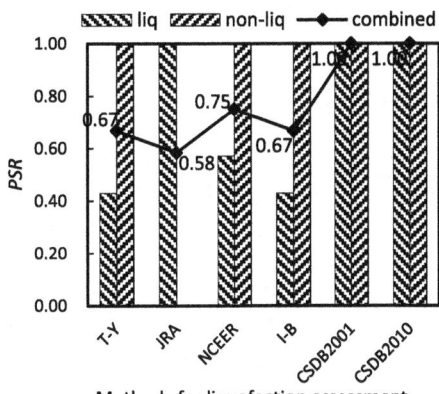

Fig. 1. Liquefaction evaluation results of China's original liquefaction database.

Fig. 2. Liquefaction evaluation results in the 1975 Haicheng earthquake.

The results from the Tangshan earthquake data are similar to those in China's original liquefaction database except for the numerical differences, as shown in Fig. 3. In the Tangshan earthquake, the CSDB methods perform better with higher success rates of 0.82 and 0.84, which is mainly because that nearly 60% of the database used in the CSDB methods is derived from the Tangshan earthquake. Seven liquefaction sites in the Tangshan earthquake are adopted by Seed's database [18], and the T-Y, NCEER and I-B methods perform quite well with no obvious underestimation of the liquefaction potential. On account of the large earthquake magnitude and the large acceleration, the JRA method is still over-conservative.

For the Chi-chi earthquake, all methods are conservative due to large local acceleration and high fines content of the earthquake, as shown in Fig. 4. The T-Y and JRA method, which are established based on Japanese liquefaction data, performed

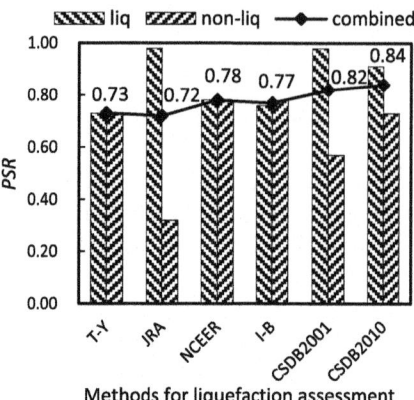

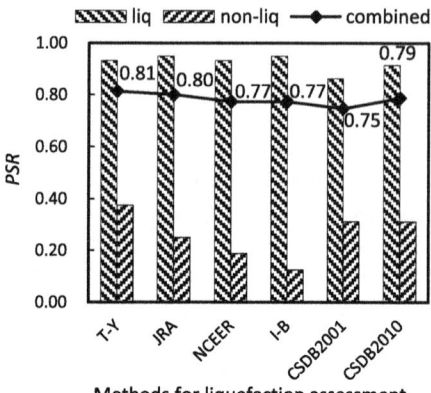

Fig. 3. Liquefaction evaluation results in the 1976 Tangshan earthquake.

Fig. 4. Liquefaction evaluation results in the 1999 Chi-chi earthquake.

marginally better than others. This could be due to Japan and Taiwan being close on the pacific seismic zone [19]. In the Bachu earthquake, liquefaction occurred even in a small earthquake magnitude, and all methods underestimate the liquefaction potential except the JRA method, as shown in Fig. 5. Moreover, the JRA method has the highest success rate though being conservative, while the NCEER method has the lowest successful assessment rate.

The results of the combined liquefaction data in China collected in this study are plotted in Fig. 6. Due to poor performance in the Chi-chi and Bachu earthquakes, the successful assessment rates of the CSDB methods are lower in the combined data compared with those of the original database. Likewise, the T-Y, NCEER and I-B methods tend to underestimate liquefaction potential, and the JRA method performs

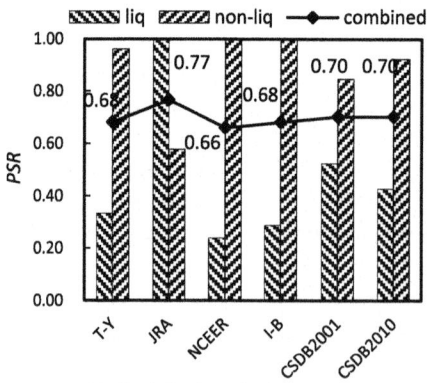

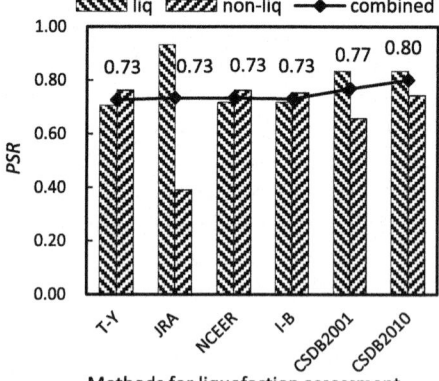

Fig. 5. Liquefaction evaluation results in the 2003 Bachu earthquake.

Fig. 6. Liquefaction evaluation results of the existing liquefaction data in China.

conservatively. Interestingly, the judgement results of the four non-Chinese methods are the same with a success rate of 0.73.

In the liquefaction assessment in China, the JRA method overestimates the liquefaction potential while the T-Y, NCEER and I-B methods underestimate liquefaction potential. Comparing the difference in these four SPT-based liquefaction assessment methods and the evaluation results, we can find that the factors taken into consideration are similar, effects of the evaluation procedure and the parameters selection are similar, and the most obvious difference is the database used in different methods. In Seed's database, the *CSR* value is generally higher than that in Chinese liquefaction database, for example, the *CSR* value of liquefaction data in the Haicheng earthquake adopted by Seed's database ranges from 0.172–0.260 compared with 0.061–0.160 in the Chinese liquefaction data [18]. Thus, there is an underestimation in the liquefaction assessment in China using the T-Y, NCEER and I-B methods, which are more applicable to sites with a relatively large magnitude or acceleration. The liquefaction data in the 1995 Kobe earthquake is mainly used in the JRA method, and the critical liquefaction depth and groundwater table can reach up to 11.5 m and 7.7 m respectively, which are larger than those in China's liquefaction data [18]. Therefore, the JRA method overestimates the liquefaction potential in China, and it's more applicable to liquefaction sites with large sand depth and deep groundwater table.

In conclusion, the CSDB methods are most applicable in China with the CSDB 2010 method marginally better than the CSDB 2001 method, because they are both established based on Chinese liquefaction data. However, these two methods don't perform well when applying to liquefaction sites in the 1999 Chi-chi earthquake and the 2003 Bachu earthquake due to the lack of high fines content and small earthquake magnitude data during the calibration of the methods, and further modifications are needed to expand the liquefaction database.

5   Conclusions

In this paper, data from most recorded existing earthquake liquefaction events in China is collected and analyzed, and several major SPT-based methods for liquefaction assessment are evaluated using the data. It should be pointed out that the liquefaction data in China used in this paper is still incomplete, for example, the data on fine content is lacking except in the Chi-chi earthquake. The main conclusions drawn from this study are as follows:

(1) In the Chinese liquefaction data, earthquake magnitude ranges from 6.3 to 7.8; peak ground acceleration ranges from 0.1 g to 0.8 g; maximum liquefaction depth is 25.2 m; maximum groundwater depth is 5 m; maximum SPT-N value is 22.2; and liquefied soil mainly consists of silt and fine sand. During the 1999 Chi-chi earthquake the greatest peak ground acceleration is recorded, while the liquefied soil occurs at relatively large depths and generally consists of high fines content. In the 2003 Bachu earthquake, although the magnitude is relatively small, and the SPT-N value is relatively large, liquefaction was still observed.

(2) Commonly adopted SPT-based methods take many factors into account, including seismic intensity, ground motion duration, groundwater depth, sand depth, unit weight, particle size, fines content, influence of overburden pressure, and influence of SPT procedure. The differences of these methods are mainly reflected in the assessment procedure, the selection of parameters, and the database used.

(3) The effectiveness of the CSDB2010 method for liquefaction assessment in China is the best, then followed by the CSDB2001 method, and then the I-B, NCEER and T-Y methods, and finally the JRA method. The CSDB methods are in general most applicable in China, but are not suitable for the recent two earthquakes and require further modifications based on the expanded liquefaction database. The NCEER, I-B and T-Y methods tend to underestimate liquefaction potential in China, and are more applicable to sites with large magnitude or great acceleration. The JRA method is very conservative when used for China's liquefaction data, and is more applicable to sites with large sand depth and deep groundwater table.

References

1. Seed, H.B.: Simplified procedure for evaluating soil liquefaction potential. J. Soil Mech. Found. Div. **97**(SM9), 1249–1273 (1971)
2. Seed, H.B.: Soil liquefaction and cyclic mobility evaluation for level ground during earthquakes. J. Geotech. Eng. Div. **105**(2), 201–255 (1979)
3. Seed, H.B.: Evaluation of liquefaction potential using field performance data. J. Geotechn. Eng. **109**(3), 458–482 (1983)
4. Seed, H.B.: Influence of SPT procedures in soil liquefaction resistance evaluations. J. Geotech. Eng. **111**(12), 1425–1445 (1985)
5. Tokimatsu, K.: Empirical correlation of soil liquefaction based on SPT-N values and fines content. Soils Found. **23**(4), 56–74 (1983)
6. Japan Rail Association (JRA): Design code and explanations for roadway bridges, Part V-seismic resistance design, Japan (1990)
7. Japan Rail Association (JRA): Design code and explanations for roadway bridges, Part V-seismic resistance design, Japan (1996)
8. Youd, T.L.: Liquefaction resistance of soils: summary report from the 1996 NCEER and 1998 NCEER/NSF workshops on evaluation of liquefaction resistance of soils. J. Geotech. Geoenviron. Eng. **127**(10), 817–833 (2001)
9. Idriss, I.M., Boulanger, R.W.: Soil liquefaction during earthquakes. No. MNO-12. EERI, Oakland, California, USA (2008)
10. Construction Department of the People's Republic of China: Code for Seismic Design of Buildings. GB50011-2001. China Building Industry Press, Beijing (2001)
11. Construction Department of the People's Republic of China: Code for Seismic Design of Buildings. GB50011-2010. China Building Industry Press, Beijing (2010)
12. Xiaoming, Y.: The development of China's normative liquefaction analysis method. Rock Soil Mech. **32**(2), 351–358 (2011)
13. Junfei, X.: Some opinions on modification of soil liquefaction evaluation method in seismic code. Earthq. Eng. Eng. Vib. **4**(2), 95–109 (1984)
14. Soil Liquefaction in the 1999 Chi-Chi, Taiwan, Earthquake. https://cecas.clemson.edu/chichi/TW-LIQ/Homepage.htm. Accessed 11 Sept 2017

15. Zhaoyan, L.: Research on the liquefaction evaluation method based on field investigations in the Bachu earthquake. Institution of Engineering Mechanics, CEA, Harbin (2012)
16. Cetin, K.O.: Field case histories for SPT-based insitu liquefaction potential evaluation. Geotechnical Engineering Research Report No. UCB/GT-2000/09. Department of Civil and Environmental Engineering, University of California, Berkeley (2000)
17. Guoxing, C.: Research on reliability of liquefaction prediction. Earthq. Eng. Eng. Vib. **11**(2), 85–96 (1991)
18. Guoxing, C.: The determinate and probabilistic method for liquefaction assessment of sand soil based on the standard penetration test. Rock Soil Mech. **36**(1), 9–27 (2015)
19. Muhsiung, C.: Comparison of SPT-N-based analysis methods in evaluation of liquefaction potential during the 1999 Chi-chi earthquake in Taiwan. Comput. Geotech. **38**, 393–406 (2011)

Experimental Research on Dynamic Characteristics of Rubber Model Soil

Rongji Xia[1], Haohao Wei[2], Guobo Wang[2(✉)], Hang Xiao[1], and Shuo Peng[1]

[1] School of Civil Engineering and Architecture,
Wuhan University of Technology, Wuhan 430070, Hubei, China
[2] Hubei Key Laboratory of Roadway Bridge and Structure Engineer,
Wuhan University of Technology, Wuhan 430070, Hubei, China
wgb16790604@126.com

Abstract. In order to meet the bearing capacity of shaking table and reduce the distortion of similarity of soil-structure stiffness, remolded soil such as rubber model soil consists of rubber particles and the intact clay has been widely used in the process of shaking table model test. Different proportion of rubber powder soil and rubber particle soil were arranged to conduct a resonance column test to determine their dynamic characteristics and compare with the dynamic characteristics of the intact clay. The experimental results show that the two models' changing regulation of the dynamic shear modulus and dynamic damping ratio is consistent with that of pure soil, which indicates that both models can reflect the dynamic characteristics of pure soil and can be used in model test. And the dynamic shear modulus and dynamic damping ratio of the rubber model soil decreased with the increase of rubber particle size. The test results can provide reference for similar test in the future.

Keywords: Rubber model soil · Dynamic characteristics
Resonance column test · Dynamic shear modulus · Dynamic damping ratio

1 Introduction

Shaking table test is a powerful method to study the dynamic interaction between soil and underground structures. In order to make the test reflect the seismic response of underground structures reasonably, it is very important to select the model soil reasonably during the shaking table model test. On the other hand, soil structure stiffness ratio is a key factor in soil structure dynamic interaction system [1]. In the shaking table model test, the distortion of the soil-structure stiffness ratio should be reduced as much as possible. At the same time, the carrying capacity of the vibration table which is made in the early time is insufficient. In combination with the above factors, sawdust soil, sawdust sand, rubber model soil and other remolded soil have been used widely. It could reduce the density of the model soil, and also can adjust the shear modulus and the damping ratio of the model soil. By adding sawdust, rubber and other materials into the undisturbed soil, we could reduce the distortion of the soil-structure stiffness ratio.

© Springer Nature Singapore Pte Ltd. 2018
T. Qiu et al. (Eds.): GSIC 2018, *Proceedings of GeoShanghai 2018 International Conference: Advances in Soil Dynamics and Foundation Engineering*, pp. 358–367, 2018.
https://doi.org/10.1007/978-981-13-0131-5_39

Because of the above advantages, many scholars have done some research on model soil. Kitada et al. [2] used silicon rubber as model soil in the shaking table test. Turan et al. [3] made a kind of model soil which included bentonite, glycerin and water. Moss et al. [4] used kaolin, bentonite, fly ash and water to make the model soil, which was adopted in the soil-underground structure shaking table test. Fatahi et al. [5, 6] designed a synthetic clay mixture which contains kaolinite clay, active bond bentonite, class F fly ash, lime, and a bender element test was performed to measure the shear-wave velocity of the mix over the cure age. And finally they found the most appropriate soil mix for the shaking table test. Based on Buckingham-Π theorem, Yan et al. [7–12] mixed the sawdust and sand with a certain proportion. And this model soil was used in a series of shaking table tests which were under non-uniform excitations. Kim et al. [13] investigated the small-strain and zero-lateral strain responses of mixtures of small rigid sand particles and large soft rubber particles. Results show that the sand skeleton controls the mixture response when the volume fraction of rubber particles is $V_{rubber} \leq 0.3$, while the rubber skeleton prevails at $V_{rubber} \geq 0.6$. Lee et al. [14] studied the stress-deformation and small-strain shear wave characteristics of rubber-sand particle mixtures. They reported that the size ratio and volumetric sand fraction are the important factors that determine the behavior of rigid and soft particle mixtures. Deng and Feng [15] studied the effect of rubber particles on the shear properties of sand and they founded that by adding 10%–20% of the tire rubber particles can improve the shear modulus of sand, and reduce material density. Shang et al. [16] did the cyclic shear tests on the mixtures of rubber particles and sand with different contents. The experimental results showed that the dynamic shear modulus of the mixture decreased with the increase of the proportion of rubber particles. Hu et al. [17] studied the influence of rubber powder on the dynamic mechanic properties through the dynamic tri-axial test. The results show that the maximum dynamic modulus (E_{dmax}) and dynamic shear modulus (G_{dmax}) decrease significantly with the increment of rubber powder percentage, the maximum damping ratio of manipulated loess increases with the increment of rubber powder percentage.

On the basis of previous studies, rubber particle model soil and rubber powder model soil with different mass ratio are prepared in this paper. Then the model soils are tested by the GDS resonant column to find out the influence of the rubber particles size on the dynamic characteristic of the mixed soil.

2 The Preparation of Model Soils

The intact clay is taken from the Wuhan subway station site, in consideration of the clay has been disturbed. In this paper, we use the name of pure soil instead of intact soil. The water content of pure soil is 17.78%. And the density is 1.89 g/cm^3. The particle size of rubber is range from 2 mm to 4 mm (Fig. 1a), and the particle size of rubber powder is range from 40 mesh to 60 mesh (Fig. 1b). The clay was placed on the playground for drying. Then the clay was crushed and sieved with a 3 mm sieve to remove the larger particles (Fig. 1c). Finally, according to the proportion, put the clay and rubber into the container and mix it evenly, then pour water into the container in three times, and stir it until homogeneous. The mixed soils were placed into the sample preparation device in three times and compacted with a jack. The samples are shown in Fig. 2.

a. Rubber particles（Particle size: 2~4mm） **b.** Rubber powders （Particle size: 40~60mesh） **c.** Sieved clay（Particle size≤3mm）

Fig. 1. Rubber particles, powders, and sieved clay

 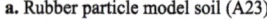  

a. Rubber particle model soil (A23)   **b.** Rubber powder model soil (B23)   **c.** Pure soil (Y3)

Fig. 2. Model soil samples

There are three kinds of soils in this experiment, which are pure soil, rubber powder model soil and rubber particle model soil. There are three samples of the pure soil, and name them with Y1, Y2 and Y3. Three mass ratios of the model soils are designed in this experiment, and three samples for each mass ratio. The experimental schemes are shown in Tables 1 and 2.

Table 1. The mass ratio of rubber particles model soil

Group	Mass ratio	Density (g/cm^3)	Sample number
A1	1.00:0.75:0.28	1.30	A11, A12, A13
A2	1.00:1.00:0.34	1.40	A21, A22, A23
A3	1.00:1.50:0.52	1.45	A31, A32, A33

The mass ratio is: rubber particles: sieved clay: water.

Table 2. The mass ratio of rubber powder model soil

Group	Mass ratio	Density (g/cm^3)	Sample number
B1	1.00:0.75:0.74	1.30	B11, B12, B13
B2	1.00:1.00:0.62	1.40	B21, B22, B23
B3	1.00:1.50:0.66	1.45	B31, B32, B33

The mass ratio is: rubber powder: sieved clay: water.

3   Test Instruments and Principles

This dynamic test is done using GDS resonance column test system. As shown in Fig. 2. When the sample is installed, the bottom is fixed, and the top is free. At the top, there is a drive system that can generate torque and operate on the top of the soil sample. The response of the sample is measured by the acceleration sensors which are at the bottom of the drive system. The damping ratio is obtained by a free vibration decay curve. And the shear modulus needs to be calculated. The parameters in the calculation process are determined as follows.

Moment of inertia of the sample:

$$I = md^2/8 \tag{1}$$

m—The quality of the sample (kg)
d—The diameter of the sample (m)

Moment of inertia of the GDS resonant column test system: $I_0 = 0.0037$. Then the value of β can be obtained according to the value of the table which is in the GDSRCA Handbook. The shear wave velocity can be obtained by formula (2).

$$V_S = 2\pi f l/\beta \tag{2}$$

f—The resonant frequency of the sample
l—Sample height

Finally, the modulus of shearing could be obtained by formula (3).

$$G = \rho V_S^2 \tag{3}$$

ρ—Sample density

The parameters of this test are set as follows: frequency: 0.01–150 Hz, voltage: 0–0.05 V, confining pressure: 0–300 kPa. Specimen geometry size: $\Phi 50 \times H100$ mm (Fig. 3).

Fig. 3. GDS resonant column test system

4 Test Results and Analysis

In this experiment, undrained-unconsolidated of the resonant column tests were carried out on rubber particle model soil and rubber powder model soil under three kinds of confining pressure (50 kPa, 100 kPa, 200 kPa). And four kinds of confining pressure (50 kPa, 100 kPa, 200 kPa, 300 kPa) for the pure soil.

4.1 Analysis of the Dynamic Characteristics of the Soil Under Different Confining Pressures

Figures 4, 5 and 6 show the dynamic characteristics of the three kinds of soils under different confining pressures, which include the curve of dynamic shear modulus (G_d), the curve of dynamic shear modulus normalized (G_d/G_{dmax}) and the curve of dynamic damping ratio with the change of dynamic shear strain(r_d). Due to the limited layout, this paper takes the date of A2, B2 and Y3 groups to analyze the dynamic characteristics of these three kinds of soils.

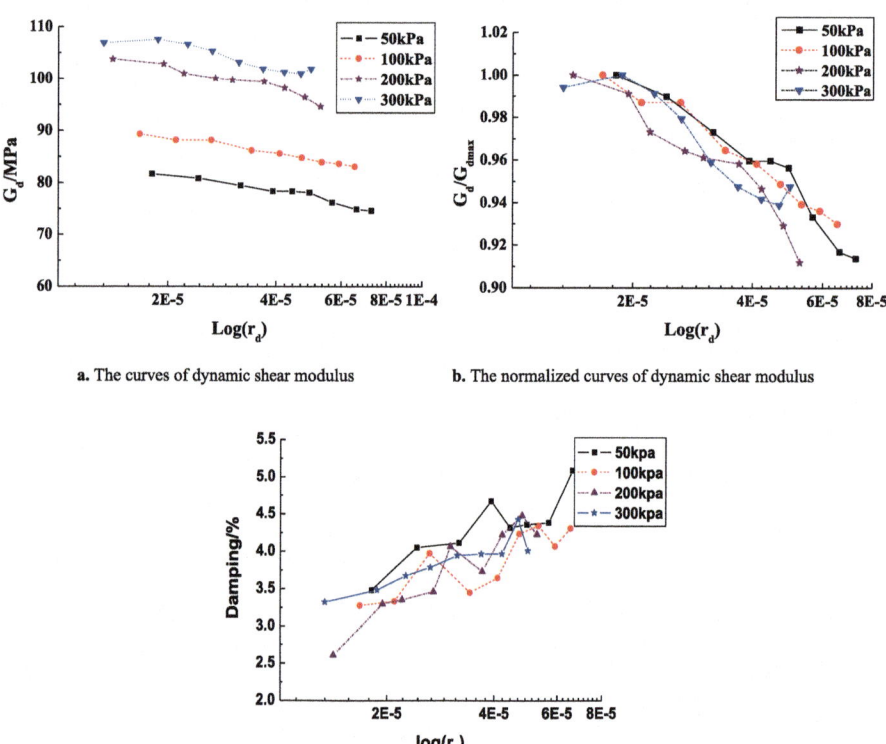

a. The curves of dynamic shear modulus   **b.** The normalized curves of dynamic shear modulus

c. The curves of dynamic damping ratio

Fig. 4. Dynamic properties of pure clay under different confining pressure

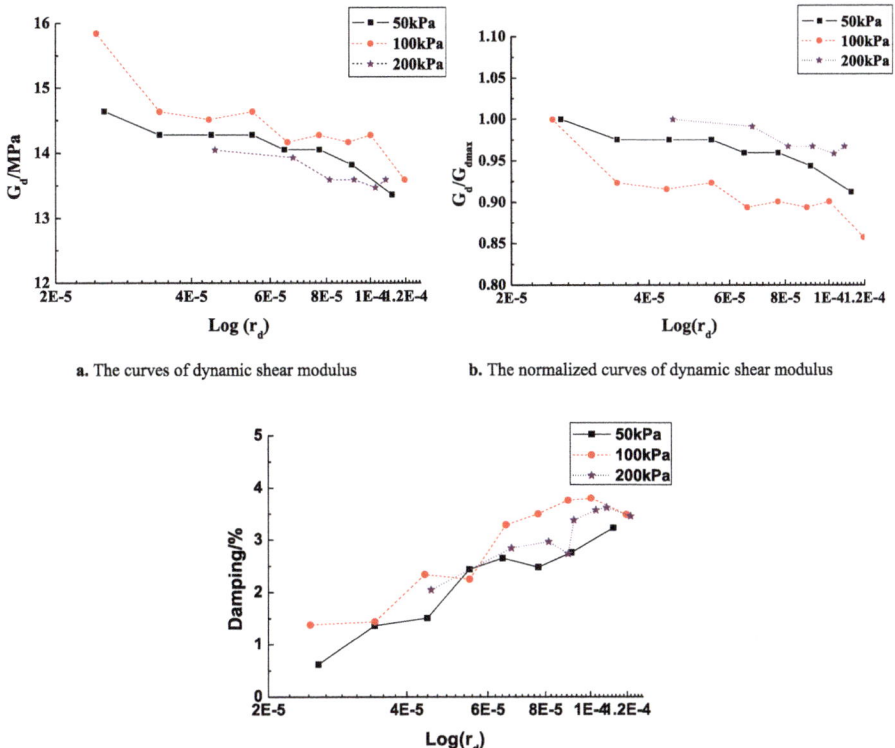

a. The curves of dynamic shear modulus

b. The normalized curves of dynamic shear modulus

c. The curves of dynamic damping ratio

Fig. 5. Dynamic properties of rubber particle model soil under different confining pressure

As shown in Figs. 4, 5 and 6:

(1) The dynamic shear modulus of rubber particles and rubber powder model soil decreases with the increase of dynamic shear strain, and the damping ratio increases with the increase of and dynamic shear strain, which are the same as the pure soil.

(2) The curves of dynamic shear modulus of three kinds of soils increase with the increase of confining pressure, but the confining pressure has little effect on the curves of G_d/G_{dmax}. As can be seen from the data in the diagram, the dynamic shear modulus of the rubber power model soil is about 2–3 times of the rubber particles model soil. It can be seen that the dynamic shear modulus of rubber model soil decreases with the increase of rubber particle size.

(3) The curves of damping ratio of the three soils have no obvious change characteristics with the increase of confining pressure, which also reflects the complexity of damping.

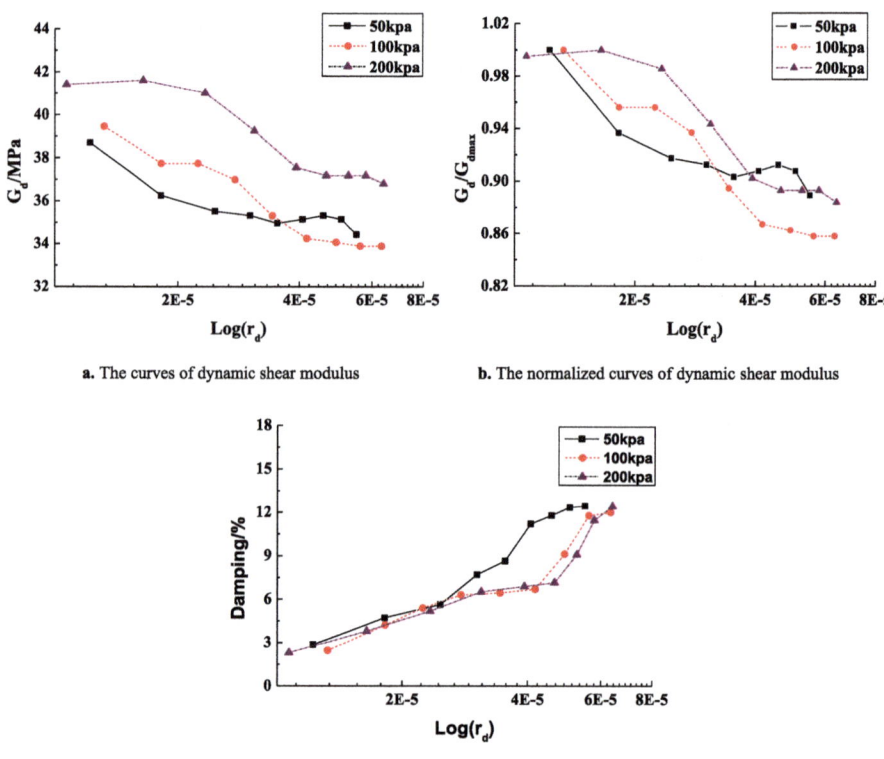

a. The curves of dynamic shear modulus   **b.** The normalized curves of dynamic shear modulus

c. The curves of dynamic damping ratio

Fig. 6. Dynamic properties of rubber powder model soil under different confining pressure

4.2   Comparative Analysis of Model Soil and Pure Soil

Figures 7, 8 and 9 shows the comparison curves of the dynamic characteristics of pure soil, rubber particle model soil and rubber powder model soil under three confining pressures (50 kPa, 100 kPa and 200 kPa). As shown in Figs. 7, 8 and 9:

(1) The three soils have a consistent trend of the curves of dynamic damping ratio and G_d/G_{dmax}, which indicates that the two model soils can simulate the dynamic characteristics of pure soil.

(2) The decrease rate of dynamic shear modulus of rubber powder is faster than that of the pure soil, while the rate of decrease of the dynamic shear modulus of rubber particles is slower than that of the pure soil.

(3) It can be seen from the figures that the dynamic damping ratio of the rubber powder model soil is larger than that of the rubber particle model soil. So the dynamic damping ratio of the soil is reduced with the increase of the particle size of the rubber particles.

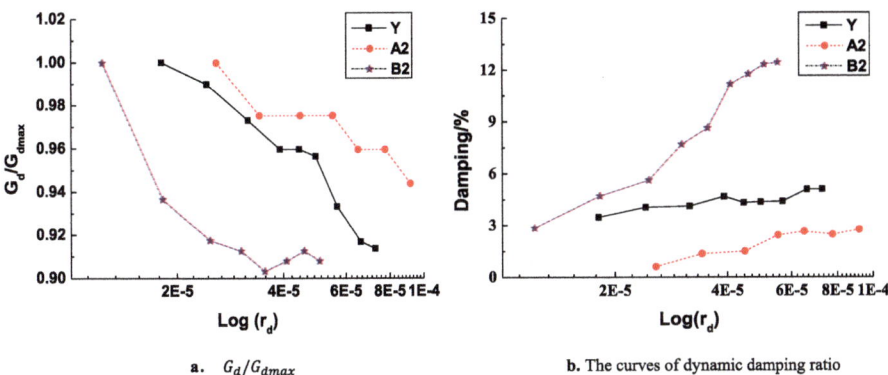

a. G_d/G_{dmax} **b.** The curves of dynamic damping ratio

Fig. 7. Testing results of model soil with confining pressure 50 kPa

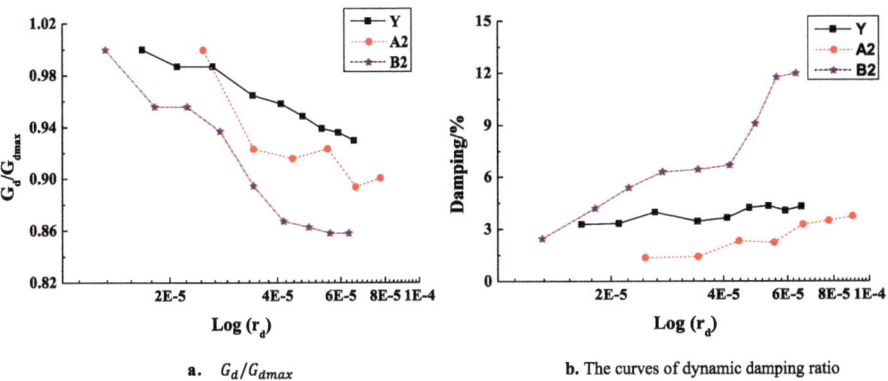

a. G_d/G_{dmax} **b.** The curves of dynamic damping ratio

Fig. 8. Testing results of model soil with confining pressure 100 kPa

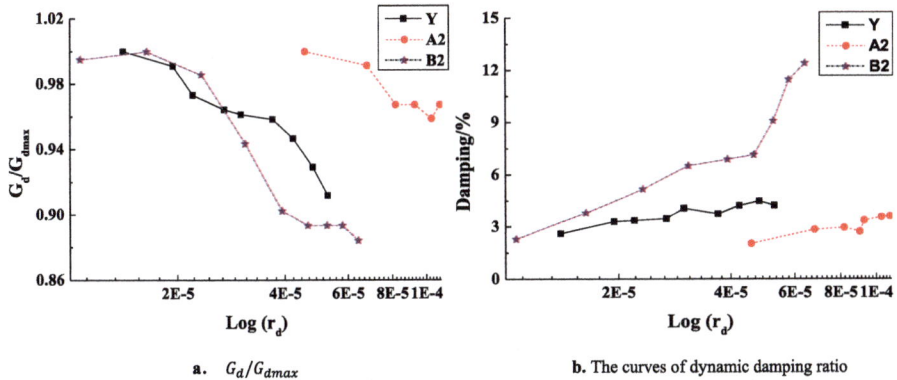

a. G_d/G_{dmax} **b.** The curves of dynamic damping ratio

Fig. 9. Testing results of model soil with confining pressure 200 kPa

5 Conclusions

In this experiment, undrained-unconsolidated resonant column tests were carried out on rubber particle model soil, rubber powder model and pure soil. And the results are as follows:

(1) The experimental results show that the two models' changing regulation of the dynamic shear modulus and dynamic damping ratio is consistent with that of pure soil, which indicates that both models can reflect the dynamic characteristics of pure soil and can be used in model test to replace the pure soil to regulate the weight and stiffness of soil.

(2) Under different confining pressure, the dynamic shear modulus of the rubber powder soil is about 2–3 times that of the rubber particle soil. It can be seen that the dynamic shear modulus of rubber model soil decreases with the increase of rubber particle size. Under the different confining pressure, the change rate of soil shear modulus is consistent with that of pure soil.

(3) Under different confining pressure, the dynamic damping ratio of the rubber powder model soil is larger than that of the rubber particle model soil. So the dynamic damping ratio of the model soil decreases with the increase of the particle size of the rubber particles.

Due to the large elasticity of rubber particles and powder, the soil samples have certain resilience, which makes the soil sample difficult to prepare, resulting in a certain dispersion of the final test data, In addition, the range of the rubber particle size of the experiment is too small. These are problems in this experiment, which need to be further improved.

References

1. Wang, J.N.: Seismic Design of Tunnels: A Simple State of the Art Design Approach. Parsons Brinckerhoff Inc., New York (1993)
2. Kitada, Y., Hirotani, T., Iguchi, M.: Models test on dynamic structure–structure interaction of nuclear power plant buildings. Nucl. Eng. Des. **192**(2), 205–216 (1999)
3. Turan, A., Hinchberger, S.D., Naggar, H.E.: Design and commissioning of a laminar soil container for use on small shaking tables. Earthq. Eng. Struct. Dyn. **29**(2), 404–414 (2009)
4. Moss, R., Crosariol, V., Kuo, S.: Shake table testing to quantify seismic soil-structure-interaction of underground structures. In: Fifth International Conference on Recent Advances in Geotechnical Earthquake Engineering and Soil Dynamics, pp. 24–29 (2010)
5. Tabatabaiefar, R., Fatahi, B., Samali, B.: Numerical and experimental investigations on seismic response of building frames under influence of soil-structure interaction. Adv. Struct. Eng. **17**(1), 109–130 (2014)
6. Hokmabadi, A.S., Fatahi, B., Samali, B.: Physical modeling of seismic soil-pile-structure interaction for buildings on soft soils. Int. J. Geomech. **15**(2), 04014046-1–04014046-18 (2015)
7. Yan, X., Yuan, J., Yuan, Y., et al.: Study on model soil of large-scale shaking table test. Struct. Eng. **31**(5), 116–120 (2015). (in Chinese)

8. Yan, X., Haitao, Y., Yuan, Y., et al.: Multi-point shaking table test of the free field under non-uniform earthquake excitation. Soils Found. **55**(5), 986–1001 (2015)
9. Yuan, Y., Yu, H., Li, C., et al.: Multi-point shaking table test for long tunnels subjected to non-uniform seismic loadings – part I: theory and validation. Soil Dyn. Earthq. Eng. **108**, 177–186 (2018)
10. Yu, H., Yuan, Y., Xu, G., et al.: Multi-point shaking table test for long tunnels subjected to non-uniform seismic loadings - part II: application to the HZM immersed tunnel. Soil Dyn. Earthq. Eng. **108**, 187–195 (2018)
11. Yan, X., Yuan, J., Yu, H., et al.: Multi-point shaking table test design for long tunnels under non-uniform seismic loading. Tunnel. Undergr. Space Technol. Inc. Trenchless Technol. Res. **59**, 114–126 (2016)
12. Yu, H., Yan, X., Bobet, A., et al.: Multi-point shaking table test of a long tunnel subjected to non-uniform seismic loadings. Bull. Earthq. Eng. **16**, 1–19 (2017)
13. Kim, H.-K., Santamarina, J.C.: Sand–rubber mixtures (large rubber chips). Can. Geotech. J. **45**(10), 1457–1466 (2008)
14. Lee, C., Truong, Q.H., Lee, W., et al.: Characteristics of rubber-sand particle mixtures according to size ratio. Mater. Civ. Eng. **22**(4), 323–331 (2010)
15. Deng, A., Feng, J.: Influence of tire rubber particles on shear properties of sand. J. PLA Univ. Sci. Technol. (Nat. Sci. Ed.) **10**(5), 483–487 (2009). (in Chinese)
16. Shang, S., Sui, X., Zhou, Z., et al.: Experimental study on the dynamic shear modulus of rubber granule - sand mixture. Rock Soil Mech. **31**(2), 377–381 (2010). (in Chinese)
17. Hu, X., Liu, Z., Zhang, Z., et al.: Test of the influence of rubber powder on dynamic characteristics of remolded loess. J. Chang'an Univ. (Nat. Sci. Ed.) **33**(4), 62–67 (2013). (in Chinese)

Deep Excavations and Retaining Structures

Limit Analysis of the Transient Stability of Slopes During Pile Driving with Nonlinear Failure Criterion

Junwei Zhao[1], Yunwei Shi[2], and Qingsheng Chen[3(✉)]

[1] Quality Supervision Station of Highway of Hulun Buir City in Inner Mongolia,
Hulun Buir 021008, China
[2] Department of Civil Engineering, University of Shanghai for Science
and Technology, Shanghai 200093, China
[3] Research Academic, Department of Civil and Environmental Engineering,
National University of Singapore, Singapore 117576, Singapore
ceecq@nus.edu.sg

Abstract. The transient stability of slopes during pile driving at the slope crest was analyzed using the upper-bound limit analysis method based on a nonlinear failure criterion. The pile-driving force was regarded as a concentrated force and the work rate done by the concentrated force was considered. The rotational logarithmic spiral discontinuity mechanism was adopted for calculation. The influences of nonlinearity coefficient, pile diameter and pile location were investigated through the study of a calculation example. From the numerical results, it is found that the safety factor decreases during the earlier stage of pile driving, and then significantly increases upon the pile tip reaching the sliding surface. After the pile tip passes through the failure surface, the safety factor remains unchanged. Nonlinearity coefficient, pile diameter and pile location have significant effects on slope stability. The results of this study may provide suggestions and guiding references for addressing engineering problems in practice.

Keywords: Limit analysis · Nonlinear failure criterion
Slope transient stability · Pile driving

1 Introduction

The use of piles to stabilize landslides has become one of the most important innovative slope reinforcement techniques in recent decades. Thus the influence of pile-driving on the slope stability has received much attention from researchers over the past years. Some achievements have been made for the analysis of piled-driving adjacent to slope [1–3].

Limit analysis has an advantage of the lower and upper bound theorems to bracket the true solution. Chen [4] employed the upper bound limit analysis method to calculate the critical height of slopes. Michalowski [5] analyzed slope stability using the translational failure mechanism which is in the form of rigid blocks that are analogous to slices in traditional slice method. Ausilio *et al.* [6] adopted the kinematic approach of

© Springer Nature Singapore Pte Ltd. 2018
T. Qiu et al. (Eds.): GSIC 2018, *Proceedings of GeoShanghai 2018 International Conference: Advances in Soil Dynamics and Foundation Engineering*, pp. 371–379, 2018.
https://doi.org/10.1007/978-981-13-0131-5_40

limit analysis to analyze stability of slopes reinforced with piles by assuming a lateral force and a moment applied at the depth of the potential sliding surface.

In practical applications, the linear failure criterion is widely used to analyze geotechnical problems. However, substantial experimental evidence shows that geo-materials tend to obey nonlinear failure criterion [7, 8]. As observed from the experiments, under the condition of low confining pressure, the stress-strain relationship of geo-materials is approximately linear. However, as the confining pressure increases, the increment rate of the geo-material' stress decreases gradually, presenting a distinct nonlinear character. The application of the nonlinear failure criterion on the upper bound solution has been addressed previously [9–11].

However, the majority of existing work of slope stability analysis is limited to slopes without the presence of piles. And the transient stability of slopes during pile driving at the slope crest has been rarely studied. In view of above context, in this paper, the work done by the pile driving force is considered. The rotational logarithmic spiral discontinuity mechanism is adopted and the transient stability of slopes during pile driving at the slope crest is studied by the upper bound limit analysis method. The results obtained by the proposed method are presented and discussed.

2 Nonlinear Failure Criterion

The nonlinear Mohr-Coulomb failure criterion can be expressed as

$$f(\sigma, \tau) = \tau - c_0(1 + \sigma/\sigma_0)^{\frac{1}{m_0}} \tag{1}$$

where c_0 is the initial cohesion, σ_0 is the absolute value of initial tensile stress, and m_0 is the nonlinearity coefficient.

As shown in Fig. 1, the tangential line to the curve at the point $G(\sigma, \tau)$ can be expressed as

$$\tau = c_t + \sigma \tan \varphi_t \tag{2}$$

where c_t and $\tan \varphi_t$ are the intercept and the slope of the tangential line, respectively, they can be expressed as

$$c_t = \frac{m_0 - 1}{m_0} c_0 \left(\frac{m_0 \sigma_0 \tan \varphi_t}{c_0} \right)^{\frac{1}{1-m_0}} + \sigma_0 \tan \varphi_t \tag{3}$$

$$\tan \varphi_t = \frac{\partial \tau}{\partial \sigma} = \frac{c_0}{m_0 \sigma_0} \left(1 + \frac{\sigma}{\sigma_0} \right)^{\frac{1-m_0}{m_0}} \tag{4}$$

The tangential line varies as σ changes. Herein, the tangential line that passes through the point $(0, c_0)$ is adopted. The equation for the line [12] is

$$\tau = c_0 + \sigma \tan \varphi_0 = c_0 + \frac{c_0}{m_0 \sigma_0} \sigma \tag{5}$$

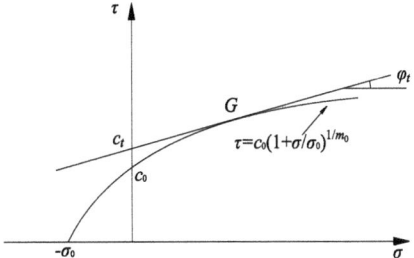

Fig. 1. Diagram of a nonlinear failure criterion

The solution hinge • on operating with a linear yield condition that exceeds the actual nonlinear condition. The increase in the yield strength of the material in any region cannot strengthen or weaken the body. Any upper bound solution of a yield condition that exceeds the true yield condition gives an upper bound to the limit load that corresponds to the true yield condition. Note that, in the calculations of current work, the cohesion $c = c_0$ and internal friction angle $\varphi = \varphi_0$ are adopted.

3 Upper Bound Limit Analysis

3.1 Basic Model

In this paper, the soil is considered infinite, homogeneous and perfectly plastic. The soil mass satisfies the nonlinear Mohr-Coulomb failure criterion and associated flow rule. A rotational logarithmic spiral discontinuity mechanism is adopted herein, as shown in Fig. 2. The height of the slope is H, the slope angle is β, the distance between the pile-driving location and the slope shoulder is s, the intersection of the discontinuity surface and the slope crest has a distance of L from the slope shoulder, the pile depth is z, and the velocity of a certain point on the surface of discontinuity is v. The region ABC rotates as a rigid block around the center of rotation O with the angular velocity Ω. The failure surface is defined as

$$r(\theta) = r_0 \exp[(\theta - \theta_0) \tan \varphi] \tag{6}$$

where θ_0 is the inclination of the chord OB, r_0 is the length of OB.

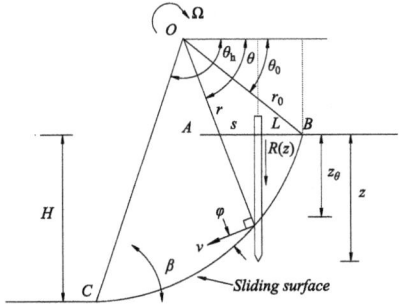

Fig. 2. Basic model for calculation

3.2 Upper Bound Limit Analysis

Limit analysis finds the range within which the true solution falls by using the static and kinematic theorems of plasticity theory. The kinematic theorem require that the rate of dissipation within the plastic zone be equated to the rate of external work for any assumed strain field that is governed by the normality rule and is compatible with the velocities at the boundary of the failing mass.

In the present work, the rate of external work includes the rate of work done by both of the soil weight and the pile-driving force. According to the rigid body assumption, there is no energy dissipation within the soil mass and only the dissipation along the sliding surface is considered.

3.3 Safety Factor

In this paper, the safety factor is employed to assess the slope stability. The safety factor is analytically defined as

$$F_s = \frac{c}{c_m} = \frac{\tan \varphi}{\tan \varphi_m} \tag{7}$$

where c_m and φ_m are the soil strength parameters necessary only to maintain the structure in limit equilibrium. The strength parameters c and φ should be reduced by F_s when calculating.

To obtain the minimum F_s, an optimization procedure is carried out with respect to the unknown parameters θ_0 and θ_h.

3.4 Rate of External Work and Internal Work

Rate of Work Done by Soil Weight. The rate of woke done by soil weight can be obtained by superposing the rate of work of regions OBC, OAB and OAC, which can be found in literature [4].

Rate of Energy Dissipation. Before the pile tip reaches the failure surface, the total rate of energy dissipation W_2 can be expressed as

$$W_2 = \int_{\theta_0}^{\theta_h} c(v \cos \varphi) \frac{r d\theta}{\cos \varphi} = \frac{c r_0^2 \Omega}{2 \tan \varphi} \{\exp[2(\theta_h - \theta_0) \tan \varphi] - 1\} \tag{8}$$

In view of the defection of the two-dimensional model, a coefficient η is adopted herein as a reduction factor. The expression for η is

$$\eta = \frac{\pi d}{4t} \tag{9}$$

where d is the pile diameter, and t is the pile spacing.

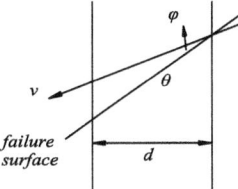

Fig. 3. Diagram of the replacement of soil by pile

The formulation of W_2 is then modified in consideration of the anti-slide effect of piles shown in Fig. 3 as

$$W_2' = W_2 + \frac{\eta \cdot d}{\sin(\theta - \varphi)} \cdot r \cos \varphi \cdot \Omega(c' - c) \tag{10}$$

where c' is the shear strength of the pile.

Rate of Work Done by Pile-driving Force According to Fan [13], the pile-driving resistance can be expressed as

$$R = R_s + R_p = \left(0.5 + \frac{z}{100}\right) \cdot P_s \cdot d + 2 \cdot (0.025P_s + 25) \cdot \frac{z}{S_t} \tag{11}$$

where R_s and R_p are the side and tip resistance of the pile, respectively, P_s is the specific resistance of static penetration. The expression for R_s and R_p are

$$R_s = 2 \cdot (0.025P_s + 25) \cdot \frac{z}{S_t} \tag{12a}$$

$$R_p = \left(0.5 + \frac{z}{100}\right) \cdot P_s \cdot d \tag{12b}$$

where S_t is the soil sensitivity.

The same coefficient η for the reduction of shear strength of piles is used herein for describing the pile-driving resistance. The reduced pile-driving resistance can be expressed as

$$R' = R_s' + R_P' = \eta(R_s + R_p) \tag{13}$$

Here the force R' is seen as a concentrated force. The rate of work done by the concentrated force is

$$W_3 = r \cos \theta \cdot \Omega \cdot R'(z) \tag{14}$$

where $r \cos \theta = r \cos \theta_0 - L + s$.

Only the pile side resistance above the failure surface is considered after the pile tip passes through the failure surface. The rate of work can be modified as

$$W_3' = r \cos \theta \cdot \Omega \cdot R_s'(z_\theta) \tag{15}$$

where z_θ is the depth of the failure surface at the pile location.

4 Results and Discussions

A parametric study is carried out to illustrate the effect of the following factors (i.e., Nonlinearity coefficient, pile diameter and pile location). The parameters for the pile and pile properties are shown in Table 1.

Table 1. Parameters of soil and pile

Initial cohesion (c_0)	Absolute value of the initial tensile stress (σ_0)	Nonlinearity coefficient (m_0)	Distance between the pile location and the slope shoulder (s)
70 kPa	247.3 kPa	1.5	2 m
Unit weight of the soil (γ)	Pile spacing (t)	Slope height (H)	Specific resistance of static penetration (P_s)
19 kN/m^3	2 m	20 m	1.2 Mpa
Pile diameter (d)	Slope angle (β)	Soil sensitivity (S_t)	Shear strength of the pile (c')
0.4 m	45°	2.0	4 MPa

4.1 Effect of Nonlinearity Coefficient

Figure 4 illustrates the effect of the nonlinearity coefficient m on the slope safety factor. Noted that, this work focuses on the transient stability of slopes during pile driving, and the horizontal axis marked as z/m in Fig. 4 represents the pile-driving depth. When $m = 1$, the nonlinear failure criterion turns into the linear one. It can be seen from Fig. 4 that with the increase of the nonlinearity coefficient, the safety factor decreases considerably. This is mainly because the traditional Mohr-Coulomb failure criterion can only be applied to evaluate the strength of the geo-materials under low stresses. At the instant of slope failure, the stress in soil is rather high. The traditional Mohr-Coulomb failure criterion would overestimate the geo-materials' strength, leading to a higher safety factor than that obtained by employing the nonlinear Mohr-Coulomb failure criterion, which is not desirable in practice.

In addition, it is observed that the safety factor drops to a certain value when the pile is driven into the soil at the very onset. Then, as the pile depth increases, the safety factor decreases linearly with increasing pile-driving force. After the pile tip passes through the failure surface, the safety factor increases to a larger value and remains unchanged subsequently. This is due to the enhanced the energy dissipation attributed to the large shear strength of piles.

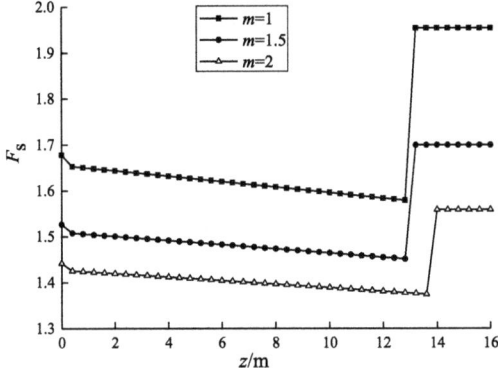

Fig. 4. Effect of nonlinearity coefficient on safety factor of slopes during pile driving

4.2   Effect of Pile Diameter

Figure 5 presents the effect of pile diameter on the safety factor. Before the pile tip reaches the failure surface, with the increase in the pile diameter, the safety factor decreases with an increasing rate. This is due to that the pile driving force increases with the increasing pile diameter. Meanwhile, the increasing pile diameter strengthens the anti-slide effect of piles at the final stage, resulting in that the corresponding safety factor increases substantially.

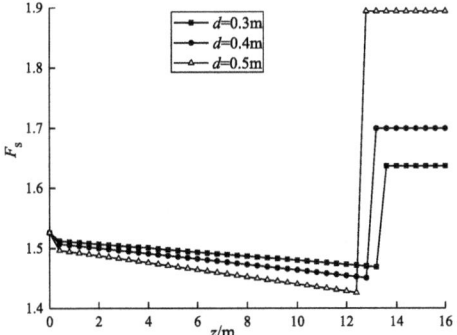

Fig. 5. Effect of pile diameter on safety factor of slopes during pile driving

4.3   Effect of Pile Location

As shown in Fig. 6, when $s = 17.5d$, the pile location goes beyond the intersection of the failure surface and the slope crest. The safety factor remains unchanged because the pile-driving force no longer affects the sliding mass.

In the cases that the pile locates inside the intersection of the failure surface and the slope crest, the mutation depth of safety factor decreases as s increases. And the anti-slide effect is more effective as s increases after the pile tip passes through the failure surface.

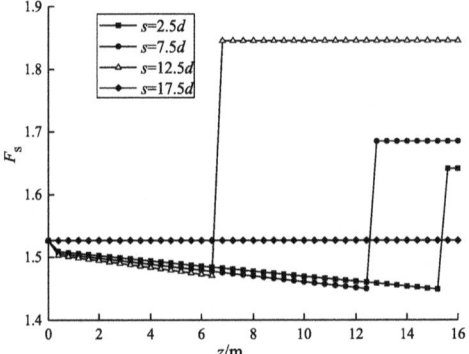

Fig. 6. Effect of pile location on safety factor of slopes during pile driving

5 Conclusions

Based on the upper bound theorem of limit analysis, the transient stability model of slopes was established. The effects of nonlinearity coefficient, slope angle, pile diameter and pile location were investigated thoroughly. From the results presented, several conclusions are drawn.

(a) During the pile-driving adjacent to slope crest: when the pile is driven into the soil in the beginning, the value of the safety factor drops suddenly; before the pile tip reaches the failure surface, the safety factor decreases constantly; the safety factor significantly increases after the pile tip reaches the failure surface; after the pile tip passes through the failure surface, the safety factor tends to be stable.

(b) Nonlinearity coefficient, pile diameter and pile location have significant effects on slope stability. The safety factor decreases as nonlinearity coefficient increases; as the pile diameter increases, the safety factor is lower and decreases more quickly at the early stage, while the higher it is during the later stage; the further the pile location is away from the slope crest, the higher the reliability of slope stability is.

(c) The results provide a reference value for engineering, but deficiencies still exist. The method searching for the most dangerous failure surface and the influence of pile-driving on the failure mechanism are future research interests.

References

1. Rao, P.P., Cui, J.F., Zhao, L.X.: Effect of pile driving adjacent to soil slopes on safety factor of specific sliding surface. Chin. J. Geotech. Eng. **38**(9), 1720–1726 (2016). (in Chinese)
2. Xu, Y., Li, T.C., Mo, J.B.: Influence of excess pore water pressure induced by pile driving on stability of wharf slopes. Rock Soil Mech. **31**(8), 2525–2530 (2010). (in Chinese)
3. Rao, P.P., Li, J.P., Cui, J.F.: Model test on squeezing effect during static pressure pile-sinking adjacent to slope. Chin. J. Highw. Transp. **27**(3), 25–31 (2014). (in Chinese)
4. Chen, W.F.: Limit Analysis and Soil Plasticity. Elsevier, Amsterdam (1975)

5. Michalowski, R.L.: Slope stability analysis: a kinematical approach. Géotechnique **45**(2), 283–293 (1995)
6. Ausilio, E., Conte, E., Dente, G.: Stability analysis of slope reinforced with piles. Comput. Geotech. **28**, 591–611 (2001)
7. Lee, K.L., Seed, H.B.: Drained strength characteristics of sands. J. Soil Mech. Found. Div., Proc. ASCE **93**(SM6), 117–141 (1967)
8. Vesic, A.S., Clough, G.W.: Behavior of granular materials under high stress. J. Soil Mech. Found. Div., Proc. ASCE **94**(SM3), 661–688 (1968)
9. Baker, R., Frydman, S.: Upper bound limit analysis of soil with nonlinear failure criterion. Soils Found. **23**(4), 34–42 (1983)
10. Zhang, X.J., Chen, W.F.: Stability analysis of slopes with general nonlinear failure criterion. Int. J. Numer. Anal. Meth. Geomech. **11**(1), 33–50 (1987)
11. Drescher, A., Christopoulos, C.: Limit analysis of stability with nonlinear yield condition. Int. J. Numer. Anal. Meth. Geomech. **12**(4), 341–345 (1988)
12. Hu, W.D., Zhang, G.X.: Plastic limit analysis of slope stability with nonlinear failure criterion. Rock Soil Mech. **28**(9), 1909–1913 (2007). (in Chinese)
13. Fan, X.Y.: Study of sinking-resistance and compaction effect of static-press pile. Tongji University, Shanghai (2007). (in Chinese)

Deformation Behavior of Foundation Pit Considering the Variation of Permeability Coefficient of Soil Mass

Yuqi Li$^{(\boxtimes)}$, Jing Fan, and Xuan Zhou

Shanghai University, Shanghai 200444, China
liyuqi2000@shu.edu.cn

Abstract. The permeability coefficient of soil mass will vary with the change of effective stress when a foundation pit is excavated in soft soil area. A three-dimensional fluid-solid coupling program was developed that can consider the change of permeability coefficient of soil with the effective stress by introducing the relationship between the permeability coefficient and the effective stress increment of soil. The foundation pit of Shanghai Bank Building in Lujiazui, Pudong District, Shanghai, China was simulated and the numerical results were compared with the ones without considering the change of permeability coefficient of soil. It is shown that the trend of the horizontal displacement of retaining structure is consistent whether or not the permeability coefficient of soil varies. The results considering the change of permeability coefficient of soil with the effective stress, however, are smaller and agree better with the measured results. As for the heave of pit base, the deformations considering the change of permeability coefficient are smaller than the results without considering it. However, the ground settlements around the pit considering the change of permeability coefficient of soil are contrary to the heave of pit base and they are bigger than the results without considering it. Therefore, the change of permeability coefficient of soil with the effective stress should be considered in excavation engineering of soft soil area.

Keywords: Excavation · Deformation · Numerical simulation
Coupling · Permeability coefficient

1 Introduction

In excavation engineering of soft soil, seepage and deformation are usually concurrent and interactive, seepage can cause the change of pore water pressure and effective stress, and further lead to the deformation of soil; and vice versa, deformation of soil also causes the change of permeability coefficient and influences stress of soil through pore water pressure. Therefore, many related researches have been made [1–4]. However, The variation of permeability coefficient of soil with the effective stress of soil mass is seldom considered. Therefore, it has important guiding significance to study the deformation behavior of an excavation considering the change of permeability coefficient with stress. In this work, to consider the variation of the permeability coefficient of soil with effective stress, a 3D finite element program [5] was further

© Springer Nature Singapore Pte Ltd. 2018
T. Qiu et al. (Eds.): GSIC 2018, *Proceedings of GeoShanghai 2018 International Conference: Advances in Soil Dynamics and Foundation Engineering*, pp. 380–387, 2018.
https://doi.org/10.1007/978-981-13-0131-5_41

developed by introducing the relationship between the permeability coefficient and the effective stress increment of soil.

2 Relationship Between Permeability Coefficient and Effective Stress

The permeability of soil influences the permeability coefficient and variation of effective stress of soil affects the permeability of soil, therefore, the variation of effective stress of soil will result in the variation of the permeability coefficient of soil. With the effective stress increasing, the porosity and permeability of soil will decrease. There are some methods to determine the relation between the permeability coefficient of soil and other parameters and here is the relation between the permeability coefficient and the effective stress increment:

$$\begin{cases} \ln\left(\frac{k}{k_0}\right) = -ap' \\ k = k_0 e^{(-ap')} \end{cases} \tag{1}$$

where k_0 is the initial permeability coefficient of soil, k is the permeability coefficient of soil considering the variation of effective stress, p' is the effective stress increment and a is a constant. As for the soil in Shanghai area, $a = 7.16$ according to the average experiment results of Wu and He [6]. Since the example is in Shanghai in this paper, $a = 7.16$ when determining the soil parameters.

3 Numerical Simulation of Excavation Considering the Variation of Coefficient of Permeability

3.1 Engineering Example

Shanghai Bank Building is located in Lujiazui, Pudong, Shanghai, China. The main structure consists of a 46-storey main building and a 3-storey podium building. The excavation depths of main building and podium building are 17.15 m and 14.95 m respectively, and excavation area is 7454 m². Due to large excavation depth, high construction difficulty and high protection requirement of the surrounding, the diaphragm wall of 1000 mm thickness was adopted as retaining structure in order to decrease the horizontal displacement of retaining structure. Three tier supports of reinforce concrete were set and their cross-sectional dimensions at the first tier, the second tier and the third tier were 900 mm by 700 mm, 1300 mm by 700 mm and 1300 mm by 700 mm, respectively. Table 1 details the construction procedure of deep excavation of Shanghai Bank Building.

Table 1. Construction process.

Excavation stage	Process	Excavation thickness: m	Total excavation depth: m	Strut setting	Duration: d
1	Excavation	2.7	2.7	Nothing	3
	Intermission	0	2.7	Nothing	7
2	Excavation	5.8	8.5	One layer	8
	Intermission	0	8.5	One layer	12
3	Excavation	5	13.5	Two layer	9
	Intermission	0	13.5	Two layer	11
4	Excavation	3.65	17.15	Three layer	2
	Intermission	0	17.15	Three layer	8

3.2   Numerical Model

Considering the influencing scope of an excavation, the symmetry about the center line and the calculation efficiency, an unit was taken in y-direction of excavation. Finite element meshes of numerical model are shown in Fig. 1 and the dimensions of the model in x- and z-direction are 100 m (length) by 80 m (depth).

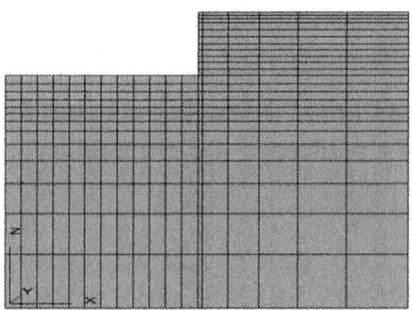

Fig. 1. Finite element meshes.

As to the hydraulic heads inside and outside the excavation, they were assumed to locate on the excavation surface and the ground surface, respectively.

The soil constitutive model adopts the revised Duncan-Chang nonlinear elastic model [7] and the model parameters are listed in Table 2. In Table 2, K is the modulus number, n is the modulus exponent, R_f is the failure ratio, K_{ur} is the unloading-reloading modulus number, D, F and G are material parameters, c' and φ' are the effective cohesion and the effective internal friction angle of soil, respectively, k is the permeability coefficient of soil (vertical and horizontal permeability coefficients are the same), and H is the thickness of soil layer.

Table 2. Soil parameters used during modelling.

Soil	Miscellaneous fill	Sandy silt	Mucky silt	Silty clay	Silty clay	Silt mixed silty sand	Silty-fine sand
K	1116	253	268	59	118	268	267
n	0.65	0.56	0.56	0.76	0.35	0.9	0.84
R_f	0.88	0.83	0.76	0.76	0.8	0.76	0.77
c': kPa	6	9	16	17	14	16	19
φ': °	38	34	38	29	31	38	37
F	0	0.075	0.112	0.049	0.054	0.112	0.147
G	0.8	0.354	0.294	0.2	0.262	0.294	0.358
D	0	2.92	8.11	3.28	0.04	8.11	7.06
K_{ur}	1500	628	1144	362	617	1144	1155
K: 1×10^{-7} cm/s	50	5.6	6540	2.36	3.44	1.75	1.75
γ': kN/m^3	5.8	8.6	9	8.6	7.9	8.5	8.7
H: m	1.7	7.45	8.85	6.6	3.7	4.7	47

Boundary conditions include displacement boundary condition and hydraulic boundary condition. As for the displacement boundary condition, the model bottom was assumed to be fixed, the model surface was free, and displacements perpendicular to the boundaries were restrained at the lateral boundaries. As for the hydraulic boundary, the model bottom, the symmetrical plane and the diaphragm wall were impermeable, whereas the model top was permeable.

Figure 2 is the comparison of horizontal displacement between the calculated results considering the variation of permeability coefficient and the measured results. It can be seen from Fig. 2 that the calculated values considering the variation of permeability coefficient and the measured values are in good agreement.

4   Result Analysis

4.1   Horizontal Displacement of Retaining Wall

Figure 2(a)–(c) also shows the comparisons of horizontal displacement considering the variation of permeability coefficient and without considering it after each excavation intermission. It can be seen from Fig. 2 that the horizontal displacements of retaining structure considering the variation of permeability coefficient are smaller than the ones without considering it. Take the fourth excavation intermission for example, the maximum displacement without considering the variation of permeability coefficient is 15.1 mm bigger than the one considering it. However, the horizontal displacements considering the variation of permeability coefficient are closer to the measured values than the ones without considering it.

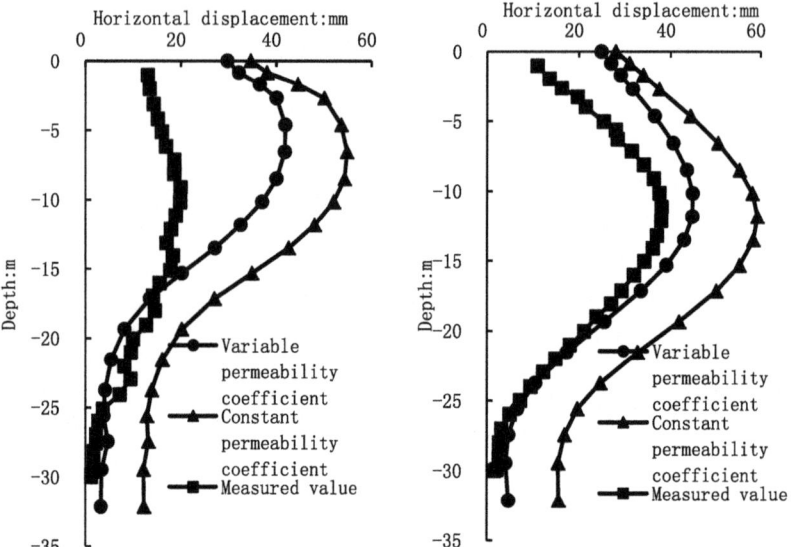

(a) After the second excavation intermission   (b) After the third excavation intermission

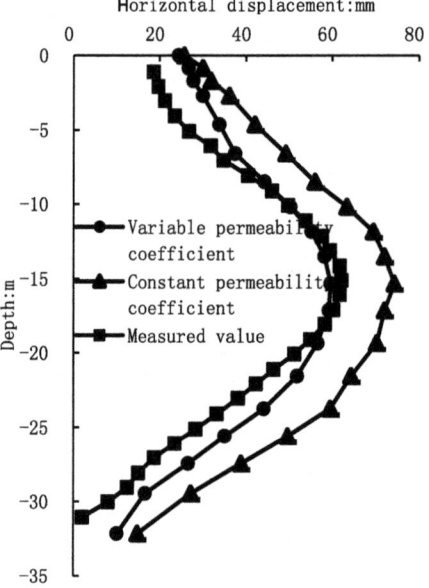

(c) After the fourth excavation intermission

Fig. 2. Comparison of horizontal displacement of retaining structure.

4.2 Heave Deformation of Pit Base

Figure 3(a)–(c) indicates the comparisons of heave deformation of pit base after each excavation intermission. It can be seen from Fig. 3 that the change trend of heave deformation is the same after each excavation intermission. However, the values without considering the variation of permeability coefficient are much bigger than the

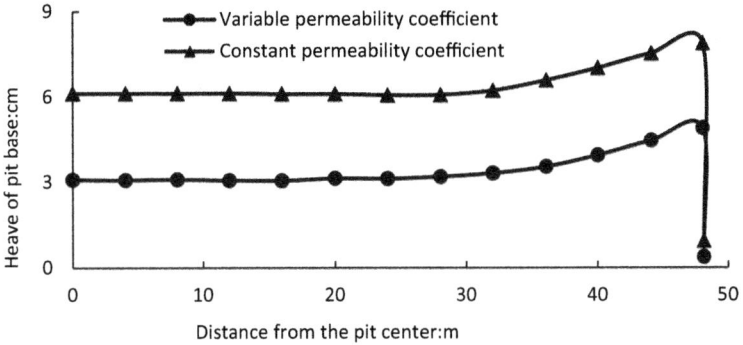

(a) After the second excavation intermission

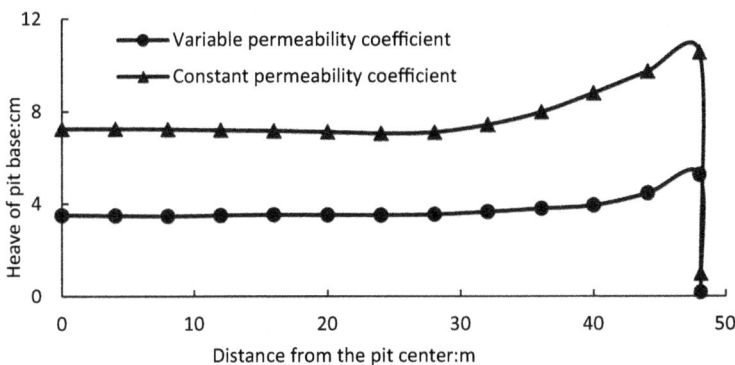

(b) After the third excavation intermission

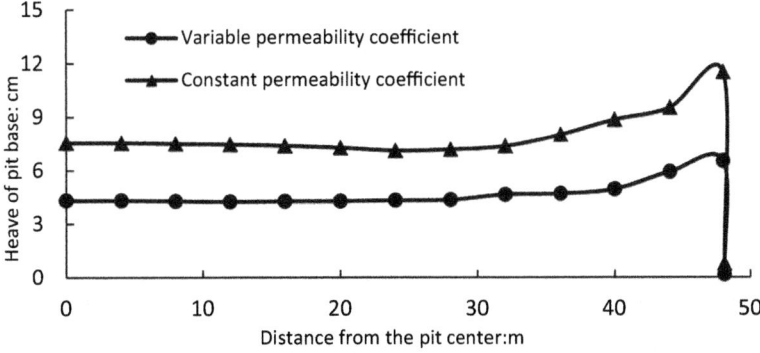

(c) After the fourth excavation intermission

Fig. 3. Heave deformation of pit base.

ones considering it. Take the fourth excavation intermission, for example, within 30 m from the center of excavation the heave value considering the variation of permeability coefficient is approximately 4.3 cm, whereas the one without considering it is approximately 7.3 cm.

4.3   Ground Settlement Around Foundation Pit

Figure 4(a)–(c) are the comparisons of ground settlement around the pit after each excavation intermission. As can be seen from Fig. 4, the trend of ground settlement around the foundation pit in two cases is the same after each excavation intermission.

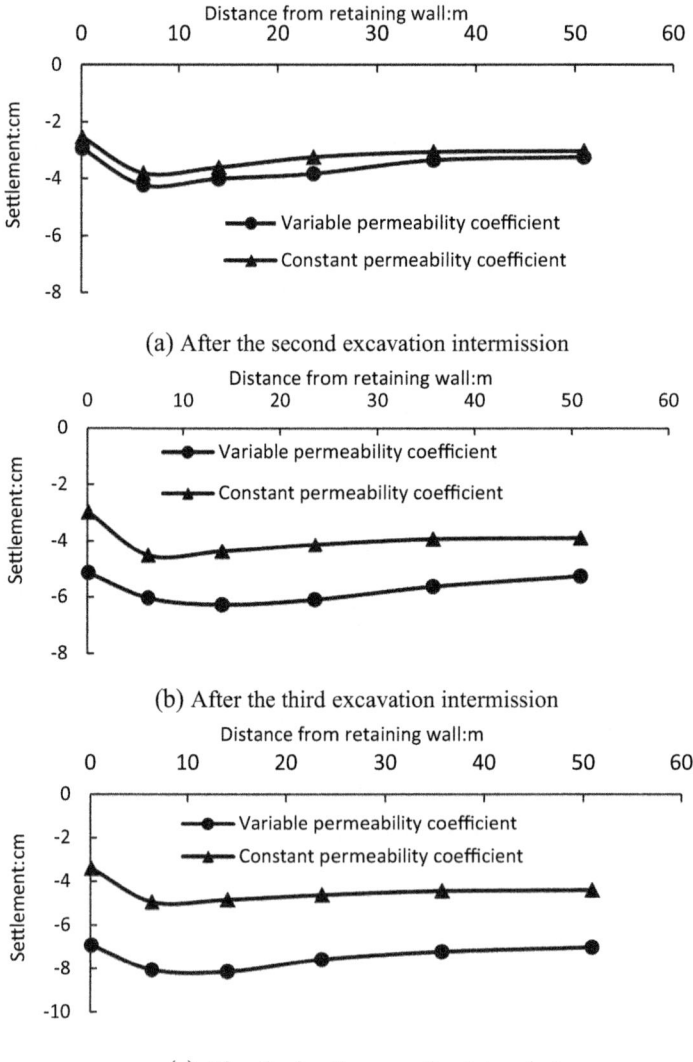

(a) After the second excavation intermission

(b) After the third excavation intermission

(c) After the fourth excavation intermission

Fig. 4. Ground settlement.

However, the result without considering the variation of permeability coefficient is smaller, the difference of settlement and the influence range in two cases both becomes bigger and bigger.

5 Conclusions

Based on the actual excavation of Shanghai Bank Building, a program was developed considering the coupling of seepage field and stress field, and was verified through the comparison of the calculated horizontal displacements of retaining wall and the measured values. The conclusions from this study can be summarized as follows:

(1) Whether or not considering the variation of permeability coefficient, the change trend of the horizontal displacement of support structure is the same. However, the horizontal displacement of support structure considering the variation of permeability coefficient with the effective stress of soil is smaller the one without considering it and the former is closer to the measured value than the latter.
(2) The change trend of heave of pit base is also the same whether or not to consider the variation of permeability coefficient. However, the result without considering the variation of permeability coefficient is bigger than the one considering it.
(3) After each excavation intermission, the trend of ground surface settlement around the foundation pit is the same whether to consider the change of the permeability coefficient or not. However, the result without considering the variation of permeability coefficient is smaller.

References

1. Meris, G., Rokhsar, A.: Theory of consolidation for clay. J. Geotech. Eng. Div. **8**, 889–904 (1974)
2. Tavenas, P., Jean, P., Leblond, P., et al.: The permeability of natural soft clays. part II: permeability characteristics. Can. Geotech. J. **20**(4), 645–659 (1983)
3. Osaimi, A.E., Clough, G.W.: Pore-pressure dissipation during excavation. J. Geotech. Eng. ASCE **105**(4), 481–498 (1979)
4. Fan, Y.Q., Liao, X.H., Li. X.K.: Transient analysis of braced excavation in elastic-plastic saturated/unsaturated soils and parametric study. In: Proceeding of 9th International Conference on Computer Methods and Advances in Geomechanics, pp. 258–264 (1997)
5. Li, Y.Q.: Studies on the behavior of foundation pit with excavation considering seepage. Zhengjiang University, Hangzhou (2005). (in Chinese)
6. Wu, L.G., He, Z.G.: Considering the groundwater seepage of aquifer parameters with stress changes. J. Tongji Univ. **23**(6), 281–287 (1995). (in Chinese)
7. Potts, D.M., Zdravkovic, L.: Finite Element Analysis in Geotechnical Engineering: Theory. Thomas Telford Limited, London (1999)

Responses of Two Buildings Located at the Corners of a Rectangular Excavation: A Case Study in Hangzhou

Yongmao Hou[1], Biao Huang[2], and Mingguang Li[2(✉)]

[1] Shanghai Tunnel Engineering Co., Ltd., Shanghai, China
[2] Department of Civil Engineering,
Shanghai Jiao Tong University, Shanghai, China
lmg20066028@sjtu.edu.cn

Abstract. Firstly, an analysis of field data of an underground passage pit in Hangzhou was carried out, which was done by dividing the monitoring points into two groups, based on their distance from corners. The results indicated that corner effects contribute to the decrease of wall deflections and ground settlements during the excavation process. Secondly, a studies with focus on the corner effects on adjacent building settlements was conducted. The results showed that, in north-south direction, the building settlements close to the walls were larger, and in east-west direction, the values near the corners were smaller. Therefore, the settlements of two buildings nearby were the accumulation in two directions and the maximum settlement occurred at the monitoring points closest to the middle span of the pit. The differential settlements occurred in two directions, therefore the mechanism of the adjacent building settlements was a complicated three-dimensional problem. The corner effects of rectangular excavations and the behavior of wall deflections, ground settlements and building settlements in the paper provide a good basis for further research and for similar future projects.

Keywords: Excavation · Corner effects · Wall deflection
Ground settlement · Building settlement · Three-dimensional problem

1 Introduction

Evaluation of the distribution and magnitude of the diaphragm wall deflections and ground settlements is a significant part of the design and construction process. However, because of the three-dimensional spatial effect of excavations, the corners of excavations have a higher stiffness than that of the middle span. As a result, smaller wall deflections and ground settlements will occur because of the corner effects. Consequently, the conventional plane strain assumption is inappropriate for studying the wall displacement and ground settlement when considering the corner effect and may lead to large errors [1–4].

Much research has been done in the past to study the corner effects of excavations and their influence on wall deflections. Tan et al. [5] analyzed the field data from rectangular underground passage pits with different aspect ratios. Their results

© Springer Nature Singapore Pte Ltd. 2018
T. Qiu et al. (Eds.): GSIC 2018, *Proceedings of GeoShanghai 2018 International Conference: Advances in Soil Dynamics and Foundation Engineering*, pp. 388–395, 2018.
https://doi.org/10.1007/978-981-13-0131-5_42

indicated that the minimum wall deflections occurred near the pit corners and the maximum occurred near the middle span of the pits. Lee et al. [6] monitored a typical-sized excavation in Singapore and the results indicated that the strengthening effect of excavation corners can lead to a significant reduction in wall deflections and ground settlements. Furthermore, a finite element back analysis was conducted to confirm that three-dimensional analysis may provide more precise predictions than two-dimensional analysis because of the corner effects. Finno et al. [7] carried out finite element simulations to study the three-dimensional effects of a supported excavation in clay. The analysis showed that the results of plane strain simulations found displacements in the center of that wall equal to that analyzed by a three-dimensional stimulation when the excavated length normalized by the excavated depth was greater than six. They concluded that the corner effects of excavations may have a large influence on adjacent diaphragm walls. Results obtained by conventional methods showed large errors in ground movements. Therefore, Fuentes and Devriendt [8] proposed a simple empirical methodology for a quick and cost-effective preliminary ground movement analysis.

However, much research has focused on corner effects in case of excavations, most studies investigated diaphragm wall deflections and ground movements in the close vicinity of excavation corners. Only few researchers studied the responses of adjacent buildings located at the corner of excavations. By analyzing the field data of a rectangular underground passage pit in Hangzhou, this paper investigated the development of wall deflections and ground settlements close to the excavation corners during the whole excavation process. Furthermore, the paper also elaborated the mechanism of corner effects on settlements of two adjacent buildings. The analysis of corner effects of rectangular excavation and the characteristics of wall deflections, ground settlements and building settlements can be a reference for other similar projects.

2 Project Information

2.1 Project Background

Hangzhou, the provincial capital of Zhejiang Province, is a famous city in China for its economy and culture. Due to the rapid economic development in Hangzhou, the city faces the land's shortage and traffic jam like other large cities in China. In order to solve the mass transportation problem, the city has invested a lot of effort to construct underground passages.

The underground passage project of Wenyi Road, stretched from Baojiao North Road to Zijingang Road, is 5.1 km. This project was divided into several sections and the pit investigated in this paper is a buried section of the first-stage project.

The considered section includes one working well and three PX parts. Figure 1 shows the layout of monitoring points and the detailed surrounding conditions. The width of the working well is 24 m, the length is 38 m and the depth of excavation is 22.5 m. The width of PX1/2 is 19 m, the width of PX3 is 26 m and the length of PX3 is 33 m. The excavation depth of the PX area is approximately 16 m. The working well has six supports, the first and fourth supports are concrete struts and others are steel pipe struts. The parts PX1, PX2, PX3 have four supports, the first support is a concrete strut and others are steel pipes. All the steel pipe struts were preloaded with 2000 kN force before propping. The working well was excavated by using the open-cut method and struts were propped after excavating per layer. The PX1, PX2, PX3 parts were excavated at first and filled after finishing the construction of structure.

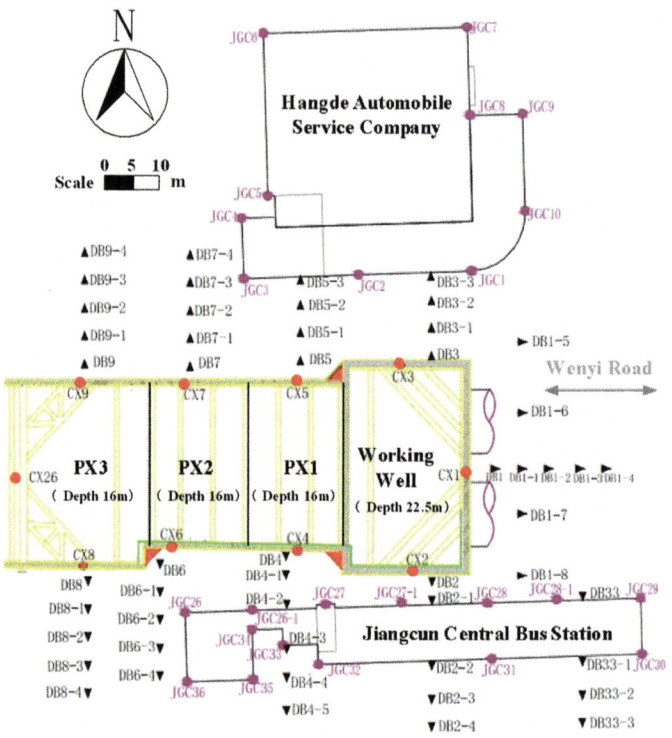

Fig. 1. Layout of monitoring points and excavation surrounding conditions

2.2 Geological Condition

Before construction, the soil conditions at the site had been explored by several tests. The soil layers of this project are mainly the Quaternary layers. The soil layers of the open-cut section contain: ① fill, ②1 sandy silt, ② silty clay, ③1 muddy clay, ③2 muddy and silty clay, ⑤ muddy clay, ⑥2 clay, ⑩2 gravelly sand. The characteristics of soil layers ③1, ③2 and ⑤ are current-formed, poorly permeable, contain a high water content, have a low strength and highly compressible.

According to the depth of underground water level, the pore water, confined water and bedrock fissure water all have a great influence on the project. The depth of pore water is 0.7–3.3 m, which is influenced by precipitation and seasonal changes of underground water level. The water-resisting layers of confined water are ③2, ⑤, ⑥2 and ⑦, which have little influence on this project.

2.3   Instrumentations

The deflections of diaphragm walls were monitored by 10 inclinometers (designated as CX1 to CX9 and CX26). In order to observe the behaviors of ground settlements during excavation, dozens of settlement points (DB1 to DB9) were placed behind the diaphragm walls. The first settlement monitoring point in per group was set 0.5–5 m away from the diaphragm walls and every other two points were approximately 5 m away from each other. Two important buildings (one is Hangde Automobile Service Company and the other is Jiangcun Central bus station) were observed by setting ten (JGC1 to JGC10) and eleven (JGC26 to JGC36) monitoring points, respectively. In order to study the corner effects on the excavation behaviors, the monitoring points were divided into two groups. The points CX2,CX3,CX8,CX9 were designated as the first group and the other points were designated as the second group. The layout of instruments is shown in Fig. 1.

2.4   Construction Procedure

The construction of the west open-cut section was divided into four parts which are PX1, PX2, PX3 and a working well. Table 1 shows the scheduling of construction of four parts in this buried section. The working well had seven construction stages and other parts had four stages. The first stage mainly contained the preparation, construction of retaining walls and concreting of the first struts. The excavation and construction of other struts were completed in the later period. The detailed information of excavation and construction of supporting structures is listed in Table 1.

Table 1. Construction scheduling of the buried section.

Events	Date (2016/month/day)			
	Working well	PX1	PX2	PX3
Concreting 1st struts	4/1–4/17			
Excavation and propping 2nd struts	4/18–5/2	4/29–5/2	4/28–5/1	4/29–5/2
Excavation and propping 3rd struts	5/8–5/19	5/10–5/12	5/7–5/11	5/1–5/5
Excavation and propping 4th struts	5/20–5/26	5/18–5/22	5/15–5/19	5/12–5/19
Excavation and propping 5th struts	5/27–6/1	\	\	\
Excavation and propping 6th struts	6/2–6/7	\	\	\
Excavation and concreting base slab	6/8–6/14	5/26–6/1	5/22–6/1	5/22–6/1

3   Observed Excavation Behaviors

3.1   Lateral Movements of Diaphragm Walls

Figure 2 presents the relationship between maximum wall deflections and depths of excavation. The monitoring data of the first group of points (CX2, 3, 8, 9), which were closer to the excavation corners, fitted the red straight line and the second group fitted the blue line. The ratio of maximum deflection and excavation depth of group 1 was $2.2\text{‰}H_e$ and the ratio of group 2 was $3.4\text{‰}H_e$, where H_e is the excavation depth. It is

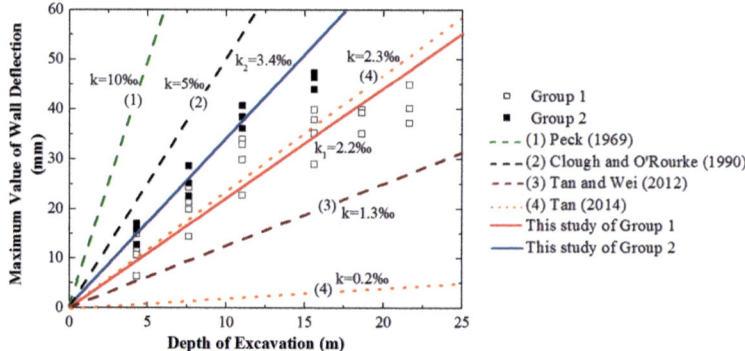

Fig. 2. Relationship between maximum wall deflections and depths of excavation

evidence that the points at the middle span of walls had larger maximum wall deflections than those close to the excavation corners. Comparing data of CX1 with CX2, CX3 and CX26 with CX8, CX9, the points far away from corners had larger deflections than those closer to corners. Therefore, the corner effects contributed to the reduction of the diaphragm wall deflections.

3.2   Excavation-Induced Ground Settlements

Figure 3 presents the relationship between maximum ground settlements and depths of excavation. In order to study the corner effects on ground settlements, the field data of monitoring points (DB2, DB3, DB8, DB9) fitted with the red straight line and the data of DB4, DB5, DB6, DB7 fitted with blue line. The ratio of maximum settlements and excavation depths of group 1 was $2.1‰H_e$ and the ratio of group 2 was $3.2‰H_e$. The blue line had larger maximum values of ground settlements than the red line with the same depths of excavation. Therefore, because of the corner effects, the range of ground closer to the excavation corners had smaller settlements than that close to the middle span of walls.

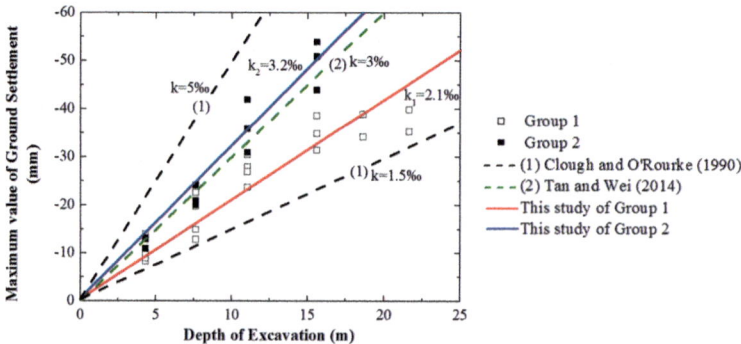

Fig. 3. Relationship between maximum ground settlements and depths of excavation

3.3 Analysis of Spatial Building Settlements

The observed data of two buildings were analyzed to study the corner effects on building settlements. Figure 4 presents the settlement plan views of Hangde Automobile Service Company from May 1st to May 26th. When considering the north-south direction, which is almost perpendicular to the diaphragm walls, the settlement values of the south side, which is closer to the walls, were larger than those of the north side. In the east-west direction, which is parallel to the walls, the settlements close to the middle span of excavation were larger than those close to the pit corners. Based on the above analysis, the final settlements of Hangde Automobile Service Company was the accumulation of the settlements in two directions and the southwest corner of the company had the largest settlement. Obviously, because of the corner effects on building settlements, using the conventional plane assumption to analyze the spatial building settlements is inappropriate.

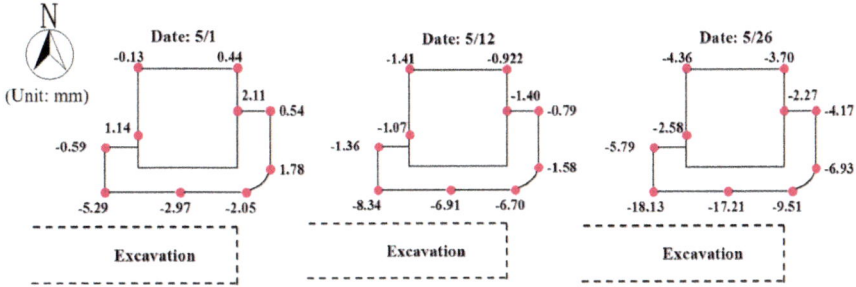

Fig. 4. Settlement plan view of Hangde Automobile Service Company

Figure 5 shows the development of building settlements at some monitoring points from JGC1 to JGC10. With increase of excavation depth, the building settlement at JGC3 increased more significantly than other monitoring points, closely followed by JGC1. This figure indicates that the point JGC26 closest to the middle span of excavation had the largest building settlement.

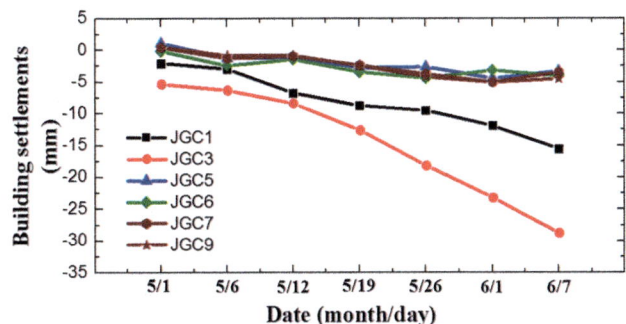

Fig. 5. Development of the settlements at special points from JGC1 to JGC10

Figure 6 presents the settlement plan views of Jiangcun Central Bus Station from May 1st to June 1st. The shape of Jiangcun Central Bus Station is a narrow rectangular, whose west side is close to the middle span of excavation and east side is close to the pit corners. Similar to the automobile service company, the maximum settlement occurred at JGC26, which was closest to the middle span. It was obvious that the differential settlements of west side were larger than those of east, which means the corner effects also contributed to reduction of building settlements.

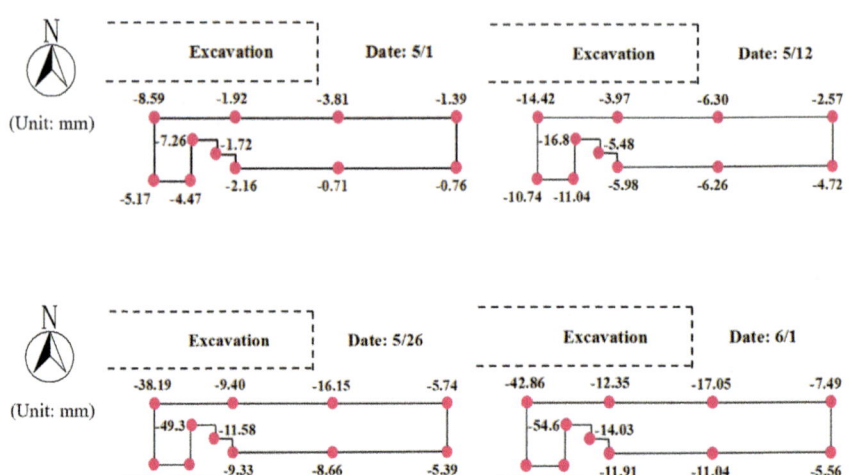

Fig. 6. Settlement plan view of Jiangcun Central Bus Station

Figure 7 shows the development of building settlements at some points from JGC26 to JGC36. The settlements at JGC26 and JGC36 increased faster than other points, which means the corner effects had a large influence on the reduction of building settlements. Based on the above analysis, the building settlements during excavation were not a simple 2D plane problem but a complicated 3D problem.

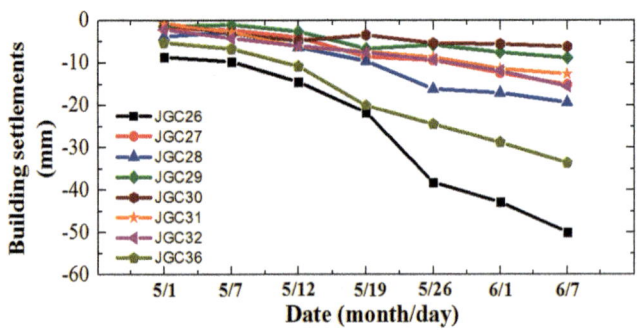

Fig. 7. Development of the settlements at special points from JGC26 to JGC36

4 Conclusions

By analyzing the field data of a rectangular underground passage pit in Hangzhou, this paper studied the influence of the corner effects on the deflections of diaphragm walls, ground settlements and building settlements during excavation process. Based on the analysis above, the following major conclusions can be drawn:

1. On the basis of analyzing the deflections of two monitoring point groups, the ratio of maximum deflections and excavation depths near corners was approximately $2.2\%_0 H_e$ and the ratio at the middle span of walls was $3.4\%_0 H_e$. Since the monitoring points close to the corners of excavation had smaller diaphragm wall deflections, the corner effects contributed to the reduction of diaphragm wall deflections.
2. With the development of excavation, the ratio of maximum settlements and excavation depths near corners was approximately $2.1\%_0 H_e$ and the ratio at the middle span of walls was $3.2\%_0 H_e$. The corner effects can effectively decrease the ground settlements nearby.
3. The final building settlements were the accumulation of the settlements in two directions and the points closest to the middle span of excavation had the largest settlement. The corner effects made the building settlements a complicated three-dimensional problem and conventional analysis method with plane assumption was inappropriate.

References

1. Ou, C.Y., Shiau, B.Y.: Analysis of the corner effect on excavation behaviors. Can. Geotech. J. **35**(3), 532–540 (1998)
2. Liu, G.B., Ng, C.W., Wang, Z.W.: Observed performance of a deep multistrutted excavation in shanghai soft clays. J. Geotech. Geoenviron. Eng. **131**(8), 1004–1013 (2005)
3. Lin, D.G., Woo, S.M.: Three dimensional analyses of deep excavation in Taipei 101 construction project. J. Geoengin. **2**(1), 29–42 (2007)
4. Wu, C.H., Ou, C.Y., Tung, N.: Corner effects in deep excavations - establishment of forecast model for Taipei basin T2 zone. J. Marine Sci. Technol. **18**(1), 1–11 (2010)
5. Tan, Y., Wei, B., Diao, Y.P., Zhou, X.: Spatial corner effects of long and narrow multipropped deep excavations in shanghai soft clay. J. Perform. Constr. Facil. **28**(4), 1–17 (2014)
6. Lee, F.H., Yong, K.Y., Quan, K.C.N., Chee, K.T.: Effect of corners in strutted excavations: field monitoring and case histories. J. Geotech. Geoenviron. Eng. **124**(4), 339–349 (1998)
7. Finno, R.J., Blackburn, J.T., Roboski, J.F.: Three-dimensional effects for supported excavations in clay. J. Geotech. Geoenviron. Eng. **133**(1), 30–36 (2007)
8. Fuentes, R., Devriendt, M.: Ground movements around corners of excavations: empirical calculation method. J. Geotech. Geoenviron. Eng. **136**(10), 1414–1424 (2010)

Braced Excavation Responses in BTG Residual Soils Under Significant Groundwater Drawdowns: DTL2 Case Study in Singapore

Wengang Zhang[1(✉)], Runhong Zhang[1], Wei Wang[1], Zhongjie Hou[1],
and Anthony Teck Chee Goh[2]

[1] School of Civil Engineering, Chongqing University,
Chongqing 400045, China
cheungwg@126.com
[2] School of Civil and Environmental Engineering,
Nanyang Technological University, Singapore 639798, Singapore

Abstract. The excavation system performance and ground movement behavior for the cut-and-cover excavation of four Mass Rapid Transit stations for the Downtown Line stage 2 in Singapore is presented. These excavations are characterized by the significant groundwater drawdown, which varies from average 2 m to 16 m. The information presented includes the maximum ground surface settlement values and the profiles, and the maximum wall deflection. Comparisons of the measured wall deflections and the ground surface settlements are also performed against the empirical methods/charts from the literature. It is hoped that these general behaviors will provide useful reference and insights for future projects involving excavation in the Bukit Timah Granite (BTG) residual soils under significant groundwater drawdowns.

Keywords: Ground movements · Braced excavation · Bukit Timah Granite
Residual soil · Groundwater drawdown · Empirical charts

1 Introduction

In a densely built-up urban environment such as Singapore, more construction of transportation tunnels and Mass Rapid Transit stations is needed to cater for the continuing population growth and urbanization. As the fifth MRT line in Singapore, the Downtown Line (DTL) is a major MRT line that links people directly from the northern and eastern parts of Singapore into the downtown area and its intent is to provide a quick, convenient, affordable and comfortable means of transport.

This paper compiled the characteristics of braced excavation responses for the cut-and-cover excavations, based on the instrumentation results of the DTL stage 2 sites: Cashew, Hillview, Bukit Panjang (BKP), and Beauty World (BTW). An overview is given, providing information including the subsurface conditions, the excavation support system, the field instrumentations, and the observed excavation responses. Construction activities and remedial measures that were undertaken are also presented. Comparisons of the measured wall deflections as well as the ground surface settlements are performed against the empirical methods/charts from the literatures.

© Springer Nature Singapore Pte Ltd. 2018
T. Qiu et al. (Eds.): GSIC 2018, *Proceedings of GeoShanghai 2018 International Conference: Advances in Soil Dynamics and Foundation Engineering*, pp. 396–403, 2018.
https://doi.org/10.1007/978-981-13-0131-5_43

2 General Project Description

2.1 Site Layout

Figure 1 provides an overview of site layouts for Cashew and Hillview. Actually the four sites were all located in densely built-up urban environments. Therefore, control of ground deformation is important in minimizing the impact on adjacent structures.

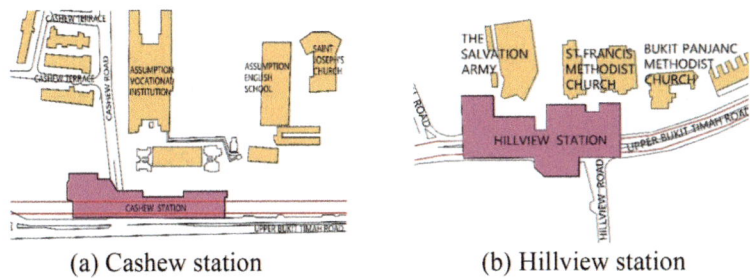

(a) Cashew station   (b) Hillview station

Fig. 1. Site layout of the 4 stations

2.2 Subsurface Conditions

The ground conditions comprise of Fill, Kallang Formation, BTG residual soil (GVI), completely weathered materials (GV), highly weathered materials (GIV) and moderately weathered (GIII) to fresh (GI) BTG rock. For the detailed properties of these geological units, Sharma et al. (1999) is referred to. Figure 2 shows the cross-sectional view and the soil profiles based on existing boreholes for Cashew site. This figure presents the typical geologic units, the generalized site stratigraphy and the relative position between each strut level and geologic unit.

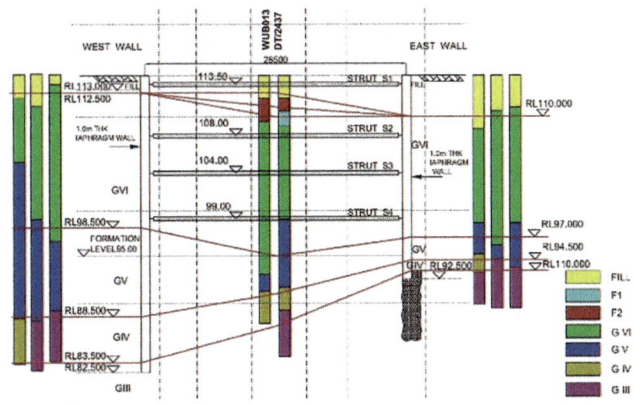

Fig. 2. Geological soil profiles and cross-sectional view of Cashew station

2.3   Excavation Support System Description

Table 1 summarized the excavation geometries and the ground conditions. Table 2 provided a summary of the retaining systems adopted for the four stations.

Table 1. A summary of excavation geometries and ground conditions for the four stations

Station	Max exca. depth (m)	Max exca. length (m)	Max exca. width (m)	Thickness of Fill/Kallang Formation (m)	Thickness of GVI (m)	Rockhead/mRL
Cashew	20	225	60	4.5–11	3–19	87.2–97
Hillview	24	190	48	0–9	3–25	87.1–105
Bukit Panjang	22	165	60	5–9	2.6–11	83.5–98.2
Beauty World	20	225	50	5.7–9	2.9–8.5	87.5–107.9

Table 2. A summary of retaining systems adopted for the four stations

Station	Wall system	Levels
Cashew	Diaphragm wall (thickness: 1 m, length: 25 m)	4
Hillview	D-wall (1 m & 1.2 m thickness, length: 29 m)	6
Bukit Panjang	Secant bored pill wall (1.2 m diameter with 1.5 m spacing, length: 21.5 m to 35.3 m)	5
Beauty World	Secant bored pill wall (1.2 m diameter with 1.6 m spacing, length: 10.3 m to 25.1 m)	4

2.4   Field Instrumentation

Monitoring data collected during the construction included diaphragm wall and secant bored pile wall deflections (using in-wall inclinometers), pore water pressures (using vibrating wire piezometers), ground settlements (using settlement markers) and strut loadings (using strain gauges and load cells). Data from the inclinometers and piezometers were obtained daily during excavation.

3   Observed Behavior of Field Instrumentations

3.1   Lateral Wall Movements

Figure 3 shows the maximum wall deflection for each excavation stage for all cross sections of four stations, respectively. The results show that from the excavation of the second layer soil onwards, the normalized maximum wall deflection is below 0.2%, which indicates that the deflection of the diaphragm wall and the secant bored pile wall during excavation is generally small.

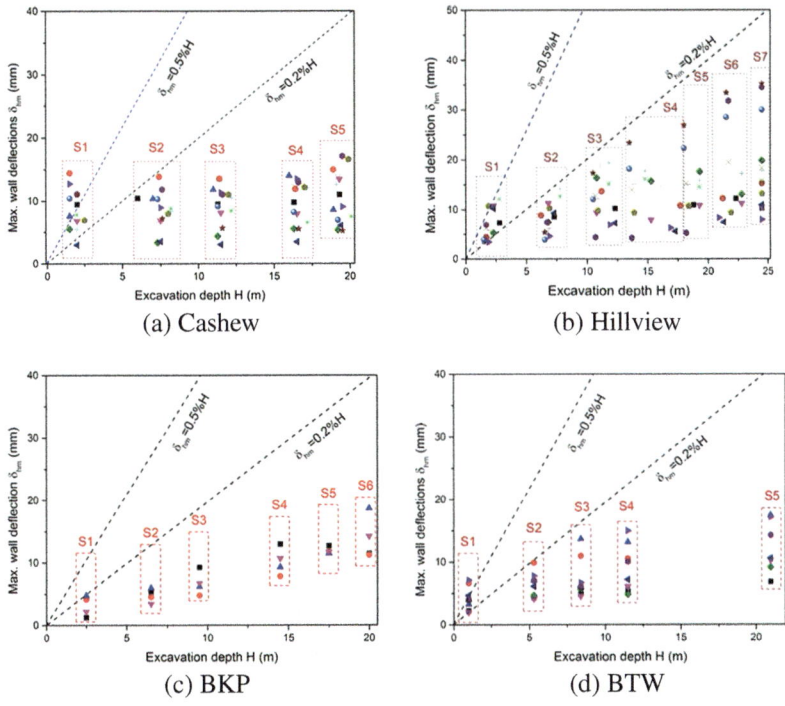

Fig. 3. Maximum wall deflection versus excavation depth

3.2 Ground Surface Settlements

Figure 4 shows the ground surface settlement profiles for different excavation stages for Cashew and Hillview stations. It is obvious that the ground settlement increases with increasing excavation depth.

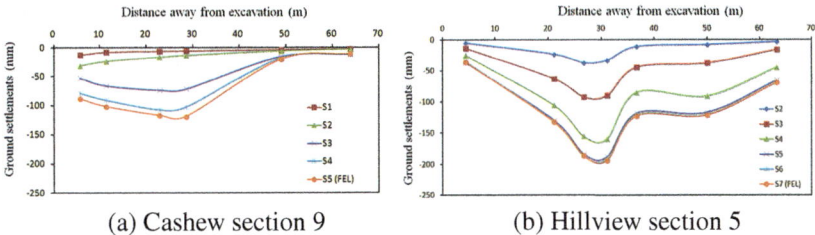

Fig. 4. Ground surface settlement profiles for different stages

Figure 5 shows the ground surface settlement profiles for the final elevation stage of all the cross sections (represented by the solid lines of different colors) of four sites, respectively. It can be observed that the ground surface settlement profiles vary significantly, even for the same site. In addition, for sections of Cashew and Hillview

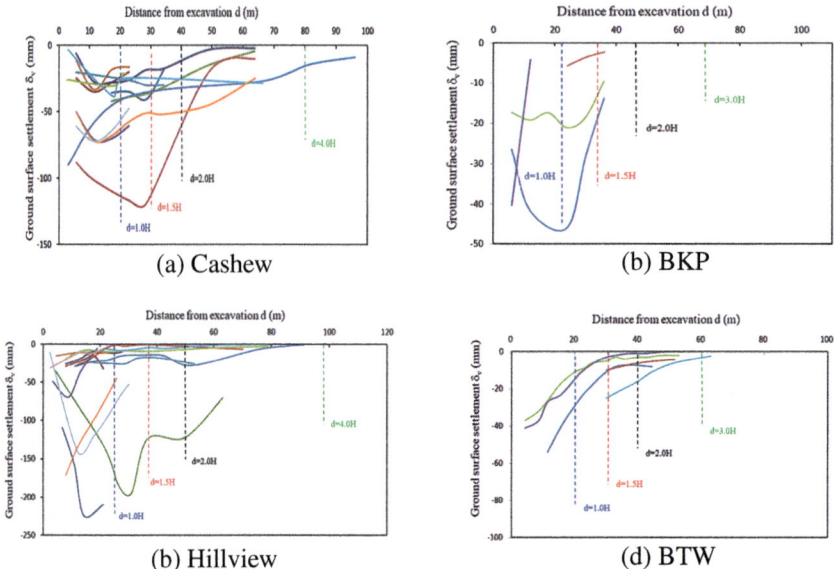

Fig. 5. Ground surface settlement profiles for FEL of the four stations

sites, most of the maximum ground surface settlements occur within the range of d = 1.0 H while for some exceptions, the maximum values exist between d = 1.0 H and d = 1.5 H. Furthermore, the settlement profiles extend beyond d = 4.0 H for Cashew and Hillview sites. For sections of BKP and BTW sites, the maximum ground surface settlements occur within the range of d = 1.0 H and the settlements occur within d = 3.0 H.

3.3   Groundwater Level Changes

Figure 6 plots the observed ground water drawdown behind the retaining wall for the four sites. Each point is the average of the standpipes readings in the sites. It is obvious that as excavation proceeds to the FEL, the ground water table decreases. The minimal

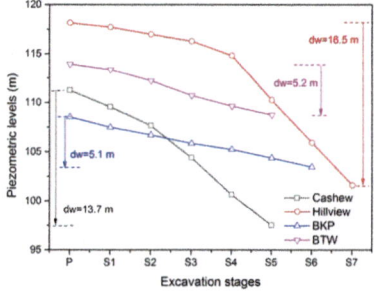

Fig. 6. The observed average ground water drawdown

piezometric levels are obtained when the last soil layer was excavation. In addition, the ground water drawdown for the Cashew and Hillview sites was more significant than that for sites of BKP and BTW, which also explains the closely related observations that the ground surface settles more for these two sites.

4   Characteristics of Braced Excavation Responses

4.1   Characteristics of Lateral Wall Movements

Table 3 shows published values of δ_{hm}/H for different soil types in different regions, including the BTG residual soil from this paper. It is obvious that the wall deflections measured in BTG residual soil were much less than those in the soft soil regions researched by Wang et al. (2005a, 2005b), Li et al. (2015), Cetin (2016). The much less wall deflections might be a result of more stiff retaining walls adopted, preloading of strut, and the observed significant ground water drawdown reported in Sect. 3.3.

Table 3. δ_{hm}/H values in different regions

Data source	Soil layer	δ_{hm}/H: %
Clough & O'Rourke (1990)	Clay of medium strength	0.20
Ou et al. (1993)	Clay, Taipei	0.20–0.50
Masuda (1993)	Clay, Japan	0.05–0.50
Carder (1995)	Hard clay	0.13–0.40
Wu et al. (1997)	Hard soil layer, Taipei	0.07–0.20
Wong et al. (1997)	Marine clay, Singapore	<0.20
Liu et al. (1999)	Soft soil region	0.20-0.90
Long (2001)	Interbed hard and soft soil	0.13–0.36
Yoo (2001)	Layered soil	0.05–0.15
Wang et al. (2005a)	Soft soil	<0.70
Wang et al. (2005b)	Soft soil, Shanghai	0.10–0.60
Li et al. (2015)	Silty clay, Nanjing	0.08–0.32
Cetin (2016)	Istanbul soft clay	0.043–0.32
Present study	BTG residual soil	0.025–0.15

4.2   Characteristics of Ground Surface Settlement Values

Table 4 shows published values of δ_{vm}/H for different soil types in different regions. It is obvious that the ground surface settlements measured in BTG residual soil were much greater than those in the soft soil regions reported by Wang et al. (2005a, 2005b), Li et al. (2015), Cetin (2016).

Table 4. δ_{vm}/H values in different regions

Data source	Soil layer	δ_{vm}/H: %
Peck (1969)	Clay, Britain	<0.70
Clough and O'Rourke (1990)	Hard clay, sandy soil	0.05–0.15
Ou et al. (1993)	Clay, Taipei	0.13–0.36
Carder (1995)	Hard clay	0.20–0.90
Fernie and Suckling (1996)	Hard clay, Britain	0.07–0.20
Wong et al. (1997)	Soft soil	0.05–0.50
Long (2001)	Hard and soft soil layer stagger	0.13–0.40
Leung and Ng (2007)	Silty sand, Hong Kong	0.20–0.50
Hashash et al. (2008)	Fill, clay	0.20
Wang et al. (2010)	Soft soil, Shanghai	<0.42
Wang et al. (2011)	Soft soil, Shanghai	0.10–0.60
Li et al. (2012)	Clay, Beijing	0.08–0.32
Li et al. (2015)	Silty clay, Nanjing	0.04–0.19
Ali and Khan (2017)	Clay	0.10–0.40
Present study	Bukit Timah Granite residual soil	0.09–0.90

5   Summary and Conclusions

Based on the observations, the characteristics of braced excavation responses in Bukit Timah Granite residual soils are summarized as follows:

1. The wall deflections are small, far below the alert level;
2. Based on the piezometric level changes, the ground water drawdown in the four sites are significant;
3. The ground surface settlements are much greater than those in other regions, most probably due to the significant ground water drawdown.

References

Ali, J., Khan, A.Q.: Behaviour of anchored pile wall excavations in clays. Geotech. Eng. Proc. Inst. Civ. Eng. (2017). https://doi.org/10.1680/jgeen.16.00071

Carder, D.R.: Ground Movements Caused by Different Embedded Retaining Wall Construction Techniques. Transport Research Laboratory, Wokingham (1995)

Cetin, D.: Performance of soil-nailed and anchored walls based on field-monitoring data in different soil conditions in Istanbul. Acta Geotech. Slov. **13**(1), 48–63 (2016)

Clough, G.W., O'Rourke, T.D.: Construction induced movements of in-situ walls. In: Proceedings of Design and Performance of Earth Retaining Structure. Geotechnical Special Publication No. 25, pp. 439–470. ASCE, New York (1990)

Fernie, R., Suckling, T.: Simplified approach for estimating lateral movement of embedded walls in U.K. ground. In: Proceedings of the International Symposium on Geotechnical Aspects of Underground Construction in Soft Ground, London, UK, pp. 131–136. CRC Press/Balkema, Leiden, The Netherlands (1996)

Hashash, Y.M., Osoulia, A., Marulanda, C.M.: Central artery tunnel project excavation induced ground deformations. J. Geotech. Geoenviron. Eng. **134**(2), 1399–1406 (2008)

Leung, E.H.Y., Ng, C.W.W.: Wall and ground movements associated with deep excavations supported by cast in situ wall in mixed ground conditions. J. Geotech. Geoenviron. Eng. **133** (2), 129–143 (2007)

Li, D., Li, Z., Tang, D.: Three-dimensional effects on deformation of deep excavations. Geotech. Eng. Proc. Inst. Civ. Eng. **168**(GE6), 551–562 (2015)

Li, S., Zhang, D.L., Fang, Q., Lu, W.: Research on characteristics of ground surface deformation during deep excavation in Beijing subway. Chin. J. Rock Mech. Eng. **31**(1), 189–198 (2012). (in Chinese)

Liu, X.W., Shi, Z.Y., Yi, D.Q., Wu, S.M.: Deformation characteristics analysis of braced excavation on soft clay. Chin. J. Geotech. Eng. **21**(4), 456–460 (1999). (in Chinese)

Long, M.: Database for retaining wall and ground movements due to deep excavations. J. Geotech. Geoenviron. Eng. **127**(3), 203–224 (2001)

Masuda, T.: Behavior of Deep Excavation with Diaphragm Wall. MS thesis, Massachusetts Institute of Technology, Cambridge, MA, USA (1993)

Ou, C.Y., Hsieh, P.G., Chiou, D.C.: Characteristics of ground surface settlement during excavation. Can. Geotech. J. **30**(5), 758–767 (1993)

Peck, R.B.: Deep excavation and tunneling in soft ground. In: Proceedings of 7th International Conference on Soil Mechanics and Foundation Engineering, pp. 225–290 (1969)

Sharma, J.S., Chu, J., Zhao, J.: Geological and geotechnical features of Singapore: an overview. Tunn. Undergr. Space Technol. **14**(4), 419–431 (1999)

Wang, J.H., Xu, Z.H., Chen, J.J., Wang, W.D.: Deformation properties of diaphragm wall due to deep excavation in Shanghai soft soil. Chin. J. Undergr. Space Eng. **1**(4), 485–489 (2005a). (in Chinese)

Wang, J.H., Xu, Z.H., Wang, W.D.: Wall and ground movements due to deep excavations in Shanghai soft soils. J. Geotech. Geoenviron. Eng. **136**(7), 985–994 (2010)

Wang, W.D., Xu, Z.H., Wang, J.H.: Statistical analysis of characteristics of ground surface settlement caused by deep excavations in Shanghai soft soils. Chin. J. Geotech. Eng. **33**(11), 1659–1666 (2011). (in Chinese)

Wang, Z.W., Charles, W.W.N., Liu, G.B.: Characteristics of wall deflections and ground surface settlements in Shanghai. Can. Geotech. J. **42**(5), 1243–1254 (2005b)

Wong, I.H.L., Poh, T.Y., Chuah, H.L.: Performance of excavations for depressed expressway in Singapore. J. Geotech. Geoenviron. Eng. **123**(7), 617–625 (1997)

Wu, P.Z., Wang, M.J., Peng, Y.R.: Discussion on deformation of diaphragm wall. In: Proceedings of the 7th Conference on Current Researches in Geotechnical Engineering, Taipei, pp. 601–608. Taiwan Geotechnical Society, Taipei (1997). (in Chinese)

Yoo, C.: Behavior of braced and anchored walls in soils overlying rock. J. Geotech. Geoenviron. Eng. **127**(3), 225–233 (2001)

The Discussion of Influential Factors on Tunnel Upheaval Deformation Induced by Deep Pit Excavations Above Metro Tunnels

Wei Gao[1(✉)], Wenhua Zhang[1], Li Li[1], Haibin Li[1], Weihong Zhao[2], Shichun Zhang[3], and Aiguo Li[4]

[1] Shenzhen Integrated Geotechnical Investigation and Surveying Co., Ltd.,
Shenzhen, China
381136989@qq.com
[2] Qianhai Development and Investment Co., Ltd., Shenzhen, China
[3] Southwest Municipal Engineering Design and Research Institute of China,
Chengdu, China
[4] Shenzhen Geotechnical Investigation and Surveying Co., Ltd.,
Shenzhen, China

Abstract. Projects of deep excavation became more common nowadays in modern city development for underground spaces exploitation, and some of them were crossing over the existing metro tunnels. Safety of metros would be threatened in case that excavation processes were not controlled considerately and unexpected deformation of tunnels was induced accordingly. The upheaval behaviors of the tunnel were actually caused by the above soil unloading and stress redistribution within the soil body. Due to the complexity of problems, some of factors, such as excavation depth, excavation manner and tunnel shape were mainly discussed in the paper. Factors study was made based on monitoring data and three-dimensional Finite Element Method. It is helpful to better understand upheaval behavior of tunnels induced by upper excavation so as to avoid excessive tunnel deformation in similar engineering cases.

Keywords: Deep excavation · Tunnels · Upheaval deformation
Influential factors · Finite Element Method

1 Introduction

With the urbanization process of cities, more underground spaces were exploited to satisfy the demands of modern city development. Deeper foundation pits were designed, and more excavations were constructed crossing over the existing metro tunnels. As a result, the original stress and strain of soil around tunnels underwent a great change due to excavation [1, 2]. Redistribution of soil stress occurred finally during the above soil unloading process. In such situation, unexpected upheaval deformation of underlying tunnels might be induced consequently.

Because few experiences to predict and protect metro structures had been obtained so far, strict regulations were declared for metro protection in cities, such as Shenzhen, Shanghai in China. The maximum displacement of metro tunnel should not exceed 20 mm.

© Springer Nature Singapore Pte Ltd. 2018
T. Qiu et al. (Eds.): GSIC 2018, *Proceedings of GeoShanghai 2018 International Conference: Advances in Soil Dynamics and Foundation Engineering*, pp. 404–411, 2018.
https://doi.org/10.1007/978-981-13-0131-5_44

Due to the complexity of this kind of problems, how to predict and assess possible deformation to ensure tunnels' safety became a significant issue. Although influences from excavations were studied by researchers [3–6], cases of upper excavations above tunnels were not usually experienced in infrastructure construction, and main factors affecting the tunnel deformation were not discussed correspondingly.

2 Project Background

The project was located in Qianhai Shenzhen-Hong Kong Modern Service Industry Cooperation Zone. A planned municipal underground expressway with the length of over 3 km and 60 m in width would be built above the current metro line 1. The tunnel length underlying excavation pit is totally 83 m, which includes left tunnel, right tunnel and entrance/exit tunnel. The tunnel shape is square with each side length of 6 m. Approximately, 3.0–4.75 m deep foundation pits will be excavated for this expressway (Fig. 1). But above entrance/exit tunnel, the maximum excavation depth can reach 6.5 m. The pit bottom is always close to the tunnel roof in the project, and the minimum distance is less than 1 m (only 0.43 m).

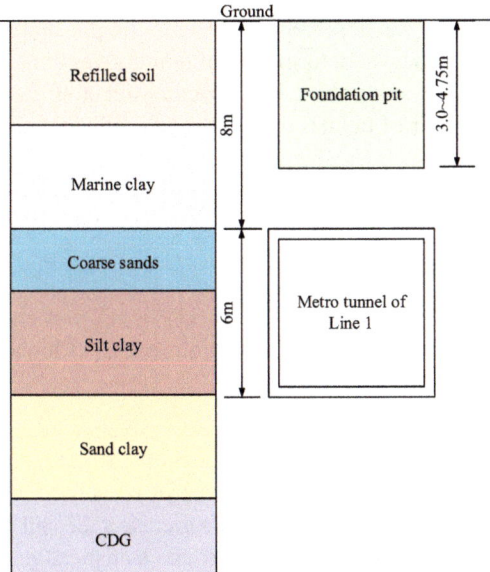

Fig. 1. A cross-section description between tunnels and the above excavation pits.

The original landform of Qianhai Zone was coastal beach with the characteristics of very thick soft marine clay and high groundwater level. The whole area was reclaimed after years, and the typical soil geological stratification (see Fig. 1), from top to bottom, was mainly composed of refilled soil, soft marine clay, silt clay, coarse sands and underlying Completely Decomposed Granites (CDG). The properties of soil strata were

Table 1. Summary of soil properties

Description	Thickness/m	Density/kN/m³	Permeability coefficient K (m/d)	Cohesion/kPa	Internal friction angle/₀
Refilled soil	4.0	19.0	0.02	12	10
Marine clay	4.0	16.5	0.001	8	2
Coarse sands	2.0	20.0	10.00	0	30
Silt clay	4.0	19.0	0.5	20	18
Sand clay	4.0	19.0	0.1	22	20
CDG	6.0	19.5	0.5	25	30

listed in Table 1, which is mainly refer to the geotechnical investigation report and statistical analysis results of by Luo et al. (2012) and Qin et al. (2013) [7, 8].

In order to reduce the negative effects on existing tunnels as possible as we can, detailed analysis of deformation prediction were conducted in advance of excavations. According to these results, countermeasures and protection principles, such as soil reinforcement, excavation manners and anti-uplift structures were considered and designed conservatively.

During excavation, it was observed the factual complicatedly although the monitoring data showed the upheaval deformation of tunnels was controlled as expected [9]. Thus, it was necessary to study those main factors affecting tunnel deformation and how the upheaval behavior of tunnels developed and influenced in different situations.

3   Finite Element Analysis

3.1   Model Establishment

In this study, the geotechnical software: MIDAS GTS NX was employed by building a three-dimensional finite element model of metro tunnel and foundation pits. The main principles of modeling were listed as follows:

(1) Three-dimensional numerical model of two different excavation manners were built to investigate how the constructing process of foundation pit excavation influence the deformation of underlying metro tunnels.

(2) The values of physical and mechanical parameters of soil layers were obtained from the geotechnical investigation report and statistical analysis results as listed in Table 1.

(3) The soil material was modeled by the modified Mohr-coulomb model and was assumed to be an ideal elastic plastic body [5]. The metro tunnel and the retaining structure of foundation pit were assumed to be linear elastic bodies.

The size of the established three-dimensional finite element model was $143 \times 93 \times 24$ m. The stress and deformation of the soil body were analyzed with three-dimensional solid element, and the metro tunnel and the retaining structure were simulated by shell element.

4   Approach and Discussion

It was observed that the upheaval behavior of tunnels was actually a very complicated process during excavation. Due to the complexity of the issues, prediction of tunnels deformation relied heavily on empirical and semi-empirical approaches established from field measurements [10]. The typical geological strata (Fig. 1) in this area would be used in the model so that the difference in geological conditions was not considered in the study. All discussions were based on the factually collected monitoring data. In this way, main influential factors: excavation depth, excavation manner, as well as tunnel shape were discussed.

4.1   Excavation Depth

In this paper, comparison was made between FEM results and the monitoring data during the entire construction process (Fig. 2). The maximum deformation value (6 mm) by FEM method was slightly less than the factually monitored data. It can be found that the predicted trend of upheaval behavior turned to be a straight line and went through the monitored values. The actual upheaval behavior of metro tunnels developed in an up-and-down accumulation manner which was possibly determined by the complex excavation procedures.

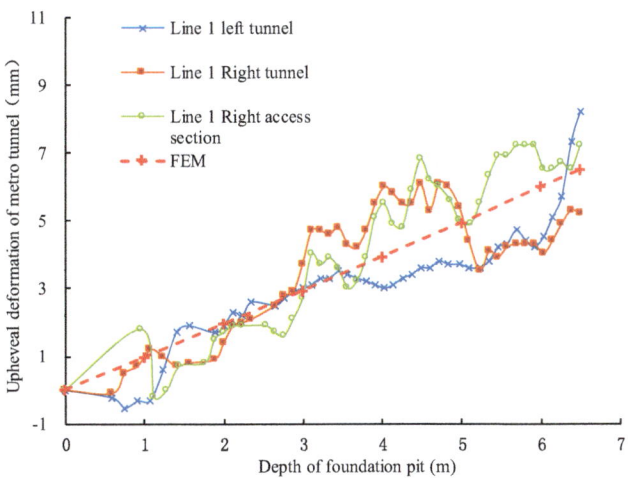

Fig. 2. Development of upheaval deformation of metro tunnels with the process of excavation.

In order to analyze how the excavation depth affected tunnels deformation, cases of different pit depth, i.e., 1.0 m, 2.0 m, 3.0 m, 4.0 m, 5.0 m, 6.0 m, 6.5 m were discussed respectively. The process of excavation was assumed in a symmetrical step-slope manner. Figure 3 demonstrates the trend of tunnel upheaval deformation with the process of excavations in the FEM calculation.

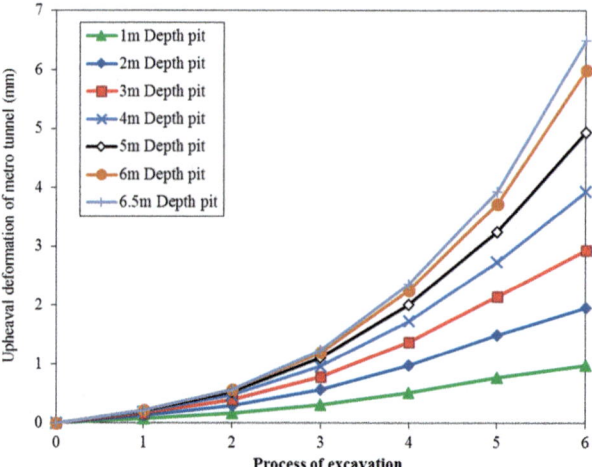

Fig. 3. Trendlines of tunnel upheaval deformation in cases of different excavation depths.

It can be found obviously that the upheaval deformation increases accordingly with the incremental of the excavation depth atop. Very small deformations (less than 2 mm) can be observed in case of a relatively shallow depth (excavation is no more than 3 m). When the pit goes deeper, larger upheaval deformation can be more easily induced. More than 6 mm of maximum deformation can be observed as the foundation pit reaches a depth of 6.5 m. Moreover, the results also indicate how the upheaval deformation developed in the condition of different depth excavation. Nonlinear behaviors can be figured out with the process of excavation, and deformation values seem to increase easily while the excavation level almost reaches the pit bottom.

4.2 Excavation Manner

The excavation manner of a foundation pit was also an essential factor to possibly ensure the safety of the existing metro tunnel. During the construction of foundation pit, two different methods of vertical shaft excavation and step-slope excavation were usually applied (Figs. 4 and 5).

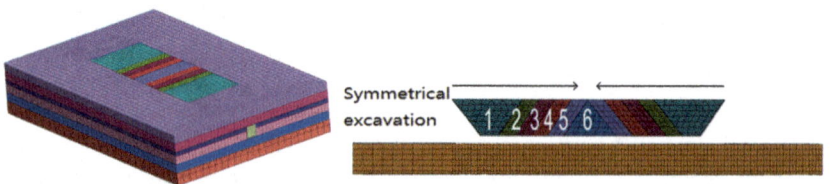

Fig. 4. Finite element model of metro tunnel and step-slope excavation manner.

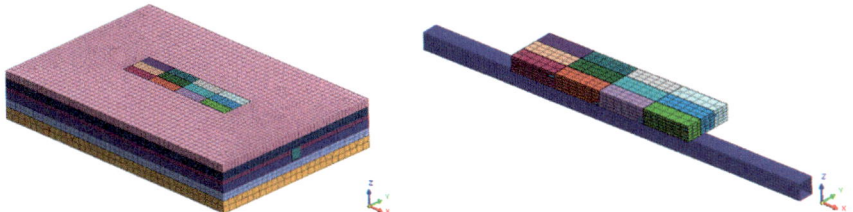

Fig. 5. Finite element model of metro tunnel and vertical shaft excavation manner.

Step-slope method is adopted to divide the whole excavation zone into several successive subzones, and dig out soil progressively and continuously in a direction from one side to another side. During the process of subzone excavation, ambient stable slopes were formed temporarily. Vertical shaft excavation, also called 'divided alternate excavation method' [11], is to separate the whole excavation zone into several isolated subzones. The divided retaining walls have to be built and excavation sequence is alternate for better utilizing the principle of the time-space effect.

Figure 6 shows the results of upheaval deformation in these two different excavation manners in numerical simulation. Vertical shaft excavation leads to less upheaval deformation of tunnel than the step-slope excavation, and nearly 20% deformation can be reduced in this case.

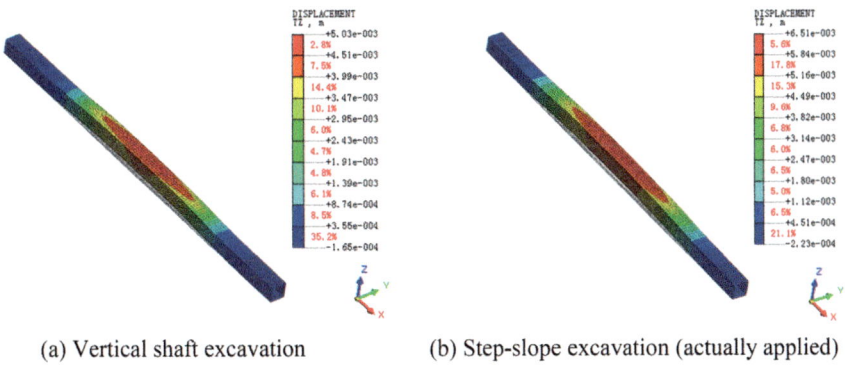

(a) Vertical shaft excavation   (b) Step-slope excavation (actually applied)

Fig. 6. Upheaval deformation of metro tunnels with two different excavation manners.

Figure 7 further demonstrates how the upheaval behavior of tunnels developed in these two different excavation manners. It shows that in the same situation, controlled alternate excavation sequence is more helpful to result in less deformation of tunnels.

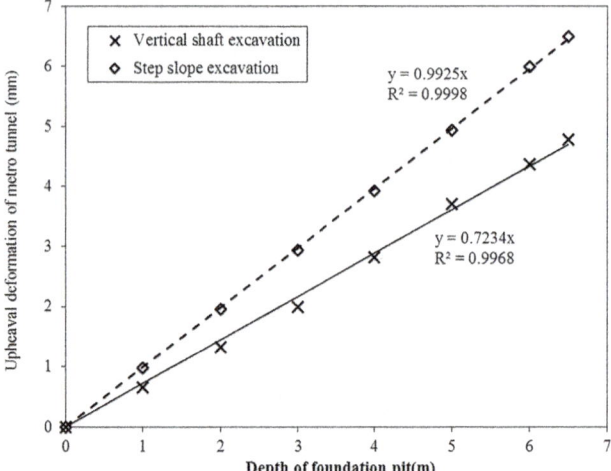

Fig. 7. Upheaval behavior of metro tunnels with different excavation manners.

4.3   Tunnel Shape

The tunnel shape or structure is believed to be another factor to possibly result in different tunnel upheaval behaviors. Larger deformation can be observed in TBM-shaped (circle) tunnels based on the monitored information comparing to the frame-structured tunnel (Three squares tunnels combined) [9].

However, when the circle tunnel and square tunnel has the same cross area, the induced maximum upheaval deformation values are almost invariant (Fig. 8). As a result, the shape of a metro tunnel may have limited effects on deformations of a metro tunnel.

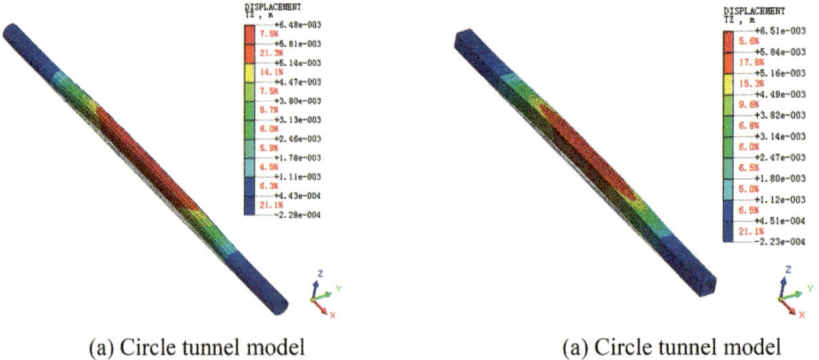

(a) Circle tunnel model   (a) Circle tunnel model

Fig. 8. Upheaval deformation of metro tunnel in two different shapes.

5   Conclusion

Based on the factually gathered data and the FEM method, factors influencing the upheaval behavior of tunnels, such as excavation depth, excavation manner and tunnel shape were discussed. Meanwhile, conclusion can be drawn as follows:

(a) The upheaval deformation of a metro tunnel usually increases accordingly with the incremental of the excavation depth. The upheaval behaviors may be induced nonlinearly as the excavation of a foundation pit.

(b) Vertical shaft excavation leads to less upheaval deformation of tunnel than the step-slope excavation. A controlled alternate excavation manner is helpful to reduce upheaval deformation of tunnels.

(c) Different shape of the metro tunnel with the same cross area may have limited effects on metro tunnel deformation.

References

1. Zeng, Y., Li, Z., Wang, Y.: Research on influencing factors of deep excavation adjacent to subway station. Chin. J. Undergr. Space Eng. 4(1), 642–645 (2005)
2. Zhang, M., Yang, X., Liu, T.: Comparison and analysis of foundation pit construction schemes near metro tunnels. Chin. J. Undergr. Space Eng. 07(6), 1203–1208 (2011)
3. Jiang, H., Hou, X.: The influence of deep excavation on adjacent metro tunnel in soft ground. Chin. Ind. Constr. 5, 53–56 (2002)
4. Zheng, G., Liu, Q., Deng, X.: Field measurement and analysis of effect of excavation on existing tunnel boxes of underlying metro tunnel in operating. Rock Soil Mech. 33(4), 1109–1116, 1140 (2012)
5. Hou, Y., Wang, J.H., Chen, J.J.: 3D FEM analysis of over size & deep excavation. Chin. J. Geotech. Eng. 52, 42–46 (2006)
6. Chinese Codes: Code for Design of Building Foundation, pp. 1465-1469. China Architecture & Building Press, Beijing (2011)
7. Luo, Y., Yang, G., Zhang, Y., et al.: Engineering characteristics and its statistical analysis of soft marine clay on Shenzhen west coast. Chin. J. Civ. Eng. Manag. 29(2), 79–86 (2012)
8. Qin, J., Lu, W., Gao, P., et al.: Structural design and deformation analysis of deep foundation pit in marine soft soil region. Chin. J. Undergr. Space Eng. 9(5), 1115–1120 (2013)
9. Gao, W., Li, H.B., Wei, X.W., Zhang, A.J.: The case study of a deep pit excavation above multiple metro tunnels and influential factors analysis on tunnel upheaval behaviors. Int. J. Eng. Technol. 9(5), 374–377 (2017)
10. Wen, S.: Study on construction technology and protection measures of the excavation above operating metro tunnel. Chin. J. Undergr. Space Eng. 53–56 (2011)
11. Chen, J., Zhu, Y.I., Li, M., Wen, S.: Novel excavation and construction method of an underground highway tunnel above operating metro tunnels. J. Aerosp. Eng. 28(6), A4014003 (2015)

Numerical Analysis of Multi-tunnel Interaction in Clay

Bing-Qing Zhu[1,2], Yin-Fu Jin[3], Zhen-Yu Yin[1,2,3(✉)],
Dong-Mei Zhang[1,2], and Hong-Wei Huang[1,2]

[1] Department of Geotechnical Engineering, College of Civil Engineering,
Tongji University, Shanghai, China
zhenyu.yin@gmail.com
[2] Key Laboratory of Geotechnical and Underground Engineering of Ministry
of Education, Tongji University, Shanghai 200092, People's Republic of China
[3] Research Institute of Civil Engineering and Mechanics (GeM),
UMR CNRS 6183, Ecole Centrale de Nantes, Nantes, France

Abstract. Constructing a new tunnel close to an existing one is a concerned engineering problem since the interaction between tunnels at a close range could lead to ground settlement, resulting in damage of buildings. This paper presents the particular interest in ground settlement induced by tunnel excavations and the configurations of twin-tunnel. For this purpose, the numerical model is established, in which the S-CLAY1 model is implemented. A series of numerical simulations on twin-tunnel at spacing of 1.5D, 3D and 4.5D are carried out. The simulated results and centrifuge measurements are compared for different ground loss rates and tunnel configurations. All the results demonstrate that the numerical analysis in conjunction with the S-CLAY1 model can well predict the ground settlement induced by multi-tunnel excavation, which is ready for practical engineering in tunnel groups.

Keywords: Multi-tunnels · Anisotropy · Ground loss · Numerical simulate
Clay

1 Introduction

Utilizing the underground space is an important measure solving the major problems in large cities, such as population, resources, and environment. With increasing the underground constructions, it inevitably leads to tunnel group in a close distance, forming multi-line parallel, overlapping, cross or other complex relations [1, 2]. Tunnel group excavation disturbs the soil, which can cause the ground settlement, strata deformation, and affect the safety of surrounding facilities [3]. The construction of a new tunnel will lead to large deformation and change the internal force of the tunnel to the adjacent tunnel [4–9]. The ground settlement will lead to local damage of existing structures or other underground facilities [10]. As the constructions continuing, tunnel group will be more common and needs to be paid more attention. The ground movements due to tunnelling is a result of many factors, such as ground loss, lining deformation, and long-term consolidation [11]. Among them, the ground loss

© Springer Nature Singapore Pte Ltd. 2018
T. Qiu et al. (Eds.): GSIC 2018, *Proceedings of GeoShanghai 2018 International Conference: Advances in Soil Dynamics and Foundation Engineering*, pp. 412–419, 2018.
https://doi.org/10.1007/978-981-13-0131-5_45

contributes the most. Thus, the settlement induced by ground loss during the multi-tunnel excavation is worth to be studied.

Numerical analysis as a useful tool has been adopted by many researchers and engineers. In terms of the multi-tunnel excavation, the ground displacement can be predicted by the numerical simulation by introducing an appropriate boundary conditions and a suitable constitutive model [1]. Among previous studies, a variety of soil models have been adopted in the analysis, e.g., nonlinear elastic model with cross-anisotropic elasticity [12], Mohr–Coulomb model [13], nonlinear Mohr–Coulomb model with small-strain [14], modified cam-clay model [12], two-surface kinematic hardening soil model [15] and Hypoplastic model with low friction angles [12]. However, the consideration of soil anisotropy on the ground settlement induced by the multi-tunnel excavation is rarely reported. Therefore, the numerical analysis in conjunction with a soil model considering the anisotropy is a good way to predict the ground loss.

This paper aims to investigate the ground settlement induced by multi-tunnel excavation. For this purpose, a series of centrifuge tests on twin-tunnels excavation are selected and the corresponding numerical models are established. The anisotropic elastoplastic model S-CLAY1 is briefly presented and implemented into FEM as a user defined material. Thus, the selected centrifuge tests on twin tunnels of aligned-horizontally at spacing of 1.5D, 3D, and 4.5D are simulated. Based on the simulations, the effects of ground loss rate and tunnel space on the ground settlements are analyzed.

2 Numerical Modeling of Tunnel Tests

2.1 Overview of the Centrifuge Tests

An eight-group plane twin tunnel centrifuge model test is documented in Divall et al. [16], investigating the ground deformation affected by twin-tunnel excavation. The size of centrifuge tank is 500 mm × 200 mm × 180 mm, and the radius of the tunnel is 40 mm. The soil in the test is kaolin clay, with a moisture content of 120%. The centrifuge test focuses on different spacing of the twin-tunnel and ground loss rate affection during excavation. The configurations of the tunnel are shown in Fig. 1.

2.2 Numerical Model

The commercial finite element code ABAQUS was employed to simulate the centrifuge test. The meshed model is shown in Fig. 2. The size of the numerical model is the same as the centrifuge test. The ideal two-dimensional simulation can also draw some practical conclusions of the ground deformation caused by the excavation [17]. The soil was simulated by the quadrangle CPE4 element and the tunnel lining was by beam element. The lining concrete is classified as grade C55. The elastic modulus and the Poisson's ratio are 3.55×10^4 MPa and 0.2, respectively. In this study, no water is involved in the simulation, which is reasonable according to previous studies [18].

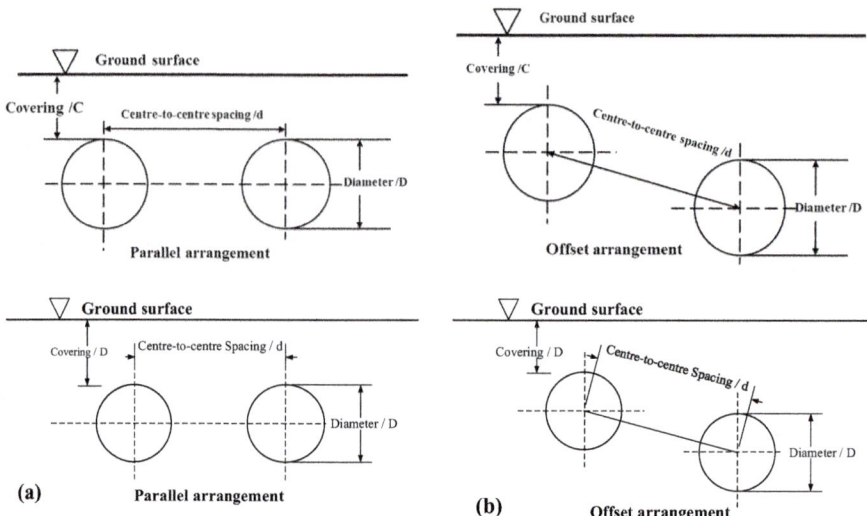

Fig. 1. Illustration of the twin-tunnel arrangements (a) parallel arrangement (b) offset arrangement

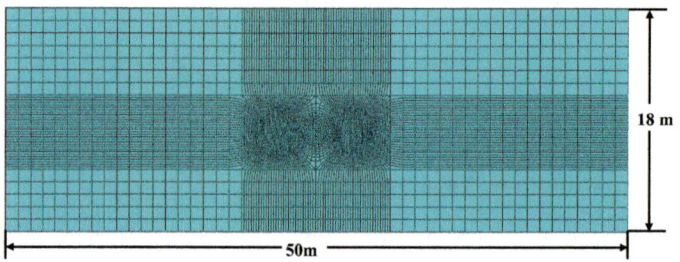

Fig. 2. Illustration of the whole meshed model

2.3   Constitutive Model and Implementation

S-CLAY1 is an extension of the critical state models, with anisotropy of plastic behaviour represented through an inclined yield surface and a rotational hardening of yield surface to model the development or erasure of fabric anisotropy during plastic straining. More information about the S-CLAY1 can be found in Wheeler et al. [19] The parameters of S-CLAY1 for the Kaolin clay employed in the simulation are summarized in Table 1 according to Atkinson [20] using the same clay.

The adopted model was implemented into ABAQUS as a user-defined constitutive model via user material subroutine UMAT. The procedure of model implementation follows the way of Hibbitt et al. [21]. For the stress integration, the cutting plane algorithm proposed by Ortiz and Simo [22] was adopted.

Table 1. Values of model parameters of Kaolin clay

Parameter	υ	M	κ	e_0	λ	α_{K0}	ω	ω_d
Value	0.3	0.89	0.035	1.27	0.18	0.35	31.0	0.37

Where υ is Poisson's ratio; M is the slope of critical state line; κ is the swelling index; e_0 is the initial void ratio; λ is the compression index; α_{K0} is the stress ratio corresponding to the K_0 condition; ω and ω_d are two constants controlling the evolution of rotation hardening.

2.4 Method to Simulate the Ground Loss

Ground loss model affects the convergence behaviour and the calculation efficiency. In this case, the non-uniform distributed ground loss method proposed by Lee et al. [23] was adopted. To achieve the given ground volume loss (VL) in the simulation, the boundary shrink method was used in the calculating process. First, the boundary of the over excavated tunnel ring was fixed, and then the boundary shrinks to the size of the designed tunnel. For each rate of ground loss, the magnitude of shrinkage is 0.03 m for 3% VL and is 0.05 m for 5% VL.

2.5 Calculation Process

Figure 3 shows the calculation steps for twin-tunnel excavation. To investigate the response of the existing tunnel affected by the excavation of new tunnel, the twin tunnels are successively excavated. Tunnel A is the first excavated tunnel, and Tunnel B is the second. The simulating process is the same as the centrifuge test.

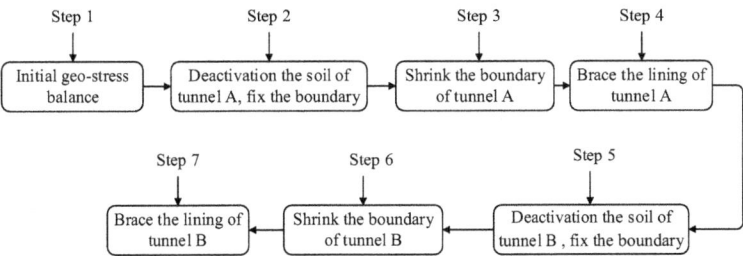

Fig. 3. Numerical simulation steps for twin tunnel excavation

3   Results and Comparisons

3.1 Cases of Parallel Arrangement

Figure 4 shows the comparisons of final ground settlement between the simulated and the experimental results for horizontally arranged parallel twin-tunnels in different spacing. As shown in Fig. 4, the two dashed lines represent the location of axis for the tunnel A and B, respectively. It can be seen that the shape of the settlement can be well captured using the numerical calculation adopting the S-CLAY1 model. All simulations indicate that the used model and its parameters associating with the FEM model are reasonable in predicting the ground settlement for twin-tunnel excavation. More situations of twin-tunnels can be studied though this numerical model. The shape of settlement curve likes a "v" with a peak point for the spacing of 1.5D. And the peak point of the settlement curve corresponding to the maximum settlement locates near the center of two tunnels. With increasing the spacing distance of two tunnels, the shape of settlement changes gradually from "v" to "w" with two peak points. For the spacing of 3.0D and 4.5D, the location of maximum settlement is on the tunnel axis. Furthermore,

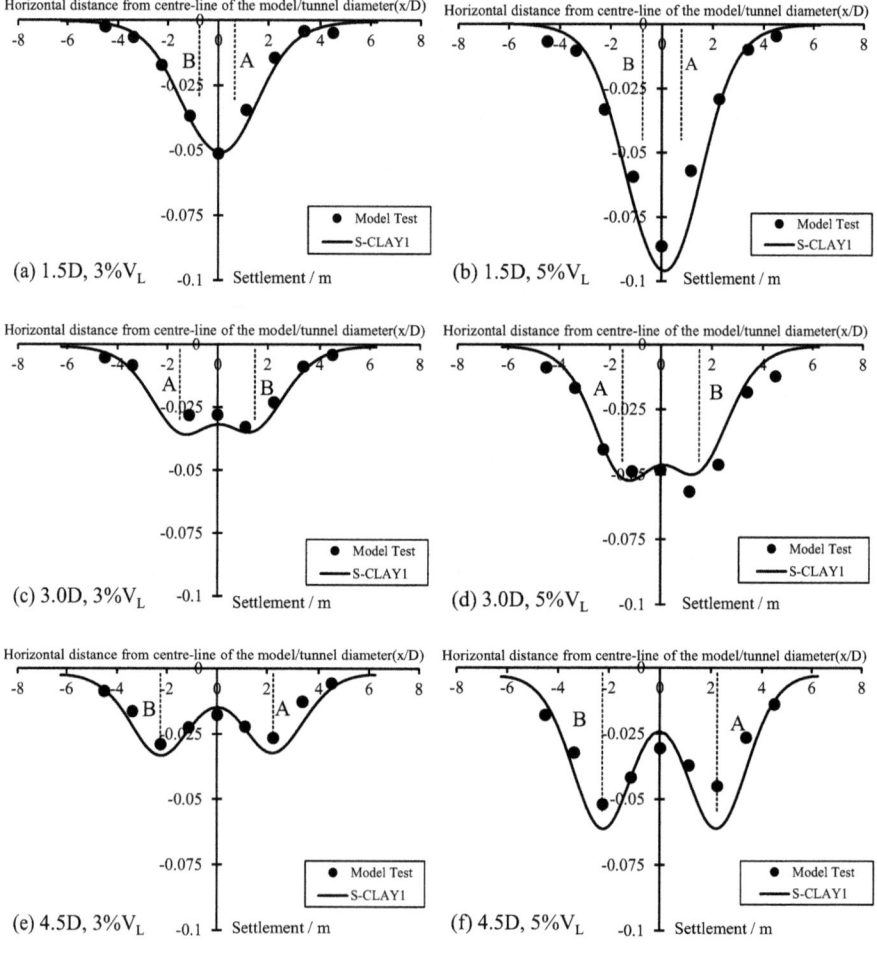

Note: A-Tunnel A; B-Tunnel B

Fig. 4. Surface settlement after excavation of parallel tunnels: (a, b) the case of 1.5 D; (c, d) the case of 3.0 D; (e, f) the case of 4.5 D

for each excavation, the settlement increases when increasing the volume loss from 3% to 5%, which is reasonable in practice.

3.2 Cases of Offset Arrangement

Figure 5 shows the ground surface settlement after excavating the second tunnel under offset arrangement condition. As shown in Fig. 5, the points of A and B indicate the relative locations of the twin-tunnel. For both situations, the good agreement between measurements and simulations demonstrates that the ground settlement under offset arrangement condition can also be well predicted. It seems that the location corresponding to the maximum settlement moves toward to the second tunnel. Furthermore, the effect caused by the construction of a new tunnel, on the existing tunnel, decreases with increasing the tunnel spacing.

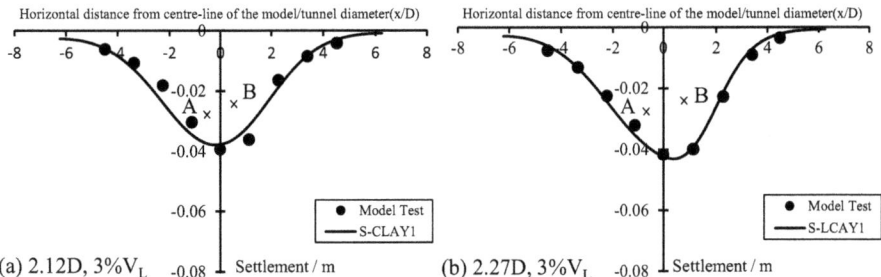

Fig. 5. Surface settlement after excavation of offset tunnels: (a) the case of 2.12D with 3% volume loss; (b) the case of 2.27D with 3% volume loss

4   Conclusion

This paper addressed the effects of twin-tunnel configurations on ground settlement induced by tunnel excavation by finite element analysis. The anisotropy of soft clay was considered by adopting the S-CLAY1 model. Eight groups of plane twin-tunnel centrifuge model tests have been simulated. All comparisons between simulations and measurements demonstrate that the numerical analysis combining the constitutive model accounting for the anisotropy of clay can well predict the ground settlement induced by the multi-tunnel excavation. The well validated numerical method can be used in practical engineering of tunnel groups in the future.

Acknowledgment. This research project is financially supported by National Natural Science Foundation of China (41372285 and 51579179).

References

1. Do, N.A., Dias, D., Oreste, P.: Three-dimensional numerical simulation of mechanized twin stacked tunnels in soft ground. Tunn. Undergr. Space Technol. **42**, 40–51 (2014)
2. Zhang, Z., Huang, M.: Geotechnical influence on existing subway tunnels induced by multiline tunneling in Shanghai soft soil. Comput. Geotech. **56**, 21–32 (2014)
3. Gang, Z., Yu, D., Xue-Song, C., et al.: Safety control of supported deep excavation and underground engineering and the impact on surrounding environment (2015)
4. Cui, Q.-L., Wu, H.-N., Shen, S.-L., Yin, Z.-Y., Horpibulsuk, S.: Protection of neighbour buildings due to construction of shield tunnel in mixed ground with sand over weathered granite. Environ. Earth Sci. **75**, 458 (2016)
5. Wu, H.-N., Shen, S.-L., Liao, S.-M., Yin, Z.-Y.: Longitudinal structural modelling of shield tunnels considering shearing dislocation between segmental rings. Tunn. Undergr. Space Technol. **50**, 317–323 (2015)
6. Jiang, M., Yin, Z.-Y.: Influence of soil conditioning on ground deformation during longitudinal tunneling. C.R. Mec. **342**, 189–197 (2014)
7. Li, P., Du, S.-J., Ma, X.-F., Yin, Z.-Y., Shen, S.-L.: Centrifuge investigation into the effect of new shield tunnelling on an existing underlying large-diameter tunnel. Tunn. Undergr. Space Technol. **42**, 59–66 (2014)
8. Shen, S.-L., Wu, H.-N., Cui, Y.-J., Yin, Z.-Y.: Long-term settlement behaviour of metro tunnels in the soft deposits of Shanghai. Tunn. Undergr. Space Technol. **40**, 309–323 (2014)
9. Jiang, M., Yin, Z.-Y.: Analysis of stress redistribution in soil and earth pressure on tunnel lining using the discrete element method. Tunn. Undergr. Space Technol. **32**, 251–259 (2012)
10. Xiao, X., Zhang, M.X., Hui-ming, W.U., et al.: Numerical simulation analysis on ground settlements caused by multi-line shield tunnel. Chin. J. Undergr. Space Eng. **07**(5), 884–889 (2011)
11. Zhang, D., Huang, Z., Yin, Z., Ran, L., Huang, H.: Predicting the grouting effect on leakage-induced tunnels and ground response in saturated soils. Tunn. Undergr. Space Technol. **65**, 76–90 (2017)
12. Masin, D., Herle, I.: Numerical analyses of a tunnel in London clay using different constitutive models. In: Geotechnical Aspects of Underground Construction in Soft Ground (2005)
13. Addenbrooke, T.I., Shin, J.H.: A numerical study of the effect of groundwater movement on long-term tunnel behaviour. Géotechnique. **52**, 391–403 (2002)
14. Zdravkovic, L.: Modelling of a 3D excavation in finite element analysis. Géotechnique. **55**, 497–513 (2011)
15. Standing, J., Potts, D., Vollum, R., Burland, J., Tsiampousi, A., Afshan, S., et al.: Investigating the effect of tunnelling on existing tunnels (2015)
16. Divall, S., Goodey, R.J.: Twin-tunnelling-induced ground movements in clay. Proc. Inst. Civ. Eng.: Geotech. Eng. **168**, 247–256 (2015)
17. Mair, R.J.: Centrifuge modelling of tunnel construction in soft clay. Cambridge University, Cambridge (1969)
18. Hang, J., Zhang, D.L.: Numerical simulation of stratum deformation above overlapping metro tunnel. Chin. J. Undergr. Space Eng. **24**(12), 2176–2182 (2005)
19. Wheeler, S.J., Näätänen, A., Karstunen, M., Lojander, M.: An anisotropic elastoplastic model for soft clays. Can. Geotech. J. **40**, 403–418 (2003)
20. Atkinson, J.H., Richardson, D., Robinson, P.J.: Compression and extension of K0 normally consolidated kaolin clay. J. Geotech. Eng. **113**, 1468–1482 (1987)

21. Hibbitt, Karlsson, Sorensen: ABAQUS/Standard: User's Manual. Hibbitt, Karlsson & Sorensen (1998)
22. Ortiz, M., Simo, J.: An analysis of a new class of integration algorithms for elastoplastic constitutive relations. Int. J. Numer. Methods Eng. **23**, 353–366 (1986)
23. Lee, K.M., Rowe, R.K., Lo, K.Y.: Subsidence owing to tunnelling. I. Estimating the gap parameter. Can. Geotech. J. **29**, 929–940 (1992)

Observed Performance of a Large-Scale Double Annular Arched Girders Structural System in Deep Excavation

Ming-yi Zhao, Shang-rong Chen, and Fa-yun Liang[(⊠)]

Department of Geotechnical Engineering, Tongji University,
Shanghai 200092, China
fyliang@tongji.edu.cn

Abstract. Large amount of field and experimental data has been devoted to excavations supported by single annular arched girders structural system thus far. In contrast, only few data were available for excavations supported by double annular arched girders structural system, especially those with large diameters in thick soft clay deposits. Therefore, it is worthwhile to study such kind of cases. This paper investigates an excavation located in Pudong New Area, Shanghai. In this project, two reinforced concrete supports are arranged vertically, and the excavation can be divided into three areas. Area A is supported by double cylindrical arched girders structural system, the diameter of each arched girders is 86 m. Area C is supported by side girders and angle diagonal brace. Area B just uses angle brace to support it. This project is constructed by bottom-up method. Based on site measured data, this paper investigated items included: groundwater level changes, lateral wall settlements, subsurface horizontal movements, metro pier settlements, and annular concrete support force changes. As the result, we find that the double cylindrical excavation behaves much differently with single cylindrical excavation. This study provided valuable data and result for double cylindrical excavation design and construction. And we found the internal force distribution of the double annular support structure is complicated and is sensitive to the boundary conditions and the construction progress. In addition, casting of the base slabs is particularly important for the safety of the double annular support structure.

Keywords: Double annular arched girders · Excavation · Field data
Case study

1 Introduction

The deep foundation pit projects in soft soil area usually utilizes certain kind of enclosure system such as bored cast-in-place pile, steel sheet pile, SMW method pile and diaphragm wall, and the cast-in-place reinforced concrete or steel as horizontal support set in the pit. The support system is usually arranged in the form of perpendicular-coupled, angle-coupled and side-truss-coupled [1]. These forms of support have been widely used, which accumulated much engineering experience as well as a set of systematic design and construction method. However, for the

© Springer Nature Singapore Pte Ltd. 2018
T. Qiu et al. (Eds.): GSIC 2018, *Proceedings of GeoShanghai 2018 International Conference: Advances in Soil Dynamics and Foundation Engineering*, pp. 420–428, 2018.
https://doi.org/10.1007/978-981-13-0131-5_46

foundation pit whose size is too large, the supporting form listed above may encounter some stability or economic difficulties. In addition, the excavation efficiency will reduce if the support beams are set too tight. The annular supporting system can utilize the foundation pit space effectively, reducing the project cost and speeding up the project progress. However, we can consider using double annular support system, if the length and width of the pit have a great large difference. At present, some projects have successfully applied double annular supporting system for foundation pit construction, which shows great economic and time costing advantage. Similar kind of pit case record is still less, so this article presents a double annular foundation pit project in Shanghai soft soil for reference.

2 Project Overview

2.1 Foundation Pit Overview

The foundation pit area is 44232 m^2, the excavation depth is 11.85–13.15 m. There is a metro viaduct on the north side of the pit, and the viaduct is about 34.5 m from the outer edge of the foundation pit, which is the key protection object of this project. This foundation pit is divided into three areas (area A, B and C) for excavation. Area A is located on the northern side of the foundation pit, and the double annular arched girders structural supporting system is used at this area. Area B is used as separation area, which is supported by angle brace. The southern part of the foundation pit is area C, which is supported by cross brace and angle brace. Supporting system of area A and area B is as Fig. 1 shows (area C is not related in this paper, so we don't present it in Fig. 1). Area A is mainly discussed and analyzed because this paper mainly investigates the characteristics of foundation excavation of double annular arched girders structural supporting system. Figure 2 is a typical front view of supporting system in area A and area B. The cross-section. 1-1 is the supporting system presents in Fig. 1.

2.2 Enclosure System Overview

The supporting system of the foundation pit is conducted by ∅900 mm, ∅950 mm and ∅1000 mm cast-in-place bored pile with Φ850@600 tri-axial stirring pile for sealing up. The main structure of supporting brace is made up of two layer reinforced concreted annular beam, which are set at 2.8 m and 8.8 m below the ±0.000 elevation.

2.3 Site Subsurface Conditions

The site subsurface conditions are listed in Table 1. Where γ is unit weight, w is moisture content, e is void ratio, c is cohesion, φ is friction angle of soil.

2.4 Monitoring System

To collect and analyze the surrounding environment elements in the construction, the monitoring points mentioned above are detailed in Fig. 3.

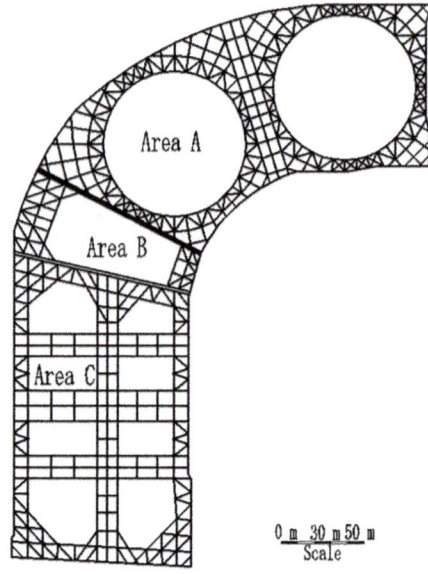

Fig. 1. Sketch of supporting system in area A and B

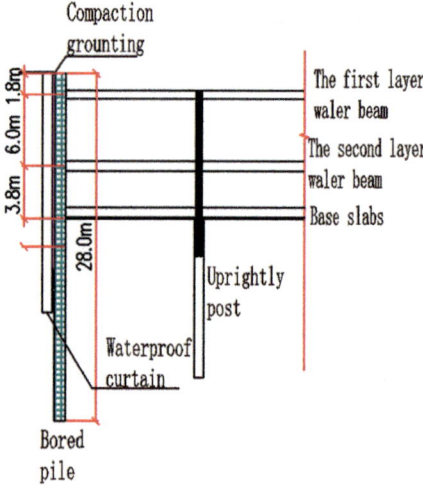

Fig. 2. Front view of supporting system in area A and B

2.5   Construction Process

The project construction lasted for 322 days. The construction sequence is detailed in Table 2. Table 2 only presents the construction sequence in area A for this paper focus on the double annular arched girders structural system only.

Table 1. Site stratum overview

Stratum No.	Stratum property	γ (kN/m³)	w (%)	e	c (kPa)	φ (°)	Thickness (m)
①	Loose miscellaneous fill	–	–	–	–	–	1.00–3.20
②	Gray brown yellow clay	18.0	36.1	1.026	19	15.5	1.30–1.60
③	Gray muddy silty clay with organic matter	17.6	39.3	1.118	11	19.0	0.60–1.80
④	Grey muddy clay	16.6	50.8	1.446	14	10.5	9.80–11.00
⑤1	Gray clay	17.2	43.1	1.245	16	11.5	4.00–5.70
⑤2-1	Gray silty clay with clayey silt	18.0	34.2	0.999	18	20.0	6.60–11.40
⑤2-2	Gray sandy silt with clay	18.3	32.0	0.925	9	26.5	2.40–18.50
⑥	Dark green silty clay	19.4	24.4	0.716	45	15.0	3.40–14.70
⑦	Grass yellow to gray sandy silt	18.6	28.4	0.824	3	31.5	1.80–7.00
⑧1	Gray clay	39.2	39.2	1.134	24	16.0	1.20–14.30
⑧2-1	Gray clay silt	28.9	18.6	0.841	12	25.5	17.70–21.10
⑧2-2	Gray silty clay	30.7	18.5	0.884	26	19.5	10.30–25.20
⑨	Gray silty-fine sand	25.8	25.8	0.746	0	32.5	5.20–19.90

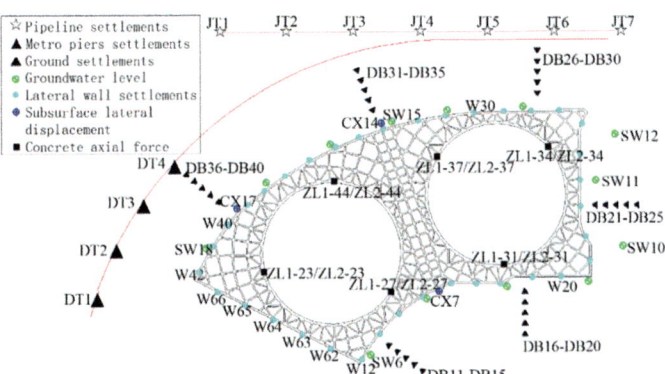

Fig. 3. The distribution of monitoring points in the foundation pit

Table 2. Construction process

Stage	Construction Content	Date	Days
0	Prepare to construct supporting system	2015.1.5-2015.1.17	13
1	Carry out supporting system construction	2015.1.17-2015.4.15	89
2	No construction	2015.4.16-2015.5.24	39
3	Construction of tower crane foundation	2015.5.25-2015.5.30	7
3a	Excavate the first layer of soil in area A	2015.6.1-2015.6.18	18
3b	Casting and conservation the first layer of support in area A	2015.6.20-2015.7.25	36
4b	Excavate the second layer of soil in area A	2015.7.31-2015.8.10	11
4c	Casting and conservation the second layer of support in area A	2015.8.11-2015.8.30	20
5b	Excavate the third layer of soil in area A	2015.9.10-2015.9.21	23
6	Base slabs casting	2015.9.22-2015.10.18	26

3   Monitoring Data and Analyze

3.1   Metro Pier Settlements

Cause there is a metro viaduct on the western side and northern side of the foundation pit, whose minimal distance from the pit is only 32 m, so we set a settlement monitor on each pier of the metro viaduct which in total of 12 monitors and marked as DT1–DT12. In this article, we choose 4 typical monitors' data to analyze (DT1–DT4), which is presented in Fig. 4.

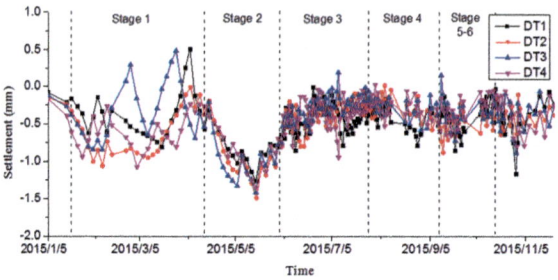

Fig. 4. Distributions of settlement of metro pier along time (DT1–DT4)

During the whole process of project, the foundation pit gives limited influence on the metro piers plotted above near the foundation pit. The maximum settlement appeared on DT3 at May 12th, whose magnitude is about 1.5 mm. At stage 1, all the piers elevated about 1 mm. We believe the construction of cast-in-place piles influence the pier. Especially, there is a settlement increase around May 25th and later the settlement return to the normal condition. According to the construction logs, stage 1 is carrying out tower crane construction, many construction tracks and equipment is parked and placed at nearby ground, which caused elastic settlement. However, the construction at other stages only gives limited influence on the piers settlements.

3.2 Lateral Wall Settlements

Figure 5(a) plots typical measured lateral wall settlements monitors' data variation along time.

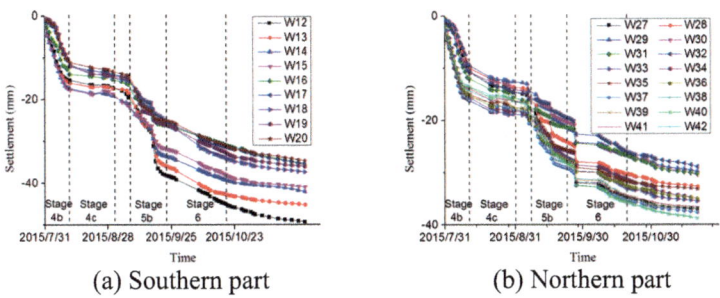

(a) Southern part   (b) Northern part

Fig. 5. Distributions of lateral wall settlement of the foundation pit along time

There are three times large variation in this graph. The first decrease happens at stage 4b, and the second decrease is at the earlier period of stage 5b, which is executing the soil excavation project. These decreases are directly related to the soli excavation. The third decrease is at the later period of stage 5b. At this period, all the 3rd layer of soil has been excavated, but because construction arrange is not so perfect, the base slabs didn't been cast until stage 6. After the base slabs cast, the speed of vertical wall settlements slows down a lot. We can see it from the graph that the third decrease is the fastest one among the three decreases, which means the absence of base slabs, will greatly lead to lateral wall settlements.

If we analyze from the magnitude of each vertical piles' settlements, it is not difficult to find that the settlements of W12–W15 are larger than W16–W20. This is because monitors W16–W20 are set near the corner of area A, which is easier to constrain the nearby soil. Monitors W12–W15 is further to the corner and located at the long edge of this foundation pit.

Figure 5(b) plots the changes of lateral wall settlements measured by some typical monitors on the northern side of foundation pit. Generally, we found the changes of lateral wall settlements on this side is very similar to that of the northern side.

In this section, we found that the foundation pit excavation contributes to the lateral wall settlements most, so it's important to monitor the lateral wall settlements during the double cylindrical excavation construction. Besides, the base slab is critical to lateral wall settlements, so the base slab should be casted as soon as possible.

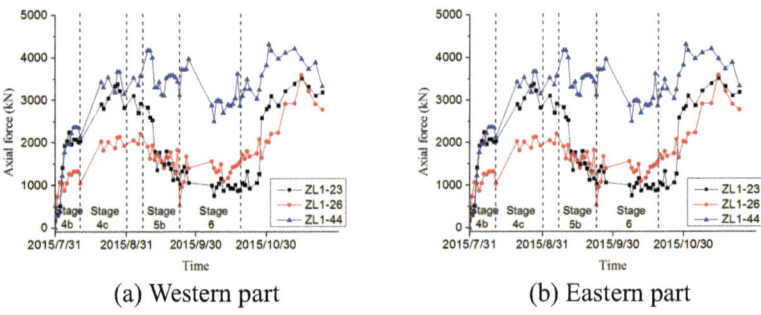

(a) Western part   (b) Eastern part

Fig. 6. Distributions of axial force of the first layer concrete annular support along time

3.3   Annular Structure Axial Force

Figure 6 plots the axial force of the first layer concrete annular support along time.

Figure 6 represent that the axial force starts to increase sharply at stage 3a, and keep increase until the end of stage 4b. At stage 4c, the axial force keeps increasing at low speed. At stage 5b, the axial force drops sharply, this is probably because the second layer's supporting system begin to work and the force from the lateral wall have transferred to it. Later, with the process of stage 5b, the axial force on the first layer support annular start to decrease, this is because more force has been transferred to the second layer.

In October, the project demolished the second layer support system, so we can see there is a sharp increase at the graph.

The variation trend and the range of the axial force distribution of the east side annular structure is like that on the west side. It is worth noting that, for the west side of the annular structure, the force on ZL1-44 is the largest one, but on the east side, ZL1-31 is the largest one, showing an asymmetric distribution of the force. It is clear that boundary conditions and the progress of the construction of the double annular structure of the force have a significant impact. We believe it's meaningful to enhance the monitoring frequency on the first layer supporting braces when the first stage excavation is finished but the second layer supporting braces is still under construction. Also, the monitoring on the first layer braces should be enhanced when the second layer braces is demolishing for a double cylindrical excavation foundation pit.

Figure 7 is the development of axial force on the second layer concrete annular support along time. At stage 5b, the axial force on both west and east side concrete annular structure increase a lot until the end of stage 5b. At the process of base slabs casting, the axial force on the second layer concrete annular structure keep decrease at low speed. This is probably because the base slab connects the whole support system, and more force transferring route established. When the base slabs complete, the axial force on second layer concrete annular structure keep stable until they are demolished.

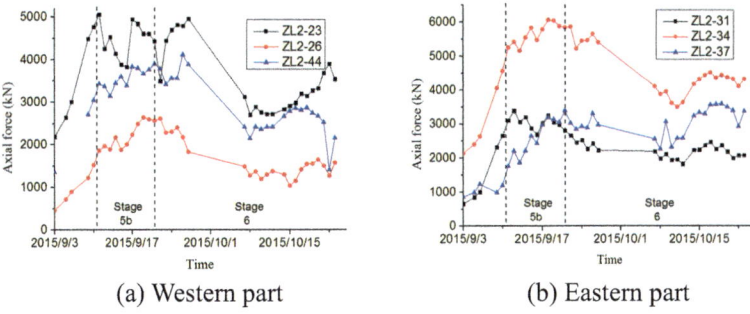

(a) Western part   (b) Eastern part

Fig. 7. Distributions of axial force of the second layer concrete annular support along time

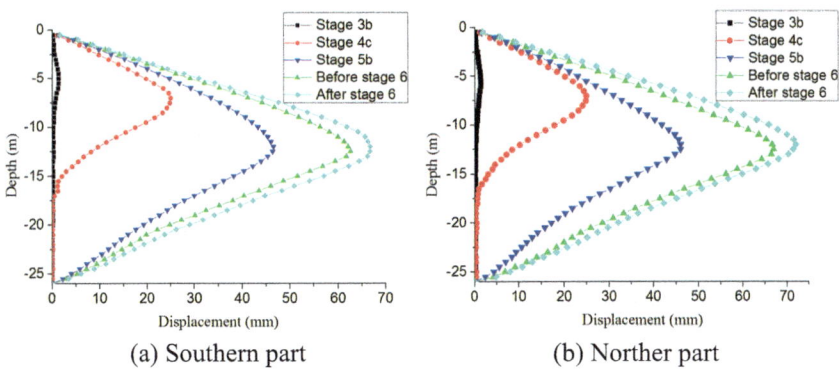

(a) Southern part   (b) Norther part

Fig. 8. Subsurface horizontal movements around Area A

3.4 Subsurface Horizontal Displacement

In this part, we set the inclinometer at July 26th, i.e., the zero-graduation displacement represents the displacement on July 26th'.

Figure 8 plots the subsurface horizontal movement around area A, which shows that each monitor represents similar trend and magnitude variation.

The second layer soil excavation completed at August 30th (the red dot marked curve in Fig. 8), and about to start the third layer of soil excavation, for now, the excavation depth is 6.05 m. The maximum displacement is about 25 mm, and the maximum displacement happens at 7 m below the ground, which is about 1.1–1.2 times of the excavation depth. The excavation construction completed at September 21st (the blue triangle marked curve in Fig. 8), for now, the excavation depth is about 12 m, and the maximum horizontal displacement increases to about 45 mm, and the maximum displacement happens at about 12.5 m below the ground, which influences the soil 16 m below the ground. Area A is preparing to cast the base slabs from September 21st to October 9th, which exposes the base soil for about 19 days. In this period, the subsurface horizontal soil displacement increases rapidly to 12–15 mm, which shows the importance of base slabs to a foundation pit safety. The base slabs completed at October 18th (the green triangle marked curve in Fig. 8), the subsurface horizontal displacement increased in low speed, which is much slower than before. For now, the maximum subsurface displacement is about 70 mm. The movement happened from stage 5b to stage 6 is almost the same with the movement happen in stage 3b and 4b, which represent that the exposure of the foundation pit base is bad for the subsurface horizontal movements control. Base slab plays a critical role on controlling the subsurface horizontal displacement, so it should be casted as quickly as possible.

4 Conclusions

(1) The data analysis results show that a reasonable design of a double annular support can meet the requirements of large foundation pit support.
(2) The internal force distribution of the double annular support structure is complicated and is sensitive to the boundary conditions and the construction progress.
(3) The casting of the base slabs is particularly important for the safety of the double annular support structure. After excavation to the base, the bottom plate should be applied as soon as possible. Otherwise, the displacement of the soil around the foundation pit will increase rapidly and affect the stability and safety of the foundation pit.
(4) The influence of the double-annular support structure on the surrounding soil is similar to that of the general support. The maximum horizontal displacement of the soil occurs at 1.0–1.2H and is symmetrical along the excavation surface.

References

1. Liu, J.H., Hou, X.Y.: Excavation Engineering Handbook, pp. 75–138. China Architecture & Building Press, Beijing (1997). (in Chinese)

Deformation of Subway Tunnels Affected by Adjacent Excavation: In-situ Monitoring and Centrifugal Model Test

Jifei Cui[1,2], Jingpei Li[1,2(✉)], Lin Li[1,2], and Gaowen Zhao[1,2]

[1] Key Laboratory of Geotechnical and Underground Engineering of Ministry of Education, Tongji University, Shanghai 200092, China
lijp2773@tongji.edu.cn
[2] Department of Geotechnical Engineering, Tongji University, Shanghai 200092, China

Abstract. The basement excavation will change the initial stress equilibrium of the surrounding soil and hence result in surface subsidence and strata movement. The adjacent subway tunnels would be inevitably influenced by the basement excavation in the dense urban environment. Based on the excavation engineering of Wufangyuan project in Shanghai, the effects of excavation on adjacent subway tunnels are studied by monitoring data and centrifugal model test in this paper. Detailed discussions are performed to investigate the different excavation partitions and excavation sequences. The deformations of an adjacent tunnel under different construction conditions are obtained and the comparisons between the tests are conducted. The results indicate that the excavation of the soil in the areas closer to the tunnel induced greater additional settlements and additional convergence. Some of the construction technologies were employed to reduce the deformation of adjacent tunnels including setting a small partition in the foundation excavation close to the tunnel, excavating the small partition after the completion of the main body of foundation excavation as well as enhancing the support in the direction perpendicular to the tunnel

Keywords: Basement excavation · Adjacent subway tunnels
Field monitoring · Centrifugal model test

1 Introduction

The rapid development of urban construction calls for growing excavation engineering in an urban area, and the metro system has been a lifeline of city transport. A lot of deep foundation pits are dug adjacent to existing tunnels. Effects on ground movement induced by deep excavations would cause a significant influence on the stress and the movement of tunnels (Wei et al. 2013). The structural cracks might lead to leakage and the rail differential displacements, which may seriously affect the safety of subway (Zhang et al. 2013a, b). Therefore, the deformations of the tunnel should be controlled within the acceptable level during excavation.

Lots of literature has reported the effects of deep excavations on the adjacent metro tunnel nearby. Zhang et al. (2013a, b) presented a semi-analytical method to evaluate

© Springer Nature Singapore Pte Ltd. 2018
T. Qiu et al. (Eds.): GSIC 2018, *Proceedings of GeoShanghai 2018 International Conference: Advances in Soil Dynamics and Foundation Engineering*, pp. 429–438, 2018.
https://doi.org/10.1007/978-981-13-0131-5_47

the heave of underlying tunnel induced by adjacent excavation and verified by field measurement results. Many previous studies (Dolezalova 2001; Sharma et al. 2001; Ge 2002; Hu et al. 2003; Zhang et al. 2007; Kog 2010) focus on numerical simulation approach. In their works, surrounding unloading soils and existing tunnels are modeled as a whole during the numerical discretization process. Birth and death element technology is used to simulate the soil excavation. Wei et al. (2014) analyzes the rules and characteristics of the vertical displacement, horizontal displacement, and convergence of the shield-driven tunnel at different construction stages based on the monitoring data. Zheng et al. (2010) study the influence of a basement excavation on an existing tunnel by carrying two centrifuge model tests.

In order to obtain a better understanding of the effects of foundation pit excavation on adjacent tunnels, the monitoring data of Wufangyuan project is analyzed in this study. Two centrifuge model tests were carried out to study the influence of excavation on an existing adjacent tunnel and the excavation scheme is compared and optimized.

2 Engineering General Situation

2.1 Site Condition

Shanghai is located near the front fringe of Yangtze River Delta in China. The Wufangyuan excavation is situated in Huangpu District, and the site plan is shown in Fig. 1. The area of the foundation pit is about 12 000 m^2, and the depth is 9.2–9.65 m. The foundation pit is divided into three areas, and the excavation sequence is A, B, and C. The south side of the foundation pit is the Shanghai Rail Transit Line 9 which has been completed and operated. The distance between the tunnel and the foundation pit is 15–20 m and the buried depth of the top of the tunnel is about 10 m.

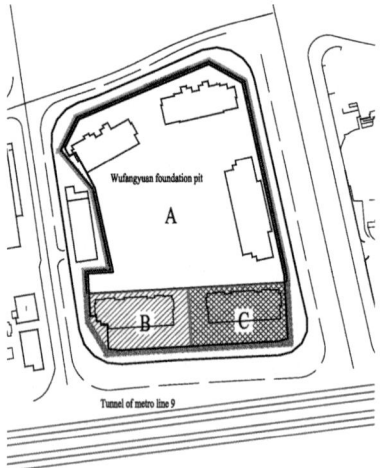

Fig. 1. Site plan of the Wufangyuan foundation pit

The retaining structure of district A is bored piles (Ø850@600, L = 20.1 m), while the retaining structures of districts B and C are concrete diaphragm walls with a thickness of 800 mm (L = 25 m). The middle wall is concrete diaphragm wall with a thickness of 600 mm (L = 21.6 m). Figure 2 shows a cross section of the support system of B and C district, which consist of concrete diaphragm wall, two concrete struts, piles and steel columns.

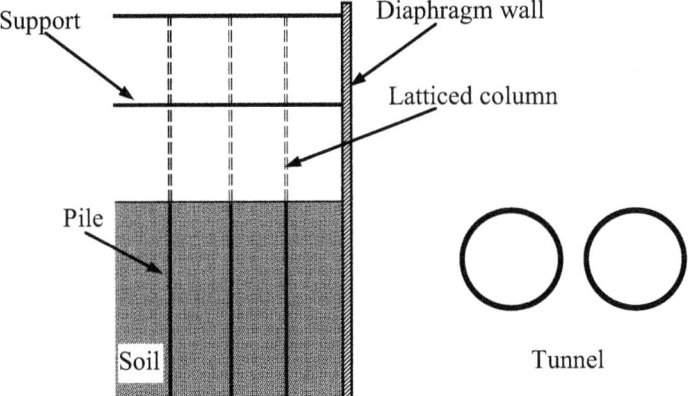

Fig. 2. Cross section of the retaining structure and tunnel

2.2 Geology

The strata of Shanghai are thick soft soils comprising Quaternary alluvial and marine deposits. A high water content, low shear strength, high compressibility and low ground bearing capacity are typical characteristics of the soft soil in Shanghai. Typical soil profiles from up to down obtained from site investigation are shown in Table 1. The groundwater conditions are approximately hydrostatic 1.0 m below ground level.

Table 1. Soil parameters

No.	Layer name	Thickness /m	Unit weight (γ)/ (kN/m^3)	Cohesion (c)/kPa	Internal friction angle (φ)/°	Poisson ratio (v)
①	Miscellaneous fill	2.67	17.5	12	11	0.40
②	Silty clay	1.93	17.9	18	14	0.35
③	Muddy silty clay	5.1	17.4	12	17.5	0.41
④	Muddy clay	8.1	16.5	13	10.5	0.42
⑤₁	Clay	10.2	17.2	17	15	0.36
⑤₂	Silty clay	5.9	18.1	17	21.5	0.27

2.3   Construction Procedure of the Excavation

The foundation pit is divided into three areas named A, B and C. The excavation sequence is first A, then B, and finally C. The soil of each area is divided into three layers, and the excavation depth is 2 m, 4.5 m, 3.5 m respectively. Table 2 summaries the main construction stages of the excavation.

Table 2. Construction stages

Stage	Construction operation	Day
1	Excavation of the first layer in A district, construction of the first support	36
2	Excavation of the second layer in A district, construction of the second support	30
3	Excavation of the third layer in A district, construction of the Base Slab	57
4	Excavation of the first layer in B district, construction of the first support	32
5	Excavation of the second layer in B district, construction of the second support	30
6	Excavation of the third layer in B district, construction of the Base Slab	33
7	Excavation of the first layer in C district, construction of the first support	33
8	Excavation of the second layer in C district, construction of the second support	34
9	Excavation of the third layer in C district, construction of the Base Slab	37

3   Observed Tunnel Deflections

3.1   Additional Settlement

Along the tunnel paralleling to the edge of the foundation pit, 28 measuring points are arranged on the upper and lower tunnel segments, respectively, to monitor the settlement of the tunnels. Figure 3 shows the additional settlement of the tunnel induced by the excavation from stage 1 to stage 9.

Figure 3 shows that the construction of the pre-excavation stage, such as diaphragm wall construction and pile foundation construction, induce slight rise of the tunnel. With the continuous excavation of the soil, the vertical displacement of the tunnel has changed from rising at the beginning to sinking, then increasing gradually. This is mainly because of the soil settlement out of the pit caused by the excavation. As seen from Fig. 3, the tunnel settlement caused by excavation after the third step is obviously greater than before. It means that the excavation of the soil in B and C areas induced greater influence on tunnel settlement than area A, which is closer to the tunnel seen in Fig. 1. Comparing the settlement of the segment along the length of the tunnel, it can be seen that the settlement of the segment beyond the boundary of the foundation pit is much smaller than the settlement at the midpoint of the pit edge.

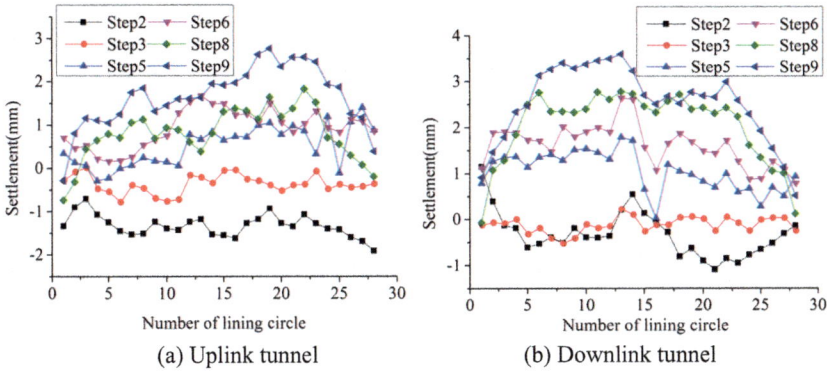

Fig. 3. Additional tunnel settlements induced by excavation

3.2   Additional Convergence

For determining additional convergence of the tunnel, the convergence was surveyed by 17 measuring points on the upper and lower tunnel segments, respectively. Four of these 34 measuring points are analyzed here for the additional convergence of the tunnel. Two of them named NO.1 and NO.4 are beyond the boundary of the foundation pit, and the other two measuring points named NO.2 and NO.3 are located at the tunnel corresponding to the midpoint of the boundary. Figure 4 shows the additional convergence of the four measuring points.

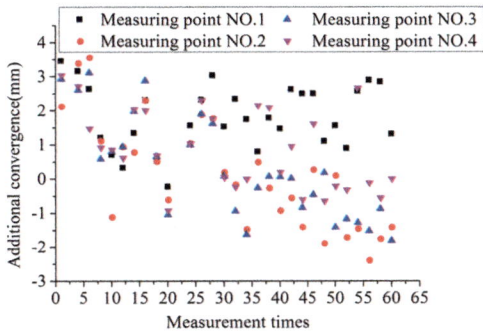

Fig. 4. Additional convergences induced by excavation

It can be seen from Fig. 4 that the additional convergence of measuring points NO.2 and NO.3 changed to the negative value which means that the excavation-induced inward compression of the tunnel. The additional convergence of the tunnel has been basically negative when the foundation pit is excavated in B and C areas. But the additional convergence of measuring points NO.1 and NO.4 vary near the averaged value and remain essentially unchanged. This shows that the excavation has a great impact on the tunnel within the foundation pit boundary, especially on the tunnel near the middle of the edge, but has little influence on the tunnel out of the foundation pit boundary.

4 Centrifugal Model Test

4.1 Apparatus

The geotechnical centrifuge in the key laboratory of geotechnical and underground engineering, at Tongji University, is a 150 gt machine with 3 m platform radius able to achieve accelerations up to 200 g. It has an automatic in-fight balancing system and a data acquisition system. A swinging model platform is located at the end of the centrifuge arm and is the base where the model container will be fixed. The soil and foundation pit model was installed on the model container (Fig. 5), 900 mm long, 700 mm wide, and 700 mm height. One side of the model container was equipped with transparent organic glass through which the deformation of the soil could be observed.

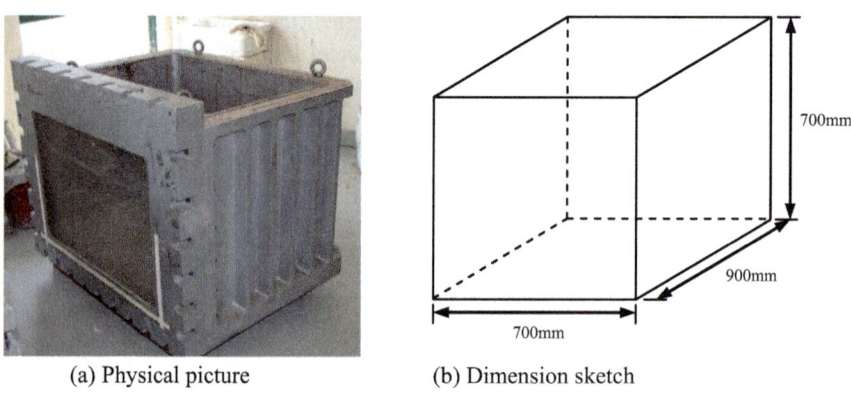

(a) Physical picture   (b) Dimension sketch

Fig. 5. Model container

4.2 Model

The centrifugal acceleration in this test is 100 g, which means the similarity ratio of this test is 1/100. There are two group of tests were conducted, one of which was in accordance with the excavation partitions and sequence of the Wufangyuan foundation pit. The second group is a parallel test with different excavation partitions and sequence of Wufangyuan foundation pit to study the impact on the tunnel of the excavation partitions and sequence, as shown in Fig. 6.

The soil used in the tests is a type of silty clay taken from $⑤_2$ layer of Wufangyuan foundation pit. The soil parameter is shown in Table 1. Before the test, the soil obtained from the site was dehydrated and crushed, configured the same moisture content as the field, and placed in a vacuum mixer, consolidated on the centrifugal under acceleration of 100 g to recovery its original state.

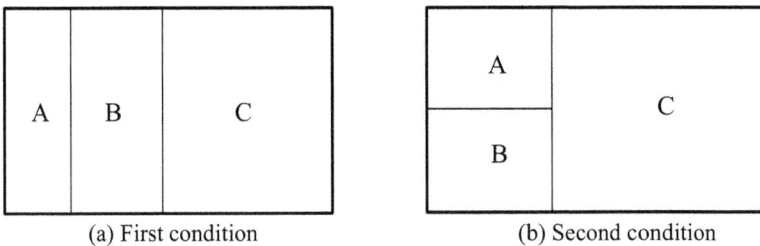

| (a) First condition | (b) Second condition |

Fig. 6. Excavation subarea

It is difficult to make a concrete foundation pit model and support system with a similar ratio of 1/100. Therefore, the aluminum alloy is used to simulate the retaining structure and support. The size of the model is calculated according to the similarity of the bending stiffness (retaining structure) and the compressive stiffness (support):

$$\left[\frac{Et^3}{12(1-v^2)}\right]_{m1} = \left[\frac{Et^3}{12(1-v^2)}\right]_{m2} \tag{1}$$

$$E_{m1}A_{m1} = E_{m2}A_{m2} \tag{2}$$

Where E is Elastic Modulus of retaining structure; t is thickness of diaphragm wall; v is Poisson ratio; m1 representative prototype material and m2 represents alternative materials.

The tunnel is simulated by PVC tube, and the size of the model is calculated by the similarity of longitudinal bending stiffness:

$$\eta\left[\frac{E\pi(D^4-d^4)}{64}\right]_{m1} = \left[\frac{E\pi(D^4-d^4)}{64}\right]_{m2} \tag{3}$$

where η is Longitudinal stiffness efficiency; E is Elastic Modulus of the tunnel; D is the outer diameter of the tunnel; d is the inner diameter of the tunnel; m1 representative prototype material and m2 represents alternative materials.

Four groups of strain gauges are pasted on the surface of the tunnel model to measure the deformation caused by excavation, as shown in Fig. 7.

4.3   Procedure

After completing the soil, the retaining structure and tunnel were installed and the model container was put into the centrifuge. The centrifuge runs for 5 min to simulate 35 days in actual operation after the foundation pit is excavated artificially. Table 3 summaries the main stages of the two tests and Fig. 8 shows the excavation process of the two tests.

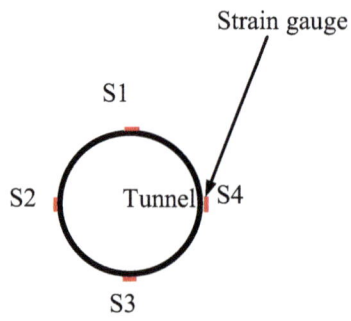

(a) Physical picture   (b) Distribution of strain gauge on tunnel surface

Fig. 7. Excavation subarea

Table 3. Test stages

Step	Construction operation	Time/min
1	First dewatering, excavate the first layer in A district, first support	5
2	Second dewatering, excavate the second layer in A district, second support	5
3	Third dewatering, excavate the third layer in A district	5
4	Excavate the first layer in B district, first support	5
5	Excavate the second layer in B district, second support	5
6	Excavate the third layer in B district	5
7	Excavate the first layer in C district, first support	5
8	Excavate the second layer in C district, second support	5
9	Excavate the third layer in C district	5

(a) First condition   (b) Second condition

Fig. 8. Excavation process

4.4   Results

The strain gauge data is recorded through the data acquisition system during the operation of the centrifuge. Figure 9 shows the additional strain of measuring points S1 and S2 on tunnel surface.

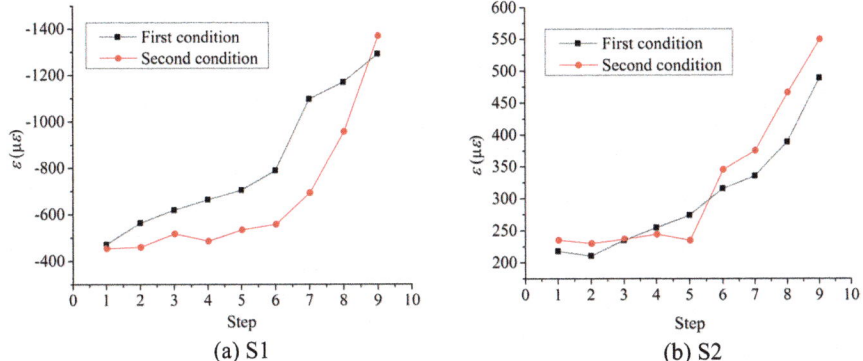

(a) S1   (b) S2

Fig. 9. Additional strain on tunnel surface induced by excavation

With the arrangement of strain gauge shown in Fig. 8, it is found that the excavation-induced downward and inward bending of the tunnel located at the side of the foundation pit, and this is consistent with the measured results shown in Fig. 3. The additional strain produced during the step 1–3 is very small, which means the excavation of district A in the far distance has little effect on the deformation of the tunnel. The additional strain of tunnel surface increases a certain extent during the excavation of district B and C. The blocked excavation can effectively control the tunnel deformation caused by excavation. The first condition can effectively restrain the deformation of the tunnel by the further partition of the districts C and add the diaphragm wall in the vertical direction, which shows that the excavation scheme adopted in the actual project is reasonable. Since the actual tunnel is joined by discontinuous segments, the dislocation between the segments should be noticed.

5   Conclusions

The field measured values of additional settlements and additional convergence induced by excavation were analyzed and two sets of centrifugal model tests were conducted to simulate the excavation near the subway tunnel. The following conclusions are drawn:

(1) The excavation of the soil in the areas closer to the tunnel induced greater additional settlements and additional convergence of the tunnel. The additional settlements and convergence value of the segment beyond the boundary of the foundation pit is much smaller than the settlement at the midpoint of the pit edge.

(2) The excavation induced downward and inward bending of the tunnel located at the side of the foundation pit. The blocked excavation can effectively control the tunnel deformation.

(3) Further partition of the district near the tunnel and using the diaphragm wall in the vertical direction can significantly reduce the influence of excavation on adjacent tunnels.

References

Wei, S., Liao, S., Zhu, Y., Li, X.: Parametric study on the effect of deep excavation on the adjacent metro station in Suzhou. In: International Conference on Geotechnical and Earthquake Engineering, pp. 223–230 (2013)

Zhang, J.F., Chen, J.J., Wang, J.H., Zhu, Y.F.: Prediction of tunnel displacement induced by adjacent excavation in soft soil. Tunn. Undergr. Sp. Technol. **36**(2), 24–33 (2013a)

Zhang, Z.G., Huang, M.S., Wang, W.D.: Evaluation of deformation response for adjacent tunnels due to soil unloading in excavation engineering. Tunn. Undergr. Space. Technol. **38**, 244–253 (2013b)

Dolezalova, M.: Tunnel complex unloaded by a deep excavation. Comput. Geotech. **28**(3), 469–493 (2001)

Sharma, J.S., Hefny, A.M., Zhao, J., Chan, C.W.: Effect of large excavation on deformation of adjacent MRT tunnels. Tunn. Undergr. Space. Technol. **16**(2), 93–98 (2001)

Ge, X.W.: Response of a shield-driven tunnel to deep excavations in soft clay. Doctor of Engineering thesis, Hong Kong University of Science and Technology, China (2002)

Hu, Z.F., Yue, Z.Q., Zhou, J., Tham, L.G.: Design and construction of a deep excavation in soft soils adjacent to the Shanghai metro tunnels. Can. Geotech. J. **40**(5), 933–948 (2003)

Zhang, Z.G., Zhang, X.D., Wang, W.D.: Numerical modeling analysis on deformation effect of metro tunnels due to adjacent excavation of foundation pit. J. Wuhan Univ. Technol. **29**(11), 93–97 (2007). (in Chinese)

Kog, Y.C.: Buried pipeline response to braced excavation movements. ASCE J. Perform. Constr. Facl. **24**(3), 235–241 (2010)

Wei, G., Li, G., Su, Q.: Analysis of the influence of foundation pit construction on an operating metro tunnel based on field measurement. Mod. Tunn. Technol. **51**(1), 179–185 (2014). (in Chinese)

Zheng, G., Wei, S., Peng, S., Diao, Y., Ng, C.W.: Centrifuge modeling of the influence of basement excavation on existing tunnels. In: Physical Modelling in Geotechnics, Two Volume Set (2010)

Evaluation of the Applicability of FEM to Simulate the Behaviour of Curtain Pile Walls in Tropical Soils

Andrea J. Alarcon[1(✉)], Renata P. Cunha[1], Juan C. Ruge[2],
and Sarah Jaccaz[3]

[1] University of Brasilia, Brasilia, Brazil
julianaalarcon2@hotmail.com
[2] Nueva Granada Military University of Colombia, Bogota, Colombia
[3] Blaise Pascal University, Clermont-Ferrand, France

Abstract. The design of retaining walls is increasingly present in engineering projects for urban areas, given their continuous development. In Brasilia and in the Federal District, many retaining walls are built in order to optimize space, such as in underground parking lots. These excavations need retaining works in order to maintain the terrain in place and to avoid any collapse. This paper has the aim of analyze numerically the behavior of "curtain piles" as retaining walls founded in tropical soils of the Federal District. The construction under study is characterized by a 13 m deep excavation, three levels of passive anchors, piles of 60 cm of diameter and 18 m long. It is founded in what is called "porous clay" of Brasilia and has instrumentation, which has yielded data displacements and stresses acting on the anchors to have the overall behaviour of the structure. Many laboratory tests were carried out to obtain the necessary soil parameters for the model. Plaxis software was selected as a numerical tool in order to simulate excavation steps, construction process, structure deformation and the presence of structural elements as piles and nails. The numerical model was developed and compared with the instrumentation data showing that the Mohr-Coulomb model used did not allow a reasonable prediction of the study case structure behavior.

Keywords: Retaining structure · Soil nailing · Numerical simulation

1 Introduction

Brazil, where the majority of the territory is situated close to the tropics, has about 75% of its area covered by lateritic soils. This type of soils has a high porosity close to 55% and is unsaturated. Meteorological conditions play an important role in their behaviour. These conditions lead to a high variation of the water content in the soil and to variations of the mechanical characteristics, which can make a structure sensitive to collapse [1]. Nowadays, in this type of soils, the containment works are more often in engineering projects like underground parking lot due to the growing occupation of urban areas.

Excavations are usually led vertically and reinforcement is needed to keep it equilibrium in the new setting. The retaining structures of the cantilever type called

© Springer Nature Singapore Pte Ltd. 2018
T. Qiu et al. (Eds.): GSIC 2018, *Proceedings of GeoShanghai 2018 International Conference: Advances in Soil Dynamics and Foundation Engineering*, pp. 439–446, 2018.
https://doi.org/10.1007/978-981-13-0131-5_48

"curtain pile" wall is used when the excavation level does not exceed 5 m. Exceeding this limit, the length of the embedded pile, to guarantee the safety factor required, becomes longer and consequently not economic. Thus, it is necessary to use reinforcement with structural elements. Passive or passive plus active anchors in the retaining structures is a technique commonly used in big cities. In general, the retaining structure calculation aims first to the estimation of it stability. For the displacements, only the methods of the finite element method and the coefficient of reaction allow estimation [2].

The objective of this research is to simulate and analyze numerically, in order to compare with instrumentation data, a retaining structure of the "curtain pile wall" type reinforced with nails founded in the porous soil of the Federal District studied in a previous research work [3]. The research was conducted by using 2D Finite Element Method (FEM) analysis with the software Plaxis 2D.

2   Study Case

The construction site used in this research is located in the SHN Federal District of Brazil called Commercial Plaza and corresponds to Block 1 project - Building Fusion as shown in Fig. 1. The retaining structure considered consists of a cantilevered pile curtain wall of 18 m length, 60 cm in diameter and 5 m embedment length reinforced with three lines of passive anchors or nails (Fig. 2).

Fig. 1. Panoramic picture of the section instrumented [3]

2.1   Soil Tests

Aforementioned, the work was based on a previous research work where were performed laboratory tests in order to characterize the soil [2]. It was thus used some of the data obtained from these tests to determinate the different parameters necessary for the modeling. A triaxial test in saturated and drained conditions has been run in Charles University (Prague, Czech Republic) at 6 and 9 m at three different confinement pressures, respectively 80, 200 and 400 kPa; and 120, 200 and 400 kPa. Also was led a triaxial test in undrained conditions here in the laboratory of the University of Brasilia (UnB). From the results, it was obtained the effective cohesion (C') and friction angle (Φ).

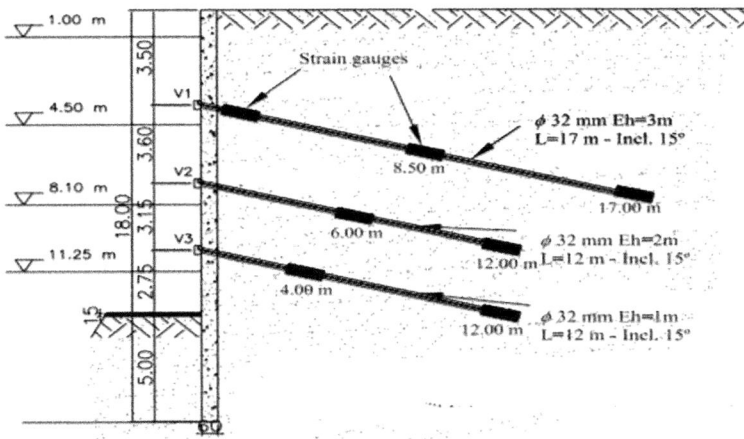

Fig. 2. Section 1. Instrumented retaining structure piles/anchors Ø = 0.60 m. [3]

Another step of characterization of the porous clay of Brasilia was to perform a direct shear stress test in the laboratory of soil mechanics of the UnB. In this case, the tests were done with a normal stress of 100, 200 and 400 kPa at a depth of 3, 6 and 9 m. With a SPT test performed at the site location, the soil modulus tangent and secant (E_i, E_{50}) and effective friction angle (Φ') can be determined with correlations proposed by Liao and Whitman [4], Skempton [5], Hatanaka and Uchida [6] and Bowles [7]. As well, it can be identified three distinctive layers: from 0 to 6 m; from 6 to 13 m and then from 13 to 25 m. In order to obtain closer parameters, at least for design purposes, it was decided to do an average of all the lab tests parameters for each depth. The results are in the Table 1 below :

Table 1. Final parameters obtained from the average of all the test results

Z (m)	C' (kPa)	Φ' (°)	E_i (kPa)	E_{50} (kPa)
3	17	29	5329	2270
6	18	28	2828	2264
9	21	30	4703	4764
13	21	30	4798	4833
18	21	30	4798	4833

2.2   Instrumentation Results

Instrumentation was installed with an inclinometer servo accelerometer probe located one meter in the upper surface of the containment, with the aim of monitoring the displacements of the structure during the phases of excavation. The maximum displacement measured at the top of the pile and after the last excavation is around

20.70 mm. Two weeks after this last excavation the displacement is stabilized at 24 mm of horizontal displacement.

3 FEM Modeling

The use of computational tools is becoming more frequent due to the excellent results obtained in studies conducted in recent decades. To study anchored containment structures and nailed it becomes necessary to use a tool that includes: analysis of layered construction to simulate the construction process, displacement limit of the structures, presence of structural elements such as piles and nails (or passive anchors). The finite element method and finite difference are the most used in solving mechanical equations involving a continuous medium. The finite element method is based on a discretization of the medium under study [8]. The software Plaxis 2D was selected for the modeling. It is one of the most efficient program that applies the Finite Element Method (FEM) with the purpose of analyze deformation and stability in geotechnical engineering problems [9].

3.1 Properties of Piles and Nails

Hence, here are the parameters of the piles and nails used in the computational model (Table 2):

Table 2. Parameters of piles and nails in the simulation

	Parameters	Value	Unit
Pile properties (Ø = 0.60 m)			
Material	Concrete	Elastic	–
Modulus of axial rigidity	EA	7.069×10^6	
Modulus of rigidity in flexion	EI	1.59×10^5	[kN/-m^2/m]
Weight	W	4.98	[kN/m/m]
Poisson's ratio	ν	0.2	–
Nails properties			
Material	Grout-Nail	Elastic	–
Modulus of axial rigidity: nail 1	EA	8.2×10^4	[kN/m]
Modulus of axial rigidity: nail 2	EA	1.2×10^5	[kN/m]
Modulus of axial rigidity: nail 3	EA	2.5×10^5	[kN/m]

3.2 Soil Properties for the Mohr Coulomb Model

In order to simulate the study case with the model of Mohr-Coulomb, it was required to obtain the 5 parameters necessary for the computation: E, ν, φ, c, ψ. As it was available the laboratory test at 6 and 9 m, and according to the SPT it was decided to divide the soil modeled into three layers: from 0 to 6 m where the parameters seems to be constant, from 6 to 12 m and from 12 to the end. All the parameters have been chosen from the average laboratory tests results that have been done above. All these information are summarized in the Fig. 3.

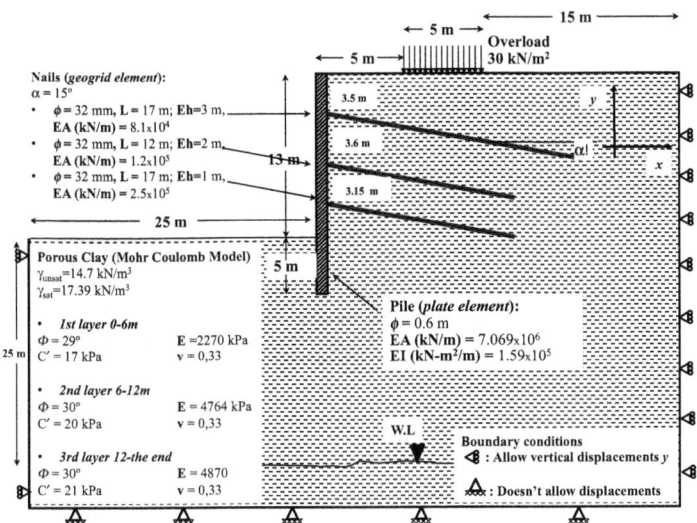

Fig. 3. Computation of the Mohr-Coulomb model.

4 Results and Discussions

The Fig. 4 presents the deformations given by the Mohr-Coulomb model after the last excavation. We can see that the results do not seem to be realistic. Indeed, we can observe that the pile displacements are the greatest at the bottom of it and not at the top of it, like it was expected, as presented in the Fig. 5.

Plaxis tutorial explains that the Mohr-Coulomb model is giving "unrealistic deformations", and gives some limitations of the model: overestimation over bottom heave, often heave of soil behind the wall and occasionally excavation widens spontaneously (even without anchors). That is what happened in our case, it can be seen a clear heave of soil behind the wall and an expansion of the excavation.

This is due in part perhaps to the fact that this model is very conservative, i.e., the linear elastic perfectly plastic Mohr-Coulomb model is a first order model that includes only limited number of features that soil behavior shows in reality, and doesn't follow the same way of stress-strain deformations as the real soil, as one can see in the Fig. 6.

The geotechnical problems must be analyzed according to the type of deformations that occur, however, a retaining structure is a complex geotechnical structure to simulate since various points of the structure behave differently on the stress-strain. In Fig. 6, the point A of the 'real' behavior of a soil mass plotted on the stress-strain (excavation in active condition), represents a place where the soil is still far from

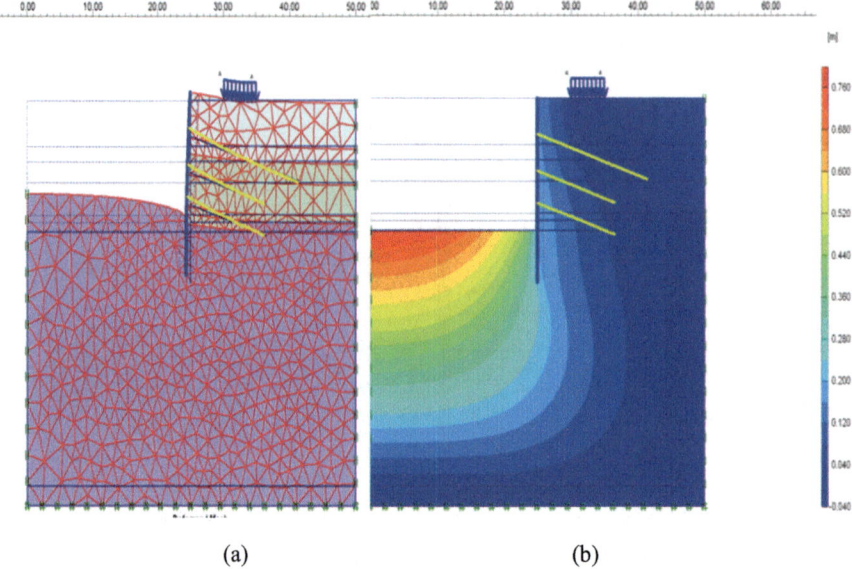

(a)   (b)

Fig. 4. Deformed mesh, extreme total displacement 766.86 * 10^{-3}m; (b) total displacement (Utot), extreme Utot 766.86 * 10^{-3}m. MC model, (Plaxis)

breaking; at point B the soil is close to the break, and the point C on softening condition [3].

The curve that reproduces the response of the MC model can hardly simulate "real" soil behaviour at these points. This by both the choice of the constitutive model, which is not only an issue of simplicity and rapidity in geotechnical design but is also subject to good judgment and experience, in order to opt for c' and φ' parameters that adequately represent the range of deformations in a prudent manner. Unfortunately, most of the time it ends in oversized geotechnical structures. For this reason, the simplest model requires an experienced designer to select appropriate value for the MC model parameters of the problem handled. In contrast, more complex models that are calibrated "appropriately" with laboratory tests and show real parameters for each type of soil response, can better represent the behaviour of the soil with respect to the range of deformations in different places of the geotechnical structures, such as happens to the retaining structures.

5   Conclusions

In terms of the numerical predictions with finite element model, it has been shown, that the Mohr-Coulomb model didn't allow a reasonable prediction of the wall behavior. Indeed, the results found were not matching the experimental data and are certainly due to the fact that the model may not recommended by the Plaxis tutorial to be used for excavations cases.

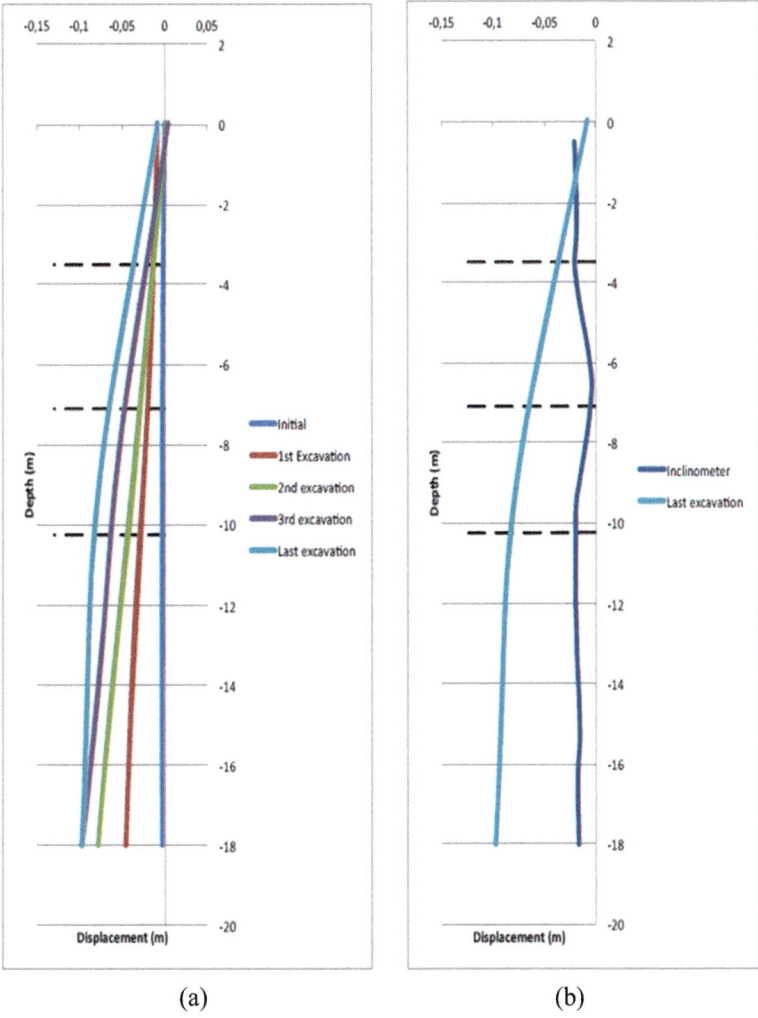

(a)   (b)

Fig. 5. (a) Displacements at different stage of excavation (MC model); (b) Comparison between numerical/experimental results of the last excavation.

It is difficult to get the real parameters of a soil and to be certain of the reliability of the tests, as long as they are not performed in the same place, in the same depth and in the same laboratory. Soil has an uncertain variable and site dependent behaviour, and can be different from one place to another one even if is close together.

Simulating using the finite element numerical code from the "Plaxis" package, and adopting the Mohr-Coulomb as rupture soil model, it was not feasible the representation of the experimental behavior of the pile curtain. Indeed, the model considers an elastic perfectly plastic curve (see Fig. 6); however as an exercise into this work, it was useful to learn and calibrate this numerical program.

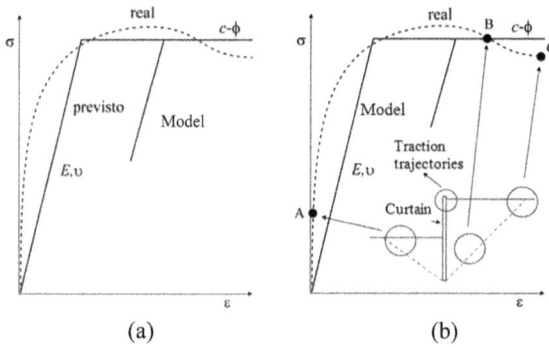

Fig. 6. (a) Stress-strain curve of the typical behavior of the MC model; (b) Stress-strain points on a typical soil behavior curve [1]

Besides, this difficulty illustrates the importance of a final civil engineering project as this one, where a real case design project and simulation is taken place with all available alternatives of parameters and data in laboratory, field and theoretical literature methods. Most probably, in practice, the designer will not have time to go through all the simulations carried out in the present work, and, moreover, will not go through the gain of experience and enrichment provided by the "sensibility" of the techniques that a work, like this one, produces in its author.

References

1. Nogami, J.S., Villibor, D.F.: Peculiarities of lateritic soils in low volume traffic paving. In: Proceedings of Ist International Symposium on Low Volume Roads, Rio de Janeiro, Brazil, pp. 542–560 (1997). (in Portuguese)
2. Marten, S., Delattre, L., Magnan, J.P.: Experimental and methodological study on the behavior of retaining screens, 300p. Laboratoire Central des Ponts et Chaussées (2005)
3. Ruge Cardenas, J.C.: Analysis of urban excavations supported by means of juxtaposed pile-type structures based on soils using a hypoplastic constitutive model (Dissertação de Doutorado em Geotecnia. Universidade de Brasília, Brasília (2014). (in Portuguese)
4. Liao, S.S.C., Whitman, R.V.: Overburden correction factors for SPT in sand. J. Geotech. Eng. ASCE **112**(3), 373–377 (1986)
5. Skempton, A.W.: Standard penetration test procedures and the effects in sands of overburden pressure, relative density, particle size, aging and over-consolidation. Geotechnique **36**(3), 425–447 (1986)
6. Hatanaka, M., Uchida, A.: Empirical correlation between penetration resistance and internal friction angle of sandy soils. Soil Found. **36**(4), 1–10 (1996)
7. Bowles, J.: Foundation Analysis and Design, 5th edn, 1133p. The McGraw-Hill Companies, Inc., New York (1997). Chapters 11, 13, 14
8. Briançon, L., Haza-Rosier, E., Thorel, L., Damiel, D., Combarieu, O.: Recommendations for design, construction and control of rigid inclusion ground improvements. IREX's Soil Specialist Cluster, 317p. (2011)
9. PLAXIS 2D Software: Tutorials (2011)

Effect of Pre-excavation Construction on Surrounding Historic Building Based on Field Measurements in Shanghai Soft Ground

Hang Li[1(✉)], Shao-ming Liao[1], Yong-jing Tang[1],
and Mingliang Shen[2]

[1] Department of Geotechnical Engineering College of Civil Engineering,
Tongji University, Shanghai, China
zgnydxlh@163.com
[2] Shanghai Construction Group Co., Ltd., Shanghai, China

Abstract. The settlement control of the building adjacent to excavation is always the important and critical issue in the excavation engineering, and the ratio of the settlement caused by the construction of pre-excavation can reach more than half of the total settlement of the excavation construction. This paper took an excavation engineering in Shanghai as an example, based on the monitoring data of the retaining structure construction stage, the effect of each stage, including bored piles construction, compaction grouting, soil treatment and diaphragm wall construction, on the historic building settlement adjacent to excavation was investigated, and some conclusions can be drawn: the construction of pre-excavation led to substantial settlement of the historic building that reached about 28 mm, and caused some cracks to exterior wall of the historic building; the settlement caused by the soil treatment and the diaphragm wall construction accounted for 30% to 50% and 40% to 60% of the total settlement in pre-excavation construction respectively. This project served as a special case study about the protection of adjacent historic building in group excavation which could also provide a reference for similar engineering in Shanghai.

Keywords: Pre-excavation construction · Historic building
Protection measures · Settlement

1 Introduction

Diaphragm walls are often used as temporary and permanent retaining structure for deep excavation in central urban areas. The construction of diaphragm wall will cause the stress relief of soil which results in the ground movements. The effects of diaphragm wall construction on the ground movements have been investigated by a number of researchers, such as Clough and O'Rourke (1990), Poh et al. (1998, 2001), Thorley and Forth (2002), Ding and Wang (2008), Chen et al. (2013). The aspects of their studies mainly included the effects of diaphragm wall construction on the ground

T. Qiu et al. (Eds.): GSIC 2018, *Proceedings of GeoShanghai 2018 International Conference: Advances in Soil Dynamics and Foundation Engineering*, pp. 447–454, 2018.
https://doi.org/10.1007/978-981-13-0131-5_49

movements based on the field test and changes in lateral earth and water pressure using numerical modeling method. In recent years, the effect of diaphragm wall construction on the performance of adjacent building have received extensive attention. Combined with the measured data accumulated in the diaphragm wall construction in Hongkong, Thorley and Forth (2002) argued that the construction activities before the excavation have obvious influence on the settlement of the adjacent building with pile foundation. According to the field data, Liu and Xie (2014) reported that the settlement of the adjacent building caused by the construction stage before the excavation could reach half of the total settlement of the building in the whole construction process, even exceeding the settlement at the excavation stage.

In Shanghai, for some sensitive buildings next to deep excavation, such as historic persevered buildings which generally adopted shallow foundation and has a weak structural integrity, the pre-excavation activities may lead to substantial settlement of the adjacent buildings. In this paper, a case of a deep excavation in Shanghai was presented. The settlement of the adjacent historic building caused by the pre-excavation activities including construction of bored piles, grouting reinforcement, ground treatment and diaphragm wall construction was monitored. From the monitored results, the effect of each construction mentioned above on the historic building was investigated respectively according to the construction sequence.

2 Project Background

2.1 The Position Relation of the Foundation Pit and the Historic Building

This excavation engineering project is located in Huangpu district in Shanghai. Figure 1 shows the plan of construction site, the dimension is about 495 m long and 225 m width. There are nine blocks that represents nine foundation pits, specifically including four smaller ones (i.e. G2, J2, A2, and F2) located on the east, west and north side of the historic building and the others five larger foundation pits (i.e. A1, F1, J, T1, and GL). The depth of the diaphragm wall ranges from 35 m to 42 m, and in order to

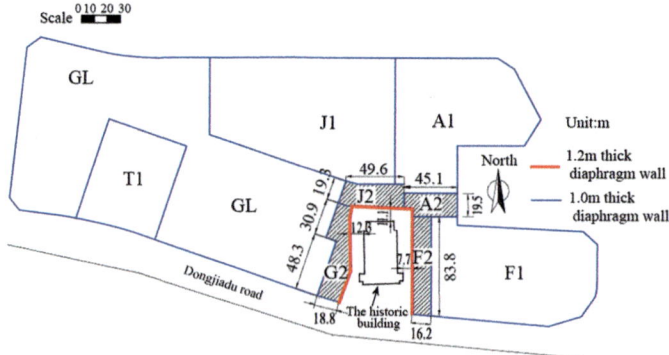

Fig. 1. The plan of the excavation engineering and location relation between the historic building and four smaller foundation pits

reduce the effect of excavation on the historic building the thickness of the diaphragm wall around it was designed to 1.2 m and other diaphragm walls was 1 m thick. The distance between the historic building and foundation pit ranges from 7.7 m to 12.8 m.

The historic building adjacent to the excavation group is a Catholic Church built in 1853, as was shown in Fig. 2 and it was listed as historical and cultural sites of Shanghai in the light of its high artistic and historical value. The church is rectangular in shape, with 51.8 m in length and 39.3 m in width. Its superstructure is brick-wood that located on the strip footing and square spread footing, as shown in Fig. 3. According to the character of the church and preliminary examination, it could be concluded that the building structure is weak and has a poor integrity, and is sensitive to adjacent construction, which should be given more attention.

 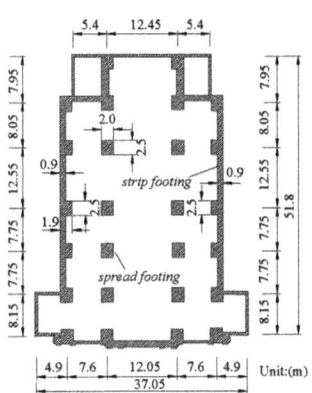

Fig. 2. Elevation of the church   **Fig. 3.** The floor plan of the foundation

2.2 Ground Conditions

The excavation site reported in this paper is near the Huangpu river. The ground water table is about 0.7 m to 2.2 m below the ground. Figure 4 shows the succession of soil layers and the variations of some geotechnical properties at the site.

The site is underlain by thick clay silt and very soft to medium clay. The upper clay silt has a lower water content but a higher shear strength (φ_u) than the marine deposits (i.e. very soft clay and soft clay) under it. The marine clay with high water content has a high compressibility coefficient and very low shear strength. In the construction of trenching, the maximum lateral soil movement will probably occurred in this soil layer as Poh et al. (2001) reported, and excessive deformation of soil may lead to the collapse of the trench, which further cause settlement of the adjacent building.

2.3 Protection and Control Measures

The protection and control measures taken around the church was shown in Fig. 5. There are four barriers to reduce the effect of excavation and other construction activities on the historic building. First of all, there were 216 bored piles to be applied

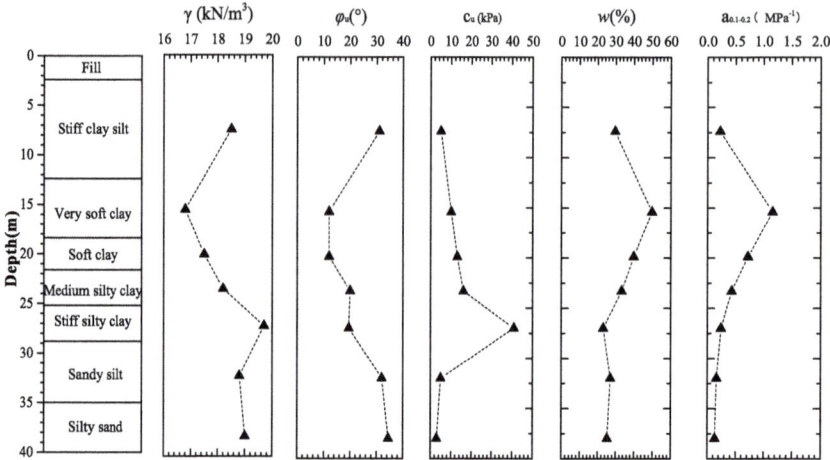

Fig. 4. Soil profile and geotechnical parameters

as partition wall between the building and foundation pits, and the nearest distance between them was only 2.1 m. Then compaction grouting between the piles was carried out to prevent the water and soil loss. After that 55 m trench with 800 mm thickness cutting re-mixing deep wall, which were used as waterproof curtain as well as groove wall reinforcement, was executed on the outer side of diaphragm wall around

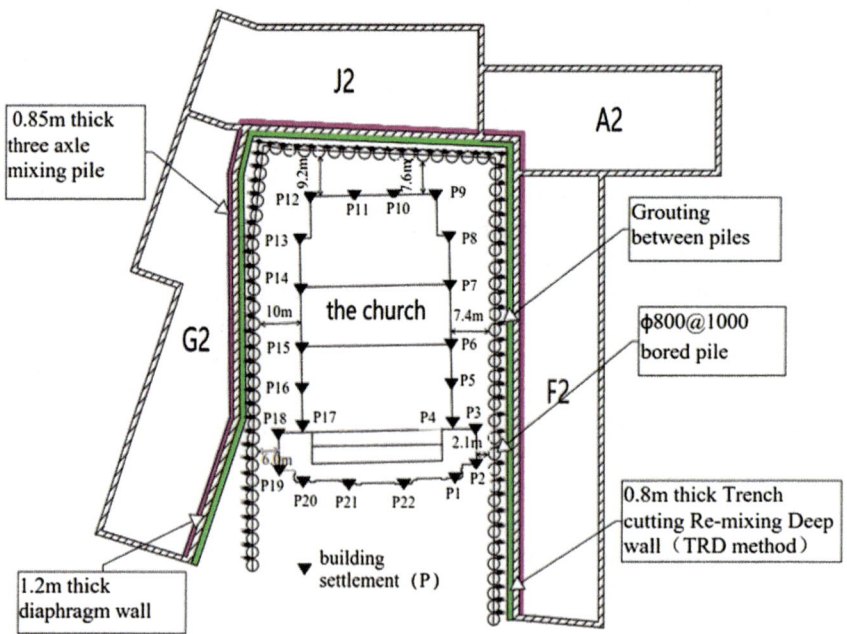

Fig. 5. Protection measures taken around the church

the church. Inner side of diaphragm wall were reinforcement as well. Finally, the diaphragm wall was constructed under the protection of these measures. Although these reinforcement measures were aimed at reducing the effect of foundation pit construction on the church, but the construction stage of retaining structure itself (as mentioned above) would make certain influence on it.

Figure 5 also showed the monitoring points layout, which were arranged on the exterior wall of the building, and these 22 monitoring points were used to record the settlement of the building during the construction.

3 Monitoring Data Analysis

Figure 6 showed the settlement of the building during the pre-excavation construction, and the settlement of each measuring point was between 12 mm and 28 mm. According to the construction consequence, it can be divided into four stages, the stage 1 was the construction of bored pile around the church and compaction grouting between piles. At this stage the settlement appeared to fluctuate with a certain range due to the squeezing effect produced by the construction of bored pile and grouting. There were a few monitoring points were uplifted and the settlement of most points were generally within 2 mm.

The stage 2 showed that settlement of each measuring point has risen obviously and ranged from 5 mm to 10 mm. Besides, the larger settlement occurred on the east wall of the building which is more closed to the diaphragm wall, as was shown in Fig. 6 and Table 1. It indicated that the historic building was probably sensitive to the construction of the second stage in the case.

The stage 3 was the construction of diaphragm wall that mainly included three steps: the construction of the guide wall, the trenching of the diaphragm wall and the concrete placement. In general, the construction of guide wall has a slight effect on the surrounding environment, and the most part of ground movements in this stage mainly occurred in the trenching process. And with the completing of concreting, the lateral movement reduced greatly, but soil settlement was influenced slightly as Poh et al. (2001) observed. Hence it could be assumed that the settlement of the historic building was caused by the construction of trench. In this project, it can be seen that the settlement of each measurement point in this stage were significant, but the settlement rate of this stage was smaller than the previous stage under the protection of the completed reinforcement measures. The data of the four monitoring points on the north wall were picked up from the date of November 1st to November 10th shown in the enlarged plot of settlement time, which showed that the settlement of the four monitoring points increased sharply by about 10 mm in five days and then remained stable at this level for several days. According to the construction records, four diaphragm wall panels on the north side of the historic building were under construction during the period, and the superimposed effect of the ground movement resulted in substantial settlement of the building. However, other monitoring points were not the case, and it was suggested that the construction quality of the diaphragm wall was an important factor in the settlement control of the adjacent historic building with shallow foundation in soft soil area. With the completion of the retaining structure around the historic

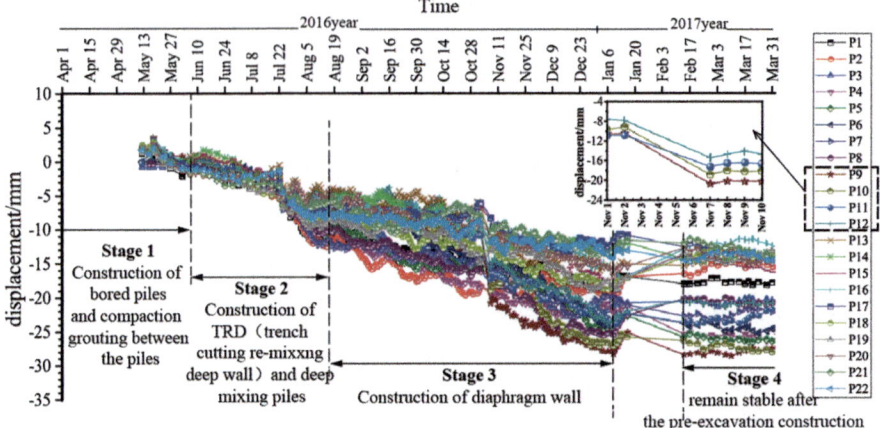

Fig. 6. Plot of settlement time during the pre-excavation construction

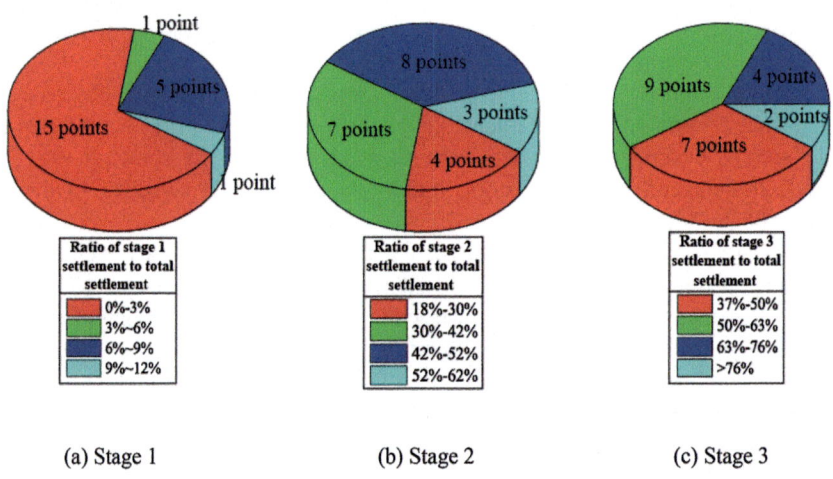

(a) Stage 1   (b) Stage 2   (c) Stage 3

Fig. 7. The ratio of each stage settlement to the total settlement in pre-excavation construction

building, the settlement of the church remains stable in a certain range as was shown in stage 4.

The monitoring data in Table 1 showed the settlement produced in each stage and its total settlement. Figure 8 which gained from the statistical analysis based on the Table 1 illustrated that the effect of each stage on the settlement of the historic building was different. For stage 1, settlement of 15 measuring points were within 3% of the total settlement. In general, settlement of and stage 1 accounted for the proportion of total settlement was less than 10%, and most settlement was caused by the soil treatment in stage 2 and construction of diaphragm wall in stage 3. The ratio of the settlement during these two stages to total settlement were respectively 30% to 50%

Table 1. Building settlement during the pre-excavation construction

Monitoring point number	Settlement at different stage/mm			Total settlement/mm
	Stage1	Stage2	Stage3	
P1	−2.1	−9.1	−7.8	−19.0
P2	−1.3	−10.4	−7.3	−19.0
P3	−0.6	−8.1	−13.3	−22.0
P4	0	−10.6	−10.7	−21.3
P5	−0.1	−10.4	−13.6	−24.1
P6	−0.1	−8.5	−15.4	−24.0
P7	0	−12.3	−8.0	−20.3
P8	−0.3	−11.4	−13.7	−25.4
P9	0.8	−8.5	−20.1	−27.8
P10	0.2	−6.4	−20.3	−26.5
P11	−0.3	−5.6	−16.7	−22.6
P12	−0.5	−4.4	−17.9	−22.8
P13	−0.1	−4.0	−11.5	−15.6
P14	0.2	−6.1	−6.9	−12.8
P15	−0.5	−5.9	−10.8	−17.2
P16	0.1	−7.0	−8.4	−15.3
P17	−0.2	−7.2	−4.4	−11.8
P18	−1.1	−7.6	−10.0	−18.7
P19	−1.3	−7.9	−8.9	−18.1
P20	−1.1	−5.2	−9.0	−15.3
P21	−1.0	−6.6	−5.7	−13.3
P22	−1.0	−7.2	−5.0	−13.2

and 40% to 60%. In the meantime, due to differential settlement of the historic building caused by the construction activities, the east wall and the south wall appeared cracks and the width of the crack was evolving during the stage 2 and stage 3.

4 Conclusion

Based on the monitoring data and analysis, some conclusions can be drawn for the project and may provide a reference for other similar projects:

(1) In the pre-excavation construction, soil reinforcement surrounding the historic buildings may lead to significant building settlement that accounted for 30% to 50% of the total settlement, but the diaphragm wall construction contributed to the most part and its construction quality was critical factor in pre-excavation construction.

(2) The settlement of the historic building reached about 28 mm when pre-excavation construction was completed, and larger uneven settlement at the East and North

wall caused some damage to the historic building. Considering the subsequent construction, the control measures applied to the historic building were not enough from the perspective of the bearing capacity of historic building.

(3) The effect of the construction of each stage on the historic building was evaluated, from which some useful experience could be gained. In fact, the protection measures taken in this case were effective in control of the settlement of the adjacent historic building. However, the settlement control of the historic building is difficult and it's hard to evaluate the damage of the building precisely and the mechanism of building response in excavation group required further study.

Acknowledgements. The writers would like to thank The National Natural Science Foundation of China provided financial support for this project (NSFC Grant No. 51278359). The writer are also grateful to the Shanghai Construction Group Co., Ltd. and following persons of the company for providing related information in this study: X. P. Wu and M. L. Shen of Shanghai Construction Group Co., Ltd.

References

Clough, G.W., O'Rourke, T.D.: Construction induced movements of in-situ wall. Geotech. Spec. Publ. **1**(25), 439–470 (1990)

Poh, T.Y., Wong, I.H.: Effects of construction of diaphragm wall panels on adjacent ground: field trial. J. Geotech. Geoenviron. Eng. **124**(8), 749–756 (1998)

Poh, T.Y., Goh, A.T.C., Wong, I.H.: Ground movements associated with wall construction: case histories. J. Geotech. Geoenviron. Eng. **127**(12), 1061–1069 (2001)

Thorley, C.B.B., Forth, R.A.: Settlement due to diaphragm wall construction in reclaimed land in Hong Kong. J. Geotech. Geoenviron. Eng. **128**(6), 473–478 (2002)

Ding, Y.C., Wang, J.H.: Numerical modeling of ground response during diaphragm wall construction. J. Shanghai Jiaotong Univ. **13**(4), 385–390 (2008)

Chen, J.J., Lei, H., Wang, J.H.: Numerical analysis of the installation effect of diaphragm walls in saturated soft clay. Acta Geotech. **9**(6), 981–991 (2013)

Liu, F.Z., Xie, X.Y.: Effect of trench construction of retaining structure in metro foundation pit on adjacent building settlement and monitoring data analysis. Chin. J. Rock Mech. Eng. **33**(s1), 2901–2907 (2014)

Countermeasures to Retaining Wall Deflection Induced by Pre-excavation Dewatering

Chao-Feng Zeng[1,2,3](✉), Zhi-Cheng Yuan[1], Xiu-Li Xue[1],
Gang Zheng[2], and Guo-Xiong Mei[3]

[1] Hunan Provincial Key Laboratory of Geotechnical Engineering for Stability
Control and Health Monitoring, Hunan University of Science and Technology,
Xiangtan 411201, Hunan, China
cfzeng@hnust.edu.cn
[2] Key Laboratory of Coast Civil Structural Safety of the Ministry of Education,
Tianjin University, Tianjin 300072, China
[3] Key Laboratory of Disaster Prevention and Structural Safety of Ministry
of Education, Guangxi University, Nanning 530004, Guangxi, China

Abstract. Pre-excavation dewatering is an important stage in deep excavation
in soft ground with high groundwater level. The deformations of retaining
structure and surrounding environment induced by pre-excavation dewatering
can reach centimeter level. However, the methods to control the pit deformation
induced by pre-excavation dewatering have not been yet proposed by the traditional pit design theory and relevant research. In this paper, some countermeasures, such as installing struts before dewatering and segmental dewatering
(i.e., letting dewatering wells operate segmentally rather than simultaneously)
are proposed. The effects of these methods on the retaining wall behaviors are
investigated based on in situ tests and numerical calculations. As to installing
struts before dewatering, because the first level of struts is installed at wall top
before dewatering, the deflections at wall top can be reduced apparently, but
great wall deflections still arise at the deep location. This method is not suitable
for the condition that deep-buried underground structures exist around excavation. The segmental dewatering can be more effective in controlling wall
deflection in the segment where dewatering wells operate at last. The segmental
dewatering can be reasonably arranged according to the location of protected
environment around excavation. It is suggested that the foundation pit segment
far away from the protected environment be firstly pumped, and the pit segment
close to the protected environment be pumped at last.

Keywords: Deep excavation · Environmental impact
Pre-excavation dewatering · Deformation control

1 Introduction

The whole processes of excavation can cause environment deformation. As to deep
excavation in soft ground with high groundwater level, as shown in Fig. 1, the whole
construction processes of excavation mainly consist of 7 stages. The test pumping (also
called as pre-excavation dewatering) is a necessary and important construction activity

© Springer Nature Singapore Pte Ltd. 2018
T. Qiu et al. (Eds.): GSIC 2018, *Proceedings of GeoShanghai 2018 International Conference:
Advances in Soil Dynamics and Foundation Engineering*, pp. 455–463, 2018.
https://doi.org/10.1007/978-981-13-0131-5_50

before excavation to check the completion quality of dewatering wells [1–6]. Typically, the discharge flow rate of dewatering wells and the drawdown in dewatering wells need to be tested and should meet the design requirement before excavation [1–4]. The results of pre-excavation dewatering (PED) can provide reference to engineers when they determine the dewatering scheme for official construction. Considering that the retaining wall already existed before PED, the retaining wall is expected to be affected by PED and appear deformation response.

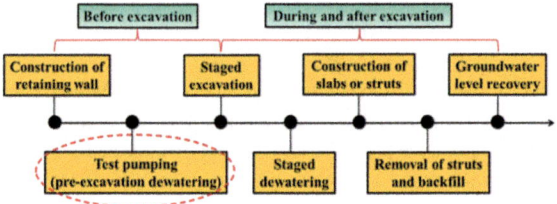

Fig. 1. Deformation induced by the whole process of excavation

Based on field measurements, Zheng et al. [5] and Zeng et al. [6] found that the deformation of retaining structure and surrounding environment induced by PED can reach centimeter level. Zeng et al. [7, 8] found that the excavation width, dewatering depth and the soil permeability have great influence on the deformation caused by PED, and the responses of retaining wall totally differ under different parameters. However, the current codes for excavation [9, 10] and most relevant research are focused on the environmental impact during and after excavation. The deformation caused by PED has not been widely concerned, and the methods to control deformation during PED has not been systematically studied.

In this paper, some countermeasures, such as installing struts before dewatering and segmental dewatering are proposed. The effects of these methods on the wall behaviors are investigated based on in situ tests and numerical calculations. On this basis, some suggestions about the best use of each countermeasure to reduce the wall deflection induced by PED are proposed for reference to designers.

2　In Situ Dewatering Tests: Installing Struts Before PED

Typically, the first level of struts is installed after PED and even after excavation to 1–2 m below the ground surface (BGS) for some projects [11–13]. As discussed above, the retaining wall will inevitably develop considerable deflection during PED if this traditional construction sequence is adopted. In this section, an in situ dewatering test considering installation of struts before PED is introduced, and the effect of this method on retaining wall behavior during PED is studied.

Figure 2 presents the plan view of the investigated excavation for the Beizhan Station of Metro Line 3 in Tianjin, China. This metro excavation mainly consisted of two sections (i.e., Excavation A and Excavation B). The excavation depth and width are shown in Fig. 2. The retaining structure was a reinforced concrete diaphragm wall that was 0.8–1.0 m thick (1.0 m for the end shafts and 0.8 m for the remainder) and 33–42 m deep (42 m for the interchange shaft and 33 m for the remainder). The allowable wall deflection for this project is 0.14% of the final excavation depth (i.e., about 27 mm).

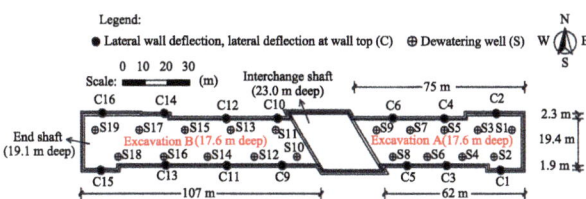

Fig. 2. Site plan and instrumentation layout

According to the geotechnical investigation report, there are one phreatic aquifer and three confined aquifers within the depth of 50 m BGS. The corresponding soil type and depth for the aquifers and aquitards are presented in Table 1. The long-term phreatic water level was observed at depths of 0.8–2.9 m BGS. The hydrostatic equilibrium for the three confined aquifers were reached at 3.26, 4.12 and 5.0 m BGS, respectively from top down.

As shown in Fig. 2, there were a total of 9 and 10 wells in Excavations A (S1–S9) and B (S10–S19), respectively. The wells screen was arranged from 2 m BGS to approximately 3 m below the final excavation depth. The deflections of the diaphragm walls were monitored by 14 inclinometer tubes. After the diaphragm walls and dewatering wells were installed on site, PED tests were carried out in Excavation A and B (labeled as T1 and T2, respectively) to check the completion quality of the wells. T1 lasted 10 days, during which all of the dewatering wells in Excavations A were operating and the pumps were submerged at different depth (i.e., 4 m, 8 m, 12 m and 16 m BGS) in four stages. Each stage lasted 2.5 days. Figure 3 presents the development of wall deflections during T1. Considerable inward wall deflections occurred. The maximum wall deflection was approximately 10 mm.

In order to reduce the wall deflection in Excavation B during T2, the first level of struts, with an interval of 3 m, was installed before T2, and segmental dewatering is adopted during T2. As shown in Fig. 4, the Excavation B was divided into three segments, and the dewatering wells operated successively from Segment 1 to 3. The dewatering time in each segment was 2 days. The dewatering depth for all dewatering wells average 17 m BGS. Figure 5 shows the comparison of wall deflections in T1 and T2. The deflections at wall top were reduced apparently, but great wall deflections still arise at the deep location, and the maximum wall deflection can reach 5 mm. Anyway, the application of installing struts before PED is useful to reduce the wall deflection development induced by PED. The reduction effect can reach 84% and 56% as to the deflection at wall top and the maximum wall deflection, respectively.

Table 1. Mechanical parameters of soil layers and soil distribution in model.

Soil classification	Hydrogeology	Depth (m, BGS)	Unit weigh, γ (kN/m³)	K_0	K_H (m/d)	K_V (m/d)	e_0	λ	κ	M
Silty clays	Phreatic aquifer	0.0–5.5	19.35	0.49	0.1	0.1	0.811	0.055 3	0.006 5	0.979
Clayey silts	Phreatic aquifer	5.5–11.0	19.30	0.43	0.5	0.5	0.792	0.031 2	0.003 6	1.192
Silty clays	Aquitard	11.0–19.0	20.10	0.50	5.0×10^{-4}	1.0×10^{-4}	0.696	0.044 5	0.005 2	0.979
Sandy silts	Confined aquifer	19.0–24.0	20.15	0.42	1.0	1.0	0.640	0.029 3	0.003 4	1.202
Clays	Aquitard	24.0–27.0	19.75	0.55	0.5×10^{-4}	0.1×10^{-4}	0.764	0.039 7	0.004 6	0.800
Sandy silts	Confined aquifer	27.0–33.0	20.65	0.35	1.0	0.7	0.583	0.028 3	0.003 3	1.202
Silty clays	Aquitard	33.0–37.0	20.50	0.39	5.0×10^{-4}	3.0×10^{-4}	0.611	0.032 0	0.003 7	0.900
Silty fine sands	Confined aquifer	37.0–42.0	20.05	0.30	2.5	1.5	0.585	0.019 1	0.002 2	1.382
Silty clays	Aquitard	42.0–50.0	19.30	0.39	5.0×10^{-4}	2.0×10^{-4}	0.676	0.030 5	0.003 5	0.900

Note: K_0 = coefficient of the earth pressure at rest; K_H = horizontal permeability; K_V = vertical permeability; e_0 = initial void ratio; λ = slope of the normal consolidation line; κ = slope of the elastic swelling line; M = frictional constant.

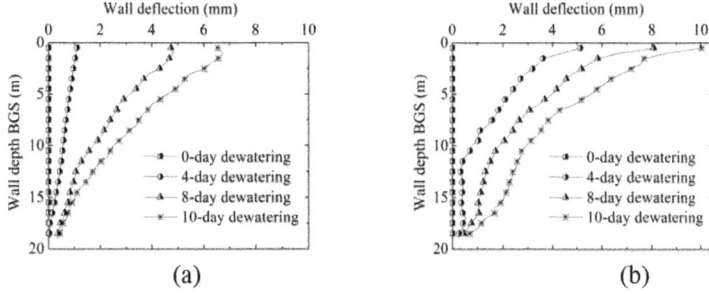

Fig. 3. Observed wall deflections at C1 and C3 during T1: (a) C1 and (b) C3.

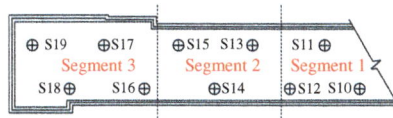

Fig. 4. 3 segments of dewatering tests.

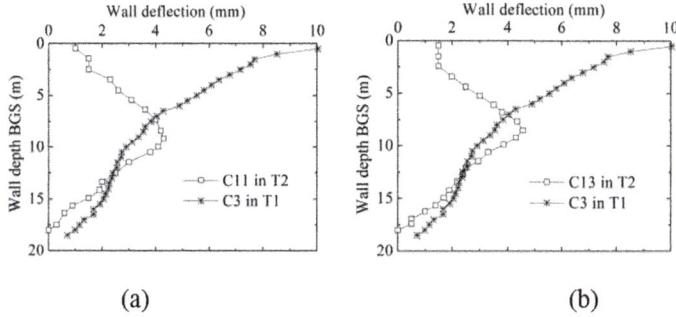

Fig. 5. Comparison of wall deflections in T1 and T2 at the end of PED: (a) C11 vs. C3, (b) C13 vs. C3

3   Numerical Model and Verification

In this section, a three-dimensional (3D) finite element (FE) model is developed in a FE software (ABAQUS), which considers the fluid-mechanical interaction using the Biot consolidation theory. The T2 is simulated. The computed wall deflections are analyzed and compared with the test results to verify the reliability of the model.

Figure 6 shows the mesh of the FE model. Considering the excavation with in-plane dimensions of 200 by 20 m, the dimensions of the model are set to 400 m by 230 m in-plane. Hence, the distance of the lateral soil boundary to retaining wall is 100 m, which exceeds the zone of influence of the dewatering well estimated by Sichardt's formula [2]. In the vertical direction, the model is set to 50 m in depth.

According to the actual field conditions of the excavation, the strata in the model are divided into 9 layers.

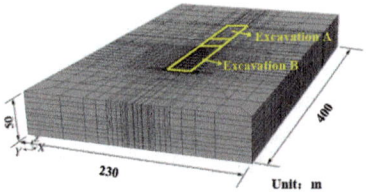

Fig. 6. Mesh of the FE model.

The soil behavior is assumed to obey the constitutive theory of modified Cam-Clay (MCC) during PED. Table 1 summarizes the main parameters of each soil layer. The linear elastic model is adopted for the retaining wall, dewatering well and struts. The Young's moduli of the diaphragm wall, dewatering wells and struts are 30, 210 and 210 GPa, respectively. The initial water level is assumed to be located at the ground surface. The lateral soil boundaries are supported by rollers and recharged by a constant head at the ground surface. The bottom of the soil is pinned and impermeable. The function of dewatering well is simulated by applying a seepage boundary on a zone where the screen of the dewatering well is located. In this paper, the drainage-only flow (DOF) seepage boundary will be set on the screen zone [5]. The DOF seepage boundary assumes that the pore fluid velocity (v_n), which is in the direction outward from the soil element surface, is equal to $k_s p_w$ when p_w is positive, where k_s is termed the seepage coefficient. v_n will be equal to 0 when p_w is negative. The initial value of k_s can be calculated using Eq. (1) based on the discharge flow rate via a well (Q) reported by Zheng et al. [5] and Zeng et al. [6], where D is the well diameter; and L_w is the length of screen range. The derivation process of Eq. (1) can be referred to literature [5].

$$k_s = \frac{Q}{\pi D L_w p_w} \tag{1}$$

The FE modeling is conducted according to T2 presented in Sect. 2. Figure 7 compares the wall deflection at C11-C14 from the field observations and calculations. It can be seen that the calculated wall deflections are close to the measured values, and the calculated deformation pattern of the wall is basically the same as that observed.

4   Effect of Segmental Dewatering

In this section, numerical simulations are carried out to specialize in the effect of segmental dewatering on wall behavior during PED. The scheme of segmental dewatering is the same to Fig. 4. Besides, the traditional dewatering method (i.e., letting all the dewatering wells operate simultaneously) is also simulated for

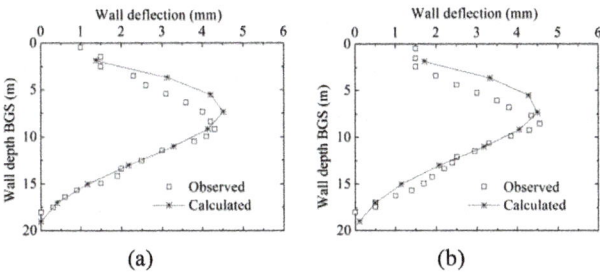

Fig. 7. Comparison between calculated and observed values of wall deflection at (a) C11 and (b) C13 at the end of PED.

comparison with segmental dewatering. The first level of struts is not installed in both segmental dewatering and the traditional dewatering method.

Figure 8 shows the deflection distribution at wall top along the longitudinal direction under the segmental dewatering and the traditional dewatering method. With the dewatering wells operating successively from Segment 1 to 3, the wall deflections grow gradually from Segment 1 to 3. When the dewatering wells in one segment operate, the retaining wall within that segment can develop apparent deflections, which can occupy more than 50% of the final wall deflection. However, the retaining wall outside the dewatering segments just develop a little deflection. As to the traditional dewatering, because all the dewatering wells operate simultaneously, relatively larger and basically uniform wall deflections occur at most locations along the longitudinal direction except around the corners. Apparently, the wall deflection is smaller under segmental dewatering compared to that under traditional dewatering method.

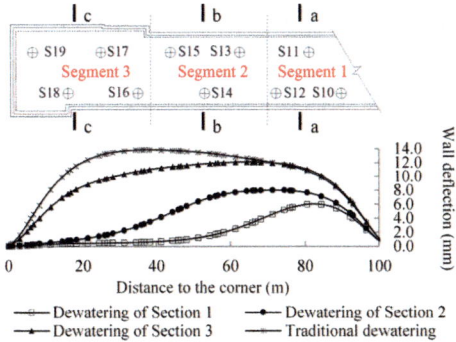

Fig. 8. Deflection distribution at wall top along the longitudinal direction under the segmental dewatering and the traditional dewatering method

In order to quantify the control effect of segmental dewatering on the wall deflection induced by PED, the control index (Δ), which means the decreased magnitude of wall deflection by using segmental dewatering, is defined by the following equation:

$$\Delta = \frac{\delta_{hm,tro} - \delta_{hm,sd}}{\delta_{hm,tro}} \times 100\% \tag{2}$$

Where $\delta_{hm,tro}$ and $\delta_{hm,sd}$ are the maximum wall deflections under traditional dewatering and segmental dewatering, respectively. Figure 9 shows Δ at three cross sections (i.e., Section a-a, b-b and c-c) which are corresponding to the center of Segment 1, 2 and 3, respectively. At Section a-a, the difference of wall deflection between the two dewatering methods is relatively smaller, and Δ is 8.8%, which means the wall deflection at Section a-a can be reduced by 8.8% when the segmental dewatering is adopted. At Segment 3, the difference of wall deflection between the two dewatering methods is relatively larger, and Δ can reach 42.3%. This result indicates that the segmental dewatering can be more effective in controlling wall deflection in the segment where dewatering wells are activated at last. Therefore, the segmental dewatering can be reasonably arranged according to the location of protected environment around excavation. It is suggested that the segment far away from the protected environment be firstly pumped, and the segment close to the protected environment be pumped at last, making the best of segmental dewatering to restrict the deformation development of protected environment.

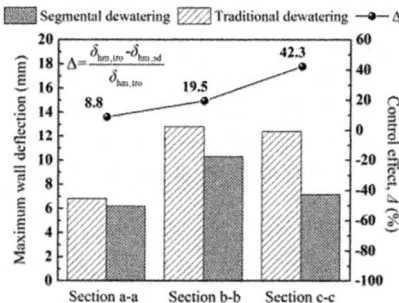

Fig. 9. Maximum wall deflection and Δ at Section a-a, b-b and c-c under segmental dewatering and traditional dewatering

5   Concluding Remarks

(1) As to the method of installing struts before dewatering, the deflections at wall top were reduced apparently, but great wall deflections still arise at the deep location. In this study, this method reduces the deflection at wall top by 84%, the maximum wall deflection by 56%.

(2) Segmental dewatering can be more effective in controlling wall deflection in the segment where dewatering wells are activated at last. Therefore, the segmental dewatering can be reasonably arranged according to the location of protected environment around excavation. It is suggested that the segment far away from the protected environment be firstly pumped, and the segment close to the protected environment be pumped at last.

Acknowledgments. The research is funded by the National Natural Science Foundation of China under Grant No. 51708206 and 11602083, the Research Project of Key Laboratory of Coast Civil Structural Safety of the Ministry of Education under Grant No. 2017-KF03. These financial supports are gratefully acknowledged.

References

1. Yao, T.Q., Shi, Z.H., Cao, H.B.: Handbook of the Dewatering of Foundation Pit. China Architecture and Building Press, Beijing (2006)
2. Cashman, P.M., Preene, M.: Groundwater Lowering in Construction: A Practical Guide to Dewatering, 2nd edn. CRC Press, Boca Raton (2012)
3. Powers, J.P.: Construction Dewatering and Groundwater Control: New Methods and Applications, 3rd edn. Wiley, Hoboken (2007)
4. Wang, J.X., Liu, X.T., Wu, Y.B., Liu, S.L., Wu, L.G., Lou, R.X., et al.: Field experiment and numerical simulation of coupling non-Darcy flow caused by curtain and pumping well in foundation pit dewatering. J. Hydrol. **549**, 277–293 (2017)
5. Zheng, G., Zeng, C.F., Diao, Y., Xue, X.L.: Test and numerical research on wall deflections induced by pre-excavation dewatering. Comput. Geotech. **62**, 244–256 (2014)
6. Zeng, C.F., Zheng, G., Xue, X.L.: Wall deflection induced by pre-excavation dewatering in large-scale excavations. Chin. J. Geotech. Eng. **39**(6), 1012–1021 (2017)
7. Zeng, C.-F., Zheng, G., Xue, X.-L.: A parametric study of wall deflection induced by pre-excavation dewatering in soft ground. Rock Soil Mech. **38**(11), 3295–3303+3318 (2017)
8. Zeng, C.-F., Zheng, G., Xue, X.-L.: Effect of soil permeability on wall deflection during pre-excavation dewatering in soft ground. Rock Soil Mech. **38**(10), 3039–3047 (2017)
9. China Academy of Building Research: Technical specification for retaining and protection of building foundation excavations. JGJ 120-2012. China Architecture and Building Press, Beijing (2012). (in Chinese)
10. Technical Committee B/526 Geotechnics. Eurocode 7: Geotechnical design—Part 1: General rules. BS EN 1997-1:2004 + A1:2013. British Standards Institution, London (2013)
11. Tan, Y., Wang, D.: Characteristics of a large-scale deep foundation pit excavated by the Central-Island technique in Shanghai Soft Clay. II: top-down construction of the peripheral rectangular pit. J. Geotech. Geoenviron. Eng. **139**(11), 1894–1910 (2013)
12. Tan, Y., Wang, D.: Characteristics of a large-scale deep foundation pit excavated by the Central-Island technique in Shanghai Soft Clay. I: bottom-up construction of the central cylindrical shaft. J. Geotech. Geoenviron. Eng. **139**(11), 1875–1893 (2013)
13. Tan, Y., Wei, B.: Observed behaviors of a long and deep excavation constructed by cut-and-cover technique in Shanghai Soft Clay. J. Geotech. Geoenviron. Eng. **138**(1), 69–88 (2012)

Estimation of Pile Set-up Effect in Clay Using Direct Shear Test

Feng Yu[1], Hai-lei Kou[2(⊠)], and Yao-bo Gou[1]

[1] Institute of Foundation and Structure Technologies, Zhejiang Sci-Tech University, Hangzhou 310018, Zhejiang, China
[2] College of Engineering, Ocean University of China, Qingdao 266100, China
hlkou@ouc.edu.cn

Abstract. A series of large-scale direct shear tests for the study of interface behavior between concrete piles and silty clay are presented in this paper. Concrete platform was used to model concrete pile in the test. The shear forces and displacements were measured. The distribution of soil partials of silty clay before and after shearing was also compared using single-lens-reflex digital camera. Test results show that the shear forces between concrete platform and soil samples are closely related to the vertical loading implemented on the soil sample and the interval times. The larger the vertical loading, the larger the shear forces. With the increase of interval times, the shear forces between concrete platform and soil samples gradually increased. It is also indicated that the vertical loading has little effect on the time effect of shear forces. As the interval time lasts, the particles of soil samples gradually upheaved and the deformation area increased. Based on the predicted three-phase model with respect to the log of time, the time points of 1.0 day and 109 days can be taken as the first and second time points respectively in this test.

Keywords: Direct shear test · Pile capacity · Time effect · Silty clay
Interface shear

1 Introduction

Soil/pile set-up is defined as an increase in pile capacity over time after installation. This phenomenon was first reported in the literature by Wendel [1], and has been recognized as occurring in most kinds of piles. The exact mechanism of pile setup has not been fully understood. The majority of setup is likely related to the dissipation of excess pore pressure within around soil. The around soil is displaced and disturbed as the pile is driven and subsequent remolding and reconsolidation of that soil [2]. The consideration of set-up effect can efficiently improve the estimation of pile capacity, so that the length, diameter or number of production piles may be economically reduced [3]. The rate and magnitudes of setup are functions of several parameters such as pile type, soil type, pile size and so on [4, 5]. Setup is recognized as occurring for virtually all driven pile types, in saturated clay, loose to medium dense silt, silty or fine sand, and is related to both soil and pile properties [6–8]. The phenomenon is referred to a pile set-up by geotechnical engineers and such effect is definitely favorable to engineering designs [9, 10].

© Springer Nature Singapore Pte Ltd. 2018
T. Qiu et al. (Eds.): GSIC 2018, *Proceedings of GeoShanghai 2018 International Conference: Advances in Soil Dynamics and Foundation Engineering*, pp. 464–477, 2018.
https://doi.org/10.1007/978-981-13-0131-5_51

Over the years, two major mechanisms have been postulated to explain the set-up effect of pile capacity [11–13]: (1) dissipation of excess pore water pressure; and (2) soil aging. The first mechanism originates from the dissipation of excess pore water pressure which is generated due to soil remolding or disturbance during pile driving. The associated increase in lateral effective stress with increasing time gives rise to an increase in shear strength and thus the axial capacity of the pile. The duration of this reconsolidation depends on the permeability of the soil. It ranges from days in coarse-grained soils to months or years in fine-gained soils [12]. The second mechanism is related to practical rearrangement around the pile shaft. This rearrangement may be accompanied by the collapsing of temporary arches formed around the shaft which increases the lateral stress on the pile and thus the soil's shear strength and pile's axial capacity [12, 14].

Many studies have been conducted on the set-up effect of concrete piles as onshore foundation [15–17]. Nowadays, the increased exploitation of offshore energy resources make concrete piles are preferred as offshore foundation. The smaller ratio of height to thickness of this kind pile makes the performance different from that of steel pipe piles during and after installation [18, 19]. Thus, the set-up effect in concrete pipe piles as offshore foundation can be significantly different from of steel pipe piles and thus needs to be studied. To further investigate and demonstrate concrete pipe pile set-up in silty clay, a series of large-scale direct shear tests are conducted. The shear behavior between concrete plate and soil samples was presented. The setup effect and soil particle distributions after shearing were also observed and described. Based on the mechanisms of set-up, a three-phrase model is proposed. In addition, a thorough review of previous studies, including a compiled database of case histories and empirical formulas, is also undertaken in this paper.

2   Set-up Mechanism

As a pile is driven, soil is displaced predominately radially along the shaft. Some vertical displacement along the shaft may occur, and soil is vertically and radially compressed beneath the toe [12]. As soil around and beneath the pile is displaced and disturbed, excess pore water pressures are generated, decreasing the effective stress of the affected soil [11, 20]. As excess pore water pressures dissipate, the effective stress of the affected soil increases, and set-up predominately occurs as a result of increased shear strength and increased lateral stress against the pile.

The soil/pile set-up mechanisms can be divided into three phases according to the dissipation process of excess pore water pressure [9]. That is

Phase I: Logarithmically nonlinear rate of excess pore water pressure dissipation;
Phase II: Logarithmically linear rate of excess pore water pressure dissipation;
Phase III: Independent of effective stress.

The physical process is schematically shown in Fig. 1 and described in detail as follows.

Phase I: Because of the highly disturbed state of the soil, the rate of dissipation of excess pore water pressure is not constant (nonlinear) with respect to the log of time for

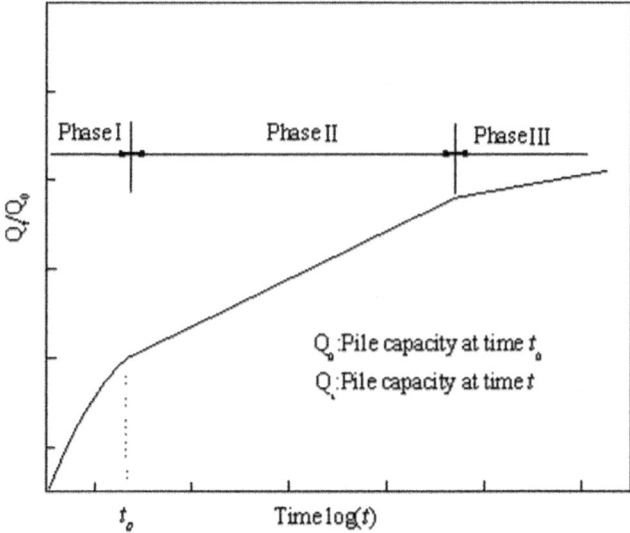

Fig. 1. Schematic idealized set-up phases [9]

some period after driving. During this phase of variable rate of dissipation of excess pore pressure, the affected soil experiences an increase in effective and horizontal stress, consolidated, and gains strength in a manner which is not well understood and is difficult to be modeled or predicted. The duration of the logarithmically nonlinear rate of excess pore water pressure dissipation is a function of soil and pile properties. In clean sands, the logarithmic rate of dissipation may become linear almost immediately after driving. In cohesive soils, the logarithmic rate of dissipation may remain nonlinear for several days.

Phase II: In this phase, at some time after driving, the rate of excess pore water pressure dissipation becomes constant (linear) with respect to the log of time. In a number of empirical set-up predictive models, this time after driving at which the rate of excess pore water pressure dissipation becomes logarithmically linear is referred to as the initial time, t0. In fine-grained granular or mixed soils, logarithmically linear dissipation may continue for several hours, several days, or several weeks. In cohesive soils, logarithmically linear dissipation may continue for several weeks, several months, or even years [20]. Skov and Denver [20] suggested that t0 = 1.0 and 0.5 in cohesive soils and sand, respectively.

Phase III: During this third phase of set-up, set-up rate is independent of effective stress. This is related to the phenomenon of aging. Aging effects increase the soil's shear modulus, stiffness, and dilatancy, and reduce the soil's compressibility [21, 22]. Aging effects also bring an increased friction angle at the soil/pile interface [23].

For cohesive soils, the majority of set-up is related to the dissipation of excess pore pressure (i.e. Phase I and Phase II), including consolidation effect, recover effect and shell effect. For a given soil type at a given elevation along the pile shaft, there is likely some overlap between successive phases, so, more than 1 phase may be contributing to set-up at a time. In addition, unless soil conditions are uniform along the entire length of the shaft and beneath the toe, different soils at different elevations will be in different phases of set-up at a given time.

A number of empirical equations have been proposed to quantify the magnitude of set-up, most of which are summarized in Table 1. Amongst those proposed formulas, the logarithmic relationship proposed by Skov and Denver [20] has been commonly adopted for the prediction of the set-up.

Table 1. Empirical formulas for predicting pile capacities with time

Number	Source	Equation	Soil type
1	Skov and Denver [20]	$Q_t = Q_0 [A\log(t/t_o) + 1]$	Fine grained soil and clay
2	Huang [24]	$Q_t = Q_{EOD} + 0.236[1 + \log(t)$ $(Q_{max} - Q_{EOD})]$	Soft soil
3	Long et al. [25]	$Q_t = 1.1Q_{EOD}t^{0.13}$ upper limit $Q_t = 1.1Q_{EOD}t^{0.05}$ lower limit	Sands, clay and mixed soil
4	Hu et al. [26]	$P_{ut} = [a \times \ln(t) + b] \times P_{uo} + P_{uo}$	Soft clay
5	Zhang [27]	$q_t = q_0[0.3\log(t) + 2.8]$	Soft soil

Note: Q_t: the capacity at time t; Q_0: the capacity at time t_o; A: time factor, suggesting between 0.2–0.8; Q_{EOD} is the bearing capacity after driving; Q_{max}: maximum capacity; P_{ut} is the bearing capacity at time t; P_{uo} is the initial bearing capacity; q_t is the shaft resistance at time t; q_0 is the shaft resistance after driving.

3   Test Setup and Procedures

3.1   Direct Shear Apparatus

The shear apparatus used in this test is illustrated in Fig. 2. This apparatus is reconfigured based on conventional shear box apparatus, including shear box, loading system, force and displacement acquisition system. A square shear box is used as the soil container with an inside length of 30 cm and a height of 15 cm. In the above shear box, a concrete platform with an area of 30 cm × 30 cm and height of 5 cm was placed to simulate the concrete pile.

Different vertical loads were implemented on the concrete sample through the top plate to simulate different soil stiffness surrounding the pile. Two-way horizontal movement is applied to the concrete platform by a motor working through a variable speed gearbox. The horizontal displacement between the soil sample and concrete platform is measured by a dial gauge. The horizontal load is measured through a load cell, fixed on the top shear box, as shown in Fig. 3.

3.2   Test Materials

The tests were conducted on a silty clay sample, which was obtained from drilled holes at a construction site. The properties of the soil sample are summarized in Table 2. The cohesion c and friction angle φ were determined using the quick shear tests. The

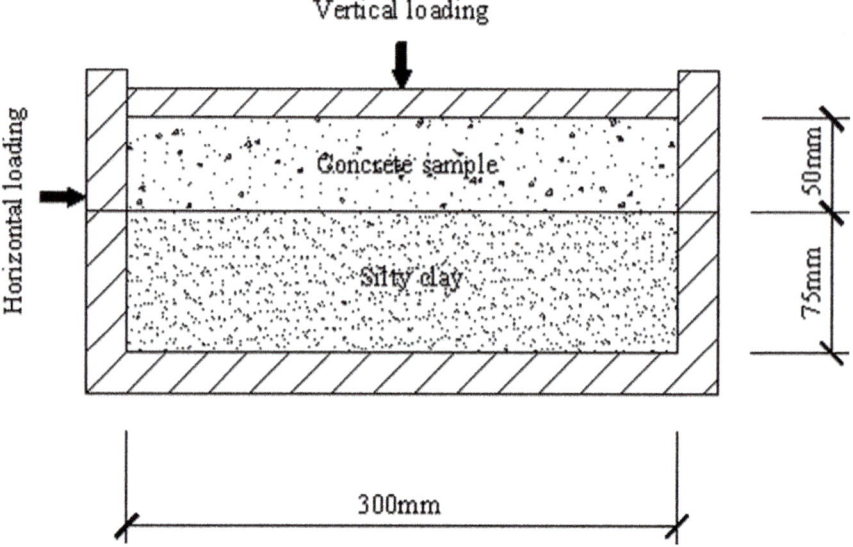

Fig. 2. Schematic view of the shear apparatus

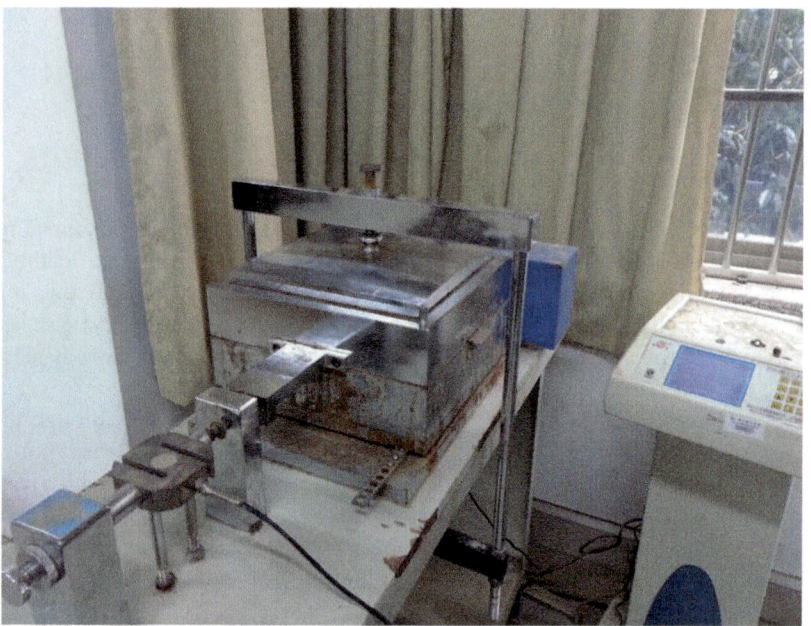

Fig. 3. Shear apparatus used in the test

Table 2. Soil properties

Soils type	w (%)	ρ (g/cm^3)	ρ_d (g/cm^3)	c (kPa)	φ (°)
Silty clay	25.0	1.81	1.45	24.0	22.0

Note: w is the natural water content; ρ is the density; ρ_d is the dry density; c is the cohesion; φ is the friction angle.

concrete plate is made of Grade C30 concrete and the mass ratio of water, cement, sand and gravel is 0.76: 2.00: 2.22: 5.44 according to the American Society for Testing and Materials standard ASTM C39. The cube compression strength of such concrete is specified to be not less than 30 MPa with 95% confidence, while its design strength is 14.3 MPa. The elastic modulus of Grade C30 concrete is routinely adopted to be 3×10^4 MPa.

3.3 Test Procedures

The test procedure in this paper is similar with conventional direct shear test and was conducted in accordance with the American Society for Testing and Materials standard ASTM D3080. After placing the soil sample in bottom shear box, vertical loadings of 50 kPa and 100 kPa are applied through a metal platen resting on top of the concrete platform, respectively. Using the displacement control system, the displacement rate is set to be 1.0 mm/min, which ensures the drained condition of silty clay during test. The shear forces can be measured by means of force measuring system. The displacement can be thus calculated by recording the time. After the shear test, single-lens-reflex digital camera is used to observe the distribution of soil particles. Several shear tests were performed at different interval times.

4 Test Results and Discussion

4.1 Relationship Between Shear Forces and Displacement

The measured shear forces versus displacement are illustrated in Fig. 4. It should be noticed that the horizontal axis in Fig. 4 shows the time interval between tests. It is indicated that the shear forces would reach the peak value at the displacement of 2 mm. This observation is similar with the findings of Zhang et al. [27] and McVay et al. [23]. With the increase of interval times, the measured shear forces will increase. It is also clear that the larger the vertical loading, the larger the ultimate shear force.

4.2 Development of Ultimate Shear Forces with Interval Time

The development of shear forces with interval time is calculated by comparing the ultimate shear forces illustrated Fig. 4(a) and (b). In the figure, τ_0 is the shear stress at time t_0 and τ is the shear stress at time t. It seems that the increment of shear stress appears in a three-phase manner as proposed by [20]. That is

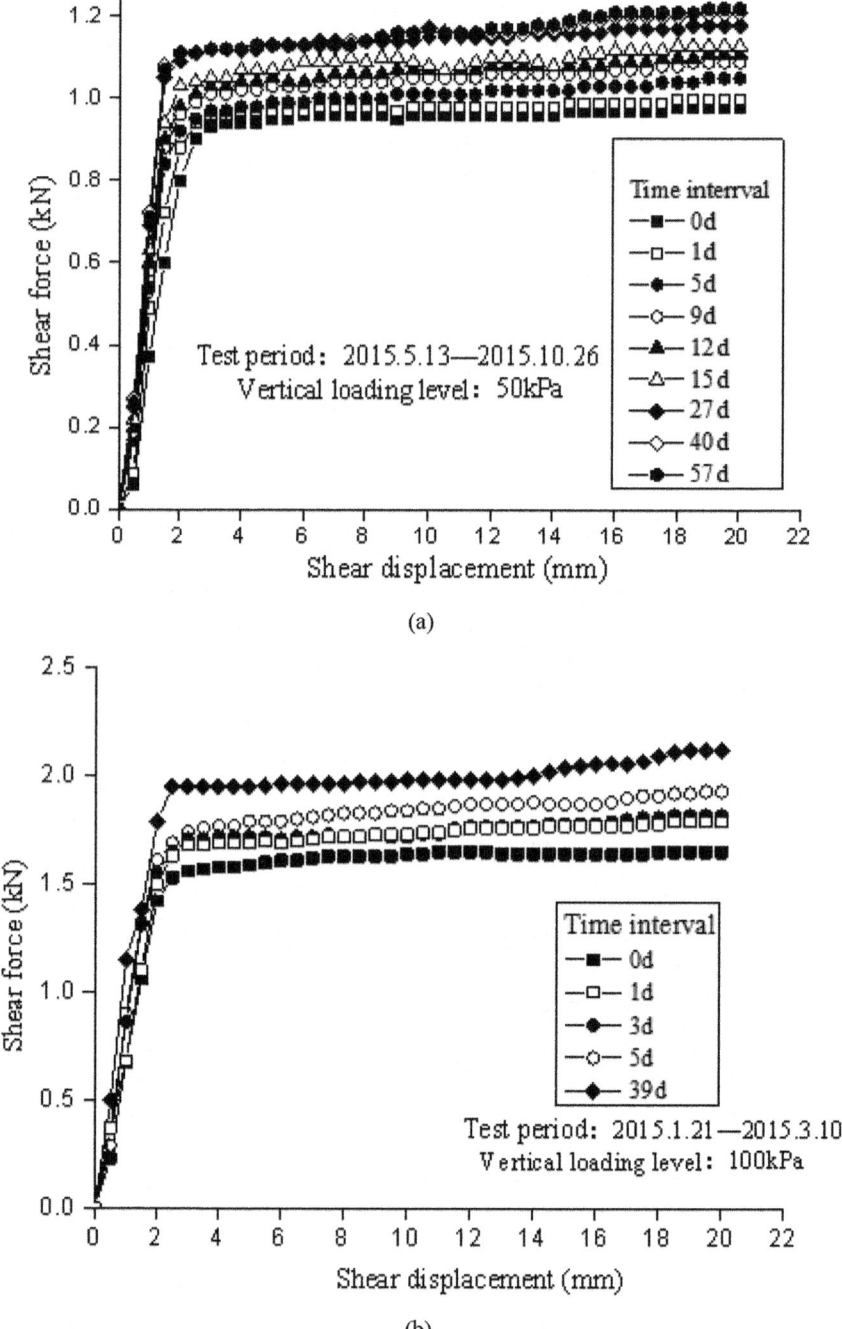

(a)

(b)

Fig. 4. Time-related displacement-forces behavior under vertical loading of: (a) 50 kPa; (b) 100 kPa

Phase I: Logarithmically nonlinear rate of excess pore water pressure dissipation. t_0 can be taken as 1.0 day;

Phase II: Logarithmically linear rate of excess pore water pressure dissipation. The whole process lasts relatively long time. In this test, the process is about 109 days when the vertical loading is 50 kPa;

Phase III: Independent of effective stress. The dissipation of excess pore water pressure is completed in this phrase and the aging effect plays the main role.

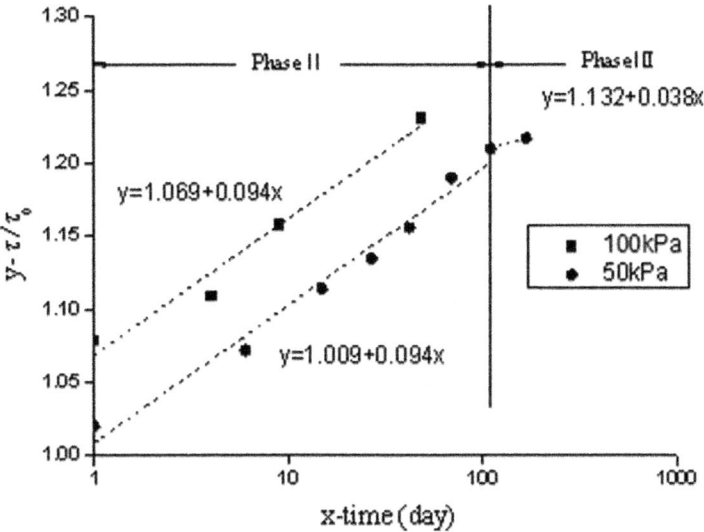

Fig. 5. Relationship between $\tau/\tau 0$ and log $(t/t0)$

It can be seen from Fig. 5 that the two fitting straight lines are almost parallel at phase II. This indicates that the vertical loading only affects the line intercept. If assuming the intercept increase linearly, the relationship between τ and τ_o at phase II can be induced as

$$\tau/\tau_0 = 1.009 + \frac{0.03P}{50} + 0.094 \log(t/t_0) \qquad (1)$$

where P is the vertical loading in kPa. Equation (1) can be rewritten as

$$\tau = \tau_0 \left[0.094 \log(t/t_0) + 1.009 + \frac{0.03P}{50} \right] \qquad (2)$$

It is indicated that the above formula is similar with the one for predicting the set-up of pile capacity proposed by Skov and Denver [20] as follow

$$Q_t = Q_0[A \log(t/t_0) + 1] \tag{3}$$

The difference between Eqs. (2) and (3) lies in the slope and the intercept. The similarity in expression form indicates that the two models are adaptive in methodology.

(a) Initial surface

(b) Initial shear

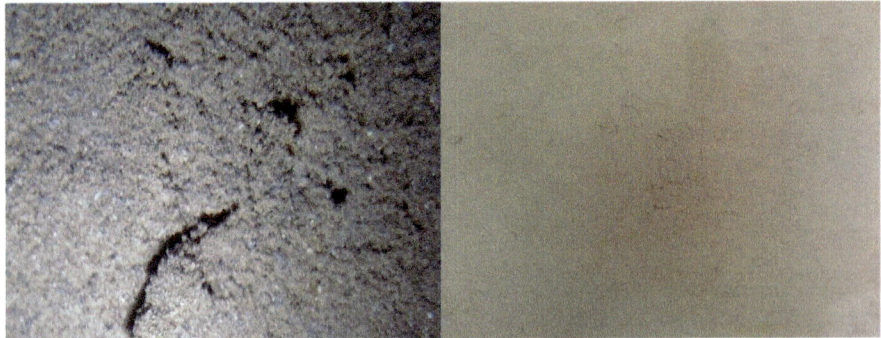

(c) Shear after 1 day

Fig. 6. Distribution of soil particles after shear tests

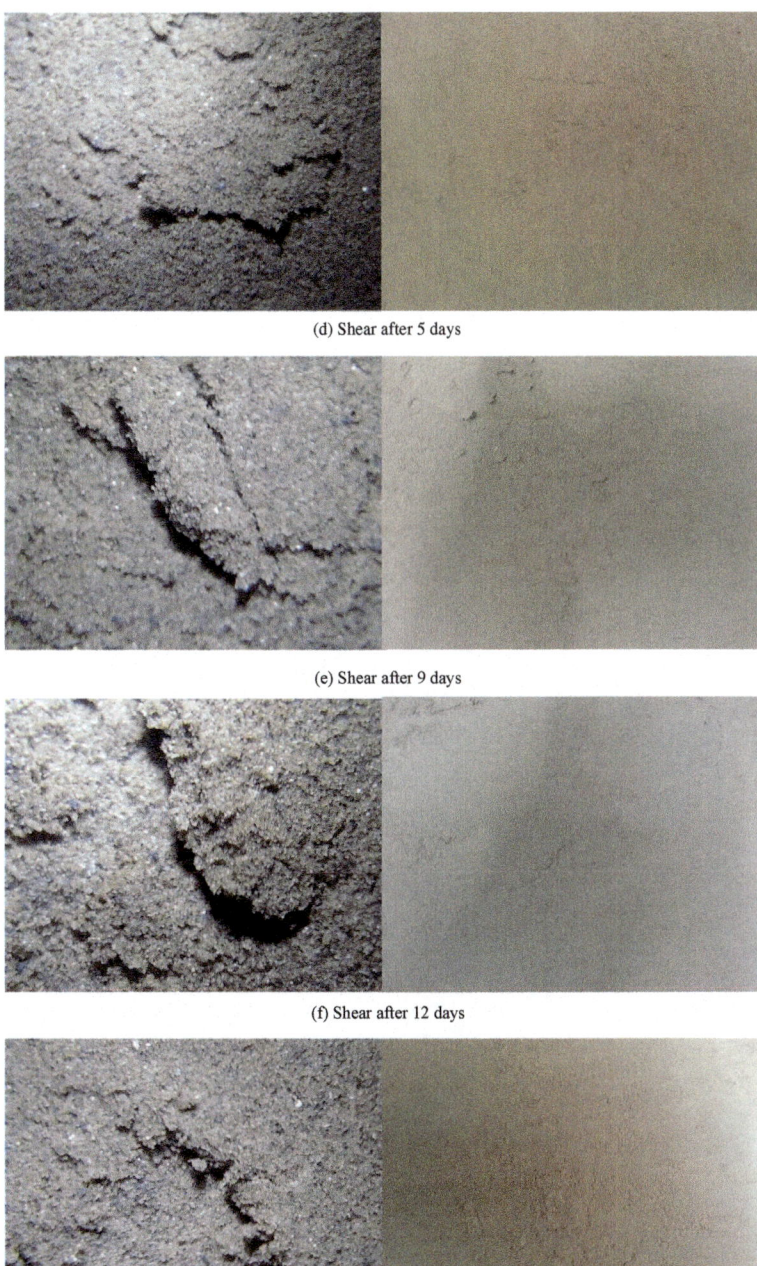

(d) Shear after 5 days

(e) Shear after 9 days

(f) Shear after 12 days

(g) Shear after 15 days

Fig. 6. (*continued*)

(h) Shear after 27 days

(i) Shear after 40 days

(j) Shear after 57 days

Fig. 6. (*continued*)

The increment of bearing capacity can be divided into three phases according to the interval time. The time points are 1.0 day and 109 day, respectively. Equation (2) is taken to calculate the bearing capacity when the interval time is between 1.0 day and 109 days.

4.3   Distribution of Soil Particles

The distributions of soil particles after shear tests with the vertical loading of 50 kPa are shown in Fig. 6. The left is the partial enlarged detail with amplification of 80. The right is macrograph. It can be seen from the figures that after initial shear, the surface of the soil becomes irregular and deflection and movement occur in the soil particles. As the interval time increases, the soil particles are gradually upheaved and the deformation areas also gradually increase. This will cause some heave like hills and stripping at the soil surface. The heave of soil particles is most significant at the interval time of 15 days and the heave of soil particles gradually evolves from the surface to around area. The following shearing surfaces illustrate that the heave height is decreasing and the stripping area is gradually extending. In this test, the stripping part can be taken as "soil crust", namely the relevant "shear zone". It is induced from the relative displacement of pile-soil caused by the modulus difference of pile-soil. This phenomena is closely related to pile set-up. The larger the shearing surface, the bigger the pile set up. Similar observations are reported by references of [28, 29].

5   Conclusions

A series of large-scale direct shear tests were conducted to study the pile set-up effect. The following conclusions could be drawn.

(1)   The shear forces can reach the peak value at the relative displacement of 2 mm for silty clay. The measured shear forces increase with the interval time. The larger the vertical loading, the larger the measured shear force.

(2)   The evolvement of increased shear forces can be divided into three phases with logarithm of time. For vertical loading of 50 kPa, three phases are: 1-day Phase I with logarithmically nonlinear dissipation rate of excess pore water pressure, 108-day Phase II with logarithmically linear dissipation rate of excess pore water pressure, and a following effective stress independent Phase III governed by the aging effect.

(3)   With interval time increase, soil particles are gradually upheaved and the deformation area also gradually increases. This will cause some heave in terms of hills and stripping generated from the soil surface. The heave of soil particles gradually evolves to deeper area as the interval time proceeds.

Acknowledgments. The funding supports from the National Natural Science Fund (Nos. 41472284 & 51408439) and Shandong Housing and Urban Construction Department (No. KY048) of China are gratefully acknowledged.

References

1. Wendel, E.: On the test loading of piles and its application to foundation problems in Gothenburg. Tekniska Samf Goteberg Handl. **7**, 3–62 (1900)
2. Kou, H.L., Zhang, M.Y., Yu, F.: Shear zone around jacked piles in clay. J. Perform. Constr. Facil. **29**(6), 04014169 (2014)
3. Liu, J.W., Zhang, Z.M., Yu, F., Lin, C.G.: Effect of soil set-up on the capacity of jacked concrete pipe piles in mixed soils. J. Zhejiang Univ. Sci. A **12**(8), 637–644 (2011)
4. Chow, F.C., Jardine, R.J., Naroy, J.F., Brucy, F.: Effects of time on capacity of pipe piles in dense marine sand. J. Geotech. Geoenviron. Eng. ASCE **124**(3), 254–264 (1998)
5. Bullock, P.J., Schmertmann, J.H., McVay, M.C., Townsend, F.C.: Side shear setup i: test piles driven in Florida. J. Geotech. Geoenviron. Eng. ASCE **131**(3), 292–300 (2005)
6. Bond, A.J., Jardine, R.J.: Effects of installing displacement piles in high OCR clay. Geotechnique **41**(3), 341–363 (1991)
7. Bullock, P.J., Schmertmann, J.H., McVay, M.C., Townsend, F.C.: Side shear setup ii: results from Florida test piles. J. Geotech. Geoenviron. Eng. ASCE **131**(3), 301–310 (2005)
8. Lim, J.K., Lehane, B.M.: Time effect on the shaft capacity of jacked piles in sand. Can. Geotech. J. **52**(11), 1830–1838 (2015)
9. Komurka, V.E., Wagner, A.B., Edil, T.: Estimating soil/pile set-up, Report No 03-05. Wisconsin Highway Research Program (2003)
10. Kou, H.L., Chu, J., Guo, W., Zhang, M.Y.: Field study of residual forces developed in pre-stressed high-strength concrete (PHC) pipe piles. Can. Geotech. J. **53**(4), 696–707 (2015)
11. Axelsson, G.: Long-term set-up of driven piles in non-cohesive soils, Ph.D. thesis, Royal Institute of Technology, Stockholm (2000)
12. Pra-ai, S., Boulon, M.: Soil–structure cyclic direct shear tests: a new interpretation of the direct shear experiment and its application to a series of cyclic tests. Acta Geotech. **12**(1) 1–21 (2016)
13. Kou, H.L., Guo, W., Zhang, M.Y.: Pullout performance of GFRP anti-floating anchor in weathered soil. Tunn. Undergr. Space Technol. **49**, 408–416 (2015)
14. York, D.L., Brusey, G.B., Clemente, F.M., Law, S.K.: Set-up and relaxation in glacial sand. Int. J. Rock Mech. Min. Sci. Geomech. Abs. **32**(6), 265A (1995)
15. York, D.L., Brusey, W.G., Clemente, F.M., Law, S.K.: Set-up and relaxation in glacial sand. J. Geotech. Eng. **120**(9), 1498–1513 (1994)
16. Shek, L.M.P., Zhang, L.M., Pang, H.W.: Set-up effect in long piles in weathered soils. Proc. ICE-Geotech. Eng. **159**(3), 145–152 (2006)
17. Chen, X., Zhang, J., Xiao, Y., Li, J.: Effect of roughness on shear behavior of red clay-concrete interface in large-scale direct shear tests. Can. Geotech. J. **52**(8), 1122–1135 (2015)
18. Brucy, F., Meunier, J., Nauroy, J.F.: Behavior of pile plugs in sandy soils during and after driving. In: Proceeding of 23rd Annual Offshore Technology Conference Houston 1 145-154 (1991)
19. Paik, K., Salgado, R., Lee, J., Kim, B.: Behavior of open-and closed-ended piles driven into sands. J. Geotech. Geoenviron. Eng. **129**(4), 296–306 (2003)
20. Skov, R., Denver. H.: Time-dependence of bearing capacity of piles. In: Proceeding of the 3rd International Conference on the Application of Stress-Wave Theory to Piles, 25–27 May 1988, Vancouver, BC, pp. 879–888 (1988)
21. Schmertmann, J.H.: The mechanical aging of soils. J. Geotech. Eng. **117**(9), 1288–1330 (1991)

22. Axelsson, G.: Long-term set-up of driven piles in sand: evaluated from dynamic tests on penetration rods. In: Proceedings of the 1st International Conference on Site Characterization, Atlanta, U.S.A, April 1998

23. McVay, M., Zhang, L., Molnit, T., Lai, P.: Centrifuge testing of large laterally loaded pile groups in sands. J. Geotech. Geoenviron. Eng. **124**(10), 1016–1026 (1988)

24. Huang, S.: Application of dynamic measurement on long H-pile driven into soft ground in Shanghai. In: Proceedings of 3rd International Conference on the Application of Stress-Wave Theory to Piles, Ottawa, Ontario, Canada, 25–27 May 1988

25. Long, J.H., Bozkurt, D., Kerrigan, J.A., Wysockey, M.H.: Value of methods for predicting axial pile capacity. Transp. Res. Rec. **1663**(99–1333), 57–63 (1999)

26. Hu, Q., Jiang, J., Yan, X.S., Chen, Y.M.: Regression analysis of the time effect of ultimate bearing capacity of single reinforced pile. J. Harbin Inst. Technol. **38**(4), 602–605 (2006). (In Chinese)

27. Zhang, M.Y., Liu, J.W., Yu, X.X.: Field test study of time effect on ultimate beaing capacity of jacked pipe pile in soft clay. Rock Soil Mech. **30**(10), 3005–3008 (2009). (In Chinese)

28. Randolph, M.F., Carter, J.P., Wroth, C.P.: Driven piles in clay-the effects of installation and subsequent consolidation. Geotechnique **29**(4), 361–393 (1979)

29. Lehane, B.M., White, D.J.: Lateral stress changes and shaft friction for model displacement piles in sand. Can. Geotech. J. **42**(4), 1039–1052 (2005)

Modified Pseudo-dynamic Method for Seismic Passive Earth Thrust of Submerged Backfill

Huang Rui[1,2(✉)]

[1] Center of Rock Mechanics and Geohazards,
Shaoxing University, Shaoxing 312000, Zhejiang, China
zjuhr@hotmail.com
[2] College of Civil Engineering, Shaoxing University,
Shaoxing 312000, Zhejiang, China

Abstract. The accurate assessment of seismic earth pressure acting on a retaining wall is an important problem in earthquake geotechnical engineering. The existing calculations of pseudo-dynamic method mainly focus on the dry soil condition. To investigate the seismic passive earth thrust of submerged backfill, a general modified pseudo-dynamic method is established based on the limit equilibrium analysis. The derivation aims at a vertical gravity wall with a planar rupture surface retaining a horizontal, cohesionless and fully submerged backfill. Meanwhile the method assumes that the amplitude of the seismic acceleration increases linearly along the wall and the backfill is divided into two extreme cases of free water and restrained water conditions according to the permeability difference. Through the comparison with the previous work, the trend of seismic passive earth thrust for submerged backfill is basically consistent with that of the dry soil, but the submerged condition has a reducing effect on the passive earth thrust. Then a parametric study is carried out to investigate the influences of soil friction angle, wall friction angle, horizontal seismic action and vertical seismic action on the intensity distribution of seismic passive earth pressure. The results indicate that the passive earth pressure increases with the increase of soil friction angle and wall friction angle, but the impact of soil friction angle is more significant. The horizontal seismic acceleration is still the determining factor affecting the magnitude and distribution of the passive earth pressure rather than the vertical one.

Keywords: Passive earth pressure · Earthquake · Submerged backfill

1 Introduction

The determination of soil pressure acting on a retaining wall under earthquake actions is a fundamental problem in geotechnical seismic design. Many theoretical and experimental studies [1–3] on the seismic earth thrust have developed since the pseudo-static method was first introduced to solve this problem in the mid-1920s.

Supported by the National Natural Science Foundation of China (Grant No. 51708354).

T. Qiu et al. (Eds.): GSIC 2018, *Proceedings of GeoShanghai 2018 International Conference: Advances in Soil Dynamics and Foundation Engineering*, pp. 478–487, 2018.
https://doi.org/10.1007/978-981-13-0131-5_52

Although the pseudo-static method is widely used in different codes of foreign countries, Steedman and Zeng [4] proposed a pseudo-dynamic method to overcome the drawbacks of the assumptions in the pseudo-static method. Then Choudhury and Nimbalkar [5, 6] extended the pseudo-dynamic method for both active and passive earth pressure conditions considering the effects of horizontal earthquake actions as well as vertical seismic action. Based on the pseudo-dynamic method, the seismic stability of retaining wall has also been improved [7, 8].

However, the existing calculations of pseudo-dynamic method mainly aim at dry soil condition, and a little literature on the seismic earth pressure considering the submerged condition by pseudo-dynamic method is found, although Bellezza et al. [9] tried to propose a more rational approach for this problem. On the other hand, during the last decades the effects of submerged or seepage conditions on seismic earth thrust have been paid attention to by some scholars. Matsuzawa et al. [10] improved the pseudo-static method for seismic active earth thrust of submerged backfill considering a restrained or free water condition exists in the soil. Ebeling and Morrison [11] used empirical coefficients to represent the ratio of the seismic excess pore pressure and the vertical effective stress, and then put forward the expressions of seismic earth and water pressure. Afterwards, Wang et al. [12, 13] calculated the seismic passive earth pressure with steady seepage conditions by using limit equilibrium analysis and pseudo-static method. As mentioned before, the pseudo-static method has some drawbacks, such as cannot describe the dynamic characteristics of earthquake actions and hardly considers the influence of the seismic acceleration magnitude on the earth resistance.

In order to improve the pseudo-dynamic method of submerged condition, a modified formula to calculate the seismic passive thrust of fully submerged backfill is proposed based on the limit equilibrium analysis in this work. The derivation takes into account the seismic acceleration amplification and the effect of phase change. Then the results of passive earth thrust between the submerged backfill and the dry soil are compared to verify this proposed method. Besides, a parametric study is performed to investigate the influences of soil friction angle, wall friction angle, horizontal seismic action and vertical seismic action on the distribution of seismic passive earth pressure.

2   Method of Analysis

2.1   Calculation Hypothesis

The interaction mechanism of the soil and water under earthquake action is complicated. In order to simplify the calculation model, some basic hypothesis should be made in this study. The gravity wall is vertical, retaining a horizontal submerged backfill, as shown in Fig. 1. The cohesionless soil behind walls is assumed to be homogeneous, isotropic and saturated. The friction angles, porosity, permeability, Poisson's ratio and elastic modulus of the soil are constant. Besides, the groundwater level is maintained on the surface of the backfill, and the seepage condition and excess pore pressure are ignored [9].

The retaining wall moves towards the backfill until the soil reaches the passive limit equilibrium state, and the planar rupture surface inclined at angle θ to the horizontal is assumed like the Coulomb's earth pressure theory. During the earthquake, the shear wave and the primary wave propagate through the backfill, and the soil vibrates in harmonic state on the basis of pseudo-dynamic method [5, 6]. The maximum amplitude of seismic acceleration increases linearly along the wall. The directions of horizontal and vertical seismic inertial forces are supposed as shown in Fig. 1.

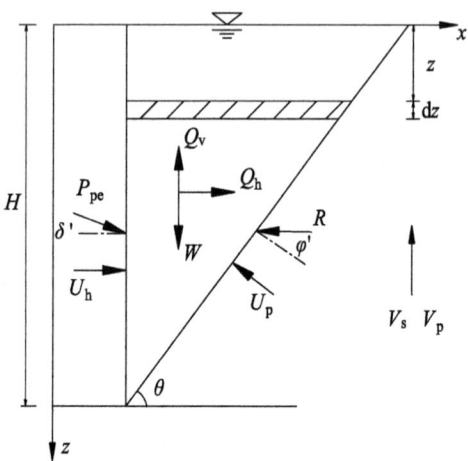

Fig. 1. Model of seismic passive earth thrust

2.2   Modified Pseudo-dynamic Method

Seismic Acceleration
Due to considering the amplification effects of seismic acceleration amplitudes, the relationship of the seismic acceleration coefficients between the top and bottom of retaining wall should be given as

$$k_{h,0}(z=0) = f_a k_h(z=H), \; k_{v,0}(z=0) = f_a k_v(z=H) \tag{1}$$

where k_h and k_v are the horizontal and vertical seismic acceleration coefficients at the wall base which represent the ratio of seismic acceleration to gravitational acceleration, z is the vertical depth, H is the wall height, and f_a is the amplification factor of seismic action.

Therefore, in the view of these previous assumptions, the seismic accelerations at any depth z and time t below the surface can be expressed as [14]

$$a_h(z,t) = \left[1 + \frac{H-z}{H}(f_a - 1)\right] k_h g \sin \omega \left(t - \frac{H-z}{V_s}\right) \tag{2}$$

$$a_v(z,t) = \left[1 + \frac{H-z}{H}(f_a - 1)\right]k_v g \sin \omega \left(t - \frac{H-z}{V_p}\right) \tag{3}$$

where g is the gravitational acceleration, ω is the angular frequency of vibration, $\omega = 2\pi/T$, T is the period of vibration, V_s and V_p are the wave velocity of the shear and primary waves, respectively.

Seismic Inertial Force

The mass of the soil element at any depth z in Fig. 1 is given as

$$dm = \frac{\gamma^* }{g} \frac{H-z}{\tan \theta} dz \tag{4}$$

where γ^* is the unit weight of backfill soil, the value of which depends on the seismic inertial forces and the soil permeability as illustrated in the following.

The total horizontal seismic inertial force on the soil wedge is given by the integral [14]

$$\begin{aligned}
Q_h &= \int_0^H a_h(z,t)dm \\
&= \frac{\lambda \gamma^* k_h}{4\pi^2 \tan \theta}[2\pi H \cos \omega\zeta + \lambda(\sin \omega\zeta - \sin \omega t)] \\
&+ \frac{\lambda \gamma^* k_h(f_a - 1)}{4\pi^3 H \tan \theta}\left[2\pi H(\pi H \cos \omega\zeta + \lambda \sin \omega\zeta) + \lambda^2(\cos \omega t - \cos \omega\zeta)\right]
\end{aligned} \tag{5}$$

where λ is the wavelength of the shear wave, $\lambda = TV_s$, $\zeta = t - H/V_s$.

Similarly, the total vertical seismic inertial force on the soil wedge is given by the integral

$$\begin{aligned}
Q_v &= \int_0^H a_v(z,t)dm \\
&= \frac{\eta \gamma^* k_v}{4\pi^2 \tan \theta}[2\pi H \cos \omega\psi + \eta(\sin \omega\psi - \sin \omega t)] \\
&+ \frac{\eta \gamma^* k_v(f_a - 1)}{4\pi^3 H \tan \theta}\left[2\pi H(\pi H \cos \omega\psi + \lambda \sin \omega\psi) + \eta^2(\cos \omega t - \cos \omega\psi)\right]
\end{aligned} \tag{6}$$

where η is the wavelength of the primary wave, $\eta = TV_p$, $\psi = t - H/V_p$.

With the consideration of the submerged soil condition in this study, the value of γ^* is somewhat different from that of the dry soil condition. Base on Matsuzawa et al. [10], the submerged backfill can be divided into two extreme types according to the permeability difference. For the highly permeable soil, 'free water' condition is defined in which the horizontal inertial force only acts on the solid portion of the soil element. Therefore, the value of γ^* in Eq. (5) equals the dry unit weight γ_d of soil, and Q_h is proportional to the dry weight W_d of the soil wedge. In the limit state without the seismic amplification, Eq. (5) can be given as

$$\lim_{V_s \to \infty} (Q_{h,F})_{max} = k_h \frac{\gamma_d H^2}{2 \tan \theta} = k_h W_d \tag{7}$$

For the low permeable soil, it is supposed that the solid portion and the water portion of the soil element behave as a unit subjected to the horizontal seismic action, which is defined as 'restrained water' condition. Thus Q_h is proportional to the saturated weight W_{sat} of the soil wedge. Similarly, it is easy to conclude that

$$\lim_{V_s \to \infty} (Q_{h,R})_{max} = k_h \frac{\gamma_{sat} H^2}{2 \tan \theta} = k_h W_{sat} \tag{8}$$

Meanwhile, as Matsuzawa et al. [10] suggested, it is important to note that the hydrodynamic water pressure should be added separately to the backfill side in the stability analysis for 'free water' condition, but there is no need to take this step for 'restrained water' condition. In addition, the calculation of vertical seismic inertial force depends on the submerged weight W' of the soil wedge, regardless of the situation in 'free water' or 'restrained water' condition. Consequently, the value of γ^* in Eq. (6) is the effective unit weight γ', and in the limit case Eq. (6) can be derived as

$$\lim_{V_p \to \infty} (Q_v)_{max} = k_v \frac{\gamma' H^2}{2 \tan \theta} = k_v W' \tag{9}$$

Seismic Passive Thrust
When the soil wedge reaches the passive limit state, all the forces acting on the wedge include the total seismic passive earth thrust (P_{pe}), the self-weight of the soil wedge ($W = \gamma_{sat} H^2 / (2 \tan \theta)$), the horizontal and vertical seismic inertial forces (Q_h, Q_v), the hydrostatic pressure on both sides of the earth wedge (U_h, U_p) and the force from the backfill soil (R). The directions of all the forces in this derivation are illustrated as Fig. 1. The horizontal and vertical force equilibrium conditions of the soil wedge are expressed as

$$P_{pe} \cos \delta' + Q_h - R \sin(\theta + \varphi') + U_h - U_p \sin \theta = 0 \tag{10}$$

$$W + P_{pe} \sin \delta' - Q_v - R \cos(\theta + \varphi') - U_p \cos \theta = 0 \tag{11}$$

where φ' is the soil friction angle and δ' is the wall friction angle.

Substituting Eq. (10) into Eq. (11), the hydrostatic pressure on the both sides of the soil wedge is offset, therefore the total self-weight W of the soil wedge is replaced by the submerged weight W'. Then the seismic passive earth thrust of submerged condition is given as

$$P_{pe} = \frac{W' \sin(\theta + \varphi') - Q_h \cos(\theta + \varphi') - Q_v \sin(\theta + \varphi')}{\cos(\theta + \varphi' + \delta')} \tag{12}$$

Substituting Eqs. (5), (6) into Eq. (12), the modified pseudo-dynamic expression of seismic passive earth pressure can be expressed as

$$
\begin{aligned}
P_{pe} = \frac{1}{2}\gamma^* H^2 \Bigg\{ & \frac{1}{\tan\theta}\frac{\sin(\theta+\varphi')}{\cos(\theta+\varphi'+\delta')} - \frac{\cos(\theta+\varphi')}{\cos(\theta+\varphi'+\delta')} \cdot \left[\frac{k_h}{2\pi^2\tan\theta}\frac{\lambda}{H}m_1 + \frac{k_h(f_a-1)}{2\pi^3\tan\theta}\frac{\lambda}{H}m_2 \right] \\
& - \frac{\sin(\theta+\varphi')}{\cos(\theta+\varphi'+\delta')} \cdot \left[\frac{k_v}{2\pi^2\tan\theta}\frac{\eta}{H}m_3 + \frac{k_v(f_a-1)}{2\pi^3\tan\theta}\frac{\eta}{H}m_4 \right] \Bigg\}
\end{aligned}
\tag{13}
$$

where the value of γ^* is defined according to the seismic inertial action and the permeability,

$$
m_1 = 2\pi\cos\omega\zeta + \left(\frac{\lambda}{H}\right)(\sin\omega\zeta - \sin\omega t)
$$

$$
m_2 = 2\pi\left[\pi\cos\omega\zeta + \frac{\lambda}{H}\sin\omega\zeta\right] + \left(\frac{\lambda}{H}\right)^2(\cos\omega t - \cos\omega\zeta)
$$

$$
m_3 = 2\pi\cos\omega\psi + \left(\frac{\eta}{H}\right)(\sin\omega\psi - \sin\omega t)
$$

$$
m_4 = 2\pi\left[\pi\cos\omega\psi + \frac{\eta}{H}\sin\omega\psi\right] + \left(\frac{\eta}{H}\right)^2(\cos\omega t - \cos\omega\psi)
$$

The intensity distribution of the passive earth pressure is $p_{pe} = \partial P_{pe}/\partial z$, and the passive earth pressure coefficient of submerged backfill is $K_{pe} = 2P_{pe}/(\gamma' H^2)$. Through the derivation, it is demonstrated that K_{pe} is a function of the dimensionless expressions H/λ, H/η, t/T and the rupture surface angle θ. H/λ and H/η describe the ratio of the wall height to the seismic wavelength, which can be taken as constant when the material parameters of the backfill are determined. As a result, the minimum value of K_{pe} is obtained by optimizing it with respect to t/T and θ. The range of t/T is from 0 to 1, and the range of θ is from 0 to $\pi/2$.

3 Comparison

To illustrate the validity of the present method, the results calculated by the proposed formula are compared with those calculated by pseudo-static method and pseudo-dynamic method [14] in dry soil condition as shown in Fig. 2, where $k_h = 0.2$, $k_v = 0.1$, $\varphi' = 30°$, $\delta' = 15°$. From the figure, it can be seen that the distribution of passive earth pressure obtained by pseudo-dynamic method is a nonlinear form, while the results by pseudo-static method is linear. This discrepancy is mainly due to the difference of the basic hypothesis between the two kinds of methods. Meanwhile, it can also be drawn that the earth pressure intensity of the submerged backfill is lower than that of the dry soil, at the same time the nonlinear characteristic of the pressure distribution curve of the submerged condition is relatively more obvious than the dry one under the same earthquake.

Table 1 lists the comparison of the seismic passive earth pressure coefficients K_{pe} between the submerged backfill and dry soil conditions. It can be observed that the trends of K_{pe} between the two conditions are basically consistent, but the value of the submerged condition is somewhat smaller and the difference gets greater with the increase of seismic action. Thus, the groundwater has a certain role in reducing the seismic passive earth pressure.

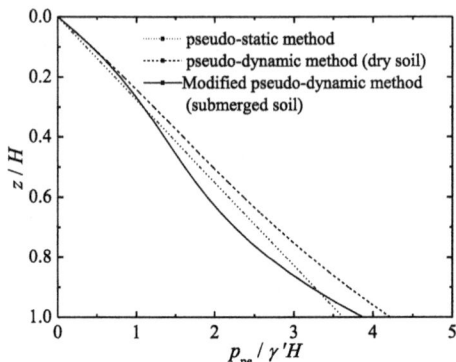

Fig. 2. Comparison of seismic passive earth pressure

Table 1. Comparison of K_{pe} between submerged soil and dry soil

$\varphi'/°$	$\delta'/°$	k_h	k_v	K_{pe} (submerged soil)	K_{pe} (dry soil)
20	10	0.05	0.025	2.292	2.441
		0.1	0.05	1.906	2.241
		0.15	0.075	1.344	2.032
25	12.5	0.05	0.025	3.118	3.301
		0.1	0.05	2.657	3.048
		0.15	0.075	2.130	2.787
30	15	0.05	0.025	4.402	4.638
		0.1	0.05	3.809	4.300
		0.15	0.075	3.175	3.957

4 Results and Discussion

In this section, the effects of seismic amplification factor, soil friction angle, wall friction angle, horizontal and vertical seismic actions on the distribution of seismic passive earth pressure of submerged backfill are analyzed. Only the restrained water situation is considered in this discussion. The values of the basic parameters are listed as follows: $\gamma_{sat} = 19$ kN/m3, $\gamma' = 9$ kN/m3, $H/\lambda = 0.3$, $H/\eta = 0.16$.

The influence of seismic amplification factor f_a on the distribution of passive earth pressure is shown in Fig. 3, where $\varphi' = 30°$, $\delta' = \varphi'/2$, $k_h = 0.2$, $k_v = k_h/2$, f_a ranges from 1 to 1.3. From the figure it can be seen that the distribution of passive earth pressure calculated by the modified pseudo-dynamic method is curved. The seismic amplification has a certain increase effect on the passive earth pressure, and the growth rate of the pressure strength increase with the vertical depth. The magnification of the earthquake loading in the soil is objective and cannot be neglected in the seismic design, although a linear assumption is taken to simplify the effect. Accordingly f_a takes the value of 1.2 in the following calculation.

Figure 4 presents the influence of soil friction angle φ' on the distribution of earth pressure, in which φ' ranges from 15° to 30° with a step length of 5°, $\delta' = \varphi'/2$, $k_h = 0.1$, $k_v = 0.05$. It is seen from the figure that the soil friction angle has a significantly increasing effect on the seismic passive earth pressure, and the enlargement rate also gets higher with larger soil friction angle. The gap among the curves of p_{pe} becomes more remarkable with the wall depth. Besides, under the same earthquake action, the curve of earth pressure distribution tends to be linear with the increase of φ', which is demonstrated that the effect of soil friction angle will play a dominant role in seismic passive earth pressure with larger value.

Figure 5 depicts the effect of wall friction angle δ' on seismic passive earth pressure with $\varphi' = 24°$, $\delta' = 0$, $\varphi'/3$, $\varphi'/2$, $2\varphi'/3$, $k_h = 0.1$, $k_v = 0.05$. From the plot, it is seen that the increasing influence of wall friction angle on p_{pe} is similar to that of soil friction angle, but the growth rate of p_{pe} with δ' is inferior to that with φ'.

Figure 6 illustrates the variations of p_{pe} with z/H under different horizontal seismic acceleration coefficient k_h with $\varphi' = 30°$, $\delta' = \varphi'/2$, $k_h = 0$, 0.05, 0.1, 0.15, 0.2, $k_v = k_h/2$. From the figure, it is shown that the passive earth pressure decreases with the increase of k_h. For example, the value of p_{pe} at the bottom of wall decreases about 12.8% when k_h changes from 0 to 0.1, and about 10.5% when k_h increases from 0.1 to 0.2. Moreover, as the backfill is assumed to vibrate in a harmonic form in the pseudo-dynamic analysis, the passive earth pressure distributes in a curve form under seismic action while the pressure distribution is linear without earthquake. The non-linearity of the distribution curve becomes more obvious as the earthquake gets stronger.

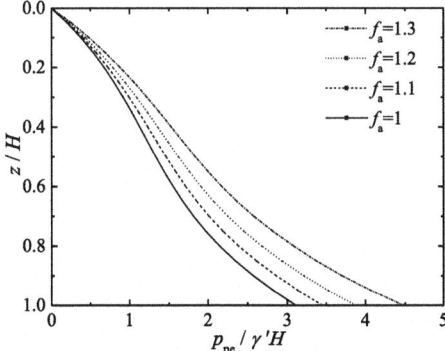

Fig. 3. Effects of f_a on seismic passive earth pressure distribution

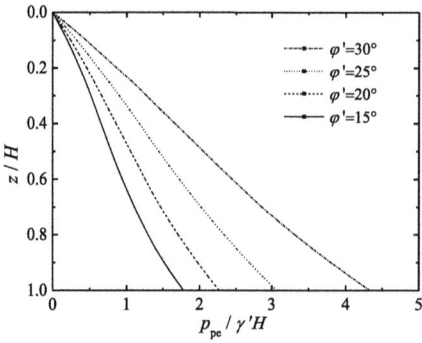

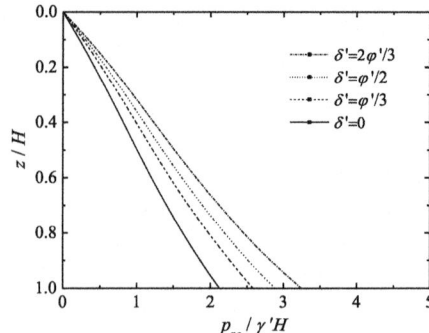

Fig. 4. Effects of φ' on earth pressure distribution

Fig. 5. Effects of δ' on earth pressure distribution

Figure 7 shows the seismic passive earth pressure under different vertical seismic acceleration coefficient k_v with $k_h = 0.15$, $k_v = 0$, $k_h/3$, $k_h/2$, $2k_h/3$. From the figure, the effect of vertical seismic force on passive earth pressure is much smaller than that of horizontal seismic force. In our calculation, the vertical seismic inertial force is supposed to be directed upward, thus the intensity of passive earth pressure decreases slightly with the increase of k_v.

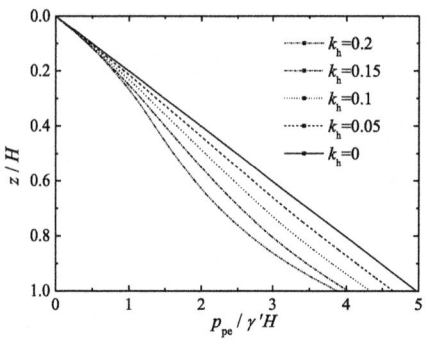

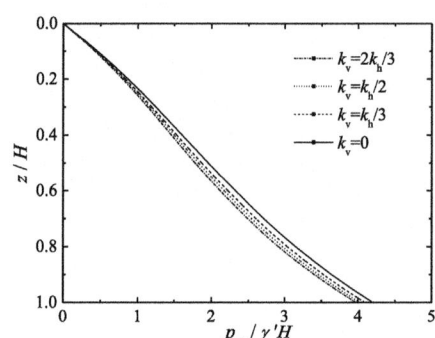

Fig. 6. Effects of k_h on earth pressure distribution

Fig. 7. Effects of k_v on earth pressure distribution

5   Conclusion

A modified pseudo-dynamic method to compute the seismic passive earth pressure for the submerged backfill is proposed in this study. The derivation is based on the limit equilibrium method with the assumption of planar failure surface, and it considers the influence of the groundwater condition and the earthquake amplification in the backfill on the seismic inertial forces.

The results manifest that the distribution of the seismic passive earth pressure by the proposed method is a non-linear form. Compared with the previous work, the trend of the earth pressure distribution of submerged backfill calculated by the proposed method is consistent with those obtained by the traditional pseudo-static and pseudo-dynamic approach of dry soil. Furthermore, it is worth noting that the submerged condition will reduce the passive earth thrust and the decreasing effect becomes greater with stronger earthquakes, which should be paid attention to in the seismic stability analysis of retaining walls.

Through the parametric study, the seismic acceleration amplification has a significantly increasing effect on the earth thrust, which cannot be neglected in the seismic design. The intensity of passive earth pressure increases with the increase of soil friction angle and wall friction angle where the former plays a more important role. The horizontal seismic action is still the key factor affecting the magnitude and distribution of the passive earth pressure rather than the vertical earthquake force.

References

1. Zhang, J., Shamoto, Y., Tokimatsu, K.: Seismic earth pressure theory for retaining walls under any lateral displacement. Soils Found. **38**(2), 143–163 (1998)
2. Soubra, A.H.: Static and seismic passive earth pressure coefficients on rigid retaining structures. Can. Geotech. J. **37**(2), 463–478 (2000)
3. Nouri, H., Fakher, A., Jones, C.: Evaluating the effects of the magnitude and amplification of pseudo-static acceleration on reinforced soil slopes and walls using the limit equilibrium horizontal slices method. Geotext. Geomembr. **26**(3), 263–278 (2008)
4. Steedman, R.S., Zeng, X.: The influence of phase on the calculation of pseudo-static earth pressure on a retaining wall. Geotechnique **40**(1), 103–112 (1990)
5. Choudhury, D., Nimbalkar, S.: Pseudo-dynamic approach of seismic active earth pressure behind retaining wall. Geotech. Geol. Eng. **24**(5), 1103–1113 (2006)
6. Choudhury, D., Nimbalkar, S.: Seismic passive resistance by pseudo-dynamic method. Geotechnique **55**(9), 699–702 (2005)
7. Ruan, X., Sun, S., Liu, W.: Effect of the amplification factor on seismic stability of expanded municipal solid waste landfills using the pseudo-dynamic method. J. Zhejiang Univ. Sci. A **14**(10), 731–738 (2013)
8. Huang, R., Xia, T.D.: Seismic stability of gravity retaining structures by pseudo-dynamic method. Marine Georesour. Geotechnol. **34**(1), 1–9 (2016)
9. Bellezza, I., D'Alberto, D., Fentini, R.: Pseudo-dynamic approach for active thrust of submerged soils. Proc. ICE-Geotech. Eng. **165**(5), 321–333 (2012)
10. Matsuzawa, H., Ishibashi, I., Kawamura, M.: Dynamic soil and water pressures of submerged soils. J. Geotech. Eng. **111**(10), 1161–1176 (1985)
11. Ebeling, R.M., Morrison Jr., E.E.: The Seismic Design Of Waterfront Retaining Structures. Office of Navy Technology & Department of the Army, Washington DC (1993)
12. Wang, J.J., Zhang, H.P., Chai, H.J., et al.: Seismic passive resistance with vertical seepage and surcharge. Soil Dyn. Earthq. Eng. **28**(9), 728–737 (2008)
13. Wang, J.J., Zhang, H.P., Liu, M.W., et al.: Seismic passive earth pressure with seepage for cohesionless soil. Mar. Georesour. Geotechnol. **30**(1), 86–101 (2012)
14. Nimbalkar, S., Choudhury, D.: Sliding stability and seismic design of retaining wall by pseudo-dynamic method for passive case. Soil Dyn. Earthq. Eng. **27**(6), 497–505 (2007)

Effects of Foundation Pit Width on the Anti-overturn Stability of Its Support Structure Under Ground Load

Shengfeng Zhao[1]([✉]), Zhiyang Chen[2], and Guanglong Huang[1]

[1] College of Transportation Science and Engineering,
Nanjing Tech University, Nanjing, China
442877591@qq.com
[2] Nanjing Institute of Surveying, Mapping & Geotechnical
Investigation Co., Ltd., Nanjing, China

Abstract. In this study, the anti-overturn stability of excavation support structure is firstly analyzed based on Rankine's earth pressure theory and the static equilibrium method. Then the row piles plus inner support structure is taken as the research project, and the criteria for classification of excavations are proposed based on the anti-overturn stability calculation with the consideration of the effects of both pit width and ground load and derived a formula for calculating anti-overturn safety coefficient of the row pile support structure. At last, the formula is applied to analyze the effects of both pit width and ground load on the anti-overturn stability of two examples in the JGJ120-2012 technical specification and literature. The results showed that the calculation formulas of the anti-overturn stability of excavation support system used in both the technical specification and literature are the special cases of the formula derived in this study under conditions of broad excavation and without ground load, respectively, indicating that our formula is more feasible.

Keywords: Excavation classification · Excavation width · Ground load
Anti-overturn stability

1 Introduction

Rapid economic development has promoted and is promoting the construction of a large amount of metropolitan subways, civil air defense works, and underground integrated pipe galleries, resulting in the excavation of more and more trenches for foundation [1]. However, most of these works lie in big cities and run through the residential areas. Small construction sites make foundation trench excavation more disadvantageous. For these projects, adopting soil nails or anchor supports will affect dwelling houses, pipelines and underground facilities outside the trenches. Therefore, it is necessary using combined support form of both row stakes and internal supports.

Many construction practices adoption of row pile support structure with same embedded depth of support piles in excavation engineering often results in satisfaction in overall stability test, but the failure to meet the specifications in anti-overturn stability examinations. When other conditions are the same, the embedded depth of the

© Springer Nature Singapore Pte Ltd. 2018
T. Qiu et al. (Eds.): GSIC 2018, *Proceedings of GeoShanghai 2018 International Conference: Advances in Soil Dynamics and Foundation Engineering*, pp. 488–502, 2018.
https://doi.org/10.1007/978-981-13-0131-5_53

narrow excavation support piles could be appropriately shortened. With deeper, but narrower excavation, the effect of foundation trench width on the stability should be taken into consideration [2]. Therefore, it is inevitable to cause waste to design foundation trench supports without considering the effect of trench width on the anti-overturn stability and safety coefficients in accordance with the formulas in "Technical Specification for Retaining and Protection of Building Foundation Excavations" (JGJ120-2012) [3] (hereinafter shorted as the technical specification).

In recent years, the effects of excavation width on its stability have been explored by some researchers. He et al. [4] analyzed the arching characteristics and forming mechanism of the bottom of wide and large soft-soil foundation trenches, divided the bottom soil body into the strong action zone, weak action zone, and free deformation zone, and put forward a method for dividing the boundary of soil's strong and weak action zones with comprehensive consideration of the excavation depth, support structure insertion ratio and the soil characteristics. Zhu et al. [5] used a soft soil excavation project as an example and proposed an approach using the pressure of passive soil on the coupling interface to increase the anti-sliding torque and decrease the support structure depth with consideration of the coupling effect of the bilateral sliding faces of narrow foundation trenches in the middle of the trenches. Ying et al. [6] established a theoretical model for computing the passive earth pressure on the rigid retaining wall of narrow excavations based on the traditional Coulomb plane soil wedge hypothesis and showed that when the width of the passive earth mass is smaller than the critical value satisfying the semi-infinite soil body and the friction angle of the wall soil is larger than zero, the inclination angle of the passive slip-fracture surface is larger than that of the traditional Coulomb passive slip-fracture surface and the passive earth pressure is greater than the classical Coulomb solution. In addition, the narrower the excavation is, the greater the wall soil friction angle and difference between the passive earth pressure and the classical Coulomb solution. According to Terzaghi's theory of foundation bearing capacity, Wang et al. [7] deduced a formula for calculating the safety coefficient against heave-resistant stability, proposed the concept of the critical width of the minimum safety coefficient and presented the theoretical solution of the critical width of homogeneous foundations. Wang [8] derived a calculation formula of the anti-overturn stability safety coefficient based on the classical earth pressure theory with consideration of the effect of the excavation width (hereinafter referred as the literature method), but not the effect of ground load on the overturn stability in construction process of the excavation. However, in actual construction, the ground load induced by the stacked construction materials and machinery is non-negligible. Aiming at the ground load against the excavation, Li et al. [9] studied the effects of external load and its applied area on the horizontal support structure displacement, earth surface subsidence, and bottom soil arching based on Biot's three dimensional consolidation theory and found that load amount has a greater effect on horizontal support structure displacement, earth surface subsidence, and bottom soil arching, while its applied area has a greater effect on the horizontal displacement and surface subsidence of the supporting structure, but a slighter effect on bottom soil arching. Wang and Cao [10] studied the effects of pit margin overload and excavation on pit deformation and failure using the foundation pit of subway station as example and found that the horizontal displacement of the wall top increased with pit-margin overload constantly increasing.

Based on the previous researches, in this study, the impacts of both pit width and ground load on the anti-overturn stability of the supporting structure are considered, a new calculation formula of the anti-overturn stability is deduced according to Rankine's earth pressure theory, and the calculation results with those calculated are compared based on the technical specification method and literature method using engineering examples.

2 Analysis of Pit Support Structure Overturn Disability

When there is a larger ground load outside the pit, using the row pile support structure is necessary but corresponding, the stress applied on the soil mass behind the support piles will increase. Therefore, insufficient embedded depth of the retaining piles will lead to severe deformation even collapse of the support system. According to Rankine's soil pressure theory, when the overturn torque is greater than the anti-overturn torque, the support structure will turn around a point and capsize. When the soil mass on the outer side of the excavation undergoes active soil pressure failure, it will result in the formation of a fracture surface with an angle of $45° - \varphi/2$ against the vertical direction. When the soil mass of the inner side of the pit undergoes passive fails, it will results in a fracture surface with an angle of $45° + \varphi/2$ against the vertical direction, where φ is the soil's internal friction angle [11]. Figure 1 shows the schematic diagram of the overturn destabilization and failure of the foundation pit.

The cantilever support structure, pile-anchor support structure with a single layer anchor, and pile-strut support structure with a single row pile layer will undergo destabilization and failure. Assuming that the support pile end, anchor head and strut fulcrum all will undergo overturn destabilization and failure, the formula for calculating the anti-overturn safety coefficient K_c in the technical specification follows [3]:

$$K_c = \frac{E_p \times a_{pl}}{E_a \times a_{al}} \quad (1)$$

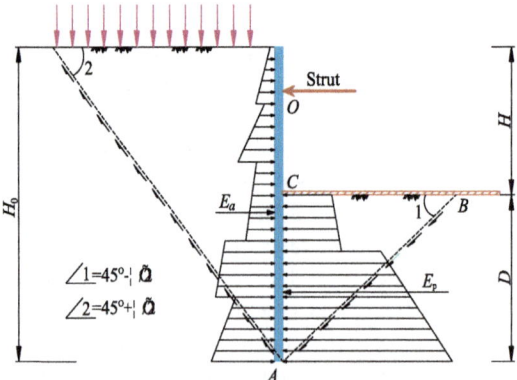

Fig. 1. Schematic of excavation destabilization and failure

where E_p and E_a are the passive and active soil pressures on the outer side and inner side of the pit respectively; a_{pl} and a_{al} are the distances from the support pile bottom to the acting point of the resultant force of both passive and active soil pressures on the outer and inner sides of the pit, respectively; K_c is anti-overturn safety coefficient.

3 Effects Analysis with Consideration of Pit Width

3.1 Classification of Pit Width Based on Earth Pressure Principle

This study uses Rankine's earth pressure theory for calculation and analysis due to its simple hypothetical conditions [12]. According to its basic principle, when excavation width $B \geq D\ tan(\pi/4 + \varphi/2)$, the soil mass of the pit bottom in the passive zone undergoes sliding wedge failure. Otherwise, the support structure will impose a constraint upon the sliding wedge of soil in the passive zone, preventing the soil from falling into the wedge failure. Therefore, the ratio of the excavation width B to the embedded depth of the support pile D can be used to classify foundation trenches into three types. Type 1 trenches are the so-called narrow foundation trenches, meeting $B/D \leq tan\varphi$. The angle θ formed by both the rupture surface and the vertical direction is smaller than the internal friction angle φ of soil of the passive zone along the pit width direction, thus the passive zone is safe. As long as the support piles meet the strength requirement, they will satisfy the requirement for anti-overturn stability. Type 2 trenches are the so-called ordinary pits, meeting $tan\varphi < B/D < tan(\pi/4 + \varphi/2)$. The support piles on both sides of the foundation trench exert their constraints on the sliding wedge of soil of the passive zone. Their width affects the anti-overturn stability. The narrower the width is, the better the anti-overturn stability. Type 3 trenches are the so-called wide trenches, meeting $B/D \geq tan(\pi/4 + \varphi/2)$. As the soil of the passive zone collapses, the support piles on both sides of the excavation lose their restraints on the sliding wedge of soil. Thus, the pit width no longer has impact on the anti-overturn stability. Figure 2 shows the schematic diagram of pit widths classification.

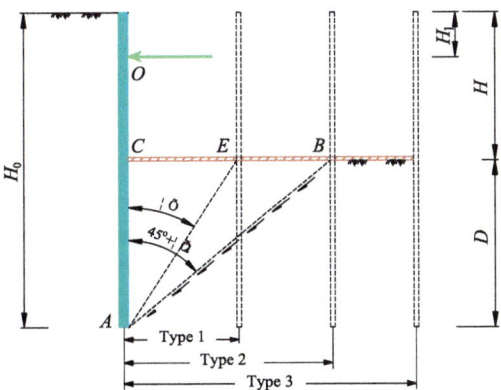

Fig. 2. Schematic of classification of excavation width

3.2   Classification of Pits with Consideration of Ground Load

In the design of pit supports, the embedded depth of the support structure is inversely calculated according to the safety coefficient given by the technical specification. With the consideration of ground load, the active earth pressure E_a must increase compared to that without considering the ground load. Accordingly, the active earth pressure torque also increases. Therefore, when the same anti-overturn safety coefficient is satisfied, the embedded depth of support piles inevitably increases to D_1 which changes the ratio of the excavation width to the embedded depth of the support pile. In turn, the pit width classification also changes accordingly. Figure 3 shows the comparison of pit width classifications, where H_1 is the burial depth of the lowest support.

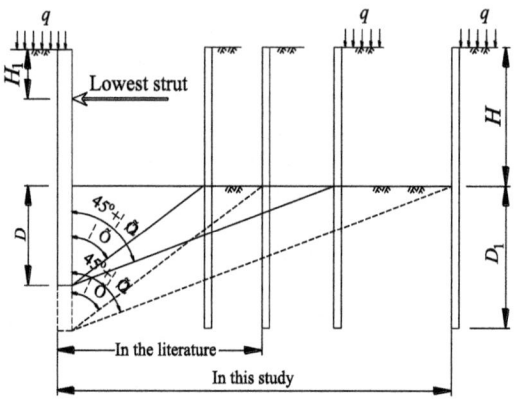

Fig. 3. Schematic of comparison in pit width classification

Once the anti-overturn stability factor $K_c > a$, a constant, is determined, under the ground load, the torque induced by the active earth pressure on the outside of the pit increases with ground load increasing. To satisfy the requirement of K_c, the torque generated by the passive earth pressure on the inside the pit also increases accordingly. However, increasing of the torque generated by the passive earth pressure can only be achieved by increasing the embedding depth of the support structure.

With the same anti-overturn safety coefficient K_c and the same soil internal friction angle φ, as the embedded depth increases, the volume of the sliding wedge of soil enlarges accordingly, resulting in a greater range of damage in the horizontal direction. If the pit width is still classified using the above method shown in Fig. 2, the division range is obviously smaller than that done with consideration of ground load. Therefore, considering the effect of ground load is more in line with the engineering reality.

4 Calculation of Anti-overturn Stability of the Support Structure

4.1 Calculation of Passive Soil Pressure of the Support Structure

According to the basic assumptions of Rankine's soil pressure principle, when the passive zone falls into failure, the collapsed wedge ΔABC is formed as shown in Fig. 4, where the gravity of the wedge is W and the force produced by the soil friction and applied on the passive rupture surface is R, whose normal angle against the fractured surface is Φ. When the soil layer is cohesive, on the fracture surface is still applied cohesion-induced cohesive force C with magnitude of $c \cdot \overline{AB}$.

According to theoretical mechanics, when the forces suffered by the wedge reaches the equilibrium, they meet the condition of force balance, namely,

$$\sum F_{xi} = 0, \sum F_{yi} = 0, \sum M_o(F) = 0 \tag{2}$$

According to the vertical force balance equilibrium of the collapsed wedge ΔABC and the law of Sine in the trigonometry $\frac{a}{\sin A} = \frac{b}{\sin B} = \frac{c}{\sin C}$, one has:

$$R = \frac{W + c \cdot D}{\sin(\theta - \phi)} = \frac{\gamma \cdot D^2 \tan\theta + 2c \cdot D}{2\sin(\theta - \phi)} \tag{3}$$

The anti-overturn safety coefficient of the support structure can be worked out by using Eqs. (2) and (3) under the joint consideration of both pit width and ground load effects, and the passive earth pressure can be found by using Rankine's earth pressure hypothesis when the excavation suffers destabilization and failure:

$$E_p = \left(\frac{1}{2}\gamma D^2 \tan\theta + cD\right)\cot(\theta - \varphi) + cD\tan\theta \tag{4}$$

When the support structure under the action of stress has to shift to the direction of the soil mass in the horizontal direction, and the stress to which the soil is subject in the horizontal direction adds up to a certain value, the foundation soil will enter into the passive failure state. When the soil in the passive zone suffers failure, the earth pressure E_p is at its minimum, thus the following relation holds:

$$\begin{aligned}
\frac{\partial E_p}{\partial \theta} &= \frac{\partial E_1}{\partial \theta} + \frac{\partial E_2}{\partial \theta} \\
&= \left[\frac{1}{2}\gamma D^2 \tan\theta \times \cot(\theta - \varphi) + cD\cot(\theta - \varphi)\right]d\theta + cD\tan\theta d\theta
\end{aligned} \tag{5}$$

$$\frac{\partial E_p}{\partial \theta} = -\frac{1}{\sin^2(\theta - \varphi)} \times \left(\frac{1}{2}\gamma DL - cD\right) = 0 \tag{6}$$

Solving Eq. (6) can find the passive soil pressure E_p.

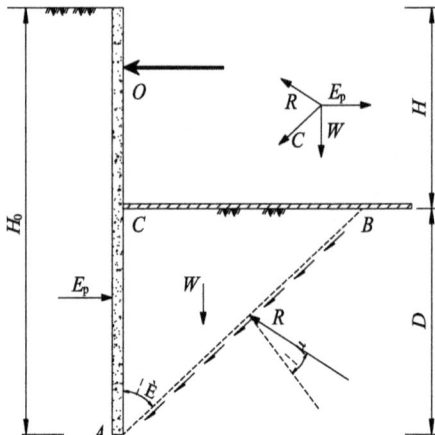

Fig. 4. Schematic of forces acting on the passive zone and its resultant failure

4.2 Calculation of the Overturn Torque of the Support Structure

In considering the ground load, in accordance with the "Technical specification for Retaining and Protection of Building Foundation Excavation" (JGJ 120-2012), the additional load on the surface diffuses downward at the angle of 45° from the vertical direction. In accordance with the principle of classical earth pressure, the diffusion angle of the additional load on the surface is $(45° - \varphi/2)$ [13], as shown in Fig. 5. The classic soil pressure principle is used in this study for calculating the diffusive angle of the additional load, and the overturn torque is obtained according to the following formulas:

$$M_a = M_{a1} + M_{a2} \tag{7}$$

$$M_{a1} = \left(K_a\gamma H_1 - 2c\sqrt{K_a}\right)\frac{(H + D_1 - H_1)}{2} \tag{8}$$

$$M_{a2} = \frac{1}{2}K_a\gamma\left(H + D_1 - H_1\right)^2 \times \frac{2}{3}(H + D_1 - H_1) \tag{9}$$

where K_a is the active soil pressure coefficient. If there is the surface load, the method to calculate the overturn torque induced by the surface load is given as follows:

(1) As the uniformly-distributed load is present,

$$M_q = K_a\Delta\sigma_k(H + D_1) \times \left(\frac{1}{2}H + \frac{1}{2}D_1 - H_1\right) \tag{10}$$

where $\Delta\sigma_k$ is the value of uniformly-distributed load.

(2) As the local load is present,

$$M_q = K_a[\frac{P_0b}{b + 2a \div \tan(45° - \varphi/2)}] \times \min[b + \frac{2a}{\tan(45° - \frac{\varphi}{2})}, H + D_1 - a]$$
$$\times |Z_a - H_1| \tag{11}$$

where P_0 is the value of uniform pressure q, Z_a is the Vertical distance from the top of the retaining structure to the additional vertical stress calculation point in the soil.

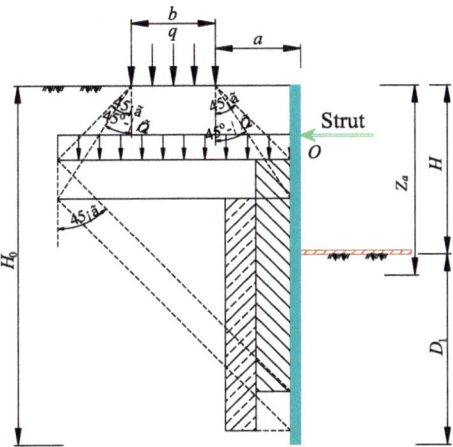

Fig. 5. Schematic of the acting range and related diffusive angle of additional load on the earth

4.3   Calculation of Anti-overturn Torque of the Support Structure

The minimum passive earth pressure of the ordinary excavation is calculated using Eq. (4), where the passive earth pressure E_{p1} includes both self-gravity and cohesion of soil and is related to the embedded depth of supports into soil. The acting point of the joint passive earth pressure force due to soil's self-gravity is $2/3D_1$, while that due to soil cohesion is $1/2D_1$.

$$E_{p1}^1 = \frac{1}{2}\gamma D_1 B \cot\left(\arctan\frac{B}{D_1} - \theta\right) \tag{12}$$

$$E_{p1}^2 = c\left[D_1\cot\left(\arctan\frac{B}{D_1} - \theta\right) + B\right] \tag{13}$$

$$E_p = E_{p1} = E_{p1}^1 + E_{p1}^2 \tag{14}$$

When the excavation depth of the trench advances to the depth H, and the burial depth of the lowest strut (the distance from the surface to the strut position) is H_1, the anti-overturn torque of the general excavation is

$$M_{P1} = \frac{1}{2}\gamma D_1 B \cot\left(\arctan\frac{B}{D_1} - \theta\right)\left(H - H_1 + \frac{2}{3}D_1\right) + c\left[D_1\cot\left(\arctan\frac{B}{D_1} - \theta\right) + B\right]\left(H - H_1 + \frac{1}{2}D_1\right)$$

(15)

For broad foundation pit, the calculations of both passive earth pressure and Rankine's earth pressure are similar, that is, $E_p = E_{p2} \cdot E_{p2}$, and the anti-overturn torque M_{P2} are respectively listed as follows:

$$E_{p2} = \frac{1}{2}\gamma D_1^2 \tan^2\left(\frac{\pi}{4} + \frac{\varphi}{2}\right) + 2cD_1\tan\left(\frac{\pi}{4} + \frac{\varphi}{2}\right)$$

(16)

$$M_{p2} = \frac{1}{2}\gamma D_1^2 \tan^2\left(\frac{\pi}{4} + \frac{\varphi}{2}\right)\left(H - H_1 + \frac{2}{3}D_1\right) + 2cD_1\tan\left(\frac{\pi}{4} + \frac{\varphi}{2}\right)\left(H - H_1 + \frac{1}{2}D_1\right)$$

(17)

For narrow foundation pit, the anti-overturn torque M_p tends to be infinite, thus no failure occurs.

4.4   Calculation of Anti-overturn Safety Coefficient

From the above analyses, the calculation methods of the anti-overturn safety coefficients of the pit support structure under the joint consideration of ground load and pit width are summarized as follows: (1) For the narrow foundation pit, K_c tends to positive infinite. Therefore, there is no need for check; (2) For the ordinary foundation pit, $K_c = M_{p1}/(M_a + M_q)$, where M_{p1} is calculated from Eq. (15) and M_a from Eq. (7); (3) For the broad foundation pit, $K_c = M_{p2}/(M_a + M_q)$, where M_{p2} is calculated according to Eq. (17), and M_a according to Eq. (7). This study considers the effect of the ground load around the pit and makes the active earth pressure calculation more conformable to the reality, thus the safety coefficient is more accurate.

5   Comparison and Analysis of the Results Calculated Using the Above-Mentioned 3 Methods

Suppose that some foundation trench project adopts the row-piles-plus-internal-struts support structure, the excavation depth is $H = 15.5$ m, the burial depth of the lowest support is $H_1 = 11.0$ m, the embedding depth is $D_1 = 16.8$ m, the whole soil layer in the influential range of the foundation fit is the cohesive soil with the cohesion is $c = 20.0$ kPa, the internal friction angle is $\varphi = 14.0°$, and the soil buck density $\gamma = 19$ kN $\cdot$ m^{-1}.

5.1   Analysis of Overturn Stability of Soil Subject to Different Ground Loads

Figure 6 shows the relationships of the anti-overturn safety coefficient of the excavation to its width obtained by using the three methods under semi-infinite loads of three ground actions of 20, 30 and 40 kPa. It can be seen from Fig. 6 that (1) the effect of the excavation width is not taken into account by the technical specification, thus the

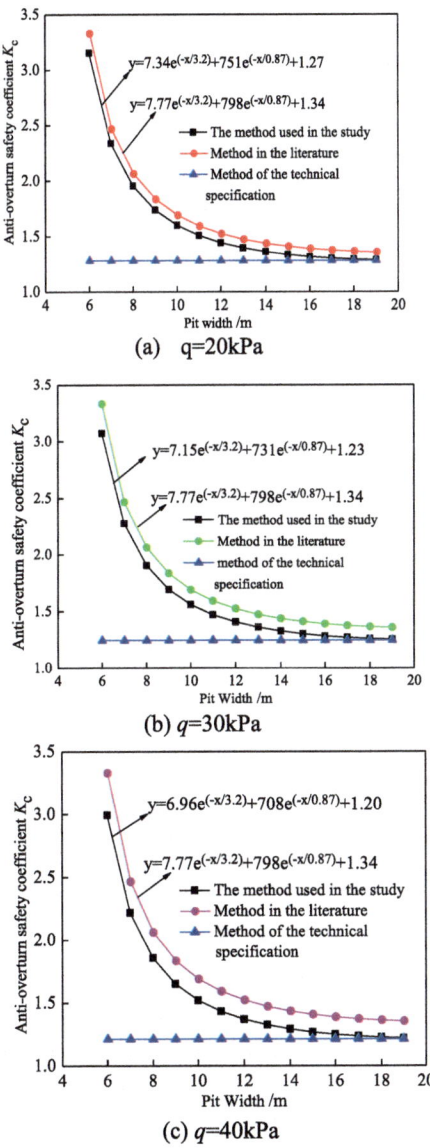

Fig. 6. Relationships of the anti-overturn safety coefficient are calculated using the three methods to the foundation pit width at tri ground force of 20, 30 and 40 kPa, respectively

calculated safety coefficient under the action of the ground load is a constant and does not change with the pit width varying; (2) The literature method[8] introduces the effect of pit width on the anti-overturn stability but doesn't consider the action of ground load. Its calculated safety coefficient of the excavation decays exponentially with the pit width increasing; (3) Our method considers both the effect of pit width and the action of ground load and the calculated safety coefficient changes consistently with that of the literature method but with a slightly lowering extent.

5.2   Analysis of Anti-overturn Stability Under Different Load Acting Distances

Figure 7 shows the relationships between the anti-failure safety coefficient of the excavation to its width calculated with the ground load $q = 20$ kPa as the semi-infinite load and the distances from the load piled site to the edge of the pit of 10.0 m and 20.0 m, respectively. From Fig. 7, clearly, with the increase of the distance from load

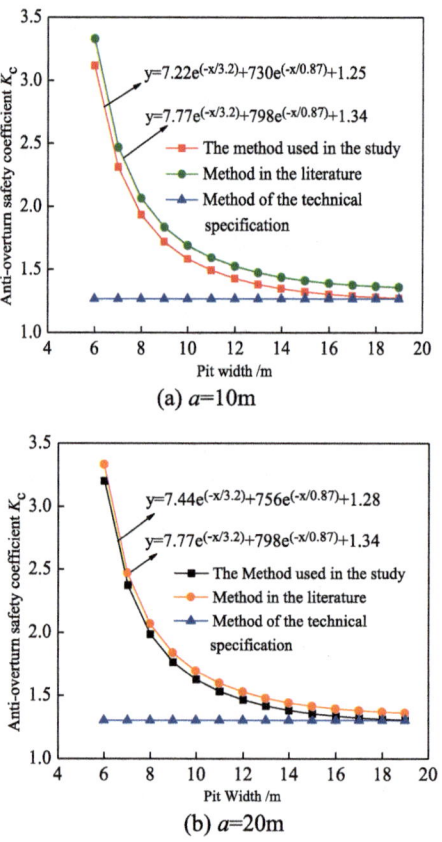

Fig. 7. Relationships of anti-overturn safety coefficient are calculated at $q = 20$ kPa and the distances from load site to pit edge of 10.0 m and 20.0 m, respectively, to the foundation pit width

site to pit edge, the effect of the pit width on the anti-overturn stability decreases gradually. When the pit width exceeds one times as the excavation depth, the effect of the pit width on the anti-overturn stability tends to zero gradually, in consistence with the case where the ground load diffuses downward at the diffusion angle of $45° - \varphi/2$. This indicates that as the row pile plus strut structure is used, the ground load with the range of the one-fold excavation depth from the pit edge should be analyzed in detail and valuated reasonably so as to enable us to avoid the insufficient embedded length of the support pile and the risk of pit failure. According to the exponential fitting of the calculated results, the corresponding anti-overturning safety factor for different pit widths corresponds to the equation as follows:

$$y = Ae^{(-x/3.2)} + Be^{(-x/0.87)} + C \tag{18}$$

Where A, B, C are constants.

5.3 Analysis of Anti-failure Stability Under Actions of Different Pit Width

Figure 8 shows the relationships between the anti-overturn safety coefficient of the excavation to its width calculated with the ground load $q = 30$ kPa and three strip-shaped loads of 10.0 m, 20.0 m, and 30.0 m respectively. From the figure, obviously, when the pit width is 0.6 times as the excavation depth, the strip-shaped ground load has a greater effect on the pit's anti-overturn stability. When the pit width is > the excavation depth, the load has a gradually lowering effect on the anti-overturn stability. When the pit width is over 1.2 times as the excavation depth, the load has little effect on the stability of the excavation. These conclusions are in a rough agreement with the effects of the semi-infinite load on the pit's anti-overturn stability, revealing that the effect of strip load on the pit's anti-overturn stability also conforms to the downward diffusing situation with diffusion angle of $45° - \varphi/2$. Therefore, for the case that roads or strip-shaped buildings are present around the designed foundation trench, if the distance from the load to the edge of the pit is over 1.2 times of the excavation depth of the pit, it is reasonable not to consider the effect of the road load on the pit's anti-overturn stability in designing row-piles-plus-internal-struts support structure.

5.4 Comparative Analysis of Embedded Depth

The same soil layer status was used for our analysis. In this case, $tan\varphi$ is about 0.249 and $tan(45° + \varphi/2)$ is about 1.28. Therefore, foundation pits with $0.249 \leq B/D \leq 1.28$ are considered as the ordinary pits. Because narrow foundation pits do not undergo failure and the calculation of the broad foundation pits excavation is similar to that of Technical Specification, the following analysis focuses only on the ordinary foundation pits.

Assuming that the excavation depth is $H = 11.65$ m, the burial depth of the lowest support is $H_1 = 6.5$ m, the ground load as the semi-infinite load is $q = 30$ kPa, the

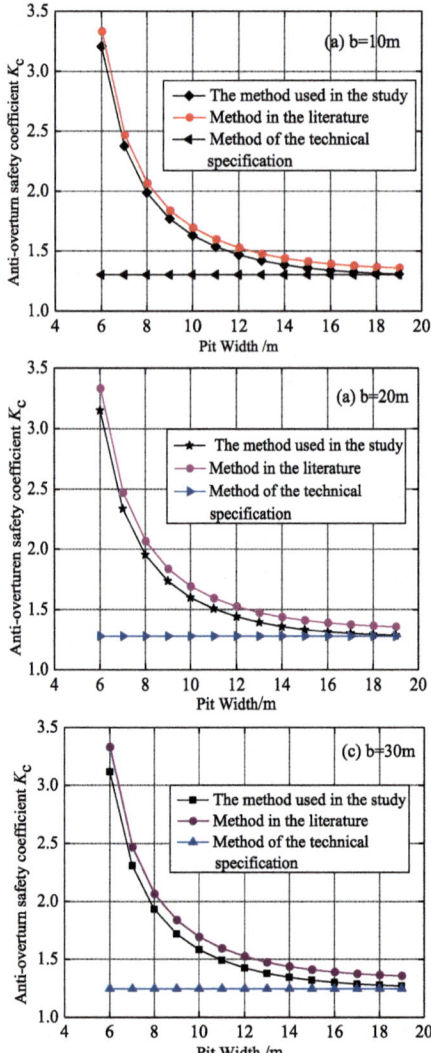

Fig. 8. Relationships between anti-overturn safety coefficient calculated at ground load $q = 30$ kPa and three strip-shaped loads of 10.0 m, 20.0 m, and 30.0 m, respectively, to the foundation pit width

excavation width, B, to be 6.0, 9.0 and 12.0 m, respectively, when the safety coefficient K_c reaches 1.3, the embedded depth of the row piles is calculated and listed in Table 1.

As can be seen from Table 1, it is clear that (1) the impact of pit width is not taken account by the technical specification and has the same anti-overturn safety coefficient, the embedded depth of the support piles keeps unchanged; (2) the literature method only considers the effect of pit width but not ground load, the calculated total lengths of the support piles are about 19.30, 20.70 and 21.40 m, respectively, and the embedding depth shortens by about 39.42%, 29.20% and 24.09%, respectively, compared with

Table 1. Comparison in embedded depth among three calculation methods

Pit width/m	$B = 6.0$			$B = 9.0$			$B = 12.0$		
	Specification	Literature	This study	Specification	Literature	This study	Specification	Literature	This study
Embedded depth/m	13.7	8.3	9.4	13.7	9.7	11.5	13.7	10.4	12.8
Safety coefficient K_c	1.3	1.3	1.3	1.3	1.3	1.3	1.3	1.3	1.3

those calculated in the technical specification; (3) the total length of the support pile calculated in this study are about 20.40, 22.50 and 23.80 m and the corresponding embedding depth shortens by about 31.39%, 16.06% and 6.57%, respectively, compared with those done in the technical specification. These results reveal that (1) the results obtained using the technical specification are too conservative and will result in waste in preparing support piles because it does not consider the effect of pit width on anti-overturn stability, (2) the results obtained using the literature method are too aggressive because it only considers the impact of pit width but not the ground load, and (3) the results obtained using the method in this study are more suitable to the engineering reality because it accommodates the former two.

To more clearly compare these three calculation methods, the anti-failure safety coefficients computed using the three methods for of a case with pit width $B = 5.0$ m and embedded depth $D_1 = 13.85$ m under the actions of different semi-infinite ground loads are given in Table 2. It can be seen from Table 2 that (1) when the ground load $q = 0$ kPa, the result calculated using the method in this study is consistent with that calculated using the literature method; (2) when the ground load $q > $ kPa and increase gradually, the safety coefficients calculated using the methods in this study and the technical specification reduce slowly, indicating that our calculation method is more representative; (3) under the conditions with same pit anti-overturn safety coefficient, but different ground loads, the corresponding embedded depths of the pit support pile also vary, which is very important for the determination of excavation support pile length in projects using row piles.

Table 2. Comparison in safety coefficients among three calculation methods under different ground loads

Ground load/kPa	This study	Literature	Specification	Ground load/kPa	This study	Literature	Specification
0.0	3.660	3.660	1.494	30	3.214	3.660	1.312
20	3.350	3.660	1.367	40	3.089	3.660	1.261

6 Conclusions

This study investigated the effect of excavation width of the support structure on anti-overturn stability with joint consideration of the ground load around the pit, and obtains the following main conclusions:

(1) Under the same excavation depth, the smaller the foundation pit width is, the smaller the embedded depth and under the same pit width, the greater the ground load on the pit is, the deeper the embedment depth of the support pile.

(2) In design and calculation of the row-piles-plus-internal-struts support structure, it is necessary to comprehensively analyze the ground load in the range of a $\leq$ H so as to avoid the risk of pit failure due to the insufficient depth of support piles. When the distance from the pit edge to the previously constructed road surrounding the pit is greater than 1.2 H, there is no need to consider the impact of road load on the anti-overturn stability.

(3) The effects of both pit width and ground load are simultaneously considered by the pit anti-overturn calculation method developed in this study, thus the result obtaining is more feasible because the method of the technical specification is only a special case of the study with broad pit width and the literature method is only a special case of the study without ground load.

References

1. Zhang, X.C.: Behavior of narrow-deep excavation in soft clay ground. Ph.D. thesis, Zhejiang University, Hangzhou (2014). (in Chinese)
2. Xu, Y.L.: Effects analysis of excavation width on retaining structure and soil nearby. Ph.D. thesis, Zhejiang University, Hangzhou (2014). (in Chinese)
3. Technical specification for retaining and protection of building foundation excavations (JGJ120—2012). China Architecture & Building Press, Beijing (2012). (in Chinese)
4. He, C., Chen, P., Zhou, S.H.: Influence of width effect of soft soil excavation on basal heave. J. East China Jiaotong Univ. 32(6), 82–87 (2015). (in Chinese)
5. Zhu, D.F., Wang, G.M., Liu, W.S.: Influence of excavation width on overall stability of pits. Port Waterw. Eng. 12, 188–191 (2013). (in Chinese)
6. Ying, H.W., Zheng, B.B., Xie, X.Y.: Study of passive earth pressures against translating rigid retaining walls in narrow excavations. Rock Soil Mech. 32(12), 3755–3762 (2011). (in Chinese)
7. Wang, C.H., Lu, Q., Sun, P.: Critical width method for analyzing stability of foundation pits against basal heave failur. Chin. J. Geotech. Eng. 28(3), 295–300 (2006). (in Chinese)
8. Wang, H.X.: Influence of excavation width on enclosure-structure stability of foundation pits. China Civ. Eng. J. 44(6), 120–126 (2011). (in Chinese)
9. Li, Y.Q., Zhou, X., Xie, K.H.: Influence of outside load on deformation behavior of foundation pit excavation in soft soil area. J. Archit. Civ. Eng. 33(4), 97–102 (2016). (in Chinese)
10. Wang, B.X., Cao, X.S.: Finite element analysis on design of long and narrow deep foundation pit. J. East China Jiaotong Univ. 32(1), 65–70 (2015). (in Chinese)
11. Wang, H.L., Song, E.X., Song, F.Y.: Analysis of earth pressure on the retaining structure due to surcharge near the excavation. Ind. Constr. 42(9), 90–95 (2012). (in Chinese)
12. Wang, S.C., Sun, B.J., Shao, Y.: Modified computational method for active earth pressure. Rock Soil Mech. 36(5), 1375–1379 (2015). (in Chinese)
13. Zhao, S.D.: Soil Mechanics. Higher Education Press, Beijing (2001). (in Chinese)

Study on the Force and Deformation Characteristics of Prefabricated Pile-Wall Compound Structure and Its Application in Foundation Pit Engineering

C. H. Xiu[1(✉)], H. Li[2], W. X. Wang[2], and Q. W. Xu[2]

[1] Jinan Rail Transit Group Co., Ltd., Jinan 250101, China
[2] Key Laboratory of Road and Traffic Engineering of Ministry of Education,
Tongji University, Shanghai 200092, China

Abstract. The combination of supporting structure and permanent structure is an effective way to build multi-story underground structure, and this new support form has been developed and applied rapidly in recent years. During the construction of the foundation pit of the West Yanmazhuang Station of Ji'nan Metro line 1, a new form of prefabricated pile-wall compound structure was proposed as the retaining structure. In order to investigate the force and deformation characteristics of this new structure, a three-dimensional 'load-structure' calculation model of the pile-wall compound structure is established, and the soil is simplified as a spring to simulate the lateral earth pressure acting on the structure. According to the connection performance between the pile and the wall, three cases are simulated, i.e. (1) the pile and wall are connected with shear lugs instead of filling concrete; (2) the pile and wall are bonded with jet concrete entirely into a whole, and the deformation of these two structures is coordinated; and (3) the strength of jet concrete is between that of the previous two cases, which means that both the pile and the wall can slide along the interface. Based on the calculation results, it proves that the structural measures in case 2 are more suitable to improve the bearing capacity of force and deformation of the new structure. Such findings can be used to guide the optimal structure design.

Keywords: Pile-wall compound structure · Foundation pit
Numerical analysis

1 Introduction

With the rise of urban rail transit construction, more and more deep foundation pits of subway station have emerged in recent years. In the process of foundation pit design and construction, it is a new trend of combining the foundation pit retaining structure and the station main structure as a whole, which can not only save project investment, but also do good to reduce the deformation and stress of the structure (Zhang et al. 2011; Li 2006; Wang et al. 2012; Xu et al. 2011; Li 2014). Therefore, a few scholars are studying the theory and practice of how to apply the composite wall and compound wall to the

© Springer Nature Singapore Pte Ltd. 2018
T. Qiu et al. (Eds.): GSIC 2018, *Proceedings of GeoShanghai 2018 International Conference: Advances in Soil Dynamics and Foundation Engineering*, pp. 503–510, 2018.
https://doi.org/10.1007/978-981-13-0131-5_54

foundation pit engineering. For example, Ma (2016) carried out the study on the shear performance of concrete composite walls with precast ribbed panels, and the FEA results show that the shear performance of composite walls reduces with the increase of ratio of wall height to width, but increases with the rise of the axial compression ratio. Zhang et al. (2010) presented a novel kind of partially prefabricated reinforced concrete composite walls and investigated its failure pattern, deformation mode and the bonding capacity by conducting tested on the typical specimens under reversed cyclical loading. Zhang et al. (2013) analyzed the mechanical behavior of reinforced concrete composite shear walls and then put forward a simplified calculation method.

In this paper, a new pile-wall compound structure is proposed as the retaining structure of the West Yanmazhuang Station foundation pit. Subsequently, numerical simulations are carried out to investigate the influence of different connection mode between pile and wall on the deformation and stress of the new structure, so as to provide suggestion and reference for the design and construction of foundation pit.

2 Project Overview

The Metro line 1 is located in the west of Jinan city. It is one of the main line which runs from the north to the south of the city. As shown in Fig. 1, the west Yanmazhuang Sation of Metro line 1 is located in the Qilu Avenue between the Qingdao Road and Liaocheng Road in Huaiyin district. As an underground two-story platform station, it is 356.6 m long. The section size of the station foundation pit is 16 m deep and 19.7 m wide. The thickness of overburden soil above station structure is 2.5 to 4.0 m, and the elevation of the main structure bottom is 11.50 m. The foundation pit is constructed with open excavation method.

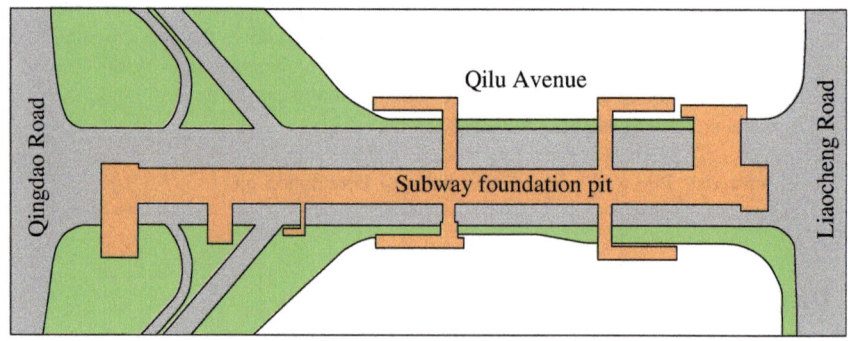

Fig. 1. Plan view of the subway foundation fit.

In this project, a new pile-wall compound structure is proposed as the retaining structure of the foundation pit, and Fig. 2 shows the schematic diagram of the station main structure and the retaining pile. As a new form of support structure of foundation pit, the prefabricated concrete square pile and the concrete wall of the station main structure are combined together with a special connecting device. This device is

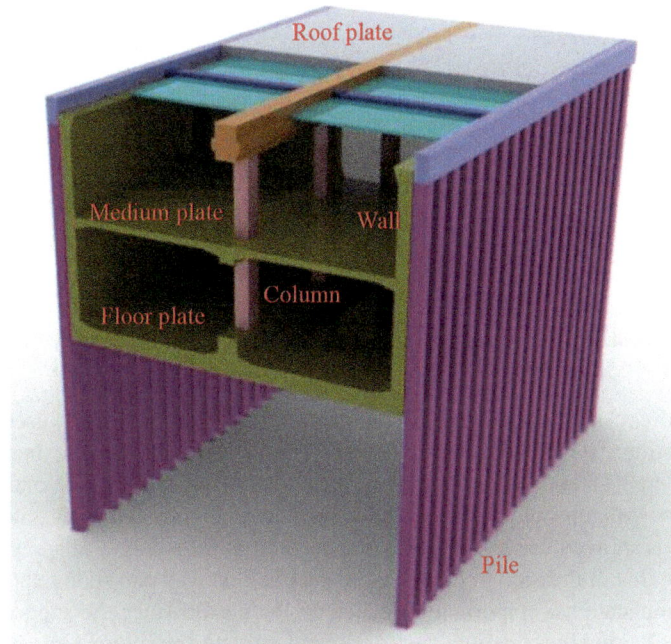

Fig. 2. Schematic diagram of the main structure of subway station.

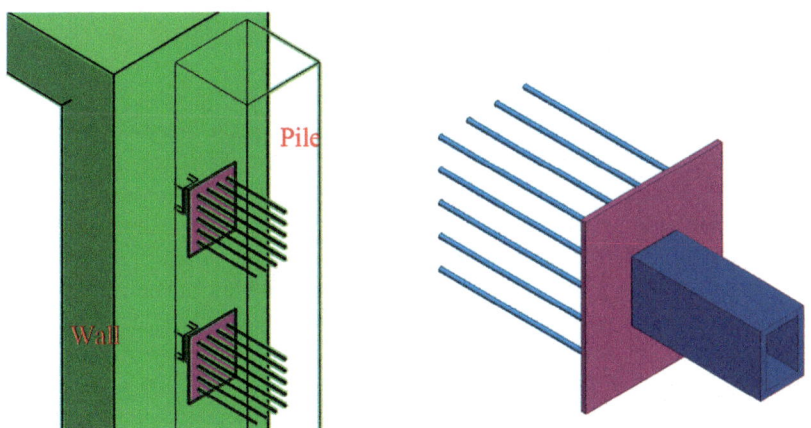

Fig. 3. Steel connecting device.

composed of embedded steel plate (purple), anchorage bar (wathet) and the shear key (dark blue). As shown in Fig. 3, along the pile body, there are 9 steel connecting devices arranged with interval of 1.4 m from the top to the bottom.

Figure 4 shows the size of embedded plate and anchorage bar of the steel connecting device. The section size of the shear key is 200 × 150 mm, with thickness of 20 mm.

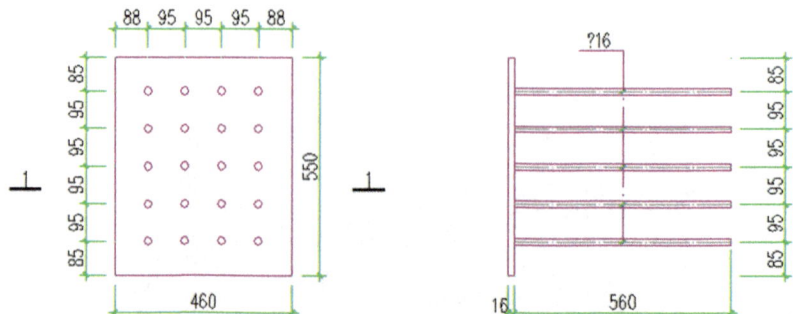

Fig. 4. Dimension diagram of embedded steel plate and anchor bar.

In actual construction process, if the above connection device is used, there are three connection state existed between the pile and the wall, i.e. (1) no concrete is filled, the pile and the wall are only connected by steel connecting device; (2) concrete infilled between the structures completely combines the pile and the wall into a whole; (3) concrete is sprayed between the pile and the all, but the strength of sprayed concrete is lower than that in the second case, that is, the pile, the wall can slide relatively. Since the connection state has important influence on the performance the support structure, the numerical simulations are carried out to investigate the internal force and deformation characteristics of the compound structure so as to provide suggestion and reference for the design and construction of the foundation pit.

3 Numerical Model

To calculate the composite wall structure, three-dimensional load-structure model is established, and the soil is simplified as spring to simulate the effect of stratum on the structure. In order to facilitate calculation, only part of the structure along the longitudinal direction of foundation pit is selected to establish the calculation model, that is, the model length along the longitudinal direction is 1.5 m, which is equal to the pile spacing. The thickness of the roof plate, medium plate and floor plate are 900, 500 and 900 mm respectively, while the side wall is 500 mm thick. The section size of the column is 1500 × 760 mm, and the side length of the square pile is 700 mm. The thickness of the filled concrete between the pile and the wall is 200 mm.

As shown in Fig. 5, according to the connection state between pile and wall, three calculation modes are established. The main structure and the prefabricated pile are simulated with the solid element. As for the special connection device, the embedded steel plate is simulated with shell element, the shear lugs are simulated with beam element, and the anchorage bar is simulated with implantable truss element which is free of bending moment and only subjected to axial force. All the structure elements adopt the ideal linear elastic model. The mechanical parameters of calculation model are shown in the Table 1.

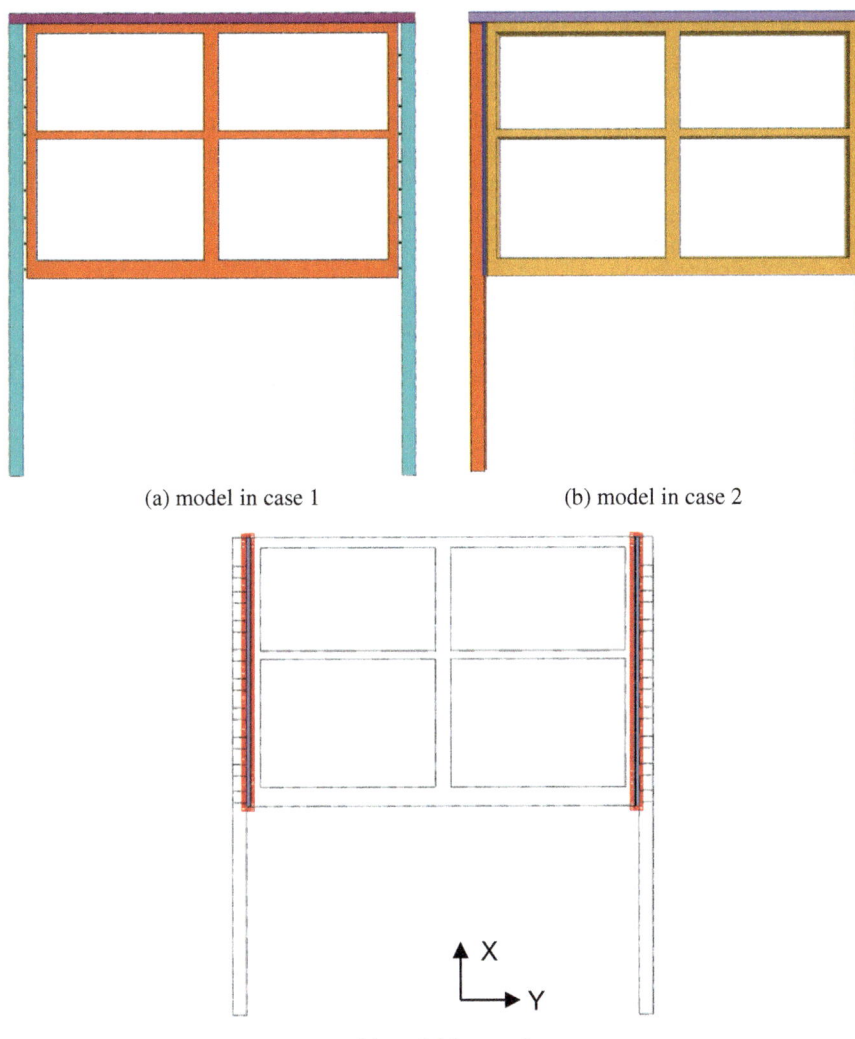

(a) model in case 1   (b) model in case 2

(c) model in case 3

Fig. 5. No concrete filled.

Table 1. Mechanical parameters for calculation model.

Structure type	Elastic modulus (MPa)	Poisson ratio	Density (kg/m³)
Wall	36600	0.2	2500
Pile	40170	0.2	2500
Steel plate	200000	0.3	7697
Anchorage bar	200000	0.3	7697
Shear key	200000	0.3	7697

In order to simulate the resistance of stratum on the structure, the elastic support constraints are arranged at the lateral sides and the bottom of pile and the bottom of the main structure floor. The horizontal springs acted on the pile side under the floor plate can not only bear the compression but also bear the tension, while the other springs can only bear the compression. The Elastic resistance coefficient of soil layer is 350 MPa/m. The loads in the calculation model mainly include two situations, namely, the construction stage and the service stage, and the corresponding load values can be found in Tables 2 and 3.

Table 2. Loads in construction stage.

Load	Load type	Load value	Load direction
Earth pressure on roof	Uniform	75 kN/m	Negative Z axis
Ground overload	Uniform	30 kN/m	Negative Z axis
Pressure on medium plate	Uniform	12 kN/m	Negative Z axis
Lateral water pressure	Trapezoidal	28.5–215.3 kN/m	X axis
Lateral earth pressure	Trapezoidal	29.3–125.6 kN/m	X axis
Earth pressure on pile under floor	Trapezoidal	125.6 kN/m	X axis

Table 3. Loads in service stage.

Load	Load type	Load value	Load direction
Earth pressure on roof	Uniform	75 kN/m	Negative Z axis
Floating pressure	Uniform	215.3 kN/m	Negative Z axis
Pressure on medium plate	Uniform	12 kN/m	Negative Z axis
Lateral water pressure	Trapezoidal	28.5–215.3 kN/m	X axis
Lateral earth pressure	Trapezoidal	29.3–125.6k N/m	X axis
Earth pressure on pile under floor	Trapezoidal	125.6 kN/m	X axis

4 Calculation Results Analysis

4.1 Deformation Analysis

The calculation results show that the structure deformation in the construction stage is higher than that in the service stage, so the deformation results in the construction stage is analyzed as below.

In case 1, 2 and 3, the maximum vertical displacements of main structure are 6.867, 4.483 and 4.493 mm respectively, which all appear in the middle wall, while the maximum horizontal displacements are 1.076, 0.71 and 0.75 mm respectively appeared at the right wall. It can be seen that, the deformation in the case 2 is relatively small. In addition, whether the pile and the wall are completely combined by concrete or not has little effect on the deformation value.

4.2   Analysis on the Internal Force of Structure

The internal force of the pile and the wall at the positions of nine embedded steel connectors are selected to analyze as below. Tables 4 and 5 show the maximum internal force of the nine sections. For the prefabricated piles, 1, 8 and 9 planes are dangerous, while for the side wall, 1 and 4 planes are dangerous. Therefore, the deformation and internal force of the dangerous plane should be monitored in the construction process. In addition, the internal force in case 2 is the smallest.

Table 4. Maximum internal force in different sections of prefabricated piles.

	F_x (kN)		F_y (kN)		F_z (kN)		M_x (kN m)		M_y (kN m)	
	Value	Plane	Value	Plane	Value	Plane	Value	Plane	Value	Plane
Case 1	−1220	1	−5.28	4	408	2	5.29	8	−159	1
Case 2	−955	9	−6.03	8	−819	9	2.33	8	−105	9
Case 3	−972	9	−6.81	8	−826	9	2.92	8	−107	9

Table 5. Maximum internal force in different sections of the side wall.

	F_x (kN)		F_y (kN)		F_z (kN)		M_x (kN·m)		M_y (kN·m)	
	Value	Plane	Value	Plane	Value	Plane	Value	Plane	Value	Plane
Case 1	−1110	1	−50.8	8	−1120	4	42.6	4	−110	1
Case 2	−1290	9	−12	1	−614	4	−8.27	9	56.7	7
Case 3	−1250	9	43.3	2	−865	4	34.3	5	−76.3	1

The axial force of anchor bar is shown in Table 6. It is obvious that the maximum tensile force and the maximum compression of the anchor bar in case 1 is higher than that in case 2 and 3. In addition, the axial force of the anchor bar in case 2 is slightly smaller than that in the case 3.

Table 6. Anchor bar axial force table.

	F_x (kN)		F_y (kN)	
	Value	Connecting device number	Value	Connecting device number
Case 1	−1110	1	−50.8	8
Case 2	−1290	9	−12	1
Case 3	−1250	9	43.3	2

4.3   Analysis on Structure Stress

The maximum horizontal and vertical stresses of the pile-wall structure are shown in Table 7. If the structures are fully bonded by the concrete, except for the vertical compressive stress is large, the other maximum stresses are less than that of the other

Table 7. Stress of the structure.

	Horizontal stress (kPa)		Vertical stress (kPa)	
	Tensile stress	Compressive stress	Tensile stress	Compressive stress
Case 1	21854	34311.1	4963.94	8234.3
Case 2	8574.51	15842.3	3466.31	10578.8
Case 3	17374.9	35962.7	6324.56	15589.2

two connections. That is to say, when concrete completely combines the pile and the wall, the structure is subjected to the least stress, which is favorable to the structure.

5 Conclusions

Based on the calculation results analysis, the following conclusions can be drawn:

(1) Filling concrete between the pile and the wall can effectively reduce the deformation and stress of the structure.
(2) If the concrete can completely combine the pile and wall, the deformation and stress is less than that in the case of spraying concrete between the pile and wall
(3) As a new type of support structure, the pile-wall composite structure can not only save resources, simplify construction steps, but also ensure structural safety.

Acknowledgments. The authors appreciate the support of the Natural Science Foundation of P.R. China (No. 41672360).

References

Zhang, H.M., Lu, X.L., Duan, Y.F., et al.: Experimental study and numerical simulation of partially prefabricated laminated composite RC walls. Adv. Struct. Eng. **14**(5), 967–980 (2011)

Li, Y.: Construction technology of composite external wall of prefabricated concrete floor with finish. Archit. Technol. **37**(11), 821–824 (2006). (in Chinese)

Wang, Z.J., Liu, W.Q., Lu, J.S.: Nonlinear finite element analysis on composite reinforced concrete shear walls. Concrete **267**, 18–21 (2012). (in Chinese)

Xu, F., Lu, X.L., Zhang, H.M.: Computational simulation for nonlinearity reinforced concrete composite walls. Struct. Eng. **27**(1), 34–39 (2011). (in Chinese)

Li, X.C.: Study on long-term and short-term working behavior of composite wall of super-deep and super-large working shaft. Tunn. Constr. **34**(11), 1024–1030 (2014). (in Chinese)

Ma, J.: Finite element analysis on shear performance of concrete composite walls with precast ribbed panels. Lanzhou Jiaotong University, Lanzhou (2016). (in Chinese)

Zhang, H.M., Lu, X.L., Duan, Y.F., et al.: Nonlinear behavior of partially prefabricated reinforced concrete composite walls (PPRC-CW). China Civ. Eng. J. **43**, 93–100 (2010). (in Chinese)

Zhang, L.H., Wan, Y.X., Chen, Y., et al.: Mechanical behaviors and application of reinforced concrete composite shear walls. Build. Struct. **43**(11), 19–23 (2013). (in Chinese)

Deformation Prediction of Excavated Slopes with a Neural Network Model Based on Nonlinear Numerical Analyses

Chenghua Wang[✉] and Xiaoxuan Wang

School of Civil Engineering, Tianjin University, Tianjin 300354, China
Chwang@tju.edu.cn

Abstract. The control and predication of deformations of excavated slopes is one of the most important problems in foundation engineering, but the evaluation of the deformation of excavated slopes has been lack of adequate methods, to which special attentions need to pay for the sack of protection of engineering environment. In order to make prediction of deformations of excavated slopes, an artificial neural network model was set up based on nonlinear finite element analyses of excavated slopes. Firstly, the pattern of the deformation of excavated slopes is generalized through a large amount of finite element analyses. Secondly, a practical and fast algorithm for predicting the deformations of excavated slopes—a neural network predication model based on numerical analysis was set up. The neural network for predicting deformations of excavated soil slopes contains four layers of neural elements, i.e. the input layer, the first and second hidden layers and the output layer. Totally 70 sets of data from finite element analyses were used for training the network, while deformation prediction were conducted with other 20 sets of dada, which given a good accuracy with errors within 10% for practical applications. The result of predictions with the neural network model demonstrates that the combination of the numerical methods and neural networks is a feasible way of deformation predication.

Keywords: Excavated slope · Deformation prediction · Neural network
Nonlinear numerical analysis

1 Introduction

With regard to the stability aspect of excavated slopes, according to China's design standards, there are many of practical methods for analyzing slope stability [1]. However, for analyzing the deformations of excavated slopes, very few simple and practical methods are available, though the deformation aspect as been always taken seriously in engineering practice. Most of works in control of the deformation and the surrounding buildings and facilities can only rely on engineering monitoring, which resulted severe weaknesses in designing excavated slopes. Theoretical researches on the evaluation or prediction of deformation of excavated soil slopes are very limited [2], and the development of fast and convenient methods is badly needed.

Finite element methods are widely used in analyzing the behavior of slopes in scientific researches, but they are suffering from the complexity and time consuming in

© Springer Nature Singapore Pte Ltd. 2018
T. Qiu et al. (Eds.): GSIC 2018, *Proceedings of GeoShanghai 2018 International Conference: Advances in Soil Dynamics and Foundation Engineering*, pp. 511–519, 2018.
https://doi.org/10.1007/978-981-13-0131-5_55

practical design works. Artificial neural networks have been demonstrated powerful tools in prediction in geotechnical engineering for decades [3, 4], but the development of a neural network often faces with the limitation of lacking enough measured data in practice.

The researches in combining both advantages of finite element methods (FEM) and artificial neural networks in fast prediction of the behaviors of geotechnical projects have been made in recent years. In order to make fast prediction of deformations of excavated soil slopes, a large amount of nonlinear finite element analyses were conducted to investigate the deformation behavior of excavated soil slopes and to provide the training data for a back propagation (BP) neural network. The theoretical backgrounds and techniques of the nonlinear finite element analyses and the back propagation neural network are described briefly in this paper, followed with introduction of the prediction results of deformation of excavated soil slopes with the neural network, which is of a good accuracy of prediction comparing with finite element analyses.

2 Finite Element Method for Analyzing Excavated Soil of Slopes

2.1 Constitutive Relations for Soils

In order to reflect the basic nonlinearity of soils, the well-known Duncan-Chang's model was used in the finite element analysis of deformation of excavated soil slopes [5]. Due to the limited space, only the key features of the model are introduced herein. The relation of stress and strain of soils is given in Eq. (1) as follows:

$$\sigma_1 - \sigma_3 = \frac{\varepsilon_1}{a + b\varepsilon_1} \tag{1}$$

Where, the σ_1 and σ_3 are the maximum and minimum principle stresses, the ε_1 is the maximum principle strain, while a and b are parameters that determine the nonlinear feature of the equation.

The Mohr-Coulomb's law for describing the failure state of soils in the model is shown by Eq. (2).

$$(\sigma_1 - \sigma_3)_f = \frac{2c \cos \varphi + 2\sigma_3 \sin \varphi}{1 - \sin \varphi} \tag{2}$$

Where, the $(\sigma_1 - \sigma_3)_f$ is the derivative stress in failure state of soil, c and φ are shear strength parameters of soil.

The tangent modulus of soil E_t is given in Eq. (3), which represents the stress dependent stress-strain relationship.

$$E_t = \left[1 - \frac{R_f(1 - \sin \varphi)(\sigma_1 - \sigma_3)}{2c \cos \varphi + 2\sigma_3 \sin \varphi}\right]^2 K p_a (\frac{\sigma_3}{p_a})^n \tag{3}$$

In where, K is the coefficient of modulus; n is the power index for the nonlinearity of soil modulus E_t with surrounding pressure σ_3; R_f is the failure ratio, and p_a the pressure of atmosphere, while others are of meanings ditto.

2.2   The Method for Simulation of Excavation

The analysis of deformation under working conditions is based on the initial state of foundation soils. The excavation process is simulated by applying the opposite equivalent loadings at nodes, which are on the designed surfaces at each excavation stages as shown in Fig. 1. Figure 1(a) shows the state of initial stress $\{\sigma^*\}$ in a soil mass with a leveled ground; Fig. 1(b) represents the exerting the opposite equivalent loadings $\{p\}$ at nodes to simulation the process of excavation, and (c) shows the formation stress state $\{\sigma_0\}$ of an excavated slope.

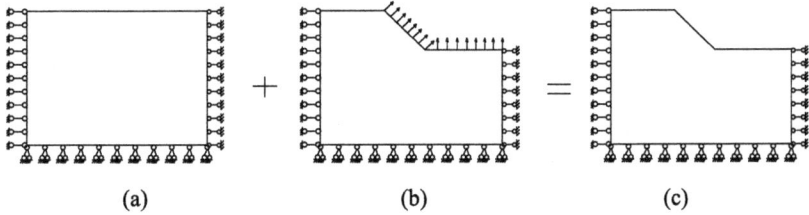

Fig. 1. Simulation of excavation of a soil slope: (a) initial stress fields before excavation $\{\sigma^*\}$; (b) equivalent node loading due to excavation $\{p\}$; (c) stress fields after excavation $\{\sigma_0\}$

2.3   Determination of Boundary Limits and Mesh Format

According to experiences, from both the authors and others, in numerical analyses of slope stability problems, the boundary limits or dimensions of a computation domain for an excavated soil slope are selected as follows: (1) The distance of toe of a slope to the right boundary is 1.5 times of the slope height; (2) The distance of top of a slope to the left edge is 2.5 times of the slope height; (3) The distance of upper to lower boundary is more than 2 times of the slope height; (4) The distance of upper to lower boundary takes 2 times the slope height.

Before normal calculations, the deformation behavior of excavated slopes with different dimensions and different soil conditions and surface loadings were tentatively examined for checking the suitability and efficiency of the finite element mesh format. Both tentative and standard calculations were conducted to verify the suitability of the setup of boundary limits and their results were quite favorable.

In normal calculations, the displacement distribution of slope in different soil conditions, of which the slope height is $H = 2\text{--}7$ m and the slope gradient is $m = 1:1$ to $1:4$. Figure 2 shows the initial and deformed finite element mesh and displacement distributions of a slope with height $H = 7$ m and the slope gradient $m = 1:1$.

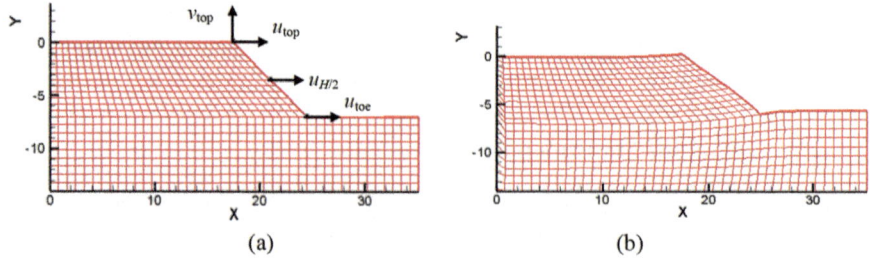

Fig. 2. The boundary limits and mesh format of an excavated slope: (a) dimensions and finite element mesh; (b) deformed mesh.

2.4   Key Features of Program FEASLOPE

The analyses of deformations of excavated soil slopes had been made by application of a 2-dimensional finite element program called FEASLOPE that is developed at Tianjin University [6]. In the program, a type of plane four-node isoparametric element is adopted, and Duncan-Chang's model and Mohr-Coulomb failure criterion are used to describe the nonlinear soil behavior, and the above mentioned excavation technique is implemented and used in FEASLOPE.

3   Parametric Analyses of Deformation of Excavated Slopes

The neural network for predicting displacements of slopes is established by using the data calculated by the finite element analysis program FEASLOPE, so it is important to learn the influence of the change of each parameter on the results of finite element analyses. Four indexes including the horizontal and vertical displacements at the top point of the slopes are compared and quantitative analyzed for the incidence of soil parameters and geometric parameters on slope displacements.

3.1   Influence of Soil Parameters on Results of Finite Element Analyses

In the analysis of influence of soil parameters, a standard condition for a slope without ground surface loadings was first set by selection of geometric parameters as following: The height of the slope $H = 7$ m, the slope gradient $m = 1:1$; the material parameters for a given soil were set as $K = 200$, $n = 0.5$, $R_f = 0.6$, $c = 20$ kPa, $\varphi = 20°$, $\gamma = 19$ kN/m^3, $\mu = 0.25$. The influence of soil parameters on the four displacement indexes were computed and examined by varying the values of each parameter in a rate of change with respect to its original value in the standard case. The horizontal and vertical displacement at the top point of the slopes are given in Fig. 3, the results of other indexes are omitted due to limited space.

As Fig. 3 shows, different soil parameters have different influence on the results of displacements and the same value of soil parameter has different influence on different displacement indexes. This has a guiding significance for the subsequent establishment of neural networks and the selection of parameters.

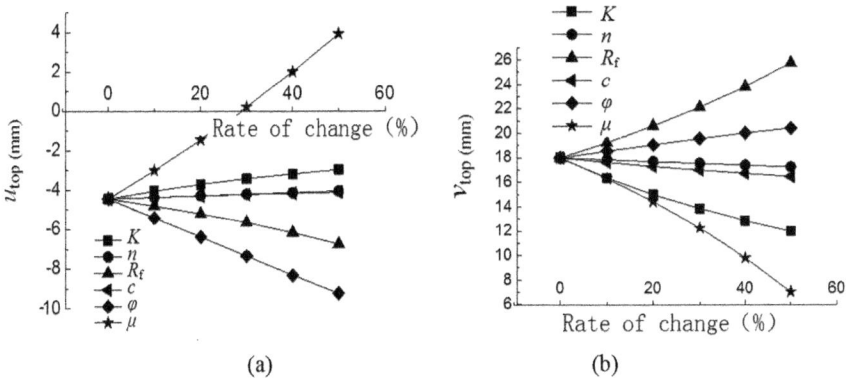

Fig. 3. The influence of soil parameters on displacement indexes: (a) u_{top}; (b) v_{top}

3.2   The Influence of Geometry Parameters on Results of Finite Element Analyses

The influence of the geometrical parameters, i.e. the height H and the gradient m of a slope, on the displacements is also checked in a standard case without ground surface loadings and whose material parameters were selection as following: $K = 200$, $n = 0.6$, $R_f = 0.8$, $c = 20$ kPa, $\varphi = 25°$, $\gamma = 19$ kN/m^3, and $\mu = 0.3$. For the displacement indexes for the top point of the slopes, the analysis results corresponding to the slope height varying from 3 to 7 m and the slope gradient ranges from 1:1 to 1:4 are shown in Fig. 4. The results of other indexes are omitted again due to limited space.

By examining the slope or derivative of a curve for a soil parameter on a displacement index, it can be sure that the greater the slope is, the stronger the parameter influences the index. It can be seen from Fig. 4, both the gradient m and the height H influence significantly all the displacement indexes and should be considered in setting the neural network for predicting the deformation of excavated slopes.

4   A BP Neural Network for Deformation Prediction

4.1   Back Propagation Neural Networks

Artificial neural networks have found their uses in geotechnical engineering since about 1990's all over the world, for their good capabilities in many prediction and decision making issues. Back propagation neural network is a kind of multilayer feed-forward network based on an error back propagation algorithm for training, which was proposed by Rumelhart and McCelland in 1986 [7]. The transfer function used in the BP network neuron is usually the Sigmoid's differentiable function, which can realize any nonlinear mapping between input and output information. At present, it has become the most widely used neural network model, and nearly 90% of applications of neural network are based on BP algorithm [8].

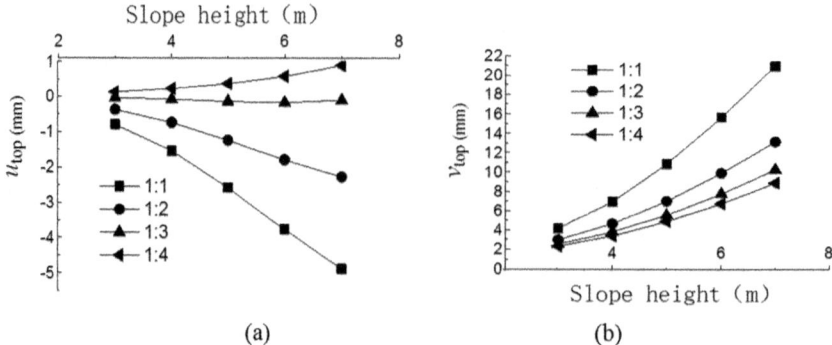

Fig. 4. The influence of geometrical parameters on displacement indexes: (a) u_{top}; (b) v_{top}

4.2   Key Features of Program BPSLOPE

A back propagation neural network had been designed the authors in recent years at Tianjin University and a corresponding program called BPSLOPE was set and used for prediction of deformation of excavated soil slopes. The basic structure of the BP network is shown in Fig. 5. There are totally four layers of neural elements, i.e. the input layer, the first and second hidden layers and the output layer, in BPSLOPE.

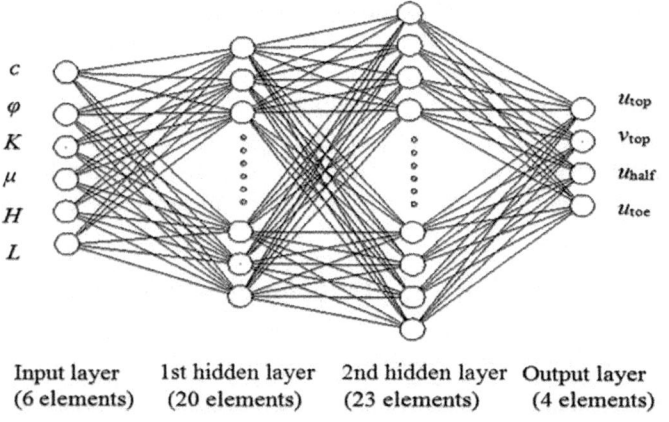

Fig. 5. The basic structure of the back propagation neural network BPSLOPE

There are six neural elements in the input layer, which inputs the key influencing factors including soil parameters such as the internal friction angle φ, the soil cohesion c, the coefficient of soil modulus K, Poisson's ratio μ, and the height H and the horizontal length $L(L = H/m)$ of a slope in the deformation analyses by the nonlinear finite element method. The six factors were selected for their relatively stronger influences to the others as analyzed based on the sensitivity analysis of main parameters in prior sections.

The first and second hidden layers contain 20 and 23 neural elements, respectively, in order to achieve good effect of simulation of the high nonlinearity between influencing factors and the behavior of slope deformations.

Through the trial and for simplicity, four deformation variables, including the horizontal displacement u_{top}, vertical displacement of the top of the slope v_{top}, the horizontal displacement at the middle of a slope $u_{H/2}$ and horizontal displacement of the toe of slope u_{toe}, as shown in Fig. 2(a), are considered suitable for representing the deformation behavior of excavated soil slopes and selected as the output parameters that formed the output layer of the BPSLOPE network.

5   Prediction of Results from BPSLOPE

At the beginning of the running of the BPSLOPE, totally 70 sets of sample data from a large number of finite element analyses were used for training the network. After 5 to 7 iterations, the study process of neural network finished with excellent accuracy of all the four deformation indexes, as the u_{top} and v_{top} in Fig. 6 for examples.

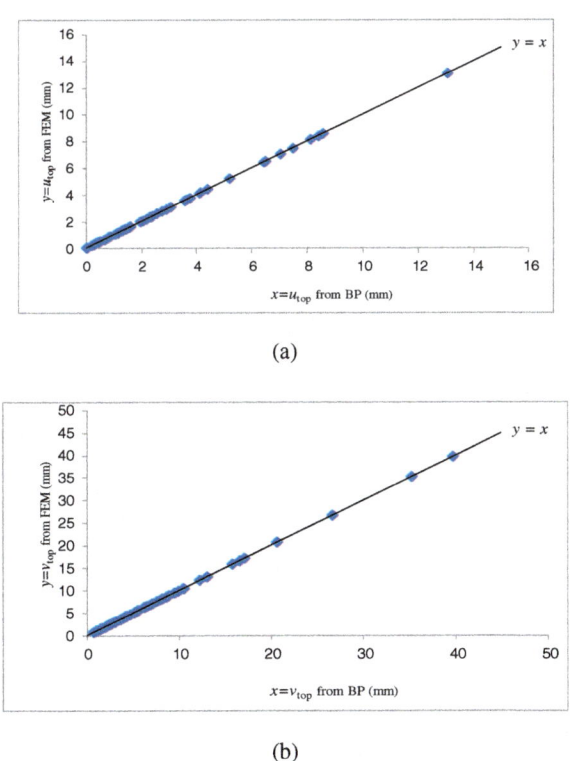

(a)

(b)

Fig. 6. Training results of displacement by neural network prediction: (a) u_{top}; (b) v_{top}

After the training of the BPSLOPE network, totally 20 sets of input data were used to predict the deformations of the slopes. The predicted results of the u_{top} and v_{top} and comparisons with that from finite element analyses are given in Fig. 7. The prediction results of other indexes are of similar accuracy, but omitted due to limited space.

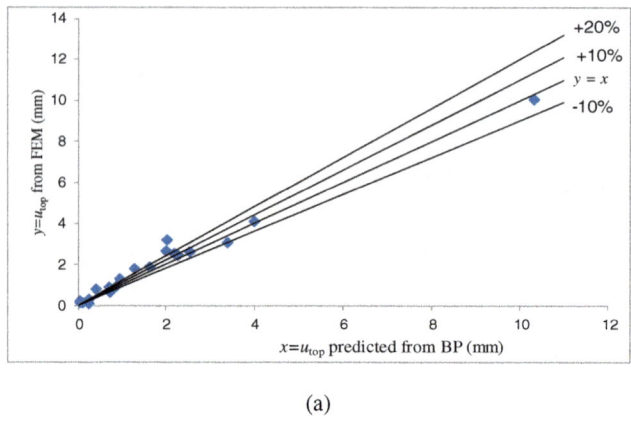

(a)

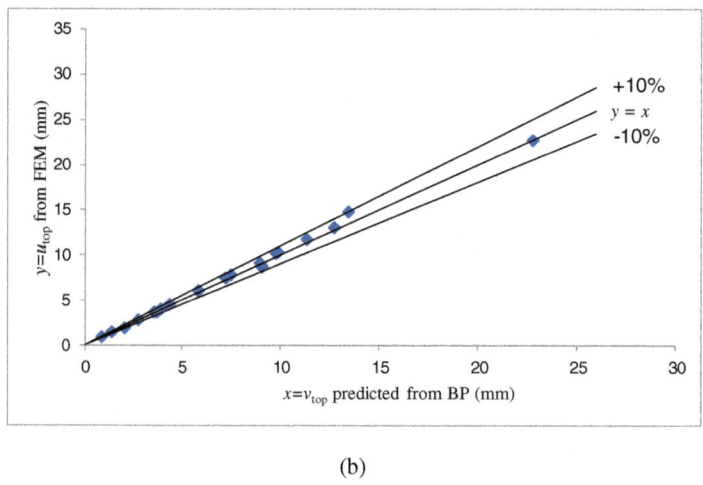

(b)

Fig. 7. Prediction of displacement by neural network prediction: (a) u_{top}; (b) v_{top}

From the comparison results above in Fig. 7, it can be find that the accuracy of neural network predictions based on finite element analyses is pretty good, and most of the comparison data are allocated within the lines that of error ±10%. Only in the prediction of the horizontal displacement of the top of the slope, the accuracy is little poorer when predicting the smaller displacement value, but beyond that, the prediction error of the other three displacement indexes can be controlled within 10%, which is reasonable in geotechnical engineering practice.

6 Conclusions

Combining the advantages of neural networks and finite element methods, a neural network for prediction of displacements of excavated slopes based on finite element analyses is suggested.

In the finite element analyses, the influence of factors of material parameters and geometric parameters on the deformation behavior of excavated slopes were analyzed, which provided the possibilities for selection of layers and neural elements for the neural network.

In the setup of the BPSLOPE network, 70 sets of data from finite element analyses were used for training the network, which yield very high degree of accuracy. Deformation prediction were conducted with other 20 sets of dada, which given a good accuracy with errors within 10% for practical applications.

The works described herein are limited to cases of excavated soil slopes in which soils are homogenous and isotropic media. The loadings on ground surface were also not considered. The deformation perditions of inhomogeneous and/or anisotropic soil slopes need further investigation and so are the cases where surface loadings are to be considered.

The capability of the BPSLOPE network need to be further verified with field observations or results in case histories of excavated slopes, which needs massive investigation but is out of the scope of present study.

Acknowledgement. The financial support of this work by the National Natural Science Foundation of China (Contract No. 51478313) is gratefully acknowledged.

References

1. GB 50007–2011: Code for Design of Building Foundation. China Architecture and Building Press, Beijing (2010). (in Chinese)
2. Wei, H.W.: Theory calculation method of slope displacement. Hydrogeol. Eng. Geol. **02**, 75–79+83 (2006). (in Chinese)
3. Sun, H.T.: Study on neural networks method of deformation prediction of foundation pit based on artificial. Rock Soil Mech. **4**(12), 63–68 (1998). (in Chinese)
4. Xu, Y.W.: Rock slope deformation prediction by artificial neural network method in Three Gorges Project. Rock Soil Mech. **20**(2), 27–31 (1999). (in Chinese)
5. Li, G.X.: Advanced Soil Mechanics, 2nd edn. Tsinghua University Press, Beijing (2016). (in Chinese)
6. Shi, H.T., Wang, C.H.: Some problems in finite element analysis of slope stability. Rock Soil Mech. **21**(2), 152–155 (2000). (in Chinese)
7. Rumelhart, D.E., McClelland, J.L., et al.: Parallel Distributed Processing: Explorations in the Microstructure of Cognition, vol. I. MIT Press, Cambridge (1986)
8. Yang, Y.B.: Prediction of slope displacement based on gray model and neural network. J. Nat. Disasters **17**(2), 138–143 (2008). (in Chinese)

Experimental Study on the Mechanical Properties and Deformation Characteristics of Prefabricated Concrete Square Pile

L. H. Lu[1(✉)], W. T. Wang[2], G. F. Wang[1], Q. W. Xu[2], and Y. J. Wang[1]

[1] Jinan Rail Transit Group Co., Ltd., Jinan 250101, China
[2] Key Laboratory of Road and Traffic Engineering of Ministry of Education, Tongji University, Shanghai 200092, China

Abstract. Prefabricated concrete square piles are more and more widely used as retaining structure in deep foundation excavation. In the actual construction process, special joint is usually used to connect short piles to meet the design requirements, and the structural performance of the prefabricated pile joint will directly affect the bearing capacity of the retaining structure. During the foundation pit construction of the West Yanmazhuang Station of Ji'nan Metro line 1, the improved joints with pin rolls are designed for connecting prefabricated concrete square piles. The structural performance of the prefabricated piles with ten pin rolls and eight pin rolls under the external force are analyzed contrastively by means of 1:1 full scale model test. The test results show that the structural performance of the two forms of joint can satisfy the design requirements, and the flexural capacities of the two kind of specimen are improved 35% and 45.14% respectively than the designed value. Meanwhile, the mid-span deflection deformation and the concrete cracks can also satisfy the design requirements. After comparison and analysis, it comes to the conclusion that the joints with ten pin rolls can ensure the stability of structure system more efficiently.

Keywords: Pile · Model experiment · Bearing capacity

1 Introduction

Due to the advantages of high body quality, easy construction, low environment pollution and low engineering cost etc., the prefabricated concrete square piles are being widely used in deep foundation pit engineering. Due to the limitation of pile length, short piles are connected with certain joints to meet the design requirements of the required pile length in actual engineering. At present, the studies of structure performance of the precast concrete pile joint are mainly focused on the joint flexural behavior of the prefabricated concrete square pile and the PHC pipe pile, which are mostly carried out by means of experiment and theoretical calculation. For example, Zhang (2004) proved that the simple welded joint of the prefabricated concrete pile with small section was practical in engineering projects through the indoor test. Xu et al. presented that there are two failure forms of the prefabricated concrete pile joint in

© Springer Nature Singapore Pte Ltd. 2018
T. Qiu et al. (Eds.): GSIC 2018, *Proceedings of GeoShanghai 2018 International Conference: Advances in Soil Dynamics and Foundation Engineering*, pp. 520–527, 2018.
https://doi.org/10.1007/978-981-13-0131-5_56

the flexural behavior experimentation of the full-scale prefabricated concrete square pile with composite reinforcement. Li et al. (2010) proposed a method of pasting steel plate to the welded PHC pipe piles which is validated to be effective to improve the structural performance and the construction technology of the joint. Liu et al. (2008) conducted the flexural behavior experiment of the prefabricated concrete hollow square pile to prove that the section stiffness can be lost when the cracks appear in the pile body, which is extremely harmful. Guo et al. (2013) proposed the hold-hoop connection between piles is proved to be good for improving the uplift bearing capacity of the pipe piles by means of theoretical analysis and calculation.

In this paper, a new form of reinforced joint with pin rolls is proposed to connect the prefabricated concrete square piles in the construction of Yanmazhuang station foundation pit of Jinan Metro line 1. In order to verify the flexural performance of pile joint under horizontal load, full-scale model test was carried out to investigate the internal force, the deformation behavior and the bearing capacity of two forms of joints. In the tests, the lateral water and earth pressure is simulated as single force applied on the simply supported beam which is the represent of the prefabricated concrete square pile. According to the test results, the reasonable prefabricated pile connector type is determined to ensure the normal performance of the foundation pit retaining structure system.

2 Form of Pile Joints

At present, there are basically three kinds of prefabricated reinforced concrete pile joints in practical engineering, i.e. the angle steel welded joint, steel plate butt welded joint and sulfur glue mud anchor joint. Since the bending strength of the pile with the sulfur glue mud anchor joint is lower, the welded joint is the main form of square pile connection. One typical form of the welded joint is that produced according to the Standard Cartography (97G361), such as the angle steel welded joint and steel plate butt welded joint. The other forms are the joints designed by the manufacturer. To be frankly, the joint produced according the Standard Cartography need more embedded components. It is not only labor cost, time-consuming and expensive, but also makes the concrete perfusion not dense enough (Ding and Cao 2014).

As for the structure design of Yanmazhuang station foundation pit of Ji'nan Metro line 1, a new type of supporting structure is proposed as the retaining structure. This new structure consists of the external prefabricated pile and the inner main structure continuous wall, and several shear keys are used to connect the pile and the wall as a whole from top to bottom. Therefore, it is required that the prefabricated concrete pile not only has better structural performance as support structure during the construction phase, but also meets the design requirements as part of the main structure in the service stage. In this project, the length of prefabricated concrete square piles is 24 m which is the combination of the 15 m and the 9 m short piles connected by the improved joints with pin rolls. The new type of joint is composed of different number of pin rolls. The schematic diagram of the joint is shown in Figs. 1 and 2.

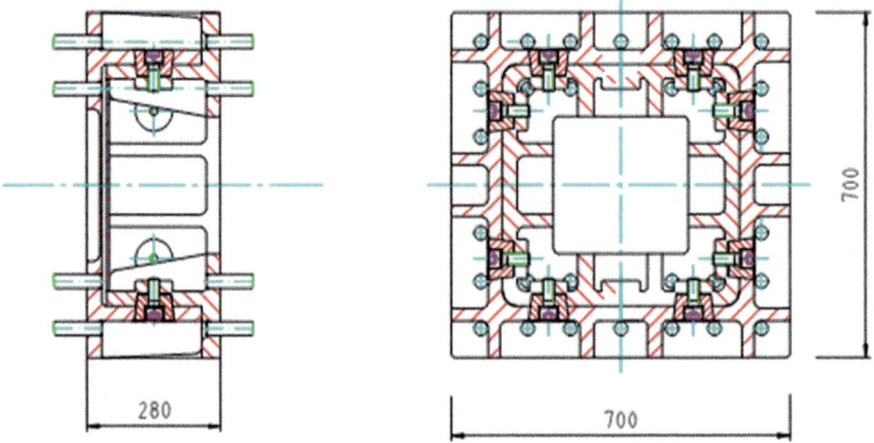

Fig. 1. Schematic diagram of the joint with eight pin rolls (unit: mm).

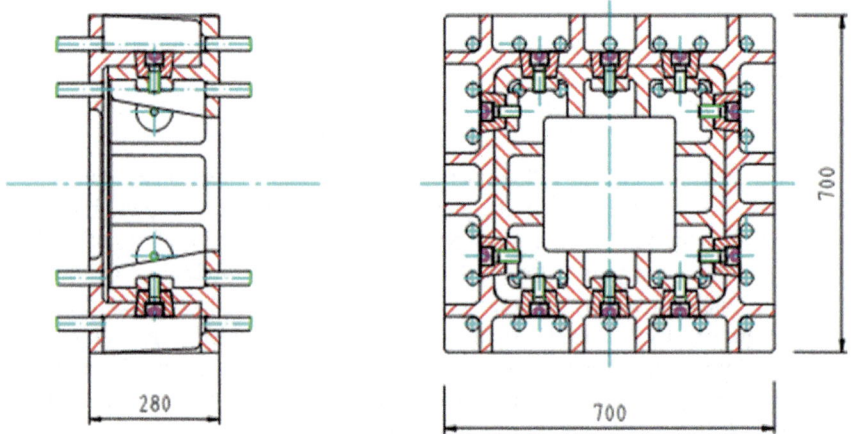

Fig. 2. Schematic diagram of the joint with ten pin rolls (unit: mm).

3 Model Test Design

3.1 Design of Test Specimen

In order to precisely simulate the deformation characteristics of the prefabricated pile joint, part of the prototype full-scale pile contained joint is selected as the test specimen. As shown in Fig. 3, the length of model pile is 1/4 of the prototype pile. Two kinds of joint are comparatively analyzed in this test. One is the joint with eight pin rolls, and the other with ten pin rolls. The detailed design of the test pieces are shown in Table 1.

Fig. 3. Picture of the pile joint with eight pin rolls.

Table 1. Design specification and quantity of the test pieces.

Specimen number	Connector type	Specimen length (mm)	Section size (mm)	Main bar (mm)	Stirrup (mm)
YJT1	8 pin rolls	9	700 * 700	16φ28, 2φ20, 13φ28	φ10@150
YJT2	10 pin rolls	9	700 * 700	16φ28, 2φ20, 13φ28	φ10@150

3.2 Design of Test Load and Monitoring Scheme

The composite support structure composed of pile and wall is subjected to the lateral water and soil pressure, which may cause large internal force in the pile. Therefore, the test load applied on the pile body is determined according to the most unfavorable combination. In this experiment, the pile used as the retaining structure of foundation pit is simplified as the simply supported beam, as shown in Fig. 4.

Fig. 4. Loading device of the model test.

In order to facilitate the test and reduce the number of loading, the uniform load is equivalent to several concentrated load. Figure 5 shows the schematic diagram of test load design, i.e. the lateral water-earth pressure is converted to several single forces, and the concentrated load is applied in incremental steps with hydraulic jacks. Four dial indicators are arranged at the beam bottom to measure the pile body deformation. In addition, the strain gauges are attached to the pile joints to measure the force of the structural joints.

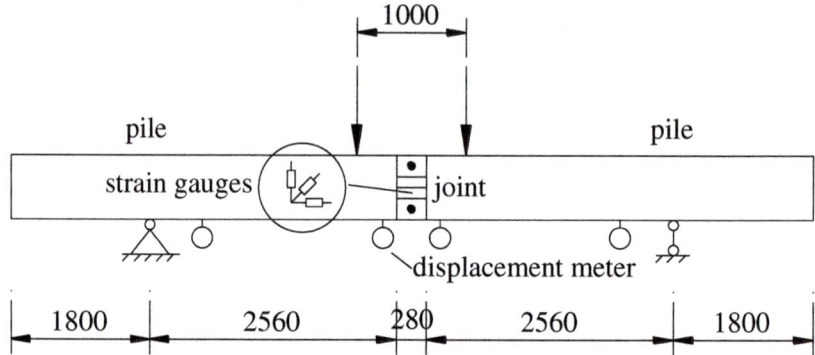

Fig. 5. Schematic diagram of test load design (unit: mm).

As shown in Fig. 6, the cracks in the pile body are monitored by the crack detector instrument during the test. It should be pointed out that the deflection deformation and cracks distribution of the test piece need to be recorded when it comes to the serviceability limit state. Thereafter, the load remains unchanged until the specimen enters the ultimate limit state (Zhang 2003; Chen et al. 2002; Wan 2015).

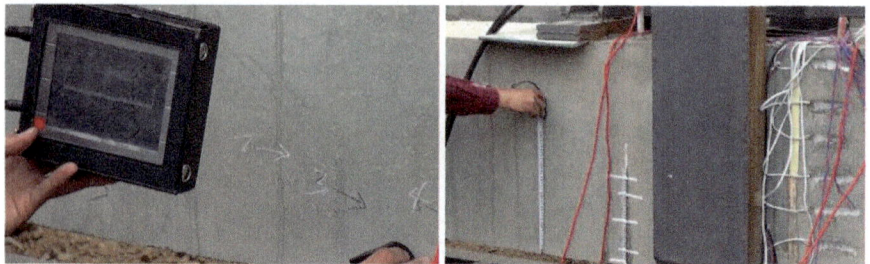

Fig. 6. Photos of the crack detection.

4   Test Results Analysis

4.1   Test Results Description

The test results of prefabricated pile with two kinds of joints are shown in Table 2, which contains the bending moment, mid-span deflection and crack width. The result shows that both types of piles can meet the design requirements of bearing capacity, while the deflection and fracture can be controlled. In comparison, the pile joint with ten roll pins with has better structural performance.

Table 2. Results of model tests.

	Serviceability limit state				Ultimate limit state		
	Design M_u (kN * m)	Trial M_u^t (kN * m)	Deflection (mm)	Crack width (mm)	Design M_u' (kN * m)	Trial $M_u'^t$ (kN * m)	Failure state
YJT1	500.01	559.4	5.920	0.08	816.83	836.0	Yield
YJT2	500.01	618.8	6.404	0.08	816.83	898.17	Yield

4.2 Mechanical Performance Comparison of Different Joints

As shown in Fig. 7, the positive strain of the pile side has the same trend with that of the pile bottom in the loading possess. The strain speed increases very slowly at the initial phrase because of the good performance of the pile in the normal service condition. Thereafter, the strain speed increase gradually due to the failure of the pile body. It is very clear that the load required for the specimen YJT2 begins to produce a larger strain is much higher than that of the specimen YJT1.

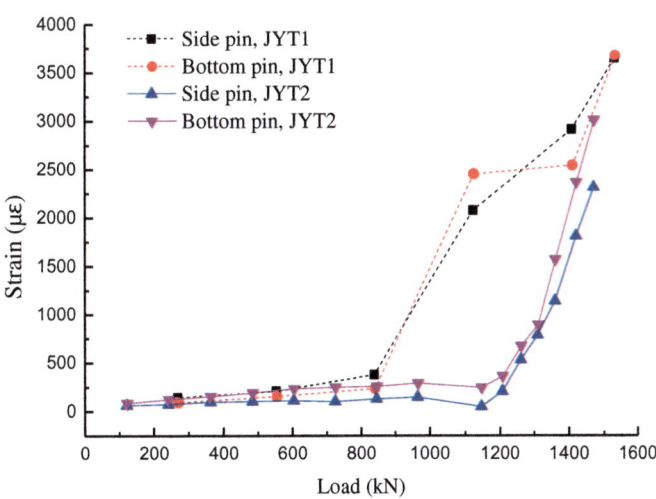

Fig. 7. Pin roll strain of YJT1 and JYT2 under different applied loads.

Figure 8 shows the mid-span deflection deformation in two test cases. It can be seen that the deformation of the mid-span grows gradually with the increase of external load, and the measured values are higher than the calculated values, especially when it comes to the ultimate limit state. In addition, when the applied the load is same, under the serviceability limit state, the measured deflection of the specimen YJT1 is a little larger than that of the specimen YJT2; while under the ultimate limit state, the specimen YJT2 can bear even larger deformation value.

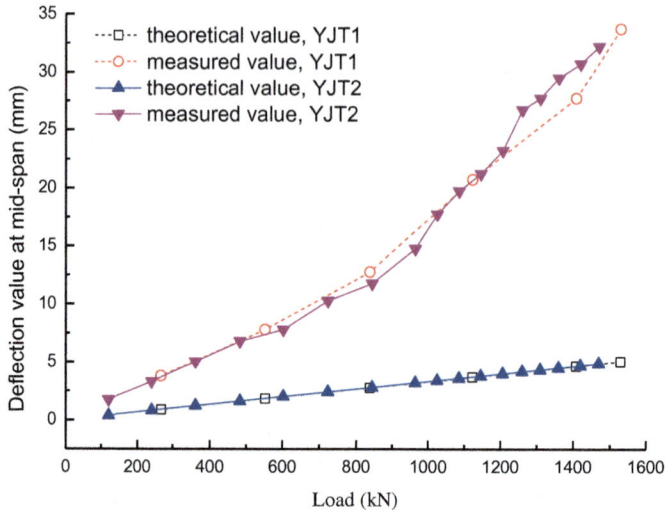

Fig. 8. Mid-span deflection of YJT1 and YJT2 under different applied loads.

As revealed in Fig. 9, in the process of loading, the crack width increases continuously, which has a mutation when it comes to the ultimate limit state. In the initial phrase, the width of the cracks is basically the same. In addition, under the condition of same load, the crack width of the specimen YJT1 is relatively large.

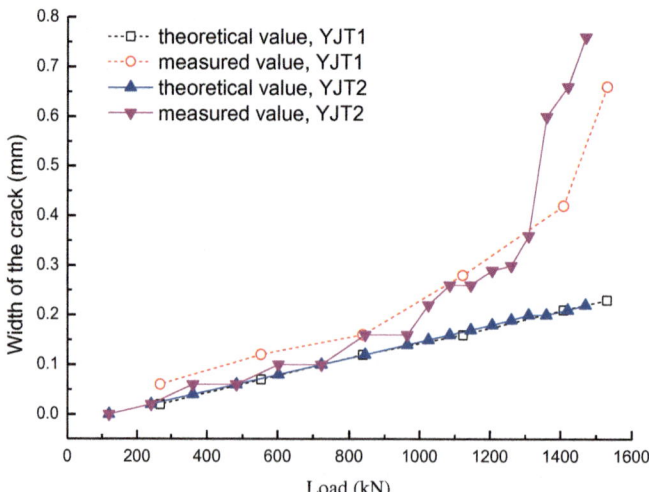

Fig. 9. Crack width curve of the specimen YJT YJT1 and YJT21.

Comparatively speaking, the pile joint with ten roll pins has even better structural performance, which is then recommended as the proper form to connect short piles as a whole.

5 Conclusions

In this paper, full-scale model tests were carried out to investigate the internal force, the deformation behavior and the bearing capacity of two forms of joints. The test results at different phrases show that both two forms of joint can meet the design requirements of bearing capacity, while the deflection and fracture can be controlled. In comparison, the pile joint with ten roll pins has better structural performance, so it is recommended as the proper pile connection form.

Acknowledgments. The authors appreciate the support of the Natural Science Foundation of P.R. China (No. 41672360).

References

Zhang, J.: The study of simple welding joint tests of small section static pressed prefabricated pile. J. Fujian Univ. Technol. **2**(2), 219–221 (2004)

Li, W., Wan, Y., Liu, Q.: Design and experimental study on a new connection between PHC uplift pile segments of the Expo Theme Pavilion. J. Build. Struct. **31**(5), 86–94 (2010)

Liu, F., Jia, L., Li, C.: The test study on welding joint flexural bearing capacity of prestressed concrete hollow square pile. J. Wuhan Univ. Technol. **30**(5), 105–108 (2008)

Guo, Y., Cui, W., Chen, F.: Analysis and experimental study of a PHC uplift pile with hold-hoop connection. Chin. J. Geotech. Eng. **35**(s2), 1007–1010 (2013)

Ding, J., Cao, Y.: Select and improvement of diaphragm wall connector in special condition. Constr. Technol. **12**(43), 50–52 (2014)

Zhang, J.: Results analysis of dead load experiment on static pressure precast pile in some project. J. Fujian Univ. Technol. **1**(2), 49–51 (2003)

Chen, F., Jian, H., Xu, W., et al.: An in-situ test and analysis of static pressed precast piles with small section. Rock Soil Mech. **32**(2), 213–216 (2002)

Wan, M.: Experimental study on flexural behavior of segment joints strengthened with high strength fiber and mortar-filled steel tubes composite. China Civil Eng. J. **48**(2), 75–80 (2015)

Influence Analysis of Frame Buildings Induced by Foundation Pit Excavation Considering Arbitrary Angle Between Longitudinal Wall and Foundation Pit

Zhi-guo Zhang[✉] and Sheng-nan Li

School of Environment and Architecture, University of Shanghai
for Science and Technology, Shanghai, China
zgzhang@usst.edu.cn

Abstract. The excavation of deep-foundation pit can impact the deformation and stability of adjacent buildings greatly. In this paper, shallow-foundation frame building affected by excavation of the pit, on the condition that the building has formed different angles with the side of the pit, is analyzed by establishing a 3D finite element numerical model. The distribution rules for the inclined value and horizontal displacement of the walls, the internal force of the framework were given high attention when majoring in this project. The results show that the walls will tilt when there is an angle between the building and the side of the pit, causing a horizontal displacement which is paralleled to the pit transversally and longitudinally. When $\alpha = 90°$ and also the building structure is located just above the lowest point of settlement trench outside the pit, the inclination of vertical wall tilt is the largest. When $\alpha = 30°$ the horizontal wall tilts the most significantly. When $\alpha = 30°$ and also $D = 5$ m, the peak value of the horizontal displacement of the horizontal wall and the vertical wall occurs. As the building gradually moves away from the excavation surface of the foundation pit, though relative deflection which occurs in the longitudinal wall of the façade that corresponds to $\alpha = 30°$ will present in prominent concave deformation shape, the relative deflection of the longitudinal wall of the façade of the building will gradually change from the concave deformation to the "⌣" deformation and the upward convex deformation.

1 Introduction

The balance of the original stress field was broken due to soil excavation in the construction site. The relative distance and different angle between the buildings and the foundation pit side will significantly change the displacement field of soil around the foundation pit, Thus it would lead to the redistribution of the internal stress in the buildings and the large deformation caused by uneven settlement, which results in local crack and damage. For historical buildings, the structure of the buildings have already deteriorated quickly due to historical reasons. Therefore, taking the relative distance and angle between the building and the foundation pit into consideration really matters significantly when referring to the study on the deformation of the adjacent frame structure in the excavation of the foundation pit.

© Springer Nature Singapore Pte Ltd. 2018
T. Qiu et al. (Eds.): GSIC 2018, *Proceedings of GeoShanghai 2018 International Conference: Advances in Soil Dynamics and Foundation Engineering*, pp. 528–535, 2018.
https://doi.org/10.1007/978-981-13-0131-5_57

Recently some scholars all around the world have studied the impact on buildings caused by soil excavation. Among them, Kung et al. made a thorough inquiry on the settlement and deformation of the walls in the soft soil layer through simplifying the nonlinear model of small strain of soil. Son and Cording, through numerical simulation, proposed that the shear deformation was dominant among all the deformation of buildings caused by soil excavation.

Among all the existing research achievements of the interaction between pit and buildings, only a few projects took relative distance and angle between buildings and foundation pit into consideration at the same time. Even few research targeted on the tilt of the wall, relative flexure deformation in wall, distribution trend of internal force of walls.

In this paper, the three-dimensional finite element method is applied to simulate the interaction between buildings and pit, considering the different angles between structures and excavation flat. Aimed at shallow foundation structures, this paper focused on the inclination of walls, horizontal displacement, and relative flexure deformation characteristics as well as the main variation features of tensile strain while the pit is being excavated.

The Establishment of Numerical Model
Excavation Parameters

Taking the excavation of a foundation pit in Shanghai as an example, its excavation size is 80 m × 80 m × 15 m. The elastic modulus and Poisson ratio of the underground continuous wall are 2.5×10^7 kN/m^2 and 0.25, and its thickness is 0.7 m and the penetration depth is 5 m. In numerical simulation, the annular support was simulated by plate element to avoid the influence of uneven effect on the deformation of buildings. The equivalent compressive stiffness of the plate was 5×10^5 kN/m, and poisson's ratio was 0.2. The center elevation of the concrete slab support section is −1 m, −7 m, −10.5 m. The three-dimensional finite element model of foundation pit-building is shown in Fig. 1.

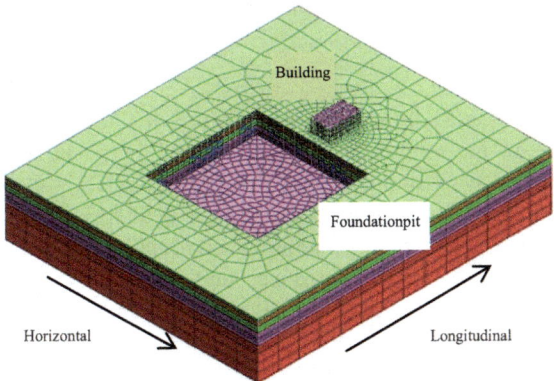

Fig. 1. Finite element mesh of excavation-building model

Table 1. Physical and mechanical parameters of soil

Soil name		Plain fill	Brown powder clay	Gray silty powder clay	Gray mucky silty clay	Gray powder clay
Thickness d	m	2.0	3.0	4.0	8.0	23.0
Weight γ	kN·m^{-3}	18.0	18.4	17.4	16.8	18.1
Compression modulus E_s	MPa	2.55	4.48	2.54	2.09	4.66
Poisson ratio v		0.33	0.32	0.32	0.33	0.29
Cohesion C	kPa	0	19	11	10	15
Friction angle φ		20.0	19.0	16.0	12.0	18.0
Secant stiffness E_{50}^{ref}	MPa	23.0	40.0	20.0	16.0	30.0
Unload stiffness E_{ur}^{ref}	MPa	69.0	120.0	60.0	48.0	90.0
Break ratio R_f		0.9	0.9	0.9	0.9	0.9

The field is composed of five soil layers, and the constitutive relation of soil is adopted to modify Mohr-Coulomb model, and the specific physical and mechanical indicators are shown in Table 1. Main parameters include: cohesion, internal friction Angle, poisson ratio, compression modulus, and R_f, E_{50}^{ref}, E_{oed}^{ref}, E_{ur}^{ref}.

Building Parameters

The construction site of foundation pit is near a historic building, which is with striped shallow foundation framework. The base width, thickness, the dimensions of the 3-story building is 0.75 m, 0.9 m, 22.5 m × 13.5 m × 9 m respectively, and the height of each floor keeps the same. The measurement of doors and window openings are 2.0 m × 1.5 m, 1.8 m × 1.5 m. The elastic modulus of the walls and floors of the building are 220 Mpa and 20 GPa, while poisson ratio are 0.1 and 0.2 respectively. The thickness are 0.24 m and 0.1 m. Considering the influence of beam column on the structural integrity of the building, C30 concrete materials were adopted as the beam, and C40 concrete materials were chosen as the column. The size of beam and column is 250 mm × 600 mm, 500 mm × 500 mm. The layout of the building is shown in Fig. 2.

Numeric Simulation Scheme

In order to further study the impact on structure of the building caused by excavation disturbance in the circumstances that the building and foundation pit have formed arbitrary angle, we setpoint O, which is on the vertical wall of building façade and near the foundation pit, as fixed point, shown in Fig. 3. Horizontal distance D, which is the distance between the fixed point O and foundation pit, in turn, is 1 m, 5 m, 9 m and 12 m. In the case of the same distance, angle α, which is the angle between the structure and edge of the pit is respectively 30°, 45°, 60° and 90° (building longitudinal wall and foundation pit side vertical).

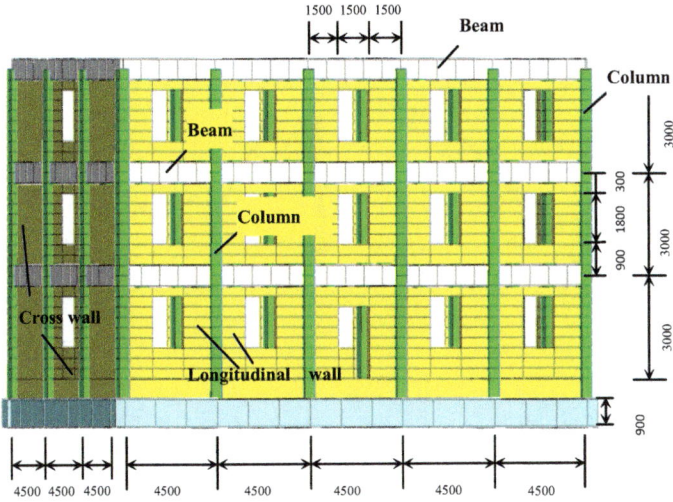

Fig. 2. Building dimensions (unit: mm)

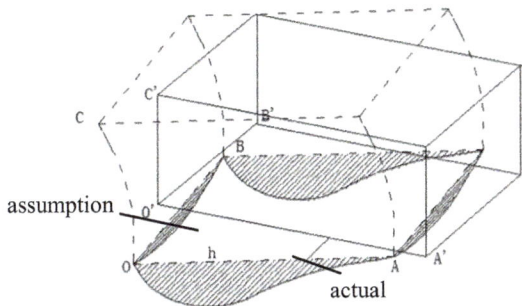

Fig. 3. Tilt calculation of longitudinal wall

Deformation of Buildings Affected by the Angle

Relative Deformation Law of Building Longitudinal Wall

Based on Fig. 3, curve OHA is the actual settlement curve of the longitudinal wall of the building. Dotted line OHA is the nominal subsidence curve of building longitudinal wall (a straight line formed by two end-point connections). The relative flexural deformation value of the building longitudinal wall is based on the difference of the actual settlement curve of the building and the nominal settlement curve of the longitudinal wall of the building.

As shown in Fig. 4, in the situation that the angle α is an arbitrary value, the relative flexural deformation of the longitudinal wall is significantly different with the change of the horizontal distance D.

When D = 1 m, the longitudinal wall shows the relative deflection of the concave shape. When $\alpha = 30°$, the relative deflection of the longitudinal wall is minimal, the

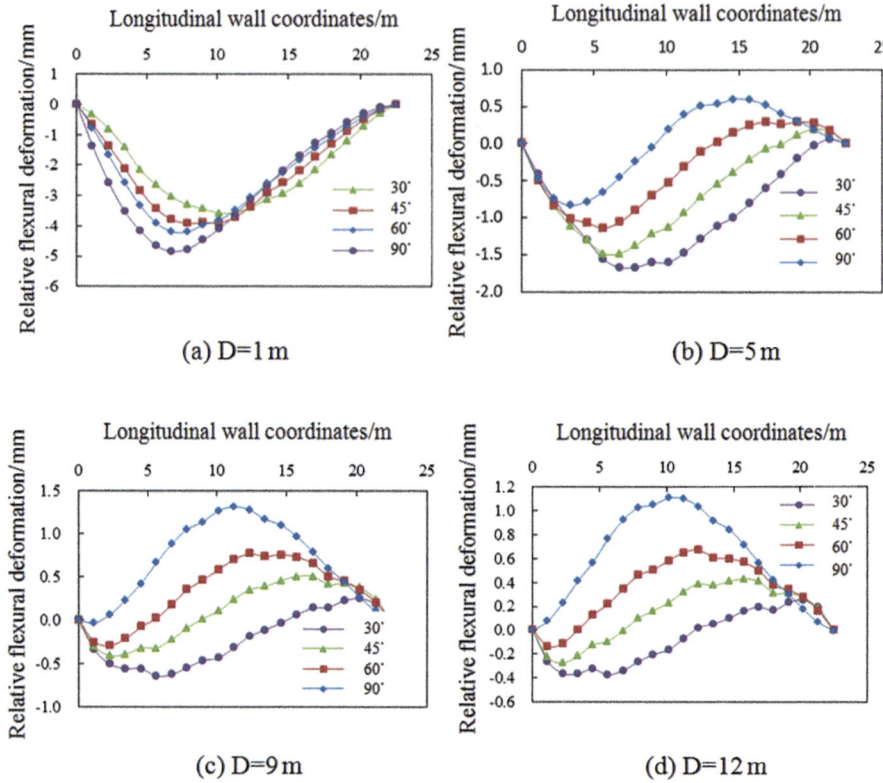

Fig. 4. Relative deformation curves of longitudinal wall

peak value is only 3.58 mm. When $\alpha = 45°$ and $\alpha = 60°$, the relative deflection of the longitudinal wall of the building gradually increases, its value is 3.92 mm and 4.21 mm. When $\alpha = 90°$, the relative deflection of the building is most obvious, the peak reaches 4.85 mm, at this point the buildings suffers a lot from instability.

When D = 9 m and angle α is any value, the relative deflection of the longitudinal wall is significantly different. When $\alpha = 30°$, the relative flexure of the facade longitudinal wall of the building is mainly in concave shape, the maximum of relative flexural deformation is 1.67 mm. When $\alpha \geq 30°$, facade vertical wall are characterized by the relative flexure deformation "⌣" form, and as the angle grows, "⌣" form of relative flexure deformation becomes more and more obvious. The characteristics of the "⌣" form of relative flexure deformation are shown as follows. The concave flexure deformation occurs mainly in the side of the longitudinal wall adjacent to the excavation surface, the upper convex flexural deformation mainly occurs in the longitudinal wall away from the excavation side. The maximum deflection of the flexural deformation decreases with the increase of angle, it occurs in the relative deflection of the concave of the lateral longitudinal wall of the foundation pit, the maximum value is 1.49 mm, 1.14 mm, 0.83 mm.

When D = 9 m, with the increase of angle α, the relative deflection of the longitudinal wall of the building by "⌣" form gradually transformed into a convex shape.

When $\alpha = 30°$, the concave relative flexural deflection which occurs in the longitudinal wall of the façade is prominent. The maximum of relative flexural deformation, which is 0.65 mm, occurs at the side of the wall which is adjacent to the excavation surface. When $\alpha = 45°$ and $\alpha = 60°$, the relative flexural deflection of the building is characterized by showing "⌣" form. And relative flexural deflection of the longitudinal wall, with the maximum being 0.51 mm and 0.77 mm respectively, mainly includes the type of convex. When $\alpha = 90°$ the convex relative flexural deformation happening in the facade longitudinal wall of the building will reach its biggest value 1.31 mm. The degree of the relative flexural deflection of the longitudinal wall will be the maximum when the building is perpendicular to the foundation pit and the center of the building is located in the natural surface subsidence trough.

When D = 12 m, the deformation value of the longitudinal wall is significantly lower than that of other horizontal distance D.

When $\alpha = 30°$, the concave relative flexural deflection of the longitudinal wall of the building shows its significance with the maximum deflection value being 0.37 mm. When $\alpha \geq 30°$ convex relative flexural deformation is prominent, and the deformation value increases gradually with the increase of angle, and its value is 0.41 mm, 0.67 mm and 1.1 mm respectively.

The Deformation Law of Buildings Affected by Stiffness

Rule of Subsidence Deformation of Longitudinal Wall

In Fig. 5, comparing the longitudinal wall settlement curve with the natural surface settlement curve, it can be seen that although the stiffness of the building restricts its settlement deformation, the trend of the two changes is roughly the same. With the change of the distance between the excavation of the structure and the soil, the constraint and coordination of the deformation of the soil in the foundation pit are significantly different. The coordination function reaches the highest value when the building is located at the maximum of both sides of the settlement and the maximum curvature of the upper convex region. And also, compared with natural surface subsidence, longitudinal wall deflection tends to be more gradual and smoother.

As the stiffness of the beam and column declines by 90% and 50%, the general stiffness of the building decreases. The settlement coordination effect of soil, which is produced by the building and happens where the foundation is located, is not obvious enough. When the longitudinal wall is adjacent to the excavation side of the foundation pit, the settlement trend is obviously concave and the maximum settlement value is 21.13 mm. When the foundation center is located in the maximum convex relative deformation of the soil, the structure has a prominent convex shape. In the case that the loss of stiffness of beam and column is both 30% and the other case that their stiffness are all in good condition, with the beam and column stiffness increasing gradually, the overall flexural subsidence of the building will gradually decrease due to the restriction which comes from the stiffness of the building itself. And finally the distribution of flexural subsidence will present linearly.

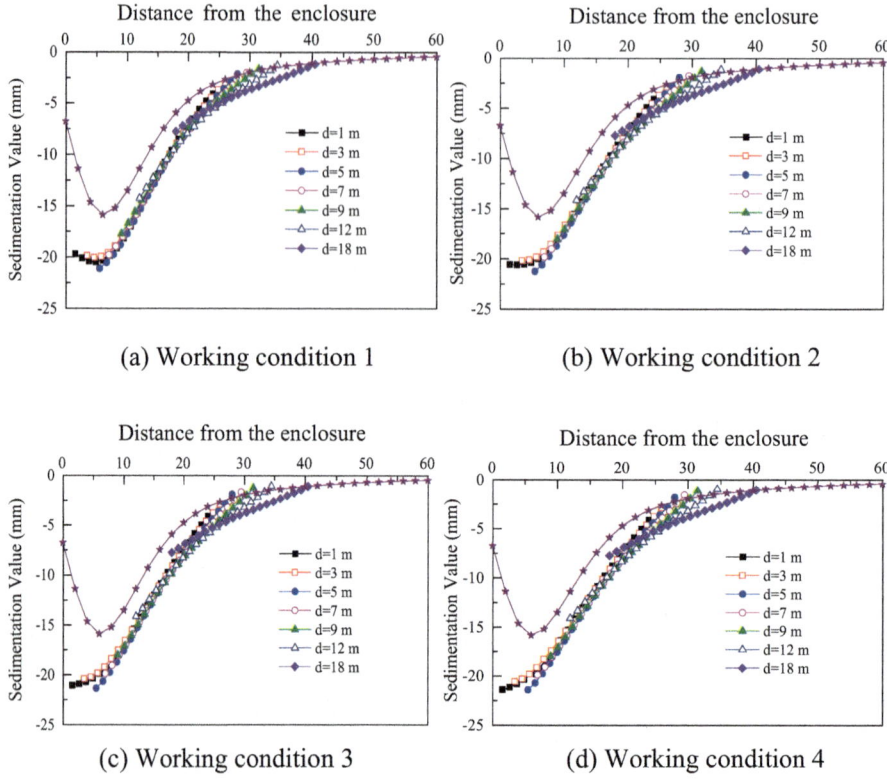

Fig. 5. Settlement curves of longitudinal wall

2 Conclusions

When D = 1 m, the relative flexure of the vertical wall is a simple concave shape. In addition to the case that the corresponding vertical wall of the $\alpha = 30°$ has obvious deflection deformation, with the increase of and horizontal distance D, longitudinal walls of building façade will be gradually transformed into "~" form and convex bending deformation. The maximum of the relative flexural deformation of the concave shape under the longitudinal wall and the maximum of the relative flexural deformation of the upper convex shape occurs at $\alpha = 90°$ and the maximum point of the settlement and the maximum point value of the convex flexure curvature.

The stiffness of the building restricts its settlement deformation, the trend of the two changes is roughly the same. With the change of the distance between the excavation of the structure and the soil, the constraint and coordination of the deformation of the soil in the foundation pit are significantly different.

The coordination function reaches the highest value when the building is located at the maximum of both sides of the settlement and the maximum curvature of the upper convex region. And also, compared with natural surface subsidence, longitudinal wall deflection tends to be more gradual and smoother.

Acknowledgments. Many people have made invaluable contributions, both directly and indirectly to my research. I would like to express my warmest gratitude to Zhang Zhi-guo, my supervisor, for his instructive suggestions and valuable comments on the writing of this thesis. Without his invaluable help and generous encouragement, the present thesis would not have been accomplished. At the same time, I am also grateful to Fei Si-yi for providing me with valuable advice and access to the related resources on my thesis. Besides, I wish to thank my colleagues ZhangRui, who helped me search for reference. My heart swells with gratitude to all the people who helped me.

Effect of Passive Zone Improvement on the Stability of a Double-Row Pile-Retained Excavation

Zhiyan Zhou, Dongqing Nie$^{(\boxtimes)}$, and Wei Zhang

Shanghai Municipal Engineering Design General Institute (Group) Co., Ltd.,
Shanghai, China
ndqasd@163.com

Abstract. The Shanghai Bailonggang Underground Wastewater Treatment Plant is located alongside the Yangtze River, Shanghai, China and features a deep and large excavation project. A retaining system consisting of a two-step berm and double-row piles is adopted to support the excavation, and the passive zone soil of the double-row pile structure is improved. This paper presents an analysis of the effects of the improved passive zone soil on the stability of the double-row pile-retained excavation based on a finite element method. Parametric studies were conducted with different widths and heights of the improvement zone. The analysis results demonstrate that the factor of safety (FOS) of the excavation increased with an increase in the improvement zone width, and two different failure mechanisms were observed with different improvement zone widths. The FOS first increased and then became constant with an increasing improvement zone height. Meanwhile, the FOS continued to increase if the improvement zone was embedded deeper than the retaining piles.

Keywords: Excavation · Double-row piles · Ground improvement
Finite element · Stability

1 Introduction

The lateral stiffness of double-row piles is much larger than that of single-row piles, when the same piles are used to construct them, and thus, the lateral displacement of double-row piles will be smaller than that of single-row piles when they are applied within an excavation project. Double-row piles have been widely used in deep and large excavations throughout China. Many studies on double-row piles have been carried out, and several methods, including those based on the limit-equilibrium theory [1, 2], elastic foundation beam method [3–5] and soil arch theory [6, 7], have been proposed to represent double-row piles.

Double-row pile-retained excavations are applicable in soft soils, and ground improvement is often used to reduce displacements and improve the stabilities of double-row piles. However, ground improvements applied to double-row pile-retained excavations have rarely been reported. This paper reports on a large area of deep excavation retained by double-row piles, and the effect of the ground improvement used in this excavation project is analyzed using a finite element method.

© Springer Nature Singapore Pte Ltd. 2018
T. Qiu et al. (Eds.): GSIC 2018, *Proceedings of GeoShanghai 2018 International Conference: Advances in Soil Dynamics and Foundation Engineering*, pp. 536–542, 2018.
https://doi.org/10.1007/978-981-13-0131-5_58

2 Project Description

The Bailonggang Municipal Wastewater Treatment Plant is located alongside the Yangtze River, Shanghai, China. The project was designed and constructed to improve the wastewater treatment capacity of the plant, and a new underground wastewater treatment plant is planned. The basement excavation of the planned underground wastewater treatment plant is approximately 575 m × 340 m in plan view, and the excavation depth ranges from approximately 12.8 m to 15.8 m. The excavation area is exceedingly large, and thus, the braced excavation method would be costly. However, as there are no important buildings or structures around the excavation site, a retaining system consisting of a two-step berm and double-row piles is adopted to support the excavation.

Figure 1 shows the classical profile of the excavation. The height of the two-step berm is 7 m, and the berm slope is approximately 35°. The double-row pile consists of two rows of bored piles and a capping beam connecting the caps of the two rows of piles. The bored piles have a diameter of 1 m and a length of 25 m, and the distance between the piles is 1.2 m. The shape of the capping beam is shown in Fig. 2. As there are thick, mucky clay layers in the ground, soil cement columns are used to improve the soil between the two rows of bored piles and the passive zone soil. The improved zone lengths are 18 m and 6 m for the soil between the two rows of bored piles and the passive zone soil, respectively, and the improved zone widths are 4 m and 5 m, respectively. Soil cement columns are also constructed 5 m from the top of the berm to prevent seepage.

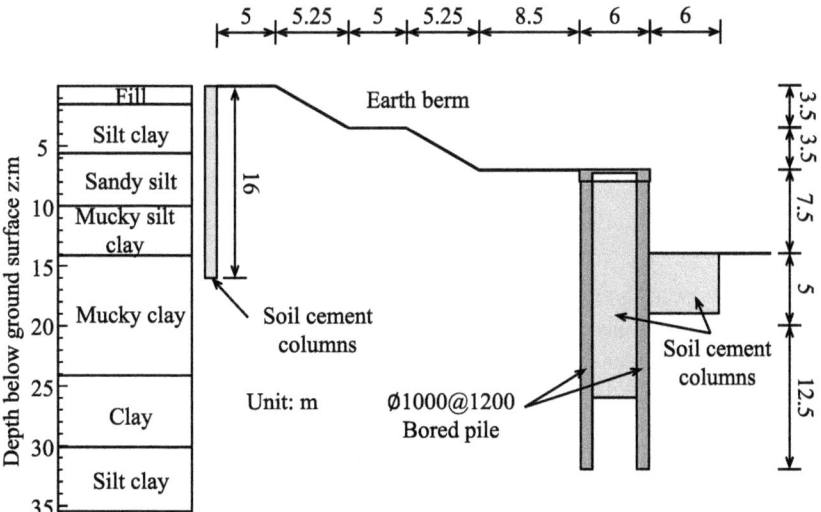

Fig. 1. A-A profile of the Bailonggang excavation project.

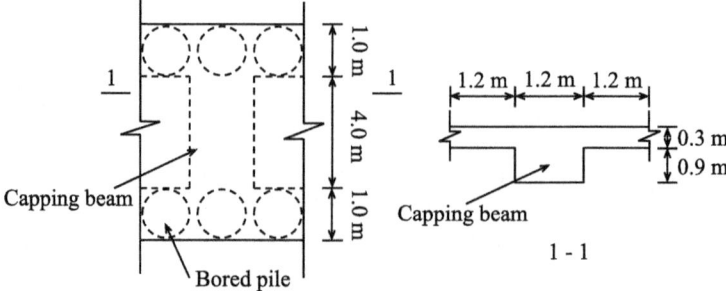

Fig. 2. Capping beam.

3 Finite Element Model

For the purpose of understanding the effects of ground improvement on the excavation stability, a two-dimensional plane strain finite element model was established based on the Bailonggang deep excavation project. The finite element model is shown in Fig. 3. The model is 160 m × 50 m (width × height). The left and right boundaries of the model were fixed with roller supports, and the bottom boundary was fixed with pin supports.

In this model, the bored piles and capping beam were both simulated using plate elements. The bending stiffness (*EI*) and normal stiffness (*EA*) of the bored piles were 1.23×10^6 kN · m²/m and 1.47×10^7 kN/m, respectively, and those of the capping beam were 4.1×10^6 kN · m²/m and 2.05×10^7 kN/m, respectively. The Poisson's ratio of these structures was 0.2.

Considering the low permeability coefficient of the ground soil and the short excavation period, the total stress method was adopted in this analysis. The Mohr-Coulomb constitutive model was employed to simulate the behaviors of the soil layer and soil cement columns. It can be assumed that the undrained shear strength of the soil S_u^s (unit: kPa) increases linearly with depth [8], and it can be calculated by

$$S_u^s = 20 + z$$

where z is the depth below the ground surface.

In an undrained analysis, the Poisson's ratio should be 0.5. However, the stiffness matrices will be singular during the calculation if the Poisson's ratio is 0.5. Hence, an approximate value of 0.495 was adopted in this analysis to avoid singular matrices. The soil shear modulus G was assumed to be 100 S_u^s [9], and the Young's modulus of the soil was $3G$. The density of the soil was 1900 kg/m³, and the lateral stress coefficient was 0.65.

In general, the unconfined compressive strength of a soil cement column should be no less than 0.8 MPa [10], and this value was accepted for this analysis. The Young's modulus of the soil cement columns was 60 MPa, and their density was 2400 kg/m³.

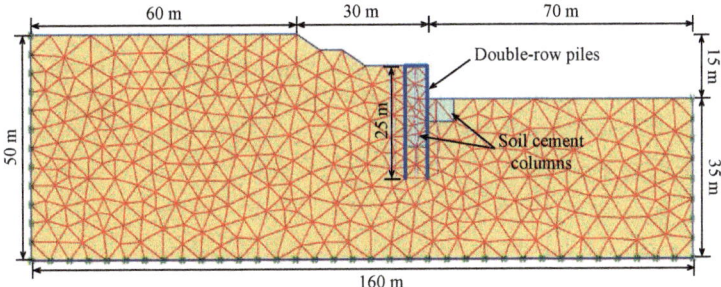

Fig. 3. Finite element model.

The modeling procedure includes the following steps: (1) construct the retaining system and initialize the deformation to zero while neglecting the effects of the construction of the retaining system structure; (2) excavate the soil above the double-row piles; (3) excavate the soil in front of the double-row piles; (4) evaluate the stability of the excavation using the shear strength reduction (SSR) method.

4 Result Analysis

The factor of safety (FOS) for the excavation calculated using the SSR method was 1.35, and the computed slip surface was observed mainly in the first step of the berm. To further analyze the stability of the double-row pile-retained excavation, the berm was converted into an equivalent distributed load before performing the SSR calculations. The FOS of the double-row pile-retained excavation was 1.66 under this condition. The shear strain of the excavation at the critical state is shown in Fig. 4. Shear bands formed around the passive improvement zone, while the soil between the two rows of piles did not fail. This indicates that the passive improvement zone was relatively effective at improving the stability of the double-row pile-retained excavation. Thus, the height H and the width B of the passive improvement zone were mainly studied in this paper.

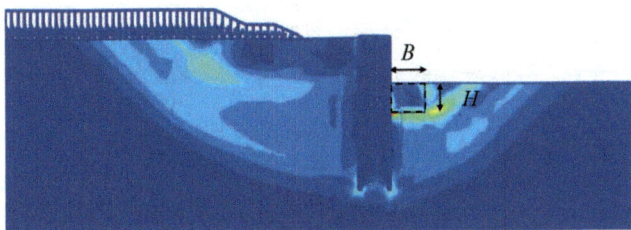

Fig. 4. Distribution of shear strain throughout the double-row pile-retained excavation.

4.1   Effect of the Width of the Passive Improvement Zone

To account for the effect of B, H was set to 6 m in this analysis. As shown in Fig. 5, the FOS exhibited a large increase when the ground improvement was applied to the passive zone, and the FOS increased almost linearly with B when B was larger than 5 m.

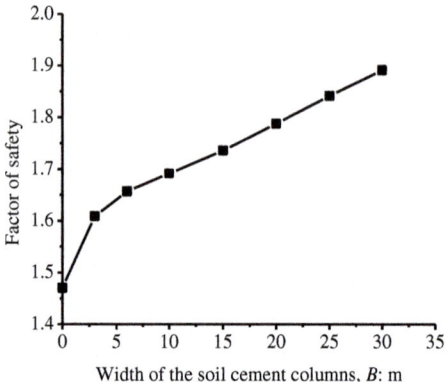

Fig. 5. Effect of the width of the passive improvement zone to FOS.

Figure 6 shows the failure mechanisms of the excavation for different B at the critical state. The widths of the active and passive shear bands were almost identical for excavations with small or zero passive improvement zone widths ($B < 30$ m). Furthermore, their failure mechanisms were all the same; that is, they failed with the overturning of the double-row piles, as shown in Fig. 6(a)–(c). For a large improvement zone width ($B = 30$ m), which was larger than the width of the passive shear band for the excavation without passive improvement, the new passive shear band surrounded the improvement zone and the width of the shear band became larger. In this case, the double-row piles almost translated horizontally, as shown in Fig. 6(d).

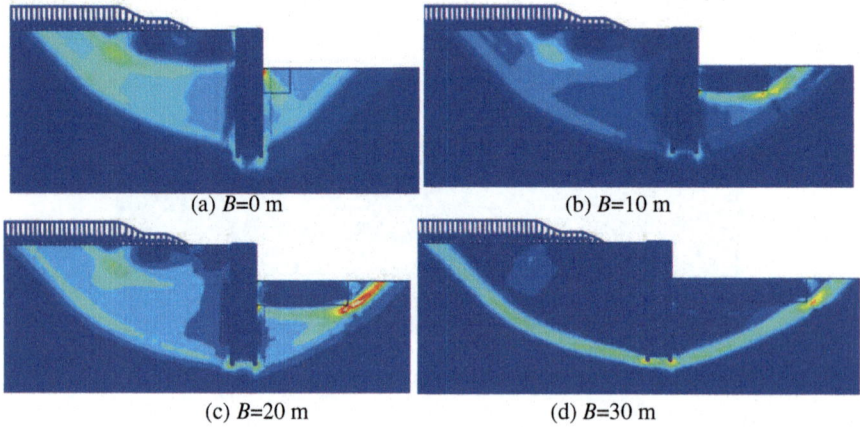

Fig. 6. Distribution of shear strain for different widths of the passive improvement zone.

4.2   Effect of the Height of the Passive Improvement Zone

To study the effect of H, B was maintained at 5 m in this analysis. Figure 7 shows the effect of H on the FOS. When H increased from 0 m to 10 m, a shear band appeared to the right of the improvement zone; in addition, the length of the shear band increased with an increase in H, as is shown in Figs. 6(a) and 8(a). Therefore, the FOS increased with an increase in the value of H. However, the FOS was constant when H varied from 10 m to 17.5 m. This was because the vertical shear band on the right side of the improvement zone had been connected to the shear band below the piles in Fig. 8(a), so the length of the shear band did not change with H, as is shown in Fig. 8(b), (c). When the improvement zone height continued to increase (>17.5 m), the bottom of the improvement zone became deeper than the bottom of retaining piles (although this is

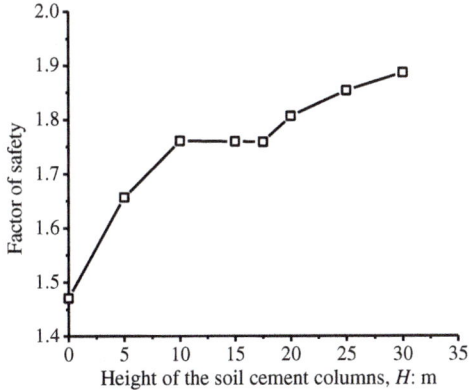

Fig. 7. Effect of the height of the passive improvement zone to FOS.

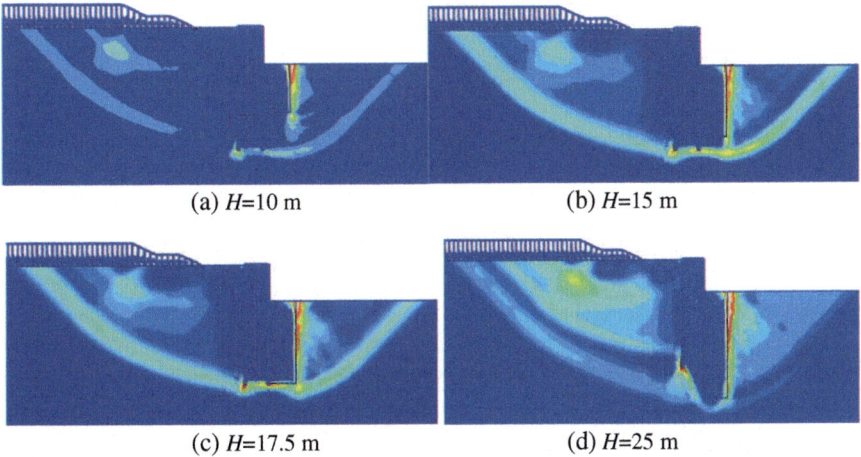

(a) H=10 m   (b) H=15 m

(c) H=17.5 m   (d) H=25 m

Fig. 8. Distribution of shear strain for different heights of the passive improvement zone.

rare in practice). The shear band was forced to surround the bottom of the improvement zone, as shown in Fig. 8(d). This made the FOS increase with H again.

5   Conclusions

The effects of passive improvement on the stability of a double-row pile-retained excavation are studied in this paper. Some conclusions and suggestions can be drawn accordingly.

1. The existence of a passive improvement zone improved the FOS of this double-row pile-retained excavation.
2. The FOS increased with an increase in the width of the passive improvement zone B. When B was larger than the width of the passive shear band for the excavation without improvement, the failure mechanism of the double-row pile-retained excavation changed from overturning failure to translation failure.
3. When the height of the improvement zone H was less than 10 m, the FOS increased with an increase in H. The FOS was constant when H was in the range of 10 m– 17.5 m. When the bottom of the improvement zone was deeper than the bottom of the piles ($H > 17.5$ m), the FOS continued to increase.

References

1. He, Y.H., Yang, B., Jin, B.S., Li, R.R., Tan, Y.J., Wang, T.H.: A study on the test and calculation of double-row fender piles. J. Build. Struct. 17(2), 58–66 (1996)
2. Huang, Q.: The simplified calculation method for slope protection pile space force. Build. Struct. 1(8), 43–45 (1995)
3. Zheng, G., Li, X., Liu, C., Gao, X.F.: Analysis of double-row piles in consideration of the pile-soil interaction. J. Build. Struct. 25(1), 99–106 (2004)
4. Wu, G., Bai, B., Nie, Q.K.: Research on calculation method of double-row piles retaining structure for deep excavation. Rock Soil Mech. 29(10), 2753–2758 (2008)
5. Ying, H.W., Chu, Z.H., Li, B.H., Liu, X.W.: Study on calculation method of retaining structure with double-row piles and its application. Rock Soil Mech. 28(6), 1145–1150 (2007)
6. Zeng, G.Q.: Design and analysis of the forces of retaining structure with double-row piles of deep foundation pit. Master thesis. Xi'an University of Architecture and Technology (2005)
7. Wan, Z., Wang, Y.S., Li, G.: Analysis and calculation of retaining structure with double-row piles. J. Hunan Univ. 28(3), 116–120 (2001)
8. Dong, Y.P., Burd, H.J., Houlsby, G.T.: Finite-element analysis of a deep excavation case history. Géotechnique 66(1), 1–15 (2015)
9. Ladd, C.C.: Stress-strain modulus of clay in undrained shear. J. Soil Mech. Found. Div. 90 (5), 103–132 (1964)
10. Liu, G.B., Wang, W.D.: Excavation Engineering Manual, 2nd edn. China Building Industry Press, Beijing (2009)

Permanent Displacement Based Seismic Design Chart for Cantilever Retaining Walls

Prajakta Jadhav[1(✉)], Mohit Singh[2], and Amit Prashant[1]

[1] Indian Institute of Technology Gandhinagar, Gandhinagar, Gujarat, India
prajakta.ramesh.jadhav@iitgn.ac.in
[2] Indian Institute of Technology Patna, Patna, Bihar, India

Abstract. The primary aim of this paper is to suggest a suitable design procedure using a seismic sliding displacement based chart for the design of cantilever retaining walls. Permanent displacement based design chart have been proposed by Franklin and Chang for slopes and have been used by Richards and Elms for the seismic design of gravity retaining walls. Engineers have been using the same chart for seismic design of cantilever retaining wall without understanding the implications of such considerations and hence may result in unreliable design. The suitability of design procedure to be adopted would depend upon its closeness with the actual mechanism taking place. Experimental investigations have witnessed the formation of a V-shaped wedge in the backfill of cantilever retaining walls. The Double wedge model computes displacements considering the formation of this wedge and its relative movement with the wall during seismic loading. The upper bound curve has been developed on the basis of Double wedge model by analyzing 153 earthquakes for four different heel-length to height ratio and compared with Franklin and Chang's chart. The procedure followed for the development of this chart has been explained in the paper. This study has been performed to understand the suitability of Franklin and Chang's chart for the design of cantilever retaining walls. A suitable design procedure has been suggested on the basis of the V-shaped mechanism for the seismic design of cantilever retaining wall.

Keywords: Design chart · Permanent sliding displacement
Cantilever retaining walls · Double wedge model · V-shaped wedge
Seismic loading

1 Introduction

Seismic design philosophies are gaining confidence in the direction of deformation based approach as failure can be measured in terms of displacements. Newmark's sliding block model [1] has been widely used to compute the permanent displacements for slopes. Franklin and Chang analyzed 169 earthquake motions using Newmark's sliding block theory with unsymmetrical resistance for earth fill dams and have proposed a design chart to be used by engineers [2]. Richards and Elms [3] proposed methodology for the seismic design of gravity retaining walls on the basis of Franklin and Chang's chart. According to their design philosophy, yield acceleration of wall to be designed can be obtained from the chart for the allowable displacements. This yield acceleration

© Springer Nature Singapore Pte Ltd. 2018
T. Qiu et al. (Eds.): GSIC 2018, *Proceedings of GeoShanghai 2018 International Conference: Advances in Soil Dynamics and Foundation Engineering*, pp. 543–550, 2018.
https://doi.org/10.1007/978-981-13-0131-5_59

can be used to compute the required weight of wall using pseudostatic force equilibrium equations. The pseudostatic forces are computed according to the M-O theory which considers equilibrium of V-shaped wedge, exerting thrust on the predefined back face of gravity retaining wall. However, the applicability of Franklin and Chang's chart for cantilever retaining walls is yet to be established. There is a need to understand if the existing charts can be used directly for the design of cantilever retaining walls. In order to use such charts reliably, one needs to consider a suitable mechanism that can occur in the backfill. There is an ambiguity regarding what mechanism to be taken place for the design of cantilever retaining walls. Some of the current design philosophies suggest to adopt ad hoc arrangement wherein soil above the heel is considered to be a part of the wall and analyze cantilever retaining walls as gravity retaining walls [4–8]. The implications of this assumption may result in the over-conservative design of cantilever retaining walls. Experimental investigations for cantilever retaining walls have shown that there is a formation of inverted V-shaped rupture planes in the backfill [9, 10]. A Double wedge model has been proposed which simulates the formation of these rupture planes and computes yield acceleration and hence displacements separately for the wall and sliding soil wedge considering its relative movement with respect to the wall [11]. Seismic design procedure has been proposed on the basis of formation of the V-shaped wedge in the backfill for cantilever retaining walls.

2 Double Wedge Model

Double wedge model has been proposed by Jadhav and Prashant [11] to compute seismic sliding displacements of a cantilever retaining wall. This model computes dynamic yield accelerations for wall and soil wedge separately. The computed yield accelerations depend not only on the geometry and soil properties but also on the ground acceleration at that time instant. During ground motion, two rupture planes, AB and BC as shown in Fig. 1 would be developed from the heel of the wall.

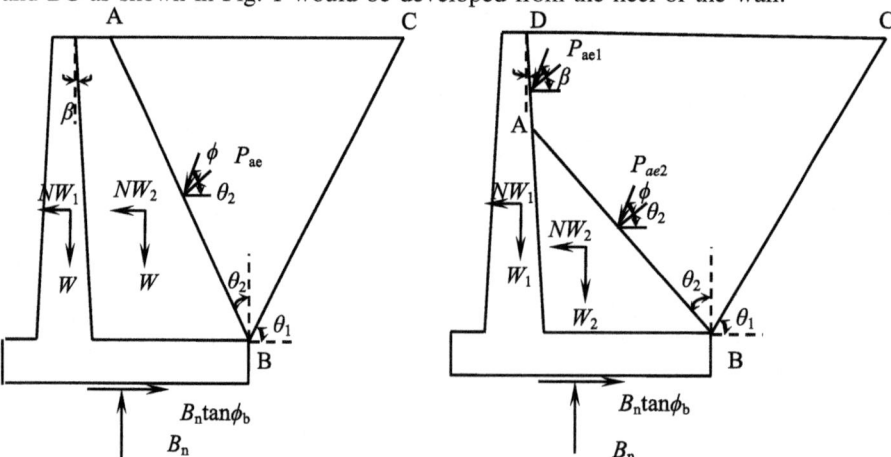

Fig. 1. Schematic diagram showing outer rupture plane not intersecting wall

Fig. 2. Schematic diagram showing outer rupture plane intersecting wall

The inner rupture plane BC would develop from the heel towards the backfill and the outer rupture plane AB would develop from the heel towards the back face of the wall. The critical angles of inclination of rupture planes are determined at which the wall has minimum yield acceleration. The model computes displacement for two cases, i.e., when the outer rupture plane does not intersect the back face of the wall as shown in Fig. 1 and the other when this plane intersects the back face of the wall as shown in Fig. 2. The formulation has been accordingly developed to compute the yield acceleration of wall. The model assumes that the developed soil wedge slides tangentially with respect to the wall with locked soil mass along AB as shown in Figs. 1 and 2, and it remains in contact with it throughout the motion. The model ensures both acceleration and velocity compatibility to be maintained during the motion. Double wedge model has been validated for centrifuge tests performed by Sitar et al. [12, 13] at UC Davis on a geometry of wall subjected to Takatori, Kobe earthquake motion [14] shown in Fig. 3. The rigid body translation measured was 0.0165 m for the considered geometry and PGA 0.64 g. The Double wedge model estimated 0.018 m which is quite close to the measured value thus imparting enough confidence on the Double wedge model.

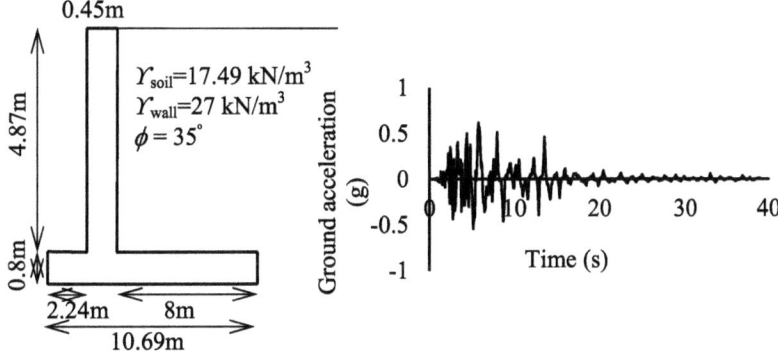

Fig. 3. Geometry of wall and ground motion used in case study

2.1 Verification of the Earthquake Data Used for Developing Franklin and Chang's Chart

The upper bound envelope using Newmark's sliding block has been redeveloped in this study. This chart predicts the displacements computed using Newmark's sliding block theory for a N/A value wherein, N represents yield acceleration, and A represents expected PGA at the site. The yield acceleration considered in the chart is computed at the factor of safety with respect to sliding equal to one which is a function of geometry and soil properties and hence remains constant throughout the ground motion. This exercise has been performed to gain confidence in the earthquake data to be used for the development of curve using Double wedge model which would be compared with Franklin and Chang's [2] curves. It was possible to obtain 153 earthquakes data from PEER [13] out of 169 earthquakes used by Franklin and Chang. All the ground motion accelerations have been scaled to 0.5 g, and computed displacements have been scaled with respect to scaling velocity of 0.762 m/s. The displacement curves from Franklin and Chang have been digitized to get the data points for comparison. The digitized curves and the computed upper bound envelope have been plotted in Fig. 4, which showed a good match for 153 earthquakes.

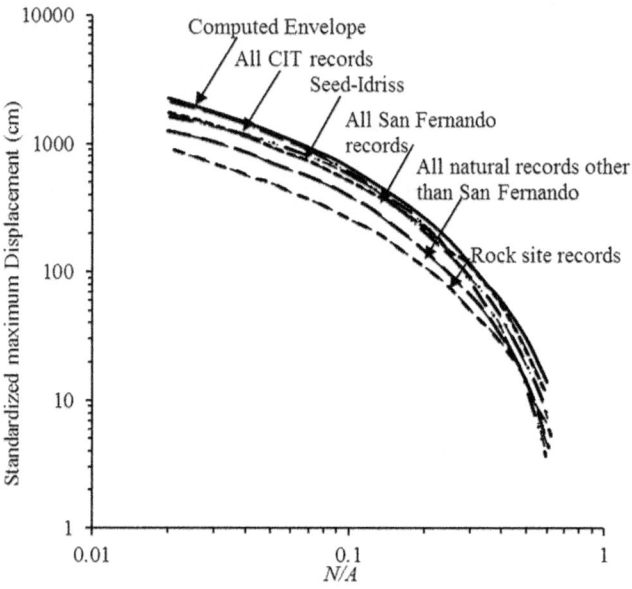

Fig. 4. Comparison of obtained curve with Franklin and Chang's upper bound curves

2.2   Methodology to Develop Design Chart Using Double Wedge Model for Cantilever Retaining Walls

A design chart has been developed for cantilever retaining walls using Double wedge model. The implementation of Double wedge model involves two-step calculation of yield acceleration. The first step involves computation of cutoff yield acceleration, N which is a function of geometry and soil properties only. During ground motion, if this cutoff yield acceleration is exceeded, dynamic yield acceleration of wall which is a function of ground acceleration at that time instant, is computed. If dynamic yield acceleration coefficient is also exceeded, then corresponding velocity and displacement of the wall is computed. As dynamic yield acceleration would vary at each time instant, the chart has been developed for different cutoff yield acceleration coefficient values but displacements have been computed only when the dynamic yield acceleration is exceeded. This assumption is in convergent with the physical situation, wherein, the wall-soil wedge would not undergo any movement, unless the cutoff yield acceleration is exceeded. Once cutoff yield acceleration is exceeded, the system would undergo displacements with respect to the dynamic yield acceleration of wall. In this paper, 12 m height, H, of the wall and four different heel lengths, L, viz., 3 m, 4 m, 4.5 m and 6 m, supporting a backfill with friction angle 30° and unit weight 16.67 kN/m³ have been considered. These configurations with L/H ratios 0.25, 0.33, 0.375 and 0.5 have been considered with an idea that there would be significant change in the developed mechanism and hence the magnitude of displacements with different L/H ratios. Following steps have been executed for developing chart.

Step 1: Consider N/A values same as Franklin and Chang chart, where N is the cutoff yield acceleration coefficient, and A is the scaled peak ground acceleration value which has been considered as 0.5 g. Using N/A values from the chart and A as 0.5 g, the values of N have been computed.

Step 2: Compute pseudostatic earth pressure force using N at different values of inclination of outer rupture plane, θ_2 using M-O equation. The maximum value of pseudostatic earth pressure force, P_{ae} and the corresponding value of θ_2 has been chosen for further calculations as shown in Fig. 5.

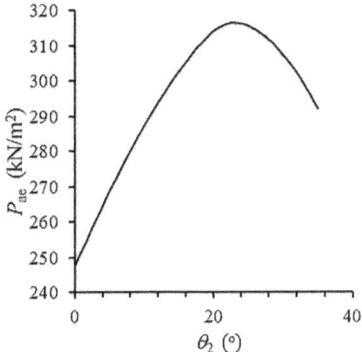

Fig. 5. Variation of lateral earth pressure force with inclination of outer rupture plane

Step 3: The weight of wall, W_1 has been computed by solving the equilibrium equation for both the cases when outer rupture plane is not intersecting and intersecting the back face of the wall as shown in Figs. 1 and 2. The Eqs. (1) and (2) have been derived to compute the weight of wall, W_1 using pseudostatic earth pressure force and inclination for outer rupture plane from step 2 for outer rupture not interesting and intersecting wall respectively. In these equations, β represents the inclination of back face of the wall and W_2 represents the weight of locked soil mass which can be calculated from the geometry. ϕ and ϕ_b represent friction angles for backfill soil and foundation soil respectively, and δ represents interface friction angle between wall and soil. δ and ϕ_b have been considered two-third ϕ in this analysis.

$$W_1 = \frac{P_{ae}(\cos(\delta + \beta) - \sin(\delta + \beta)\tan(\phi_b))}{(\tan(\phi_b) - N)} - W_2 \tag{1}$$

$$W_1 = \frac{P_{ae1}K_1 + P_{ae2}K_2}{N - tan_b} - W_2 \qquad (2)$$

$$K_1 = sin(\delta + \beta)tan\phi_b - cos(\delta + \beta)$$

$$K_2 = sin(\phi + \theta_2)tan\phi_b - cos(\phi + \theta_2)$$

Step 4: Compute seismic sliding displacement for cantilever retaining walls using the obtained values of weight of wall following the steps in Double wedge model [11] for different N/A values. According to the formulation in Double wedge model, dynamic yield accelerations are obtained for wall and soil wedge for critical wedge by iterating the inclination of rupture planes. The displacements are computed only when the yield acceleration of wall is exceeded by the considered ground motion at that instant by using numerical integration. These steps have been repeated for 153 different earthquakes, and the upper bound curve has been plotted for four L/H ratios as shown in Fig. 6. The upper bound curve for Franklin and Chang has also been plotted in Fig. 6 with the curve for cantilever retaining walls on the log-log scale and normal scale to understand the difference in magnitudes of displacement.

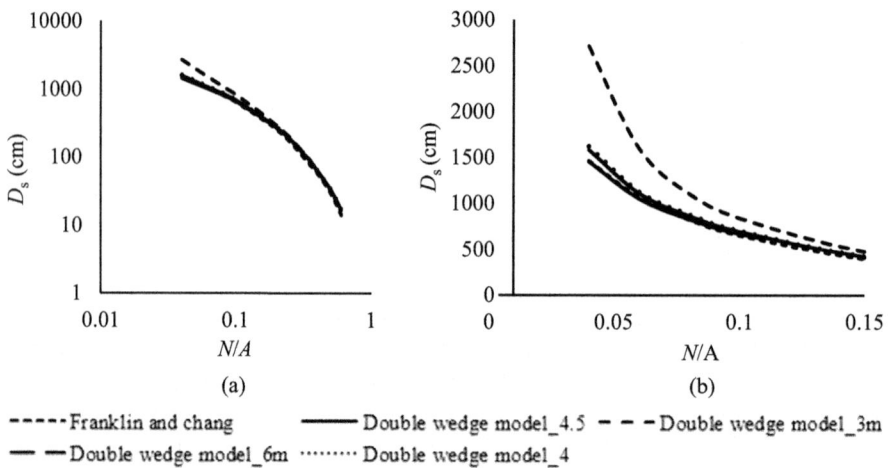

Fig. 6. Upper bound curves representing standardized maximum displacement, D_s versus N/A ratio obtained from Double wedge model and Franklin and Chang in (a) log-log scale, and (b) normal scale

3   Results and Discussions

It can be observed that the displacements computed by Double wedge model for L/H ratio greater than 0.3 matches well with the displacements obtained from Franklin and Chang. The upper bound curve for lower N/A value shows significant variation for L/H ratio less than 0.3. This shows that wall with smaller N/A and lesser heel length is susceptible to undergo more sliding displacement than the wall with same N/A and greater heel length. With the decrease in heel length, the mechanism gets switched from case1 to case2 as shown in Figs. 1 and 2. When the outer rupture plane is not intersecting the back face of the wall, i.e., $L/H > 0.3$, the wall undergoes nearly same magnitude of displacement owing to the soil-soil friction angle ϕ along the outer rupture plane. For L/H less than 0.3, there is reduced wall-soil interface friction angle along the outer rupture plane which is in contact with wall thus resulting in an increased magnitude of displacements at lower N/A. At higher N/A values, which is quite uneconomical, the wall would undergo nearly same displacements irrespective of the heel length due to the sufficient resistance by the geometry of wall. Thus, Franklin and Chang's curve can be directly used for the estimation of sliding displacements for the wall geometries with L/H ratio greater than 0.3.

4 Suggested Design Procedure for Cantilever Retaining Walls

The upper bound curve for cantilever retaining walls with L/H ratio greater than 0.3 has been observed to match well with the Franklin and Chang's curve. The Franklin and Chang's curve with this limitation can be used directly for the design of cantilever retaining walls. Use of this chart in design would require assumption of suitable mechanism in the backfill. Owing to the experimental observations, formation of inverted V-shaped wedges has been assumed in this paper. Accordingly, following steps can be followed to compute the required weight of wall using the proposed chart.

- Estimate the maximum seismic sliding displacement that the wall would be allowed to undergo during ground shaking. Consider height of the wall according to the site requirement and assume heel length such that L/H greater than 0.3.
- Depending upon the site location, estimate a suitable PGA for the ground motion.
- Determine the N/A ratio from the proposed chart for the corresponding expected displacement. Using the value chosen for PGA, determine the value of cutoff acceleration, N, for the wall.
- Compute the lateral earth pressure force values from the M-O equation using the obtained value of N for different values of θ_2. The value of earth pressure force and θ_2 at which earth pressure is maximum would be used for further calculations.
- Depending upon the value of θ_2, either of Eqs. (1) or (2) can be used to compute the required weight of the wall.
- If the weight of the considered geometry is greater than the required weight of the wall, then the geometry would undergo less than or equal to expected displacement. The charts are applicable to walls resting on soil with properties taken in the range of this study. The authors are further continuing to perform the required computations to improve the design recommendations.

5   Conclusion

The permanent displacement based seismic design chart has been developed for cantilever retaining walls for different heel-length to height ratio using Double wedge model. The developed chart can be used for walls with backfill properties lying in the range of this study. The upper bound curves for cantilever retaining walls matched closely with the Franklin and Chang's curve with $L/H > 0.3$ whereas the curve for $L/H < 0.3$ showed significant variation at lower N/A ratio. The proposed design procedure considers the formation of V-shaped mechanism in the backfill and is suitable for geometry with $L/H > 0.3$.

References

1. Newmark, N.M.: Effects of earthquakes on dams and embankments. Geotechnique **15**(2), 139–160 (1965)
2. Franklin, A.G., Chang, F.K.: Permanent displacements of earth embankments by Newmark Sliding block analysis, Report 5, miscellaneous paper S-71-17, U.S Army Corps of Engineers Waterways Experiment Station, Vicksburg, MS (1977)
3. Richards Jr., R., Elms, D.G.: Seismic behavior of gravity retaining walls. J. Geotech. Geoenviron. Eng. ASCE **105**(14496), 447–464 (1979)
4. Earth retaining structures: British Standard (1994)
5. Earthquake Engineering Sectional Committee: Criteria for Earthquake Resistant Design of Structures (Part 3) Bridges and Retaining Walls, India (2002)
6. EN 1998-5:2004 - Eurocode 8: Design of structures for earthquake resistance - Part 5: foundations, retaining structures and geotechnical aspects, Eurocode 8 (2004)
7. Anderson, D., Martin, G., Lam, I., Wang, J.: Seismic Analysis and Design of Retaining Walls, Buried Structures, Slopes, and Embankment, NCHRP (2008)
8. U.S. Department of Transportation Federal Highway Administration: LRFD Seismic Analysis and Design of Transportation Geotechnical Features and Structural Foundations, FHWA-NHI-11-032 (2011)
9. Watanabe, K., Munaf, Y., Koseki, J., Tateyama, M., Kojima, K.: Behaviors of several types of model retaining walls subjected to irregular excitation. Soils Found. **43**(5), 13–27 (2003)
10. Penna, A., Scotto, A., Kloukinas, P., Taylor, C.A., Mylonakis, G., Evangelista, A., Simonelli, A.L.: Advanced measurements on cantilever retaining wall models during earthquake simulations. In: 20th IMEKO TC4 International Symposium and 18th International Workshop on ADC Modelling and Testing Research on Electric and Electronic Measurement for the Economic Upturn Benevento, Italy, pp. 127–132 (2014)
11. Jadhav, P., Prashant, A.: Double wedge model for computing seismic sliding displacements of cantilever retaining walls. Soil Dyn. Earthq. Eng. (2017, under review)
12. Al Atik L, Sitar N,: Experimental and analytical study of the seismic performance of retaining structures. Pacific Earthquake Engineering Research Center (2009)
13. Geraili Mikola, R., Candia, G., Sitar, N.: Seismic earth pressures on retaining structures and basement walls in cohesionless soils. J. Geotech. Geoenviron. Eng. **142**(10), 04016047 (2016)
14. Pacific Earthquake Engineering Research Center (PEER). NGA Database (2017). https://ngawest2.berkeley.edu/

Three-Dimensional Numerical Parametric Investigation of the Influence of Large-Scale Overexcavation on Deep Excavations

Jian-yong Han[1], Wen Zhao[1(✉)], Ling-hou Miao[2], Peng-jiao Jia[1], and Yong-ping Guan[3]

[1] School of Resource and Civil Engineering, Northeastern University, Shenyang 110819, People's Republic of China
[2] Economic and Technology Research Institute, State Grid Shandong Electric Power Company, Ji'nan 250000, People's Republic of China
[3] China Railway Design Corporation, Tianjin 300142, People's Republic of China

Abstract. Using a numerical analysis program based on the finite difference method (FDM), a study is conducted to reveal the mechanical behaviors of deep excavations influenced by a large-scale overexcavation, retained by a tieback pile wall. Based on an overexcavated deep excavation, the parameters used in the numerical models are verified using field observation. A parametric investigation is performed in order to analyze the influence of three factors, length of the wall, distance between the edge of overexcavation and the pile wall, and ovexcavation depth. According to the simulation results, the pile wall bottom is significantly influenced by the overexcavation and a deep overexcavation can cause the excavation collapse and retaining structure failure. Special emphasis is given to the analysis of horizontal pile wall deformation, ground settlement and potential slip surface. Conclusions obtained from the numerical study can provide guidelines on good practice in construction of anchored pile wall retained excavation in sandy soil.

Keywords: Deep excavation · Overexcavation · Tieback pile wall
Numerical simulation · Sandy soil

1 Introduction

Deep excavations are widely employed in the construction of subway stations, high-rise buildings and underground commercial centers. The study and construction of deep excavations in urban districts have significantly increased in recent years [1–4]. For a large-scale excavation projects, overexcavation easily occurs caused by many factors, such as geotechnical conditions, project requirements and construction mistakes. In recent years, a large number of deep excavations have collapsed with severe consequences due to overexcavation. Figure 1 shows a collapsed excavation for subway station in soft clay in China. Investigations for this accident found that the structure failures probably caused by overexcavation and improperly popped steel

© Springer Nature Singapore Pte Ltd. 2018
T. Qiu et al. (Eds.): GSIC 2018, *Proceedings of GeoShanghai 2018 International Conference: Advances in Soil Dynamics and Foundation Engineering*, pp. 551–560, 2018.
https://doi.org/10.1007/978-981-13-0131-5_60

struts. The overexcavation can not only increase the deformation of retaining structures, but also reduce the excavation stability, even lead to a collapse. However, the issues about the behaviors of retaining structures and surroundings of overexcavated excavations have not been resolved. Therefore, a deeper understanding of the influence of overexcavation is required to avoid excessive deformations and mitigate losses.

Until now, lots of researchers and engineers have contributed many achievements to the research of excavation (e.g., Peck [5]; Clough and O'Rourke [6]; Ou et al. [7]). However, available literatures rarely study related to the behaviors of an overexcavated deep excavation. Finno et al. [8] first mentioned overexcavation of a deep excavation in clay and indicated that overexcavation during construction could result in large movements. Based on a deep excavation case constructed by using cut-and-cover technique in soft clay, Tan and Wei [9] studied the performance of the excavation according to field observations obtained during soil excavated. Analyses of field observations showed that overexcavation was an important factors influencing the retaining wall deformation. Other related efforts are described by Gue and Tan [10], Huo et al. [11], Xu et al. [12], and Shen et al. [13]. Although some conclusions were obtained from the available literatures, the responses of excavations and retaining structures to overexcavation still remain an issue, especially for a large-scale overexcavation. Therefore, this issue requires further researches.

In this paper, a series of numerical models were established based on an overexcavated deep excavation case, to analyze the mechanical characteristics and deformation responses of retaining structures after excavation overexcavated. Parametric study about the primary influence factors was conducted using finite difference method (FDM). This study provides valuable information for design and construction of deep excavations.

Fig. 1. Excavation collapse of overexcavated subway station in China [14]

2   Project Overview and Numerical Model

2.1   Project Overview

The deep excavation studied was located at the center district of Shenyang, China, which is shown in Fig. 2. The excavation consisted of two rectangular parts, Area A and Area B. As shown in Fig. 2, many buildings and busy roads existed adjacent to the deep excavation. The original excavation depth of Area A was 23.1 m and the area was 218 m × 140 m. In order to control deformations of pile walls and surrounds during the excavation was under construction, many monitoring devices were installed around the excavation, including inclinometers in piles, force sensors in anchors and ground settlement monitoring points.

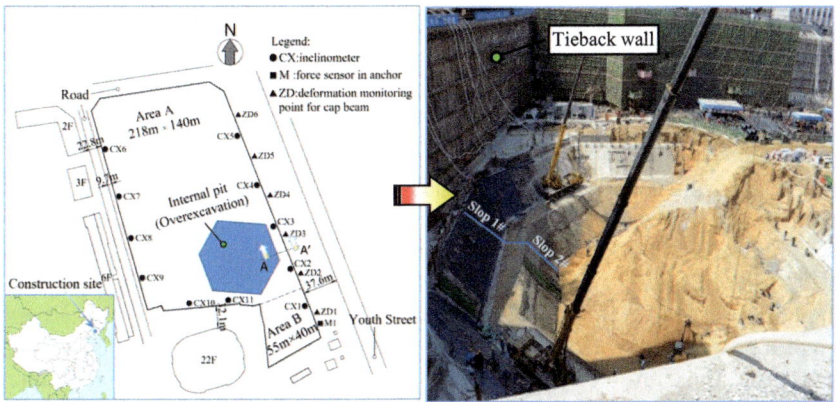

Fig. 2. Plan and construction site of deep excavation.

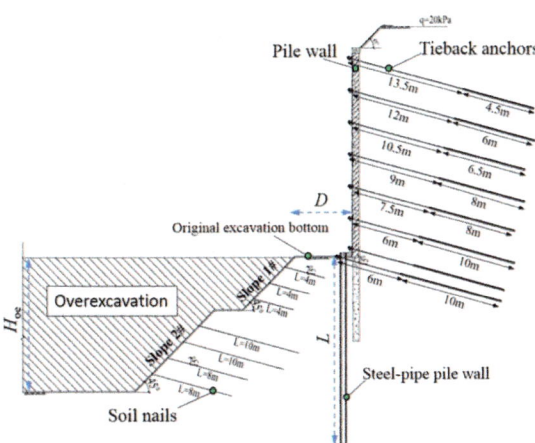

Fig. 3. Profile of retaining structures and overexcavation (section AA' in Fig. 2).

The deep excavation was supported by a seven-row tieback pile wall originally (see Fig. 3). The pile wall consisted of reinforced cast-in-situ piles, with 27 m in length, 0.6 m in diameter and 1.1 m in horizontal spacing. A concrete cap beam was constructed to joining the heads of piles, with an area of 0.5 m × 0.8 m. The tieback anchors consists of free length and fixed length, which are shown in Fig. 3. Three or four strands were installed in one anchor hole and each strand consisted of seven steel wires with 5 mm in diameter and 1860 MPa in strength.

After the excavation was constructed to design depth, a foundation pit was designed in the excavation for a high-rise building, which needed deeper excavation with a depth of 14 m below the original excavation bottom (see Fig. 3). The overexcavation region was excavated with two-stage slope, Slope 1# and Slope 2#, with depths of 6 m and 8 m, respectively. In order to reinforce the overexcavated deep excavation, double-row steel-pipe pile wall was installed at the edge of the excavation bottom, with 18 m in length and 0.4 m in diameter. The steel-pipe pile wall was also reinforced by tiebacks at the head of pile wall. Slope 1# and Slope 2# were reinforced by three rows soil nails and four rows soil nails, respectively. Each soil nail consisted of one steel reinforcement with a diameter of 25 mm.

The soils in construction site underneath Shenyang city mainly contains medium-coarse sand and gravelly sand with a little clay. The mechanical parameters of soils were obtained from a series of geotechnical investigations, which are illustrated in Table 1. The filled soil and silty clay have low friction angles. The cohesion and friction angle of silty clay was obtained from the direct shear tests and the test results are uniform. However, the results of the direct shear tests for the filled soil show discreteness. Thus, the average value of the friction angles for the filled soil was selected. The average depth of groundwater level is about 6.4 m to 8.3 m, and the groundwater was kept below the excavation bottom during excavation construction.

Table 1. Mechanical parameters of soils.

Num.	Soils	Depth/m	Unit weight/(kN·m^{-3})	c/kPa	Es/MPa	ϕ/(°)
①	Filled soil	4	16.66	-	15	10
②	Silty clay	1.5	18.13	22	4.5	7.3
③	Medium-coarse sand	4	19.11	1	18	30
④	Gravelly sand	10	19.60	1	30	36.4
⑤	Medium-coarse sand	8	19.11	1	27	33.4
⑥	Pebble	21	20.58	1	40	37.8
⑦	Gravelly sand	53	19.60	1	33	37

2.2 Numerical Model

At present, many methods can be used for design and investigation of deep excavation, including theory of beam on elastic foundation, numerical method and robust geotechnical design method [15]. However, some of them can only accommodate two-dimensional analysis. Thus, the finite difference method (FDM) was selected to investigate the behavior of overexcavated deep excavation, because of its excellence performance in solving three-dimensional problems and large deformation problems [16, 17]. Three-dimension finite difference models were established using the FDM program FLAC 3D [18]. In accordance with the symmetry of the problem, only a quarter of the excavation was modeled, as shown in Fig. 4. The size of the model was assumed to be 85 m in width and 67 m in length, which was large enough to avoid the influence of the boundaries. The four lateral boundaries of the mesh were restrained in normal direction, and the bottom boundary was restrained in all directions.

The soil stratum in the model was a simplification of typical soil conditions in Shenyang area with two soil layers, medium-coarse sand and gravelly sand. The behaviors of all soils were simulated using the Mohr-Coulomb (MC) model. Because the Mohr-Coulomb model has difficulty in simulating the unloading behavior of soil, the Young's modulus of soil used in the model was set to three times of that obtained from the investigation [19]. The Poisson's ratio was equal to 0.26 and the coefficient of earth pressure at rest $K_0 = 1 - \sin\varphi$.

In the model, the original pile wall and the steel-pipe pile wall were both modeled using linear elastic liner elements based on equivalent of lateral stiffness, with Young's modulus of 32.5 GPa and 210 GPa, and Poisson's ratio of 0.2 and 0.3. The tieback anchors were modeled using cable elements with a Young's modulus of 195 GPa, a Poisson's ratio of 0.3, and an average prestress of 200 kN. The soil nails were modeled using rockbolt elements without prestress and the slopes was reinforced by shotcrete modeled by linear elements.

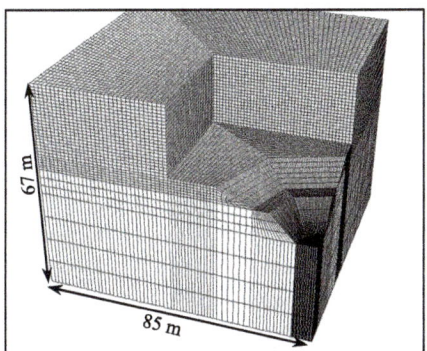
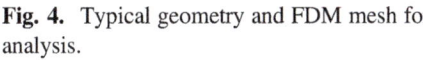

Fig. 4. Typical geometry and FDM mesh for analysis.

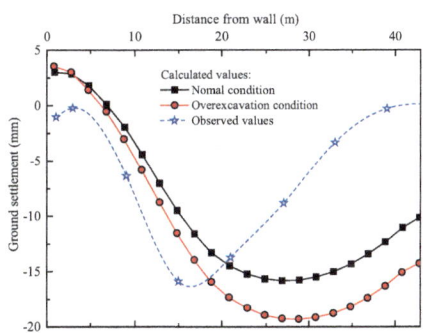

Fig. 5. Comparison between observed and calculated ground settlement.

The observed and calculated ground settlements are plotted in Fig. 5, and the calculated values are in good agreement with the field observations, especially for the peak values. However, the calculated settlements seem to be slight larger than the observed ones at place far from 20 m to the retaining wall, of which the reason is considered to be soil loss between retaining piles during construction and the simplifications of the model. Based on the comparisons, the accuracy of the FDM model for predicting the excavation behavior was verified.

3 Parametric Study of the Overexcavated Excavation

3.1 Length of the Steel-Pipe Pile Wall

For an overexcavated excavation, the type of additive retaining structure showed important for keeping the stability of the excavation. Figure 6 shows horizontal deformations of pile wall reinforced by steel-pipe pile wall with different lengths. As shown in Fig. 6, the maximum horizontal deformation (δ_m) was found to decrease with the length of steel-pipe pile wall (L) increasing, but at a reduced rate. The δ_m was 32.02 mm when $L = 15$ m, which was slightly larger than the one (30.16 mm) under condition without overexcavation. The pile wall deformation decreased with the L increasing above the excavation bottom, while the opposite response occurred below the excavation bottom. It is mainly because that as the L increased, the earth pressure acting on the steel-pipes increased, causing a larger deformation below the excavation bottom.

The pile wall deformation under the condition of overexcavation without reinforcement is shown in Fig. 6. The δ_m of unreinforced overexcavation was 33.75 mm, and a potential slip surface occurred around the slope shown in Fig. 6. It shows that the ground below the excavation bottom has a large movement under the condition of unreinforced overexcavation.

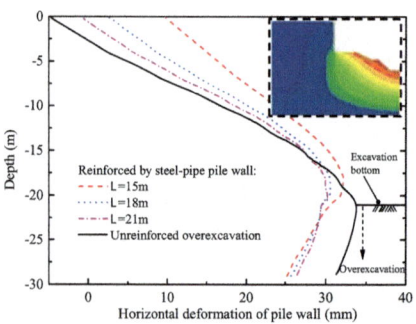 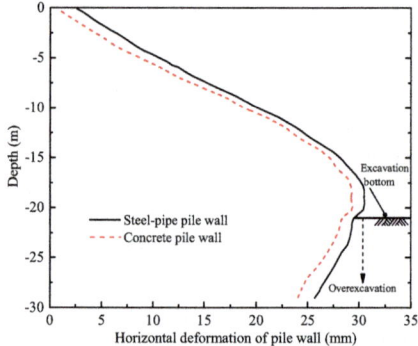

Fig. 6. Horizontal deformation of pile wall reinforced by different retaining structures.

Fig. 7. Comparison of horizontal deformation of pile wall induced by two retaining structures.

In order to compare the effects of the type of additive retaining structure on the pile wall, the numerical models reinforced by steel-pipe pile wall and concrete pile wall with same length were conducted, respectively. Figure 7 shows the horizontal pile wall deformations of overexcavated excavation reinforced by two types pile walls with a length of 18 m. A concrete pile wall restrained the deformation caused by overexcavation better than a steel-pipe pile wall. Compared with the steel-pipe pile wall, the concrete pile wall caused a reduction of 3.92% in terms of pile wall deformation, which was because of a higher lateral stiffness of concrete piles than the steel-pipe piles. For the construction conditions mentioned above, the position of δ_m was near the excavation bottom.

3.2   Distance Between the Edge of Overexcavation and the Pile Wall

Figure 8 shows relationship between horizontal pile wall deformation and depth from ground surface. Distance between the edge of overexcavation and the pile wall (D) varies from 6.5 m to 14.5 m. The δ_m values decreased with the D increasing, but at a reduced rate. The trend was predicted as the obtained relationship between these two variables when the D was larger than 15.4 m (see Fig. 8). The relationship between the δ_m value and the D shown in Fig. 8 indicated that, the overexcavation slightly influenced horizontal pile wall deformation when the D was larger than 14.5 m. For the pile wall deformation above excavation bottom, it has almost no changes when the D varies from 14.5 m. Compared with the pile wall deformation above excavation bottom, the pile deformation below the excavation bottom increased gradually when the D was reduced. It is because that the overexcavation unloads the earth pressure before the wall, causing soil moves forward, and the influence of overexcavation on the pile wall deformation is reduced when the D gets larger.

Figure 9 shows variations of ground settlement behind the pile wall. As shown in Fig. 9, the ground settlements decreased with increases in the D. Furthermore, when the D varies from 14.5 m to 10.5 m, significant settlement increases occurred at a distance of 25 m to 45 m from the wall, while the incremental settlement can be negligible within 25 m from the wall. When the D = 6.5 m, the maximum of ground settlement located at the position with a distance of 25 m far from the wall, and was 24.7 mm, 56.3% larger than that of normal condition.

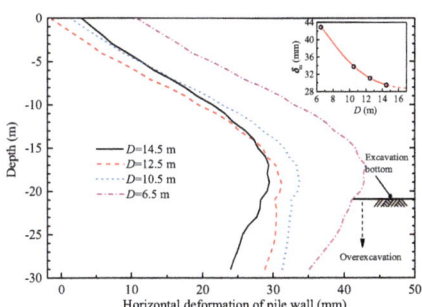

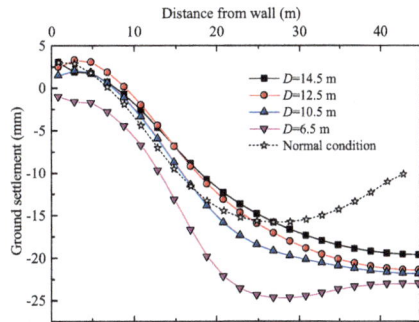

Fig. 8. Effects of distance between the edge of overexcavation and the wall on pile wall deformation.

Fig. 9. Effects of distance between the edge of overexcavation and the wall on ground settlement.

According to above analysis, a clear trend of decreasing pile wall deformation and ground settlement with the D increasing was observed. The stiffness of soils behind the wall which were reinforced by tiebacks increased, so the pile wall deformation and ground settlement around this region were less effected by the overexcavation. The soils behind the wall moved around the embedded part of the wall because of the unloading induced by the overexcavation. When the overexcavation move closer to the pile wall, the whole pile wall and soils behind the wall can be influenced.

3.3 Overexcavation Depth

In order to analyze the effects of the overexcavation depth, three models with overexcavation depths (H_{oe}) of 13 m, 18 m and 22 m were established, respectively. The models of H_{oe} = 13 m and H_{oe} = 18 m were excavated with two-stage slope, while the model of H_{oe} = 22 m was excavated with three-stage slope. The parameters of three models are listed in Table 2.

Table 2. Slope parameters of different models.

H_{oe}	Depth (m)			Slope 2#
	Slope 1#	Slope 2#	Slope 3#	
13 m	5	8	-	1:1
18 m	5	13	-	1:0.6
22 m	5	13	4	1:0.6

For various overexcavation depths, the pile wall deformation was influenced significantly (see Fig. 10). The pile bottom moved towards to the overexcavation, especially for the model of H_{oe} = 22 m. When the H_{oe} = 22 m, pile wall rotation failure occurred and the slip surface appeared at the slopes shown in Fig. 10. Figure 11 shows relationships between ground settlement and distance from wall. It showed that excessive ground settlement occurred when the H_{oe} = 22 m. The numerical results indicated that as the H_{oe} increased, the structure deformation and excavation stability was influenced significantly. It was dangerous if the overexcavation bottom was deeper than the pile bottom, causing lack embedded length of pile.

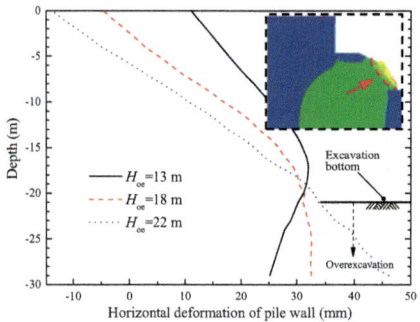

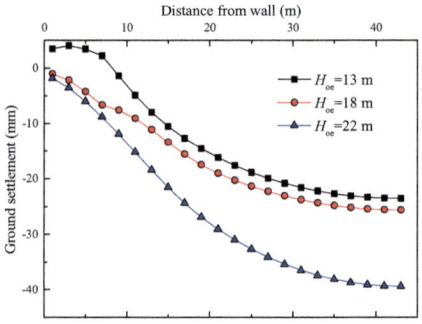

Fig. 10. Effects of overexcavation depth on pile wall deformation.

Fig. 11. Effects of overexcavation depth on ground settlement.

4 Conclusions

(1) The length of additive retaining structure influence the pile wall deformation significantly, especially for the embedded part of pile. A higher lateral stiffness of piles can reduce the pile wall deformation.
(2) The pile wall deformation and ground settlement increased as the D decreased, but at a reduced rate. When the D was larger than 14.5 m, the influence of the ocverexcavation can be negligible.
(3) As the overexcavation depth increased, the pile bottom moved towards to the overexcavation and the ground settlement increased. When the overexcavation bottom was deeper than the retaining structure bottom, a rolling failure of the structure tended to occur.

References

1. Arai, Y., Kusakabe, O., Murata, O., Konishi, S.: A numerical study on ground displacement and stress during and after the installation of deep circular diaphragm walls and soil excavation. Comput. Geotech. 35(5), 791–807 (2008)
2. Khoiri, M., Ou, C.Y.: Evaluation of deformation parameter for deep excavation in sand through case histories. Comput. Geotech. 47, 57–67 (2013)
3. Zhao, W., Han, J.Y., Li, S.G., Guan, Y.P.: Stresses and deformations in pile-anchor support system of deep foundation pit in sandy layers. J. Northeast. Univ. (Nat. Sci.) 4, 26 (2015). (in Chinese)
4. Cheng, X.S., Zheng, G., Diao, Y., Huang, T.M., Deng, C.H., Lei, Y.W., Zhou, H.Z.: Study of the progressive collapse mechanism of excavations retained by cantilever contiguous piles. Eng. Fail. Anal. 71, 72–89 (2017)
5. Peck, R.B.: Deep excavation & tunneling in soft ground. State-of-the-art-report. In: Proceedings of 7th International Conference of Soil Mechanics and Foundation Engineering, ISSMGE, Mexico City, pp. 225–281 (1969)
6. Clough, G.W., O'Rourke, T.D.: Construction induced movements of in situ walls. In: Design and Performance of Earth Retaining Structures (GSP 25), ASCE, Reston, VA, pp. 439–470 (1990)
7. Ou, C.Y., Liao, J.T., Lin, H.D.: Performance of diaphragm wall constructed using the top-down method. J. Geotech. Geoenviron. Eng. 124(9), 798–808 (1998)
8. Finno, R.J., Atmatzidis, D.K., Perkins, S.B.: Observed performance of a deep excavation in clay. J. Geotech. Eng. 115(8), 1045–1064 (1989)
9. Tan, Y., Wei, B.: Observed behaviors of a long and deep excavation constructed by cut-and-cover technique in Shanghai soft clay. J. Geotech. Geoenviron. Eng. 138(1), 69–88 (2011)
10. Gue, S.S., Tan, Y.C.: Prevention of failures related to geotechnical works on soft ground. In: Special Lecture, Malaysian Geotechnical Conference, vol. 714, pp. 21–28 (2004)
11. Huo, J., Gong, Q., Chen, J.: Analysis of the deformation of retaining structure of pit-in-pit excavation. J. Civ. Archit. Environ. Eng. 33(S1), 139–142 (2011). (in Chinese)
12. Xu, W., Tu, Y.: Landslide analysis and reinforcement design of the pit-in-pit. Rock Soil Mech. 31(5), 1555–1563 (2010). (in Chinese)
13. Shen, M., Liao, S., Zhou, X., et al.: Parametric analysis on stress field of pit in pit excavation. Chin. J. Geotech. Eng. 32(S1), 187–191 (2010). (in Chinese)

14. Tan, Y., Wei, B.: Performance of an overexcavated metro station and facilities nearby. J. Perform. Constr. Facil. **26**(3), 241–254 (2011)
15. Ou, C.Y., Lin, Y.L., Hsieh, P.G.: Case record of an excavation with cross walls and buttress walls. J. GeoEng. **1**(2), 79–86 (2006)
16. Ou, C.Y., Teng, F.C., Seed, R.B., Wang, I.W.: Using buttress walls to reduce excavation-induced movements. Proc. Inst. Civ. Eng.-Geotech. Eng. **161**(4), 209–222 (2008)
17. Lim, A., Hsieh, P.G., Ou, C.Y.: Evaluation of buttress wall shapes to limit movements induced by deep excavation. Comput. Geotech. **78**, 155–170 (2016)
18. Itasca Consulting Group, Inc.: FLAC Volume 3D (Fast Lagrangian Analysis of Continua in 3 Dimensions) Version 4.0 Manual (2009)
19. Ladd, C.C.: Stability evaluation during staged construction. J. Geotech. Eng. **117**(4), 540–615 (1991)

Mechanical Analysis of Supporting Struts for Special-Shaped Foundation Pits

Kun Wang[1], Mengbo Liu[2(✉)], Shaoming Liao[2], and Jun Wu[1]

[1] Troops 63926 of People's Liberation Army, Beijing 102202, China
[2] Department of Underground Engineering, Tongji University,
1239 Siping Rd., Shanghai 200092, China
100683@tongji.edu.cn

Abstract. The method of vertical beam on elastic foundation (BEF) was widely used in the design of supporting system of foundation pit. When it was applied to unsupported cylindrical excavation, the arching effect of the circular structures was treated as an equivalent brace. The reasonability of this simplification was that the pit shape and the load pattern were both centrosymmetric. However, more and more special-shaped foundation pits emerged, which were composed of circles and rectangles. The shapes of this kind of pit, as well as the loads pattern are non-centrosymmetrical, so the equivalent stiffness of structures is impaired and difficult to be determined. By means of 2D bar system finite element method (FEM) and the force method, this paper analyzed the mechanical characteristics of the horizontal struts of one type of special-shaped pit on the assumption that the loads is uniform. The results showed that the stiffness varied along the struts. At arch dome, the equivalent stiffness of the struts was decreased to half of that of full-circle struts, while the stiffness of arch foot was much larger than that of arch dome. Moreover, there is no significant reduction on the arching effect of structure. 3D FEM and in-situ tests showed that the displacements of arch dome was much smaller than that of the arch foot resulting from the distribution of lateral earth load. As a result, it was strongly recommended that the variation of structure stiffness and the earth load distribution should be considered comprehensively in special-shaped struts design.

Keywords: Special-shaped struts · Arching effect · Equivalent stiffness

1 Introduction

Cylindrical foundation pits are preferred in practice, which are believed to have strong capabilities of resisting deformation because of the arching effect (hoop effect) of circular struts and retaining walls without braces or ground anchors (Tan and Wang 2013; He et al. 2017). As a result, this type of excavation is advantageous to achieve less wall deformation, smaller internal forces.

The design methods for circular pits include 2D model of vertical beam on elastic foundation (BEF) taking the arching effect of structures into account, 3D spring-slab method and 3D FEM (Zhou and Luo 2003). The stress status and deformation are determined by the vertical flexural rigidity and circumferential stiffness (Tan and Wang 2013), so the key point to the design of circular pits is to determine the equivalent

© Springer Nature Singapore Pte Ltd. 2018
T. Qiu et al. (Eds.): GSIC 2018, *Proceedings of GeoShanghai 2018 International Conference:*
Advances in Soil Dynamics and Foundation Engineering, pp. 561–569, 2018.
https://doi.org/10.1007/978-981-13-0131-5_61

stiffness of the arching effect. Many contributions and codes have been made in this field (*JTJ303-2003* (Chinese design code); Zhou and Luo 2003; Li et al. 2013; Wang et al. 2017; Li and Ge 2017; He et al. 2017). Zhou and Luo (2003) derived a simplified method for the calculation of deformation of equivalent retaining structure for arching effect based on the solution of cylinder under uniform distributed loading in mechanics of elasticity. Wei described an axisymmetric beam on a nonlinear elastic foundation method derived from thin-shell theory to analyze circular excavations taking into account the nonlinear characteristics of soil-structure interactions.

However, with the rapid development of underground engineering, more and more special-shaped excavations emerged, which are termed as the foundation pits that consist of circles and rectangles in this paper. The mechanical characteristics of special-shaped vertical excavations have not been examined extensively.

In this paper, the stiffness and deformation characteristics of one type of special-shaped underground structures were analyzed, which was used as underground garage as shown in Fig. 1. The layout of the structure was non-centrosymmetrical, consisting of two isometric circles linked by a rectangular passage. Consequently, the ring stiffness could be decreased and difficult to be determined. By means of analysis methods and bar system FEM, the equivalent stiffness and internal forces of this type of struts were analyzed. For reference and comparison, a circumferential strut of a full-circle foundation pit was firstly analyzed in Sect. 2. Lastly, the 3D FEM model was established to analysis the influence of the interaction between the ground and the structure.

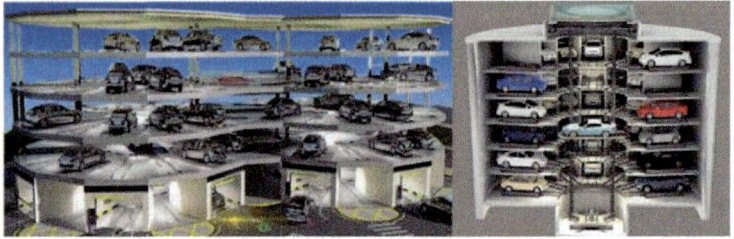

Fig. 1. Special-shaped underground garage.

2 Design Methodology and Model of Full-Circle Foundation Pit

2.1 Determining the Equivalent Stiffness of Struts by Codes

2D model of vertical beam on elastic foundation (BEF) can be applied in cylindrical foundation pits. The design model was revised as shown in Fig. 2. The key point of this model was treating the arching effect of waler beam (ring beam) and retaining wall as an equivalent bracing strut by determining the equivalent stiffness (k_{hi} & k_d) reasonably, where k_{hi} is the equivalent stiffness of ring beam and k_d is the equivalent stiffness of circular retaining wall.

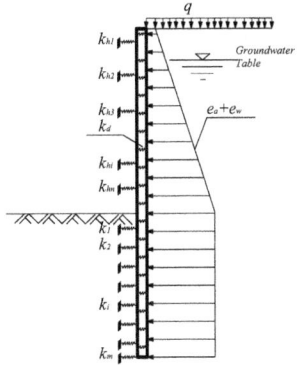

Fig. 2. 2D model of vertical BEF for full-circle foundation pit.

The *Design and construction technical code for diaphragm wall structure of port engineering (JTJ303-2003)* put forward the methods of determining k_h & k_d as following:

$$k_h = \frac{E_h A_h}{R_{h0}^3} \quad k_d \frac{\alpha t E_d}{R_0^2} \tag{1}$$

Where, E_h and E_d are the elastic module of the circular waler beam and retaining wall, A_h is the cross section area of the waler beam, t is the thickness of retaining wall; R_{h0} and R_0 are the initial radius of the waler beam and retaining wall, $\alpha = 0.5–0.7$. In this paper, the stiffness of circular waler beam k_h is the main issue of concern.

2.2   Bar System Finite Element Methods

The bar system finite element method (FEM) is another useful method analyzing the circular pits and it needs no equivalent calculating. The calculating parameters of full-circle struts was as shown in Table 1; The FEM model and results were as show in Fig. 3(a)–(d); The results obtained by the above two methods compared in Table 2.

According to the calculating results of FEM, it could be found that the full-circle struts (ring beam) had strong arching effect. The bending moment M was zero while the axil force N was near 900 kN. The total displacement was uniform and about 1 mm. Since the design code was conservative to some degree, the calculating displacement was relatively larger than that of bar system FEM. At the same time, there is no comparable methods for special-shaped struts, so the results of FEM was termed as the reference for the special-shaped struts.

Table 1. Calculating parameters of full-circle struts.

R/m	t/m	h/m	E/GPa	v	q/kPa
6.0	0.35	0.5	30	0.3	300

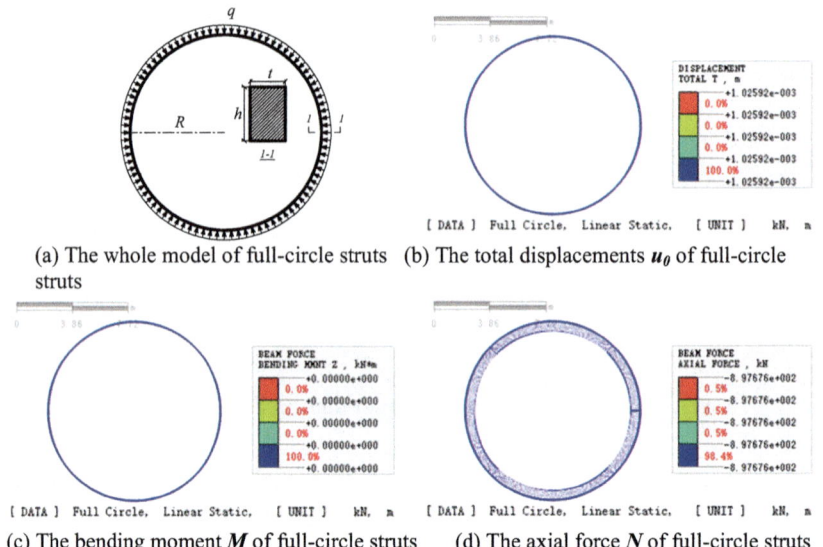

(a) The whole model of full-circle struts   (b) The total displacements u_0 of full-circle
    struts

(c) The bending moment M of full-circle struts   (d) The axial force N of full-circle struts

Fig. 3. The analysis model and results of the full-circle struts

Table 2. Calculating results of full-circle struts.

Bar system FEM			Codes results (JTJ303-2003)	
u_0/mm	M/kN·m	N/kN	k_h/MPa·m^{-1}	u_0/mm
1.05	0.00	897.7	24.3	6.17

3   Analysis of the Special-Shaped Struts

3.1   Analysis Method and Bar System FEM

The mechanic model of special-shaped struts was as shown in Fig. 4. The dimension of cross section was the same with that of the full-circle struts aforementioned and the load applied on it was assumed uniform along the struts.

According to the geometrical symmetry properties, as shown in Fig. 5(a)–(b), a quarter of this strut can be decomposed into an arc with elastic supports and a truss subjected to support reaction. The former part is the main issues of concern in this paper. Based on the principles of force method in structural mechanics, the basic structure of the arc can be obtained as shown in Fig. 5(c).

The internal forces of the arc at the position of α, including bending moment $M(\alpha)$, axial forces $N(\alpha)$ and shear forces $Q(\alpha)$, can be obtained by Eq. (2) according to the principle of force method.

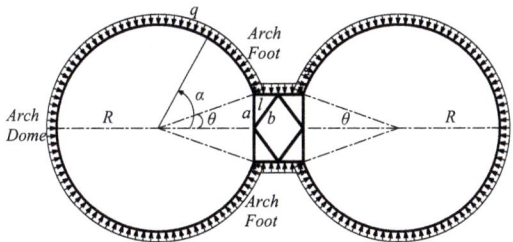

Fig. 4. Schematic diagram of the special-shaped struts

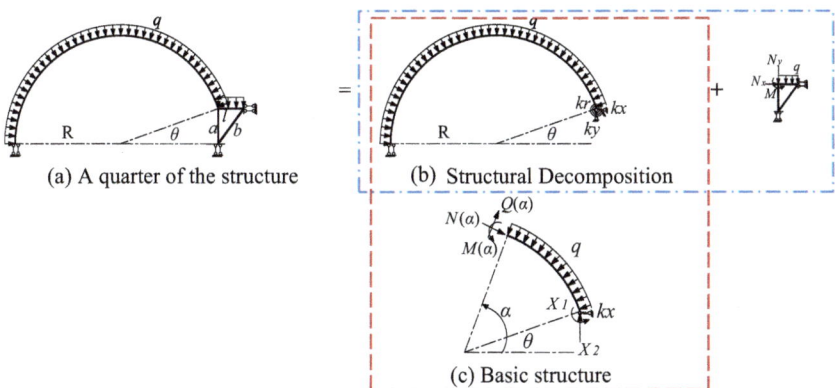

(a) A quarter of the structure

(b) Structural Decomposition

(c) Basic structure

Fig. 5. Decomposition of the mechanic model and the basic structure

$$
\begin{cases}
M(\alpha) = M(\alpha)_q + \overline{M(\alpha)_1} \cdot X_1 + \overline{M(\alpha)_2} \cdot X_2 \\
N(\alpha) = N(\alpha)_q + \overline{N(\alpha)_2} \cdot X_2 \\
Q(\alpha) = Q(\alpha)_q + \overline{Q(\alpha)_2} \cdot X_2
\end{cases}
\tag{2}
$$

Where, $M(\alpha)_q$, $N(\alpha)_q$, $Q(\alpha)_q$ are the internal forces induced by the external loads q (kPa) and they can be obtained by Eq. (3); X_1 and X_2 are the supports reaction and can be worked out by Eq. (4).

$$
\begin{cases}
M(\alpha)_q = qR^2(\cos(\alpha - \theta) - \cos^2\theta - \sin\theta\sin\alpha) + qlR(\cos\alpha - \cos\theta) - \dfrac{1}{2}ql^2 \\[2mm]
N(\alpha)_q = F_B\sin\alpha + ql\cos(\pi - \alpha) + \int_\theta^\alpha qR\sin(\alpha - \varphi)d\varphi = qR\sin\theta\sin\alpha - ql\cos\alpha - qR\cos(\alpha - \theta) \\[2mm]
Q(\alpha)_q = F_B\cos\alpha + ql\cos(\dfrac{\pi}{2} - \alpha) + \int_\theta^\alpha qR\cos(\alpha - \varphi)d\varphi = qR\sin\theta\cos\alpha + ql\sin\alpha + qR\sin(\alpha - \theta) \\[2mm]
\text{where, } F_B = \int_0^\theta qR\cos\theta d\theta = qR\sin\theta
\end{cases}
\tag{3}
$$

$$\begin{cases} \Delta_{1P} + \delta_{11}X_1 + \delta_{12}X_2 = -1/k_r \\ \Delta_{2P} + \delta_{21}X_1 + \delta_{22}X_2 = -1/k_y \end{cases} \tag{4}$$

Where,

$$\begin{cases} \delta_{11} = \dfrac{R(\pi - \theta)}{EI} \\[2mm] \delta_{22} = \displaystyle\int_\theta^\pi \dfrac{\overline{M(\alpha)_2} \times \overline{M(\alpha)_2}}{EI} Rd\alpha = \dfrac{R^3}{EI}\left[(\pi - \theta)(\tfrac{1}{2}\cos 2\theta + 1) + \tfrac{3}{4}\sin 2\theta\right] \\[2mm] \delta_{12} = \delta_{21} = \displaystyle\int_\theta^\pi \dfrac{\overline{M(\alpha)_1} \times \overline{M(\alpha)_2}}{EI} Rd\alpha = \dfrac{R^2}{EI}\left[(\pi - \theta)\cos\theta + \sin\theta\right] \end{cases} \tag{5}$$

$$\begin{cases} \Delta_{1P} = \displaystyle\int_\theta^\pi \dfrac{M(\alpha)\cdot\overline{M(\alpha)}}{EI} Rd\alpha = \tfrac{qR}{EI}\left[(-R^2\cos^2\theta - lR\cos\theta - \tfrac{1}{2}l^2)(\pi - \theta) - \tfrac{1}{2}R^2\sin 2\theta - LR\sin\theta\right] \\[2mm] \Delta_{2P} = \displaystyle\int_\theta^\pi \dfrac{\overline{M(\alpha)_2} \times M(\alpha)}{EI} Rd\alpha = R\cos\theta \displaystyle\int_\theta^\pi \dfrac{M(\alpha) \times \overline{M(\alpha)_1}}{EI} Rd\alpha - R\displaystyle\int_\theta^\pi \dfrac{\cos\alpha M(\alpha)}{EI} Rd\alpha \\[2mm] = \dfrac{qR^2}{4EI}\left[\begin{array}{l} R^2(1 + 2(\pi - \theta)(\cos\theta - 2\cos^3\theta) - 2\sin\theta - \cos 2\theta - 3\cos^2\theta\sin\theta) \\ + lR(-2\cos 2\theta(\pi - \theta) + \sin 2\theta - 4\sin\theta) + l^2(\sin\theta - 2\pi + 2\theta) \end{array}\right] \end{cases} \tag{6}$$

$$\begin{cases} \overline{M(\alpha)_1} = 1 \\ \overline{N(\alpha)_1} = 0 \,, \\ \overline{Q(\alpha)_1} = 0 \end{cases} \quad \begin{cases} \overline{M(\alpha)_2} = R(\cos\theta - \cos\alpha) \\ \overline{N(\alpha)_2} = \cos\alpha, \\ \overline{Q(\alpha)_2} = -\sin\alpha \end{cases} \tag{7}$$

The calculating results of bar system FEM were as shown in Fig. 6. Comparing to the full-circle struts, the displacements of the special-shaped struts showed a completely different pattern. The displacements varied along the struts and at the position of arch dome, the deformation was the largest, which is about twice larger than that of the full-circle struts. In this sense, the equivalent stiffness of the strut at this position was decreased to half. In contrast, the deformation at the position of arch foot was near zero, which accounted for that the stiffness at this position was much larger than that at arch dome, which resulted from the strengthen effects of the truss on the arch.

From the perspective of internal forces, the axial force and bending moments was basically identical with that of the full-circle struts. That means there is no significant reduction on the arching effect of this type of special-shaped strut.

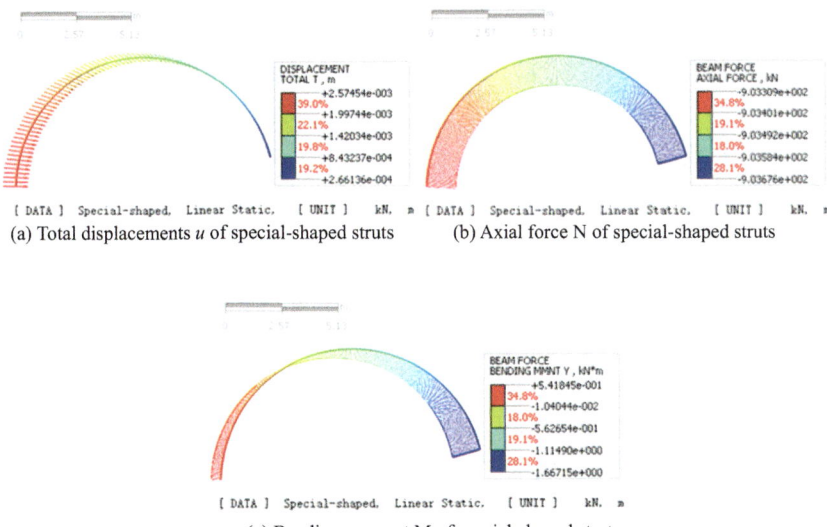

[DATA] Special-shaped, Linear Static,   [UNIT]   kN, m
(a) Total displacements u of special-shaped struts

[DATA] Special-shaped, Linear Static,   [UNIT]   kN, m
(b) Axial force N of special-shaped struts

[DATA] Special-shaped, Linear Static,   [UNIT]   kN, m
(c) Bending moment M of special-shaped struts

Fig. 6. Results of the special-shaped struts used bar system FEM.

3.2 Methods Considering the Interaction Between the Ground and Structures

In this section, a 3D FEM model was established considering the interaction between the ground and structures as shown in Fig. 7. Modified Mohr-Coulomb model was used to describe the behavior of the ground for it could take the rebound modulus into consideration and the critical parameters were as shown in Table 3. The cross section and the material model of the circumferential strut was the same as the bar system FEM.

Table 3. Material parameters of the ground in 3D FEM

Elastic modulus E/GPa	Rebound modulus E_{ur}/GPa	Poisson's ratio v	Cohesive strength c/kPa	Friction angle φ/°
5	20	0.3	40	20

The cross section of the circumferential strut was the same as the bar system FEM. Figure 8 showed the results of the total displacements of the diaphragm wall and the circumferential struts. The maximum displacements of the diaphragm wall occurred near the level of the basal. The displacements of struts obtained by 3D FEM showed a different pattern compared with the bar system FEM. According to the results of this model, the maximum displacements occurred near the arch foot, or to be precise, at the position that $\alpha = 40$–$45°$. The displacement at the position of arch dome was relatively minor, which was about zero.

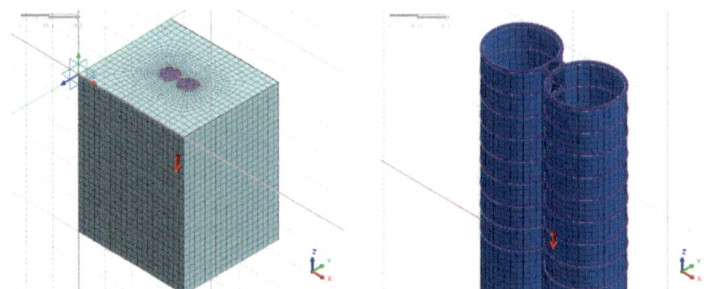

Fig. 7. 3D FEM model considering the interaction between the ground and structures

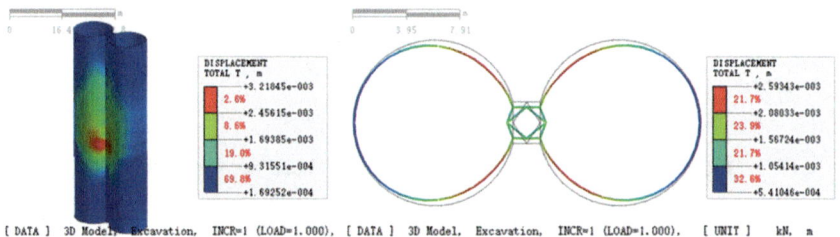

Fig. 8. Displacements of the diaphragm wall and ring struts

The results of 3D FEM agreed with the conclusions obtained by the in-situ measures of similar cases. The in-situ measures also showed that the displacements of the arch dome was much smaller than that of the arch foot. 3D FEM can simulate the actual situation more reasonably than bar system FEM since the latter was based on the hypothesis that the load is uniform. The distribution of lateral earth load is the critical factor that influences the deformation of the struts. The earth near the arch dome has the self-stability to some degree for the arching effect of the earth, while earth near the arch foot applied larger loads on the structures. In practice, designer should consider the variation of structure stiffness and the earth load distribution comprehensively to determine the critical position of the special-shaped structure.

4   Conclusions

This paper analyzed the characteristics of one type of special-shaped foundation pits by means of analysis methods, bar system FEM and 3D FEM. The main conclusions were as following:

(1) 2D model of vertical BEF can be applied in the design of special-shaped foundation pits. The key point was treating the arching effect of ring beam as an equivalent bracing strut by determining the equivalent stiffness reasonably.

(2) On the assumption that the earth load is uniform, the stiffness vary significantly along the struts. The equivalent stiffness of the arch dome was half of that of the full-circle struts, while the stiffness of the arch foot was relatively large.

(3) Considering the interaction between soil and structure, 3D FEM and in-situ test showed that the displacement at arc dome is much smaller than that at arc spring resulting from the non-uniform distribution of earth load.

(4) From two perspectives above, it was recommended that the variation of structure stiffness and the earth load distribution should be considered comprehensively in special-shaped struts design.

Acknowledgements. This work was supported by National Basic Research Program of China (973 Program, Grant No. 2015CB057806) and projects (Grant No.16DZ1200202) supported by Shanghai Committee of Science and Technology, China.

References

Wang, C.Y., Zhang, F.T., Ma, Y., et al.: A method for calculating horizontal stiffness coefficient of ring supporting system for foundation pit. Rock Soil Mech. **38**(3), 840–846 (2017)

Li, J., Gu, K.Y., Lin, J.: Design method for bracing of asymmetric cylindrical foundation pits. Chin. J. Geotech. Eng. (Supp. 2) **35**, 888–891 (2013)

JTJ303-2003: Design and construction technical code for diaphragm wall structure of port engineering. China Communications Press, Beijing (2004)

Zhou, J., Luo, X.B.: Structural computation of circular retaining structure using method of equivalent retaining of arch effect. Rock Soil Mech. **24**(2), 169–172 (2003)

Li, S.L., Ge, Y.X.: Calculation method of retaining piles with annular beams elastic support stiffness coefficient for circular foundation pit. Chin. J. Undergr. Sp. Eng. **13** (z1) (2017)

He, W., Luo, C., Cui, J., et al.: An axisymmetric BNEF method of circular excavations taking into account soil-structure interactions. Comput. Geo-tech. **90**, 155–163 (2017)

Tan, Y., Wang, D.: Characteristics of a large-scale deep foundation pit excavated by the central-island technique in shanghai soft clay. I: bottom-up construction of the central cylindrical shaft. Geotech. Geoenviron. Eng. **139**(11), 1875–1893 (2013)

Basal Stability Analysis of Braced Excavations with Embedded Walls in Non-homogeneous Clay by a Kinematic Approach

Zhen Tang[1,2(✉)], Mao-song Huang[1,2], and Ju-yun Yuan[1,2]

[1] Department of Geotechnical Engineering, Tongji University,
Shanghai 200092, China
tangzhen9@126.com
[2] Key Laboratory of Geotechnical and Underground Engineering of Ministry
of Education, Tongji University, Shanghai 200092, China

Abstract. The embedded wall has beneficial effects on the basal stability of braced excavation in clay. This paper investigates the basal stability of excavations with wall embedment in undrained clay using the kinematic approach of limit analysis, with the consideration of non-homogeneous undrained shear strength of clay. The proposed failure mechanism consists of three rigid blocks and one shear zone, and the least upper bound solutions and the corresponding critical failure mechanisms are obtained through optimization with respect to the geometric parameters. Comparison of the factor of safety from the proposed method with the limit equilibrium predictions based on the modified Terzaghi method was performed. Parameter study showed that the basal stability of excavations is significantly influenced by the wall embedment and the non-homogeneous strength of clay. Finally, the proposed mechanism was validated by five field excavation cases near or at failure and a numerical excavation case. The results show that the proposed method could reasonably describe the actual stability of excavations in both homogeneous and non-homogeneous clay.

Keywords: Basal stability · Limit analysis · Wall embedment
Anisotropic shear strength

1 Introduction

For deep excavations in soft clay, the factor of safety against base heave failure plays a key role in assessing the amount of ground movement generated by the plastic deformation of clay around the excavation. The retaining walls are used to penetrate the excavation base with certain depth to prevent the base heave failure and limit the ground movements. For calculations of basal stability, it is desirable to consider the effects of wall embedment on the basal stability. The study of the basal stability of excavations in soft clay has been investigated by several authors in literature. Most authors have performed numerical or analytical approaches.

The most widely used numerical approach for basal stability analysis may the finite element method (FEM). The FEM provides a comprehensive framework to evaluate

© Springer Nature Singapore Pte Ltd. 2018
T. Qiu et al. (Eds.): GSIC 2018, *Proceedings of GeoShanghai 2018 International Conference:
Advances in Soil Dynamics and Foundation Engineering*, pp. 570–579, 2018.
https://doi.org/10.1007/978-981-13-0131-5_62

multiple facets that affect the basal stability and thus results in an accurate estimation. The non-linear finite element analysis conducted by Hashash and Whittle showed the wall embedded depth and support conditions can significantly affect the failure models and the stability of excavation base [1]. Goh and Faheem et al. used the FEM with shear strength reduction technique to study the basal stability of excavations [2, 3]. Parametric study showed that the basal stability of excavations increases with the embedded depth of the wall. Ukritchon et al. used the numerical limit analysis to study the stability the braced excavations [4], whose results agreed well with those by Goh and Faheem et al. [2, 3]. The results of numerical analysis show the close dependency of the basal stability on wall embedment.

Most analytical approaches for calculating stability are through limit equilibrium analysis or upper bound limit analysis. Terzaghi and Bjerrum and Eide assumed that unloading caused by an excavation is analogous to the failure of a wide footing [5, 6]. This assumption provided a convenient way to calculate the basal stability factor by introducing the bearing capacity expressions. Obviously, these two methods neglect the effects of wall penetration. To investigate the effect of wall embedment, Eide modified the method of Bjerrum and Eide by considering the adhesion between the soil and the buried parts of the wall [7]. Wong and Goh modified Terzaghi's method by considering the resistance caused by the soil between the buried parts of the wall [8]. Another widely used limit equilibrium method is the slip circle method [9], which defines the ratio of resisting moment to the driving moment of the lowest support as the factor of safety for basal stability. By limit analysis, Su et al. and Liao and Su studied the effect of anisotropic shear strength on the basal stability of excavations by upper bound approach [10, 11]. Chang revised Terzaghi's method by upper bound limit analysis [12]. Faheem et al. studied the effects of aspect ratio, the thickness of the soft soil layer and the embedded depth of the wall by adopting the Terzaghi-type and Prandtl-type mechanisms [3]. Huang et al. introduced the multi-block mechanism into the basal stability problems [13].

In this paper, the basal stability of excavations in undrained clay is investigated by using the kinematic approach of limit analysis. A failure mechanism consisting of three rigid blocks and a shear zone is proposed, with the consideration of non-homogeneous shear strength. Least upper bound solutions are obtained by optimizing the geometry of the failure mechanism with respect to the geometric variables. Then the proposed method was compared with the modified Terzaghi method to study the effect of wall embedment and the non-homogeneous strength on the basal stability. Finally, the applicability of the proposed is validated five field excavation cases and a numerical excavation case.

2 Basal Stability Analysis with Upper Bound Method

2.1 Upper Bound Theorem

The upper bound theorem shows that collapse will occur under the smallest values of the surface tractions for which it is possible to find a kinematically admissible velocity field. Under the condition of Tresca's yield criterion. Shield and Drucker demonstrated that the upper bound solution for the surface traction $T = \{T_x, T_y, T_z\}$ can be calculated from the following equation [14]:

$$\int_S T^T v dS = \int_V 2c_u |\dot{\varepsilon}|_{max} dV + \int_{S_D} c_u |\Delta v| dS \tag{1}$$

where v is the kinematically admissible velocity field; $|\dot{\varepsilon}_{max}|$ is the absolutely largest principal component of the plastic strain rate; Δv is the velocity jump across any discontinuity; S is the surface that bounds the body or the assemblage of the bodies; V is the volume of the assemblage of the bodies; and S_D is the surface of all discontinuities.

2.2   Failure Mechanism and Upper Bound Calculation

Due to one dimensional deposition and subsequent K_0 consolidation, the undrained shear strength of clay is not uniform with depth, this study assumed the undrained shear strength of clay increases linearly with depth according to:

$$c_u(z) = c_{u0} + \eta z \tag{2}$$

in which $c_u(z)$ is the undrained shear strength at any given depth z, c_{u0} is the undrained shear strength at the ground surface, and $\eta = dc_u/dz$ is the rate of shear strength increase with depth. $\eta = 0$ in Eq. (2) results in uniform shear strength. The failure mechanism shown in Fig. 1 is considered. It consists of three rigid translational blocks abcd, dce and hmdfg and a slip fan def. As shown in Fig. 1, H and B are the excavated depth and width of the excavation, D is the embedded depth of the wall, q is the uniformly distributed surcharge pressure. Three geometric parameters are required to determine the mechanism. These are the angle α of right triangle dce, the angle β of slip fan def and the width B_2 of pentagon hmdfg. The width of the failure mechanism outside the excavation can be calculated as

$$B_1 = \frac{-B_2}{2\cos(\alpha + \beta)\cos\alpha} \tag{3}$$

Under the kinematically admissible condition, the rigid block abcd moves vertically with a velocity v_0, and the velocities on the respective discontinuities are shown in Fig. 1. The work done by the internal stress for this mechanism can be expressed as follows.

Work done by internal stress along velocity discontinuity bc, cd, ce, ef, fg, fd, gh and dm are

$$E_{bc} = v_0 \int_0^{H+D} c_u(z) dz \tag{4}$$

$$E_{cd} = v_0 \tan\alpha c_u(H+D) \tag{5}$$

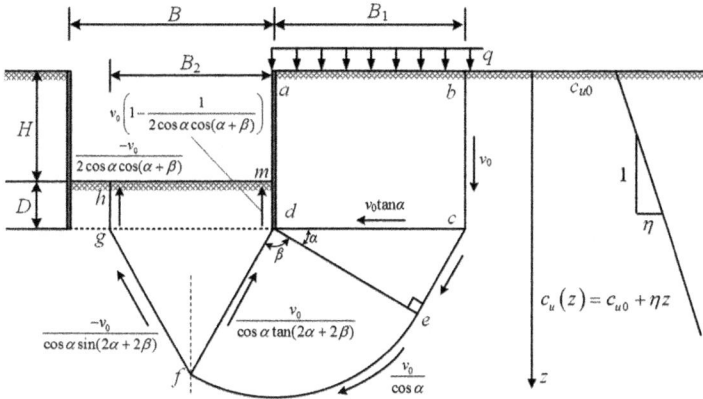

Fig. 1. Failure mechanism

$$E_{ce} = \frac{v_0}{\cos^2 \alpha} \int_{H+D}^{H+D+B_1 \cos \alpha \sin \alpha} c_u(z)dz \tag{6}$$

$$E_{ef} = \frac{v_0}{\cos \alpha} \int_{\alpha}^{\alpha+\beta} c_u(z)B_1 \cos \alpha d\theta \tag{7}$$

$$E_{fg} = \frac{-v_0 \csc^2(\alpha+\beta)}{2 \cos \alpha \cos(\alpha+\beta)} \int_{H+D}^{H+D+B_1 \cos \alpha \sin(\alpha+\beta)} c_u(z) \, dz \tag{8}$$

$$E_{fd} = \frac{v_0 \cot(2\alpha+2\beta)}{\cos \alpha \sin(\alpha+\beta)} \int_{H+D}^{H+D+B_1 \cos \alpha \sin(\alpha+\beta)} c_u(z)dz \tag{9}$$

$$E_{gh} = \frac{-v_0}{2 \cos \alpha \cos(\alpha+\beta)} \int_{H}^{H+D} c_u(z)dz \tag{10}$$

$$E_{dm} = - \frac{v_0 \sin(\alpha+\beta)}{\cos \alpha \sin(2\alpha+2\beta)} \int_{H}^{H+D} c_u(z)dz \tag{11}$$

Work done by internal stress in the slip fan *fcd*

$$E_{ef} = \frac{v_0}{\cos \alpha} \int_{\alpha}^{\alpha+\beta} \int_{0}^{B_1 \cos \alpha} c_u(z)rdrd\theta \tag{12}$$

The work done by external forces q, the gravity forces of block *abcf*, *dec* and *hmdfg* and slip fan *def* are expressed as follows.

$$W = v_0 q B_1 + v_0 \gamma B_1 (H+D) \tag{13}$$

By adopting the strength reduction technique, the factor of safety F_s of the proposed mechanism is defined as

$$F_s = \frac{\sum E}{W} \tag{14}$$

Equation (14) can be solved for any given combination of α, β and B2. The combination of these parameters that minimizes the solution defines the critical kinematic mechanism and gives the least upper-bound solution. The minimization can be performed with Optimization Tool Global Search in Matlab.

The following limitations are applied to the solution

$$\alpha + \beta \leq \frac{3\pi}{4}; \, 0 \leq \alpha < \frac{\pi}{2}; \, 0 < \beta \leq \frac{3\pi}{4} \tag{15}$$

so that the optimized mechanism is kinematically admissible.

Another limitation applied is

$$B_2 \leq B \tag{16}$$

This ensures that the rigid block (C) does not extend beyond the excavation.

3   Results and Comparison

In order to validate the proposed upper bound method, the modified Terzaghi method [8] is used to assess the basal stability of braced excavations with embedded walls in undrained clay. The failure mechanism of modified Terzaghi method is shown in Fig. 2, and the factor of safety is defined as the ratio of total resisting force to the total driving force, which is expressed as:

$$F_{smT} = \frac{5.71 c_{ub2} + \sqrt{2} c_{uu2}(H+D)/B + 2c_{up}D/B}{\gamma H + q} \tag{18}$$

where c_{ub2} is the undrained at $z = H + D + B/(2\sqrt{2})$, c_{uu2} is the undrained strength at $z = (H+D)/2$ and c_{up} is the undrained strength at $z = H + D/2$.

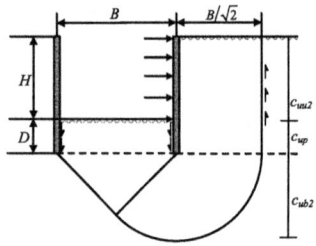

Fig. 2. Failure mechanism of the modified Terzaghi method

3.1   Effect of Wall Embedment

The proposed is first used to analyze excavation cases in isotropic and homogeneous clay with wall embedment D/H ranges from 0 to 1.5. For excavations in homogeneous clay, the stability of the excavation can be most conveniently expressed in terms of the stability number N, which is given by $N = F_s\gamma H/c_u$. The stability number of the proposed method and the modified Terzaghi method are shown in Fig. 3 for two different excavation aspect ratios $H/B = 0.5$, and $H/B = 2$. As seen, the stability number N increases with the increase of wall embedment D/H, increasing by 35.5% for wide excavation ($H/B = 0.5$) and by 100% for narrow excavation ($H/B = 2$) from $D/H = 0$ to $D/H = 1.5$, respectively. The N values of the proposed method are close to that of the modified Terzaghi method for wide excavation ($H/B = 0.5$), and are lower than that of the modified Terzaghi method for narrow excavation ($H/B = 2$).

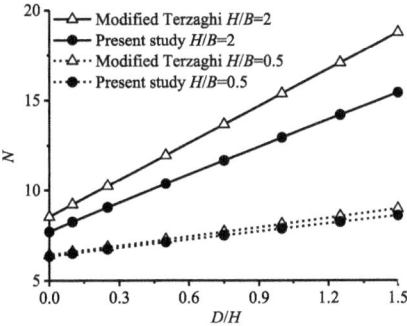

Fig. 3. Comparisons of stability numbers vary with wall embedment for different aspect ratios

3.2   Effects of Non-homogeneous Strength

In order to validate the reliability of the proposed method that considers the

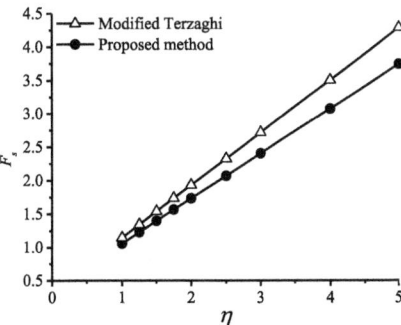

Fig. 4. Comparisons of factor of safety for non-homogeneous clay

non-homogeneity and anisotropy of clay, an excavation case with depth $H = 12$ m, width $B = 24$ m and wall embedded depth $D = 6$ m is investigated. The unit weight of soil γ is equal to 20 kN/m^3, and the undrained strength at the ground surface cu0 is equal to 12 kPa. For the case of non-homogeneous soil, the non-homogeneous ratio η ranges from 1 to 5. The safety of factors of the proposed method are presented in Fig. 4.

It can be seen from Fig. 4 that the safety of factor increase with the increase of non-homogeneous ratio as expected. Moreover, the safety of factors of the proposed method are lower than that of the modified Terzaghi method, the discrepancy of the two solutions increase from 8.4% at $\eta = 1$ to 14.5% at $\eta = 5$. The critical failure mechanisms of the proposed method and the modified Terzaghi method are plotted in Fig. 5. It can be seen from Fig. 5 that the failure range of the proposed method decreases with

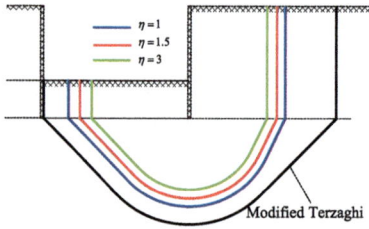

Fig. 5. Critical failure mechanism of the proposed mechanisms and the modified Terzaghi method for different non-homogeneous ratios

the increase of non-homogeneous ratio, while the modified Terzaghi method defined a constant failure mechanism.

4 Cases Study

4.1 Field Excavation Cases in Isotropic and Homogeneous Clay

In this section, five field excavation cases shown in Table 1 are adopted to validate the proposed method, all those cases are known to have been at failure or near failure. The basic geometry, soil properties and the factor of safety calculated from the proposed method are listed in Table 1. For comparison, the factor of safety calculated from modified Terzaghi method are listed in Table 1. For the four failure excavation cases (from Table 1), the safety factor calculated from the proposed method are ranging from 0.967 to 1.078, which are realistic and reasonably to the real failure condition. Meanwhile, the safety of factor of the modified Terzaghi method are slightly larger than that from the proposed method. However, the basal stability of braced excavation in undrained clay is affected by many factors. The difference between the safety of factor calculated from the proposed method and the actual ones are less than 10%, which verified the applicability of the proposed method.

Table 1. Validation of the proposed method with field excavation cases at or near failure

Parameters		Excavations cases				
		Davidson Ave.1	Washington DC	Pumping station, Fornebu, Oslo	Sewerage tank, Drammen	Excavation, Grey Wedels Plass, Oslo
H (m)		9.1	9.1	3	3.5	4.5
B (m)		7.6	21.3	5	5.5	5.8
D (m)		5.4	6.7	0.7	1	0.5
c_u (kPa/m^2)		19.2	23	7.5	10	14
γ_{sat} (kN/m^3)		16.3	17	17.5	18	18
F_s	Modified Terzaghi	1.27	1.10	1.01	0.99	1.09
	Present study	1.17	1.08	0.99	0.97	1.06
Observed behavior		No failure	Failure	Failure	Failure	Failure
Reference				Bjerrum and Eide [6]		

The Davidson Ave. 1 excavation refers to the general conditions surveyed along the braced cut of Davidson Ave. excavation. Although there is no visible failure observed, the measured maximum wall deflection varied from 100 mm to 250 mm, which is fairly high for a 9.1 m deep cut. The Fs obtained from the proposed method is 1.173, which is close to unit and consistent with observed behavior of the excavation.

4.2 Numerical Excavation Cases in Non-homogeneous Clay

The MIT-E3 model which describes the anisotropic strength behavior of soil was used by Hashash and Whittle [1] for the nonlinear finite element analysis of excavation in Boston blue clay. The excavation models investigated here are long braced excavations with a wide of 40 m, retained by 0.9 m thick concrete diaphragm walls. Three numerical cases were studied with different wall lengths ($L = 12.5$ m, 20 m, 40 m). Failure in the finite element analysis was defined when the numerical convergence is not achieved within a specified excavation step, the failure depth H_f predicted by the finite element analysis are 10–12.5 m, 15–17.5 m and 22.5–25 m for wall length $L = 12.5$ m, $L = 20$ m and $L = 40$ m, respectively.

For normally consolidated Boston blue clay, the normalized undrained shear strengths c_{uDSS}/σ'_v is 0.21, total unit weight of soil is 18 kN/m^3, and vertical effective consolidation pressure σ'_v (kPa) is equal to $8.19z + 24.5$. Figure 6 show the safety of factor calculated from the proposed method and the modified Terzaghi method with excavation depth for wall length $L = 12.5$ m, 20 m, and 40 m. As shown, the predicted failure depth H_f of the proposed method are 12 m and 18 m for wall length $L = 12.5$ m

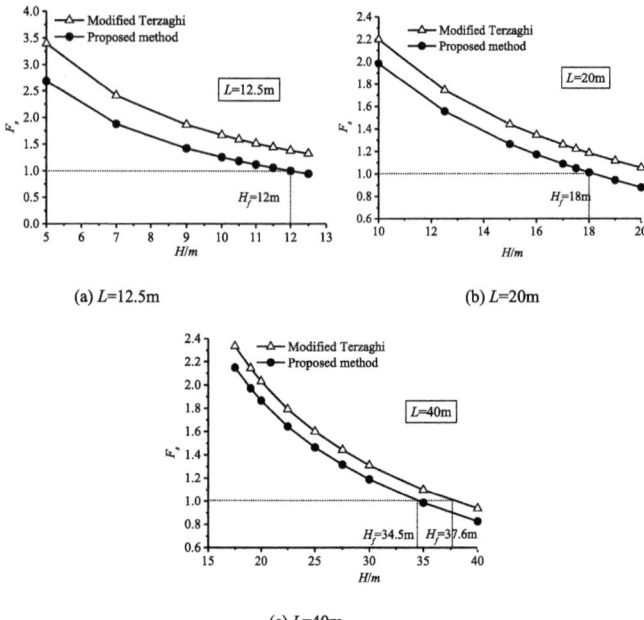

Fig. 6. Factor of safety vary with excavation depth for different wall length in normally consolidated Boston blue clay: (a) L = 12.5 m, (b) L = 20 m, (c) L = 40 m

and 20 m, respectively, which are close to that of the finite element analysis method. While the modified Terzaghi method predict no failure of the excavation for wall length L = 12.5 m and 20 m.

It should be noticed that for the case wall length L = 40 m, the predicted failure depth of the proposed method and the modified Terzaghi method are 34.5 m and 37.6 m, respectively, which is inconsistent with the finite element analysis. The discrepancy may attribute to that the failure mechanisms of proposed method and the modified Terzaghi method don't involve the failure of the wall, while the failure mechanism of the finite element method does involve the failure of the wall for wall length L = 40 m.

5 Conclusions

According to the kinematic approach of limit analysis, this paper presents a failure mechanism to estimate the basal stability of excavations with wall embedment in soft clay, with the consideration of non-homogeneous undrained shear strength. The least upper-bound solutions and the critical failure mechanism can be obtained through the optimization with respect to geometric parameters of the failure mechanism. The following conclusions can be drawn on the basis of the work presented herein:

1. For excavations with wall embedment in isotropic and homogeneous clay, the proposed method is capable to evaluate the effect of wall embedment on the basal

stability by comparison with the modified Terzaghi method. The stability number of the proposed method are close to that of the modified Terzaghi method for wide excavations, but lower than that of the modified Terzaghi method for narrow excavations. Parameters study showed that the stability of the excavation increase with the increase of non-homogeneous ratio.

2. The study of the field excavation cases showed that the proposed method could reasonably describe the actual stability of excavation base in homogeneous clay, which had been at or near failure. The difference between predicted and field case in factor of safety against base heave are within ±10% for the four failure cases.

3. The results of numerical excavation cases in non-homogeneous clay indicate that the predicted failure depth obtained by the proposed method were consistent with that obtained by the finite element analysis for short walls, but inconsistent with that from the finite element analysis for long walls.

Acknowledgement. The authors acknowledge the financial support of the National Key Research and Development Program (Grant No. 2016YFC0800202).

References

1. Hashash, Y.M.A., Whittle, A.J.: Ground movement prediction for deep excavations in soft clay. J. Geotech. Eng. **122**(6), 474–486 (1996)
2. Goh, A.T.C.: Assessment of basal stability for braced excavation systems using the finite element method. Comput. Geotech. **10**(4), 325–338 (1990)
3. Faheem, H., Cai, F., Ugai, K., Hagiwara, T.: Two-dimensional base stability of excavations in soft soils using FEM. Comput. Geotech. **30**(2), 141–163 (2003)
4. Ukritchon, B., Whittle, A.J., Sloan, S.W.: Undrained stability of braced excavations in clay. J. Geotech. Geoenvironmental Eng. **129**(8), 738–755 (2003)
5. Terzaghi, K.: Theoretical Soil Mechanics. Wiley, New York (1943)
6. Bjerrum, L., Eide, O.: Stability of strutted excavations in clay. Geotechnique **6**(1), 32–47 (1956)
7. Eide, O., Aas, G., Josang, T.: Special application of cast-in-place walls for tunnels in soft clay in Oslo. In: Proceedings of 5th European Conference on Soil Mechanics and Foundation Engineering, Madrid, Spain, pp. 485–498 (1972)
8. Goh, A.T.C., Kulhawy, F.H., Wong, K.S.: Reliability assessment of basal-heave stability for braced excavations in clay. J. Geotech. Geoenvironmental Eng. **134**(2), 145–153 (2008)
9. Hsieh, P.G., Ou, C.Y., Liu, H.T.: Basal heave analysis of excavations with consideration of anisotropic undrained strength of clay. Can. Geotech. J. **45**(6), 788–799 (2008)
10. Su, S., Liao, H., Lin, Y., Lin, Y.H.: Base stability of deep excavation in anisotropic soft clay. J. Geotech. Geoenvironmental Eng. **124**(9), 809–819 (1998)
11. Liao, H.J., Su, S.F.: Base stability of grout pile-reinforced excavations in soft clay. J. Geotech. Geoenvironmental Eng. **138**(2), 184–192 (2011)
12. Chang, M.F.: Basal stability analysis of braced cuts in clay. J. Geotech. Geoenvironmental Eng. **126**(3), 276–279 (2000)
13. Huang, M.S., Yu, S.B.: Upper bound analysis of basal stability in undrained clay based on block set mechanism. Chin. J. Geotech. Eng. **34**(8), 1440–1447 (2011). (in Chinese)
14. Shield, R.T., Drucker, D.C.: The application of limit analysis to punch indentation problems. J. Appl. Mech.-Trans. ASME **20**(4), 453–460 (1953)

Reliability-Based Assessment of Internal Stability for MSE Walls in Heavy Haul Railway

Biao Hu$^{(\boxtimes)}$, Zhe Luo, and Youwen Wang

The Key Laboratory of Road and Traffic Engineering of Ministry of Education,
Tongji University, Shanghai 201804, China
{biaohu,zluo,wangyouwen}@tongji.edu.cn

Abstract. An efficient reliability-based approach for the design of reinforced earth retaining walls in heavy haul railway is proposed in this paper. This approach addresses explicitly the effect of uncertain parameters in the design against internal instability for reinforced earth retaining walls. The first-order reliability method is integrated into the spreadsheet to assess the internal stability of retaining wall against rupture and pullout under various external loadings. The effect of uncertainties in the backfill unit weight, friction angle of soils, width, thickness and tensile strength of steel bars as well as the coefficient of friction for soil-reinforcement interface is presented in this study. The results generated by the proposed reliability-based approach agree well with those by Monte Carlo Simulation. The proposed approach is easy to use and this efficient tool can be adapted for other design considerations for the retaining walls in railway engineering.

Keywords: Reinforced earth retaining walls · Heavy haul railway
First-order reliability method · Rupture · Pullout

1 Introduction

In recent years, the reinforced earth retaining walls (or mechanically stabilized earth (MSE) walls) have been widely used as an economical infrastructure in the transportation engineering such as road, bridge, railway, etc. This technology has many advantages over other retaining structure including the lower cost, space-conservative, improvement of soil qualities, well-controlled deformation as well as easier construction. Nowadays, the designs for reinforced earth retaining wall are mainly based on the safety coefficient method in the manual of several countries. The minimum factors of safety (*FS*) in regard to sliding and overturning under static loading are recommended respectively as 1.5 and 2.0 by the American Association of State Highway and Transportation Officials [1]. The *FS* against local pullout and pullout is taken as 2.0 by Iron in the Second Survey and Design Institute [2]. In the deterministic safety coefficient method, it is not logical to apply the same *FS* to the conditions with various degrees of uncertainty, while the reliability-based design provides a means of quantitatively evaluating combined effect of uncertainties.

© Springer Nature Singapore Pte Ltd. 2018
T. Qiu et al. (Eds.): GSIC 2018, *Proceedings of GeoShanghai 2018 International Conference: Advances in Soil Dynamics and Foundation Engineering*, pp. 580–589, 2018.
https://doi.org/10.1007/978-981-13-0131-5_63

In the reliability-based analysis of reinforced earth retaining wall, several efforts have been made considering the internal and (or) external stability under various loadings. For example, a reliability-based design method for external stability of MSE walls in static conditions is developed using a two-phase approach [3]. More attentions are paid to the retaining wall stability under earthquake condition. A reliability-based design was presented considering the external seismic stability of reinforced soil structures [4–6]. Other examples include the reliability analysis of internal stability of reinforced soil structures against rupture, local pullout and pullout by Basha and Babu [7].

In this paper, an efficient reliability-based assessment tool for the internal stability of reinforced earth retaining walls for heavy haul railway is developed in this paper. The axle load for heavy haul railway is significantly larger than traditional railway, which typically ranges from 28.0 ton to 32.5 ton and can even reach 39 ton. Hence, an efficient tool is required to ensure the various stability requirements of retaining walls. The first-order reliability method (FORM) is integrated into spreadsheet and then the probability of failure can be calculated with the given design parameters. The effect of uncertainty in the backfill unit weight, friction angle of soils as well as the coefficient of friction for soil-reinforcement interface is investigated explicitly. The veracity of this developed FORM-based assessment method is examined by comparing with Monte Carlo simulation (MCS). This design tool requires much less computational effort and can be applied to other aspects of railway engineering.

2   Design for Reinforced Earth Retaining Wall Stability

There are multiple failure modes involved in the stability design of reinforced earth retaining walls, including but not limited to (1) overturning failure, (2) sliding failure, (3) slip circle failure of embankment, (4) bearing capacity failure, (5) excessive settlement, (6) pullout failure, and (7) rupture failure [2]. These individual failure modes can be grouped into external failure modes [(1)–(5)] and internal failure modes [(6)–(7)]. Due to the scope limit, this paper emphatically explores the effect of uncertainty in the material property of the backfills, coefficient of friction for soil-reinforcement interface and steel erosion on the internal stability of reinforced earth retaining walls for heavy haul railway. Thus, the internal stability against pullout and rupture failures is investigated, rather than covering the external failure modes.

Figure 1 illustrates the example section of a reinforced earth retaining wall subjected to heavy haul railway load. This example wall is located in DK283 + 523.70– DK283 + 628.24, Tianlin Station, of Nanning-Kunming Railway Section in China. The design parameters used in the design example are shown in Tables 1 and 2.

The factor of safety for retaining walls against local pullout, overall pullout and rupture can be determined with Eqs. (1)–(3) [2]:

$$FS_{Local\ Pullout} = \min(\frac{S_{f_i}}{E_{x_i}}) = \min(\frac{2\sigma_{vi}aL_{bf}}{\sigma_{hi}S_xS_y}) \tag{1}$$

$$FS_{Pullout} = \frac{\sum S_{f_i}}{\sum E_{x_i}} = \frac{\sum 2\sigma_{vi}aL_{bf}}{\sum \sigma_{hi}S_xS_y} \tag{2}$$

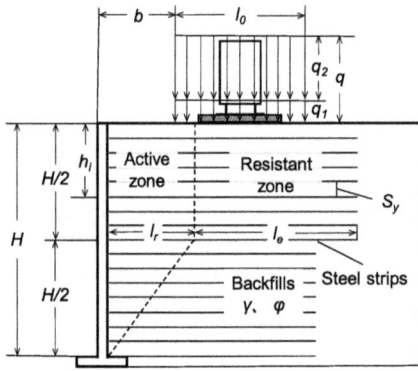

Fig. 1. Example design section of reinforced earth retaining walls for heavy haul railway.

Table 1. Parameters as random variables in the design of reinforced earth retaining wall design against rupture and pullout.

Input parameters	Notation	Mean	Unit	COV
Unit weight of backfill	γ	19	kN/m^3	10% [12]
Friction angle of soil	φ	35	deg	20% [13]
Coefficient of friction for soil-reinforcement interface	f	0.3 [2]	–	10%
Tensile strength of steel strips	$[\sigma]$	200	kN/m^2	10%
Width of steel strips	a	0.05	m	5%
Thickness of steel strips	e	0.0022	m	5%

$$FS_{Rupture} = \min(\frac{[\sigma]}{T_i}) = \min(\frac{[\sigma]}{K\sigma_{h_i}S_xS_y}) \tag{3}$$

where $FS_{Local\ Pullout}$ = minimum factor of safety against local pullout, $FS_{Pullout}$ = factor of safety against overall pullout, $FS_{Rupture}$ = minimum factor of safety for single steel strips against rupture, S_{fi} = total friction forces along both top and bottom surfaces of steel strips, E_{xi} = lateral earth force on the block, σ_{vi} = vertical pressure on the steel strips at the plate i, a = width of steel strips, f = coefficient of friction for soil-reinforcement interface, σ_{hi} = horizontal pressure on the plate i, S_x and S_y = horizontal and vertical spacing between steel strips, $[\sigma]$ = tensile strength of steel strips, T_i = tensile force on the steel strips, K = amplification factor of tensile force of reinforcement and L_b = efficient reinforcement length, which can be categorized with Eq. (4)

$$L_b = \begin{cases} 0.7H & \text{for } h_i \leq H/2 \\ 0.4H + 0.6h_i & \text{for } h_i > H/2 \end{cases} \tag{4}$$

where H = overall height of the retaining walls.

Table 2. Parameters as constants in the design of reinforced earth retaining wall design against rupture and pullout.

Basic calculation parameters	Notation	Value	Unit
Foundation load	q_1	13.91 [11]	kN/m^2
Train load	q_2	54.11 [11]	kN/m^2
Total load	q	68.02 [11]	kN/m^2
Load distribution width	l_0	3.3 [11]	m
The distance from the inner edge of the load to the panel	b	1.9	m
Overall height of retaining wall	H	5.56	m
Initial zinc coat thickness	e_i	86	μm
Amplification factor of tensile force of reinforcement	K	1.5 [2]	–
Horizontal spacing between steel strips	S_x	0.4	m
Vertical spacing between steel strips	S_y	0.4	m
Length of reinforcement	L	6	m

2.1 Horizontal Pressure on Wall Facing Blocks

The horizontal pressure caused by backfill materials and surcharge load can be computed with

$$\sigma_{h_i} = \sigma_{h_{1i}} + \sigma_{h_{2i}}$$
$$= \lambda_i \gamma h_i + \frac{q_1 + q_2}{\pi} \left(\frac{bh_i}{b^2 + h_i^2} - \frac{h_i(b + l_0)}{h_i^2 + (b + l_0)^2} + \arctan\frac{b + l_0}{h_i} - \arctan\frac{b}{h_i} \right) \quad (5)$$

in which $\sigma_{h_{1i}}$ = horizontal earth pressure at the depth of h_i, $\sigma_{h_{2i}}$ = horizontal pressure caused by surcharge load at the depth of h_i, γ = unit weight of backfills, h_i = depth of steel strips, q_1 = track load, q_2 = train load, b = distance from the inner edge of the load to the panel, l_0 = surface load distribution width, and λ_i = coefficient of earth pressure at the depth of h_i, and can be categorized as follows

$$\lambda_i = \begin{cases} (1 - \sin\varphi)(1 - h_i/6) + \tan^2(45° - \varphi/2)(h_i/6) & \text{for } h_i \leq 6m \\ \tan^2(45° - \varphi/2) & \text{for } h_i > 6m \end{cases} \quad (6)$$

where φ = internal friction angle of backfills.

2.2 Vertical Pressure on the Steel Strips

The vertical pressure induced by backfill materials and surcharge load can be determined with Eq. (7):

$$\sigma_{vi} = \gamma h_i + \frac{\gamma h_0}{\pi} \left(\arctan X_1 - \arctan X_2 + \frac{X_1}{1 + X_1^2} - \frac{X_2}{1 + X_2^2} \right) \quad (7)$$

where $X_1 = (2x + l_0)/(2h_i)$, $X_2 = (2x - l_0)/(2h_i)$, x = distance from the calculation point to the load centerline.

3   Conventional FORM Procedure for Reliability-Based Design Considering Single Failure Mode

Figure 2 illustrates the uncertainty propagation in the probabilistic design of retaining walls. In this study, six major parameters are modeled as random variables, including the backfill unit weight (γ), friction angle of soils (φ), width (a), thickness (e) and tensile strength ([σ]) of steel strips as well as the coefficient of friction for soil-reinforcement interface (f). These six random parameters are assumed to follow normal distributions and Table 1 shows their mean values and coefficient of variations (COVs). All other input parameters involved in this retaining wall design are treated as constants. Table 2 shows the constant parameters. Due to the effect of uncertain parameters, the resulting *FS* also includes uncertainty and the *FS* distribution is illustrated in Fig. 2. Given a limiting *FS* of 1.0, the probability of failure is the ratio of the shaded area over the entire area of the *FS* distribution.

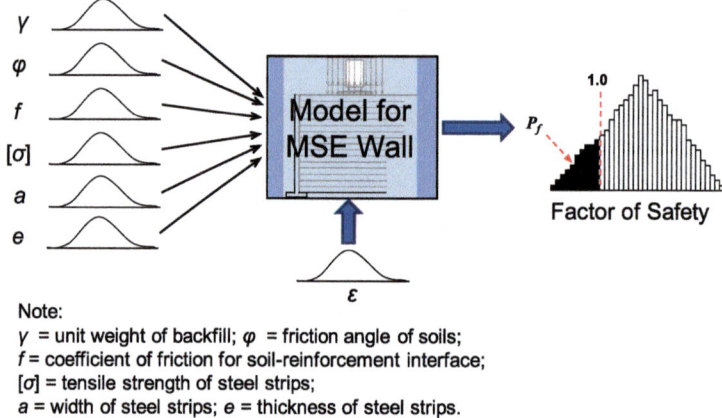

Note:
γ = unit weight of backfill; φ = friction angle of soils;
f = coefficient of friction for soil-reinforcement interface;
[σ] = tensile strength of steel strips;
a = width of steel strips; e = thickness of steel strips.

Fig. 2. Uncertainty propagation in geotechnical probabilistic analysis of reinforced earth retaining wall of heavy haul railway.

It is known that the probability of failure can be calculated by a few methods such as Monte Carlo Simulation (MCS), First-Order Second-Moment Method (FOSM), First-Order Reliability Method (FORM), Point Estimate Method (PEM), Sub-Set Method (SSM), etc. In this study, we choose the efficient FORM to reduce the computational effort. The results obtained using FORM will be verified by comparing with those obtained using MCS. Similar to other reliability-based approached, for FORM the reliability index (β) is first evaluated and then the homologous probability of failure can be computed with Eq. (8) [8]:

$$P_f = 1 - \phi(\beta) \tag{8}$$

in which ϕ represents the standard normal cumulative distribution function. In FORM, β is defined as the shortest distance between the limit state surface and the origin in the normalized variable space [9]:

$$\beta = \min_{x \in F} \sqrt{\left[\frac{x_i - \mu_i}{\sigma_i}\right]^T \mathbf{R}^{-1} \left[\frac{x_i - \mu_i}{\sigma_i}\right]} = \min_{x \in F} \sqrt{n^T \mathbf{R}^{-1} n} \qquad (9)$$

where x_i = random variable in the vector x of input parameters in the reliability-based analysis, μ_i and σ_i = the corresponding mean value and standard deviation of valuable x_i, respectively, R = the correlation matrix for random variables (γ, φ, f, $[\sigma]$, a and e), and n = dimensionless equivalent standard normal vector.

Next, the FORM procedure for the reliability-based assessment of reinforced earth retaining walls is implementing into spreadsheet. The layout of this spreadsheet is shown in Fig. 3. The aforementioned formulations [Eqs. (1)–(9)] are implemented into this spreadsheet. Interested in readers are referred to Low et al. for more details of this procedure. As will be shown in this paper, this spreadsheet-based FORM procedure is computationally efficient and robust comparing with Monte Carlo simulation

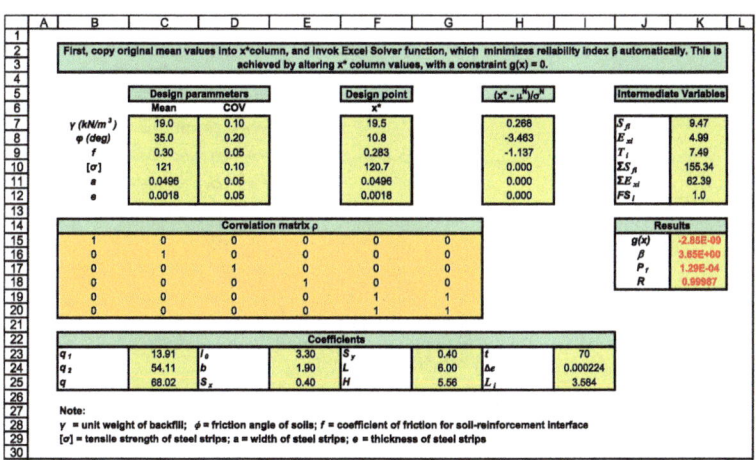

Fig. 3. Spreadsheet layout of first-order reliability analysis for single failure mode.

4   Reliability-Based Assessment of Reinforced Earth Retaining Walls Subjected to Heavy Haul Railway Load

Using the developed spreadsheet tools as illustrated in Fig. 3, the reliability-based assessment of reinforced earth retaining wall for heavy haul railway is conducted in this study. Focusing on the internal stability of retaining wall, several combinations of the mean and COV for the unit weight of backfills (γ), friction angle of soils (φ) and coefficient of friction for soil-reinforcement interface (f) is adopted in the analysis. The effect of uncertainty in width, thickness and tensile strength of steel strips

$(a, e$ and $[\sigma])$ due to construction error is also considered. The detailed parameters used in the reliability-based assessment are shown in Tables 1 and 2.

4.1   Effect of Uncertain Material Property of Backfills

First, the internal stability of retaining walls is investigated by taking the uncertainty in backfill material property into account. Using the spreadsheet-based FORM tool as illustrated in Fig. 3, the probability of failure against local pullout failure, overall pullout failure and rupture failure for several combinations of the mean and COV for unit weight of backfill (γ) is estimated and shown in Fig. 4. It is obvious that the probability of failure (P_f) for all three failure modes increases slightly with mean γ regardless of COV level. At the same mean value of γ, the effect of COV on P_f is negligible for overall pullout and local pullout failure modes, while larger COV results in larger probability of failure (P_f) against rupture failure. In Fig. 4, the magnitude of P_f for pullout failure and rupture failure is less than 10^{-6}. Among the three failure modes, the magnitude of P_f for local pullout failure is largest and slightly exceeds 10^{-4}. Nevertheless, the effect of uncertain γ is insignificant in the internal stability analysis of reinforced earth retaining walls.

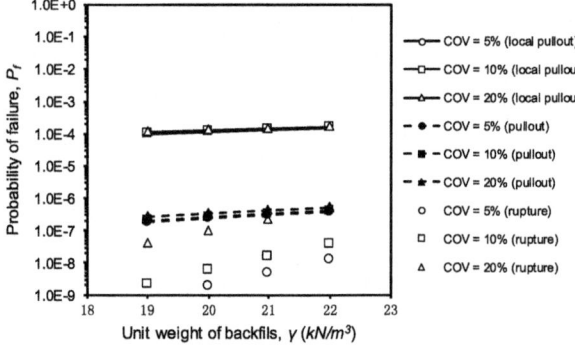

Fig. 4. Effect of uncertainty in unit weight of backfills (γ) on probability of failure (P_f) against local pullout failure, overall pullout failure and rupture failure.

Figure 5 shows the effect of uncertainty in friction angle of soils (φ) on the probability of failure for the three failure modes. The mean value for φ ranges from $30°$ to $45°$, and three levels of COV are considered: 10%, 20% and 40%. In Fig. 5, P_f decreases slightly with φ. For the same friction angle (φ), larger COV leads to much higher P_f, regardless of the failure mode. Taking $\varphi = 35°$ and local pullout as an example, the value of P_f increases from 7.32×10^{-11} to 2.70×10^{-2} as COV increases from 10% to 40%. Thus, the estimated P_f can differ by several orders of magnitude if the level of variability for φ is not properly assessed. This also points to the importance of quality control during the backfilling and compacting of the backfill materials.

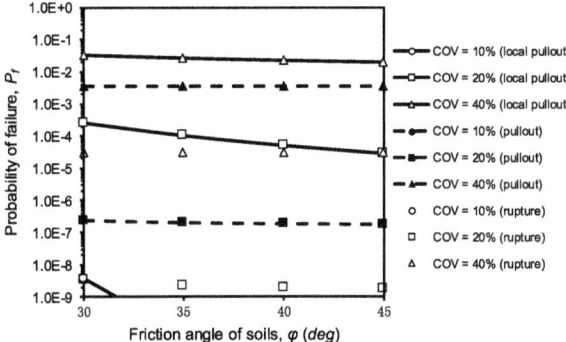

Fig. 5. Effect of uncertainty in friction angle of backfills (φ) on probability of failure (P_f) against local pullout failure, overall pullout failure and rupture failure.

4.2 Effect of Uncertain Soil-Reinforcement Interface Friction

It is known that due to various factors such as inconsistent prefabricated steel strips, imperfect backfill compaction, and backfill erosion, there exists uncertainty in the friction along soil-reinforcement interfaces. In this regard, the internal stability of retaining wall is further investigated by considering the uncertainty in coefficient of friction for soil-reinforcement interface (f). Following the aforementioned procedure and using the spreadsheet tool in Fig. 3, a series of reliability-based analysis of internal failures is conducted by entering several combinations of mean and COV values for f in Cells C9 and D9, respectively, in the spreadsheet tool in Fig. 3. The resulting relationship between P_f and f is shown in Fig. 6. It is shown in Fig. 6 that for rupture failure, P_f is comparable and smaller than 10^{-8} for all mean and COV combinations for f. This observation is consistent with the rupture model as in Eq. (3), in which f is not involved in the factor safety against rupture. For overall pullout and local pullout failures, it is shown in Fig. 6 that P_f increases significantly with COV at the same mean f; for the same level of COV, P_f decreases with mean f. For example, when COV = 10% for local

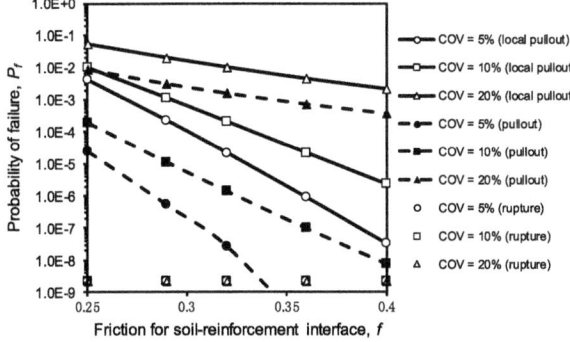

Fig. 6. Effect of uncertainty in friction along soil-reinforcement interface (f) on probability of failure (P_f) against local pullout failure, overall pullout failure and rupture failure.

pullout, P_f decreases from 1.02×10^{-2} to 2.30×10^{-6}, as f increases from 0.25 to 0.40. Similar to the observation in Figs. 4 and 5, Fig. 6 also shows that local pullout failure has the largest P_f and dominates the internal stability analysis. Hence, soil-reinforcement interface friction f has a significant impact on P_f due to pullout failures. It is critical to ensure the full development of the friction between the backfill materials and reinforcement.

5   Comparison Between FORM and Monte Carlo Simulation

To validate the developed spreadsheet tools, the results generated by the proposed efficient FORM-based approach are compared with those by Monte Carlo Simulation (MCS). The parameters (γ, φ, f, [σ], a and e) in MCS are assumed to be normally distributed multivariate variables, with mean and COV values following those in Table 1. As illustrated in Fig. 2, the probability of failure in MCS is calculated as the ratio of the number of cases with $FS < 1.0$ over the total number of simulation. It is recommended that the minimum number of MCS can be taken as the number of random variables multiplied by ten times of the reciprocal of corresponding probability of failure [10]. Following this rule of thumb, the simulation number of 1.0×10^7 is determined for each scenario in this study. The comparison between FORM and MCS is shown in Fig. 7. It is shown that the FORM-based P_f agrees well with those computed with MCS. However, the spreadsheet tools require much less computational effort.

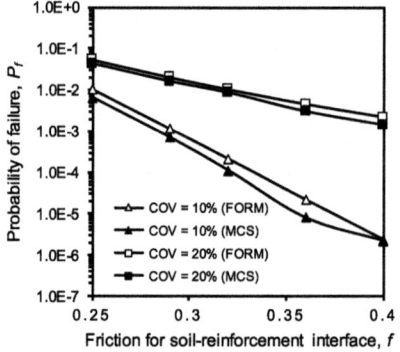

Fig. 7. Effect of uncertainty in friction for soil-reinforcement interface (f) on probability of failure (P_f) against local pullout by First-Order Reliability Method (FORM) and Monte Carlo Simulation (MCS).

6   Conclusion

By integrating the first-order reliability method into spreadsheet, this paper develops an efficient reliability-based analysis method of reinforced earth retaining walls for heavy haul railway. The step-by-step procedure for the developed approach is illustrated in the internal stability analysis of retaining wall against local pullout, pullout and rupture. Effect of material property of backfills and soil-reinforcement interface on internal stability of retaining wall is discussed in this paper. The accuracy of this developed reliability-based assessment approach is demonstrated by comparing with Monte Carlo simulation.

The effect of uncertainty in unit weight of backfills on retaining wall stability against local pullout and overall pullout is quite little, whereas higher for rupture failure. Yet the friction angle of soils and coefficient of friction for soil-reinforcement interface are more significant on the internal stability of retaining walls. Compared with Monte Carlo simulation, the developed FORM-based reliability method is more computationally efficient, which can be easily applied to other aspects in the railway engineering.

Acknowledgment. The study on which this paper is based was supported by the National Natural Science Foundation of China through Grant No. 41702296.

References

1. AASHTO: Standard specifications for highway bridges, 16th edn. American Association of State Highway and Transportation Officials, Washington, DC (1996)
2. Iron in the Second Survey and Design Institute: Code for design on retaining structures of railway subgrade [S]. TB10025—2006, J127—2006, Beijing (2006)
3. Chalermyanont, T., Benson, C.: Reliability-based design for external stability of mechanically stabilized earth walls. Int. J. Geomech. **5**(3), 196–205 (2005)
4. Basha, B.M.: Optimum design of retaining structures under static and seismic loading: a reliability based approach. Ph.D. thesis, Department of Civil Engineering, Indian Institute of Science Bangalore (2009)
5. Sayed, S., Dodagoudar, G.R., Rajagopal, K.: Reliability analysis of reinforced soil walls under static and seismic forces. Geosynthetics Int. **15**(4), 246–257 (2008)
6. Basha, B.M., Babu, G.L.S.: Seismic reliability assessment of external stability of reinforced soil walls using pseudo-dynamic method. Geosynthetics Int. **16**(3), 197–215 (2009)
7. Basha, B.M., Babu, G.L.S.: Earthquake resistant design of reinforced soil structures using pseudo-static method. Am. J. Eng. Appl. Sci. **2**(3), 565–572 (2009)
8. Melchers, R.E.: Structural Reliability: Analysis and Prediction. Ellis Horwood Ltd., Chichester (1987)
9. Ditlevson, O.: Uncertainty Modeling with Applications to Multidimensional Civil Engineering Systems. McGraw-Hill Book Co., Inc., New York (1981)
10. Ang, H.S., Tang, W.H.: Probability Concepts in Engineering: Emphasis on Applications to Civil and Environmental Engineering, 2nd edn. Wiley, New York (2007)
11. The first survey and design institute of railway: Code for Design of Railway Earth structure. TB 10001—2016, J 447—2017, Beijing (2017)
12. Duncan, J.: Factors of safety and reliability in geotechnical engineering. J. Geotech. Geoenvironmental Eng. **126**(4), 307–316 (2000)
13. Phoon, K.K., Kulhawy, F.H.: Evaluation of geotechnical property variability. Can. Geotech. J. **36**(4), 625–639 (1999)

Numerical Modelling of the Excavation Damaged Zone in Hollow Cylinder Experiment of Boom Clay

Shiyi Liu[1(⊠)], Longtan Shao[2], and Xiangling Li[3]

[1] School of Resources and Civil Engineering, Northeastern University,
Shenyang 110819, China
Liu-shiyi@hotmail.com
[2] State Key Laboratory of Structural Analysis for Industrial Equipment,
Dalian University of Technology, Dalian 116024, China
[3] European Underground Research Infrastructure for Disposal of Nuclear Waste
in Clay Environment (EIG EURIDICE), 2400 Mol, Belgium

Abstract. Boom clay is currently under investigation as a potential host rock for deep geological disposal of nuclear waste in Belgium. The excavation-induced damaged zone (EDZ) around galleries could have a negative effect on long-term safety. Therefore, it is necessary to properly predict mechanical behavior and failure mechanism of the clay during excavation. For this purpose, small scale thick-wall hollow cylinder simulation experiments were performed to study the EDZ in the Boom clay. In this paper, the focus is made on the simulation of the EDZ in the hollow cylinder experiment through finite element analyses with a modified Drucker-Prager model considering hardening and softening plasticity and elasto-plastic cross-anisotropy. Furthermore, it presents an aided method used to predict fractures in the EDZ by simulating slip lines. The modelling results agree with experiments quantitatively and confirm the feasibility of predicting the EDZ according to the strain-softening zone.

Keywords: EDZ · Boom clay · Anisotropy · Hollow cylinder experiment

1 Introduction

In Belgium, Boom clay is selected as a potential host rock formation for deep geological disposal of high-level waste thanks to its favorable properties, among which is very low hydraulic conductivity. Excavation of galleries leads to both mechanical and hydraulic disturbances, and creates the excavation damaged zone (EDZ) in which major hydro-mechanical modifications occur [1]. The EDZ could have a negative effect on long-term safety due to the degree of permeability increase. Therefore, processes creating fractures in the EDZ need to be evaluated. And a better understanding of mechanical behavior and failure mechanism of the clay during excavation will be benefit to the safety of the disposal.

The underground research facility HADES (High-Activity Disposal Experimental Site), at a depth of about 223 m, has been expanded several times [2]. Within the TIMODAZ (thermal impact on the damaged zone around a radioactive waste disposal

© Springer Nature Singapore Pte Ltd. 2018
T. Qiu et al. (Eds.): GSIC 2018, *Proceedings of GeoShanghai 2018 International Conference: Advances in Soil Dynamics and Foundation Engineering*, pp. 590–597, 2018.
https://doi.org/10.1007/978-981-13-0131-5_64

in clay host rocks) project, at the LMR-EPFL laboratory, hollow cylinder experiments were carried out to study at reduced scale the EDZ around openings in Boom clay samples by decreasing the inner confining pressure aiming at mimicking a gallery excavation [3]. Before and after the mechanical unloading, X-ray computed tomography scans of the hollow cylinder samples were carried out and a particle tracking technique was used to perform quantitative analyses of the displacements inside the Boom clay. Eye-shape damaged zone around the central hole was inferred by the slope discontinuity observed in the displacement profiles along the diameters. The hydro-mechanical constitutive interpretation of the displacement in this experiment was made through finite element modelling with a constitutive model including elastic and plastic cross-anisotropy [4]. But only qualitative comparison of the shape of the EDZ was made between numerical prediction and the experiments.

The overconsolidation ratio (OCR) of the Boom clay is about 2.4 [5]. In general, peak strength is observed in this overconsolidated clay where the natural fabric of the soil must be destroyed before shear stress reach residual strength. The decrease of the strength (termed "strain softening") would make the clay cannot sustain. To properly describe such mechanical behavior, a modified Drucker-Prager model considering hardening and softening plasticity was proposed. This work contributes to simulate the EDZ in the Boom clay and understand mechanical behavior and failure mechanism of the clay during excavation. Emphasis is place on the application of the model adopted for predicting the EDZ and slip lines (potential failure surfaces) in hollow cylinder experiment.

2   Constitutive Model and the Aided Method

2.1   Constitutive Model

A hardening/softening Drucker-Prager model considering elastic and plastic cross-anisotropy has been developed to improve understanding on specificities of Boom clay [4, 6]. Based on this model, a new hardening and softening process was proposed to describe failure of the clay. The model was implemented in a FORTRAN subroutine as a user-defined material model (UMAT) in the ABAQUS finite element software [7]. And a fully implicit backward Euler stress update algorithm was used [8].

Elasticity. The elasto-plasticity principle allows that the total strain rate, $\dot{\varepsilon}_{ij} = \dot{\varepsilon}_{ij}^e + \dot{\varepsilon}_{ij}^p$, split into elastic, $\dot{\varepsilon}_{ij}^e$, and plastic, $\dot{\varepsilon}_{ij}^p$. $\dot{\varepsilon}_{ij}^e$ is linked to stress rate through the Hooke law:

$$\dot{\varepsilon}_{ij}^e = D_{ijkl}^e \dot{\sigma}_{ij}', D_{ijkl}^e = \begin{bmatrix} 1/E_p & -v_{tp}/E_t & -v_p/E_p & 0 & 0 & 0 \\ -v_{pt}/E_p & 1/E_t & -v_{pt}/E_p & 0 & 0 & 0 \\ -v_p/E_p & -v_{tp}/E_t & 1/E_p & 0 & 0 & 0 \\ 0 & 0 & 0 & 1/G_t & 0 & 0 \\ 0 & 0 & 0 & 0 & 1/G_p & 0 \\ 0 & 0 & 0 & 0 & 0 & 1/G_p \end{bmatrix} \quad (1)$$

The D^e_{ijkl} matrix considers cross-anisotropic elasticity, where subscript p and t stand for "in-plane" and "transverse", respectively.

Yield Surface and Flow Rule. In terms of the stress invariants, the Drucker-Prager yield criterion can be expressed as:

$$Y = q - p' \tan \beta - d = 0, \tan \beta = 6 \sin \varphi'/(3 - \sin \varphi'), d = c' \tan \beta/\tan \varphi' \quad (2)$$

p' and q are the mean effective stress and the deviatoric stress, and c' and φ' are the cohesion and the internal friction angle respectively.

Plastic potential function is expressed as:

$$G = q - p' \tan \beta', \tan \beta' = 6 \sin \psi/(3 - \sin \psi) \quad (3)$$

where ψ is the angle of dilation.

Hardening and Softening. The model allows hardening and softening process during plastic flow. They are introduced via two hyperbolic variations of the internal friction angle and cohesion between initial (φ'_i and c'_i), peak (φ'_p and c'_p) and residual (φ'_r and c'_r) values as a function of the equivalent plastic strain ε^p_{eq}:

$$\begin{cases} x' = x'_i + 2\left(x'_p - x'_i\right)\varepsilon^p_{eq} \Big/ \left(B_h + \varepsilon^p_{eq}\right), \left(\varepsilon^p_{eq} \leq B_h\right) \\ x' = x'_p + \left(x'_r - x'_p\right)\left(\varepsilon^p_{eq} - B_h\right) \Big/ \left(B_s + \varepsilon^p_{eq} - B_h\right), \left(\varepsilon^p_{eq} > B_h\right) \end{cases} \quad (4)$$

where x is φ or c and the equivalent plastic strain ε^p_{eq} is obtained by integration of the equivalent plastic strain rate, $\varepsilon^p_{eq} = \int_0^t \sqrt{2\dot{\varepsilon}^p_{ij}\dot{\varepsilon}^p_{ij}/3}dt$, B_h and B_s are hardening and softening parameters.

The hardening parameter is defined as Eq. (5) and could represent the hardening or softening process. Hence the softening zone would be easily determined by this parameter. This article does not consider pathological mesh-sensitivity. If material is considered as elastic-perfectly plasticity, $H = 0$.

$$H = -\left(\partial Y/\partial\varepsilon^p_{eq}\right)\sqrt{(2/3)\left(\partial G/\partial\sigma'_{ij}\right)\left(\partial G/\partial\sigma'_{ij}\right)}, \{H > 0, \text{hardening}/H < 0, \text{softening} \quad (5)$$

Anisotropic Plasticity. The plastic cross-anisotropy is taken into account through the material cohesion that depends on the angle between major principal stress and the normal to the bedding orientation [4]. The mathematical expression of the cohesion is as follows (Fig. 1):

$$c' = \max\left[\left(\left(c'_{(45°)} - c'_{(0°)}\right)\Big/45°\right)\alpha_{\sigma_1} + c'_{(0°)}; \left(\left(c'_{(90°)} - c'_{(45°)}\right)\Big/45°\right)\left(\alpha_{\sigma_1} - 45°\right) + c'_{(45°)}\right] \quad (6)$$

where three cohesion values are defined, for major principal stress parallel ($\alpha_{\sigma_1} = 0°$), perpendicular ($\alpha_{\sigma_1} = 90°$) and with an angle of 45° ($\alpha_{\sigma_1} = 45°$) with respect to the normal to bedding plane. α_{σ_1} is the angle between the normal to the bedding plane $\vec{n}$ and the major principal stress $\vec{\sigma}_1$, $\alpha_{\sigma_1} = \cos^{-1}((\vec{n} \cdot \vec{\sigma}_1)/(\|\vec{n}\| \|\vec{\sigma}_1\|))$.

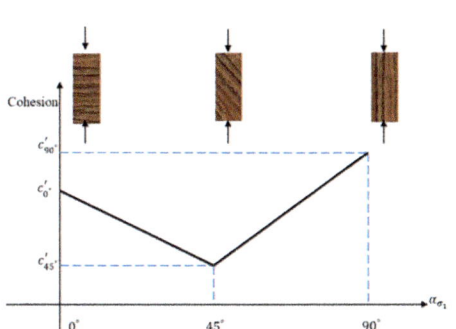

Fig. 1. Schematic view of the cohesion evolution as a function of the angle between the normal vector to bedding plane and the direction of major principal stress [4].

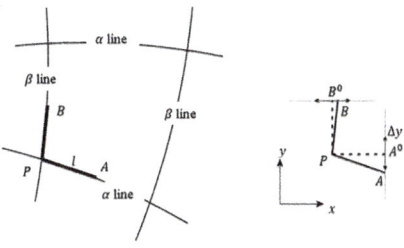

Fig. 2. Pairs of intersecting α and β lines and local slip lines.

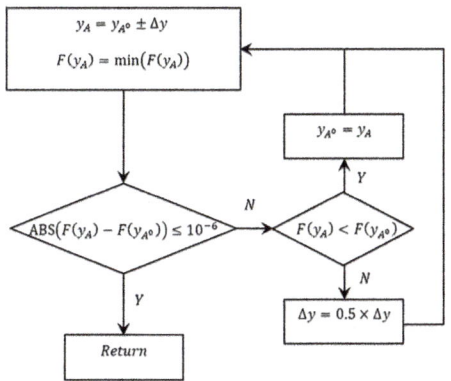

Fig. 3. Flow chart of the searching process.

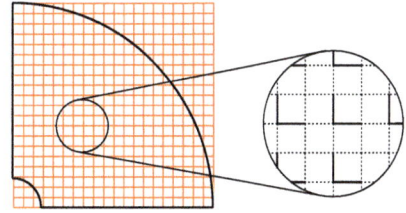

Fig. 4. Initial local slip lines (solid lines) in the whole modelling region.

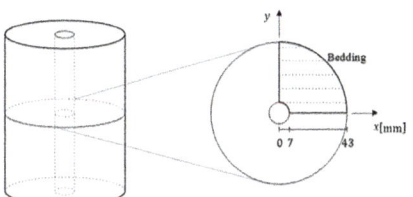

Fig. 5. Modelling of the hollow cylinder under 2D plane strain conditions [4].

2.2 A Aided Method for Locating Slip Lines

Slip-line field theory is a powerful mathematical technique that can be used to solve plane strain boundary value problems in plasticity. It gives exact solutions for rigid plastic solids [9]. The slip lines are considered to be potential failure surfaces which are helpful to properly predict mechanical behavior and failure mechanism of geotechnical materials [10]. In practice, it is difficult to solve a particular problem. Because it is not easy to find the slip-line field and not suitable to consider geotechnical material as rigid plasticity. At present, the method has been largely superseded by finite element modelling.

A geotechnical stability approach, named the "enhanced limit" finite element method [11, 12] or the finite element limit equilibrium method (FELEM) [13, 14], based on finite element stress analysis with searching the critical failure surface has been proposed to provide the factor of safety and critical slip surface to evaluate safety of slope. Based on the above method, a new method was proposed to search slip lines at fixed location in the strain softening plastic zone which is calculated by the FEM. Slip lines in the strain softening zone are used to predict fractures in the EDZ.

Formulation of the Aided Method. As a hardening and softening plastic cohesive-frictional material, in the finite element stress analysis, it is impossible to obtain continuous and smooth slip lines that can be expressed by some mathematic function. But we could assume a slip line composed by finite small segments. An effective optimization algorithm is necessary to be used to search small segments called local slip lines at fixed location. And for a slip line segment l, objective function is defined as the safety factor in FELEM:

$$F = \int_l \tau_f dl \Big/ \int_l \tau dl \tag{7}$$

where τ is the shear stress and τ_f is the shear strength. When $|F - 1.0| \leq \varepsilon$, ε is the calculation parameter (in this paper, $\varepsilon = 0.01$), slip lines are supposed to be reaching the state of failure. Considering the cohesive-frictional material, yield function or failure criterion is $\tau_f = \sigma' \tan \Phi' + C'$, where σ' is the normal stress. Parameters Φ' and C' have to be matched with the Drucker-Prager model that we used in the FE analysis for the nondilatant flow [7], i.e., $\sin \Phi' = \tan \beta \sqrt{3}/3, C' = d/\cos \Phi'$.

Implementation of the Searching Process. A simple and effective search strategy for searching local slip lines is proposed. In Fig. 2, initial lines PA^0 and PB^0 are given to locate local slip lines PA and PB. Both lines are parallel to x direction and y direction respectively. Briefly, the purpose of the search is to find local minima of the objective function $F(y_A)$ or $F(y_B)$ as well as $|\min F(y_A) - 1.| \leq \varepsilon$ or $|\min F(y_B) - 1.| \leq \varepsilon$. It is a one-dimensional minimization problem. So pattern search is used for solving the optimization problem. The searching process is shown in Fig. 3. In the modelling region, initial local slip lines are presented in a grid form, as shown in Fig. 4.

3 EDZ in Hollow Cylinder Experiment

The objective of this simulation is to predict the EDZ in hollow cylinder model and understand its feature. Simulation of local slip lines is used to estimate the fractures in the EDZ. In this simulation, the behavior of the material has been reproduced in the mid-plane section of the hollow cylinder with the assumption of plane strain conditions in the axial direction. Because of the mechanical cross-anisotropy of the material, a quarter of the entire section has been considered, as shown in Fig. 5. The entire boundary conditions are summarized in Fig. 6. The clay is considered as homogeneous and to be fully saturated. The mechanical, hydraulic and physical parameters are reported in Table 1.

Figure 7 compares the computed displacement with the experimental measurement [4] in the three considered radial sections (parallel, at $45°$ and perpendicular to bedding). Apart from profile of displacement along diameter parallel to bedding, the model provides a very accurate estimate of profiles of total displacement. Both experiment results and numerical results show that strain localization occurs in hollow cylinder samples on the basis of the fluctuation of the profile of displacement and local deformation around inner boundary of samples, as shown in Fig. 8. Both results also reveal that the size of the EDZ in the direction parallel to bedding is larger than the direction perpendicular to bedding. The range of the predicted EDZ obtained from numerical result is similar to experimental result.

Table 1. Set of the Boom clay geomechanical, hydraulic and physical parameters in the anisotropic Drucker-Prager model.

Parameters		Cross-anisotropy	Parameters		Cross-anisotropy
Young elastic modulus (MPa)	E_p	400	Residual cohesion (kPa)	c'_r	85 ($0°$)
	E_t	200			80 ($45°$)
Poisson ratio (-)	v_p	0.125			110 ($90°$)
	v_{tp}	0.0625	Initial friction angle ($°$)	φ'_i	8
Shear modulus (MPa)	G_t	178	Peak friction angle ($°$)	φ'_p	18
Initial cohesion (kPa)	c'_i	255 ($0°$)	Residual friction angle ($°$)	φ'_r	16
		240 ($45°$)	Softening parameters (-)	B_s	0.02
		330 ($90°$)	Hardening parameters (-)	B_h	0.04
Peak cohesion (kPa)	c'_p	255 ($0°$)	Dilatancy angle ($°$)	ψ	0
		240 ($45°$)	Permeability (m/s)	k_p	4×10^{-12}
		330 ($90°$)		k_t	2×10^{-12}
			Initial porosity (-)	n_0	0.39

Contours of the equivalent plastic strain and increment of equivalent plastic strain rate obtained by the FEM with the anisotropic Drucker-Prager model are shown in Fig. 9. After the excavation, when the plastic mechanism is activated, the spatial and temporal distribution of equivalent plastic strain is not significantly different from increment of equivalent plastic strain rate on Fig. 9(a) and (b). However, localization of increment of equivalent plastic strain rate is more noticeable at 4,000 s in Fig. 9(d). With the increase of increment of equivalent plastic strain rate, the shear bands (contour of equivalent plastic strain) are clearly observed, as shown in Fig. 9(e), and the localization pattern becomes more stable.

With the assumptions of horizontal and vertical initial lines, local slip lines in strain softening zone at the end of test are plotted in Fig. 10. As expected, direction of local slip lines is same as shear bands. It is revealed that the shear stress on the local slip lines has already reached shear strength. As a result, slip lines in the strain softening zone could be used to predict fractures in the EDZ.

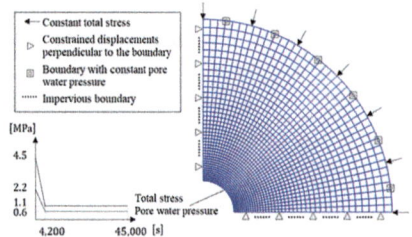

Fig. 6. Finite element mesh and hydro-mechanical boundary conditions of the 2D plane strain problem of the hollow cylinder experiment.

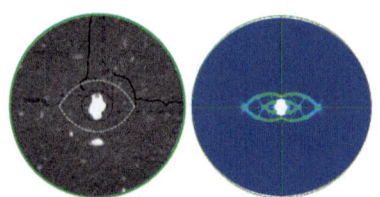

Fig. 8. Comparison of the shape of the excavation damaged zone between prediction obtained by observation in displacement profiles [4] (left-hand side) and contour of equivalent plastic strain (right-hand side) at the end of test (deformation scale factor: 1).

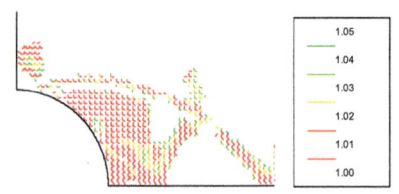

Fig. 10. Local slip lines: initial lines are parallel to the horizontal direction and vertical direction, respectively.

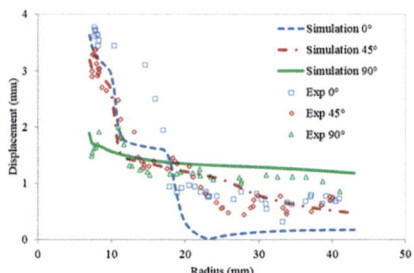

Fig. 7. Radial profile of displacements. Comparison between experiment measurements and numerical results obtained by the FEM with the anisotropic Drucker-Prager model.

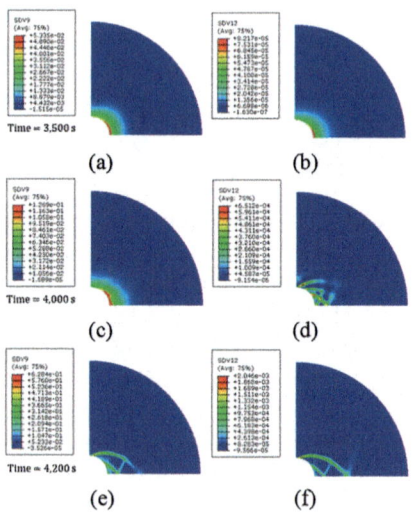

Fig. 9. Contours of equivalent plastic strain (left-hand side) and the increment of equivalent plastic strain rate (right-hand side) obtained the FEM with the anisotropic Drucker-Prager model.

4   Conclusions

Boom clay is currently under investigation as a potential host rock for deep geological disposal of nuclear waste in Belgium. But there is no proper constitutive interpretation of the EDZ induced by gallery excavation. This problem is addressed by developing a modified Drucker-Prager model considering hardening and softening plasticity and elastic and plastic cross-anisotropy. The model is adopted for predicting the EDZ in hollow cylinder experiment. From a perspective of soil strength, it is able to quantitatively analysis the EDZ in the Boom clay. Potential fractures in the EDZ are estimated by slip lines obtained by a proposed aided method. Comparison between the experiment and numerical results in hollow cylinder case reveal the anisotropic EDZ and the localization of deformation.

Acknowledgments. This work is funded by the Doctoral Scientific Research Foundation of Liaoning Province (Grant No. 20170520341).

References

1. Tsang, C.-F., Bernier, F., Davies, C.: Geohydromechanical processes in the excavation damaged zone in crystalline rock, rock salt, and indurated and plastic clays—in the context of radioactive waste disposal. Int. J. Rock Mech. Mining Sci. **42**(1), 109–125 (2005)
2. Bernier, F., Li, X.L., Verstricht, J., Barnichon, J.D., Labiouse, V., Bastiaens, W.: CLIPEX: clay instrumentation programme for the extension of an underground research laboratory: final report. Commisson of the European Communities, Luxembourg (2002)
3. Labiouse, V., Sauthier, C., You, S.: Hollow cylinder simulation experiments of galleries in boom clay formation. Rock Mech. Rock Eng. **2013**, 1–13 (2013)
4. François, B., Labiouse, V., Dizier, A., Marinelli, F., Charlier, R., Collin, F.: Hollow cylinder tests on boom clay: modelling of strain localization in the anisotropic excavation damaged zone. Rock Mech. Rock Eng. **2012**, 1–16 (2012)
5. Bernier, F., Li, X.L., Bastiaens, W.: Twenty-five years' geotechnical observation and testing in the tertiary boom clay formation. Géotechnique **57**(2), 229–237 (2007)
6. Chen, G.J., Sillen, X., Verstricht, J., Li, X.L.: ATLAS III in situ heating test in boom clay: field data, observation and interpretation. Comput. Geotech. **38**(5), 683–696 (2011)
7. HKS: ABAQUS user's manual, version 6.9. Hibbitt, Karlsson and Sorensen, Inc. Providence (2010)
8. Belytschko, T., Liu, W.K., Moran, B.: Nonlinear Finite Element for Continua and Structures. Wiley, New York (2001)
9. Bower, A.F.: Applied Mechanics of Solids. CRC Press, New York (2010)
10. Davis, R.O., Selvadurai, A.P.S.: Plasticity and Geomechanics. Cambridge University Press, New York (2002)
11. Fredlund, D.G., Scoular, R.E.G., Zakerzadeh, N.: Using a finite element stress analysis to compute the factor of safety. In: Proceedings of the 52nd Canadian Geotechnical Conference, Saskatchewan (1999)
12. Pham, H.T.V., Fredlund, D.G.: The application of dynamic programming to slope stability analysis. Can. Geotech. J. **40**(4), 830–847 (2003)
13. Shao, L., Tang, H., Han, G.: Finite element method for slope stability analysis with its applications. Chin. J. Comput. Mech. **18**(1), 81–87 (2001). (in Chinese)
14. Liu, S., Shao, L., Li, H.: Slope stability analysis using the limit equilibrium method and two finite element methods. Comput. Geotech. **63**, 291–298 (2015)

Numerical Analysis About Influence of Protecting Wall on the Displacement Field Around the Pit

Jian Guo[1(✉)], Guobin Liu[1,2], and Tao Qin[1]

[1] Department of Geotechnical Engineering, Tongji University,
1239 Siping Road, Shanghai 200092, People's Republic of China
zxct2005@163.com
[2] Key Laboratory of Geotechnical and Underground Engineering of Ministry
of Education, Tongji University, Shanghai 200092, People's Republic of China

Abstract. Nowadays, with the rapid development of the economy, the deep excavations become more and more common. However, the protection of the important architecture or pipelines such as historical buildings also becomes a main concern and problem. In order to solve such contradiction, the diaphragm wall or isolation piles acting as a protecting wall between the deep excavation and the architecture was proposed. Based on an excavation project in Shanghai, this paper investigated the influence of the protecting wall on the displacement field around the pit by using 2D numerical analysis. The results showed that protecting walls can restrain the ground subsidence behind the pit effectively but have little effect on the retaining wall behavior of the pit. On the contrary, the excavation had a great effect on the protecting wall. The length of the protecting wall along with the distance between the protecting wall and the pit are the key factors influencing the ground subsidence. However, the restraint effect is limited if such length or distance go beyond a certain range. The findings can offer some guidance for similar projects to some extent.

Keywords: Numerical simulation · Protecting wall · Parametric study
Ground subsidence

1 Introduction

With the rapid development of economy, many excavations are carried out in downtown area with congested environment. Under such circumstances, the adjacent buildings may face great risks of being damaged. Moreover, many important structures and historical buildings with shallow foundation need extra protection. Therefore, the pile wall or diaphragm wall used as protecting wall for such buildings are adopted in Shanghai. Liu et al. [1] studied the observed performance of a deep multistrutted excavation in Shanghai. Yoo and Lee [2] analyzed the characteristics of ground subsidence induced by excavation. Ou et al. [3] and Wu et al. [4] investigated the effect induced by excavations using cross walls on nearby structures. Zhai et al. [5] summarized three engineering projects using protecting wall and the promising results were obtained. Hua [6] pointed that the traditional flexible protecting wall may not reach the

© Springer Nature Singapore Pte Ltd. 2018
T. Qiu et al. (Eds.): GSIC 2018, *Proceedings of GeoShanghai 2018 International Conference: Advances in Soil Dynamics and Foundation Engineering*, pp. 598–605, 2018.
https://doi.org/10.1007/978-981-13-0131-5_65

expected protective effect so the stiff wall should be used. Generally, the studies of interaction between protecting wall and deep excavation along with its effect on surroundings in soft clay are still limited. Based on this background, this paper investigated the excavation behaviors and ground settlement between a protecting wall and a deep excavation by using finite element software ABAQUS. Different parameters such as location and length of the protecting wall were analyzed and the findings may provide some guidance for similar projects.

2 Geotechnical Conditions and Parameters of Material

Shanghai is located at the flat alluvial plain which is called Yangtze River Delta and the main stratum in Shanghai were silty clay, soft clay and silty sand. A large basement excavation was carried out in downtown area due to the limited space. However, some important communication cables were close to the excavation and a 0.6-m-thick protecting wall were constructed between them. Figure 1 shows the plan view of this project and Table 1 presents the parameter of soil in the numerical model. The Modified Cam-clay model was used in this simulation and the parameters such as unit weight of soil (γ), cohesion (c), friction angle (φ), void ratio (e), swelling index (κ), compression index (λ) and Poisson's ratio (ν) were obtained or calculated from the results of field tests and laboratory tests. The retaining wall and protecting wall were assumed to behave as a linear–elastic material and they were simulated as beam element. By taking the construction quality into consideration [7], the Young's modulus of the concrete struts, retaining wall and protecting wall were reduced to 25 GPa, 25 GPa and 6 GPa, respectively. The excavation depth was 20 m and embedded depth ratio of the retaining wall was 1.0. The detailed construction stages were listed in Table 2 and the numerical simulation model was shown in Fig. 2, where B_2 refers to the distance between the protecting wall and the foundation pit (10 m for this project). H_2 is the length of the protecting wall (40 m for this project). H denotes the excavation depth (20 m) and B is pit width (20 m) in this study. The upper boundary of the model was set as free surface while the lateral displacement at left and right side of the model was restrained and the bottom of the model was set as fixed boundary. According to

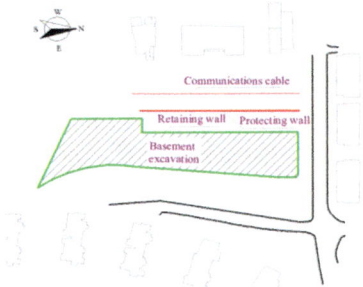

Fig. 1. Plan view of the project

Table 1. Typical soil layers and geotechnical parameters (Modified Cam-clay model)

Layer	Thickness (m)	$\gamma/kN \cdot m^{-3}$	c/kPa	φ (°)	v	λ	κ	e
Silty clay	10	18.5	16.2	16.7	0.32	0.123	0.015	0.921
Soft clay	4	16.8	11.4	13.1	0.37	0.186	0.025	1.302
Silty clay	10	18.2	15.6	15.6	0.35	0.161	0.018	1.108
Clay	6	17.7	17.6	20.6	0.32	0.101	0.011	0.858
Silty sand	16	18.7	2.0	31.2	0.27	0.088	0.004	0.768
Stiff clay	16	18.1	18.2	17.3	0.28	0.054	0.003	0.814
Fine sand	20	19.2	3.0	33.4	0.25	0.042	0.002	0.600

Table 2. Detailed information of construction stages and supporting system

Stage	Activity	Size of concrete struts	Thickness of retaining wall
1	Construct level 1 struts and excavate to 6 m below ground surface (BGS)	1 m (width) × 0.8 m (height)	0.8 m
2	Construct level 2 struts and excavate to 10 m BGS	1 m (width) × 0.8 m (height)	
3	Construct level 3 struts and excavate to 14 m BGS	1.2 m (width) × 1 m (height)	
4	Construct level 4 struts and excavate to 20 m BGS	1.2 m (width) × 1 m (height)	

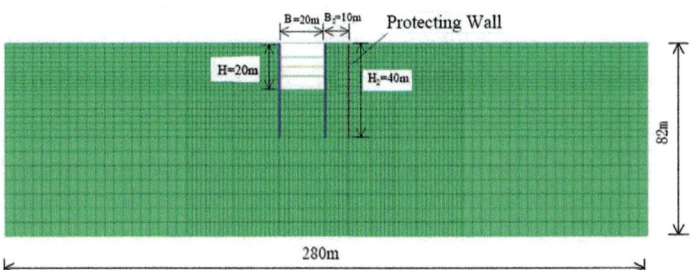

Fig. 2. Numerical simulation model

Huang et al. [7], to alleviate the boundary effect, the whole length of the model is 280 m and the total depth is 86 m. The underground water was not included in this study. Since the excavation duration was short and the overload near the excavation was negligible, the soil consolidation was not included to simplify the numerical simulation and shorten the calculation time.

3   Analysis of the Numerical Results

The results computed by ABAQUS and the field measured data were shown in Fig. 3. From Fig. 3(a) and (b), the computed results are similar to the field monitoring data. It can be obtained that a protecting wall could significantly affect the displacement field around the pit and the ground settlement behind the protecting wall can be reduced remarkably, which would restrain the settlements of pipelines and buildings. The results shown in Fig. 3 also verified the reliability of the numerical model.

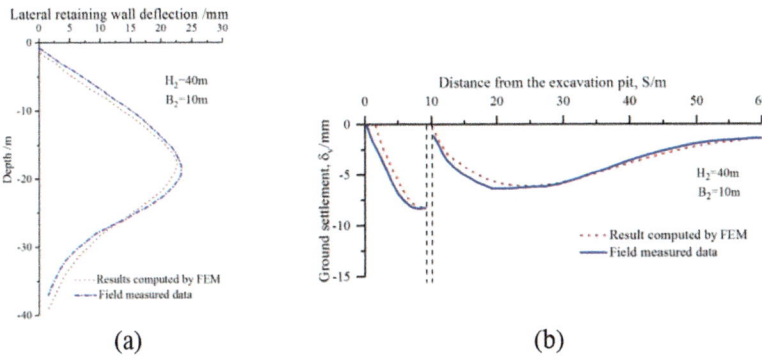

(a)   (b)

Fig. 3. Comparison between the computed results and field monitoring data: (a) retaining wall deflection; (b) ground settlement outside the pit

4   Parametric Studies

4.1   The Distance Between the Excavation and the Protecting Wall

In this section, B_2 varied from 5 m to 40 m while other parameters kept unchanged to conduct parametric studies.

Deformation of the Retaining Wall. Figure 4 plots the right sided retaining wall deflection. It can be found that when $B_2 = 5$ m and $B_2 = 40$ m, the maximum wall deflection δ_{hm}, were 21.1 mm and 24.5 mm, respectively while δ_{hm} was 25 mm if there was no protecting wall outside the pit. With the increase of the distance between the protecting wall and the pit, the lateral retaining wall deformation increased slowly and when $B_2 > 40$ m, the influence from the protecting wall can be ignored. Generally, the influence of the protecting wall on the lateral deformation of retaining wall was not obvious.

Ground Subsidence. Figure 5 plots the ground settlement outside the pit. From Fig. 5, it can be seen that the ground settlement experienced an obvious saltation at the location of protecting wall. When $B_2 = 10$ m, the maximum ground settlement between the protecting wall and the pit, δ_{vmi}, was 8.1 mm and the maximum ground settlement behind the protecting wall, δ_{vmo}, was 6.3 mm. When $B_2 = 20$ m, δ_{vmi} was

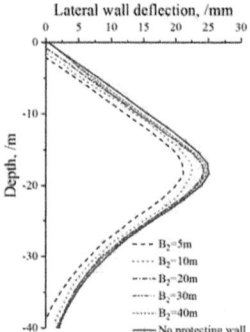

Fig. 4. Deflection of retaining wall

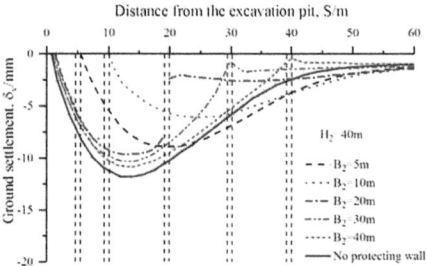

Fig. 5. Ground subsidence

9.3 mm while the δ_{vmo} was only 2.6 mm. It can be found that in different cases, the ground settlement between the pit and protecting wall was less affected than that outside the protecting wall, which indicated that the protecting wall had a great restraint on the ground settlement behind it. Therefore, the protective effect from protecting wall is evident. When B_2 reached 40 m, the protecting wall slightly decreased the ground settlement and the ground settlement curve was close to that with no existence of protecting wall.

Basal Heave. Figure 6 presents the influence of the protecting wall on the basal heave inside the pit. It can be noted that the basal heave increased to some extent if there was a protecting wall outside the pit. When $B_2 = 5$ m, the increment of basal heave was 4.7 mm while it was 3.1 mm when $B_2 = 10$ m, which were only 6.8% and 4.3% respectively of the basal heave with no protecting wall outside the pit. It can be seen that the distance between the protecting wall and pit had slight influence on the basal heave.

Lateral Deflection of Protecting Wall. During the excavation process, the construction activities inside the pit can cause some effect on the protecting wall. Figure 7 plots the lateral deformation of protecting wall with different B_2, where positive value presents the wall moves towards the pit. When B_2 was 5 m, the maximum lateral

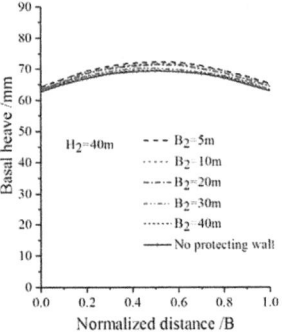

Fig. 6. Basal heave inside the pit

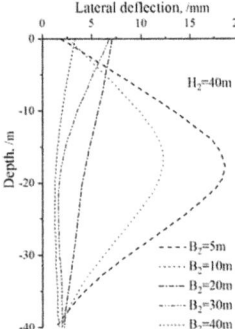

Fig. 7. Lateral deflection of the protecting wall

deformation was 18 mm and it occurred at the depth of 18.6 m. The deformation at both the top and the bottom of the wall were much small and a bulging deformation profile appeared. As B_2 increased, the location of the maximum lateral wall deformation moved upwards. When B_2 was 40 m, the maximum deflection of the protecting wall was 3.2 mm and it occurred at the wall top. Generally, when the protecting wall is close to the pit, it resists the earth pressure together with the retaining wall of the pit. But with the increase of B_2, the excavation of the pit had little influence on the soil in further area. As a result, its influence on the protecting wall decreased obviously.

4.2   The Length of the Protecting Wall

From the analysis above, the protecting wall had little effect on the deflection of the retaining wall and the basal heave inside the pit but a huge influence on the ground subsidence. Moreover, the protecting wall was usually used to restrain the ground settlement outside the pit but not to restrain the excavation behaviors of the pit. Therefore, the length of the protecting wall varied from 20 m to 60 m in this section and only ground subsidence along with the deflection of the protecting wall were analyzed.

Ground Subsidence. Figure 8(a) and (b) presents the numerical results under different conditions. It can be obtained that the protecting wall with different length had a great influence on the ground settlement. From Fig. 8(a), when H_2 went beyond 30 m, the restraint effect was limited. As shown in Fig. 8(b), when B_2 = 20 m and H_2 increased from 20 m to 60 m, the δ_{vmo} were 61%V_n, 40%V_n, 22%V_n, 20%V_n, respectively, in which V_n represents the maximum ground settlement if there was no protecting wall outside the pit. It can be noted that with the increase of the length of protecting wall, its restraint on ground settlement increased correspondingly. Under the condition of B_2 = 10 m, the restraint effect became maximum when H_2 = 30 m (1.5H). However, when B_2 = 20 m, the most significant restraint effect occurred when H_2 = 40 m (2.0H). By comparing Fig. 8(a) and (b), It can be indicated that the influence of protecting wall on ground settlement behind the protecting wall was related with length of the protecting wall (H_2) and the distance between the protecting wall and the pit (B_2).

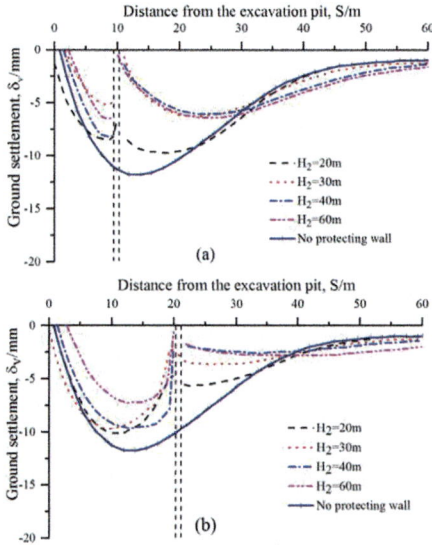

Fig. 8. Ground subsidence: (a) B_2 = 10 m; (b) B_2 = 20 m

Lateral Deflection of the Protecting Wall Figure 9(a) and (b) plots lateral deflection of the protecting wall with different length. When B_2 = 10 m, the maximum deflection of the protecting wall decreased from 13.7 mm to 11.5 mm with its length increasing. The location where the maximum deformation occurred was between 15–20 m BGS. When B_2 = 20 m, the maximum lateral deformation decreased from 11.1 mm to 6.9 mm as length of the protecting wall increased and the maximum lateral movement occurred at the top of the wall. When H_2 reached 40 m, the wall deflection became stable.

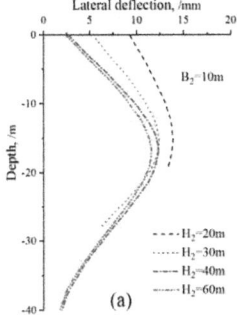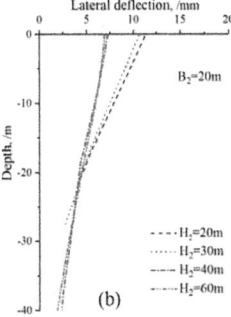

Fig. 9. Lateral deflection of the protecting wall: (a) $B_2 = 10$ m; (b) $B_2 = 20$ m

5   Conclusions

Based on a series of aforementioned numerical analysis, the conclusions can be drawn as follows:

1. The protecting wall can effectively affect displacement field behind the pit caused by excavation but hardly influence the deflection of the retaining wall and basal heave inside the pit for this project. However, the excavation inside the pit has a great effect on the protecting wall.
2. Distance between the protecting wall and the pit is the main factor that influence protecting wall's restraint on ground settlement. In this project, when B_2 is 0.5H–1.0H, the restraint effect can be very evident. However, when the distance is larger than 1.5H, the restraint effect was limited.
3. Length of the protecting wall can also be an important factor. Generally, the larger length the protecting wall has, the better restraint effect appears. But when it goes beyond some certain length, the restraint effect is not obvious.

References

1. Liu, G.B., Ng, C.W.W., Wang, Z.W.: Observed performance of a deep multistrutted excavation in Shanghai soft clays. J. Geotech. Geoenviron. Eng. **131**(8), 1004–1013 (2005)
2. Yoo, C., Lee, D.: Deep excavation-induced ground movement characteristics-A numerical investigation. Comput. Geotech. **35**(2), 231–252 (2008)
3. Ou, C.Y., Hsieh, P.G., Lin, Y.L.: Performance of excavations with cross walls. J. Geotech. Geoenviron. Eng. **137**(1), 94–104 (2011)
4. Wu, S.H., Ching, J., Ou, C.Y.: Predicting wall displacements for excavations with cross walls in soft clay. J. Geotech. Geoenviron. Eng. **139**(6), 914–927 (2013)
5. Zhai, J.Q., Jia, J., Xie, X.L.: Practice of partition wall in the building protection projects near deep excavation. Chin. J. Undergr. Space Eng. **1**(6), 102–166 (2010)
6. Hua, G.Q.: The partition wall should be carefully used in underground engineering to protect the surrounding structures. China Municipal Eng. **5**, 42–45 (2009)
7. Huang, P., Liu, G.B., Huo, R.K.: Parameter analysis of zoned excavation in soft soil area. J. Cent. South Univ. (Sci. Technol.). **46**(10), 3859–3864 (2015)

Settlement Prediction of Large LNG Tanks Built in Beaumont Formation

Anil Bhandari[1(✉)], Mahi Galagoda[1], and James T. Cameron[2]

[1] Bechtel Oil, Gas, and Chemicals, Houston, TX 77056, USA
abhandar@bechtel.com
[2] Bechtel National, Inc., San Francisco, CA 94105, USA

Abstract. This paper compares the settlement prediction of a group of large LNG tanks utilizing simple one-dimensional calculation, an axisymmetric Finite Element Method (FEM), and relatively complex three-dimensional FEM solutions, including the load interaction effect. The soil parameters used for the settlement prediction were derived from a suite of in-situ and laboratory tests. Commercial software, PLAXIS 2D and PLAXIS 3D, were utilized for the FEM analyses. Settlement predictions from these three approaches are presented and compared. The LNG tanks are currently under construction and the settlement will be monitored during the construction, hydrotest and operation phases. The measured settlement and the comparison with the estimates made during design phase will be published in the future.

Keywords: Settlement · LNG tanks · Finite Element Method
Load interaction

1 Introduction

Tanks are an integral part of a Liquefied Natural Gas (LNG) projects and are used to store the LNG products until they are shipped. The LNG tanks are typically full containment structures with diameters ranging from 70 to 95 m with storage capacities on the order of 100–200 thousand cubic meters. Because of the criticality of the LNG due to their cryogenic nature and the large quantity of stored product, the design and maintenance of LNG tanks are strictly regulated under National fire protection Agency Standard NFPA 59A (2016) and American Petroleum Institute Standard API 650 (2013). An appropriate design of the LNG tanks foundations is a key component per these regulations and important to the success of the project in providing uninterrupted storage capacity to load the LNG ships without incurring operational delays. LNG projects are permitted by Federal Energy Regulatory Commission (FERC) which provides a guidance document for permitting process for public use. The FERC guidance document includes a section on LNG tank foundation design which was followed in this design.

There are limited published design/performance case histories of these structures in conferences/journals and most of these limited publications describe the LNG tanks supported on piled foundations (Fellenius and Ochoa 2013). Large storage tanks supported on piled foundations can be analyzed as a flexible raft placed at the pile toe

© Springer Nature Singapore Pte Ltd. 2018
T. Qiu et al. (Eds.): GSIC 2018, *Proceedings of GeoShanghai 2018 International Conference: Advances in Soil Dynamics and Foundation Engineering*, pp. 606–618, 2018.
https://doi.org/10.1007/978-981-13-0131-5_66

level or closer to the toe level for settlement analysis. This conclusion was derived from the back analysis of the performance (settlement) of five large storage tanks (Fellenius and Ochoa 2013).

D'Orazio and Duncan (1987) have complied the case histories for the failure of oil storage tank foundations and identified two modes of failure: edge shear and base shear. Additionally, the importance of differential settlement and the shape of the settlement dish, and their effect on the design of the tank foundation, was discussed. The foundation design should incorporate the following checks (D'Orazio and Duncan 1987):

1. Procedure for estimating the total settlement.
2. Procedure for estimating the shape of the settlement dish.
3. Criteria of judging the acceptability of differential settlement.

The design requirements from Federal Energy Regulatory Commission (FERC) are largely in line with the philosophy discussed by D'Orazio and Duncan (1987). Based on the FERC draft guidance document (2007), design of LNG tank foundation shall meet the following settlement and planar tilt requirements.

1. Differential settlement around the periphery of the tank < 1:500 of the arc length under consideration.
2. Differential settlement from center to the edge of the tank < 1:300 of the tank radius.
3. Planar tilt angle of the tank foundation < 0.002 radians.

Note that there are additional requirements for bearing capacity in the FERC document; however, are not discussed in this paper since the objective of the paper is to discuss the settlement estimation of the large storage tanks.

Deep foundations are expensive relative to raft foundations. There is a paucity of published case studies of LNG tanks supported on raft foundations. The purpose of this paper is to present and compare different approaches for settlement estimations of raft foundations for LNG tanks. There are three adjacent tanks at various stages of planning and construction at the time this paper was prepared. These tanks will be referred as Tank 1, Tank 2, and Tank 3 in the paper. The settlement of the tanks will be monitored and will be published in the future.

2   Historical Use, Subsurface Conditions, and Design Parameters

The area of the LNG tanks is generally flat with average elevation of +25 ft above Mean Sea Level (MSL). Approximately 60% of the area was previously covered with bauxite stockpiles up to 60 ft high, imposing areal loads of approximately 7,500 psf. Stockpiling the bauxite started in 1965 and progressed with different stockpile heights until it was 60 ft high in 1985. Removal of the stockpile started around 2000 and was completed in 2005. The bauxite stockpile functioned as a preload. A portion of the LNG tank farm (Tank 3 and part of Tank 2), located outside of the bauxite stockpile area, would not have this beneficial effect of preloading. The implication of spatially variable historical loading on the settlement estimate is discussed in the subsequent sections.

At each of the LNG tank locations, a number of Cone Penetration Tests (CPTs), Seismic Cone Penetration Tests (SCPTs), soil borings and pressuremeter tests in soil borings, were performed to characterize the subsurface conditions. Maximum depth explored was on the order of 330 ft and exceeded one tank diameter. Soil conditions at the site consisted of interchanging layers of lean clays, sandy lean clays, fat clay, silty sands and clayey sands to the exploration depth. The thicknesses of these interchanging layers through the tank centerlines are depicted in Fig. 1. Light yellow shade indicates coarse-grained soils (sands) and light blue shade indicates fine-grained soil (clay and silt). The available boring logs and CPT data indicated that the fine-grained soils generally exhibited stiff to hard consistency, and the coarse-grained soils varied between loose and very dense relative density. The schematic layout of the boreholes and CPTs as tested in the field and used for 3D FEM analysis are shown in Fig. 2.

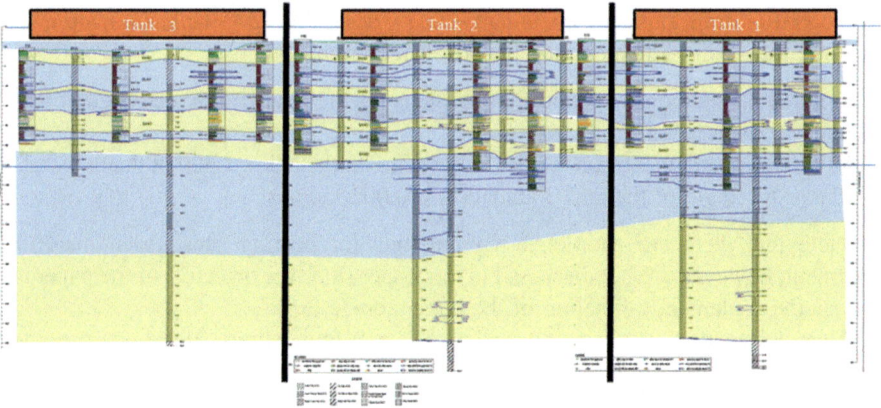

Fig. 1. Subsurface profile along the tank centerline along North-South direction (Light yellow shade indicate coarse grained soils and light blue shade indicate fine grained soil).

The interlayered clays and sands encountered at the LNG storage tank areas belong to the Pleistocene Beaumont Formation. This formation is approximately 100 ft thick at the tank site. The underlying layers belong to an older Pleistocene Lissie Formation. The Beaumont soils are over consolidated due to cyclic changes in moisture content after deposition. Over Consolidation Ratio (OCR) is generally greater than 10 for depths to 25 ft below the existing ground surface and reduces with depth. The Beaumont soils also exhibit minor cementation along grain contacts.

Soil stiffness values and other strength parameters for each tank were derived from tank-specific borehole logs, CPTs, SCPTs, pressuremeter tests, and laboratory test results. A typical summary of these properties is presented in Table 1 for Tank 1. Soil layering was determined from a consideration of descriptions from the borehole logs and analysis of the CPT/SCPT data. The effects of the previous bauxite stockpile are evident in the plots of overconsolidation ratio and preconsolidation stress versus elevation shown in Fig. 3. The two lines drawn on the preconsolidation stress plot show

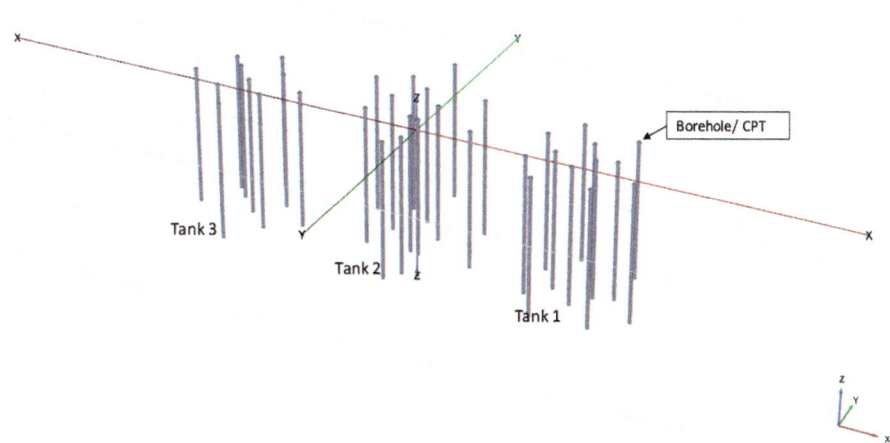

Fig. 2. CPTs and boreholes distribution in FEM model of three tanks

the effective overburden stress (thin red line in Fig. 3) and the additional effective stress the 60 ft of bauxite would have expected to cause (dark red line, which is 7.5 ksf greater than the effective overburden stress in Fig. 3) if the area was under bauxite stockpile for sufficiently long duration to achieve 100% consolidation. Note that the measured values under Tank 1 and the preloaded portion of Tank 2, nearly match the preconsolidation line expected due to bauxite stockpile preloading. Additionally, the OCR plot shows that, generally, the entire area, even without the bauxite stockpile (Tank 3 data in Fig. 3), is overconsolidated by a factor of 1.7–2.0 below Elevation −100 ft MSL.

Table 1. Parameters for settlement calculation (Tank 1 in Fig. 1).

Layer No.	Elevation (ft in Fig. 1) From	To	Soil Type	S_u ksf	OCR	E_w ksf	m	m_v	j	Cv (ft²/yr in Fig. 1)	Effective Cohesion, C', ksf	Effective Friction, φ	Ψ deg	E_{50} ref, ksf	n	Ko	k (ft/day in Fig. 1)	E_{oed} ref, ksf	E_{ur} ref, ksf
1	25	15	Sandy Clay 1	2.1	19	1575	27	270	0	339	0.1	30.0	NA	650	0.6	2.2	2.98E-06	309	1950
2	15	-2	Silty Sand 1			520	1560	0.5			0.0	35.0	5	890	0.5	0.4	2.83E-01	710	2670
3	-2	-18	Clay 2	2.7	6.5	2025	37	340	0	339	0.4	17.9	NA	260	0.7	1.8	4.57E-06	145	780
4	-18	-29	Silty Sand 2			350	1050	0.5			0.0	35.0	5	905	0.5	0.4	2.83E-01	725	2715
5	-29	-54	Clay3	2.9	5	2175	20	200	0	33.9	0.3	19.0	NA	290	0.8	1.5	4.57E-06	150	870
6	-54	-64	Silty Sand 3			285	855	0.5			0.0	35.0	5	960	0.5	0.4	2.83E-01	770	2880
7	-64	-79	Clay 4	3.5	4	2625	33	195	0	28.8	0.6	16.3	NA	385	0.8	1.4	4.57E-06	215	1155
8	-79	-89	Silty Sand 4			250	750	0.5			0.0	35.0	5	945	0.5	0.4	2.83E-01	755	2835
9	-89	-157	Clay 5	3.8	2.5	2850	22	155	0	28.8	1.0	17.5	NA	425	0.9	1.1	1.96E-06	220	1275
10	-157	-278	Silty Sand 5			182	546	0.5			0.0	35.0	5	1010	0.5	0.4	2.83E-01	810	3030
11	-278	-305	Clay 6	3.8	1.8	2850	22	155	0	16.8	1.0	17.5	NA	425	0.9	1.1	1.96E-06	220	1275
12			Concrete											6.36E+05					

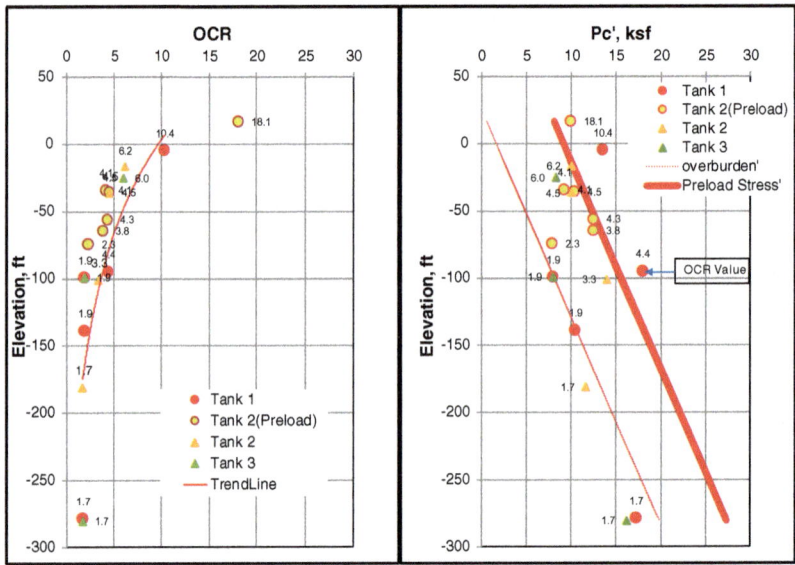

Fig. 3. Measured OCR and preconsolidation stress under tanks

3 Tank Description and Loading

The LNG tanks are full containment tanks with an external diameter of 257.5 ft and an internal diameter of 253.5 ft. The tanks are supported on 271 ft diameter concrete mat foundations at a depth of approximately 3 ft below final grade. The outer 20-ft perimeter ring of the mat is thicker than the interior portion to better distribute the weight of the tank shell. The clear spacing between adjacent tanks is 154 ft. Each tank is to be constructed over a period of about 30 months, in the tank order of 1, 3, and 2. The construction start dates for Tanks 1 and 3 were offset by about 3 months, and the planned start date for Tank 2 is not finalized yet. The design life for each tank is 20 years.

Each of the tanks has a net capacity of 160,000 m³. The total weight of concrete exterior wall and roof induces a contact pressure of 4.65 ksf over a 20 ft perimeter ring. The tanks will be hydrotested to a contact pressure over the interior mat of 5.45 ksf over a period of 4 to 6 weeks. Sustained pressure due to product loading will be 3.28 ksf for the design life. Though settlement at different stages of construction and operation were estimated, for this paper we focus primarily on the settlement during hydrotesting and production loading. The LNG tanks will witness the most extreme loading case during hydrotesting albeit for a short duration. Production load is smaller compared to hydrotest loading but will be maintained over the design life of the LNG tanks.

4   Analyses and Results

Settlements of the LNG tanks at different stages of construction and operation were evaluated using three methods: one-dimensional settlement evaluation assuming flexible foundation, an axisymmetric tank model using 2D FEM, and a three-dimensional analysis of tanks using 3D FEM. We elected to engage simplified models initially and progressed towards more complex models as the design matured. We used commercial software UniSettle 4.32 for one-dimensional analyses, PLAXIS 2D for axisymmetric FEM analyses, and PLAXIS 3D for three-dimensional analyses. In all methods, tank loads (discussed in preceding section) were applied per the planned construction schedule to estimate the settlements and settlement patterns of tanks.

UniSettle 4.32 inherently considers the LNG tank foundation as a flexible foundation while PLAXIS 2D and 3D can model the foundation and soil interaction considering the respective stiffnesses of the foundation and soils. Additionally, PLAXIS 3D can simulate the spatial variations of the soil layers and can be used to predict differential settlement along the periphery and the planar tilt of the tank foundation. The effect of Bauxite preloading can be modelled in PLAXIS 3D by incorporating variations in soil properties in both horizontal and vertical directions. All 3 models can be set up to include loading interaction of the 3 tanks. Brief descriptions of these methods are presented in the succeeding section.

4.1   One-Dimensional Analysis

UniSettle 4.32 uses Janbu's parameter for soils such as modulus number (m), recompression number (m_r) and stress components (j) to calculate the settlement under the applied load. This approach implements tangent modulus concept to calculate the settlement (Janbu 1967). Table 1 summarizes these input parameters used in UniSettle 4.32 for analyzing the settlements of Tank 1. Table 1 also includes the design input parameters used in the axisymmetric and 3D FEM analyses. Similar tables were developed for Tanks 2 and 3 based on the test results but are not included in the paper due to space limitation. Because of the layered soil profiles with varying stiffness, the Westergaard method was used to obtain the stress distribution. It is well established that Westergaard method is more appropriate to model stress distribution and settlement in layered soil than Boussinesq's or 2V:1H stress distributions. The settlement values obtained from one-dimensional analysis were corrected for three-dimensional effects as recommended by Skempton and Bjerrum (1957). We designated this analysis as "Single Tank Model (Janbu's Method)".

4.2   Axisymmetric Model in FEM Analysis

The model measured 400 ft along the radial axis and was 300 ft deep. Fifteen-node elements were used with mesh refinement around the perimeter of the mat foundation. The soil property data was tailored to meet the required input for PLAXIS numerical models. Strain hardening constitutive soil models were used that account for the non-linear behavior of soils under loading, unloading and reloading states of stress, as well as stress state history. We developed a separate FEM model for each tank using tank-specific subsurface layering and properties, and applied tank loading to establish the settlement profiles without consideration of the interaction effects. We designated this analysis as "Single Tank Model".

4.3   Three-Dimensional Model in FEM Analysis

In order to incorporate the variations in subsurface conditions and use them in three-dimensional FEM modelling, a well thought out geotechnical investigation program coupled with significant advanced laboratory tests including consolidation and triaxial strength tests are required. The detailed investigation program implemented in this project allowed us to effectively utilize the three-dimensional FEM model for the settlement interaction study.

The model measured 1600 ft along the north-south tank alignment axis and 800 ft along the east-west axis, and is about 300 ft deep. Ten–node tetrahedral elements were used with mesh refinement around the perimeter of the mat foundation. We developed a FEM model of three tanks that included tank-specific soil properties, layering, and planned construction sequences. A three-tank model was created with fully penetrating vertical subdivisions midway between the tanks. This model allows the tank stress fields to act upon one another and properly captures the interaction effects with due consideration to the measured soil properties. The effect of spatial variation of the bauxite preloading was also incorporated in this model. The premise of this approach is that the soil properties determined for LNG Tanks 1 and 2 mostly reflected the effect of bauxite stockpile. Nearly 2/3rd of the Tank 2 will be in the area which was used for bauxite stockpile though of variable heights and the zone south of bauxite stockpile (towards Tank 3) would have more compressible soil properties. The area outside the bauxite preload to midway between Tanks 2 and 3 was modelled by using the soil stiffness measured under Tank 2 reduced by 10% to account for the effects of the reduced bauxite loading in this area. Additional vertical subdivisions were modelled in the bauxite stockpile area and midway of Tanks 2 and 3 to better simulate the changing soil stiffness in this region. The remaining area towards Tank 3 and beyond was modeled using the stiffness and soil strength parameters derived for Tank 3. The complete 3D model is shown in Fig. 4. We designated this analysis as "With Interaction" in subsequent Figs. 5, 6, 7, 8, 9 and 10.

4.4   Results

During hydrotest, one-dimensional analysis using Janbu's method provided the maximum tank settlements (Figs. 5, 6 and 7). On average, tank settlements were 35% higher than those obtained from FEM analysis without considering the interaction (respective axisymmetric models). The mat was assumed to be a flexible foundation; therefore, sharp changes in settlement profiles in accordance with the sustained pressure intensities were evident in the results of one-dimensional analysis. The one-dimensional analysis uses a simple model that allows estimating the range of settlements during the initial design; however, cannot be relied on for final design since the stiffness of the foundation, subsurface variations in horizontal plane, and the effect of tank interaction cannot be accurately modelled.

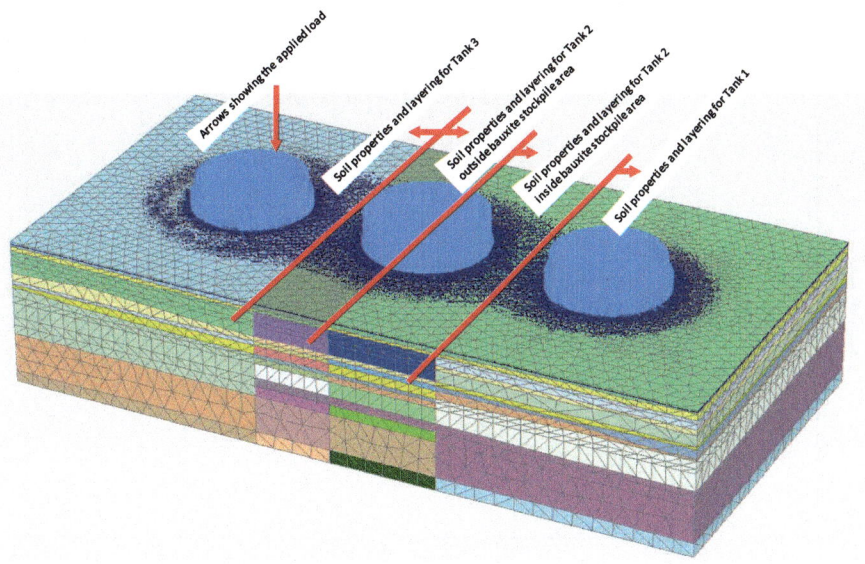

Fig. 4. Three-dimensional FEM model of three tanks

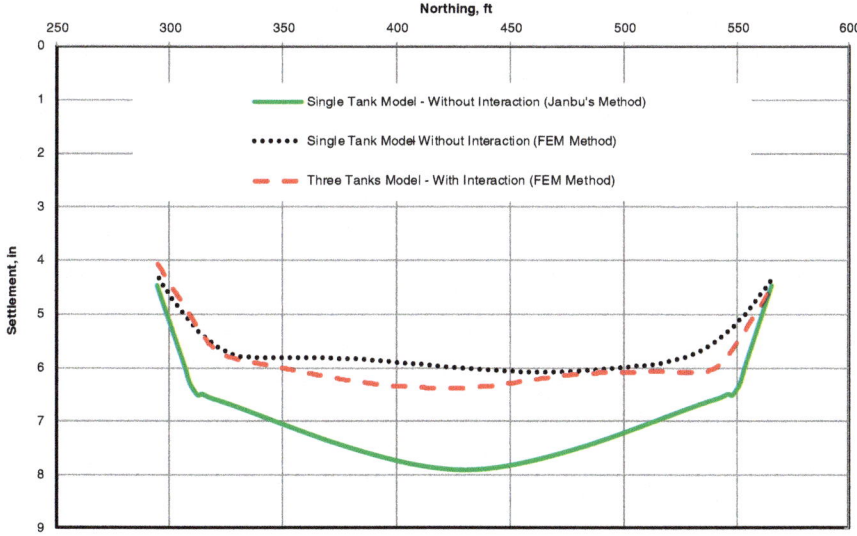

Fig. 5. Settlement of Tank 1 (right in Fig. 1) after hydrotest

The settlement profiles obtained from comparison of the axisymmetric FEM model and 3D FEM model for Tanks 1 and 3 indicate that the tank interaction effect is less than ½ in. (5% in Figs. 5 and 7) of increased settlement under hydrotest loading. Tank interaction effects are more pronounced for Tank 2 as evidenced in the settlement

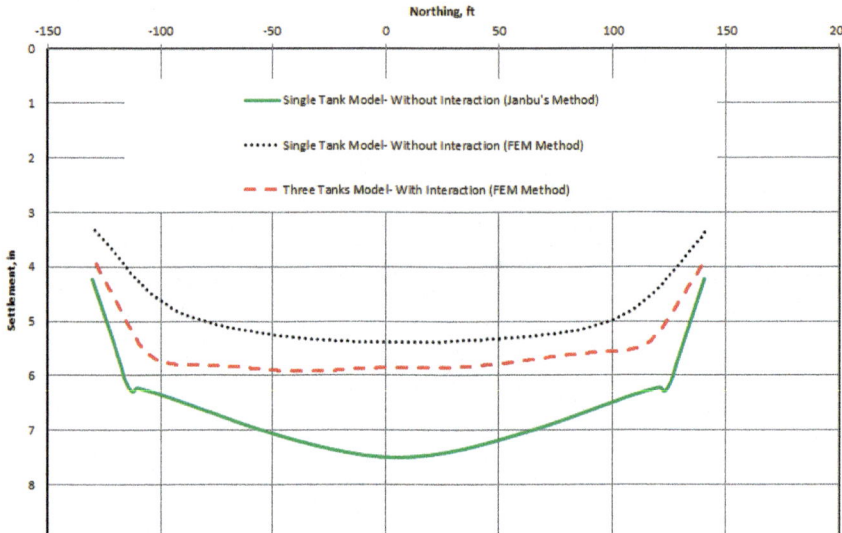

Fig. 6. Settlement of Tank 2 (middle in Fig. 1) after hydrotest

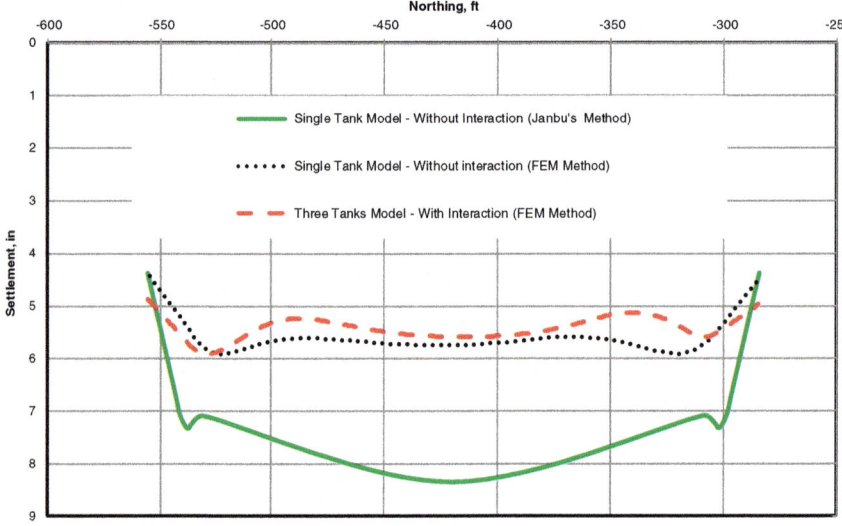

Fig. 7. Settlement of Tank 3 (left in Fig. 1) after hydrotest

profile plotted in Fig. 6. Because of the overlapping stress bulbs from Tanks 1 and 3, Tank 2 may experience ½ in. more settlement during hydrotesting compared to a "Single Tank Model" (no tank interaction). The increased settlements due to tank interactions are approximately 5% to 10% for these tanks under hydrotest loading.

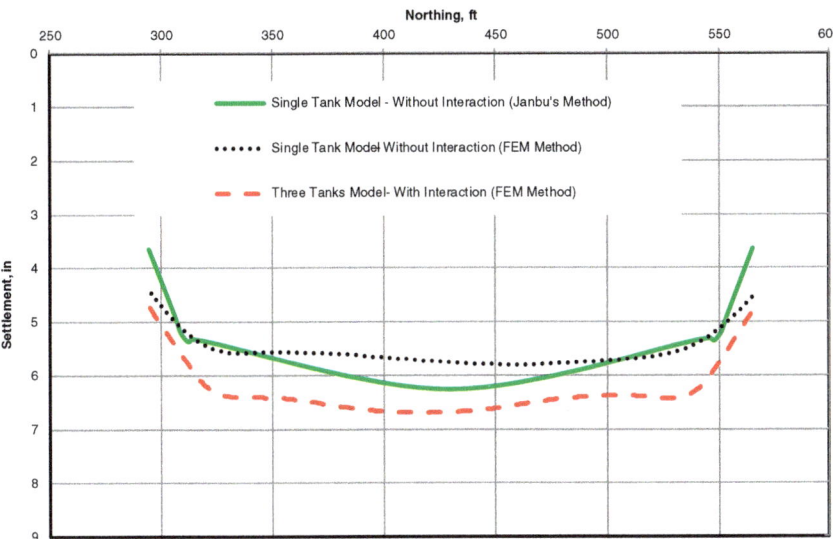

Fig. 8. Settlement of Tank 1 (right in Fig. 1) after 20-year production load

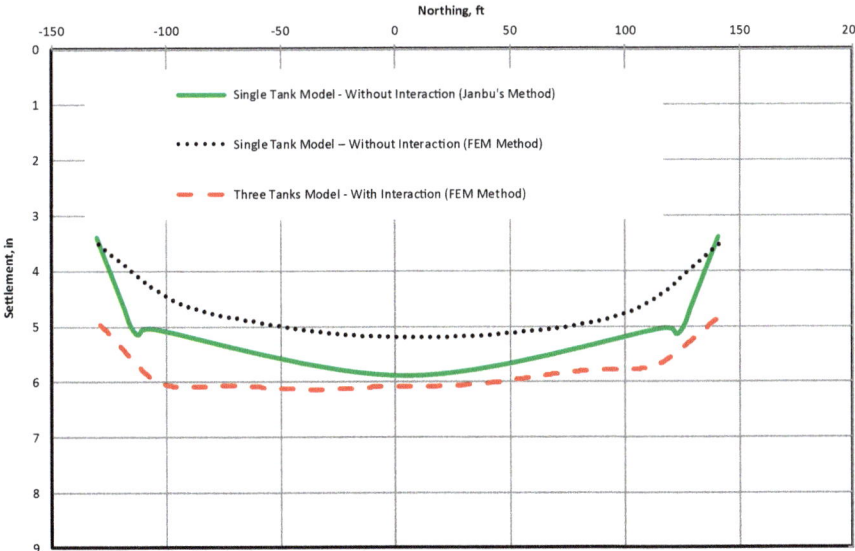

Fig. 9. Settlement of Tank 2 (middle in Fig. 1) after 20-year production load

During product loading, three-dimensional FEM analysis provided the maximum tank settlements (Figs. 8, 9 and 10). The settlement profiles obtained from comparison of the axisymmetric FEM model and the 3D FEM model for Tanks 1 and 3 indicated that interaction effects can add up to 1 in. of additional settlement (Figs. 8 and 10), or about 15% increase in settlement during product loading. Similar to the hydrotest

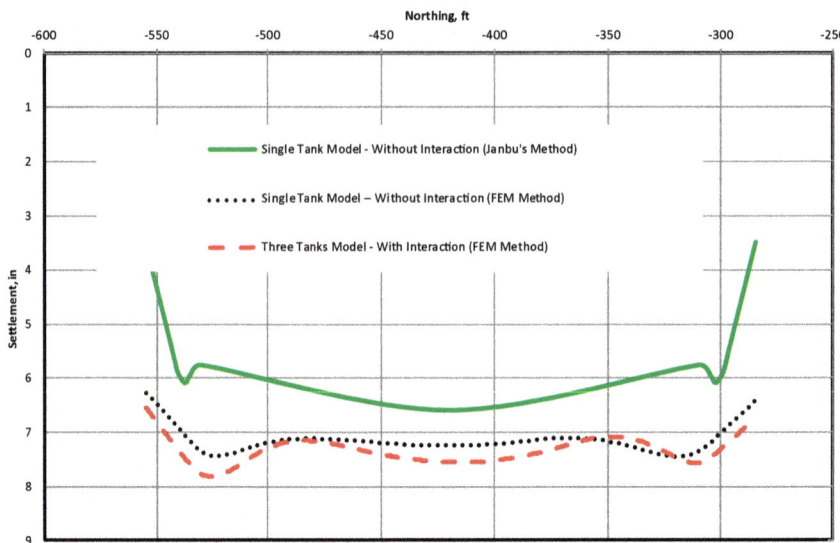

Fig. 10. Settlement of Tank 3 (left in Fig. 1) after 20-year production load

settlement results, tank interaction effects during product loading is more pronounced for Tank 2 as evidenced in the settlement profile plotted in Fig. 9. Because of the overlapping stress bulbs from Tanks 1 and 3, Tank 2 may experience up to 1½ in. more settlement compared to the corresponding "Single Tank Model". The increased settlement due to tank interactions was approximately 30% during product loading.

Three-dimensional analysis with interaction is the appropriate approach for design where tank stress fields can act upon one another, and where spatial variations of soil properties in the horizontal and vertical directions can be modelled. The results indicated that the tank interaction effect during hydrotesting is fairly insignificant. However, tank interaction effect can add up to 1½ in. settlement during the design life of Tank 2. Settlement in sand and elastic settlement in clay may be complete during hydrotesting but only a fraction of consolidation settlement can occur. On the other hand, contribution of consolidation settlement is more prominent during product loading which causes increased settlement when interaction effect is considered.

The maximum differential settlement between a tank center and its outer shell occurred for Tank 1 during hydrotesting with a 2.3-in. differential, which is less than the maximum allowable of 5.4 in. (1:300 of 135.5 ft tank radius). The maximum tilt calculated across a tank diameter occurred for Tanks 1 and 3 during hydrotesting, and Tank 3 after 20 years of product operation, of about 1×10^{-4} radians, which is well below the allowable amount of 2×10^{-2} radians. The maximum differential settlement around a tank perimeter was on the order of one to two inches, which is well below the allowable slope of 1:500. In summary, the 3D FEM analyses results indicated all three tanks satisfy the FERC requirements for allowable deformation and distortion. The evaluation of tank tilt and differential settlement around a tank perimeter would not be possible without performing the three-dimensional analysis of tank foundation.

5   Conclusions

Three different approaches adopted in a project for settlement estimations of raft foundations for LNG tanks were presented and the results from these different methods were compared. The following conclusions are drawn from the analysis results:

1. One-dimensional analysis can be used for estimating the range of settlements during the initial design of tanks. The settlement from this simplified approach can be approximately 35% higher than the respective estimates from three-dimensional analyses. The one-dimensional analysis can not be relied for the final design since the stiffness of the foundation, subsurface variations in horizontal plane, and the effect of tank interaction can not be accurately modelled. However, if the 1-D analyses show very little settlement, there may not be a need for conducting more sophisticated analyses.
2. The comparison of settlement profiles from axisymmetric FEM model and three-dimensional FEM model indicated that the tank interaction effect during hydrotesting is insignificant approaching 5% to 10% of increased settlement when tank interactions are considered.
3. The comparison of settlement profiles between axisymmetric FEM model and three-dimensional FEM model indicated that the total settlements during design life can increase by 15% to 30% when tank interactions are considered for the sub-surface conditions encountered at the project site and the loading conditions used in the design.
4. Subsurface variation in spatial and vertical directions can be effectively modelled using three-dimensional analyses. A three-dimensional finite element model can account for the stress interaction of the tanks on one another during construction, hydrotesting, and product operation.
5. In order to effectively incorporate the variations in subsurface conditions and use them in three-dimensional FEM model, a well-planned geotechnical investigation program, coupled with sufficient advanced laboratory tests, including consolidation and triaxial strength tests, are required.

The LNG tanks discussed in this paper were at various stages of planning and construction at the time of this paper preparation. The settlement of the tanks will be monitored and will be published in the future. Though the methodology discussed in this paper are applicable for different foundation designs (LNG tanks or other foundations), the conclusions derived are based on the specific problem detailed in this paper. These conclusions may not be valid for subsurface conditions other than those described in this paper.

Acknowledgement. The authors would like to express their sincere appreciation to Bechtel Oil, Gas, and Chemicals for permitting authors to share this paper.

References

API Standard 650: Welded Tanks for Oil Storage, Twelfth Edition. American Petroleum Institute (2013)

D'Orazio, T.B., Duncan, J.M.: Differential settlement in steel tank. J. Geotech. Eng. ASCE **113** (9), 967–983 (1987)

Fellenius, B.H., Ochoa, M.: Large liquid storage tanks on piled foundations. In: International Conference on Foundation and Soft Ground Engineering – Challenges in the Mekong Delta, Ho Chi Minh City, Vietnam, pp. 3–17 (2013)

FERC (Federal Energy Regulatory Commission): Draft Seismic Design Guidelines and Data Submittal Requirements for LNG Facilities. FERC, Washington D.C. (2007)

Janbu, N.: Settlement calculations based on the tangent modulus concept. University of Trondheim, Norwegian Institute of Technology, Geotechnical Institution, Bulletin 2 (1967)

NFPA 59A: Standard for the Production, Storage, and Handling of Liquefied Natural Gas (LNG). National Fire Protection Association (2016)

PLAXIS 2D: Plaxis Computer Program, Version 10.1 manual. A.A. Balkema Publishers (2010)

PLAXIS 3D: Plaxis Computer Program, Version 13 manual. A.A. Balkema Publishers (2013)

Skempton, A.W., Bjerrum, L.: A contribution to the settlement analysis of foundation on clay. Geotechnique **7**(4), 168–178 (1957)

Shafts and Deep Foundations

Centrifugal Model Tests on the Bearing Capacity of Piles with Basement Construction Under the Existing Building

Bing Ma[1], Jing-yi Zhang[2], and Fa-yun Liang[1(✉)]

[1] Department of Geotechnical Engineering,
Tongji University, Shanghai 200092, China
fyliang@tongji.edu.cn
[2] Sichuan Institute of Building Research, Chengdu 610081, China

Abstract. It is a reasonable choice to solve the problem of the limited space and the contradiction of land use in the urban areas by basement construction beneath the existing building. The excavation beneath the existing building will change the stress state of the soil around pile foundation and thus affect the post-excavation pile foundation behavior. A series of centrifuge tests are conducted to study the influence of basement excavation on the bearing capacity of single pile under typical geological conditions in Chengdu, Sichuan. Modelling the existing building by loading on the top of the pile foundation, the relation between different excavation depths and pile foundation characteristics is studied, including ultimate bearing capacity, the axial force and shaft resistance of piles. The test results show that excavation beneath existing building has great influence on the ultimate bearing capacity of both single pile and pile groups. The ultimate bearing capacity of single pile decreased by 11.8% after 4 m excavation of a 20 m long pile, while 35.3% after 8 m excavation. With the increase of excavation depth, the axial force and shaft resistance of pile beneath the excavation face gradually develop. The test results can provide some reference for the design of excavation construction beneath the existing building.

Keywords: Basement construction · Existing building · Excavation depth
Pile foundation behavior · Centrifuge test

1 Introduction

Nowadays, the car parking problem in big cities is becoming more and more serious. Urban land is too tense and expensive to add more ground parking, which results in a waste of land. In recent years, a large number of non-basement buildings have been built all over the country, and the demolition of these buildings is costly. So how to exploit the underground space without affecting the normal use of this group of buildings, and meet parking and commercial demand, has become a major issue to solve for the engineers.

Basement construction beneath the existing building method has been proved to be one of the best choices to solve the problem efficiently and economically, which creates no increase in the construction area and costs less time. This technology is of high risk

© Springer Nature Singapore Pte Ltd. 2018
T. Qiu et al. (Eds.): GSIC 2018, *Proceedings of GeoShanghai 2018 International Conference: Advances in Soil Dynamics and Foundation Engineering*, pp. 621–628, 2018.
https://doi.org/10.1007/978-981-13-0131-5_67

because involves the complex interaction between the existing structure and the new basement. Both the uneven settlement of structure of and soil deformation should be strictly controlled in the process of underpinning to decrease the effect of the function of the upper structure.

The overall level of reconstruction and underpinning technology in China is not high enough at present. There are few engineering examples of the excavation and basement construction beneath existing buildings. Few researches are conducted, and the specific engineering data is scarce. Li and Huang (2010) systematically studied the redesign of pile foundation in basement construction of existing building by model test. Li and Huang (2010) analysis the difference of Q-S curves, axial force and limit of uplift piles between before and after the excavation by centrifugal model test. Hu et al. (2013) performed centrifuge model tests to study the influence of deep excavation on the bearing capacity of uplift piles. Zheng et al. (2016) studied the effect of foundation excavation on the bearing capacity of the low friction piles and high friction piles.

This paper mainly discusses the relationship between excavation depth and the mechanism of loading characteristics of single pile during the excavation under the typical geological subsurface of Chengdu, China by the centrifuge model test. The experimental procedure and the main results of a series of centrifuge tests are shown.

2   Centrifuge Model Test

2.1   Design of Model Tests

The geotechnical centrifuge of Tongji University used for the test has a single arm with the nominal radius of 3 m. The strongbox has dimensions of 700 mm in width, 900 mm in length and 700 mm in height. The acceleration in the test is 50 g. Five pile model tests were conducted.

The centrifuge model tests were conducted to study the relationship between excavation depth and the mechanism of loading characteristics of single pile, regarding the boundary conditions and neglecting the interaction of the piles based on relevant standards. The excavation depth included three kinds: 0 cm, 8 cm and 16 cm. Preloading was used to simulate the load of the upper structure in the tests. The distance between two adjacent piles in the tests was shown in Fig. 1. Pile S1 was used to simulate the bearing capacity test and pile T0 was a preliminary test to make sure the loading by the motor was steady. T1–T3 were the main tests (Table 1).

2.2   Soil Property and Pile

According to the geological and hydrogeological background in Chengdu, the soil layers from the top downward are miscellaneous fill, silt sand layer, fine sand layer and cobble layer. Due to the fact that the thickness of soil and silt was small, this test only simulated silt sand layer and cobble layer. According to the distribution characteristics of field soil and the particle size effect, sand was considered to be the simulation of undisturbed soil in centrifugal test. The distribution of soil layers from the top to bottom were: 0–16 cm fine sand, 16–60 cm medium sand. The medium sand particle

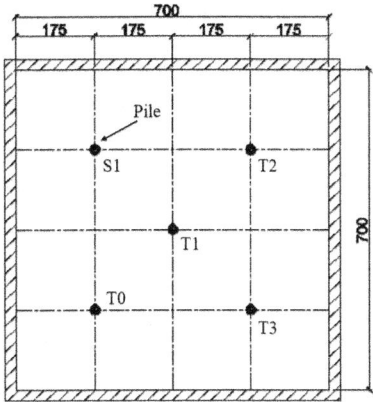

Fig. 1. Plan view of model tests (Units: mm)

Table 1. Test program.

Test number	Model pile	Excavation depth	Simulated test objective
1	S1	0 mm	Ultimate bearing capacity test
2	T0	0 mm	Preliminary test
3	T1	0 mm	Test without excavation
4	T2	80 mm	Test after excavation depth of 4 m
5	T3	160 mm	Test after excavation depth of 8 m

size was more than 50% in the range of 0.25–1 mm, namely the simulated soil particle size 12.5 mm–50 mm accounted for more than 50%. Because the particle size of the top fill was small, it was simulated by fine sand with particle size of 0.075 mm–0.25 mm. Grain size distribution of the soils was shown in Fig. 2. The groundwater level was controlled by 16 cm.

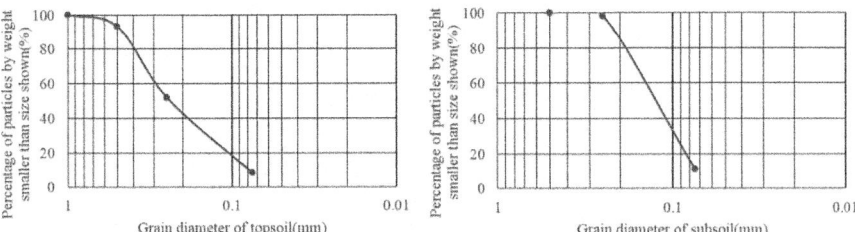

Fig. 2. Grain size distribution of the soils

Taking pile production and simulation into account in the centrifuge test model, aluminum alloy 6061 was selected as model pile material, and its elastic modulus was 70 GPa. Based on the compressive stiffness, the inner diameter (ID) of test pile was

12 mm, and the outside diameter (OD) was 16 mm, which could simulate with pile 20 m in length and 800 mm in diameter under the condition of 50 g. Specific parameters are shown in Table 2.

Table 2. Geometric parameter of model piles and prototype

Pile	Prototype	Model piles	Soil	Dry density/(kg·m^{-3})	Water content	Dr
Diameter	800 mm	ID: 12 mm	Topsoil	1610	0%	78%
		OD: 16 mm				
Length	20 m	400 mm	Subsoil	1560	13%	76%
L/D	25	25				

2.3   Instrumentation

Strain gauges were installed on the piles symmetrically beforehand to measure the axial forces along the piles. Figure 3 shows the dimensions of the 1/50 scaled-down model and strain gauges layout. A bracket was fixed on the walls of the model box, and the frequency modulation motor was set up on it. The load of the pile top was applied by the motor, and a force sensor was set between the motor and the pile so as to measure the vertical axial force.

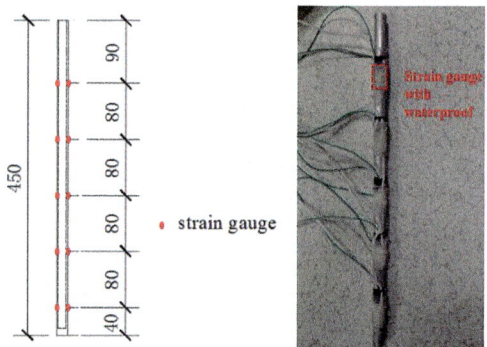

Fig. 3. Arrangement of strain gauges of piles

2.4   Testing Procedure

After calibrating the instrumentation, the following test procedure in single pile tests was adopted:

1. After being dried and sieved, the sand was added into test box with baffle, which divided the box into two parts. One was 700 mm × 700 mm for the test zone; the other was 700 mm × 200 mm for water control. The sand was laid by each level of 8 cm and then paved and compacted with a leveling bar. During sand deposition, the model piles were deployed in the model vertically to the determined position.

2. After preparation, the models were loaded on the centrifuge basket, with the devices connected to the onboard acquisition system.
3. Pre-consolidation was carried out under the acceleration of 50 g for 60 min, simulating consolidation for 3 months. To simulate the existing building, preloading the piles by half of the ultimate bearing capacity for 5 min and then loaded the pile to failure again. The ultimate bearing capacity of the original single pile was tested.
4. The excavation depth before loading in T2 was 8 cm and 16 cm in T3.

2.5   Test Results and Analyses

The effects of excavation depths on the bearing performance of piles were explored in this study. Note that all the test results, including the bearing capacity, the axial force and the shaft resistance, have been transformed into the prototype values through the similarity ratio.

Q-S **Curve.** The load and displacement of the pile top were monitored during the centrifuge test. Figure 4 shows the load-settlement curves of S1 and T0. There are no obvious inflection points on the *Q-S* curves which mean the pile was the end-support friction pile. Taking the load when the total settlement was 0.05D (i.e.40 mm) as the ultimate bearing capacity, it can be found that the ultimate bearing capacity was about 3300 kN. The *Q-S* curve in Fig. 4 shows the ultimate bearing capacity of T0 was almost the same as the bearing capacity of S1. Moreover, it can be concluded that the existing load has little influence on the bearing capacity of piles in sand.

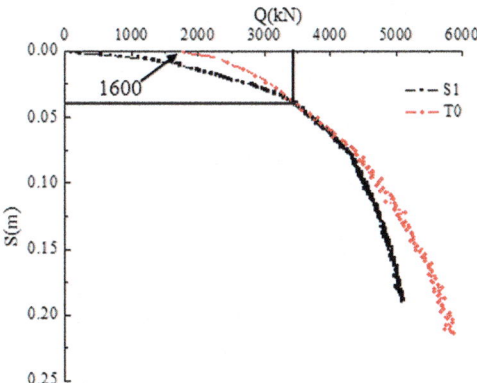

Fig. 4. *Q-S* curve in preliminary test

As is shown in Fig. 5, the Q-S curves have little difference between T1 and T2 in the initial loading stage. But as the load increased gradually, the difference between the Q-S curves is obvious. The Q-S curve of simulating 4 m in excavation depth beneath existing building shows that the ultimate bearing capacity of T2 decreased to 3000 kN after the excavation of 4 m. The Q-S curve of simulating 8 m in excavation depth

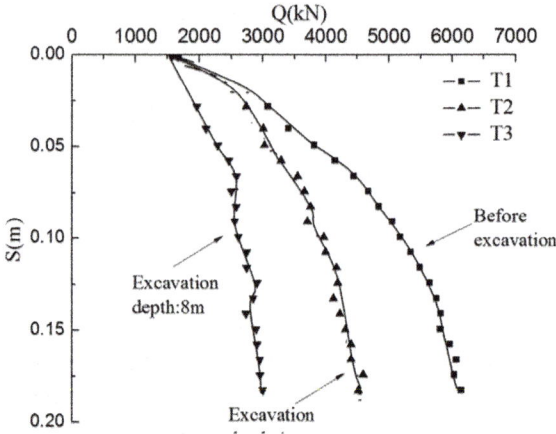

Fig. 5. *Q-s* curves of single pile before and after excavation

beneath existing building dropped steeply, when the load exceeded 2200 kN, which means the pile reaches the limit state. As a result, the ultimate bearing capacity dropped by 35.3%.

Axial Forces and Shaft Resistance. The axial forces of different depth with time was monitored, and the axial forces and shaft resistance in various stages was shown in Figs. 6 and 7. It shows that the axial force of the single pile was mainly borne by the upper part of the pile at the initial loading stage. The axial force of the pile gradually decreases along the length of the pile under the same loading. As the load increased gradually, the resistance of pile end was continuously played. The axial force of upper parts changed larger than the lower parts, meaning the side friction of upper pile developed gradually. But under the condition of small load, the lateral friction of the

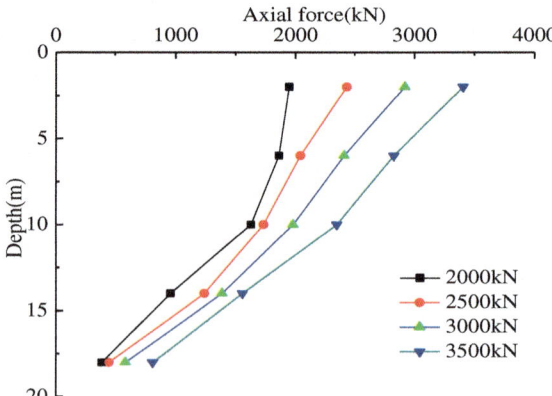

Fig. 6. Variation of axial force with loading before excavation

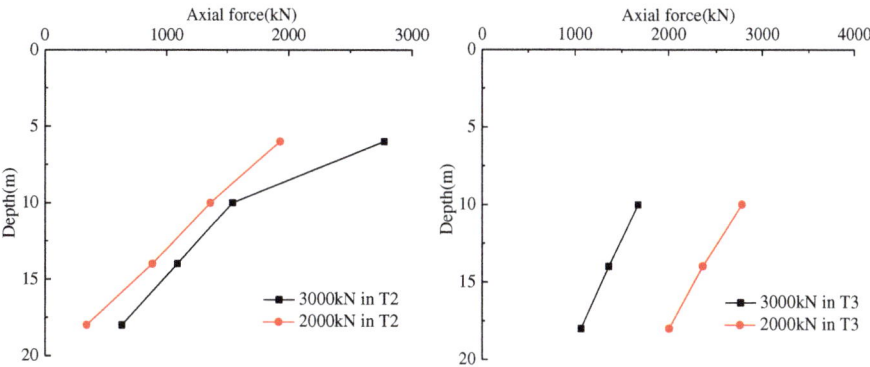

Fig. 7. Variation of axial force with loading after excavation

upper part of the soil worked out first, while the pile side friction of the lower part played with the load gradually as the slope of the upper part is almost the same.

Variation of axial force with loading after excavation was shown in Fig. 7. Due to the over consolidation state of the soil near the excavation face after excavation, axial force varied more near the excavation face, so the shaft resistance increased more. With the increase of excavation depth, the lateral friction of the pile body in the middle and lower part of the pile increased accordingly. At the same time, the pile tip resistance increased correspondingly, sharing part of the load.

3 Conclusion

A series of centrifuge tests were conducted to study the influence of basement excavation on the bearing performance of single pile. Excavation beneath existing building has great influence on the ultimate bearing capacity. The ultimate bearing capacity of single pile is decreased by 11.8% and 35.3% after excavation depth of 4 m and 8 m, respectively. The excavation depth also has significant effect on the axial forces and shaft resistance. With the increase of excavation depth, axial force varied faster near the excavation face, and the lateral friction of the pile in the middle and lower parts of the pile increased accordingly. At the same time, the pile tip resistance increased correspondingly, sharing part of the load.

Acknowledgements. The work reported herein was supported by the Science and Technology Program of Sichuan Province (Grant No. 2017JZ0035), and the Fundamental Research Funds for the Central Universities of China. The above financial support is gratefully acknowledged.

References

Li, J.J., Huang, M.S.: Centrifugal model tests on bearing capacity of uplift piles under deep excavation. Chin. J. Geotech. Eng. **32**(3), 388–396 (2010)

Zheng, G., Zhang, X., Diao, Y.: Experimental study on the performance of compensation grouting in structured soil. Geomech. Eng. **10**(3), 335–355 (2016)

Hu, Q.F., Ling, D.S., Kong, L.G.: Effects of deep excavation on uplift capacity of piles by centrifuge tests. Chin. J. Geotech. Eng. **35**(6), 1076–1083 (2013)

Analysis of Bearing Capacity of Jacked PHC Pipe Piles in Double-Layered Ground Based on Data Mining

Qingqing Zhang[1], Jun Yang[1,2(✉)], and Yan Sun[3]

[1] Guangdong Hualu Transport Technology Co., Ltd., Guangzhou 510420, China
yangjun851113@163.com
[2] Department of Geotechnical Engineering, Tongji University,
Shanghai 200092, People's Republic of China
[3] China Agricultural Engineering Construction Center, Ministry of Agriculture,
Beijing 100081, People's Republic of China

Abstract. In this paper two data-mining technologies based on different calculation principles are used to evaluate correlation between ultimate bearing capacity of jacked PHC pipe pile and its impact factors from geological investigation, pile foundation construction and in situ static load test. Major factors closely related to bearing capacity are determined and regression equation for predicting ultimate bearing capacity is derived. The result shows that data-mining technology has very attractive capabilities to carry out relevance analysis for bearing behavior of single pile. In double-layered ground as Shenyang city, ultimate bearing capacity of PHC pipe pile is mainly provided by pile tip resistance and closely related to final pressure in pile construction. The method presented can be used to evaluate ultimate bearing capacity of PHC pipe pile in similar ground and the accuracy can fill the requirements in practice.

Keywords: Jacked PHC pipe pile · Data mining · Ultimate bearing capacity
Impact factor · Relevance analysis

1 Introduction

Either vertical or horizontal ultimate bearing capacity of single pile has been affected by various factors from pile shaft and soil around pile. Because there are many uncertainties among these factors, it is difficult to determine which factors have obvious influence on bearing capacity. Jacked high strength prestressed concrete pipe pile (or PHC pipe pile) is one kind of pile foundation constructed by static pressured machine. It has been widely applied in China for more than twenty years for many advantages, such as higher construction speed, reliable quality and non-environmental pollution. Many researchers have studied the bearing mechanism of this pile foundation (Zhu et al. (2004), Zhao et al. (2005), (2008), Xie et al. (2009) and Kou et al. (2013)). However, it is difficult to find a method to quantify the relationship between ultimate bearing capacity and its impact factors comprehensively.

© Springer Nature Singapore Pte Ltd. 2018
T. Qiu et al. (Eds.): GSIC 2018, *Proceedings of GeoShanghai 2018 International Conference: Advances in Soil Dynamics and Foundation Engineering*, pp. 629–636, 2018.
https://doi.org/10.1007/978-981-13-0131-5_68

Data Mining (DM) is a nontrivial process which extracts connotative, unknown and valuable messages from database including mass data. A number of data mining technologies, such as artificial neural network, support vector machine, and genetic algorithm etc., have been successfully applied to predict the bearing capacity of pile foundation. Das and Basudhar (2006) used neural network model to analyze undrained bearing capacity of laterally loaded piles in clay and carried out sensitivity analysis on input parameters. Pooya et al. (2009) established neural network model predicting pile settlement based on large standard penetration test results and discussed selection of middle layer parameters. Zhao et al. (2010) established analysis model to predict horizontal bearing capacity of single PHC pipe piles in double-layered ground by combining grey system theory with artificial neural network, and evaluated relevance of various factors affect bearing capacity. Kordjazi et al. (2014) established a support vector machine model based on field CPT-values for predicting ultimate bearing capacity of single pile. However, most methods are based on different mathematical principles and have various adaptabilities to variety of input specimens. It brings great difficulty to understand the true working mechanism of pile foundation only through one method.

Hence, in this paper two data-mining technologies based on different calculation principles, grey correlation method (Liu 2008) and partial least squares regression method (Wang 1999), are used to evaluate correlation between vertical ultimate bearing capacity of PHC pipe pile and the impact factors in typical double-layered ground, based on data from geological investigation, pile foundation construction and static load tests. Consistence of two methods is compared and major factors affecting bearing capacity of pile are determined. The regression formula for predicting ultimate bearing capacity is proposed and the precision is evaluated with measured values.

2 Calculation Method

2.1 Grey Correlation Method

Grey correlation method comes from grey system theory, which solves shortages of traditional mathematical statistics method and uncertainty system analysis, such as limited number of specimen and poor information (Liu 2008). It is suitable for situations including less specimen number and no distinct regularity. Calculated correlation values from grey correlation analysis represent contribution of various variables to bearing capacity of single pile. Major process adopted can be seen in Liu (2008).

2.2 Partial Least Squares Regression Method

Partial least squares regression (PLSR) method is used to extract the aggregate variables with strongest explanatory to system and get rid of disturbance from multiple correlation information or irrelevant information through information decomposition and filtering, overcoming adverse effects of multiple correlation in system modelling. The procedures of data reduction, variable correlation analysis and regression modeling analysis can be realized at the same time. VIP analysis in PLSR method is to evaluate the effects of various variables on dependent variables. Detailed modelling process of VIP analysis is seen in Wang (1999).

3 Impact Factor and Relevance Analysis

Surface soil layer (thickness less than 2.0 m) is backfill or cropland, which has complex composition and poor mechanical properties. For ease of analysis, contribution of this soil layer to pile capacity is neglected. As shown in Fig. 1, the soil layer around pile shaft is mainly interlayered silty clay and clay (thickness: 13.0–18.0 m). The bearing stratum is medium dense to dense coarse sand (occasionally mixed with silty clay), which has higher bearing capacity and larger thickness. The penetrating depth of pile tip is generally more than 1.0 m or $3d$ (d is external diameter of pipe pile).

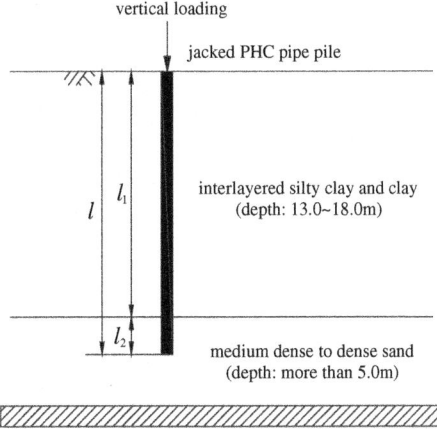

Fig. 1. Double-layered soil profile

Field tests and model experiments of open-ended pipe pile (De Nicola et al. 1999; Yetginer et al. 2006) indicated that soil profile and soil plug type have significant effects on ultimate bearing capacity of pipe pile. For open-ended PHC pipe pile with smaller inner diameter, entire plugging will appear due to soil compaction and friction effects in pile hole when passing through sand layer, which presents almost the same bearing properties as closed-end pipe piles. So for the PHC pipe pile with smaller external diameter (less than 350 mm) and pile tip penetrating into better bearing stratum, it is reasonable to treat it as wholly closed-ended pipe piles.

As the suggestion of Liu et al. (1999), the impact factors of bearing capacity of pile mainly consist of four aspects: soil parameters from pile side and pile tip, geometric parameters of pile and parameters from pile construction. According to the situation in this paper, the factors affect the ultimate bearing capacity of PHC pipe pile can be divided into three parts:

(1) Parameters related to side resistance: moisture content w, liquidity index I_L, void ratio e, cohesion c and internal friction angle φ of interlayered silty clay and clay;

(2) Parameters related to tip resistance: standard penetration test counts N and final pressure of static pile driver P_u in sand layer;

(3) Geometric parameters of pile: depth that pile tip penetrating into sand stratum l_2, pile length to diameter ratio l/d and pile length through silty clay and clay layer l_1.

It should be noted that soil parameters of pile side are weighted average considering the thickness of soil layer. For bearing stratum of pile tip, SPT-N value is selected from soil around pile tip position.

Data from site geotechnical investigation, pile construction and static load tests of 22 PHC pipe piles selected in northern region of Shenyang City is treated as data specimen in this study, which is presented in Table 1. Through two relevance analysis the construction degree of each impact factor to ultimate bearing capacity of PHC pipe pile is determined.

Table 1. Arrangement of original data

No.	$w(\%)$	e	c(kPa)	$\varphi(°)$	I_L	l_1(m)	l_2(m)	l/d	N	P_u(kN)	Q_u(kN)
1	25.42	0.71	27.61	24.14	43.15	15.0	2.0	42.5	38	2750	2100
2	27.86	0.77	55.57	20.03	64.65	15.3	1.7	42.5	42	3250	2700
3	27.92	0.82	39.81	23.15	55.23	13.6	2.4	40.0	35	2000	2000
4	27.69	0.88	26.32	14.12	58.04	13.9	3.1	42.5	48	3800	3800
5	25.29	0.80	18.62	13.31	68.22	14.4	4.4	62.6	27	2750	1800
6	30.19	0.90	28.2	13.22	64.22	13.6	4.8	61.3	24	1750	1600
7	26.31	0.84	22.64	11.23	67.63	13.8	4.7	61.6	15	1750	1600
8	26.81	0.79	17.22	8.59	68.35	13.4	2.6	40.0	40	2500	2400
9	32.23	0.91	25.16	10.94	69.59	13.8	2.2	40.0	37	2700	2200
10	24.76	0.76	15.57	7.44	62.62	14.4	5.6	50.0	45	2750	2700
11	36.85	0.98	33.05	13.87	71.82	12.8	4.2	42.5	35	2550	2000
12	34.66	0.93	29.69	12.13	73.18	14.6	2.4	42.5	30	2750	1850
13	27.45	0.89	24.64	12.94	64.79	17.5	2.5	50.0	36	2500	2400
14	28.62	0.72	19.82	9.13	94.96	17.0	1.3	45.7	47	3375	2750
15	26.32	0.87	23.61	7.89	47.15	17.9	1.4	48.2	54	3250	3000
16	25.31	0.77	18.47	10.31	53.24	17.0	1.3	45.7	45	3250	2700
17	30.78	0.96	29.50	12.76	44.77	17.5	1.0	46.2	48	3500	2850
18	32.95	0.95	25.44	15.94	79.42	17.6	0.6	45.5	49	3500	3200
19	28.41	0.81	18.82	9.45	72.14	14.0	2.0	40.0	28	2500	1755
20	29.53	0.86	47.86	20.24	64.30	14.8	1.4	40.5	42	2750	2475
21	42.09	1.07	24.10	13.16	82.81	13.2	3.8	42.5	46	2875	2445
22	30.33	0.86	21.95	10.01	70.37	13.8	1.2	37.5	43	2250	2550

Note: Q_u is measured ultimate bearing capacity of single PHC pipe pile.

From Table 2, it is illustrated that mechanical parameters (N and P_u) of bearing soil stratum have closest relationship with ultimate bearing capacity in this geological condition. The geometric parameters of pile (l_1 and l/d) have impacts on ultimate bearing capacity to some extent. Soil parameters around pile side (I_L, w, c and φ) have very limited effect on bearing capacity. In other words, ultimate bearing capacity of PHC pipe pile is mainly provided by pile tip resistance and closely related to final pressure in construction process, which is significantly different from engineering experiences of similar pile type in soft ground. Hence, jacked PHC pipe pile with medium pile length in this ground is typically frictional end bearing pile. It is important to pay enough attention to final pressure in pile construction, which is of great significance to guarantee construction quality, determine pile capacity and pile foundation design optimization.

Table 2. Results of relevance analysis

Variable	Grey correlation value	Variable	VIP-value
N	0.871441	N ($X1$)	2.06084
P_u	0.833497	P_u ($X2$)	1.94743
l_1	0.809803	l_2 ($X3$)	1.29509
e	0.786542	l_1 ($X4$)	1.22764
w	0.773218	l/d ($X5$)	0.810846
l/d	0.769565	E ($X6$)	0.493988
c	0.68533	I_L ($X7$)	0.260477
I_L	0.642293	φ ($X8$)	0.238154
φ	0.598807	w ($X9$)	0.152084
l_2	0.59433	c ($X10$)	0.134328

Meanwhile, there are still certain discrepancies between results from two methods. In grey correlation method, l_2 has minimum contribution but w and e to bearing capacity is relatively higher, which is contrary to projection sorting results from PLSR method. It is because the interaction between pile side soil and pile shaft reflects the collective effect of various parameters of pile side soil layer, which is more complex than pile tip position. Relationship among soil parameters of pile side and bearing capacity do not demonstrate distinct regularity. Hence, it is more reasonable to evaluate correlation using two methods simultaneously.

4   Derivation and Verification of Regression Equation

Based on principles of two methods, later five variables ($X6$–$X10$) and l_2 are excluded from regression analysis. Four parameters (N, P_u, l_1 and l/d) are selected to establish regression equation. In VIP analysis, 22 specimens are selected in regression modelling. The relationship between principal component t_1 (represents various impact factors) and u_1 (represents ultimate bearing capacity of single pile) is shown in Fig. 2.

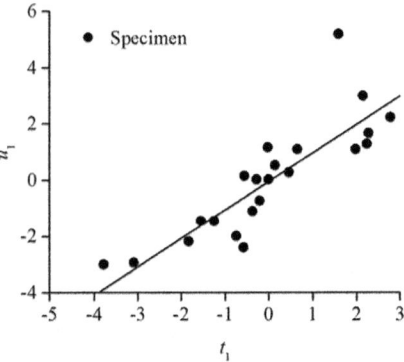

Fig. 2. Relationship of t_1 and u_1

Majority of specimens are close to a straight line. It indicates that a certain correlation does exist between independent and dependent variables. In other words, vertical bearing capacity of single PHC pipe pile does have a linear relationship with selected parameters (N, P_u, l_1 and l/d). Hence, it is reasonable to establish the relationship of the partial least squares regression between them. The prediction of bearing capacity of PHC pipe pile can be regarded as the problem of multi-independent variables to single dependent variable regression modelling. The detailed deduction of regression equation can be presented by flow chart as in Fig. 3.

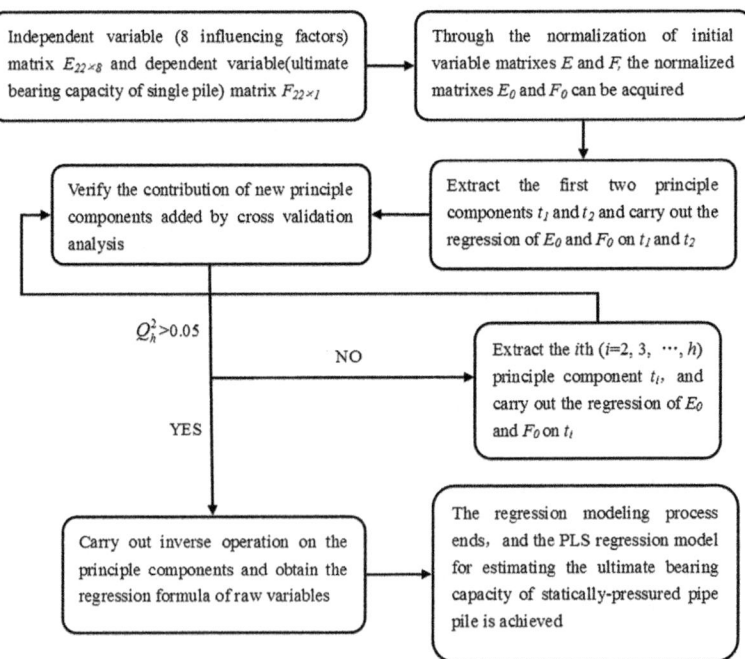

Fig. 3. Flow chart of modelling of regression equation

The regression equation for bearing capacity of single PHC pipe pile is obtained in double-layered ground, that is,

$$Y = [(\frac{X4 - 14.9}{1.6}) \times 0.03 + (\frac{X1 - 38.7}{9.1}) \times 0.51 + (\frac{X5 - 45.6}{7.1}) \times (-0.03)$$
$$+ (\frac{X2 - 2829}{561}) \times 0.47] \times 552 + 2403 \tag{1}$$

To verify the rationality of proposed equation, the calculated results are compared with measured values of other 8 PHC pipe piles under similar stratum. From Table 3, the predicted results and measured values have better consistence and maximum error is less than 20%. The calculation accuracy is satisfactory in practice. It is validated that the regression equation can be used to preliminarily predict bearing capacity of jacked PHC pipe pile in similar stratum.

Table 3. Comparison of measured and predicted bearing capacities

No.	l_1(X4)	N(X1)	l/d(X5)	P_u(X2)	Q_{meas}(kN)	Q_{pred}(kN)	Errors(%)
1	14.5	32	35	4250	3000	2874	−4.2
2	14.5	34	37	3650	2800	2653	−5.2
3	13.6	28	42	3800	2650	2516	−5.1
4	13.5	19	33	3100	2340	1934	−17.4
5	15.2	24	42	3400	2400	2223	−7.3
6	15.8	38	45	3850	2550	2864	12.3
7	14	19.3	35	3550	2400	2379	−1.0
8	13	30	36	3250	2250	2480	10.2

Note: Q_{meas} and Q_{pred} are measured and predicted ultimate bearing capacity of jacked PHC pipe pile respectively.

5 Conclusion

(1) Data mining technology (grey correlation analysis and partial least squares regression method) has attractive capabilities and advantages to analyze correlation of parameters influencing on ultimate bearing capacity of jacked PHC pipe piles in double-layered soil. Disadvantages, such as shortage of specimen number, multiple correlations among various parameters and minor factors disturb, can be overcome.

(2) For this study, standard penetration test blow count (N) of bearing sand stratum around pile tip and final pressure in pile construction(P_u) have closest relation with ultimate bearing capacity of jacked PHC pipe pile. In ultimate bearing state, pile tip resistance will develop adequately and ratio of pile side resistance to total resistance will reduce to some extent, which is totally different from that in South China.

(3) The regression equation deduced can be used to evaluate ultimate bearing capacity of similar pile types and the precision is satisfactory in engineering practice. The results provide a new way to precast pile selection and bearing capacity analysis in specific ground for early stages of pile foundation design.

References

Zhu, H.H., Xie, Y.J., Wang, H.Z.: Behavior of long PHC piles driven in Shanghai soft clay. Chin. J. Geotech. Eng. **26**(6), 745–749 (2004). (in Chinese)

Zhao, J.B., Ruan, X., Sun, C.Y., et al.: Research on the relationship between the final pressure of static pressure pile in Liaoshen area and the ultimate bearing capacity of single pile. J. Shenyang Jianzhu Univ. (Nat. Sci.) **21**(4), 302–304 (2005). (in Chinese)

Zhao, X.M., Wang, X.W., Zhao, C.R., et al.: Reliability analysis of vertical bearing capacity of PHC piles at north bank of Yangtze River, Nanjing. Rock Soil Mech. **29**(3), 785–789 (2008). (in Chinese)

Xie, Y.J., Wang, H.Z., Zhu, H.H.: Soil plugging effect of PHC pipe pile during driving into soft clay. Rock Soil Mech. **30**(6), 1671–1675 (2009). (in Chinese)

Kou, H.L., Zhang, M.Y., Bai, X.Y.: Field performance of residual stresses in jacked PHC pipe piles in layered ground. Chin. J. Geotech. Eng. **35**(7), 1328–1336 (2013). (in Chinese)

Das, S.K., Basudhar, P.K.: Undrained lateral load capacity of piles in clay using artificial neural network. Comput. Geotech. **33**(8), 454–459 (2006)

Pooya, N.F., Jaksa, M.B., Kakhi, M., et al.: Prediction of pile settlement using artificial neural networks based on standard penetration test data. Comput. Geotech. **36**(7), 1125–1133 (2009)

Zhao, J.B., Tu, J.W., Shi, Y.Q.: An ANN model for predicting level ultimate bearing capacity of PHC pipe pile. In: Gangbing, S., Ramesh, B.M. (eds.) Earth and Space 2010: Engineering, Science, Construction, and Operations in Challenging Environments, pp. 3168–3176 (2016)

Kordjazi, A., Pooya, N.F., Jaksa, M.B.: Prediction of ultimate axial load-carrying capacity of piles using a support vector machine based on CPT data. Comput. Geotech. **55**, 91–102 (2014)

Liu, S.F.: Grey System and Application. Science Press, Beijing (2008). (in Chinese)

Wang, H.W.: Partial Least-Squares Regression Method and Applications. National Defense Industry Press, Beijing (1999). (in Chinese)

De Nicola, A., Randolph, M.F.: Centrifuge modelling of pipe piles in sand under axial loads. Géotechnique **49**(3), 295–318 (1999)

Yetginer, A.G., White, D.J., Bolton, M.D.: Field measurements of the stiffness of jacked piles and pile groups. Géotechnique **56**(5), 349–354 (2006)

Liu, X.Y., Kang, J.W., Lin, W.X.: Pile Foundation Work Characteristic Analysis of Neural Network Model. China Architecture & Building Press, Beijing (1999). (in Chinese)

Full-Scale Loading Test on Pre-bored Precast Pile with Enlarged Base in Shanghai

Zao Ling[1(✉)], Wei-dong Wang[1,2], Jiang-bin Wu[2], and Ju-yun Yuan[1]

[1] Tongji University, Shanghai, China
lingzao@tongji.edu.cn
[2] East China Architecture Design & Research Institute Co., Ltd.,
Shanghai, China

Abstract. The pre-bored precast piling method is an efficient and environmental measure to construct precast piles to avoid squeezing effect, noise and vibration pollution. Although this pile has been increasingly used in China, the true ultimate bearing capacity and load transfer mechanism are still unclear. In this study, three full-scale ultimate static loading tests on instrumented pre-bored precast piles were conducted in Shanghai, China. The tests results indicated that the load-displacement curve increased slowly at the beginning, and tumbled sharply after reaching the ultimate resistance of the test piles. The pre-bored precast pile with enlarged base pertained to the type of typical end bearing friction pile. Axial force could be directly transferred to the pile toe at initial loading and the toe resistance accounted for about 25% of the pile capacity when applied ultimate load. It is concluded that the pre-bored precast pile with enlarged base in soft clay has a good axial bearing performance, and the ultimate static loading tests provide guidance and technical support for the theoretical research and engineering practice of these piles in Shanghai.

Keywords: Pre-bored precast pile · Static loading test
Ultimate bearing capacity · Load transfer mechanism

1 Introduction

The pre-bored precast piling method is an efficient, environmental and economical measure to construct precast piles to avoid squeezing effect, noise and vibration pollution. It has been widely used in Japan for last decades, and some improved versions with enlarged base have been developed to meet the demand of the higher bearing capacity piles after 2004 (Yamato and Karkee 2004; Kobayashi and Ogura 2007). To adapt to different geological and construction conditions, a modified pre-bored precast pile with flexible composition of inner precast concrete piles and special dimension of enlarged base has been increasingly popular in China (Zhang et al. 2013).

The high vertical compressive bearing capacity of such piles has been confirmed by previous studies and applications (Karkee et al. 1998; Zhou et al. 2013). The load transfer mechanism of pile shaft and enlarged base were analyzed with laboratory model tests (Kusakabe et al. 1994; Ishikawa et al. 2011) and numerical simulation (Zhou et al. 2016). A limited number of well documented static loading tests on

© Springer Nature Singapore Pte Ltd. 2018
T. Qiu et al. (Eds.): GSIC 2018, *Proceedings of GeoShanghai 2018 International Conference:*
Advances in Soil Dynamics and Foundation Engineering, pp. 637–645, 2018.
https://doi.org/10.1007/978-981-13-0131-5_69

modified pre-bored precast piles are available in early literature (Zhou et al. 2013, 2015). However, all the above-mentioned field loading tests were the proof-load tests, which were terminated before reaching the true ultimate capacity of the test piles with plunging failure. As a result, it was unable to obtain the true ultimate bearing capacity and the distribution characteristics of ultimate shaft and toe resistance.

In this study, three full-scale ultimate static loading tests on pre-bored precast piles with enlarged base were conducted in Shanghai for the first time. The main purpose of the tests is to investigate the load-displacement response and load transfer mechanism of test piles.

2 Description of Modified Pre-bored Precast Piling Method

The basic sequence of pile installation is illustrated in Fig. 1. In this method, cement soil column with enlarged high strength base should be first formed using a special helical auger, then the precast concrete pile is inserted using self-weight. With the auger going up and down continually during construction process, base-forming and side-forming cement milks are injected into the soil and mixed well with the soil to increase the toe and shaft resistance of the pile. More importantly, the piling method has been developed to provide high bearing capacity from deeper and better stratum without much reliance on the shaft resistance of the upper soft soil layers (Yamato and Karkee 2004).

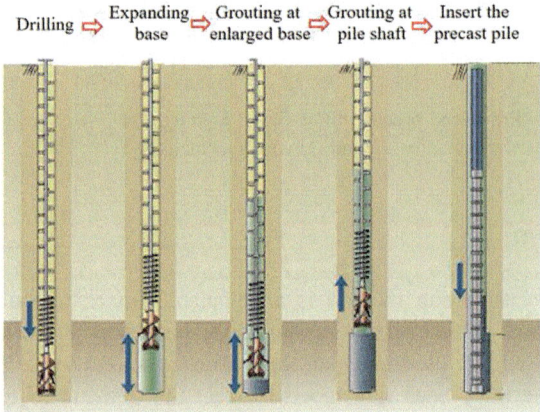

Fig. 1. Installation sequence (modified after Zhang et al. 2013)

The inner concrete precast piles can be made up of prestressed high-strength concrete (PHC) pile, PHC nodular pile, prestressed reinforced high-strength concrete (PRHC) pile and so on. The composition of inner precast piles is flexible to achieve different requirements of load performance. The water-cement ratio of cement milk of pile shaft and enlarged base is about 1.0–1.2 and 0.6–0.7, respectively. The diameter of the enlarged base should not be more than 1.6 times the diameter of the borehole, and the height of base should not be less than 3 times the diameter of the borehole.

3   Pile Test Program

3.1   Ground Conditions

The test piles were located in the central area of Shanghai, China, famous for its thick layer of marine deposits. A series of laboratory and in situ tests were performed at this site. The groundwater table was located at a depth of 1.5 m. The subsurface condition mainly consists of mucky clay, silty clay, silt and silty sand. Soil parameters of different layers from laboratory tests are summarized in Table 1. The cohesion c and friction angle φ were determined from consolidated undrained tests, and modulus of compressibility E_s worked under 100–200 kPa. The ultimate unit shaft resistance q_{sik} and unit toe resistance q_{pk} are recommended for precast piles in technical code DGJ0811 (2010).

Table 1. Summary of soil parameters

Soil layer	Depth L (m)	w (%)	γ (kN/m³)	c (kPa)	φ (°)	E_s (MPa)	q_{sik} (kPa)	q_{pk} (kPa)
Fill	0.0–3.0	25.0	17.0	/	25.0	/	10	/
Silty clay mixed silt	3.0–7.0	31.0	18.5	5	32.0	10.0	15	/
Mucky clay	7.0–19.5	50.4	16.7	10	11.5	2.0	25	/
Silty clay	19.5–28.5	33.1	18.2	15	19.5	4.3	40	/
Silty clay	28.5–39.5	32.9	18.2	16	21.0	5.0	55	/
Silty clay mixed silt	39.5–46.5	23.1	19.7	40	20.0	7.2	65	/
Silty clay	46.5–51.0	31.0	18.4	19	20.5	5.5	60	/
Silt mixed silty clay	51.0–59.5	28.2	18.7	12	26.5	9.0	80	3500
Silty sand	59.5–70.0	25.9	19.0	/	35.0	12.8	110	8500

3.2   Pile Installation and Instrumentation

Three test piles with length of 55 m and shaft diameter of 750 mm consisted of two parts of inner precast concrete piles, assembled by a 40 m long 600 mm diameter PHC pile section in the upper part and a 15 m long 650 mm (500 mm) diameter nodular pile in the lower part. The pile toes were all located in medium-dense silt mixed silty clay. The enlarged bases were 1200 mm diameter and 2750 mm long. According to the construction record, the cement ratio of pile shaft and enlarged base was 9.0% and 59.8%, respectively.

The test piles were instrumented with vibrating-wire strain gages at six different levels: 1.5, 18.0, 28.0, 39.0, 46.0, and 53.0 m from the pile head. Figure 2 shows the arrangement of the strain sensors along pile shaft. At each level, four gages were arranged in 90°. The gages in each pile were attached tightly to the spiral hoop steels of the precast concrete piles when constructed in the factory, as shown in Fig. 3. To minimize the effect of bending related strains, the average of the readings from the four strain gauges was adopted for analysis. The uppermost set of strain gages was deliberately placed at a distance of 1.5 m below the pile head to minimize the effects of localized stresses near the pile head (Lam and Jefferis 2011). To create the pile head suitable for the loading test, the upper 0.80 m of the pile was encased in a steel casing. To examine whether the strain gauges were affected during pile installation, strain values were monitored at three critical stages: in the construction site, immediately after pile forming, before the loading test.

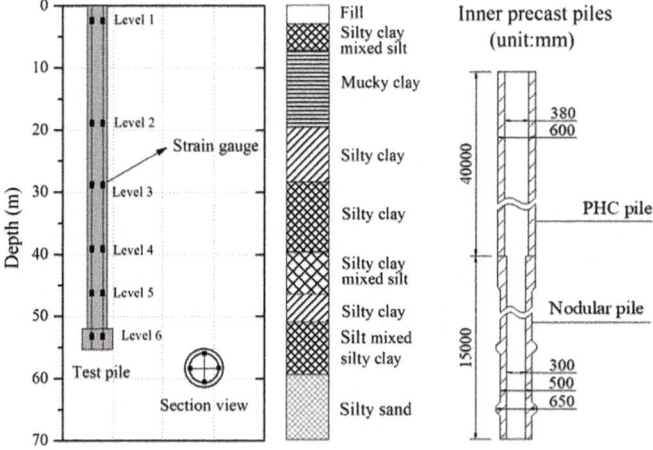

Fig. 2. Arrangement of strain gauges and inner precast concrete piles

Fig. 3. Strain gauges installation

3.3   Static Loading Test

Static loading tests were carried out 43–45 days after pile installation by the maintained load method, which was conducted in accordance with the Chinese code JGJ106 (2014). The test apparatus mainly consisted of a counterforce device using concrete blocks, a loading device using hydraulic jack, and a measuring system using force and displacement sensors, as shown in Fig. 4. All the test piles were designed to be loaded to reach the ultimate capacity of plunging failure and then fully unloaded. The load increment of 800 kN was equal to about 10% of the estimated maximum load. For each loading step, the pile head displacement was recorded at 5, 15, 30, 45, 60 min and every 30 min thereafter until the displacement rate reduced to a minimum value of 0.1 mm/h.

Fig. 4. Static loading test apparatus in the field

4   Pile Test Results

4.1   Load-Displacement Response

Figure 5 shows the measured load-displacement response of the three test piles. The load-displacement curves of TP1 and TP2 increased slowly at the beginning of loading,

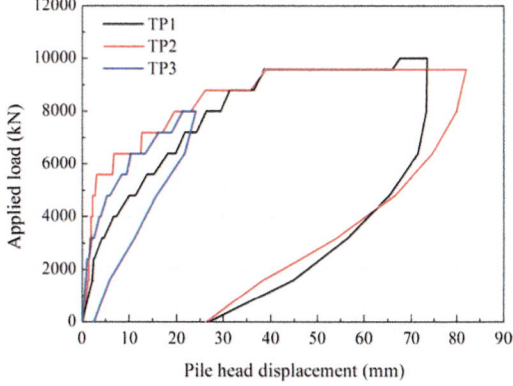

Fig. 5. Load-displacement response of three test piles

and then dropped dramatically when reaching the ultimate bearing capacity of the pile plunging failure of 8800 kN. As the pile head of TP3 was inclined during the test, the loading test on TP3 was terminated at applied load of 8000 kN. The ultimate bearing capacity of the three test piles was greater than the design requirement of 6700 kN. The pile head displacements of test piles were small during the initial stages of loading, and were less than 10 mm and 40 mm at working load of 4400 kN and ultimate load of 8800 kN, respectively. The displacement and applied load of pile head at several critical stages are summarized in Table 2.

Table 2. Displacement and applied load at several critical stages

No.	Maximum load (kN)	Maximum displacement (mm)	Ultimate load (kN)	Ultimate displacement (mm)	Working load (kN)	Working displacement (mm)
TP1	10000	73.5	8800	36.7	4400	9.3
TP2	9600	81.9	8800	35.7	4400	2.3
TP3	8000	24.0	>8000	/	4000	3.9

4.2 Load Transfer Mechanism

Load Distribution. The total axial force of test pile is equal to the axial force of the precast pile plus the axial force of the cement soil. The variation of cross sectional area and material composition along the pile axis was taken into account in computing the axial force. The Young's modulus of cement soil of pile shaft and enlarged base were assumed equal to 160 MPa and 2500 MPa from testing results with similar conditions (Zhou et al. 2016). The nonlinear relationship of Young's modulus of the inner precast concrete pile and measured strain was obtained from the strain data at the uppermost gauge level using the secant modulus method (Lam and Jefferis 2011). This relationship was applied to the other strain gage levels. In addition, the effects of residual loads were accounted for (Fellenius 2002).

The load distributions of the test piles were basically consistent. A typical load distribution of TP1 is illustrated in Fig. 6. Only very little difference of load distribution between the whole test pile and inner precast pile can be seen in Fig. 6(a) and (b). This indicated that the deformation performance of the pre-bored precast pile was mainly affected by the inner precast pile. Due to the high stiffness of the precast pile and reinforcement of pile toe with enlarged base, the axial force of the pile could be transferred directly to the pile toe at the initial stage of loading. This property was similar to that of long rock socketed pile (Wang et al. 2016). The axial force decreased with the increase of the buried depth, and the reinforcement effect of enlarged base was strengthened. The toe resistance accounted for about 25% of the applied load under the ultimate condition.

The Young's modulus of cement soil was about 0.5% of the modulus of precast concrete pile, so the axial force of the cement soil of pile shaft could be neglected without larger errors, as shown in Fig. 6(c). However, the axial force of the cement soil

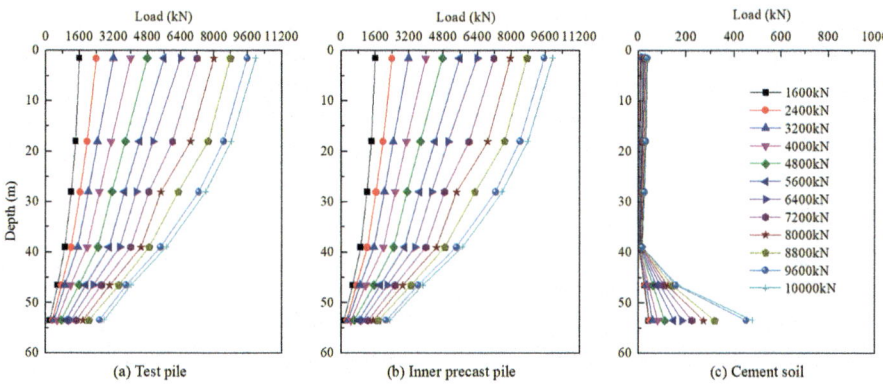

Fig. 6. Typical load distribution of TP1

at enlarged base obviously raised to 300–400 kN under ultimate conditions, which made up 15.2–19.5% of the toe resistance of test piles.

Mobilized Shaft Resistance. According to previous engineering experience and numerical analysis (Zhou et al. 2016), the shaft resistance of pre-bored precast pile with enlarged base was dominated by the interface between cement soil and surrounding soil. The unit shaft resistance (τ_s) between two instrumented levels can be back-figured from the measured axial forces at different depths. On the whole, although the total shaft resistance increased during loading test, the proportion reduced from 88.0% to 75.0% under the ultimate conditions.

The typical mobilized unit shaft resistance of TP1 is shown in Fig. 7. The measured ultimate values (τ_{max}) was about 30% higher than the recommended values in

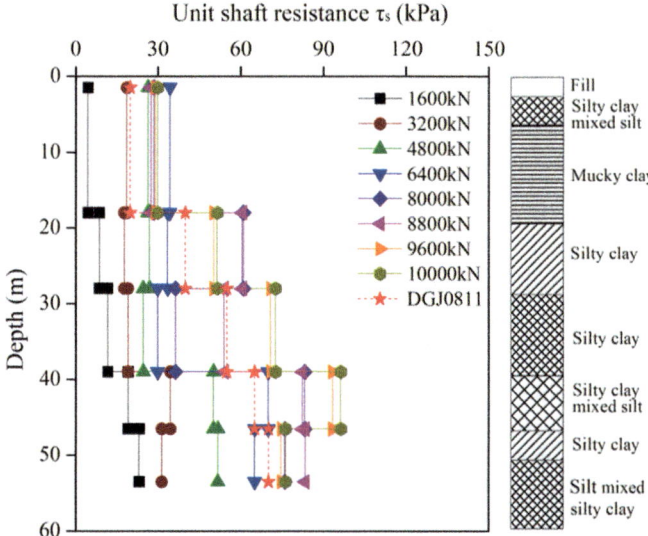

Fig. 7. Typical mobilization of unit shaft resistance on TP1

DGJ0811 (2010). The unit shaft resistance of upper and lower part of the test piles increased simultaneously. This was quite different from that of super long bored piles (Wang et al. 2011). With the increase of applied load, the upper shaft resistance between the depth of 1.5 to 28.0 m decreased to residual value (τ_{res}), with residual adhesion ratio (τ_{res}/τ_{max}) approximately ranging from 0.70 to 0.90 after reaching the peak (τ_{max}). However, the shaft resistance of lower part continued to increase until the end of the test.

Mobilized Toe Resistance. The unit toe resistance can be obtained from the readings of the strain gauges near the pile toe. Field tests indicated that pre-bored precast pile with enlarged base was a typical end bearing friction pile. The unit toe resistance of test piles reached the ultimate values when the test piles reaching pile plunging failure. Generally, the unit toe resistance increased nonlinearly with the pile head loading, as shown in Fig. 8. As the applied load exceeded the ultimate value of 8800 kN, the unit toe resistance showed an accelerated growth. The ultimate unit toe resistance of test piles was about 1950 kPa, accounting for 55% of the recommended value for traditional precast pile in technical code DGJ0811 (2010).

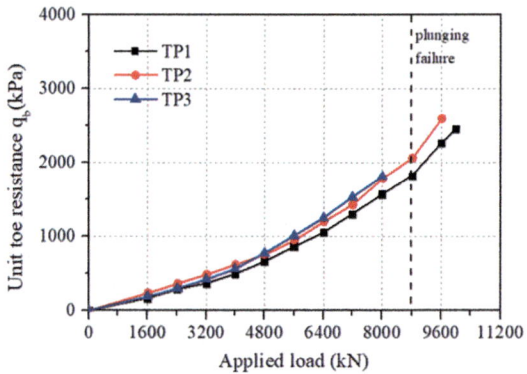

Fig. 8. Mobilization of unit toe resistance on test piles

5   Conclusions

A series of full-scale ultimate static loading tests were performed on three modified pre-bored precast piles with enlarged base installed in thick layer marine deposits in Shanghai, China. Some conclusions can be drawn as follows:

(1) The ultimate bearing capacity of three pre-bored precast piles with enlarged base NO. TP1–TP3 met the requirement of the design criteria. This type of pile with higher bearing capacity was verified in thick soft soil area, which could replace the traditional bored piles and precast piles in some cases.

(2) Field study indicated that pre-bored precast pile with enlarged base was a typical end bearing friction pile. The axial force of the pile could be transferred directly

from pile head to pile toe at the initial loading stage, and the toe resistance was about 25% of the pile capacity when applied ultimate load.

(3) The upper unit shaft resistance showed softening performance with the residual adhesion ratio ranging from 0.70 to 0.90, while the lower unit shaft resistance increased gradually until the end of the test. The unit toe resistance increased nonlinearly with the pile head loading, and the ultimate unit toe resistance accounted for 55% of the recommended value for traditional precast pile.

References

DGJ0811: Foundation Design Code. Shanghai Urban-Rural Construction and Traffic Committee, Shanghai (2010)

Fellenius, B.H.: Determining the resistance distribution in piles. Part 1: notes on shift of no-load reading and residual load. Geotech. News Mag. **20**(2), 35–38 (2002)

Ishikawa, K., Ito, A., Ogura, H., Nagai, M.: Effect of strength and tip length of enlarged grouted base on bearing capacity of nodular pile. J. Struct. Constr. Eng. **76**(670), 2107–2113 (2011). (in Japanese)

JGJ106: Technical Code for Testing of Building Foundation Piles. Ministry of Housing and Urban-Rural Development of the People's Republic of China, Beijing (2014)

Karkee, M.B., Kanai, S., Horiguchi, T.: Quality assurance in bored PHC nodular piles through control of design capacity based on loading test data. In: Proceedings of the 7th International Conference and Exhibition, Piling and Deep Foundations, Vienna, Austria, vol. 1, no. 24, pp. 1–9 (1998)

Kobayashi, K., Ogura, H.: Vertical bearing capacity of bored pre-cast pile with enlarged base considering diameter of the enlarged excavation around pile toe. In: Advances in Deep Foundations: International Workshop on Recent Advances of Deep Foundations (IWDPF07), Yokosuka, Japan, 1–2 February 2007, pp. 277–283 (2007)

Kusakabe, O., Kakurai, M., Ueno, K., Kurachi, Y.: Structural capacity of precast piles with grouted base. J. Geotech. Eng. **120**(8), 1289–1306 (1994)

Lam, C., Jefferis, S.A.: Critical assessment of pile modulus determination methods. Can. Geotech. J. **48**(10), 1433–1448 (2011)

Wang, W.W., Li, Y.H., Wu, J.B.: Field loading tests on large-diameter and super-long bored piles of Shanghai Center Tower. Chin. J. Geotech. Eng. **33**(12), 1817–1826 (2011). (in Chinese)

Wang, W.W., Wu, J.B., Nie, S.B.: Field loading tests on large-diameter super-long rock-socketed bored piles of the Wuhan Center Tower. J. Build. Struct. **37**(6), 196–203 (2016). (in Chinese)

Yamato, S., Karkee, M.B.: Reliability based load transfer characteristics of bored precast piles equipped with grouted bulb in the pile toe region. Soils Found. **44**(3), 57–68 (2004)

Zhang, R.H., Wu, L.L., Kong, Q.H.: Research and practice of JZGZ pile foundation. Chin. J. Geotech. Eng. **35**(S2), 1200–1203 (2013). (in Chinese)

Zhou, J.J., Gong, X.N., Wang, K.H., Zhang, R.H.: A field study on the behavior of static drill rooted nodular piles with caps under compression. J. Zhejiang Univ. Sci. A **16**(12), 951–963 (2015)

Zhou, J.J., Gong, X.N., Wang, K.H., Zhang, R.H., Yan, T.L.: A model test on the behavior of a static drill rooted nodular pile under compression. Mar. Georesour. Geotechnol. **34**(3), 293–301 (2016)

Zhou, J.J., Wang, K.H., Gong, X.N., Zhang, R.H.: Bearing capacity and load transfer mechanism of a static drill rooted nodular pile in soft soil areas. J. Zhejiang Univ. Sci. A **14**(10), 705–719 (2013)

Design of Drilled Shafts for Unusual High Mooring Forces

Alex Shue[(✉)] and Arthur Roesler

Langan Engineering and Environmental Services,
1818 Market Street, Suite 3300, Philadelphia, PA 19103, USA
{ashue,aroesler}@langan.com

Abstract. Most drilled shafts are treated as columns or beam-columns sub-jected to the vertical compression load of buildings or bridges. In the case presented in this paper, drilled shafts were used to take lateral and uplift forces, and were treated as circular flexural composite members. This paper presents a procedure for designing drilled shafts for very high lateral and uplift forces. Our geotechnical project involved converting a U.S. Navy base site into a cruise ship terminal. Our client requested that we design new bollards that could take a mooring lateral load of 1,500 kN and an uplift load 750 kN for the newest generation of luxury cruise ships. The subsurface conditions at the site generally consist of 24 m to 30 m of soft river mud and clay underlain by dense sand and decomposed bedrock. A reinforced concrete-filled steel tube (RCFT) drilled shaft was designed to take care of such unusually high lateral and uplift loads. Current design models adopted by three major engineering codes in the U.S. were reviewed and a Plastic Stress Distribution Model was used in our analysis. Steel shear rings inside a casing were designed to increase the bond between concrete and steel.

Keywords: Moment-resisting drilled shaft · Concrete-filled steel tube (CFT) Reinforced concrete-filled steel tube (RCFT) · Plastic stress distribution (PSD) model · Strain compatibility (SC) model · D/t slenderness ratio

1 Introduction

The project site is a new cruise ship departure point, called Cape Liberty Port, in Bayonne, New Jersey, directly across from Manhattan (see Fig. 1). The port provides a wealth of cruise sailing from Bayonne to ports in New England, Florida, Canada, and the Caribbean (see Fig. 2). The site was once a former Navy ocean terminal, but the naval mooring facilities cannot meet today's cruise line berthing requirements because today's vessels are longer, higher and heavier than the naval vessels. Our client asked Langan to design and install three additional mooring bollards, each of which could take 150 tons of line force with an inclination of 30°. The 150 ton inclined line load results in a lateral load (horizontal force) of 1,500 kN and an uplift (vertical) force of 750 kN. The 150 tons of force is an extreme loading condition that would occur during hurricane type winds and with strong currents expected to occur annually. The usual line load would be less than 50% of 150 tons. The load is kinetic and cyclic.

© Springer Nature Singapore Pte Ltd. 2018
T. Qiu et al. (Eds.): GSIC 2018, *Proceedings of GeoShanghai 2018 International Conference: Advances in Soil Dynamics and Foundation Engineering*, pp. 646–653, 2018.
https://doi.org/10.1007/978-981-13-0131-5_70

The British Standard Code for Maritime Structures (BSI) suggests that vessels usually equal to or more than 200,000 tons displacement require a mooring (bollard) load of 1,500 kN or more. The mooring load of 150 tons belongs to the highest bollard load rate of civilian oceangoing vessels. Because of the limitation of the site and difficult subsoil condition, it was not economical or practical to install a conventional mooring dolphin platform, a cluster of battered piles and vertical piles. Our feasibility study suggested that constructing a drilled shaft of reinforced-concrete-filled steel tube (RCFT) would be cost-effective and could accommodate the very high mooring load.

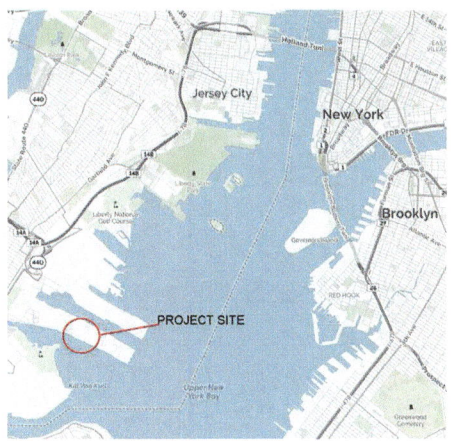

Fig. 1. Site location map

The elevation view of the drilled shaft we designed is shown in Fig. 3. The mooring-point load acting at the bollard indicates that this is a moment-resisting tensile shaft in lieu of the commonly used axial compressed drilled-shaft foundation. The kinetic and cyclic nature of the mooring load necessitated that the drilled-shaft be ductile, meaning the shaft could displace inelastically through cyclic loading without significant degradation of structural strength or stiffness. The RCFT drilled shaft combines the advantages of ductil-

Fig. 2. Cruise ship berthing and mooring at port of Bayonne NJ

ity generally associated with steel structures, with the stiffness of a concrete column. The Von Mises yield criterion states that continuum materials behave more ductilely under a high confining stress field. A wealth of research has proved that a steel casing confines the concrete core and that the confinement not only increases the strength of the concrete and prohibits excessive spalling, but also delays the local buckling of the steel tube wall.

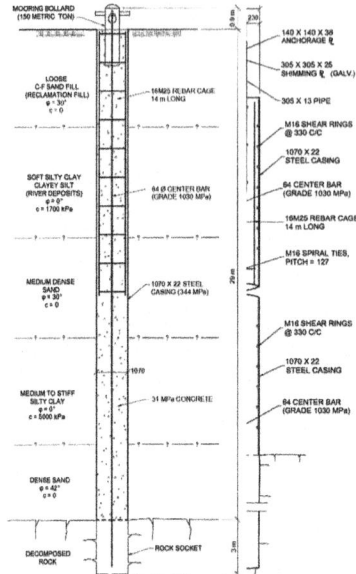

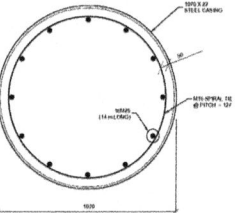

Fig. 3. 107 cm diameter drilled shaft and subsurface profile

2 Review of Design Provisions of Concrete-Filled Steel Tube (CFT)

2.1 D/t Ratio

Similar to when you design a regular column, you must first consider slenderness to avoid structural buckling. The D/t slenderness limit is employed to limit local buckling of the tube and prevent concrete spalling to ensure development of the plastic capacity of the member. Previous researchers have developed different D/t limits and expressions for three major engineering codes in the U.S. These codes are: the American Institute of Steel Construction (AISC) LRFD, American Concrete Institute (ACI), and American Associations of State Highway and Transportation Officials (AASHTO) Guide Specifications For LRFD Seismic Design. These are the different expressions:

$D/t \leq 0.15E/F_y$ (AISC, Provisions)
$D/t \leq \sqrt{8}E/F_y$ (ACI Provisions)
$D/t \leq 2\sqrt{E}/F_y$ (AASHTO LRFD Provisions).

Where, E is steel elastic modulus, F_y is the minimum yield strength of steel tube.

There are wide variations in those local slenderness limits among the codes. Using $E = 200 \times 10^3$ MPa, $F_y = 345$ MPa, the limits are 87, 68, and 48 for AISC, ACI and AASHTO design provisions, respectively.

2.2   Design Models for CFT

As we know, Mander's model (Mander et al. 1988) is the most common method of calculating the strength of confined concrete members. Mander developed a stress-strain model for concrete subjected to monotonic compression loading and confined by any type of transverse confining steel. A single equation was derived for the stress-strain relation based on one parameter: compression. However, there is no monotonic compression loading for the mooring shaft we designed; therefore, Mander's method is not applicable for this case. The drilled shaft we coped with is neither a column nor a beam-column, rather it is a flexural composite member.

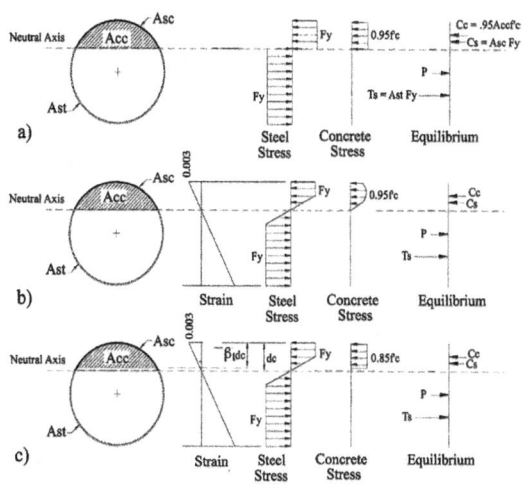

Fig. 4. Models for prediction of resistance of CFT

The current design provisions for concrete-filled steel tube (CFT) components specified in design codes in the U.S. can be divided in two categories: the plastic stress distribution (PSD) model and the strain compatibility (SC) model. These models presented in Fig. 4 were taken from a report by Lehman and Roeder (2012).

The AASHTO Guide Specification for LRFD Seismic Bridge design and AISC Specification (2005) allow the use of the plastic stress distribution (PSD) model (see Fig. 4a) for predicting the flexural and axial resistances of concrete-filled steel tube (CFT) components. The plastic stress distribution model assumes that the composite section develops a uniform compression stress of $0.95f'_c$ in the concrete and the full yield stress, F_y of the steel in tension and compression as shown in Fig. 5a. The $0.95f'_c$ in the concrete is larger than the stress of $0.85f'_c$ typically used for calculating a Whitney stress block in recognition of the beneficial effects of concrete confinement in a circular CFT. With this PSD model, the bending resistance or axial capacity can be determined by satisfying equilibrium over the cross section for each possible neutral axis to establish the axial-moment diagram.

The AISC also adopts the strain compatibility (SC) model to predict flexural strength and axial load of CFT members used in a non-seismic zone. Figure 4b shows that the model employs a linear strain distribution including an elastic-perfect plastic curve to model the steel and a parabolic curve for the concrete. By satisfying the equilibrium relations, the flexural strength is determined for a maximum compression strain in the concrete of 0.003 mm/mm ($\varepsilon = 0.003$).

Figure 4c illustrates the ACI strain compatibility (SC) model, which is similar to the AISC strain model (Fig. 4b). The ACI procedure uses an equivalent rectangular stress block with a compressive stress of $0.85f'_c$ acting over a depth $\beta_1 c$, where β_1 depends on concrete strength. In the expression, c is the depth from the location of the maximum compressive strain to the neutral axis depth.

According to previous researchers (Priestley, Park, et al.), the design strength of the strain compatibility (SC) models are quite conservative. The main reason is that the SC method is based on normal concrete strength and maximum strain $\varepsilon = 0.003$. This 0.003 strain limit provides a lower-bound estimate to spalling of the concrete cover, but concrete spalling is eliminated in CFT since all concrete is well-confined. When comparing the PSD method with the SC method, the plastic stress distribution (PSD) model is more accurate and relatively simple. It makes more sense to use the PSD model to predict the resistance of a relatively ductile CFT drilled shaft under cyclic loading.

3   Design of Mooring Shaft

The plastic stress distribution (PSD) model provides a practical and simple solution for predicting the flexure and axial resistance of a concrete-filled steel tube (CFT) drilled shaft. Often a rebar cage is placed in the concrete of the drilled shaft as we did in our design (see Fig. 3). The Washington State Department of Transportation (WSDOT) has adopted a model for predicting the resistance of reinforced-concrete-filled steel tubes (RCFT). The WSDOT's solution is essentially the same as the CFT PSD model suggested by AASHTO Guide Specifications for LRFD Seismic Bridge Design. The only difference is that the strength of the rebar cage is considered by simulating the rebar cage as an inner steel tube. (see Fig. 6). Our design was performed per WSDOT's model.

3.1   D/t Ratio

WSDOT suggests:

$D/t \leq 0.22E/F_y$ for members subjected to elastic forces
$D/t \leq 0.15E/F_y$ for members subjected to plastic forces.

The steel tube we used is 107 cm dia. $\times$ 2.2 cm thickness with $E = 200 \times 10^3$ MPa and $F_y = 345$ MPa, therefore, $D/t = 48 \leq 0.15E/F_y = 87$ as per the limit suggested by WSDOT.

3.2   Subsurface Conditions and L-Pile Analysis

According to our boring logs, the subsurface conditions generally consist of reclamation fill overlying successive strata of soft silty clay (river deposit), underlain by medium to dense silt and sand, and decomposed bedrock. In the immediate vicinity of the proposed bollard locations, rock was encountered approximately 25 m (83 ft) to 29 m (96 ft) below the existing grades. Figure 3a shows the soil profile. The Table of Soil Properties shows the shear strengths of different soil strata.

Table of Soil Properties

Soil layers	SPT N-Valves (Blows/0.3 m)	Unconfined compression strength (KPa)	Friction angle (Degree)
Reclamation sandy FILL	1 to 70	0	30
Silty CLAY river deposit	1 to 22	1,700	0
Medium dense SAND	6 to 34	0	30
Stiff silty CLAY	4 to 70	5,000	0
Dense SAND	20 to 50	0	42

L-pile analysis (also called the p-y method, per Reese and Van Impe 2001) was performed based on the parameters of the above table. The drilled shaft is a 1.07 m diameter by 29 m long (42 in. by 95 ft) CFT. Assume there is no rock socket because the overburden soil is sufficiently thick, and the computed deflection of the shaft at the rock is so small that the resistance of the rock can be neglected. Figure 5a and b show the results of L-pile analysis. Under 1,500 kN (330 kips) of lateral force and 730 kN (165 kips) of uplift, the estimated maximum bending moment M_{max} = 4,475 kN-m (3,300 kips-ft), deflection at top of shaft δ = 79 mm (3.1 in.), and maximum shear V_{max} = 1,500 kN (330 kips). The L-pile analysis indicates the maximum flexural point occurs at the depth of 5.2 m (17 ft), which is approximately 4.8 D (D is outside diameter) below the grade. The equivalent fixity for elastic displacement is located at the depth of 9 m (30 ft), which is about 8.6 D below the grade.

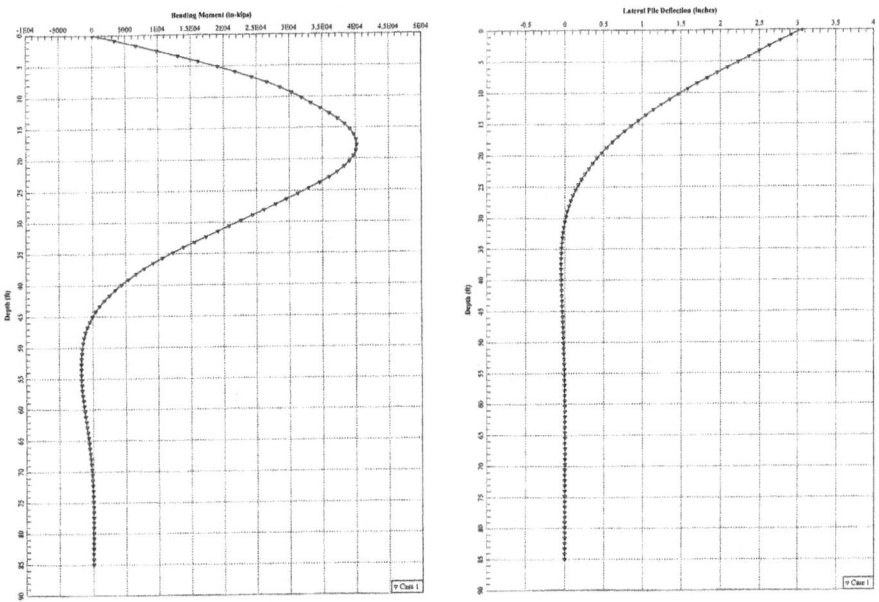

Fig. 5. (a) Moment diagram of drilled shaft (b) Deflection curve of drilled shaft

3.3   Available Moment and Shear Strength

The WSDOT manual provides the geometry for predicting stress distribution (see Fig. 6). The process begins by assuming different y and θ, and calculating the stress acting on A_{cc} and A_{sc}. The internal axial force and moment are then calculated by trial and error. The process is iterative and repeated until equilibrium is achieved. It is a simplified model, but the computation is still tedious. Because of the limit of the length of this paper, some tedious computations are not presented. Readers may refer to WSDOT's bridge-design manual for details of the approach.

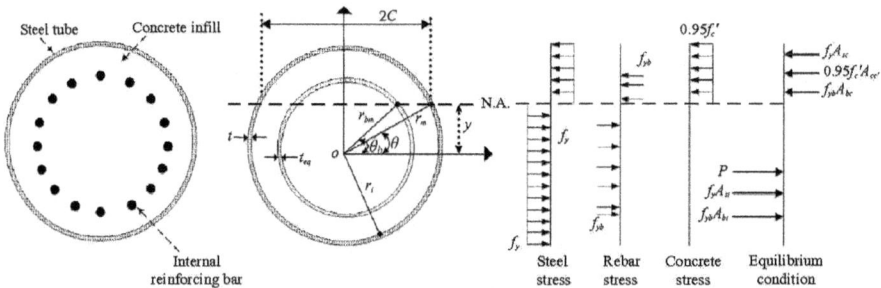

Fig. 6. Plastic stress distribution for RCFT

- Material Properties
  $f_c' = 31$ MPa, $F_y = 345$ MPa
  $E_s = 200,000$ MPa
- Loads
  Horizontal Force = 1,500 kN
  Vertical Uplift = 730 kN
  $M_u = 5,830$ kN-m
  $V_u = 1910$ kN
- Design
  Try D (OD of tube) = 1,070 mm and t (wall thickness) = 22 mm.
  Check D/t = 48 ≤ $0.15E/F_y$ = 87; therefore, it is OK.
  Assume wall thickness will be reduced by corrosion over 75 years at the rate of 0.076 mm/yr. suggested by WSDOT.
  Using the RCFT Model per WSDOT, we have
  $\Phi M_n = 0.9M_n = 7,700$kN-m. ≥ $M_u = 5,830$ kN-m as required; therefore, it is OK.
  $\Phi V_n = 9,200$ kN ≥ $V_n = 1,910$ kN; therefore, it is OK.
  Check the minimum reinforcing ratio:
  $\rho = A_s/A_c = 0.01 ≥ \rho_{min} = 0.005$ as suggested by WSDOT; therefore it is OK.

3.4   Shear Rings and Center Bar

To transfer the axial uplift force from the top to the bottom of the shaft, shear rings were installed at both ends of the tube as illustrated in Fig. 3. The rings are made of M 16 @ 330 mm with a 6 mm fillet weld all around. The ring calculations were performed based on the method recommended by Gebman of UCSD (2006). To avoid making this paper too long, the design method and calculations are not addressed.

A 63-mm-diameter high-strength threaded centralized bar was installed in the shaft and extended 3 m into bedrock. The center bar is essentially a rock anchor designed to resist the whole uplift imposed by the mooring line. The philosophy in marine engineering is: "apply redundancy as much as is practical." The threaded centralized bar is a structural redundancy other than the drilled shaft to take care of uplift.

4   Conclusions

The reinforcing-concrete-filled steel tube (RCFT) drilled-shaft bollards have been in place for more than three years, and despite having experienced several hurricanes and storms, they are performing well in this busy new cruise port on the east coast of U.S. The flexural strength of an RCFT member determined by using the plastic stress distribution (PSD) model is a practical design approach. An RCFT drilled shaft is a good solution to take care of high lateral and tensile cyclic mooring loads since it is structurally ductile and cost-effective.

References

AASHTO LRFD Bridge Design Specifications, Fifth edn. (2010)

AASHTO Guide Specifications for LRFD Seismic Bridge Design (2009)

AISC 2005 Specification for Structural Steel Buildings

ACI 318-08 Building Code Requirements for Structural Concrete

Mander, J.B., Priestley, M.J.N., Park, R.: Theoretical stress-strain model for confined concrete. J. Struct. Eng. **114**, 1804–1826 (1988)

Lehman, D.E., Roeder, C.W.: Rapid Construction of Bridge Piers with Improved Seismic Performance. Department of Civil Engineer, University of Washington (2012)

WSDOT Bridge Design Manual M23.50: Chapter 7 Substructure Design, 17 June 2017

Gebman, M., Ashford, S.A., Restropo, J.I.: Axial force transfer mechanism within cast-in-steel-shell-piles. Department of Structural Engineering, University of California at San Diego (2006)

Reese, L.C., Van Impe, W.F.: Single Piles and Pile Groups Under Lateral Loading. A.A. Balkema Publishers, Amsterdam (2001)

Prediction of Long-Term Settlements of Foundations Supporting High and Heavy Storage Tanks Based on Short-Term Field Measurements

Jianmin Hu[1] and Lianyang Zhang[2(✉)]

[1] Hebei Research Institute of Construction and Geotechnical
Investigation Co., Ltd., Shijiazhuang, Hebei, China
hbkchjm@vip.sina.com
[2] Department of Civil Engineering and Engineering Mechanics,
University of Arizona, Tucson, AZ, USA
lyzhang@email.arizona.edu

Abstract. The operation and safety of high and heavy storage tanks are very sensitive to the differential settlements of the foundations. It is therefore extremely important to precisely predict the long-term settlements so that appropriate measures can be taken if the expected settlements are too large. Researchers have proposed different methods for predicting the long-term settlements of foundations. Of these different methods, the methods based on the short-term field measurements at the beginning, such as the Asaoka method, are commonly used. Since the measurements are taken at the site, the field conditions related to the structure, foundation and soils are well reflected in the measurements. However, since the measurements are only for a short period at the beginning, they may not reflect the behavior of the whole system at a significantly later time. This paper uses the extensive data of field settlement measurements for foundations supporting high and heavy storage tanks at an alumina refinery plant to do a detailed evaluation of the Asaoka method. Specifically, the measurements within different periods of time from the beginning are used to predict the settlements at a later time and the predictions are then compared with the measurements. By doing so, the effect of the time period from the beginning with field measurements on the accuracy of long-term settlement prediction is quantified. Finally, an optimum time period from the beginning with field measurements is recommended so that the field measurements can be used to predict the long-term settlement with sufficient accuracy.

Keywords: Long-term settlement · High and heavy storage tanks
Field measurements · Prediction

1 Introduction

High and heavy storage tanks are often constructed at refinery plants to store liquid. Due to the heavy weight of the tank itself and the stored liquid, large settlement often occurs. To ensure normal operation and safety of the storage tanks, it is extremely

© Springer Nature Singapore Pte Ltd. 2018
T. Qiu et al. (Eds.): GSIC 2018, *Proceedings of GeoShanghai 2018 International Conference: Advances in Soil Dynamics and Foundation Engineering*, pp. 654–661, 2018.
https://doi.org/10.1007/978-981-13-0131-5_71

important to precisely predict the long-term settlement so that appropriate measures are taken if the expected settlements are too large.

There are different methods for predicting the long-term settlements of foundations. Of these different methods, the methods based on the short-term field measurements at the beginning, such as the Asaoka method, are commonly used [1–3]. Since the measurements are taken at the site, the field conditions related to the structure, foundation and soils are well reflected in the measurements. However, since the measurements are only for a short period at the beginning, they may not reflect the behavior of the whole system at a significantly later time [4, 5]. Therefore, it is important to consider the conditions at which a method can be used for predicting the long-term settlement of a foundation.

This paper uses the extensive data of field settlement measurements for foundations supporting high and heavy storage tanks at an alumina refinery plant to evaluate the Asaoka method. Specifically, the measurements within different periods of time from the beginning are used to predict the settlements at a later time and the predictions are then compared with the measurements. By doing so, the effect of the time period from the beginning with field measurements on the accuracy of long-term settlement prediction is quantified. Finally, an optimum time period from the beginning with field measurements is recommended so that the field measurements can be used to predict the long-term settlement with sufficient accuracy.

2 Background

The alumina refinery plant is located in Xiaoyi City, Shanxi Province, China. The subsoils at the site are mainly alternating layers of clay, silty clay, silt, silty sand and fine sand. The ground water level is 0.9–1.6 m below the ground surface. The first phase of the plant consists of 20 platforms and 40 storage tanks, with each platform supporting two tanks (see Fig. 1). The platforms are staged with height varying from 0.4 m to 8.2 m and weight from 4,600 tons to 6,300 tons. Each platform is supported by a group of reinforced concrete piles of diameter 1.0 m and length 36 m. All tanks have the same diameter of 14 m and height of 36 m. When completely filled with alkaline aluminum mineral solution, each tank weighs 18,160 tons.

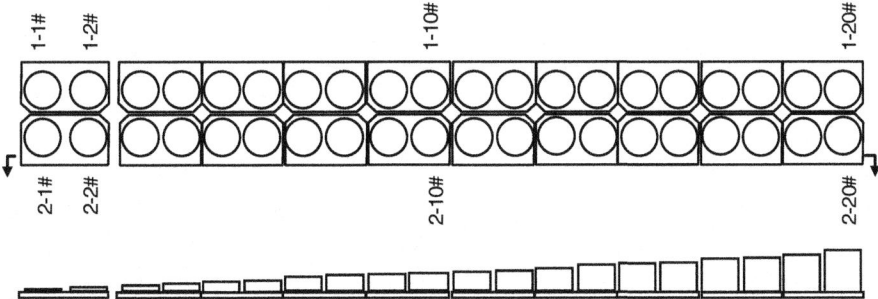

Fig. 1. Distribution of storage tanks on platforms with height varying from 0.4 m to 8.2 m (1-2# denotes storage tank 2 on row 1).

During the operation of the first phase storage tanks, it was found that large total and differential settlements occurred to the tanks based on field measurements. The field measurements were started after the construction of the storage tanks was completed. For each tank, the measurements were conducted at four monitoring locations (see Fig. 2 for the monitoring locations of tanks 2-17# to 2-20#). Figure 3 shows the measured settlements for tank 2-17#.

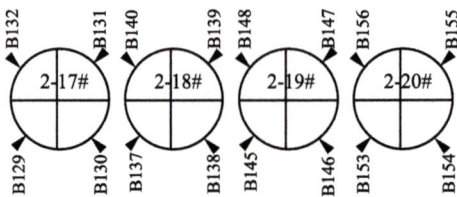

Fig. 2. Distribution of monitoring locations on storage tanks 2-17# to 2-20# for settlement measurements.

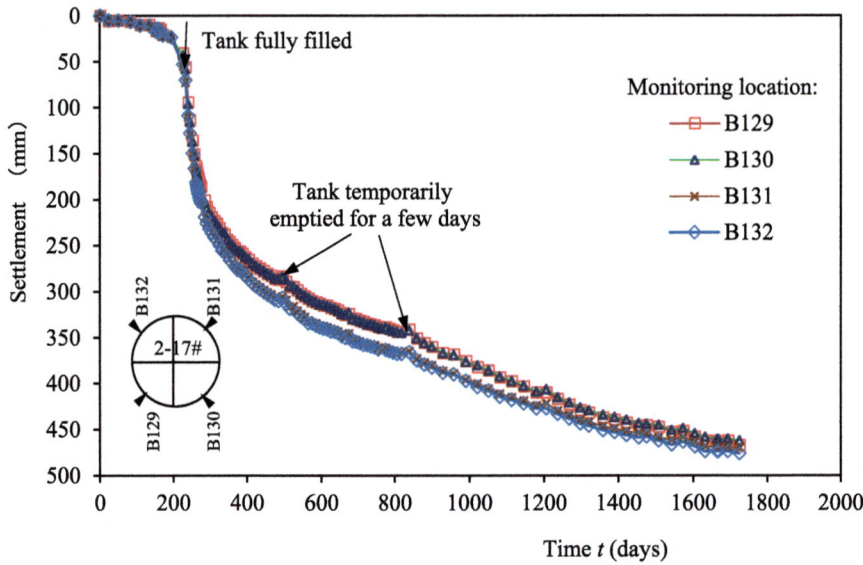

Fig. 3. Measured settlements versus time for storage tank 2-17# at four different locations.

To save space, only the measured settlements for tank 2-17# are shown in Fig. 3. At the time of 604 days, the measured maximum and minimum settlements reached 340.5 mm (B132) and 314.5 mm (B130), respectively, leading to a differential settlement of 26.0 mm. The large total and differential settlements led to tilting of the tanks, which can be clearly seen from the bending of the fence rails of the high level work platform (see Fig. 4). The owner of the alumina refinery plant was worried about the safety of the storage tanks and wanted to know what would be the long-term settlements.

Fig. 4. Bending of fence rails of high level work platform due to tilting of storage tanks caused by differential settlements.

3   Asaoka Method

Asaoka [1] proposed a technique that enables estimation of the final settlement (δ_f) from short-term settlement measurements. The method works by obtaining the settlements at a series of equally spaced times and then plotting each value δ_j at time t_j versus the value δ_{j-1} at the preceding time (see Fig. 5). The points should lie on or near a straight line with slope β_1 and intercept with the vertical axis β_0, i.e.

$$\delta_n = \beta_0 + \beta_1 \delta_{n-1} \tag{1}$$

When the consolidation is completed, $\delta_{n-1} = \delta_n = \delta_f$, and thus

$$\delta_f = \frac{\beta_0}{1 - \beta_1} \tag{2}$$

Therefore, after a linear fitting analysis of the $\delta_{n-1} - \delta_n$ data using Eq. (1) is done, parameters β_0 and β_1 can be obtained and then δ_f can be determined using Eq. (2).

4   Results and Discussion

The Asaoka method was used to predict the long-term settlement of the storage tanks based on the short-term measurements. Considering the very similar settlement versus time trend for all the monitoring locations and to save space, only the results for monitoring location B132 of storage tank 2-17# are presented here.

Since the tanks were fully filled on the 239th day (see Fig. 3), only the data after the 250th day were used in the analysis. Figure 6 shows the $\delta_{n-1} - \delta_n$ data for a time interval $\Delta t = 50$ days.

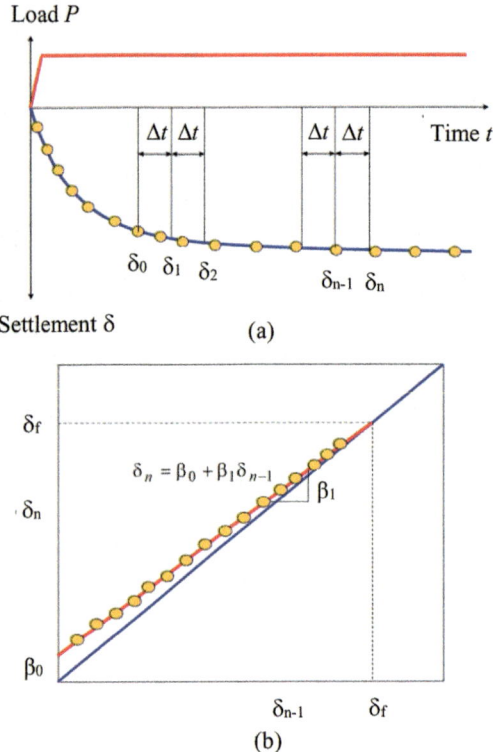

Fig. 5. (a) Settlement versus time; and (b) δ_{n-1} versus δ_n.

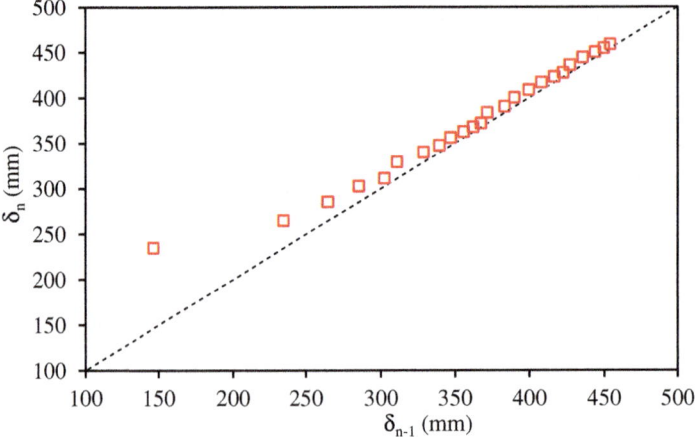

Fig. 6. $\delta_{n-1} - \delta_n$ data for a time interval $\Delta t = 50$ days.

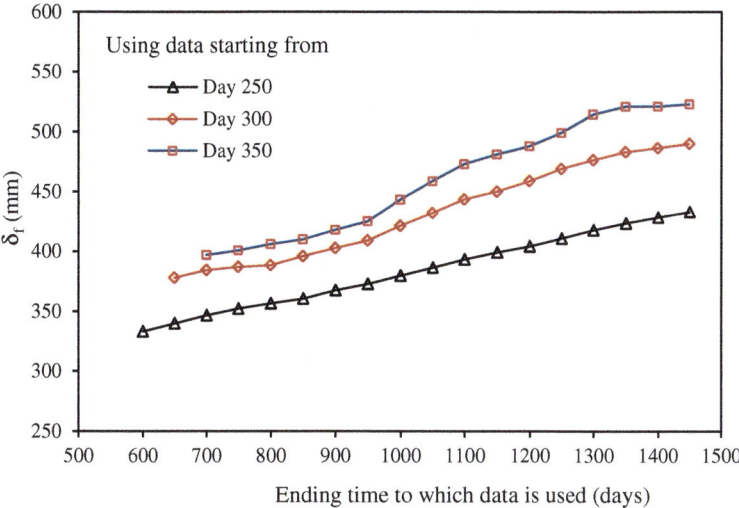

Fig. 7. Predicted long-term settlements using settlement measurement data during different time periods and at a time interval $\Delta t = 50$ days.

To evaluate the accuracy of the Asaoka method for predicting the long-term settlement based on the short-term measurements, the data within different time periods were used. Figure 7 shows the predicted long-term settlements using the measured settlement data during periods with different starting and ending times. The results show that when the data starting on the 250th day are used, the long-term settlement (δ_f) is significantly under-predicted even if the data up to the 1,450th day are used. When the data starting on the 300th day are used (i.e., the data on the 250th day is ignored), much larger δ_f is predicted. When the data starting on the 350th day are used (i.e., the data on both the 250th and 300th days are ignored), even larger δ_f is predicted. For the 250th and 300th day starting times, the predicted δ_f keeps increasing when the data during a longer period of time are used. However, for the 350th day starting time, the predicted δ_f increases up to an ending time of 1,350 days and then remains about the same. Therefore the predicted δ_f with a starting time of 350 days and an ending time of at least 1,350 days can be considered the long-term settlement, which is about 522.8 mm and reasonable based on the settlement measurement data up to 1,723 days shown in Fig. 3.

With the predicted δ_f of 522.8 mm, the settlements on the 250th, 300th and 350th days are 28.1%, 44.9% and 50.7% of the δ_f, respectively, or the degrees of consolidation on the 250th, 300th and 350th days are 28.1%, 44.9% and 50.7%, respectively. So it can be concluded that in order to accurately predict the long-term settlement, the settlement measurement data after a degree of consolidation greater than 50% should be used. This is in agreement with the observations of a number of researchers including Leroueil [3], Bergado et al. [4], Anderson et al. [5] and Urzua et al. [6]. The results also show that in order to accurately predict the long-term settlement, the settlement measurement data within a long enough period time after a degree of consolidation greater than 50% should be used. Specifically, for this studied case, the settlement measurement data up to at least 1,300 days or a degree of consolidation of 85.0% should be used.

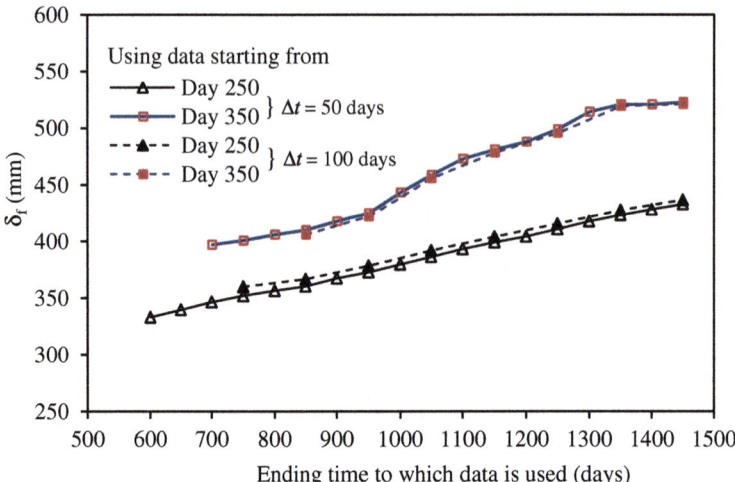

Fig. 8. Predicted long-term settlements using settlement measurement data during different time periods and at a time interval $\Delta t = 50$ days and 100 days, respectively.

The long-term settlement was also predicted by using a time interval $\Delta t = 100$ days. Figure 8 compares the results with those using a time interval $\Delta t = 50$ days. It can be clearly seen that using a different time interval only very slightly affects the predicted long-term settlement values.

5   Conclusions

The Asaoka method was used to predict the long-term settlements of foundations supporting high and heavy storage tanks at an alumina refinery plant. The following conclusions can be drawn based on the results:

(1) To accurately predict the long-term settlement, the settlement measurement data within a long enough period time (up to a degree of consolidation of 85%) after a degree of consolidation greater than 50% should be used.
(2) With the settlement measurement data at the right period of time, using different time intervals only very slightly affects the predicted long-term settlement values

References

1. Asaoka, A.: Observational procedure of settlement prediction. Soils Found. **18**(4), 87–101 (1978)
2. Magnan, J.-P., Deroy, J.-M.: (1980) Analyse graphique des tassements observés sous les ouvrages. Bulletin Liaison du Laboratoire des Ponts et Chausées **109**, 45–52 (1980). (in French)

3. Leroueil, S.: Tenth Canadian geotechnical forum: recent developments in consolidation of natural clays. Can. Geotech. J. **25**(1), 85–107 (1988)
4. Bergado, D.T., Ahmed, S., Sampaco, C.L., Balasubramaniam, A.S.: Settlements of Bangna-Bangpakong highway on soft Bangkok clay. J. Geotech. Eng. **116**(1), 136–155 (1990)
5. Anderson, L.R., Sampaco, C.L., Gilani, H., Keane, E., Rausher, L.: Settlements of highway embankments on soft lacustrine deposits. In: Proceedings of Conference on Vertical and Horizontal Deformations of Foundations and Embankments, vol. I. ASCE, Reston, VA, pp. 376–397 (1994)
6. Urzua, A., Ladd, C.C., Christian, J.T.: New approach to analysis of consolidation data at early times. J. Geotech. Geoenviron. Eng. **142**(10), 06016009-1–06016009-6 (2016)

Full-Scale Field Study on Large-Diameter Post-grouting Drilled Shafts

Guoliang Dai[✉] and Zhihui Wan

Key of Laboratory for RC and PRC Structure of Education Ministry,
School of Civil Engineering, Southeast University, Nanjing 210096, China
daigl@seu.edu.cn

Abstract. In this article, full-scale field tests were conducted to observe the field performances of large-diameter drilled shafts for three combined side-and-tip grouting shafts and one side-grouting shaft in extra-thick fine sand layer. The load-displacement response, shaft resistance, and mobilized unit tip resistance were discussed. Comparing with the test results before and after grouting shows that the ultimate bearing capacity, the total shaft resistance and the base resistance of the shaft after grouting at the shaft side alone are increased by 41.54%, 51.85%, and 1.27%, respectively, whereas the ultimate bearing capacity, the total shaft resistance and the base resistance of the shaft after grouting at the shaft tip and side are increased by 66.03–73.49%, 46.72–56.91%, and 137.87–139.17%, respectively. Consequently, the bearing behavior of the combined-grouting shaft is obviously better than that of the side-grouting shaft. Additionally, the strengthening effect of the soil improvement at the shaft tip due to tip grouting on mobilizing shaft resistance, and meanwhile, the unit tip resistance can also be enhanced by the surrounding soil improvement due to side grouting.

Keywords: Drilled shaft · Combined post-grouting at shaft tip and side
Post-grouting at shaft side · Shaft resistance · Tip resistance

1 Introduction

Drilled shaft foundations have been widely used due to their ability to resist the large load and consideration of scour for highway bridges. However, soil relaxation beneath the shaft tip due to drilling process, which brings great influence to the tip resistance, and debris remaining after cleanout will further reduce the tip resistance (Mullins et al. 2000; Safaqah et al. 2007). Meanwhile, the soil stress release and disturbance surrounding the shaft due to drilling will also reduce the shaft resistance (Zhang et al. 2009; Duan and Kulhawy 2009). Additionally, the mobilization of the tip resistance is hampered by the softened of the shaft tip condition. It can thus be found that these issues have a negative impact on the application of drilled shafts. Accordingly, post-grouting technique is considered to be an effective means to solve the above-mentioned issues and improve the bearing performance of drilled shafts.

Post-grouting at the shaft tip has been successfully employed throughout the world for the past six decades and has been to be proven an effective method to improve axial resistance and reduce settlement (Bruce 1986; Safaqah et al. 2007; Dai et al. 2010;

© Springer Nature Singapore Pte Ltd. 2018
T. Qiu et al. (Eds.): GSIC 2018, *Proceedings of GeoShanghai 2018 International Conference: Advances in Soil Dynamics and Foundation Engineering*, pp. 662–674, 2018.
https://doi.org/10.1007/978-981-13-0131-5_72

Thiyyakkandi et al. 2014). However, the application of post-grouting at the shaft side is slightly later to post-grouting at the shaft tip, and there are few reported in the literature. In recent years, post-grouting technique at the shaft side has also been widely applied to increase shaft resistance (Gouvenot and Gabaix 1975; Stocker 1983; Littlechild et al. 1998). In practice, large-diameter drilled shafts are usually installed in the soil to resist greater vertical, lateral, and uplift loads and have been widely applied in the foundation of highway bridges, high-voltage transmission towers, and offshore oil platforms. However, the shaft body in the above-mentioned foundation is often affected by the combined actions of axial force, horizontal shear and moment, whereas only the use of tip grouting technique may be difficult to meet the shaft foundation design requirements. Therefore, it is necessary to use the post-grouting technique at the shaft side or at the shaft tip and side to further improve the bearing capacity of large-diameter drilled shafts. On this basis, there is a need to analyze the response of side grouting shafts and combined grouting shafts.

In this paper, the field tests are carried out four large-diameter drilled shafts in Shishou, Hubei Province, Central China, including three combined side-and-tip grouting shafts and one side-grouting shaft. The load-displacement response, shaft resistance, and toe resistance under axial load in extra-thick fine sand layer are estimated from the strain gauge data obtained during the static load tests, and the results of combined-grouting shafts and side-grouting shaft before and after grouting are compared and analyzed. Finally, the test results of four large-diameter drilled shafts were further discussed.

2   Site and Test Shafts Conditions

Shishou Yangtze River Highway Bridge is located in Shishou, Hubei Province, Central China. The results of geological drilling in the proposed bridge site shows that there is no bedrock in the exploration of 180 m, and the regional geological conditions are relatively simple, which is mainly composed of fine sand layer. To clarify the behavior of large-diameter drilled shafts in extra-thick fine sand layer, it is necessary to test shafts in the proposed bridge site. In-situ standard penetration tests (SPT) were performed prior to the design of shaft foundations. Two SPT boreholes (BH1 and BH2) were carried out near the test shafts of the approach bridge, the locations of the site investigation are illustrated in Fig. 1. The detailed soil profiles and properties for each soil layer are given in Table 1, in which w is the natural water content; γ_{sat} is the saturated unit weight; e is the void ratio; E_{s1-2} represents the compressive modulus of each soil layer; and c and φ are the cohesion and internal friction angle of each soil layer, respectively.

Figure 5 shows the detailed soil layer distributions and corresponding SPT blow-counts (N) variations at a given site: (a) borehole BH1 near the shafts TS1 and TS2; and (b) borehole BH2 near the shafts TS3 and TS4. In Fig. 5, the SPT N values of the cohesive soil at the shallow layer varies between 2 to 11, whereas the SPT N values of the fine sand soil at the deep layer varies between 16 to 41.

Four large-diameter drilled shafts in extra-thick fine sand layer were tested to study the response of axially loaded single shaft. The test shafts TS1, TS2 and TS3, TS4 are

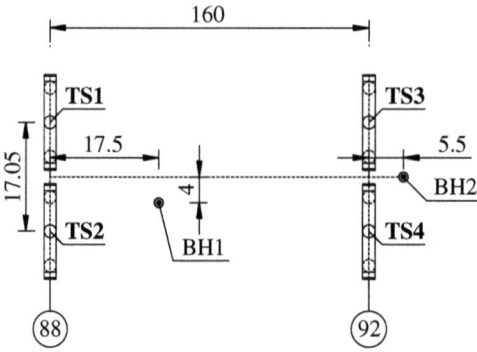

Fig. 1. The locations of SPT data near test shafts (Unit: m).

Table 1. Soil parameters.

Soil layer	w (%)	γ_{sat} (kN · m^{-3})	e	E_{s1-2} (MPa)	c (kPa)	φ (°)
Clay	31.5	18.9	0.901	4.43	16.2	13.9
Silt	31.1	18.9	0.881	4.30	14.7	13.8
Loose fine sand	20.6	19.0	0.968	8.6	15.3	27.74
Slightly dense fine sand		19.7	0.892	10.2	16.0	30.7
Medium dense fine sand		20.2	0.808	12.0	15.0	33.0
Very dense fine sand		20.5	0.672	12.4	15.0	33.9

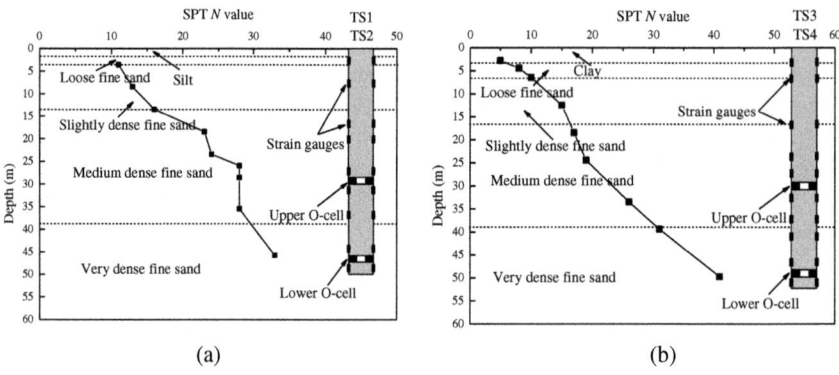

Fig. 2. Typical soil profiles, SPT N variations, and shaft instrumentation: (a) borehole BH1 near shafts TS1 and TS2; and (b) borehole BH2 near shafts TS3 and TS4.

located on pier no. 88 and pier no. 92 of the north approach bridge, respectively. The test shafts TS1 and TS2 with 2.0 m in diameter had lengths of 50 m, whereas the test shafts TS3 and TS4 with 2.0 m in diameter had lengths of 52 m. The layout of the four test shafts is shown in Fig. 1, and the geometry size of each test shaft and the subsurface profile at the location of each test shaft is shown in Fig. 2. The test shafts were installed using the reverse circulation rotary drilling method. The elastic modulus of the concrete for each test shaft was assumed to be 30 GPa. All the test shafts rested on a layer of very dense fine sand.

3 Post-grouting Technique

After the test shafts were formed, post-grouting technique was adopted to increase bearing capacity and reduce settlement of the test shafts. The test shafts TS2, TS3, and TS4 were subjected to the combined post-grouting at the shaft tip and side, and the test shafts TS1 was conducted to post-grouting at the shaft side alone. The grouting system is the same as that described by Duan and Kulhawy (2009), in which the grouting device at shaft tip is post-grouted using the straight pipe method, the grouting device at shaft side is post-grouted using the ring pipe method set to grouting pipes along the circumferential direction of the steel rebar cage. The test shafts TS2, TS3, and TS4 are placed two ring grouting pipes along the steel rebar cage at two levels, while the test shaft TS1 is placed three ring grouting pipes along the steel rebar cage at three levels. The number and location of the grouting pipe at the shaft tip and side are comprehensively determined according to the Chinese standard (Code for Design of Ground Base and Foundation of Highway Bridges and Culverts 2007; Technical Code for Building Pile Foundation 2008) and geotechnical investigations. The position of grouting ring pipes above the shafts tip for each test shaft is summarized in Table 3. The sequence of combined side-and-tip grouting should first perform grouting at the shaft side, and then grouting at the shaft tip. Additionally, when shaft side grouting pipes were attached to the reinforcement cage of the test shafts at different levels, the sequence of side grouting should be performed from shaft head to shaft bottom. It should be noted that the time interval of the side grouting and the tip grouting should not be less than 2 h. The test shafts are post-grouted using a mixture of water and ordinary Portland cement. The water-cement ratio of the grout is about 0.5:1, and the other detailed grouting parameters of the test shafts are described in Table 2.

Table 2. Grouting parameters.

Shaft no.	Quantity of cement (t)		Grouting pressure (MPa)		Position of side pipes above the shaft tip (m)	Grouting type
	Shaft side	Shaft tip	Shaft side	Shaft tip	First, second, third	
TS1	3.6	N/A	3.9	N/A	8, 23, 38	Side grouting
TS2	3.0	4.0	2.7	4.2	15, 30, N/A	Combined grouting
TS3	3.4	3.6	2.8	4.3	15, 30, N/A	Combined grouting
TS4	3.4	3.6	2.9	4.2	15, 30, N/A	Combined grouting

Note: the grouting pressure of the first cross-section of the grouting pipe at the shaft side is only given.

4 Test Method

The static load tests were conducted in two phases: before and after post-grouting. The first phase was commenced on after the pouring concrete and reached the concrete strength of the shaft body. The second phase was conducted after post-grouting and met requirements of the grout curing.

The bi-directional double-level O-cell test was employed in the field tests of the test shafts. The positions of the double-level O-cell for the test shafts are shown in Fig. 2. The bi-directional O-cell tests were conducted on the test shafts in accordance with the slowly maintained load procedure. The methods of loading and unloading were in accordance with the Chinese Technical Code for Static Loading Test of Foundation Pile-Self-balanced Method (2009). To investigate the influence of post-grouting on the bearing characteristics of large-diameter drilled shafts, strain gauges were installed on the four symmetrical reinforcing bars of each test shaft at each level before the installation of test shafts. Figure 2 presents the elevations of strain gauges along the test shaft. Two displacement transducers were fixed at the top of each test shaft for measuring the shaft head displacement, and four displacement transducers were installed at the upper and lower level of O-cell for measuring the displacement of the upper and lower O-cell.

5 Analysis on Load-Test Results

5.1 Equivalent Load-Displacement Responses

The bi-directional O-cell tests of the first phase were carried out about 20 days after pouring the concrete, and the bi-directional O-cell tests of the second phase were conducted about 30 days after grouting. The test results of the double-level bi-directional O-cell test before and after grouting were converted into the equivalent shaft head load-displacement curves for the top-down load test method. The details of the equivalent conversion of the bi-directional O-cell test were discussed by Dai et al. (2010). Therefore, the equivalent shaft head load-displacement curves of the test shafts before and after grouting are obtained, as shown in Fig. 3.

Figure 3 shows that the equivalent load-displacement curves of the test shafts before and after grouting have distinct plunging points, and the equivalent shaft head load-displacement curves for the test shafts after grouting are gentler than that for the test shafts before grouting. Prakash et al. (1990) and Zhang et al. (2014) proposed that the ultimate bearing capacity of a single shaft is taken to be the load when the shaft is plunged. According to the equivalent shaft head load-displacement curves, when the shaft head displacement s is 5, 10, 15, and 20 mm, the load value of the test shafts before and after grouting and their corresponding increased range are shown in Table 3. Furthermore, the load prior to reaching the maximum load can be regarded as the ultimate bearing capacity, the ultimate bearing capacity of each test shaft before and after grouting can be obtained. The ultimate bearing capacity, total shaft resistance, base resistance of each test shaft, and their corresponding increased range are summarized in Table 4.

It can also be observed in Fig. 3 that the equivalent displacement of the test shafts head increases with the equivalent applied load, and the equivalent shaft head load-displacement curves are almost linear for each test shaft before and after grouting during the initial load increments. There are some differences between the grouted shafts and the ungrouted shafts as the equivalent applied load increases. The equivalent displacement increase at the shaft head for grouted shafts is less than that of ungrouted shafts at the same load level. That is, the test shafts after grouting are able to bear more applied load on the shaft head at the same shaft head displacement. The test results indicate that the load-bearing capacity of grouted shafts is obviously larger than that of ungrouted shafts under the same conditions. Meanwhile, the load-bearing capacity of combined-grouting shafts is also greater than that of side-grouting shafts.

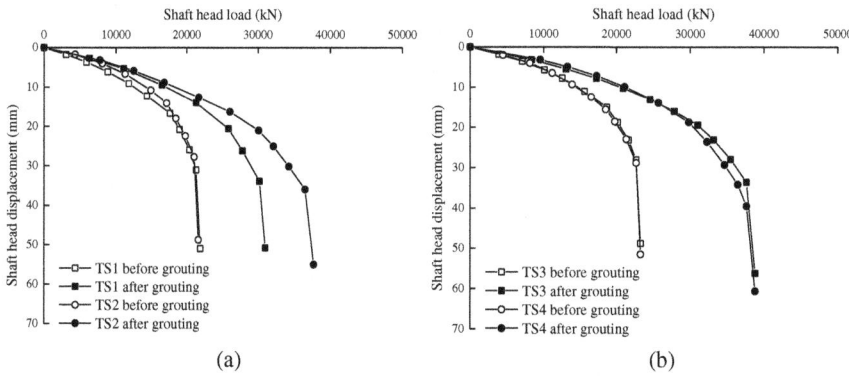

Fig. 3. Equivalent load versus displacement curves of the test shafts before and after grouting: (a) TS1 and TS2; and (b) TS3 and TS4.

It can be observed from Table 3 that the increased range of bearing capacity increases gradually with increasing of the shaft head displacement, indicating that the post-grouting shafts have a strong subsequent effect and can enhance the safe reserve capacity of the shaft foundations. When the grouted shaft reaches the ultimate state, the ultimate bearing capacity for post-grouting at the shaft side alone is increased by at least 41.54%, whereas the ultimate bearing capacity for post-grouting at the shaft tip and side is increased by at least 66.03%. It reflects that post-grouting craftwork can effectively improve the ultimate bearing capacity of the shaft foundation and its settlement can be effectively controlled.

The results in Table 4 show that compared with the test results of test shaft TS1 before grouting at the pile side, the ultimate bearing capacity of the test shaft TS1 after grouting at the pile side in extra-thick fine sand layer is increased by 41.54%, the total shaft resistance is increased by 51.85%, and the base resistance is increased by 1.27%. In addition, the ultimate bearing capacity, the total shaft resistance and the base resistance of the test shafts TS2, TS3, and TS4 after grouting at the shaft tip and side are 66.03–73.49%, 46.72–56.91%, and 137.87–139.17% larger than that of before grouting at the shaft tip and side, respectively. It can thus be concluded that the bearing behavior of the combined-grouting shaft is better than that of the side-grouting shaft, and the load transfer characteristics of the shaft are significantly improved.

Table 3. The load value and the corresponding increased range the test shafts before and after grouting at different displacements.

Shaft no.		TS1		TS2		TS3		TS4	
		Before grouting	After grouting	Before grouting	After grouting	Before grouting	After grouting	Before grouting	After grouting
s = 5 mm	Q_u (kN)	7451	10662	8977	10729	9132	12491	9609	13338
	Increased range (%)	43.09		19.52		36.78		38.80	
s = 10 mm	Q_u (kN)	12570	17167	14088	18206	14662	20329	14361	21117
	Increased range (%)	36.57		29.23		38.65		47.05	
s = 15 mm	Q_u (kN)	16447	22127	17281	24554	18635	26725	18148	26414
	Increased range (%)	34.53		42.08		43.42		45.55	
s = 20 mm	Q_u (kN)	18795	25762	19034	29101	20631	31292	20339	30396
	Increased range (%)	37.07		52.89		51.68		49.45	
Ultimate state	Q_u (kN)	21321	30178	21046	36513	22676	37648	22676	37648
	Increased range (%)	41.54		73.49		66.03		66.03	

Table 4. Ultimate bearing capacities, total shaft resistances, base resistances of the test piles and their corresponding increased range

Shaft no.	Before grouting			After grouting			$\frac{Q_{su}'-Q_{su}}{Q_{su}}$ (%)	$\frac{Q_{bu}'-Q_{bu}}{Q_{bu}}$ (%)
	Total shaft resistance Q_{su} (kN)	Mobilized base load Q_{bu} (kN)	Ultimate bearing capacity Q_u (kN)	Total shaft resistance Q_{su}' (kN)	Mobilized base load Q_{bu}' (kN)	Ultimate bearing capacity Q_u' (kN)		
TS1	16975	4346	21321	25777	4401	30178	51.85	1.27
TS2	16803	4243	21046	26365	10148	36513	56.91	139.17
TS3	17878	4798	22676	26235	11413	37648	46.74	137.87
TS4	17897	4779	22676	26258	11390	37648	46.72	138.33

5.2   Shaft Resistance

The total shaft resistance was obtained by subtracting the mobilized base load from the ultimate bearing capacity, and the shaft head displacement was obtained by the equivalent conversion of the test results of the bi-directional O-cell test method. The relationship between the total shaft resistance and the shaft head settlement of the test shafts before and after grouting are plotted in Fig. 4.

It can be observed from Fig. 4 that the changing trends of the total shaft resistance of the test shafts before and after grouting are almost identical as the shaft head displacement, but have different values of the total shaft resistance. The total shaft resistance of the test shafts after grouting is significantly larger than that of without grouting. The grout may have penetrated into the fine sand layer at the shaft side because of high pressure grouting, the boundary conditions at the shaft-soil contact may be improved and increase adhesion of the concrete to the fine sand layer, giving high strength and stiffness of the soil at the shaft side. The total shaft resistance of grouted shaft is thereby larger than that of ungrouted shaft. In addition, the pressure grout compacts the soil at the shaft side and increases horizontal stress at the shaft-soil interface, which has significant and positive impact on the improvement of the shaft resistance.

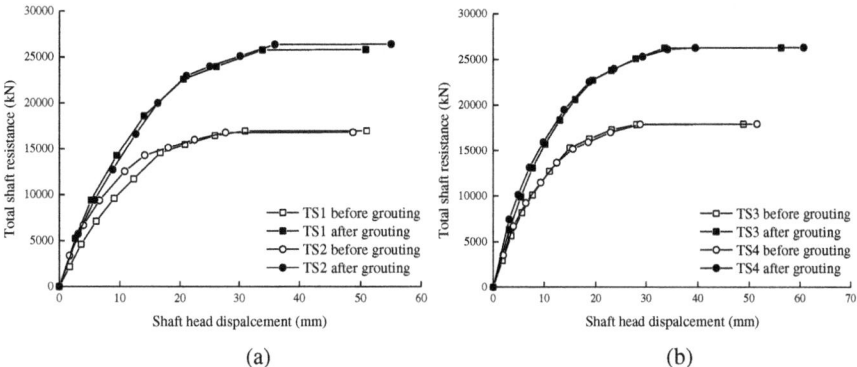

Fig. 4. Total shaft resistance versus shaft head displacement curves of the test shafts before and after grouting: (a) TS1 and TS2; and (b) TS3 and TS4.

5.3   Mobilized Base Resistance

The tip resistance of the test shaft is obtained by subtracting the shaft resistance of the lower shaft segment from the lower O-cell load, whereas the tip displacement is obtained by subtracting the compression of the lower shaft segment from the downward displacement of the lower O-cell. Therefore, the tip resistance-tip displacement curves of the test shafts before and after grouting can be obtained, as shown in Fig. 5.

It can be found from Fig. 5 that the tip resistance-tip displacement curves of the test shafts before and after grouting are similar but have different values of the tip resistance. The tip displacement after grouting is less than that before grouting at the same tip resistance, which indicates that end bearing capacity can be mobilized within smaller displacement after grouting. The pressure grout effectively strengthens the loose sediment at the shaft tip and the mudcake at the shaft side by means of compacting, permeating, and splitting, producing high strength and stiffness of the soil at the shaft tip. The tip resistance of grouted shaft is therefore larger than that of ungrouted shaft. Additionally, the tip resistance of the test shaft TS2 after grouting at the shaft tip and side is larger than that of the test shaft TS1 after grouting at the shaft side alone. It also reflects that more significant effect can be achieved compared with post-grouting at the shaft side alone, and then the tip responses of combined-grouting shaft also have a more significant impact.

Figure 6 shows the relationships between the mobilized unit tip resistance and normalized tip displacement for the four test shafts before and after grouting. The normalized tip displacement is the ratio of the tip displacement to the shaft diameter (D). As previously described, the tip displacement s_t is obtained by subtracting the

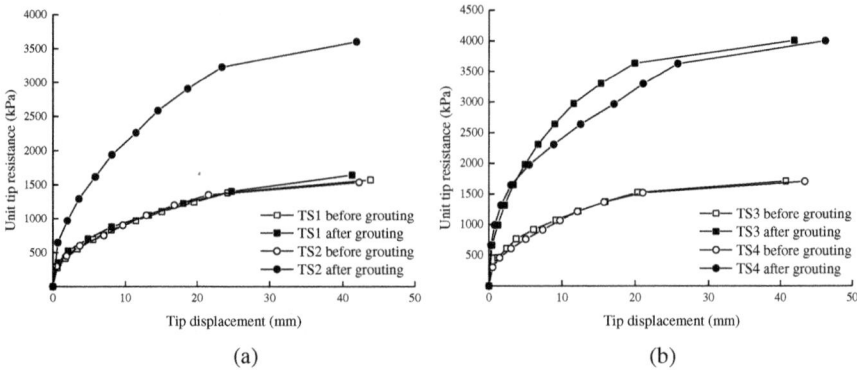

Fig. 5. The tip resistance versus tip displacment curves of the test shafts before and after grouting: (a) TS1 and TS2; and (b) TS3 and TS4.

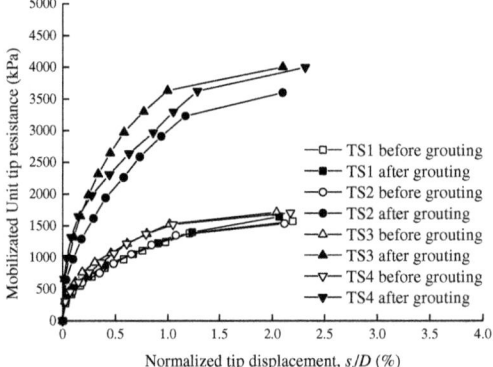

Fig. 6. Relationships between the mobilized unit tip resistance and normalized tip displacement of test shafts before and after grouting.

compression of the lower shaft segment from the downward displacement of the lower O-cell and is normalized by D. In Fig. 6, the mobilized unit tip resistance increases with increasing normalized tip displacement. The mobilized unit tip resistance for the test shafts after grouting is significantly larger than that for the test shafts before grouting at the same normalized tip settlement, except for the test shaft TS1. It should be noted that the tip displacement for the fully mobilizing tip resistance capacity of test shafts in extra-thick fine sand layer before and after grouting was about 2%–2.5% of D. This normalized tip displacement for the fully mobilizing tip resistance (i.e., 2%–2.5%) is smaller than the displacement for the fully mobilizing tip resistance (4%–5%) reported by Reese and O'Neil (1988). The tip displacement needed to mobilize ultimate tip resistance ranges may result from the different bearing stratum, the geometry size of the shaft, the construction uncertainty of each shaft, and the effect of post-grouting (Ng et al. 2001).

6 Discussion

The test results have demonstrated that the four large-diameter drilled shafts after grouting in extra-thick fine sand layer have high load-bearing capacity, and the load-bearing capacity of combined-grouting shafts is greater than that of side-grouting shafts. To further analyze the mobilized tip resistance and side resistance sharing ratio, thus, the ratios of the mobilized tip resistance or total shaft resistance to the shaft head load for the four test shafts is discussed. The ratios of the mobilized tip resistance or total shaft resistance to the shaft head load for shafts before and after grouting under different loading levels are illustrated in Fig. 7.

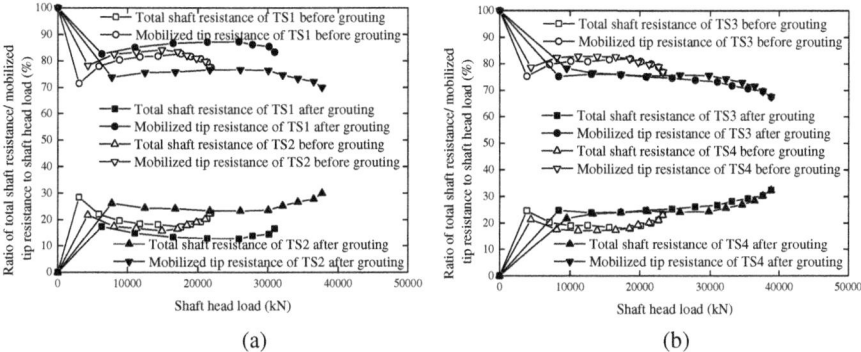

Fig. 7. Ratio of total shaft resistance/ mobilized tip resistance to shaft head load for test shafts before and after grouting at different loading levels.

It can be observed from Fig. 7 that the shaft head load for the test shafts before grouting is fully supported by the shaft resistance for small applied loads before the tip resistance was mobilized, whereas the shaft head load after grouting was first supported by the tip resistance during the initial load due to the preloading effect of tip grouting. With increasing applied load, the tip resistance was mobilized gradually, and the mobilized tip resistance sharing ratio also gradually increased. As the applied load reached the ultimate load, the mobilized tip resistance sharing ratios for the test shafts TS2, TS3, and TS4 after grouting at the shaft tip and side were 27.79%, 30.32%, and 30.25%, respectively, whereas the mobilized tip resistance sharing ratios for the test shaft TS1 after grouting at the shaft side alone was 14.58%. Hence, the large-diameter drilled shaft after grouting at the shaft tip and side functioned as an end-bearing friction shaft.

Figure 8 shows the distribution of the unit shaft resistance along the test shafts TS1 and TS2 before and after grouting under the ultimate loading. It can be seen from Fig. 8 that compared with the test results of test shafts TS1 and TS2 before grouting, the unit shaft resistance of the test shafts after grouting is obviously improved. The increased range of the unit shaft resistance for the combined-grouting test shaft TS2 is higher than that of the side-grouting test shaft TS1, especially for the unit shaft resistance near the shaft tip. It should be noted that the tip grouting for the combined-grouting shaft may

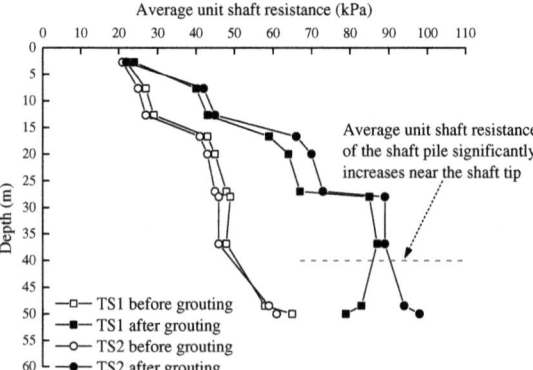

Fig. 8. Average unit shaft resistance of each section for the test shafts before and after grouting.

improve the shaft-soil contact near the shaft tip due to migration of the grout along the shaft length, producing high strength and stiffness of the soil at the shaft side, the unit shaft resistance is therefore improved. On the other hand, the tip grouting can compact and penetrate the fine sand layer at the shaft tip and mixes with the fine sand layer to effectively strengthen them, making them stronger and stiffer. Therefore, the strengthening effect of the soil improvement at the shaft tip on the unit shaft resistance. Indeed, it has been observed in the literature (Zhang et al. 2013) that enhance in the soil strength at the shaft tip puts positive impact on the mobilization of the shaft resistance.

It can also be observed from Fig. 5 that compared with the tip resistance of test shaft TS1 before grouting at the shaft side, the tip resistance of the test shaft TS1 after grouting at the shaft side is increased by 1.27%, which indicates that post-grouting at the shaft side can not only enhance the strength of the soil at the shaft side, but also improve the strength of the soil at the shaft tip. It is noted that post-grouting at the shaft side effectively solidifies the mudcake at the shaft side by means of compacting, permeating, filling, and consolidating, whereas the strengthening effect of the surrounding soil improvement on mobilizing the tip resistance.

7   Conclusions

This paper presents the results of field tests performed on three combined side-and-tip grouting shafts and one side grouting shaft constructed in extra-thick fine sand layer in the region of Shishou, China. Based on the test results and discussions, the following conclusions can be drawn.

(1) The ultimate bearing capacity of side-grouting shafts in extra-thick fine sand layer is increased by 41.54%, and the ultimate bearing capacity of combined-grouting shafts is increased by 66.03–73.49%. The results show that the grouted shafts provide larger load-bearing capacity than ungrouted shafts under the same conditions, and the combined-grouting shafts provide also greater load-bearing capacity than side-grouting shafts.

(2) Post-grouting at the shaft side effectively improves the boundary conditions between the shaft and the soil, and enhances the strength and stiffness of the soil around the shaft. Therefore, the total shaft resistances after grouting are 46.72–56.91% higher than that before grouting and have significant effects on the load transfer characteristics of the shaft.

(3) The mobilized tip resistance of grouted shafts larger than that of ungrouted shafts at the same tip displacement, and end bearing capacity can be mobilized within smaller displacement after post-grouting. Additionally, the increased range of the tip resistance for the side-grouting shaft and combined-grouting shafts is 1.27% and 137.87–139.17%, respectively, indicating that the effect of the combined-grouting shafts is more significant than that of the side-grouting shaft, and the tip responses of combined-grouting shafts also have a more significant impact.

(4) The increased range of the unit shaft resistance for the combined-grouting test shaft TS2 is larger than that of the side-grouting test shaft TS1, especially for the unit shaft resistance near the shaft tip, it reflects that the strengthening effect of the soil improvement at the shaft tip on the unit shaft resistance; the tip resistance of the side-grouting test shaft TS1 is 1.27% higher than that of ungrouted shaft, which indicates that the strengthening effect of the surrounding soil improvement on mobilizing the tip resistance. Consequently, the test results show the interaction between the shaft resistance and the tip resistance.

References

Bruce, D.A.: Enhancing the performance of large diameter piles by grouting. Ground Eng. **19**(4), 9–15 (1986)

Dai, G., Gong, W., Zhao, X., et al.: Static testing of pile-base post-grouting piles of the suramadu bridge. Geotech. Test. J. **34**(1), 34–49 (2010)

Duan, X., Kulhawy, F.H.: Tip post-grouting of slurry-drilled shafts in soil: Chinese experiences. In: Contemporary Topics in Deep Foundations, pp. 47–54. ASCE, Orlando (2009)

Mullins, G., Dapp, S.D, Lai, P.: Pressure-grouting drilled shaft tips in sand. In: New Technological and Design Developments in Deep Foundations, pp. 1–17 (2000)

Gouvenot, D., Gabaix, J.C.: A new foundation technique using piles sealed by cement grout under high pressure. In: Offshore Technology Conference, Texas, USA, pp. 645–656 (1975)

Littlechild, B.D., Plumbridge, G.D., Free, M.W.: Shaft grouted piles in sand and clay in Bangkok. In: Proceeding of 7th International Conference and Exhibition on Piling and Deep Foundations, pp. 171–178. DFI, Vienna (1998)

Ng, C.W.W., Yau, T.L.Y., Li, J.H.M., et al.: New failure load criterion for large diameter bored piles in weathered geomaterials. J. Geotech. Geoenviron. Eng. **127**(6), 488–498 (2001)

Prakash, S., Sharma, H.D.: Pile Foundations in Engineering Practice. Wiley, New York (1990)

Reese, L.C., O'Neill, M.W.: Drilled shafts: Construction and design. FHWA, Publication No. HI-88, p. 42 (1988)

Safaqah, O., Bittner, R., Zhang, X.: Post-grouting of drilled shaft tips on the sutong bridge: a case history. In: Proceeding of Geo-Denver 2007 Congress: Contemporary Issues in Deep Foundations, Denver, USA, pp. 1–10 (2007)

Stocker, M.F.: The influence of post grouting on the load bearing capacity of bored piles. In: Proceeding of 8th European Conference on Soil Mechanics and Foundation Engineering, Balkema, Helsiniki, pp. 167–170 (1983)

The Professional Standards Compilation Group of People's Republic of China. Code for Design of Ground Base and Foundation of Highway Bridges and Culverts (JTJD63-2007). China Communications Press, Beijing (2007)

The Professional Standards Compilation Group of People's Republic of China. Technical Code for Building Pile Foundation (JGJ94-2008). China Architecture and Building Press, Beijing (2008)

The Traffic Professional Standards Compilation Group of People's Republic of China. Static Loading Test of Foundation Pile-Self-balanced method (JT/T 738-2009). China Communications Press, Beijing (2009)

Thiyyakkandi, S., McVay, M., Bloomquist, D., et al.: Experimental study, numerical modeling of and axial prediction approach to base grouted drilled shafts in cohesionless soils. Acta Geotech. 9(3), 439–454 (2014)

Zhang, Z.M., Yu, J., Zhang, G.X., et al.: Test study on the characteristics of mudcakes and in situ soils around bored piles. Can. Geotech. J. 46(3), 241–255 (2009)

Zhang, Q., Li, S.C., Li, L.P.: Field study on the behavior of destructive and non-destructive piles under compression. Mar. Georesour. Geotechnol. 32(1), 18–37 (2014)

Zhang, Q.Q., Zhang, Z.M., Li, S.C.: Investigation into skin friction of bored pile including influence of soil strength at pile base. Mar. Georesour. Geotechnol. 31(1), 1–16 (2013)

A Reliability Study of a Pile Foundation Design in Soft Soils

Wenjun Dong[(✉)]

Bittner-Shen Consulting Engineers, Inc., Portland, OR, USA
wjd@bittner-shen.com

Abstract. This paper presents a reliability study of a pile foundation design in soft soils that are consolidating under a surcharge. Both limit and serviceability states are considered in the design study. The load and soil properties are assumed to possess variances that can be described by lognormal distributions. The influence of these variances on pile capacity and settlement are evaluated in terms of reliability level using the concept of reliability index of performance function that describes the design problem. Two design loads that correspond to different conventional mean factors of safety are considered, and relevant results are used to show the relationship of reliability levels of design with mean factor of safety and variance of soil properties. The results from reliability analysis for limit and serviceability states can be used in decision making process for the final design of pile foundation in soft soils.

Keywords: Reliability-Based Design · Pile foundation · Settlement Downdrag

1 Introduction

Pile foundations have been widely applied for structures over soft soils to enhance their bearing capacities by transferring applied loads to deeper competent layers or to reduce settlements of superstructures on the top (Fellenius 2015). Therefore, both limit and serviceability states need to be considered in pile designs. Usually, loads and soil/pile resistance components of such design problems possess uncertainties and they can be considered as random variables. In conventional design approach, a factor of safety was used to deal with the uncertainties of load and resistance components using their so-called characteristic values (Duncan 2000). However, the characteristic values of material properties and loads are not unambiguously defined and are usually determined based on personal judgments and past experiences of designers. A more rigorous way to assess the influence of uncertainties inherent in design problems on their safety margins is to use Reliability-Based Design (RBD) approach (Ang and Tang 1984; Harr 1987; Fenton and Griffith 2008).

The RBD has started its application to geotechnical problems to assess their safety margins (Phoon 2008). In the RBD approach, the limit state of a geotechnical design problem is described by a performance function that is conventionally assumed to have normal distribution for its merits in statistical calculations. Then, the influence of uncertainties embedded in load and resistance properties on the design is assessed by

© Springer Nature Singapore Pte Ltd. 2018
T. Qiu et al. (Eds.): GSIC 2018, *Proceedings of GeoShanghai 2018 International Conference: Advances in Soil Dynamics and Foundation Engineering*, pp. 675–683, 2018.
https://doi.org/10.1007/978-981-13-0131-5_73

the reliability index of performance function (Cornell 1969; Rackwitz 1976; Hasofer and Lind 1974; Dong 2017). Compared to conventional design approach using a factor of safety, reliability analysis provides more comprehensive insights into the influence of uncertainties on designs and its results in terms of probability of failure or reliability can be more easily incorporated with the risk-analysis of problem and decision-making process.

This paper presents a reliability study of a pile foundation design in soft soils. The pile foundation is assumed to be subjected to loads from superstructures and downdrag forces from adjacent consolidating soils. The reliability levels of pile foundation design for both limit and serviceability states are evaluated based on the statistical properties of loads and geo-materials. In the paper, the models that are used to predict the pile capacity, including skin friction and end bearing resistance, as well as soil consolidation are first reviewed. Next, a performance function is established to describe the limit state of pile foundation design, and corresponding reliability index is presented, which links the probability of reliability or failure with factor of safety and uncertainties involved in the design problem. Finally, a reliability analysis is applied to predict the reliability level of pile settlement in soft soils that consolidate under a surcharge load. A brief discussion is given on applying the results from the reliability analysis for both limit and serviceability states for final pile design.

2   A Pile Foundation Design

In this paper, we consider a pile design problem in soft soils. The soil profile is assumed to consist of an unconsolidated Holocene clay overlying a consolidated Pleistocene clay that is underlain by a sand layer, as shown in Table 1. The pile is assumed to be steel pipe pile with a diameter of 1 m and a wall thickness of 15 mm, and it penetrates 30 m in ground to the Pleistocene clay layer.

Table 1. Soil profile and parameters.

Soil	Depth (m)	γ_s (kN/m^3)	β_s	N_t	m	m_r	σ_p' (kPa)
Holocene	0	16.5	0.15	10	12	120	0
Pleistocene	20	18.5	0.25	20	18	180	10
Sand	35	20	0.30	40	250	750	10

2.1   Pile Capacity

In this example, the nominal ultimate axial capacity of pile is estimated using the following prediction formula

$$R_n = R_b + R_s = f_s A_s + q_b A_s = \beta_s \sigma_v' A_s + N_t \sigma_v' A_b \tag{1}$$

where $R_s = f_s A_s$ and $R_b = q_b A_b$ are the shaft (skin) friction resistance and the tip (base) bearing resistance of pile, respectively, $f_s = \beta \sigma_v'$ is the pile shaft resistance stress,

$q_b = N_t \sigma'_v$ is the pile tip resistance stress, β_s is the Bjerrum-Burland coefficient, N_t is the tip resistance coefficient, A_s is the pile shaft area embedded in the soils, and A_b is the pile base area. In the following discussions, the prediction formula in (1) for pile (skin and end bearing) capacities are assumed to be unbiased. Using (1) and the parameters listed in Table 1, the pile under consideration has a predicted ultimate capacity of 2446 kN. The pile is assumed to be subjected to a design working load Q_n on the top. Design codes or manuals require that the pile shall be designed to satisfy the following criterion,

$$\frac{R_n}{Q_n} \geq F_S \tag{2}$$

where F_S is the mean factor of safety that is used to deal with uncertainties involved in loads and resistances. The selection of F_S should be based on confidence levels on load and resistance variables and is usually determined using judgments and/or past experiences. In this paper, we assume the working load Q_n consists of Dead Load (DL) only. Two DL magnitudes are considered, one with 1200 kN and the other with 800 kN, resulting in two F_S of 2.0 and 3.0, respectively. Figure 1(a) shows the predicted distributions of ultimate axial capacity of the pile and negative skin frictions on the pile (due to the settlement of adjacent soils).

2.2   Settlement

In this example, the soft soils adjacent to the pile is assumed to consolidate under a surcharge load of 22 kN (which may be attributed to the placement of a gravel layer on the top). For fine-grained soils, their settlements under a surcharge can be calculated using Janbu's approach (Fellenius 2015), i.e.

$$\varepsilon = \frac{1}{m_r} \ln \frac{\sigma'_1}{\sigma'_0} \quad \text{if} \quad \sigma'_1 \leq \sigma'_p$$
$$\varepsilon = \frac{1}{m_r} \ln \frac{\sigma'_p}{\sigma'_0} + \frac{1}{m} \ln \frac{\sigma'_1}{\sigma'_p} \quad \text{if} \quad \sigma'_1 > \sigma'_p \tag{3}$$

where σ'_0 and σ'_1 are the initial and final vertical effective stress before and after consolidation, σ'_p is the pre-consolidation stress defined as the summation of pre-overburden pressure and in situ vertical effective stress, m and m_r are the tangent modulus numbers for primary loading and unloading/reloading. Using (3) and the modulus numbers as listed in Table 1, the soil settlement profile can be estimated as shown in Fig. 1(b). Figure 1 also shows two Neutral Planes (NP) corresponding to two DL conditions. In theory, the NP is defined as the depth where downdrag loads on the pile (i.e. DL plus negative skin friction) is equal to remaining pile ultimate capacity (i.e. end bearing plus positive skin friction below the NP) (Fellenius 2015). The pile under downdrag loads will settle with soils at the NP. It can be seen from Fig. 1 that the larger the DL becomes, the shallower the NP is, and the more settlement the pile has.

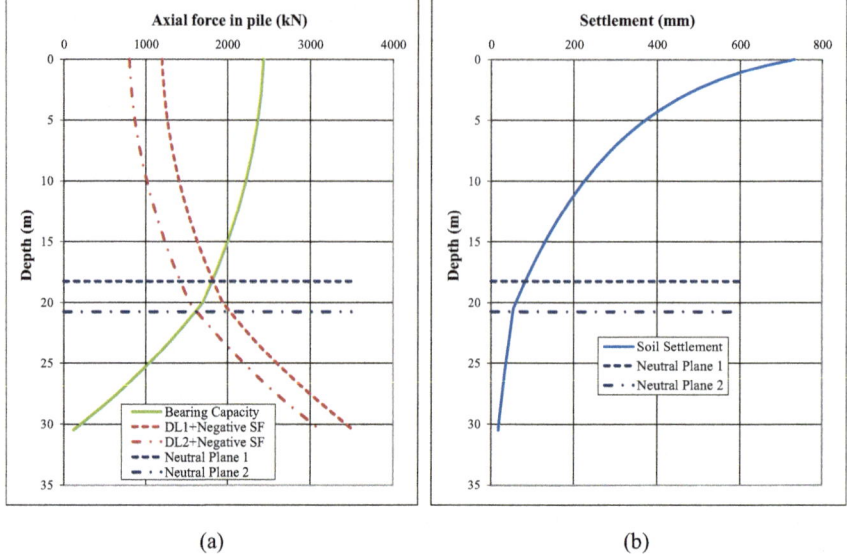

Fig. 1. Distribution curves of (a) pile capacity and downdrag load, and (b) settlement of adjacent soil under surcharge

3   Reliability of Design

3.1   Performance Function

The appropriateness of F_S as defined in (2) can be evaluated with the probability of failure or reliability of design problem using a performance function. The performance function is used to describe the limit state of geotechnical design problem and it can be developed by directly comparing lumped load Q and resistance R. The performance function is considered as random variables due to the uncertainties inherent in load Q and resistance R. Most input variables in geotechnical problems can be described by lognormal distributions. In addition, the performance function that has a logarithmic form of division operation most likely has normal probability distribution (Harr 1987; Dong 2016). Therefore, it is convenient to create a performance function g in terms of the following logarithmic form,

$$g = \ln R - \ln Q = \ln(R/Q) \qquad (4)$$

The limit state of design is represented by the failure surface of performance function g, defined as $g = 0$. The point on the failure plane is called failure point (R_f, Q_f), i.e. $g(R_f, Q_f) = 0$, implying that $R_f = Q_f = x_f$.

3.2 Reliability Index

The performance function g as defined in (4) can be considered as the linear combination of the lognormal variables, $\ln Q$ and $\ln R$. Assuming Q and R are uncorrelated, then using the Advanced First Order Second Moment (AFOSM) method (Rackwitz 1976; Dong 2017), the reliability index β of performance function g can be expressed as

$$\beta = \frac{E[g]}{\sigma[g]} = \frac{\ln\left(F_S\sqrt{\frac{1+COV_Q^2}{1+COV_R^2}}\right)}{\sqrt{\ln(1+COV_R^2) + \ln\left(1+COV_Q^2\right)}} \tag{5}$$

in which E[g] and σ[g] represent the mean and standard deviation (SD) of performance function g, $COV_Q = \sigma_Q/\mu_Q$ and $COV_R = \sigma_R/\mu_R$ are the coefficient of variation (COV) of Q and R, μ_Q and μ_R are the means of Q and R, σ_Q and σ_R are the SDs of Q and R, Fs is the mean factor of safety as defined above. Assuming a normal probability distribution to the performance function g, the probability of failure (p_f) or reliability (p_r) of the design problem described by g can be calculated by

$$p_r = 1 - p_f = P(g > 0) = 1 - \Phi\left(\frac{0 - E[g]}{\sigma[g]}\right) = 1 - \Phi(-\beta) \tag{6}$$

where Φ is the standard normal cumulative distribution function. It can be seen from (5) and (6) that the p_r of the design problem as described by g is related to F_S and COVs of load and resistance. Such relationship is shown in Fig. 2 with two different COVs.

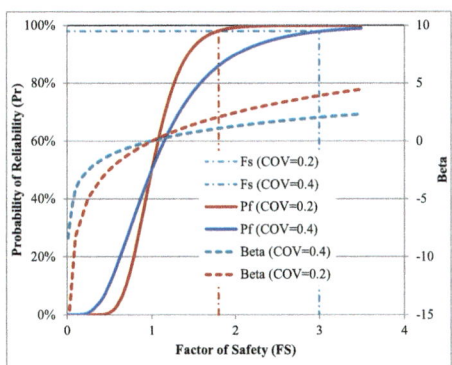

Fig. 2. Relationships of p_f and β with Fs with different COVs

It can be seen from Fig. 2 that the probability of reliability p_r is proportional to F_S, with $p_r = 50\%$ when $F_S = 1$ (for unbiased predictions of Q and R). When $F_S > 1$, p_r is greater than 50%, and it decreases with increasing COV_Q and $COVR_Q$; When $F_S < 1$, p_r is less than 50%, and it increases with increasing COV_Q and COV_R. Uncertainties reduce reliability levels when $F_S > 1$, but they increase reliability levels when $F_S < 1$, making p_r approaching 50%. The relationships as shown in Fig. 2 can provide insights on choosing F_S for different variance levels of load and resistance components. For example, if a p_r equal to 98% is set as an acceptable risk level, then F_S of 1.8 and 3.0 should be selected in conventional design approach for two levels of variance in load and resistance components, respectively.

4   Reliability of Pile Foundation

4.1   Pile Capacity

The parameters as listed in Table 1 can be considered as mean (or characteristic) values for load, soil resistance and settlement properties. In addition to the mean values, they are also assumed to possess the variations that can be described by lognormal distributions with statistical parameters as listed in Table 2. In Table 2, μ, σ and COV represent the mean, standard deviation and COV of corresponding properties. In this design example, the pile can be considered as friction pile and therefore the possible variations on pile tip resistance is ignored. Since the top Holocene clay is considered unconsolidated, only variation on its modulus number for primary loading is considered.

Table 2. Variations of load and soil properties

Soil	μ	COV	σ	$\mu - \sigma$	$\mu + \sigma$
DL-1	1200 kN	0.2	240 kN	960 kN	1440 kN
DL-2	800 kN	0.2	160 kN	640 kN	960 kN
β_{s1} (Holocene)	0.15	0.5	0.075	0.075	0.225
β_{s2} (Pleistocene)	0.25	0.5	0.125	0.125	0.375
m_1 (Holocene)	12	0.4	4.8	7.2	16.8
m_2 (Pleistocene)	18	0.4	7.2	10.8	25.2
m_{r2} (Pleistocene)	180	0.4	72	108	252

With the assumption that load and resistance properties possess lognormal distributions, the performance function (4) is used to describe the limit states of the pile design problem. As discussed above, the probability of reliability p_r or reliability index β is related to the statistical properties of loads and pile resistances. The statistical moments of load can be calculated directly using parameters listed in Table 2 and they are listed in Table 3 below. The statistical moments, means and variances of pile resistances as listed in Table 3, are calculated based on soil properties using the point estimate method (Rosenblueth 1975). The corresponding Probability Density Functions (PDF) of DLs and R are shown in Fig. 3. Using the statistical properties as listed in Table 3 and definitions as shown in (5) and (6), we can have the reliability indices and reliability probabilities for different DLs (corresponding to two different F_S) as listed in Table 4.

As discussed above, given the same COVs of loads and resistances, the larger F_S results in higher probability of reliability as expected. However, selecting an appropriate

F_S for a pile foundation, or to see if such pile foundation is suitable for a larger or smaller DL, can be evaluated by comparing corresponding p_r or p_f with the criteria that can be preset by the risk-analysis of the problem in terms of safety and cost. If p_r or p_f corresponding lower F_S satisfies the criterion, then such pile foundation is considered safe for a relatively larger DL with a potential saving on cost. If p_r or p_f corresponding to higher F_S does not satisfy the criterion (due to larger uncertainties involved, etc.), then such pile foundation needs to be improved for an increased safety margin, implying an even higher equivalent F_S.

Table 3. Statistical moments of dead load and pile resistance

	R	DL_1	DL_2	$\ln(R)$	$\ln(DL_1)$	$\ln(DL_2)$
Mean (kN)	2446	1200	800	7.747	7.070	6.665
SD (kN)	832	240	160	0.331	0.198	0.198
COV	0.34	0.2	0.2	0.043	0.028	0.030

Table 4. Reliability analysis results of pile resistance under two different loading conditions

LC	F_S	β	p_r
DL_1	2	1.755	96.0%
DL_2	3	2.806	99.7%

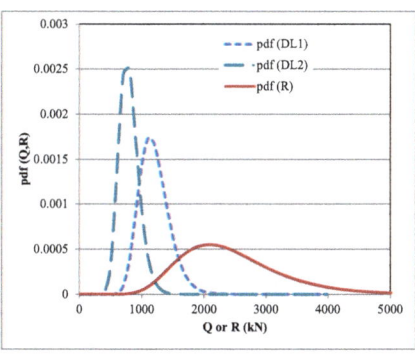

Fig. 3. Probability density functions of pile resistance and DL1 and DL2.

4.2 Pile Settlement

As mentioned above, when the soft soils adjacent to pile are subjected to a surcharge, pile will settle with soils at the NP. The location of NP is related to the magnitude of loads as well as the resistance and settlement profiles of soft soils. One illustration of such relationship is shown in Fig. 2, in which the pile resistance and soil settlement distributions are calculated using the mean values of soil properties as listed in Table 1. When soil properties possess variations, it can be expected that both NP and pile settlement will also present variations. A similar procedure to the pile capacity as discussed in the previous section can be used to estimate such variations, or probability distributions, based on the statistical parameters of loads and soil properties as listed in Table 2. Table 5 lists the calculated statistical properties of NP and pile settlement (with soil at the NP), corresponding to two DL conditions that are considered in this example. If the variation of settlement is assumed to be described by a lognormal distribution, then the PDF and Cumulated Distribution Function (CDF) of pile settlement (at the NPs) for two DL conditions can be determined as shown in Fig. 4. The CDF shown in Fig. 4(b) reflects the probability of settlement less than the corresponding value on abscissa.

Table 5. Statistical properties of neutral plane depths and pile settlement at neutral plane

	NP_1 (m)	S_1 (mm)	NP_2 (m)	S_2 (mm)
Mean	15.6	169.8	19.1	97.8
SD	6.1	137.1	3.5	54.9
COV	0.39	0.81	0.18	0.56

Table 6. Pile settlement characteristics under two different DL conditions

LC	NP_m (m)	S_m (mm)	$P(S \leq S_m)$	$S(P = 80\%)$ (mm)
DL_1	18.25	83	25.5%	240
DL_2	20.75	53	18.2%	135

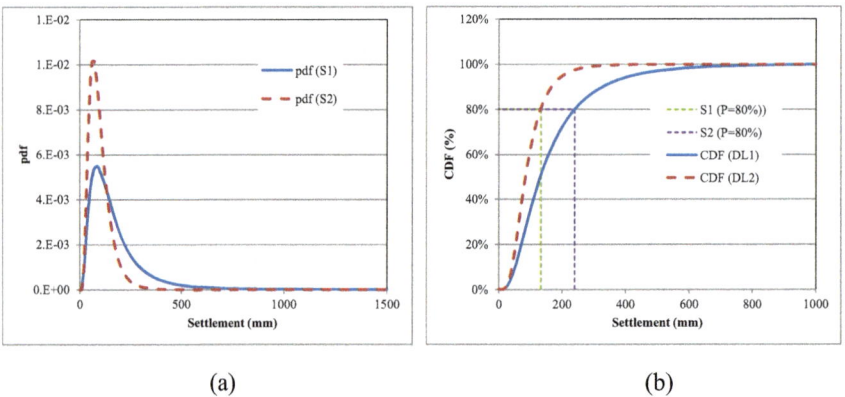

(a)   (b)

Fig. 4. Probability density functions (a) and cumulated distribution functions (b) of settlement at neutral plane for two dead load cases.

From the pile resistance and settlement distributions as shown in Fig. 2, we can have the "mean" values of neutral plane, NP_m, and pile settlement, S_m, for two DL conditions, which are listed in Table 6. They are referred as "means" because they are calculated using the mean values of loads and soil properties. However, from the CDFs as shown in Fig. 4(b), the probabilities of these mean pile settlements are only 25.5% and 18.2% for DL1 and DL2 conditions, respectively. If the required probability of settlement is preset as 80%, the corresponding settlement thread holds should be 240 mm and 135 mm for DL1 and DL2 conditions, respectively. If such thread holds do not meet requirement, the pile settlement needs to be reduced by either increasing pile embedment length (consequently lowering NP) or reducing soil settlement in general.

5 Conclusions

Reliability analysis can provide more comprehensive insights on the uncertainties of pile design in soft soils with respect to pile capacity and settlement. Compared to factor of safety adopted in conventional design approach, the safety margin expressed in terms of probability of failure or reliability is more easily incorporated with the risk-analysis and decision-making process for pile design problem. The appropriateness of selected factor of safety in conventional pile foundation design will depend not only on the magnitude of load and resistance but also on the levels of their variances, and it can be evaluated with reliability analysis. In addition to the limit state, reliability analysis can be also applied to the serviceability state of pile foundation in soft soils to investigate the statistical characteristics of its settlements, which may be quite different from one for the limit state. A robust design of pile foundation should provide acceptable reliability levels for both capacity and settlement.

References

Fellenius, B.H.: Basics of Foundation Design. Pile Buck International, Inc., Vero Beach (2015)
Duncan, J.M.: Factor of safety and reliability in geotechnical engineering. J. Geotech. Geoenviron. Eng. **126**(4), 307–316 (2000)
Ang, A.H.-S., Tang, W.H.: Probability Concepts in Engineering Planning and Design. Decision, Risk, and Reliability, vol. 2. Wiley, New York (1984)
Harr, M.E.: Reliability-Based Design in Civil Engineering. McGraw-Hill, New York (1987)
Fenton, G.A., Griffiths, D.V.: Risk Assessment in Geotechnical Engineering. Wiley, New York (2008)
Phoon, K.K.: Reliability-Based Design in Geotechnical Engineering: Computations and Applications. Taylor and Francis, London and New York (2008)
Cornell, C.A.: A probability-based structural code. J. Amer. Conc. Inst. **66**(12), 974–985 (1969)
Rackwitz, R.: Practical probabilistic approach to design. In: Bulletin 112, Comite European du Beton, Paris, pp. 13–72 (1976)
Hasofer, A.M., Lind, N.C.: Exact and invariant second-moment code format. J. Eng. Mech. Div. **100**(EM1), 111–121 (1974)
Dong, W.: An analytical approach of geotechnical reliability-based design and its application. In: Huang, J., et al. (ed.) Proceedings of the 6th International Symposium on Geotechnical Safety and Risk, Geo-Risk 2017, pp. 364–373. Denver, Colorado (2017)
Dong, W.: Reliability indices of equivalent geotechnical performance functions. In: Proceedings of the 6th Asian-Pacific Symposium on Structural Reliability and its Applications, pp. 164–169. Tongji University Press, Shanghai (2016)
Rosenblueth, E.: Point estimates for probability moments. In: Proceedings of the National Academy of Sciences, vol. 72 (10), pp. 3812–3814 (1975)

A Simplified Method for Nonlinear Analysis of Piled Raft

Jie Jiang[1,2,3], Kaiwen Hou[1,2,3], Xiaoduo Ou[1,2,3(✉)],
and Maosong Huang[4]

[1] College of Civil Engineering and Architecture, Guangxi University,
Nanning 530004, China
ouxiaoduo@163.com
[2] Key Laboratory of Disaster Prevention and Structural Safety of Ministry
of Education, Guangxi University, Nanning 530004, China
[3] Guangxi Key Laboratory of Disaster Prevention and Engineering Safety,
Guangxi University, Nanning 530004, China
[4] Department of Geotechnical Engineering, Tongji University,
Shanghai 200092, China

Abstract. Piled raft systems are often suitable as foundations for large buildings in soft soil, and the analysis of piled raft systems is very complex. On the basis of elasto-plastic analysis of single pile, the interaction relationship between pile and pile, pile and soil, soil and soil are considered to determine the stiffness matrix of pile group-soil system. The strength of soil varied along the depth is also considered. By solving the stiffness matrix equations, the vertical load-settlement relationship of piled raft foundation, axial load distribution in the piles along the depth and the load shared by piles and rafts were obtained. It needs no discretization of soil and pile along depth. Therefore, the analysis procedure is much simplified. The analyzed results show that the smaller pile length diameter ratio has a significant impact on the overall stiffness of piled raft foundation and the load shared by cap. Compared to results of boundary element method and field tests, the presented solutions are shown in good agreement. For considering the strength of soil varying along the depth and nonlinear interaction between piles and soil, the theoretical model agreed with the fact quite well. Hence the proposed method can applied to the analysis of practical engineering problems.

Keywords: Piled raft · Vertical load · Stiffness matrix · Nonlinear analysis
Interaction

1 Introduction

In the past few years, the piled foundation is usually assumed in elastic state under common working load conditions and then analyzed with elastic methods due to the complexity of this problem. Kuwabara [1] used the boundary element method to

Foundation item: Supported by the National Natural Science Foundation of China (51568006, 41372361), the Systematic Project of Guangxi Key Laboratory of Disaster Prevention and Structural Safety (2016ZDX11); Project funded by China Postdoctoral Science Foundation (2017M61 2865).

© Springer Nature Singapore Pte Ltd. 2018
T. Qiu et al. (Eds.): GSIC 2018, *Proceedings of GeoShanghai 2018 International Conference: Advances in Soil Dynamics and Foundation Engineering*, pp. 684–691, 2018.
https://doi.org/10.1007/978-981-13-0131-5_74

analyze the pile group foundation properties of the rigid raft. The three element finite element method was used to analyze the interaction between pile, raft and soil by Liang et al. [2]. Variational approach was conducted by Liang and Chen [3] to analyze piled raft foundation. Obviously, elastic analysis methods will over evaluate the interaction between pile and soil. Furthermore, it is difficult for them to estimate the deformation characteristics of piled raft foundation in plastic state.

In an attempt to overcome the limitations of the elastic analysis methods, in the past several years, many efforts have been focused on developing nonlinear analysis methods to predict the behavior of piled raft foundation (e.g., Lee and Xiao [4]; Castelli and Maugeri [5]). However, these methods ignore the significant contribution of the raft to pile foundation performance. In order to take into account the proportion of load carried by the raft, Poulos ([6, 7]) proposed some approximate nonlinear methods for pile-raft-soil interaction. The so-called "hybrid" approach, developed by Clancy and Randolph [8], was used in the analysis of piled raft foundations. Besides, a simplified nonlinear analysis method for piled raft foundation was conducted by Huang et al. [9] to analyze the pile-raft-soil interaction. However, their analysis methods need to be discretized on the raft and pile along pile shaft, which is due to the limitation of the three-dimensional properties of the piled raft foundation.

In this paper, an approximate method for nonlinear response of piled raft foundation under vertical load is analyzed. The solution can be simplified only if the deflection of the raft and the contact stress at the raft-pile-soil interface are solved. The method of analysis is similar to the method proposed by Liang and Chen [3], while some important improvements have been made. Firstly, the pile-soil interaction is calculated with a more rigorous model, which is able to simulate the slip between pile and soil. Secondly, a "load cut-off" method which was described by Hain and Lee [10] and Yang et al. [11] for ultimate load is used to simulate the ideal elastic-plasticity behavior of soil. Compared with the available published results, the present method is verified and computationally effective. Therefore, it is feasible to analyze the large piled raft foundation.

2　Method of Analysis

2.1　Nonhomogeneity of Subsoil

Guo and Randolph [12] presented a elasto-plastic solution to the vertical load settlement of a single pile in a heterogeneous foundation. The theoretical model took into account the slip between the soil and pile and the solution procedure is simple. The soil profile used in this paper is summarized as follows.

The initial soil shear modulus and the limiting shaft friction distribution along a pile are assumed as power functions of depth:

$$G = A_g z^\theta \tag{1}$$

$$\tau_f = A_v z^\theta \tag{2}$$

where A_g, A_v, θ are constants and z is the depth below the surface.

2.2   Stiffness Matrix Analysis Method for the Piled Raft Foundation

The contact surface of the raft and the soil is dispersed into a series of quadrilateral elements, and the pile head should be in a unit. As the deflection of the raft and the displacement of the pile are consistent with the soil, the reaction between soil and pile can be given by

$$\{P_{sp}\} = [K_{sp}]\{w_{sp}\} \tag{3}$$

where $\{w_{sp}\} = \{w_{s1}, w_{s2}\cdots, w_{sns}, w_{p1}, w_{p2}\cdots, w_{pnp}\}^T$, $\{P_{sp}\} = \{P_{s1}, P_{s2}\cdots, P_{sns}, P_{p1},$ $P_{p2}\cdots, P_{pnp}\}^T$, in which P_{si} and w_{si} are the equivalent concentrated forces and the average displacement of the soil unit I, respectively. P_{si} can be expressed as $P_{si} = \iint p_s dA_{si}$, in which p_s is the distribution force on the soil unit. A_{si} is the area of the soil element i, n_s is the number of the soil elements. P_{pi} and w_{pi} are the force and displacement of the head of the pile i respectively, and np is the number of piles. $[K_{sp}]$ is the stiffness matrix of piles-soil system and can be obtained from the inversion of the flexibility matrix.

$$[K_{sp}] = [F_{sp}]^{-1} \tag{4}$$

where $[F_{sp}]$, which is the $(ns + np) \times (ns + np)$ flexibility matrix of the piles-soil system and can be rewritten as a submatrix form.

$$[F_{sp}] = \begin{bmatrix} f^{ss} & f^{sp} \\ f^{ps} & f^{pp} \end{bmatrix} \tag{5}$$

in which the submatrices $[f^{ss}], [f^{sp}], [f^{ps}], [f^{pp}]$ represent the interaction matrix of soil surface to soil surface, soil surface to pile head, pile head to soil surface, pile head to pile head, respectively.

Thus the load-displacement relationship of piled raft foundation and the percentage of load carried by raft and pile are obtained easily through Eq. (3). The remaining problem is to determine the flexibility coefficients of these submatrices in Eq. (5). In order to simplify the calculation, an approximate analysis method is presented in this paper.

3   Determination of the Flexibility Matrix of Pile Group-Soil System

As described by Chow [13], the nonlinear response of the group pile is mainly determined by the nonlinear behavior of single pile, while the interaction effects remain essentially elastic. So the flexibility coefficients will be determined as:

(1) According to Chow [14], for a homogeneous soil medium, the flexibility coefficient in the submatrix $[f^{ss}]$ can be obtained by the integral form of the Boussinesq solution.

$$f_{ii}^{ss} = \frac{\sqrt{\pi}(1 - v^2)}{2E_s a_i} \quad (i = 1, 2, \ldots, n_s) \tag{6}$$

$$f_{ij}^{ss} = \frac{(1 - v^2)}{\sqrt{\pi}E_s a_i} \sin^{-1}\left(\frac{a_i}{s_{ij}\sqrt{\pi}}\right) \quad (i, j = 1, 2, \ldots, n_s; i \neq j) \tag{7}$$

where E_s is the elastic modulus of the soil, a_i is the width of the soil element i, s_{ij} is the distance between center of soil element i and soil element j at head. The flexibility coefficient f_{ii}^{ss} indicates the displacement of the soil element i caused by the unit load on itself and f_{ij}^{ss} indicates the displacement of the soil element i caused by the unit load on the soil element j. For non-homogeneous soil medium, the flexibility coefficient is determined by the axisymmetric finite element analysis which is basically two-dimensional. The details are given by Chow [14].

(2) The flexibility coefficients in submatrices $[f^{pp}]$ are given by approximate analytical solutions of the interaction factor proposed by Guo and Randolph [15].

$$f_{ii}^{pp} = \frac{w_e\left[1 + \mu k_s L^{n/2 + 1} C_v(\mu L)\right] + \frac{2\pi r_0 A_v (\mu L)^{2 + \theta}}{(2 + \theta)E_p A_p}}{w_e k_s E_p A_p L^{n/2} C_v(\mu L) + \frac{2\pi r_0 A_v (\mu L)^{1 + \theta}}{1 + \theta}} \quad (i = 1, 2 \ldots, n_p) \tag{8}$$

$$f_{ij}^{pp} = \frac{1}{G_L r_0 \pi C_{V2}} \sqrt{\frac{\zeta_2}{2\lambda}} - \frac{1}{G_L r_0 \pi C_{V0}} \sqrt{\frac{\zeta}{2\lambda}} \quad (i, j = 1, 2 \ldots, n_p; i \neq j) \tag{9}$$

In Eq. (8), f_{ii}^{pp} is the displacement at the head of pile caused by the unit load on itself, r_0 is the pile radius, $w_e = \frac{A_v}{A_g} r_0 \zeta$, $k_s = \frac{L}{r_0}\sqrt{\frac{2}{\lambda \zeta}}(\frac{1}{L})^{1/2m}$, $\zeta = \ln(r_m/r_0)$, r_m represents the maximum influence radius of the pile and is expressed with the length of the pile, as $r_m = 2.5L(1 - v)\rho$, v is the Poisson ratio of soil, $\rho = G_{L/2}/G_L$ (in homogeneous soil, $\rho = 1$), G_L and $G_{L/2}$ is shear modulus at pile base level and at depth $L/2$, respectively. E_p and A_p is the Young's modulus and cross-section area of the pile, respectively.

With the increase of pile head load, the pile-soil relative slip is assumed to start from the ground, and at any stage of the loading process, it can develop into a depth called transition depth L_1. $\mu = L_1/L$ $(0 < \mu \leq 1)$ is defined as the degree of slip. As μ approaches zero, pile-soil relative slip occurs just at the ground level, which means that the pile is in a state of transition from elasticity to plasticity. As μ is equal to 1, the slip occurs just at the pile base level. Then pile-head load reaches its limiting value and will keep invariable.

The parameter $C_v(z)$ can be expressed as $C_v(z) = \frac{C_1(z) + \chi C_2(z)}{C_3(z) + \chi C_4(z)}(\frac{z}{L})^{n/2}$, where

$$\chi = \frac{2}{\pi(1 - v)}\sqrt{\frac{2\zeta}{\lambda}} \tag{10}$$

in which $C_1(z) = -cI_{m-1}(y) + bK_{m-1}(y), C_2(z) = dI_{m-1}(y) + aK_{m-1}(y), \quad C_3(z) = cI_m(y) + bK_m(y), C_4(z) = -dI_m(y) + aK_m(y)$. I and K is the modified Bessel function of the first kind and second kind respectively. $m = 1/(n+2)$, The variable y is

$$y = \frac{2mL}{r_0}\sqrt{\frac{2}{\lambda\zeta}}(\frac{z}{L})^{1/2m} \tag{11}$$

The coefficients a, b, c, d are the values of the modified Bessel functions for $z = L$.

In Eq. (9), f_{ij}^{pp} is the displacement at the head of pile i caused by the unit load on the head of pile j. $\zeta_2 = \ln(r_{mg}/r_0) + \ln(r_{mg}/s'_{ij})$, in which r_{mg} is given by $r_{mg} = 2.5L$ $(1-v)\rho + r_g$ and s'_{ij} is the distance between center of pile i and pile j at head. More generally, r_g may be estimated by $r_g = (0.3 + 0.2n)s$, with the s being taken as the lesser of r_m and s'_{ij}. $\lambda = E_p/G_L$. The parameter C_{v0} can be expressed as $C_{v0} = \lim_{z,y \to 0} C_v(z)$, The parameter C_{v2} is defined as C_{v0}, with values of ζ_2 and ζ in Eqs. (10) and (11), respectively.

(3) The flexibility coefficient in the submatrix $[f^{sp}]$ can be obtained by the load displacement relation of the pile-soil interaction, and ignore the effect of pile bottom on soil surface (Liang and Chen [3]).

$$f_{ij}^{sp} = \frac{\ln(r_m/s''_{ij})}{\ln(r_m/r_0)} f_{jj}^{pp} \qquad (i = 1, 2 \cdots, n_s; j = 1, 2 \ldots, n_p) \tag{12}$$

where f_{ij}^{sp} is the surface displacement of soil element i due to unit load acting on the head of pile j, f''_{ij} is the distance between center of soil element i and pile j at head.

(4) The flexibility coefficients of the submatrix $[f^{ps}]$ can be obtained by the Maxwell reciprocal theorem (Hain and Lee [10]).

$$f_{ij}^{ps} = f_{ji}^{sp} \ (i = 1, 2 \cdots, n_p; j = 1, 2 \ldots, n_s) \tag{13}$$

where f_{ij}^{ps} is the surface displacement of the pile i caused by the unit load on the soil element j.

In order to simplify the analysis, it is usually assumed that the raft is rigid. Then uniform displacement w_0 occurs to the displacement of soil element and pile at head. Substituting w_0 into Eq. (3), the counterforces of soil and pile can also be easily obtained.

4   Results of Analysis

The accuracy of the elastic solution (μ approaches zero) in this paper has been verified by comparing the results with boundary element method by Kuwabara [1].

In the analysis of Kuwabara [1], 3×3 pile groups with rigid raft of dimensions $26r_0 \times 26r_0$ in homogeneous soil are subdivided into 160 elements for soil and

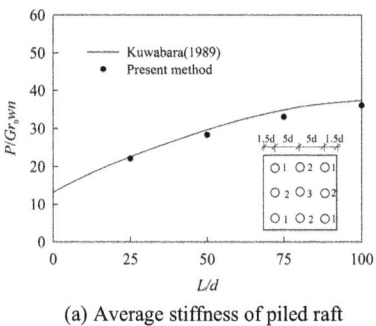
(a) Average stiffness of piled raft

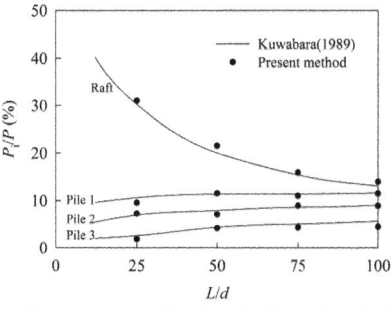
(b) Percentage of load carried by raft and piles

Fig. 1. Comparing results of 3×3 piled raft ($E_p/G = 3000$) (P_i is the load carried by raft and piles respectively)

9 elements for piles. v is taken to be 0.5. Figure 1 shows the average stiffness of pile groups, $\frac{P}{Gr_0 wn}$ and the load sharing between the piles and raft, in which P is the total external force acting on piled raft system, w is the vertical settlement of raft and n is the number of piles.

It can be seen that the present solutions are in good agreement with those rigorous methods, and the calculation of the present method is significantly reduced. It is noted that with the increase of pile length diameter ratio, the overall stiffness and load sharing ratio of piled raft foundation will increase, and the load shared by raft will decrease. The smaller pile length diameter ratio has a significant impact on the overall stiffness of piled raft foundation and the load shared by cap. The load of the angular pile under the rigid raft is the largest, the side pile is the second, the central pile is the smallest, which is in line with the actual situation of the project.

The accuracy of the elastic-plastic solution in this paper has been verified by comparing the results with those from a full-scale test on a 3×3 pile group with raft in contact with the ground by Koizumi and Ito [16]. The closed-end steel pipe piles, which embedded lengths of 5.55 m, had an outside diameter of 300 mm and a wall thickness of 3.2 mm. The pile spacing was 900 mm with a raft overhang of 450 mm. The nonhomogeneous soil profile deduced by Chow and Teh [17] were also adopted in this paper. A comparison of the calculation of the elastic-plastic load-displacement behavior of the piled raft and field test results is shown Fig. 2(a). It can be seen that load-settlement relationship of the pile-raft foundation agrees well with the measured results. A comparison of the computed and measured loads carried by the cap and the piles is shown in Fig. 2(b). In the stage of elasticity, the calculated results are in good agreement with the measured results. It is worth noting that at the plastic stage, the computed value of load carried by piles is slightly less than field test result, which results from the assumption that with the increase of applied load, the pile-head load reaches its limiting value and keeps invariable. In fact, a part of increased load will be shared by soil at the pile base level.

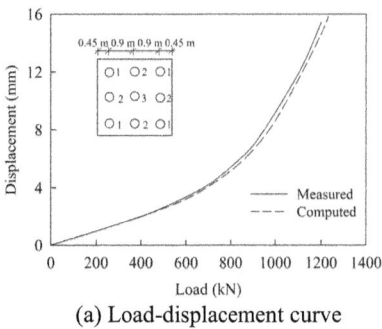
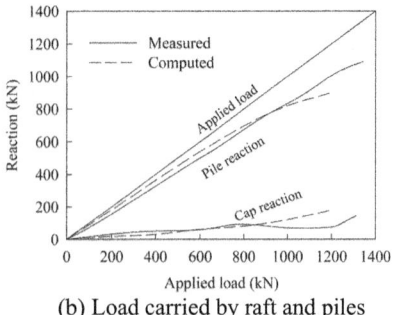

(a) Load-displacement curve　　　　(b) Load carried by raft and piles

Fig. 2. Comparison between computation and test

5　Conclusions

An analytical model was described for the nonlinear elastic and elastic-plastic analysis of vertically loaded pile groups embedded in a homogeneous and nonhomogeneous soil with the rigid raft in contact with the ground. The presented solutions are consistent with other rigorous numerical methods and case studies of field tests on the response of piled raft foundation. The main advantage of the present method is that it can describe the nonlinear response of piled raft without discretization of piles and soil at the pile soil interface as well as the raft. Therefore, the computation time and computer storage space required are much less than those required by the traditional boundary-element method or finite element method, it is feasible to use large group piles to analyze the nonlinear analysis of piled raft foundation. However, this paper uses the "load cut-off" method to simulate the ideal elastic-plasticity behavior of the soil, which causes the load to be shared by the raft is larger than the measured value. In addition, this paper takes the rigid raft foundation into consideration, and the flexible raft foundation should be further studied.

References

1. Kuwabara, F.: An elastic analysis of piled raft foundations in a homogeneous soil. Soils Found. **29**(1), 82–92 (1989)
2. Liang, F.Y., Chen, L.Z., Shi, X.G.: Numerical analysis of composite piled raft with cushion subjected to vertical load. Comput. Geotech. **30**(6), 443–453 (2003)
3. Liang, F.Y., Chen, L.Z.: A modified variational approach for the analysis of piled raft foundation. Mech. Res. Commun. **31**, 593–604 (2004)
4. Lee, K.M., Xiao, Z.R.: A simplified nonlinear approach for pile group settlement analysis in multilayered soils. Can. Geotech. J. **38**(5), 1063–1080 (2001)
5. Castelli, F., Maugeri, M.: Simplified nonlinear analysis for settlement prediction of pile groups. J. Geotech. Geoenv. Eng. **128**(1), 76–84 (2002)
6. Poulos, H.G.: An approximate numerical analysis of pile-raft interaction. Int. J. Numer. Anal. Meth. Geomech. **2**(18), 73–92 (1994)

7. Poulos, H.G.: Pile raft foundations: design and applications. Geotechnique **51**(2), 95–113 (2001)
8. Clancy, P., Randolph, M.F.: An approximate analysis procedure of piled raft foundations. Int. J. Numer. Anal. Meth. Geomech. **17**(12), 849–869 (1993)
9. Huang, M.S., Liang, F.Y., Jiang, J.: A simplified nonlinear analysis method for piled raft foundation in layered soils under vertical loading. Comput. Geotech. **38**(7), 875–882 (2011)
10. Hain, S.J., Lee, I.K.: The analysis of flexible raft–pile systems. Geotechnique **28**(1), 65–83 (1978)
11. Yang, M., Wang, S.J., Wang, B.J., Zhou, R.H.: Practical analysis of piled raft foundation considering ultimate capacity of piles. Chin. J. Geotech. Eng. **20**(5), 82–86 (1998). (in Chinese)
12. Guo, W.D., Randolph, M.F.: Vertically loaded piles in non-homogeneous media. Int. J. Numer. Anal. Meth. Geomech. **21**(8), 507–532 (1997)
13. Chow, Y.K.: Analysis of vertically loaded pile groups. Int. J. Numer. Anal. Meth. Geomech. **1**(10), 59–72 (1986)
14. Chow, Y.K.: Vertical deformation of rigid foundations of arbitrary shape on layered soil media. Int. J. Numer. Anal. Meth. Geomech. **11**(1), 1–15 (1987)
15. Guo, W.D., Randolph, M.F.: An efficient approach for settlement prediction of pile groups. Geotechnique **49**(2), 161–179 (1999)
16. Koizumi, Y., Ito, K.: Field tests with regard to pile driving and bearing capacity of piled foundations. Soils Found. **7**(3), 30–53 (1967)
17. Chow, Y.K., Teh, C.I.: Pile-cap – pile-group interaction in nonhomogeneous soil. J. Geotech. Eng. **117**(11), 1655–1668 (1991)

A Comparative Study of New Pneumatic Caisson and Underground Diaphragm Wall Used in Deep Underground Space

Kai-Hua Chen[1] and Fang-Le Peng[2(✉)]

[1] Department of Geotechnical Engineering, Tongji University,
1239 Siping Road, Shanghai 200092, China
[2] Key Laboratory of Geotechnical and Underground Engineering,
Ministry of Education, Department of Geotechnical Engineering,
Tongji University, 1239 Siping Road, Shanghai 200092, China
pengfangle@tongji.edu.cn

Abstract. It is crucial to choose the reasonable construction method for the construction of deep shaft/foundation pit in urban areas, especially close to the existing structures or facilities. For this purpose, a numerical study using a commercial software of PLAXIS was conducted to compare the new pneumatic caisson (NPC) method and the underground diaphragm wall method. A kinematic mechanical model was proposed for evaluating the influence of new pneumatic caisson method on the surrounding strata. The accuracy and reliability of this model was verified by field measurement of ground deformation. Then, on the basis of the FEM result of the two construction methods, soil-structure mechanical response and surrounding soil disturbance characteristics of deep shaft/foundation pit were discussed. Compared to the underground diaphragm wall method, the ground deformation induced by the NPC method was much smaller. The NPC method proved to be an efficient and safe construction method, which would provide reference for the development of deep underground space.

Keywords: New pneumatic caisson (NPC) · Underground diaphragm wall
Simulation

1 Introduction

The urban underground space has been developed including metro, underground road, underground commercial facilities, which still cannot meet the needs of the current urban underground space. The utilization of deep underground space solves the problem regarding shortage of shallow underground space. Deep underground space will play an increasingly significant role in the urbanization process for Shanghai.

During the construction of the deep shaft/foundation pit as most common structure for deep underground space, it is important to control the ground deformation, especially in the sensitive and unstable soft ground, which result in sharply increasing risks and adverse impacts on the surrounding buildings. Therefore, it is crucial to choose the reasonable construction method for the construction of deep shaft/foundation pit in

© Springer Nature Singapore Pte Ltd. 2018
T. Qiu et al. (Eds.): GSIC 2018, *Proceedings of GeoShanghai 2018 International Conference: Advances in Soil Dynamics and Foundation Engineering*, pp. 692–699, 2018.
https://doi.org/10.1007/978-981-13-0131-5_75

urban areas. However, there have been few research achievements for the comparative study of NPC method and underground diaphragm wall.

In this paper, a numerical study using a commercial software of PLAXIS was conducted to simulate the excavation of different depth shaft/foundation pits, used the new pneumatic caisson (NPC) method and the underground diaphragm wall method, respectively. Then, on the basis of the FEM result of the two construction methods, soil-structure mechanical response and surrounding soil disturbance characteristics of deep shaft/foundation pit were discussed.

2   Overview of the NPC Method and Diaphragm Wall Method

In Japan, there are relatively mature experiences in design and construction of NPC (Ohuchi et al. 2003). The advanced automatic system for pneumatic caisson (Fig. 1) has been developed and put into practice use (Kodaki et al. 1997). In China, NPC method is applied to one shaft construction for the first time in Shanghai and achieves great success (Li et al. 2008). These engineering practices have indicated that the NPC method is of considerable merits in deep shaft/foundation excavation.

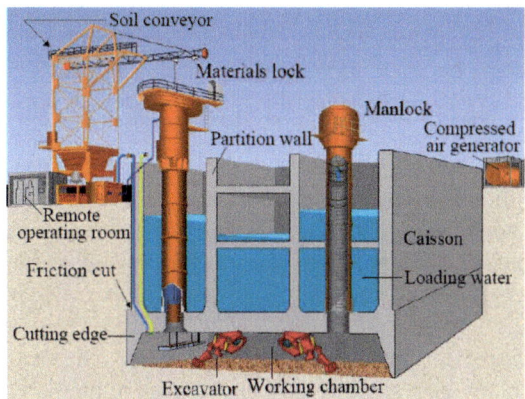

Fig. 1. Schematic view of the NPC method.

Underground diaphragm wall as a mature deep shaft/foundation pit retaining method has been widely used in the world. Currently, the diaphragm wall method is mainly adopted in deep shaft/foundation pit located in dense buildings. Some researchers pay attention to deformation characteristics for shaft/foundation pit retaining structure in the process of excavation.

However, there have been few research achievement for the comparative study of NPC method and underground diaphragm wall. In the present study, a numerical study using a commercial software of PLAXIS is conducted to compare the new pneumatic caisson (NPC) method and the underground diaphragm wall method.

3   Validation of Numerical Model

Prior to the FEM analysis of NPC method and underground diaphragm wall method, the adopted numerical model had been validated by field monitoring measurement for NPC method in this section. The shaft construction to be presented is one part of the shield tunnel project of Metro Line 7 in Shanghai (Fig. 2). The shaft is one fully-embedded four-storied reinforced concrete caisson with the outer dimensions 25.2 m × 15.6 m × 29.0 m for four sinking phases.

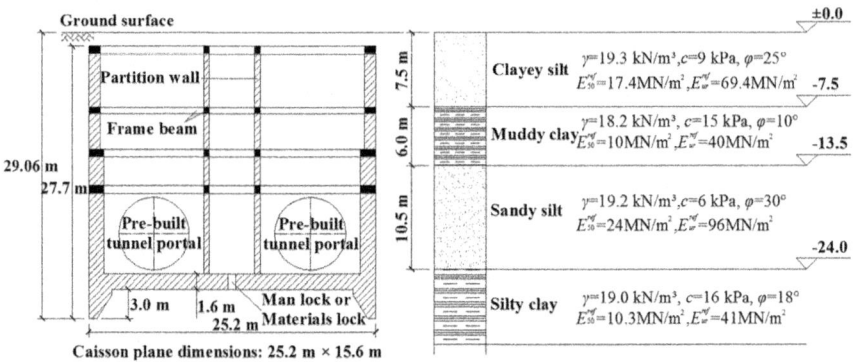

Fig. 2. Profile of the shaft and in-situ geology.

To investigate the environmental effects caused by the NPC construction, some field measurement points were laid out around the caisson before caisson sinking. The observed items included horizontal movements, surface settlement, subsurface settlement and soil stress. The plan view of in-situ measurement is shown in Fig. 3.

To model the soil behavior more accurately, the advanced elasto-plastic constitutive model Hardening-Soil (HS) model had been employed. The material parameters of drained type used in this analysis were also summarized in Fig. 2. For the parameter m, the default value 0.5 was adopted.

The distributed soil reaction equaled to 250, 250, 200 and 350 kN/m^2, respectively. And the air pressure equaled to 35, 113, 200 and 285 kN/m^2, respectively. The load increment was automatically decided with trial calculation by using the computer program. The skin friction equaled to 20 kN/m^2.

From these graphs (Fig. 4), it can be seen that, the results of numerical calculation seemed to agree with the measurements rather well in general. The difference between the calculations and measurements for Phase 1 was mainly caused by incline of caisson during construction process. The proposed numerical analysis could simulate well the whole construction process of pneumatic caisson. Hardening-soil model can be adopted to conduct non-linear analysis in the proposed numerical calculation.

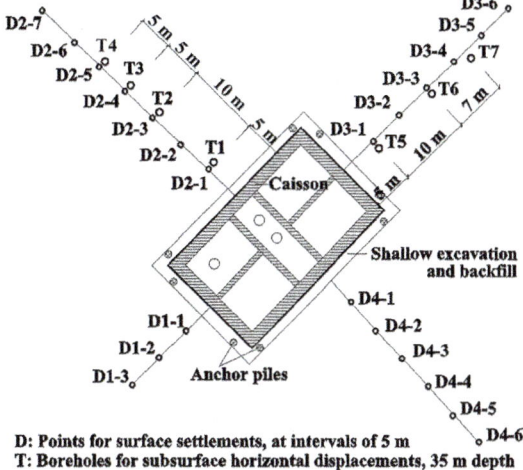

Fig. 3. Plan view of field measurements.

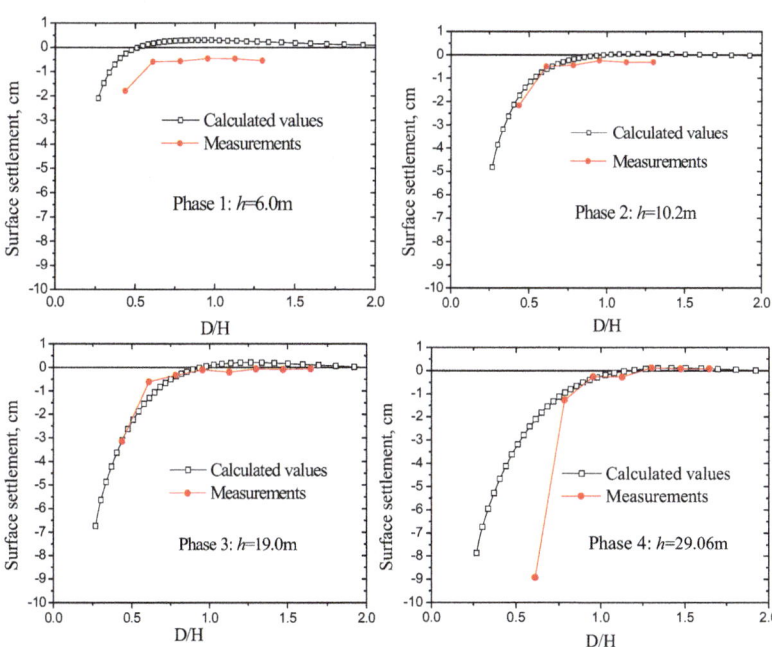

Fig. 4. Calculated values and measurements of surface settlement in D2 direction in each sinking phase.

4 Numerical Simulation

In this section, the construction process of different depth shafts/foundation pits, adopted new pneumatic caisson (NPC) method and the underground diaphragm wall method, were simulated by PLAXIS software. The simulation condition of different shafts/foundation pits adopted NPC method and diaphragm wall method included the excavation or sinking depth of 20 m, 50 m and 70 m, respectively.

The shaft was one reinforced concrete structure with the outer dimensions 30 m 30 m, and the braces were assigned in each 5 m. In addition, the insertion ratio of diaphragm wall was 1.0. Depth of sinking in each phase was 10 m, and the depth of excavation in each phase was 5 m.

As for the structure of deep shaft/foundation pit, the EA of Brace was 2.4×10^7 kN/m, and the EA and EI were 6×10^7 kN/m and 2.0×10^7 kN·m²/m. Hardening-soil model was employed to conduct non-linear analysis in the proposed numerical calculation. The soil parameters of drained type adopted in this analysis were summarized in Table 1. For the parameter m, the default value 0.5 was adopted.

Table 1. The geological parameter of People's Square in Shanghai.

Layer	Thickness (m)	c (kPa)	φ (°)	Weight (kN/m³)	E_s (MPa)	K_V (cm/s)	K_H (cm/s)	E_{50}^{ref} (KN/m²)	E_{ur}^{ref} (KN/m²)	m
① artificial fill	2.05							4000	2.00E + 04	0.5
② silty clay	1.7	21	14.5	18.2	4.67	1.66E–07	2.21E–07	7940	3.97E + 04	0.8
③ mucky silty clay	5.7	14	12.5	17.6	3.23	2.16E–07	3.57E–07	7680	3.84E + 04	0.8
④ muddy clay	6.3	14	9.5	16.8	2.25	1.77E–07	2.69E–07	6030	2.41E + 04	0.8
⑤₁₋₁ clay	3.15	15	13.5	17.6	3.71	1.31E–05	1.81E–05	7120	3.56E + 04	0.8
⑤₁₋₂ silty clay	7.5	17	16.5	17.9	4.32	3.63E–06	6.76E–07	8.64E + 03	4.29E + 04	0.8
⑤₂ silty clay with fine sand interbed	4.9	16	18	18	5.77	1.15E–05	5.37E–05	1.06E + 04	5.29E + 04	0.8
⑥ clay	3.25	50	16.5	19.7	7.78	3.96E–07	6.18E–07	1.39E + 04	6.96E + 04	0.8
⑦₁ sandy silt	6	5	29	19	10.99	1.85E–04	2.77E–04	1.15E + 04	5.73E + 04	0.7
⑦₂ silt sand	27.25	3	32.5	19.5	14.38	7.93E–04	1.37E–03	1.49E + 04	7.43E + 04	0.5
⑧₂ silty clay	8.3	26	14.5	18.2	5.55	4.48E–06	1.15E–05	1.11E + 04	5.55E + 04	0.5
⑨₁ silty-fine sand	12.9	0	34.5	18.9	14.87			1.11E + 04	5.57E + 04	0.7

For simplification, the plane strain assumption was adopted and a typical finite element mesh of the transverse section had been set up as shown in Fig. 5, in which the symmetries of the analyzed problem were also taken into consideration. The unit skin friction equals to 20 kN/m². When the caisson had sunk for the depth of 10 m, the working chamber increased with 30 kN/m² compressed air pressure.

Certainly in the meantime, the "excavated" soil elements in each phase were killed, i.e. the soil stiffness was set to a value (of) approximately zero. The ground water table within the region of the caisson was also updated in each phase. As for the numerical analysis method, after an initial stress analysis was performed, the caisson sinking process and excavation process were modeled based on the construction sequence.

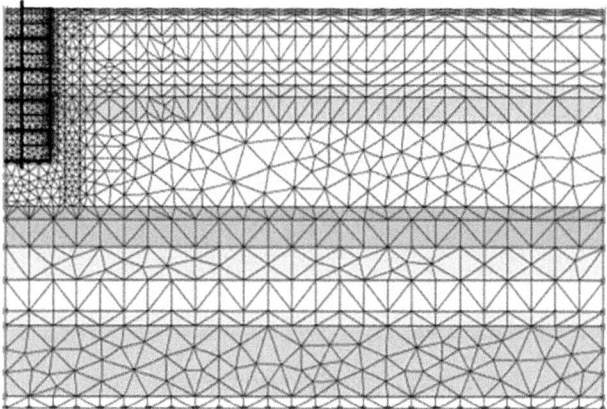

Fig. 5. FEM mesh of 50 m pneumatic caisson.

5   Results and Discussions

5.1   Surface Settlements

As shown in the Fig. 6, the final surface settlement profile obtained from NPC calculation seemed to be a parabola, the settlement decreased rapidly. As the caisson sunk deeper, the settlement was increasing and the influence zone on the ground surface was widened gradually. We also notice that, there was small upheaval at points where D/H was approximately within 0.5–0.75 for calculated values. This could be explained that, three-dimensional effects actually existed and caisson "scraping" against its surrounding strata caused more ground subsidence in the transverse section.

As the shape of grooved, the surface settlements caused by underground diaphragm wall method was more than NPC method. Differ from NPC method, the total calculated maximum settlements approximately located at 0.5 H, rather than around the wall. The surface settlement influence region of diaphragm wall method was approximately 2–3 times excavation depth for calculated values.

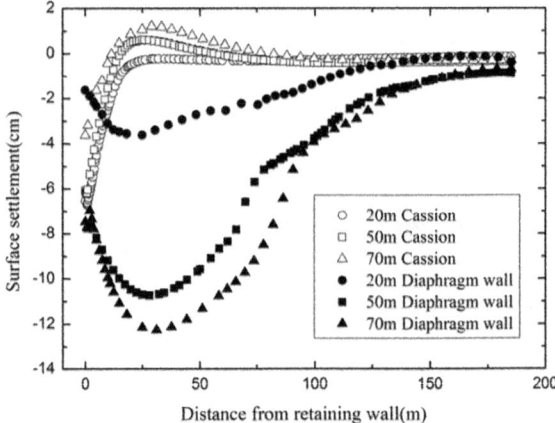

Fig. 6. Calculated values of surface settlement in the FEM analysis.

5.2 Horizontal Displacements

As shown in the Fig. 7, the maximum of horizontal movements for calculated values was no more than 20 mm for NPC method. As to the distribution pattern of the horizontal movements, soils above some critical level (assumed as H0) tended to move towards the caisson and the displacement reached the maximum at the surface, whereas soils below H0 tended to move away from the caisson and the displacement increased until the maximum then decreased to zero gradually.

The distribution pattern of the horizontal movements caused by diaphragm wall method was difference from NPC method. Soils tended to move away from the diaphragm wall and the displacement increased until the maximum then decreased to zero gradually. The total calculated maximum horizontal displacements of diaphragm wall was much more than NPC method.

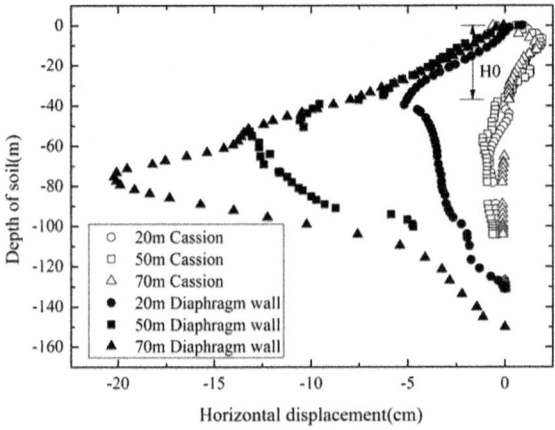

Fig. 7. Calculated values of horizontal movements in the FEM analysis.

5.3   Lateral Earth Pressure

As shown in the Fig. 8, the lateral earth pressure acting on the diaphragm wall was much closer to the active earth pressure, due to large horizontal displacement. And the lateral earth pressure acting on the caisson was closer to the static earth pressure, attributing to the small horizontal displacement.

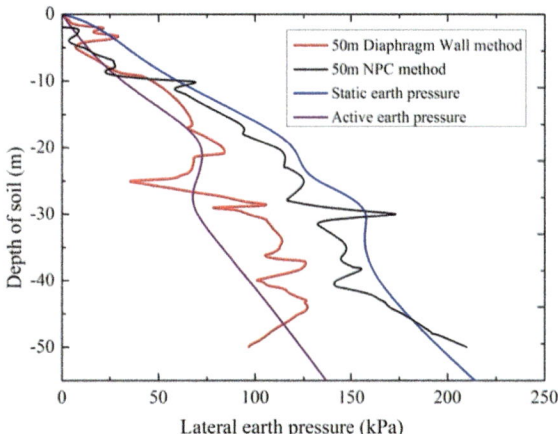

Fig. 8. Calculated values of lateral earth pressure of 50 m excavation depth in the FEM analysis.

6   Conclusions

The following conclusions can be drawn from the above:

(1) The NPC method proved to be an efficient and safe construction method, which can be adopted in the construction for deep shaft/foundation pit; (2) Compared to the underground diaphragm wall method, the soil disturbance and influence regions induced by the NPC method was much small.

Acknowledgments. The authors acknowledge gratefully the support provided by funds from Grant 2015CB057806 from The National Basic Research Program (973 Program).

References

Kodaki, K., Nakano, M., Maeda, S.: Development of the automatic system for pneumatic caisson. Automation in Construction, No. 6, pp. 241–255 (1997)

Li, Y.L., Jiang, C.H., Deng, Q.F.: Key construction techniques for modern pneumatic caisson practice with fully automatic and remote control. Build. Constr. **30**(1), 2–4 (2008). (in Chinese)

Ohuchi, M., Peng, F.L., Sun, D.X., Zhu, H.H., Liao, S.M.: New pneumatic caisson. Geotech. Eng. **15**(1), 19–22 (2003). (in Chinese)

Analysis of Load Distribution Characteristics in Piled Raft Foundation

Ruping Luo, Min Yang, and Weichao Li[✉]

Department of Geotechnical Engineering, Tongji University,
Shanghai 200092, China
WeichaoLi@tongji.edu.cn

Abstract. Piled raft have been widely used as an effective foundation for high-rise buildings for their efficiency in controlling the total and differential settlements and improving bearing capacity of foundation. To design piled raft foundation with economy and safety, and particularly to calculate with reasonable accuracy the moments in the raft, the likely distribution of loads on piles underneath the raft is of a basic importance. Based on the non-linear boundary element method (BEM), the distribution characteristics of pile loading in the piled raft foundation are investigated in this paper. The results show that the pile loading on the corner pile is significantly larger than that on the center pile, and the loading ratio is varied with the magnitude of external load and pile number. In order to provide a practical reference for industry engineers, several design charts are provided to evaluate the load distribution characteristics under different working conditions.

Keywords: Piled raft foundation · Load distribution · BEM · Design charts

1 Introduction

Piled raft has been widely employed as an effective foundation for high-rise buildings for their efficiency in controlling the total and differential settlements and improving bearing capacity, especially in the areas where the field sites are underlain by deep soft soil deposits [1, 2]. Due to the three-dimensional (3D) interactions between piles-soil-raft, the responses of the piled raft foundation under working loads are rather complicated. In order to design the piled rafts foundation with economy and safety, and particularly to calculate the moments in the rafts with reasonable accuracy, the likely distribution of load on each pile underneath the raft should be determined.

The distribution characteristic of pile loading in the piled raft foundation is heavily dependent on the rigidity of the structure-foundation-soil system [3], and based on the field observations and numerical simulations, a series of research efforts have been made on the pile loading distribution characteristic of the piled raft foundation [3–8]. For example, Cooke [5] conducted a series of model tests to investigate the pile loading distribution characteristic of piled raft foundation in London clay, and the tests results show that loads on piles at the corner are significantly larger than that at the center pile, and the pile loading ratio is greatly influenced by the pile spacing. Nevertheless, current studies are unable to provide enough data that could be extrapolated to assist the design

© Springer Nature Singapore Pte Ltd. 2018
T. Qiu et al. (Eds.): GSIC 2018, *Proceedings of GeoShanghai 2018 International Conference: Advances in Soil Dynamics and Foundation Engineering*, pp. 700–708, 2018.
https://doi.org/10.1007/978-981-13-0131-5_76

of prototype piled raft foundation, and the variation patterns of the pile loading ratio with the magnitude of external load and pile number are not fully investigated.

Based on the boundary element method (BEM), a non-linear analysis method for rigid piled raft foundation is proposed in this paper. The non-linear analysis method is validated by the field measurements, and then a series of parametric study is conducted to investigate the pile loading distribution characteristic. The variation patterns of the pile loading ratio with the magnitude of external load (foundation safety factor, F_s) and pile number are analyzed, and design charts are developed to aid the industrial design.

2   Non-linear Analysis of Rigid Pile Raft

The analysis model in this study is shown in Fig. 1, in which the rigid raft is assumed. Therefore, the settlement of each pile head under this raft is identical and the differential settlement of raft is ignored. Four interactions between the foundation elements and the soil are: (1) pile-soil interaction, (2) pile-pile interaction, (3) raft-soil interaction, and (4) pile-raft interaction. The main calculation steps of the pile-soil-raft interaction analysis method are given below.

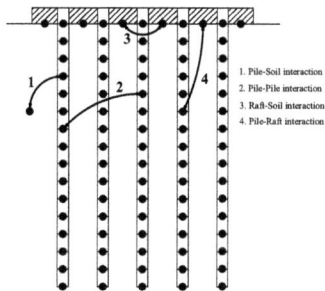

Fig. 1. Analysis model of rigid pile raft

(1) A total of n_p piles are analyzed in the model, with each pile discretized into n_{ep} nodes as depicted in Fig. 1, and it is assumed that there are a total of k and $n = n_p \times n_{ep}$ nodes for the raft and pile shaft, respectively.

(2) The soil displacement equation is given by:

$$\{w_s\} = [F_s] \cdot \{P_s\} \tag{1}$$

Where the $\{w_s\}$ and $\{P_s\}$ are column vector of soil vertical displacements and loads acting on the soil respectively, the expansions are:

$$\begin{cases} \{w_s\} = \left[w_s^{r1}, w_s^{r2}, \ldots, w_s^{rk}, w_s^{p1}, w_s^{p2}, \ldots, w_s^{pn}\right]^{\mathrm{T}} \\ \{P_s\} = \left[P_s^{r1}, P_s^{r2}, \ldots, P_s^{rk}, P_s^{p1}, P_s^{p2}, \ldots, P_s^{pn}\right]^{\mathrm{T}} \end{cases} \tag{2}$$

$[F_s]$ is the matrix of soil flexibility coefficient and the expansion is:

$$[F_s] = \begin{bmatrix} [F_{rr}] & [F_{rp}] \\ [F_{pr}] & [F_{pp}] \end{bmatrix}$$ (3)

Where:

(1) $[F_{rr}]$ is the displacement influence matrix of raft-soil interaction, and it can be calculated by the equation suggested by Kitiyodom and Matsutum [9]:

$$F_s = \frac{(1 - v_s)(1 - \exp(-H_{soil}/(2r_{equ})))}{4G_s r_{equ}}$$ (4)

Where H_{soil} is depth of compressible soil layer; r_{equ} is the equivalent radius of raft element; G_S is the soil shear modulus; vs is soil Possion's ratio. The flexibility coefficients (off-diagonal elements) of raft-soil-raft interaction are calculated by the equation suggested by Chow [10]:

$$F_{i,j} = \frac{(1 - v_s^2)}{\pi E_s r_{equ}} \sin^{-1}\left(\frac{r_{equ}}{r_{i,j}}\right) \quad i \neq j$$ (5)

Where $r_{i,j}$ is the radial distance between the centers of element i and element j of the raft. E_s is the soil Young's modulus.

(2) $[F_{pp}]$ is the displacement influence matrix of pile-soil interactions, $[F_{rp}]$ and $[F_{pr}]$ are the displacement influence matrixes of pile-soil-raft interaction. The elements of the matrixes are obtained via the Mindlin's displacement solution.

(3) According to the equilibrium and compatibility of pile-soil-raft interaction system, the relationships between the displacements and the loads are established:

$$([K_p] + [K_s]) \cdot \{w_p\} = \{Q_{top}\}$$ (6)

Where $[K_p]$ is the global stiffness matrix of the pile group system; $[K_s]$ is the soil stiffness matrix; $\{w_p\}$ is the column vector of all the node displacements of pile and raft; $\{Q_{top}\}$ is the column vector of external loads acting on the pile cap.

(4) Equation (6) is solved incrementally by considering the limiting stress when yielding at the pile-soil interface for cohesive soil, the limiting stress can be estimated using Eq. (7) [8]:

$$q_u = \begin{cases} \alpha S_u, & \text{the pile shaft nodes} \\ 9S_u, & \text{the pile base nodes} \\ 6S_u, & \text{the raft nodes} \end{cases}$$ (7)

Where S_u is the undrained shear strength of soil; α is an empirical adhesion factor, 0.35–1.25.

Meanwhile, the non-linear soil behavior is approximately considered by assuming that the soil Young's modulus E_s using the following hyperbolic function:

$$E_s = E_i(1 - R_f \cdot q/q_u)^2 \tag{8}$$

Where E_i is the initial soil modulus, R_f is a non-linear coefficient, which ranges from 0 to 1, and q, q_u are the stress acting on nodes and limiting stress respectively. The nonlinear load-displacement relation of pile raft foundation can be obtained through the above steps.

3   Case Validation

The 130 m high Messe-Torhaus was the first building in Germany designed with piled raft foundation. It consists of two separate rectangular rafts, each one is supported by 42 piles which are 20 m long and 0.9 m in diameter. The distance between the two rafts is 10 m. The raft is 17.5 m × 24.5 m in plan, 2.5 m thick and the elastic modulus of the concrete is 34,000 MPa. The elastic modulus of the concrete in the piles is 23,500 MPa and Poisson's ratio is 0.2 for both the raft and piles. The sketch and applied load on the piled raft foundation of the Messe-Torhaus is shown in Fig. 2.

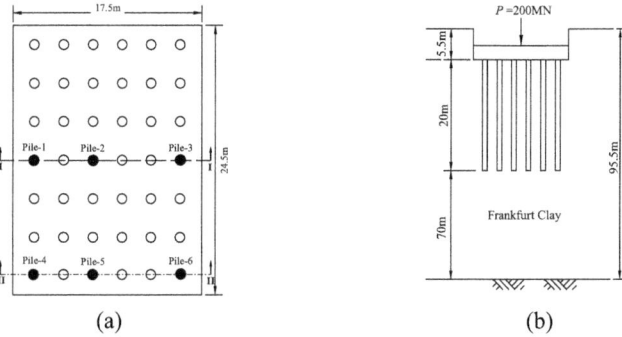

(a)   (b)

Fig. 2. Sketch and applied load of piled raft foundation of the Messe-Torhaus (Modified from Small et al. [11]) (a) plan view; (b) side view

As discussed by Franke et al. [12] and Reul et al. [7], the long term Young's modulus E_s (MPa) and shear strength S_u (kPa) of the Frankfurt clay is increased with depth linearly, and can be described by Eqs. (9) and (10). The Poisson's ratio of Frankfurt clay is $v = 0.15$ [7].

$$E_s = 45 + \left[\tanh\left(\frac{z - 30}{15}\right) + 1\right] \times 0.7z \tag{9}$$

$$S_u = 127 + 3.93z \tag{10}$$

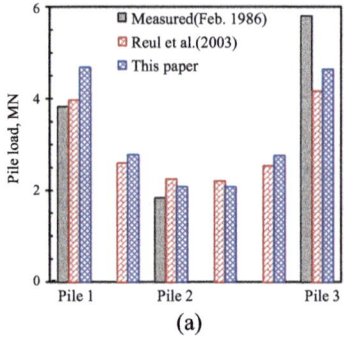
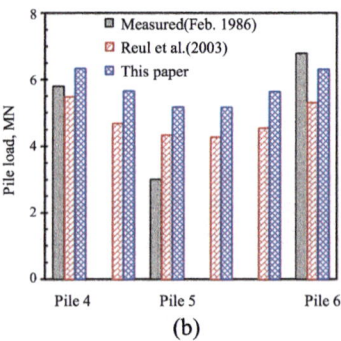

(a)   (b)

Fig. 3. Distributions of pile load after building completion (a) Section I–I; (b) Section II-II

Where: z is the depth below the ground surface (m).

Based on the raft dimensions and soil mechanical parameters, the raft in this case can be considered as rigid by calculating the raft–soil stiffness ratio K_{rs} as defined by Horikoshi and Randolph [13]. Hence, the nonlinear analysis method proposed in this paper can be reasonably applied. In this case, based on the pile material and soil strength, the empirical adhesion factor is adopted as $\alpha = 0.55$.

The distribution of pile loading after building completion is illustrated in Fig. 3. As shown in Fig. 3, for piles at various locations, due to the varying mobilization of shaft friction, the pile loading increases from a center pile (pile 2), to the side piles (pile 1, pile 3, pile 5) and finally to the corner piles (pile 4, pile 6). The comparisons of distributions characteristics of pile loading show that the pile loading calculated by this paper agrees well with the measured and that calculated by Reul *et al.* [7], which proves the validity of the non-linear analysis model proposed by this study.

4 Parametric Study

The results in the case validation show that the pile loading is varied with pile location, and the loads on the corner piles and side piles are significantly larger than that on the center piles. In order to investigate the influence of pile number on the pile loading distribution characteristic, 4 piled raft foundations with different pile number are analyzed. The pile numbers are $5 \times 5, 7 \times 7, 10 \times 10, 15 \times 15$, respectively.

The sketch of the piled raft foundation is depicted in Fig. 4. The raft is supported by piles that are 25 m long, 0.5 m in diameter and 2 m pile spacing, which are the typical pile parameters adopted in the practical engineering. The elastic modulus of the raft and piles is 25 GPa and the Poisson's ratio is 0.2. The soil is a typical soft soil, with shear strength $S_u = 40$ kPa, elastic modulus $E_s = 16$ MPa, and Poisson's ratio $v = 0.35$. For the soft soil, the empirical adhesion factor is adopted as $\alpha = 0.7$.

Figure 5 illustrates the distribution characteristic of pile loading for 7×7 piled raft foundation. As shown in Fig. 5, the pile loading ratios of corner and side piles to the center pile are varied with the foundation safety factor F_s. When the safety factor F_s is larger about 3, the applied load on the foundation is mainly resisted by the corner and

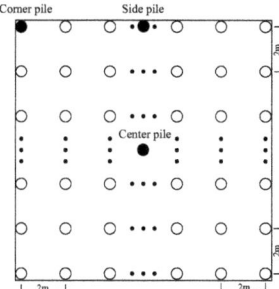

Fig. 4. Sketch of pile arrangement

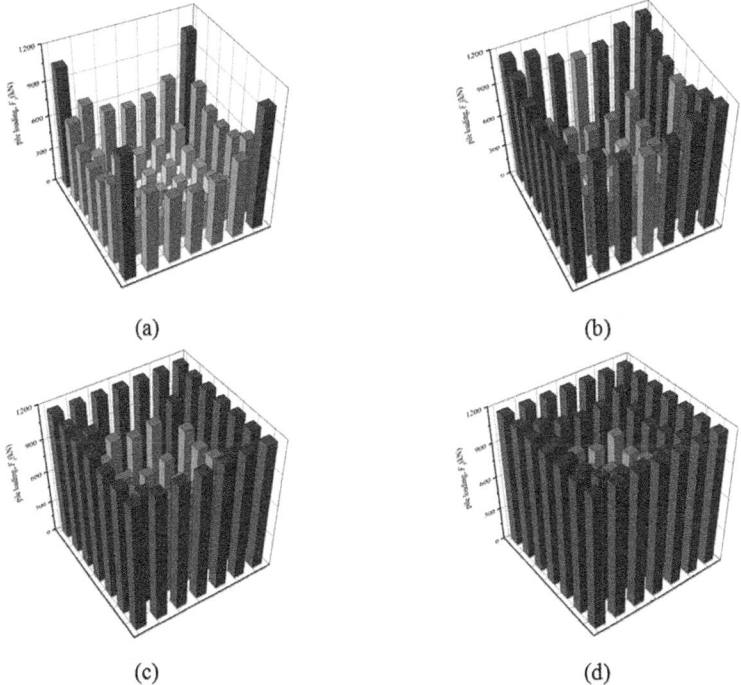

Fig. 5. Pile loading distribution characteristic under different safety factor (a) $F_s = 6.3$; (b) $F_s = 3.3$; (c) $F_s = 2.1$; (d) $F_s = 1.7$

side piles, which results in a rather small load on the center piles. With the decrease of the foundation safety factor, the shaft friction of the center pile is mobilized gradually, and the bearing percentage of the center pile is improved.

To quantify the pile loading distribution characteristic of the piled raft foundation, the variations of pile load ratio of corner pile to center pile (R_{cc}) and side pile to center pile (R_{sc}) are investigated. The variations of R_{cc} and R_{sc} with foundation safety factor F_s are illustrated in Fig. 6.

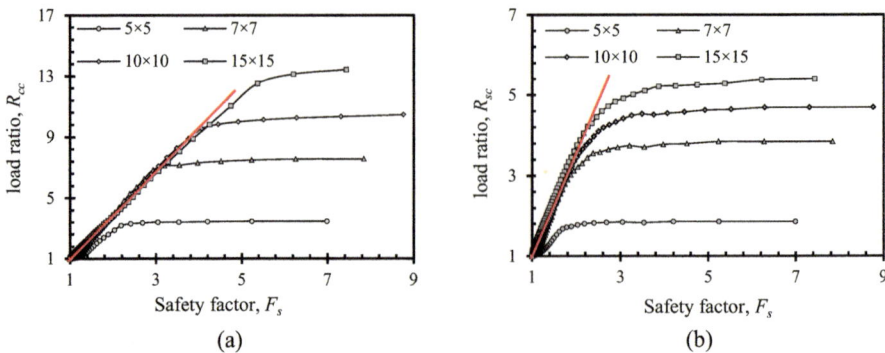

Fig. 6. Variations of pile load ratio with safety factor (a) Corner-center; (b) Side-center

As it can be seen from Fig. 6, there is a critical value of safety factor $F_{s,c}$ for the piled raft foundation. When the safety factor F_s is larger than $F_{s,c}$, the soil around the piles is in elastic condition, and the development patterns of pile loading for piles at different position are in synchronously, inducing a constant pile load ratios R_{cc} and R_{sc}. With the decreasing of safety factor F_s, the bearing capacity of corner pile and side pile has been fully mobilized, while the pile load for the center pile is increased step by step, resulting a decreasing of pile loading ratios R_{cc} and R_{sc}. As illustrated in Fig. 6, the pile loading ratios R_{cc} and R_{sc} are decreased linearly with the decreasing of the safety factor F_s, and the slops of the curves are nearly the same for the 4 piled raft foundations.

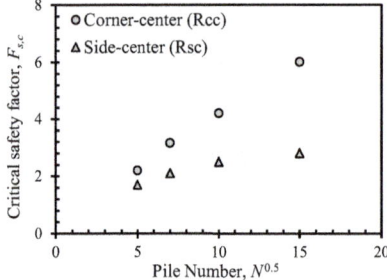

Fig. 7. Variations of critical safety factor with pile number

Fig. 8. Variations of pile loading ratio with pile number ($F_s = 3$)

The variations of critical safety factor $F_{s,c}$ with pile number are illustrated in Fig. 7, which show that the critical safety factor $F_{s,c}$ is increased linearly with $N^{0.5}$, where N is the pile number. As the corner pile is more easily to reach the ultimate bearing capacity, the critical safety factor $F_{s,c}$ for R_{cc} is larger than that of R_{sc}.

The pile loading distribution characteristic has a great influence on the bending moment of the raft. To aid the industrial design, the variation of pile loading ratio with the squared pile number ($N^{0.5}$) is developed, see Fig. 8. The safety factor F_s is 3, which is a typical value for a piled raft foundation design. For the commonly used piled raft foundation ($N \geq 100$), the pile load ratios R_{cc} and R_{sc} are about 6.8 and 5, respectively. Based on the pile load ratios R_{cc} and R_{sc}, the pile loading distribution characteristic can be roughly estimated, and then the bending moment of the raft can be calculated.

5 Conclusions

Based on the boundary element method, a practical non-linear analysis method for rigid piled raft foundation is proposed to investigate the pile loading distribution characteristic. The proposed non-linear analysis method is validated against the field measurements, and a series of parametric study is conducted to investigate the pile loading distribution characteristic. The conclusions can be drawn as followings:

1. The distribution characteristics of pile loading calculated with this paper proposed method agree well with that measured in the filed case, which proves the validity of the non-linear analysis model proposed by this study.
2. Pile load ratios of corner pile to center pile and side pile to center pile are increased with the increasing of safety factor in a linear fashion until to a critical value of safety factor, and the slops are nearly the same for piled raft foundations with different pile numbers.
3. The variation of pile load ratio with pile number is developed, and for a typical piled raft foundation with safety factor $F_s = 3$, pile number $N \geq 100$, the pile load ratios of corner pile to center pile and side pile to center pile are about 6.8 and 5, respectively.

Acknowledgement. This study is supported by National Natural Science Foundation of China (Grant No. 41372274 and No. 41502273), which is highly appreciated.

References

1. Poulos, H.G., Davis, E.H.: Pile Foundation Analysis and Design. Wiley, New York (1980)
2. Chow, Y.: Analysis of vertically loaded pile groups. Int. J. Numer. Anal. Meth. Geomech. **10**(1), 59–72 (1986)
3. Cooke, R.W.: Piled raft foundations on stiff clays—a contribution to design philosophy. Geotechnique **36**(2), 169–203 (1986)
4. Koizumi, Y., Ito, K.: Field tests with regard pile driving and bearing capacity of piled foundations. Soils Found. **7**(3), 30–53 (1967)
5. Cooke, R.W., Price, G., Tarr, K.: Jacked piles in London clay: interaction and group behaviour under working conditions. Geotechnique **30**(2), 97–136 (1980)
6. Kuwabara, F.: An elastic analysis for piled raft foundations in a homogeneous soil. Soils Found. **29**(1), 82–92 (1989)
7. Reul, O., Randolph, M.: Piled rafts in overconsolidated clay: comparison of in situ measurements and numerical analyses. Géotechnique **53**(3), 301–315 (2003)
8. Basile, F.: Non-linear analysis of vertically loaded piled rafts. Comput. Geotech. **63**, 73–82 (2015)

9. Kitiyodom, P., Matsumoto, T.: A simplified analysis method for piled raft foundations in non-homogeneous soils. Int. J. Numer. Anal. Meth. Geomech. **27**(2), 85–109 (2003)
10. Chow, Y.K.: Vertical deformation of rigid foundations of arbitrary shape on layered soil media. Int. J. Numer. Anal. Meth. Geomech. **11**(1), 1–15 (1987)
11. Small, J.C., Liu, H.L.: Time-settlement behaviour of piled raft foundations using infinite elements. Comput. Geotech. **35**(2), 187–195 (2008)
12. Franke, E., EI-Mossallamy, Y., Wittmann, P.: Calculation methods for raft foundations in Germany. In: Design Applications of Raft Foundations, pp. 283–322 (2000)
13. Horikoshi, K., Randolph, M.F.: On the definition of raft–soil stiffness ratio. Géotechnique **47**(5), 1055–1061 (1997)

Physical and Numerical Modelling of Grouting Induced Enlarged-Base Pile

Qiuyu Wang[1,2], Daokun Qi[3], Minming Zhou[1,2], Jiangu Qian[1,2(✉)], and Xin Hu[3]

[1] Department of Geotechnical Engineering, Tongji University, Shanghai, China
qianjiangu@tongji.edu.cn
[2] Key Laboratory of Geotechnical and Underground Engineering of Ministry of Education, Tongji University, Shanghai, China
[3] State Grid Henan Economic Research Institute, Zhengzhou 450052, China

Abstract. The grouting induced enlarged-base pile is an uplift pile combining conventional enlarged-base pile with post-grouting technique. This type of uplift pile has great value in engineering practice for its higher uplift bearing capacity and less material used. This paper will discuss the uplift bearing capacity and failure mechanism in depth from two aspects, including the model experiment and finite element modelling (FEM). Considering the grouting induced enlarged-base pile, experimental model was developed to investigate the uplift capacity of the enlarged base. The load-displacement curves for enlarged bases of different sizes were obtained under the overburden pressure corresponding to a certain depth. Meanwhile, the experiment was numerically investigated via FEM and the results are consistent with the experimental data. It is found that the degree of increase of uplift bearing capacity becomes greater as the length ratio of the enlarged base to the pile or the diameter ratio of enlarged base to pile increase.

Keywords: Enlarged-base pile · Grouting induced · Model test
Finite element method

1 Introduction

For soft ground, the traditional uniform-section circular pile generally has a relatively low uplift bearing capacity, due to its low skin friction or base resistance mainly related to the mechanical properties of the soil-pile interface [1–6]. In recent years, three new types of uplift piles, i.e., uniform-section piles with side-grouting, base-enlarged piles and molded grouting screwed piles have been applied to replace the uniform section piles in the soft ground in Shanghai. In this paper, concentration will be given to a newly developed grouting induced base-enlarged pile.

The non-uniform piles including base-enlarged piles and molded grouting screwed piles exerts the broaching resistance of the belled body to intensify the capacity within less materials. Although the molded grouting screw pile increases the uplift capacity by adding the length of fracture plane, its construction techniques are quite complicated [7]. In order to prevent the grouting diffusing or blocking the post grouting technique is widely adopted in soft-clay area. A newly developed grouting induced enlarged-base

© Springer Nature Singapore Pte Ltd. 2018
T. Qiu et al. (Eds.): GSIC 2018, *Proceedings of GeoShanghai 2018 International Conference: Advances in Soil Dynamics and Foundation Engineering*, pp. 709–717, 2018.
https://doi.org/10.1007/978-981-13-0131-5_77

pile put forward in this study is modified on the basis of conventional enlarged-base pile. After drilling, firstly the geotextile tubes are bound up around the base, secondly the reinforcing cage is buried prior to the cast-in-situ. Finally the cement slurry is injected with high pressure into the geotextile tubes and forms the enlarged base simultaneously accompanied by hardening. The specific procedures of the piling technology is shown in Fig. 1. This subtle modification made on the end of the reinforcing cage not only simplifies the construction technique, but the grouting can also squeeze and compact the soil around [8, 9]. Besides, the cement slurry in the unsealed tubes is more likely to diffuse to more areas, which equals the secondary hardening function. This kind of pile has been already applied to current engineering projects in Shanghai.

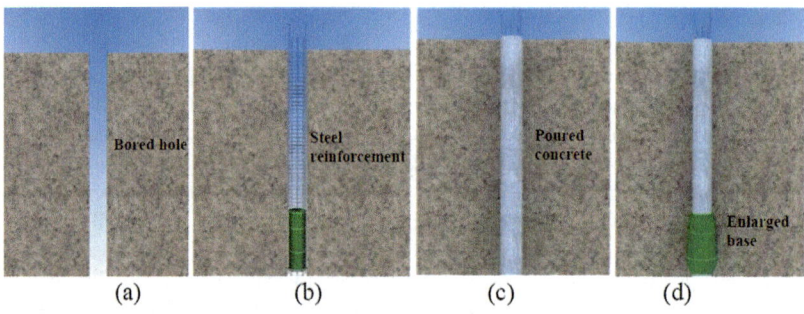

Fig. 1. Piling technology: (a) Drilling (b) Placement of reinforcing cage (c) Pouring concrete (d) Formed enlarged base.

Considerable research has been conducted on the behavior of enlarged-base piles through numerical modelling, centrifuge and field tests [10–15]. Dickin [11, 14] conducted a series of model tests on the centrifuge with respect to different pile diameter, enlarged base diameter and compactness of sand, and found that the failure displacements normalized to base diameter increased considerably with embedment ratio but decreased with increased sand density. Apart from modelling tests, some models were established to predict the uplift capacity of enlarged-base pile. Ilamparuthi [15, 16] suggested a hyperbolic Q-S curves of the geogrid-cell-reinforced belled pile. Influencing factors as the geometry or buried depth were also studied [17, 18].

It was proved the grouting enlarged base pile does increase the uplift capacity and attempts were made to predict the uplift capacity of enlarged-base pile. However, the conventional enlarged-base pile seldom accounts for the compactness of soil due to grouting and the rational dimension parameters to be determined. Accordingly, in this study, the two analyzed factors are the length ratio L/d which refers to the ratio of the length of enlarged base L to the diameter of pile d, and the diameter ratio D/d which refers to the ratio of diameter of enlarged base D to the diameter of pile d.

In this study, the pile-soil model tester designed by Tongji University was adopted for laboratory test and the uplift capacity properties were investigated of the grouting enlarged base pile. In addition, the load-displacement curves obtained by numerical model were compared with the test curves, then the influencing factors, namely, the dimension parameters D/d and L/d were analyzed and quantified based on the degree of ameliorating the uplift bearing property as well as the grouting-induced expansion. Consequently, the optimization parameter design was proposed and the developing mechanism of plastic region was simply revealed.

2 Model Test

In this study, 6 sets of scale model experiments were conducted considering the grouting induced enlarged-base pile technique. Then the numerical modelling results will be contrasted with experimental results and added groups of more sizes of enlarged base. However, one shortcoming of the loading programme in the tests is that the uniform confining pressure along the pile direction fails to reflect the in situ stress conditions, where lateral stress on piles normally increases with depth. So each set of tests simulated only a section of pile at certain depth [8]. To resolve this problem, the enlarged base was embedded at a certain depth (10 m); the uplift pile was positioned horizontally with confining pressure acted vertically on the top of model box.

2.1 Model Test Scheme

The model test material is the Yangzi river sand. Through measuring, the internal friction angle of the tested sand is 32.7°. Plus, the diameter and length of pile body are uniformly 300 mm and 400 mm. The test scheme is given in Table 1.

Table 1. Model test scheme

Test no.	Vertical confining pressure σ_v	Length of enlarged base L (mm)	Diameter of enlarged base D (mm)
1	100	75	60
2	100	75	80
3	100	100	60

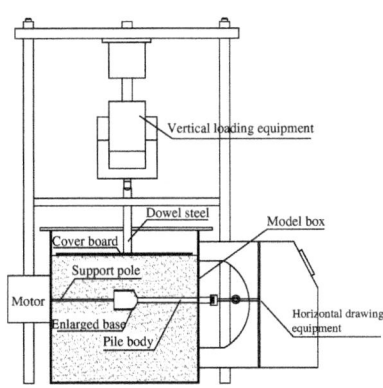

Fig. 2. Schematic of the testing system for the grouting induced enlarged-base pile.

Figure 2 is the schematic of the testing system. The test device is the pile-soil model tester designed by Tongji University including five parts: vertical loading equipment, horizontal drawing equipment, mechanical base-enlarging installation, electric motor and model test box. The vertical loading device adopts the large-scale consolidation loading centrifuge by DGJ-250. The horizontal drawing device controls the displacement through electric motor. The mechanical base-enlarging system is composed of pile body and enlarged base, controlling the internal elevation of drive pole by the rotating of the end region connected to the motor.

The enlarged base is composed of frustum drive and three arc side walls which could open up in lifting the slide frustum. Therefore, the diameter of enlarged base is increased from the pile diameter (Fig. 3).

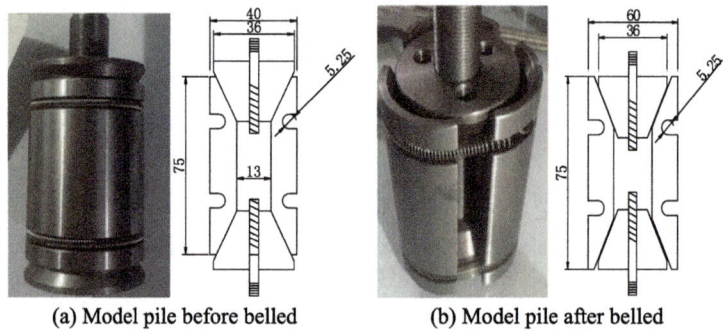

(a) Model pile before belled   (b) Model pile after belled

Fig. 3. Mechanical enlarging-base device.

2.2   Test Results

The figures in Figs. 4 and 5 present the load-displacement curves for different sizes of enlarged base considering the grouting induced expansion or not.

Enlarged bases of different D/d. Figure 4 shows the load-displacement curves of the 1st and 2nd set, of which each enlarged base has a distinct D/d, and the percentages of the increase in ultimate uplift capacity under different conditions are listed in Table 2.

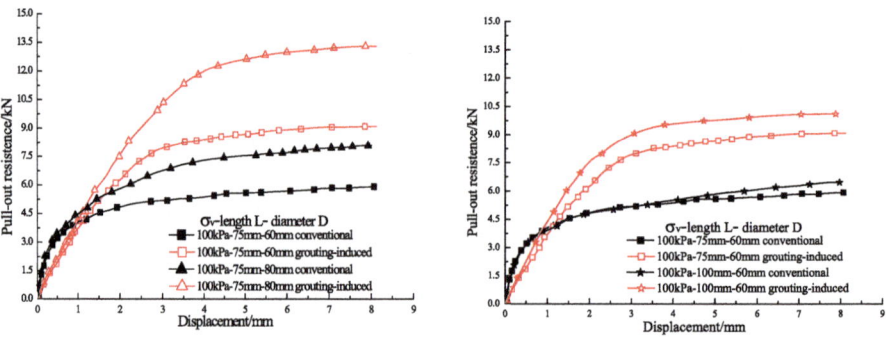

Fig. 4. Load-displacement curves of the 1st and 2nd set.

Fig. 5. Load-displacement curves of the 1st and 3rd set.

Note that the initial slopes of load-displacement curves of conventional piles are greater than that of enlarged-based pile as seen in Figs. 4 and 5, this illustrates that the enlarged base not also compacts the surrounding soil but also results in the shear failure of the upper soil, owing to which the soil around is comparably loosened. However, the relative displacement of the enlarged base and soil is not large enough so that the uplift bearing capacity is mainly generated by the upper structure. Accordingly, the initial stiffness of conventional piles is relatively higher.

Enlarged bases of different L/d. Figure 5 shows the load-displacement curves of the 1st and 3rd set, of which each enlarged base has a distinct L/d, and the percentages of the increase in ultimate uplift capacity under different conditions are listed in Table 2.

Table 2. Percentage of the increase of ultimate uplift capacity

Set number	1	2	3
Ultimate uplift capacity for conventional enlarged-base piles (N)	5907	8092	6491
Ultimate uplift capacity for grouting induced enlarged-base piles (N)	9184	13354	10147
Percentage of the increase of ultimate uplift capacity	55%	65%	56%

Figures 4 and 5 manifests the similar rules that the curves can be divided in two stages: the displacement of the first stage is linearly proportional to pull-out resistance and the soil still remains flexible with relatively greater stiffness. Subsequently the second stage represents a nonlinear relationship between pull-out resistance and displacement, the plastic region of the soil initiates around the enlarged base and develops further, and finally the pull-out resistance nearly maximizes and stabilizes.

Moreover, it can be seen that the slopes of the initial load-displacement curves of the grouting belled piles are all smaller those of the non-grouting belled piles under the same conditions, illustrating that the grouting induced expanding process disturbs the soil around so as to result in the shear failure with the stiffness decreasing. In the initial pull-out process the resistance is mainly provided by the soil near the enlarged base so the initial stiffness of grouting expansion is less than that of non-grouting expansion. Meanwhile, the ultimate lifting capacity increases with the vertical confining pressure for the enlarged bases of the same size, indicating that the increasing uplift capacity depends on the exertion of deep side friction of piles. Thus the grouting induced enlarged-base piles do remarkably improve the uplift capacity.

3 Numerical Modelling

In order to illustrate the failure mechanism of the grouting induced enlarged-base pile, the model pile is implemented by ABAQUS software through 14 sets of different diameter ratios and length ratios of model piles. The interaction of the modelling process sets the pile-soil interface to be Mohr-Coulomb interaction. The key analysis is to simulate the shear strength transmission mechanism and developing deformation on the interface. Besides, Figs. 6 and 7 show the load and boundary conditions in grouting induced expanding and pull-out process.

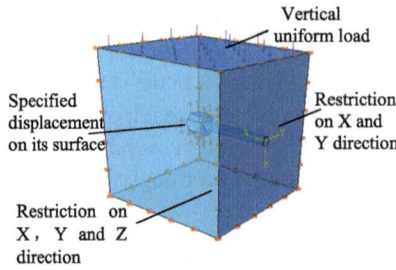

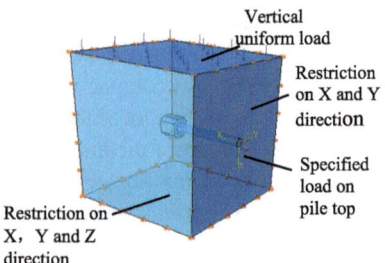

Fig. 6. Load and boundary conditions for grouting expansion.

Fig. 7. Load and boundary conditions for pull-out.

Figure 8 compares the experimental result with the numerical result of the model pile under confining pressure of 100 kPa, with diameter of 60 mm and length of 75 mm. Based on the numerical conformity with test results, added model piles of different D/d and L/d are simulated to further determine the optimal dimension parameter.

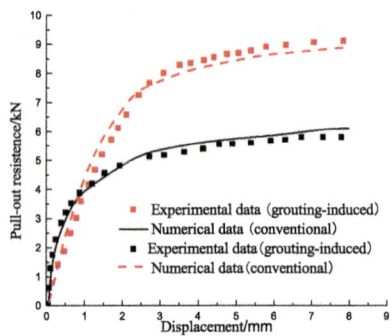

Fig. 8. Comparison between experimental data and FEM results.

Enlarged bases of different D/d. Figures 9 and 10 plot the load-displacement curves of grouting induced enlarged-base and conventional enlarged-base piles of different D/d. As seen, the numerical curves present the variant trend identical with the model test results. Plus, Fig. 11 reveals the relationship between D/d and percentage increase of ultimate uplift capacity is plotted. The figures show that ultimate uplift capacity increases with the diameter ratio and particularly the percentage increase tends to decrease slightly and flatten with a diameter ratio over 2.5.

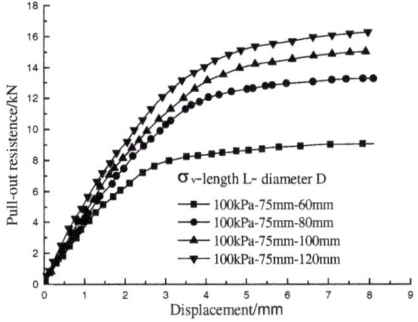

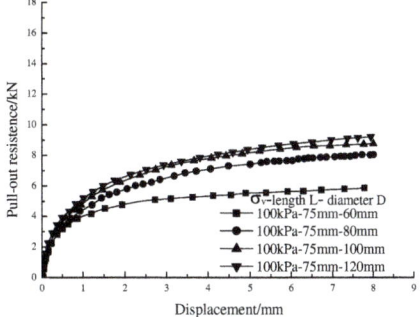

Fig. 9. Load-displacement curves of grouting induced enlarged-base piles.

Fig. 10. Load-displacement curves of conventional enlarged-base piles

Enlarged bases of different L/d. Figures 12 and 13 plot the load-displacement curves of grouting induced enlarged-base and conventional enlarged-base piles of different L/d. Figure 14 reveals the relationship between L/d and percentage increase of ultimate uplift capacity. As seen, percentage increase of ultimate uplift capacity decreases with L/d, and the ultimate uplift capacity doesn't increase infinitely when the L/d is approaching infinity. Under such a condition the enlarged-base pile is more likely to be the uniform circular pile. Accordingly, the higher uplift capacity for practical engineering use should take the length of enlarged base into account to reach the optimal diameter ratio.

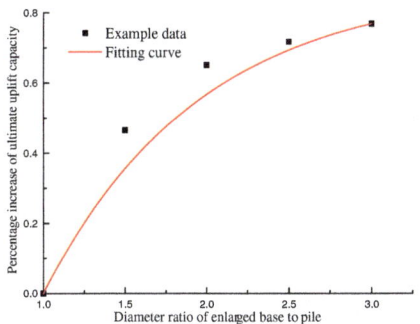

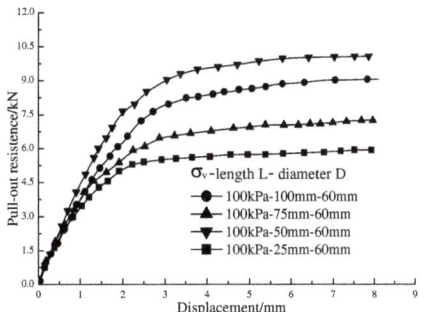

Fig. 11. Relationship between D/d and percentage increase of ultimate uplift capacity.

Fig. 12. Load-displacement curves of grouting induced enlarged-base piles.

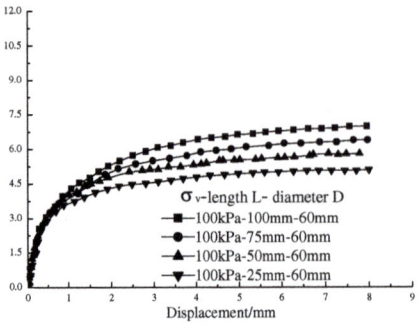
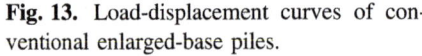

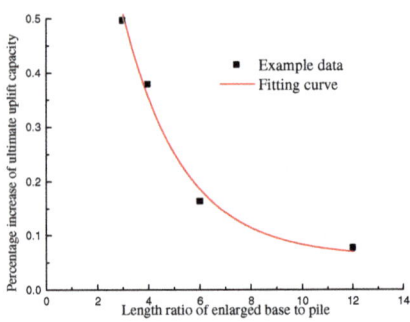

Fig. 13. Load-displacement curves of conventional enlarged-base piles.

Fig. 14. Relationship between L/d and percentage increase of ultimate uplift capacity.

4   Summary and Conclusions

In this study, the grouting induced enlarged-base pile was simulated on the pile-soil tester designed by Tongji University which transformed the grouting technique into mechanical extrusion. Furthermore, the effect of extrusion, vertical confining pressure and size of enlarged base on the load-displacement curves was studied. Based on the model test and numerical modelling, the following conclusions could be drawn:

1. The percentage of ultimate uplift capacity increase is proportional to the diameter ratio and inversely proportional to the length ratio. Moreover, when the diameter ratio reaches nearly 2.5, the increase gradually stabilize.
2. The percentage increase of ultimate uplift capacity decreases with L/d, and the ultimate uplift capacity doesn't increase infinitely when the L/d is approaching infinity.

   The extrusion change the mechanical property of soil around the enlarged base and such expanding densification transfer the pull-out load to larger regions in soil. Consequently, the uplift capacity is greater.

Acknowledgement. This work was supported by the National Grid Company Technology Project (No. 5217L0160001) and the Fundamental Research Funds for the Central Universities (No. 0200219221).

References

1. Alawneh, A.S., Malkawi, A.I.H., AlDeeky, H.: Tension test on smooth and rough model piles in dry sand. Can. Geotech. J. **36**(4), 746–753 (1999)
2. Ramasamy, G., Dey, B., Indrawan, E.: Studies on skin friction in piles under tensile and compressive load. Indian Geotech. J. **34**(3), 276–289 (2004)
3. Dash, B.K., Pise, P.J.: Effect of compressive load on uplift capacity of model piles. J. Geotech. Geoenviron. Eng. **129**(11), 987–992 (2003)

4. Ball, A.: The resistance to breaking-out of mushroom foundations for pylons. In: Proceedings of 5th International Conference on SMFE, pp. 569–576 (1961)
5. Chattopadhyay, B.C., Pise, P.J.: Uplift capacity of piles in Sand. J. Geotech. Eng. **112**(9), 888–904 (1986)
6. Parry, R.H., Swin, C.W.: Effective stress methods of calculating skin fiction on driven piles in soft clay. Ground Eng. **10**(3), 24–26 (1997)
7. Abdelghany, Y., Naggar, H.E.: Steel fibers reinforced grouted and fiber reinforced polymer helical screw piles—a new dimension for deep foundations seismic performance. In: Geo-Frontiers 2011, Advances in Geotechnical Engineering, Dallas, Texas, pp. 103–112 (2011)
8. Qian, J.G., Gao, Q., Wang, B., Xue, J.F., Huang, M.S.: Physical and numerical pull-out modelling of ribbed piles. Proc. Inst. Civ. Eng.-Geotech. Eng. **170**(1), 51–61 (2016)
9. Youn, H., Tonon, F.: Numerical analysis on post-grouted drilled shafts: a case study at the Brazo River Bridge, TX. Comput. Geotech. **37**(4), 456–465 (2010)
10. Thiyyakkandi, S., McVay, M., Bloomquist, D.: Measured and predicted response of a new jetted and grouted precast pile with membranes in cohesionless soils. J. Geotech. Geoenviron. Eng. **139**(8), 1334–1345 (2012)
11. Dickin, E.A., Leung, C.F.: Performance of piles with enlarged base subject to uplift forces. Can. Geotech. J. **27**(5), 546–556 (1990)
12. Mullins, G., O'Neill, M.: Pressure grouting drilled shaft tip a full-scale load test program. Research Report. University of South Florida, Tampa (2003)
13. Wang, W.D., Wu, J.B., Xu, L., Huang, S.M.: Full-scale field tests on uplift behavior of piles with enlarged base. Chin. J. Geotech. Eng. **29**(9), 1418–1422 (2007)
14. Dickin, E.A.: Uplift behavior of horizontal anchor plates in sand. J. Geotech. Eng. **114**(11), 1300–1317 (1988)
15. Ilamparuthi, K., Dickin, E.A.: Predictions of the uplift response of model belled piles in geogrid-cell-reinforced sand. Geotext. Geomembr. **19**(2), 89–109 (2001)
16. Ilamparuthi, K., Dickin, E.A.: The influence of soil reinforcement on the uplift behaviour of belled piles embedded in sand. Geotext. Geomembr. **19**(1), 1–22 (2001)
17. Dickin, E.A., Leung, C.F.: The influence of foundation geometry on the uplift behaviour of piles with enlarged bases. Can. Geotech. J. **29**(3), 498–505 (1992)
18. Niroumand, H., Kassim, K.A., Ghafooripour, A.: Uplift capacity of enlarged base piles in sand. Electron. J. Geotech. Eng. (EJGE) **17**, 2721–2737 (2012)

Predicting the Drained Capacity of Skirted Foundations Under Uniaxial Loads

Vali Ghaseminejad[1](✉), Mohammad Ali Rowshanzamir[2], and Amin Barari[3]

[1] Department of Civil Engineering, Isfahan University of Science and Research, Isfahan, Iran
[2] Department of Civil Engineering, Isfahan University of Technology, Isfahan, Iran
[3] Department of Civil Engineering, Aalborg University, Thomas Manns Vej 23, Aalborg Ø, 9220 Aalborg, Denmark
abarari@vt.edu

Abstract. Skirted foundations denoted as suction caissons are becoming an increasingly prevalent offshore foundation solution for either the oil and gas industry or renewable energy infrastructure. Their response to combined vertical, horizontal and moment loading must be found to ensure their stability under harsh environmental conditions. As part of this process, knowledge of uniaxial capacities is required. Previous studies have neglected effect of deformable ground by assuming that the soil within skirts behaves rigid during drained loading, but this assumption needs rigorous studies. A series of 3-D finite element analyses has been conducted to investigate directly how the skirt geometry, soil strength profile and deformable plug within skirt compartment affect the drained skirted foundation capacity under uniaxial loading. The results show that foundation embedment and soil plug placed within the skirt significantly influence the accompanying mechanisms occurring at failure and therefore the uniaxial capacities.

Keywords: Offshore wind turbines · Bucket foundations · Depth factors
Finite element analyses · Pure bearing capacity · Sand

1 Introduction

Offshore wind turbines are sensitive to deformations and tilting [1]. Compared with a monopile, the installation of bucket foundations is easier and does not need heavy installation equipment [2]. Bucket foundations are feasible under suitable soil conditions and are often used in shallow water depths from near-shore to approximately 55 m inland [3]. Wind turbines transfer a small vertical force to the bucket, but develop heavy horizontal loads and moments. Bucket foundations are circular foundations with thin skirts around their circumference consisting of a large steel cylindrical shaft of diameter D, skirt length L and skirt thickness t_s, with a closed top and open bottom. They penetrate into the seabed vertically under self-weight with a trapped soil plug underneath. Penetration discontinues when the lid of the bucket foundation comes in contact with the seafloor.

© Springer Nature Singapore Pte Ltd. 2018
T. Qiu et al. (Eds.): GSIC 2018, *Proceedings of GeoShanghai 2018 International Conference: Advances in Soil Dynamics and Foundation Engineering*, pp. 718–725, 2018.
https://doi.org/10.1007/978-981-13-0131-5_78

Gourvenec and Randolph [4, 5], Bransby and Yun [6] and Hung and Kim [7] investigated the responses of two and three-dimensional finite element analyses of the general loading of strip and circular skirted foundations in homogeneous and non-homogeneous clay and presented the ultimate limit states and failure envelopes. Gourvenec [8, 9] surveyed failure envelopes and shape effects on the capacity of shallow foundations under general loading at varying aspect ratios. Gourvenec [10] later studied the effect of embedment on the undrained bearing capacity of shallow strip footings subjected to uniaxial and combined loading through a finite element study. Barari and Ibsen [11–13] reported the experimental and numerical responses of vertical and moment loading on small-scale circular surface and suction bucket models on Baltic clay at Aalborg University. Ibsen et al. [14, 15] investigated the behavior of bucket foundations under combined static loads in dense saturated sand and conducted an extensive experiment on small-scale foundations in the laboratory.

The aim of the current report is to evaluate the effect of aspect ratios and sand relative densities on pure bearing capacitates of bucket foundations installed in saturated sands. Load-deformation behavior of suction buckets under pure loads was investigated and compared. Subsequently, the failure mechanism under pure loading based on finite element results was presented.

2 Finite Element Model

All FE analyses were performed using Plaxis 3D Foundation software package [16]. For the skirt and the surrounding soil, 15-node wedge elements were used in the 3D finite element calculations. An elastic-plastic model was used to describe the behavior between skirt and soil. Strength reduction factor in the soil-foundation interface was considered $R_{inter} = 0.7$. The effect of gaps along the bucket and surrounding soil was prevented. The skirt and lid materials were modeled as linear elastic. External boundaries were set sufficiently remotely to reach sufficient accuracy of the results. Hence the behavior of the bucket foundation is not significantly influenced by the boundary conditions. The length of the finite-element mesh boundary was set to 6 times of the bucket diameter and bottom boundary of the model was extended 3 times the bucket skirt length. In order to determine ultimate horizontal, vertical and moment capacities tangent intersection method is employed. In this methodology, ultimate bearing capacity is obtained as load corresponding to the intersection point of two tangential lines along the initial and latter part of the load-deformation curve. Figure 1 exhibits sign conventions for loads and displacements as well as the finite element mesh.

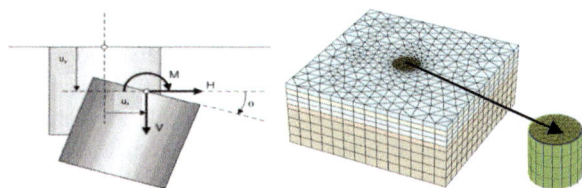

Fig. 1. Left: Sign convention for loads and displacements, Right: A Schematic view of the finite element mesh used in the analyses

Bucket foundations embedded in sands with different relative densities at $D = 12$ m, 16 m and different aspect ratios ($L/D = 0, 0.25, 0.5, 0.75, 1$) were used. D refers to the diameter of the bucket foundation and L is skirt length of the bucket. In the analysis, deformation properties of steel materials with modulus of elasticity $E = 210$ GPa and Poisson's ratio $v = 0.2$ were considered. Submerged unit weight of the steel used for the bucket body was set to $\gamma' = 68$ kN/m^3. A top plate thickness of $t_L = 0.10$ m and unit weight $\gamma' = 77$ kN/m^3 and a very large modulus of elasticity $E = 1 \times 10^9$ GPa were selected for the bucket lid. The analyses are divided into three stages. The initial step is used to consider soil normal stresses by applying unit weight of soil. In the second phase, a part of the soil is replaced by the steel bucket elements. Center top of the bucket foundation is loaded gradually during the third phase until failure. Pure horizontal, vertical and moment loads were applied separately on the bucket lid and increased gradually until the pure bearing capacity of the bucket foundation was reached. In this research, Mohr–Coulomb material model is used as constitutive model in the numerical simulation of soil behavior. To simulate non-linear soil response, stress-dependency of the oedometric modulus of elasticity was implemented through the following expression.

$$E_S = \kappa.\sigma_{at}.\left(\frac{\sigma_m}{\sigma_{at}}\right)^{\lambda} \tag{1}$$

Where σ_m is the current mean principle stress in the considered soil element and $\sigma_{at} = 100$ kN/m^2 is reference stress. Parameters κ and λ are related to soil stiffness at reference stress state. Table 1 gives the parameters of the material used for sands with different properties.

Table 1. Material properties used in the numerical analysis [17]

Property	Loose sand	Medium dense sand	Dense sand	Unit
Submerged unit weight (γ')	7	9	11	[kN/m^3]
Oedometric stiffness parameter (κ)	300	400	600	[-]
Oedometric stiffness parameter (λ)	0.65	0.6	0.55	[-]
Poisson's ratio (v)	0.25	0.25	0.25	[-]
Internal friction angle (φ')	30	35	40	[°]
Dilation angle (ψ)	2.5	5	10	[°]
Cohesion (C')	0.1	0.1	0.1	[kN/m^2]

3   Validation of the Numerical Model

In this study, a finite element simulation of bucket foundations in dense sand was utilized and validated versus the results of finite element analysis performed by Achmus et al. [17]. A bucket at $D = 12$ m, a skirt length of $L = 9$ m ($L/D = 0.75$) and a skirt thickness of $t_s = 3$ cm was analyzed. Dense soil parameters listed in Table 1, were chosen for analysis. Figure 2 compares Achmus et al. results along with the numerical simulations and loading eccentricity of h = 100 m. Moment-rotation curve represented good agreement with the literature.

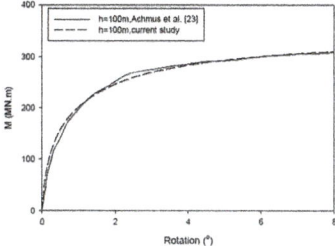

Fig. 2. Moment-rotation curve for the FE simulation compared by Achmus et al. [17] results

4 Vertical Capacity

Vertical loads on bucket foundations derive from their self-weight as well as the loading tower. Figure 3 shows view of vertical displacement contours under pure vertical loading at $D = 16$ m for $L/D = 0.25$ and 1. Failure under pure vertical loads was almost governed by pure vertical displacements. Figure 4 shows variations in pure vertical capacity depth factors ($d_{cv} = V_{ult(L/D)}/V_{ult(L/D=0)}$) as a function of aspect ratios for a range of sand profiles using $D = 12$ m and 16 m. At large aspect ratios, bucket foundations in different sands represented higher vertical bearing capacities since their sidewalls involved higher shear strengths.

There is no exact solution to determine the pure vertical capacity of a bucket foundation. However, the results of the present study showed that a linear function could be considered in V/V_{ult}- L/D curve at different diameters and sand types.

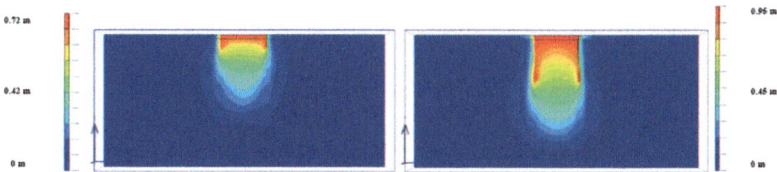

Fig. 3. Vertical displacement contours under pure vertical loading in medium dense sand at $D = 16$ m, Left: $L/D = 0.25$ Right: $L/D = 1$

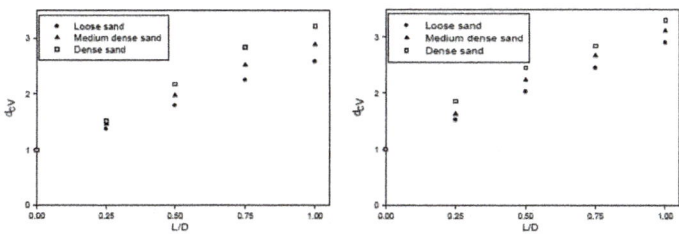

Fig. 4. Normalized pure vertical bearing capacities as a function of aspect ratios in sands with different relative densities, Left: $D = 12$ m, Right: $D = 16$ m

Additionally, to determine pure vertical bearing capacity depth factors, the following equations were found to better match with the numerical results.

$$d_{cV} = 1 + 1.77\left(\frac{L}{D}\right) \qquad \text{loose sand} \qquad (2)$$

$$d_{cV} = 1 + 2.07\left(\frac{L}{D}\right) \qquad \text{medium sand} \qquad (3)$$

$$d_{cV} = 1 + 2.41\left(\frac{L}{D}\right) \qquad \text{dense sand} \qquad (4)$$

It is worth adding that the results could be valid for large diameter buckets when $0 \leq L/D \leq 1$. At diameter of 16 m, the pure vertical capacity depth factor was a little larger than 12 m. However, in order to reach a unique response regarding both 12 m and 16 m in sands with different relative densities, the same equation was considered.

5 Horizontal Capacity

Since the analysis of horizontal bearing capacities could have a great role in offshore wind turbines, an investigation of the behavior of bucket foundations with varying soil profiles to pure horizontal load was performed. Figure 5 shows view of incremental displacement contours under pure horizontal loading at D = 16 m for L/D = 0.25 and 1. Sliding failure mechanism of surface foundations in pure horizontal loads alters from sliding behavior to a rotational mode when skirts are applied to different types of sand under investigation here. In pure horizontal loading, the coupling between the horizontal and rotational degrees of freedom played an important role. The rotational mechanism observed in terms of pure horizontal loading, was related to lateral strength of soil acting on inside and outside the bucket skirt.

Figure 6 illustrates variations in pure horizontal capacity depth factors ($d_{cH} = H_{ult (L/D)}/H_{ult(L/D=0)}$) as a function of aspect ratios in loose, medium and dense sands at D = 12 m, and 16 m. Finite element results indicated that pure horizontal capacity depth factors could be related to the linear expressions between normalized ultimate uniaxial horizontal loads and aspect ratios.

$$d_{cH} = 1 + 8.36\left(\frac{L}{D}\right) \qquad \text{loose sand} \qquad (5)$$

$$d_{cH} = 1 + 9.53\left(\frac{L}{D}\right) \qquad \text{medium sand} \qquad (6)$$

$$d_{cH} = 1 + 10.34\left(\frac{L}{D}\right) \qquad \text{dense sand} \qquad (7)$$

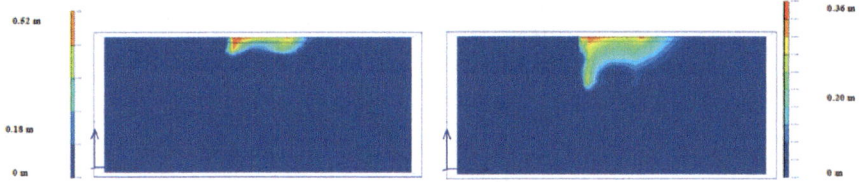

Fig. 5. Incremental displacement contours under pure horizontal loading in medium dense sand at $D = 16$ m, Left: $L/D = 0.25$ Right: $L/D = 1$

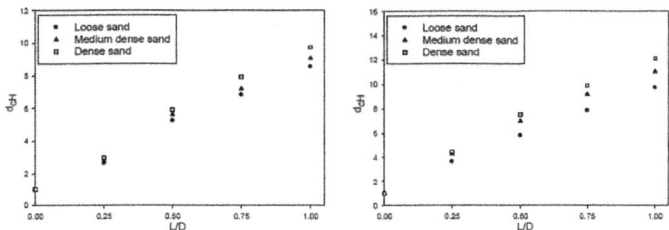

Fig. 6. Normalized pure horizontal capacities as a function of aspect ratios in sands with different relative densities, Left: $D = 12$ m, Right: $D = 16$ m

6   Moment Capacity

Moment loading is worth consideration in wind turbines, which are tall and slender structures and as a result susceptible to overturn due to eccentricity of loading. The wind at the top of the tower can produce huge moments for the bucket foundation to bear compared to vertical and horizontal loading imposed on it. Figure 7 outlines incremental displacement contours under pure moment loading at $D = 16$ m for $L/D = 0.25$ and 1. At large aspect ratios, the soil confined within the bucket foundation was observed to behave as a rigid cluster during loading, whereas in case of low aspect ratios it was affected by the failure mode. A pure rotation would not accompany large deformations under pure moment due to a combination of horizontal and rotational degrees of freedom causing horizontal and rotational translations.

Figure 8 illustrates variations in pure moment capacity depth factors ($d_{cM} = M_{ult (L/D)}/M_{ult(L/D=0)}$) as a function of aspect ratios in all types of the sands at $D = 12$ m and 16 m. At small aspect ratios, the pure moment bearing capacity was small, meaning that sidewalls could be negligible, and increases in the strength of underlying soil would induce more capacity to bucket foundations. The pure moment capacity was larger in dense sand than medium and loose sands. When the degree of embedment increases, the difference will become greater.

Finite element results indicated that pure moment capacity depth factors could be related to the square of aspect ratios and expressed by the following functions.

$$d_{cM} = 1 + 2.22\left(\frac{L}{D}\right) + 7.22\left(\frac{L}{D}\right)^2 \qquad \text{loose sand} \qquad (8)$$

$$d_{cM} = 1 + 1.73\left(\frac{L}{D}\right) + 8.94\left(\frac{L}{D}\right)^2 \qquad \text{medium sand} \qquad (9)$$

$$d_{cM} = 1 + 2.12\left(\frac{L}{D}\right) + 10.95\left(\frac{L}{D}\right)^2 \qquad \text{dense sand} \qquad (10)$$

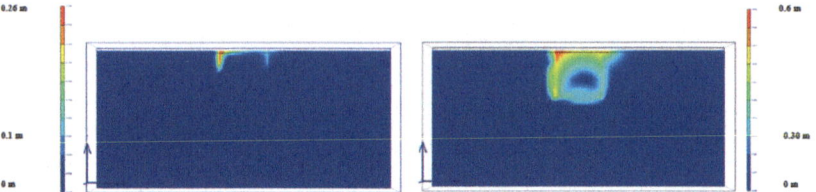

Fig. 7. Incremental displacement contours under pure moment loading in medium dense sand at $D = 16$ m, Left: $L/D = 0.25$, Right: $L/D = 1$

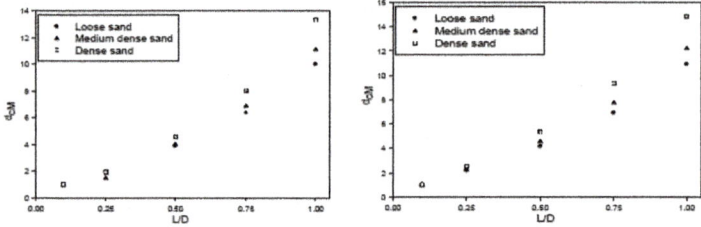

Fig. 8. Normalized pure moment capacities as a function of aspect ratios in sands with different relative densities, Left: $D = 12$ m, Right: $D = 16$ m

7 Conclusion

The study investigated the results of three-dimensional numerical analyses of bucket foundations founded in loose, medium and dense sands under pure horizontal, vertical and moment loadings. The influence of variations in the geometry of the bucket (length-to-diameter aspect ratio L/D) and soil properties on pure bearing capacity were evaluated and discussed. Bucket dimensions and types of sand would significantly affect pure ultimate capacities of bucket foundations. Sidewalls in bucket foundations proved to play an important role in soil-bucket foundation response to uniaxial loadings, being able to transfer the normal and shear stresses to the soil. The effect of embedment was examined and depth factor relationships were developed for a range of soil properties. The normalized expressions of pure horizontal and vertical bearing capacities were found to be proportional to aspect ratios linearly, while a quadratic relationship was observed between aspect ratio and pure moment capacity depth factor. As depicted above, failure under pure horizontal and moment loads may be governed by a combination of horizontal and rotational translations.

References

1. Barari, A., Bagheri, M., Rouainia, M., Ibsen, L.B.: Deformation mechanisms for Monopile foundations of offshore wind power plants considering the cyclic mobility effect. Soil Dyn. Earthq. Eng. **97**, 439–453 (2017a)
2. Barari, A.: Characteristic behavior of bucket foundations. Aalborg University, Denmark, Ph. D. thesis (2012)
3. Barari, A., Ibsen, L.B., Taghavi Ghalesari, A., Larsen, K.A.: Embedment effects on vertical bearing capacity of offshore Bucket foundations on cohesionless soil. Int. J. Geomech. (2017b). https://doi.org/10.1061/(ASCE)GM.1943-5622.0000782
4. Gourvenec, S., Randolph, M.: Effect of strength non-homogeneity on the bearing capacity of circular skirted foundations subjected to combined loading. In: Proceedings of the Twelfth International Offshore and Polar Engineering Conference, Kitakyushu, Japan (2002)
5. Gourvenec, S., Randolph, M.: Bearing capacity of a skirted foundation under VMH loading. In: Proceedings of OMAE03 22nd International Conference on Offshore Mechanics and Arctic Engineering, Cancun, Mexico (2003)
6. Bransby, M.F., Yun, G.J.: The undrained capacity of skirted strip foundations under combined loading. Géotechnique **59**(2), 115–125 (2009)
7. Hung, L.C., Kim, S.R.: Evaluation of vertical and horizontal bearing capacities of bucket foundations in clay. Ocean Eng. **52**, 75–82 (2012)
8. Gourvenec, S.: Shape effects on the capacity of rectangular footings under general loading. Géotechnique **57**(8), 637–646 (2007)
9. Gourvenec, S.: Failure envelopes for offshore shallow foundations under general loading. Géotechnique **57**(9), 715–728 (2007)
10. Gourvenec, S.: Effect of embedment on the undrained capacity of shallow foundations under general loading. Géotechnique **58**(3), 177–185 (2008)
11. Barari, A., Ibsen, L.B.: Effect of embedment on the vertical bearing capacity of Bucket foundations in clay. In: Proceedings, Pan-Am CGS Geotechnical Conference, Toronto, Ont (2011)
12. Barari, A., Ibsen, L.B.: Undrained response of bucket foundations to moment loading. Appl. Ocean Res. **36**, 12–21 (2012). https://doi.org/10.1016/j.apor.2012.01.003
13. Barari, A., Ibsen, L.B.: Vertical capacity of bucket foundations in undrained soil. J. Civ. Eng. Manag. **20**(3), 360–371 (2014)
14. Ibsen, L.B., Barari, A., Larsen, A.: Adaptive plasticity model for bucket foundations. J. Eng. Mech. **140**, 361–373 (2014)
15. Ibsen, L.B., Larsen, K.A., Barari, A.: Calibration of failure criteria for bucket foundations on drained sand under general loading. J. Geotech. Geoenviron. Eng. (2014). https://doi.org/10.1061/(ASCE)GT.1943-5606.0000995
16. Plaxis user's manual, version 1.6 (2005)
17. Achmus, M., Akdag, C.T., Thieken, K.: Load-bearing behavior of suction bucket foundations in sand. Appl. Ocean Res. **43**, 157–165 (2013)

Offshore Geotechnics

Numerical Study on the Performance of Bio-inspired Bridge Attachments as Local Scour Countermeasures with Attack Angles

Fayun Liang[1(⊠)], Chen Wang[1,2], and Xiong (Bill) Yu[2]

[1] Department of Geotechnical Engineering, Tongji University,
Shanghai 200092, China
fyliang@tongji.edu
[2] Department of Civil Engineering, Case Western Reserve University,
Cleveland, OH 44106, USA

Abstract. Scour is a natural phenomenon caused by erosion or removal of streambed or riverbank materials around bridge foundations by the stream. To reduce the scour depth around bridge foundations, local scour countermeasures are designed. During the past decades, a significant amount of countermeasure types and concepts has been proposed. Different from the traditional counter-measure concept of riverbed attachments, a new protection method named bio-inspired bridge attachments is provided in the present study. To evaluate the performance of this method, numerical simulation using Computational Fluid Dynamic (CFD) with scour initiation criteria was involved. This scour countermeasure can provide good protection of riverbed and is more economical than the traditional methods. To make full use of this concept, the preliminary design guideline of bridge attachments as countermeasures has also been carried out.

Keywords: Local scour · Numerical simulation · Countermeasures
Computational fluid dynamic

1 Introduction

Scour countermeasures at bridge piers play an important role in bridge safety and have become a major part of Federal Highway Administration's national bridge scour program. They are also considered vital in reducing the vulnerability of bridges to scour. The scour mechanism around an underwater structure is coupled to a complex three-dimensional interaction of flow, soil, and structure [1–4]. The use of local scour countermeasures has three advantages: 1. Lower the possibility of bridge failure under extreme conditions by reducing scour depth around a certain bridge pier; 2. Cut down the budget for underwater structures regarding local scour by permitting a lower embedded length or other construction requirements; 3. Fulfill some other additional functions and purposes, such as channel guiding and collision prevention.

During the past decades, various types of countermeasures have been proposed and improved to enhance the stability of the bridge. Among these methods, the concept of countermeasures can be divided into two categories, namely negative and positive countermeasures [5, 6]. Negative countermeasures include riprap and its alternatives,

© Springer Nature Singapore Pte Ltd. 2018
T. Qiu et al. (Eds.): GSIC 2018, *Proceedings of GeoShanghai 2018 International Conference: Advances in Soil Dynamics and Foundation Engineering*, pp. 729–739, 2018.
https://doi.org/10.1007/978-981-13-0131-5_79

aiming at enhancing the ability of the bed material to resist erosion. In contrast, positive countermeasures are designed to reduce the erosive power of the stream, for example, the energy associated with down flow and horseshoe vortex, and therefore mitigate the scour action on the bed material. This can be achieved by placing an extended base, use of a streamlined collar, or sacrificial piles. Negative countermeasures can also be classified as riverbed attachments, in which category bed materials are usually reinforced by units with high resistance to erosion. However, these units, like ripraps, are consumable due to the continual coming flow and need to be refilled periodically. The loss of countermeasure materials may lead to bridge unsafety and would be a heavy economic burden in bridge maintenance. Another concept is to refine the bridge pier so that the coming flow is redirected and diminished. The budget for this kind of countermeasures may be higher than riverbed attachments at the beginning due to its design, manufacture, and installation. Nevertheless, it is more economical in long term because of its stability and durability, with which the bridge attachments do not need to be repaired as frequently as riverbed materials and cost as much as resistance units refill. A significant amount of studies on various scour countermeasures have been carried out in past decades. For example, the collar is used to control scouring around the piers mainly by diverting the downward flow. The effect of collars on mitigating the local scour has been investigated by researchers [5, 7, 8]. The size and elevation of the collar are proved to be the most important factors influencing the efficiency. A slot in the pier, which permits the coming flow to pass through, has also been studied. The effect of slot height, width, and position on scour depth around a circular pile under currents and waves have been studied in previous research [9]. The similar concept also brings the ideas of sheath [10] and streamlining design [11].

However, these countermeasures have their own limitations when applied in practice. The disadvantages can be classified into three categories: 1. Hard to design and install; 2. May have a negative influence on bearing capacity; and 3. Lose its function when the flow attack angle changes. In this study, the concept of bio-inspired bridge attachments was proposed by using additional parts attached to piers for the purpose of flow power dissipation. The influence of attack angle has been studied. Numerical simulations have been carried out to investigate the performance in reducing maximum scour depth and the effect on scour process. Preliminary design criteria have been addressed to make full use of the protection method.

2   Numerical Method

A sketch of local scour around a bridge pier with and without countermeasure is shown in Fig. 1. The width of the pier is defined as D. The approaching flow velocity is V, and its velocity is also shown in Fig. 1. The idea of this bio-inspired scour countermeasure is from the sturgeon. When a sturgeon is swimming, it will swing and find the best angle towards the coming water to avoid the flow power. For a bridge pier stays still during its service, the flow pattern may change according to the attack angle. The basic mathematical formulations are summarized below.

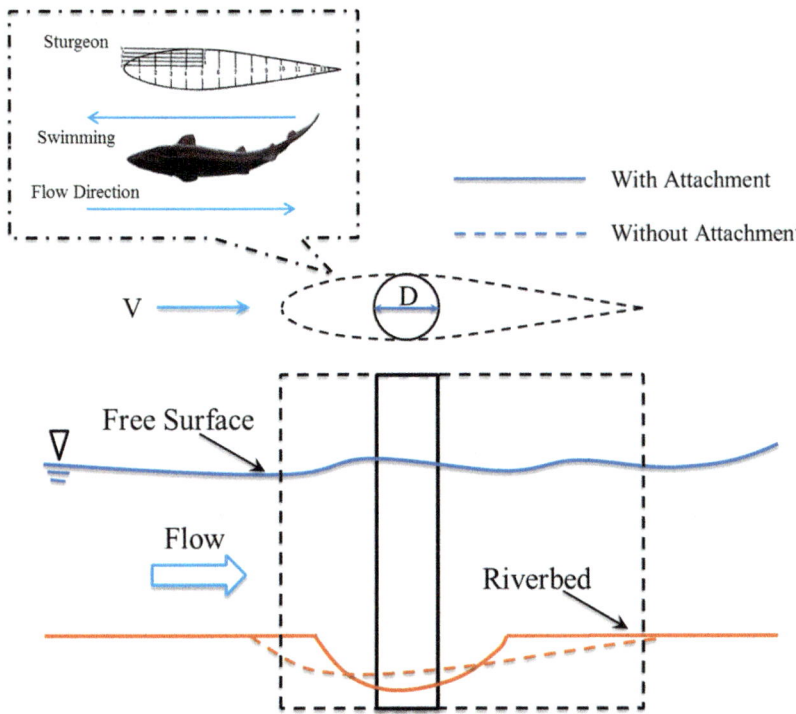

Fig. 1. Sketch of local scour around bridge pier with and without bio-inspired attachment

2.1 Governing Equations for Flow

A two-dimensional numerical model is developed for simulating local flow structure around a pier. The governing equations for the turbulent flow, which is assumed to be compressible with the Mach number Ma < 0.3, can be are the Reynolds-Averaged Navier-Stocks (RNS) equations. The equations can be expressed as

$$\begin{cases} \rho(u \cdot \nabla)u = \nabla \cdot \left[-pI + (\mu + \mu_T)\left(\nabla u + (\nabla u)^T\right) - \frac{2}{3}(\mu + \mu_T)(\nabla \cdot u)I - \frac{2}{3}\rho kI\right] \\ \nabla \cdot (\rho u) = 0 \end{cases} \tag{1}$$

where ρ is the fluid density, u is the flow velocity, μ is the dynamic viscosity, and μ_T is the turbulent viscosity, which is defined by $\mu_T = \rho C_u k^2 / \varepsilon$, k stands for the turbulent kinetic energy, which is related to turbulent dissipation rate ε. The relationship can be expressed as:

$$\begin{cases} \rho(u \cdot \nabla)k = \nabla \cdot \left[\left(\mu + \frac{\mu_T}{\sigma_k}\right)\nabla k\right] + P_k - \rho\varepsilon \\ \rho(u \cdot \nabla)\varepsilon = \nabla \cdot \left[\left(\mu + \frac{\mu_T}{\sigma_\varepsilon}\right)\nabla\varepsilon\right] + C_{\varepsilon 1}\frac{\varepsilon}{k}P_k - C_{\varepsilon 2}\rho\frac{\varepsilon^2}{k} \end{cases} \tag{2}$$

where σ_k, σ_ε, $C_{\varepsilon 1}$, and $C_{\varepsilon 2}$ are model constants with magnitude of 1.0, 1.3, 1.44 and 1.92, respectively (Wilcox 1998). P_k is a production term which can be calculated by:

$$P_k = \mu_T \left[\nabla u : (\nabla u + (\nabla u)^T - \frac{2}{3} (\nabla \cdot u)^2 \right] - \frac{2}{3} \rho k \nabla \cdot u \tag{3}$$

The Shear Stress Transport (SST) k-ω turbulence model is utilized for modeling the turbulence. The SST two-equation turbulence model was introduced by Menter [12] to deal with the strong sensitivity of the k-ω turbulence model. It was found that the SST k-ω model can give a good prediction of boundary layer flows with the adverse pressure gradient. The model can be briefly expressed as

$$\frac{D\rho k}{Dt} = \tau_{ij} \frac{\partial u_i}{\partial x_j} - \beta^* \rho \omega k + \frac{\partial}{\partial x_j} \left[(\mu + \sigma_k \mu_t) \frac{\partial k}{\partial x_j} \right] \tag{4}$$

$$\frac{D\rho \omega}{Dt} = \frac{\gamma}{\nu_t} \tau_{ij} \frac{\partial u_i}{\partial x_j} - \beta \rho \omega^2 + \frac{\partial}{\partial x_j} \left[(\mu + \sigma_\omega \mu_t) \frac{\partial \omega}{\partial x_j} \right] + 2(1 - F_1)\rho \sigma_{\omega 2} \frac{1}{\omega} \frac{\partial k}{\partial x_j} \frac{\partial \omega}{\partial x_j} \tag{5}$$

The detail of k-ω turbulence model including these parameters can be found in Menter [12] and will not be elaborated here.

2.2 Erosion Mechanisms

In erosion studies, the critical shear stress, τ_c, is identified as one of the key parameters used to judge whether the erosion occurs. When the shear stress, τ, in the sediment is lower than the critical shear stress, the erosion does not occur. The shear stress around a pier can be captured by the Computational Fluid Dynamic (CFD) method. Meanwhile, the critical shear stress for the sediments can be determined by both empiric equation [13, 14] and laboratory test [15]. The critical shear stress varies with the change of bed materials and is related to other soil characteristics. According to Shields [14], the critical shear stress can be calculated by:

$$\tau_c = K_s(\gamma_s - \gamma)d_{50} \tag{6}$$

where K_s is Shields coefficient, γ_s is the unit weight of soil particles, γ is the unit weight of water, d_{50} is the median particle size of sand.

In two-dimensional problems, the shear stress of a Newtonian fluid, at a surface element parallel to a flat plate at the point y can be calculated by

$$\tau = \mu \left(\frac{dv}{dx} + \frac{du}{dy} \right) \tag{7}$$

where μ is the dynamic viscosity of the fluid, v is the velocity of the fluid perpendicular to the long boundary, u is the velocity of the fluid along the long boundary, x and y are the distance to the long and short boundary respectively.

For local scour around a pier embedded in the same sediments, the critical shear stress is the same as well. The scour process can be regarded as the conflict between the load, which is generated by local flow, and resistance, which means the capability of sediments stability when faced with erosion. The erosion will reach the equilibrium state when

$$\alpha_i \tau_i \leq \beta_k \tau_{ck} \tag{8}$$

where τ_i is the shear stress, which is induced by flow, wave, and(or) other factors, subjected in layer i, α_i is the factor of shear stress, τ_{ck} is the critical shear stress in soil of layer k, β_k is the factor of critical shear stress. For preliminary studies, the parameters can be roughly estimated as 1.0, which means the erosion is determined by the shear stress and critical shear stress.

3 Simulation of Bio-Inspired Bridge Attachments with Attack Angle

3.1 Simulation of Free Surface Flow Around Circular Cylinder

Free surface flow around a circular cylinder has been simulated firstly. The computational parameters are the same as those used in the experiment carried out by Melville [16]. The experimental results evolving the streamline around the circular cylinder surface can be found in it as well. The experiment was conducted in a flume, which is about 19 m long, 45.6 cm wide and 44 cm deep with glass panels on each side over the entire length. The diameter of the cylinder, D_c, equals to 5.08 cm. The water depth is 0.15 m and the approaching flow velocity is 0.25 m/s with an attack angle of 0°. The streamline around the half cylinder is well predicted by the numerical results. As shown in Fig. 2, the rise of streamline before the cylinder, which may lead to the vortex ahead of the pier, has been calculated, and the drop of free surface elevation behind the obstruction has been captured as well. In addition, the vortex wake behind the obstruction has also been well predicted. The predicted flow structure agrees well with the experimental results.

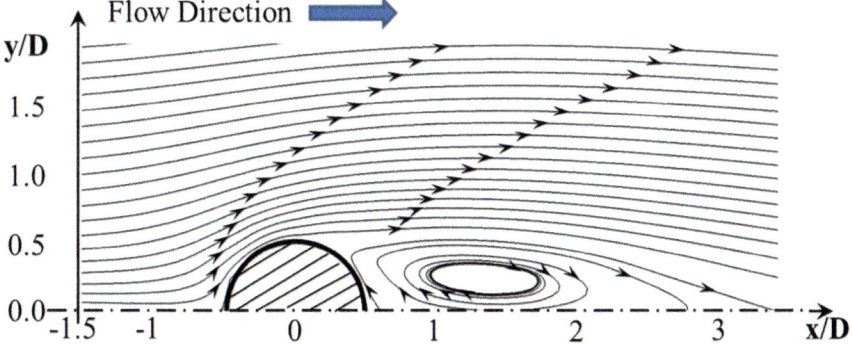

Fig. 2. Computed streamlines around the circular cylinder.

3.2 Simulation of Flow Around Bio-Inspired Bridge Attachments

A great many living beings, such as sturgeon and palm trees, are equipped with streamlined appearance to help them survive in challenging environments, especially strong wind and flow. These appearances together with their outlines are excellent sources of inspiration for developing artificial bridge attachments as scour counter-measures. The interference among flow, soil, and structure is complex and hard to be coupled in computational methods. The relationship between these factors have been unveiled comprehensively and plenty of academic achievements are accomplished every year [4]. Among these factors, flow plays an important role because it can be regarded as the beginning of the scour phenomenon (Fig. 3). It leads to the beginning of scour process, interacts with the structure to generate larger power, and rushes the initiated soil of the riverbed. To investigate the scour mitigation effect of the bridge attachment, the key point is to study on the interaction between the structure and flow.

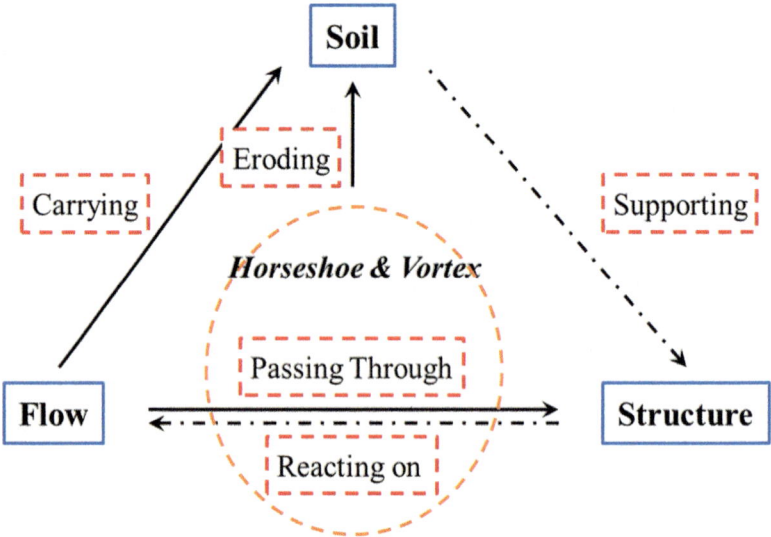

Fig. 3. The interaction between factors in scour phenomenon.

For a sturgeon swimming in the river, the streamlined body is critical during its daily life. To save energy when swimming towards flow with strong power, the sturgeon has to adjust its attack angle with the coming water. However, the bio-inspired bridge attachment cannot work so flexibly during the flood. The outline of the bridge attachment is designed regarding the streamlined sturgeon as shown in Fig. 1. The length of the bridge pier (L) is designed as 1.8 m, while the flow velocity (U_∞) is 2.5 m/s. The temperature of the water is 20 °C resulting in a Mach number of 0.0017. To simulate the scour behavior of bridge attachment with varies attack angles, the pier is inclined at an angle α to the oncoming stream as following,

$$(u_\infty, v_\infty) = U_\infty(cos\alpha, sin\alpha) \tag{9}$$

To generate a steady flow field, the upstream, top and bottom edges are located 100 L away from the pier in the computational domain. The downstream edge is 200 L away. With this large computational domain, the effect of the applied boundary conditions can be mitigated. The domain and the model can be found in Fig. 4. The computations involve a structured mesh with a high size-ratio between the elements' boundaries. A no slip condition is applied to the surface of the attachment. The model and its mesh method can be found in Fig. 4 as well. The study performs a parameter sweep with the angle of attack (α) taking the values 0°, 2°, 4°, 6°, 8°, 10°, 12°, 14° and 16°. Among these simulations, representative results have been selected to compare the flow altering behavior of bio-inspired bridge attachments.

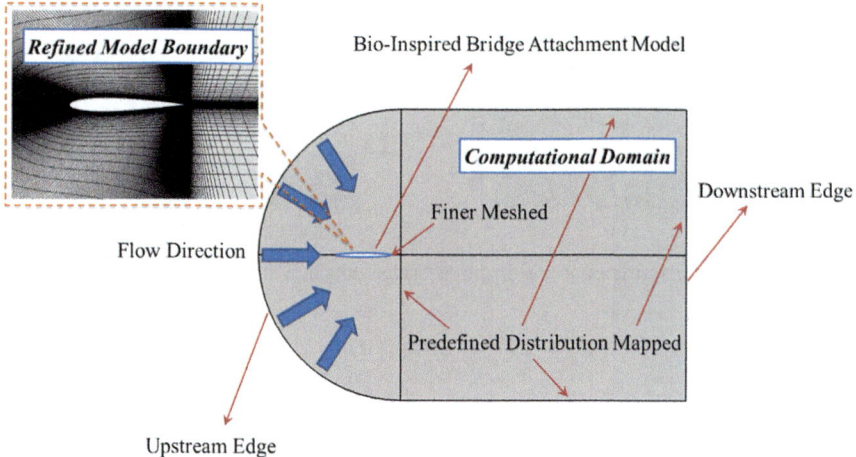

Fig. 4. Computational domain and flow-field boundary conditions.

The flow field around bio-inspired bridge attachment with the attack angle of 0°, 4°, 8° and 12° can be found in Fig. 5. When the attack angle equals to 0°, the streamlined bridge attachment can alter the flow direction well to generate a symmetric flow field to mitigate the interaction between the fluid phase and the solid phase. When compared with the bridge pier without attachment, the approaching flow is redirected gradually so that the serious horseshoe vortex and downflow stream are hard to occur. Meanwhile, due to the moderate interaction between these two phases, the wake-flow behind bridge pier is avoided as well. The critical area has been marked in Fig. 5. This area is critical because the serious fluid-solid interaction occurs in the area for each simulated group. The maximum approaching flow velocity increases when the attack angle becomes larger from 3 m/s (0°) to 7 m/s (12°). In addition, the unbalanced flow velocity on the sides of bridge attachment may lead to the incline of the pier, especially with the erosion around it. The larger velocity will result in about twice the drag force of the smaller one determined by Wen and Yu's equations [17],

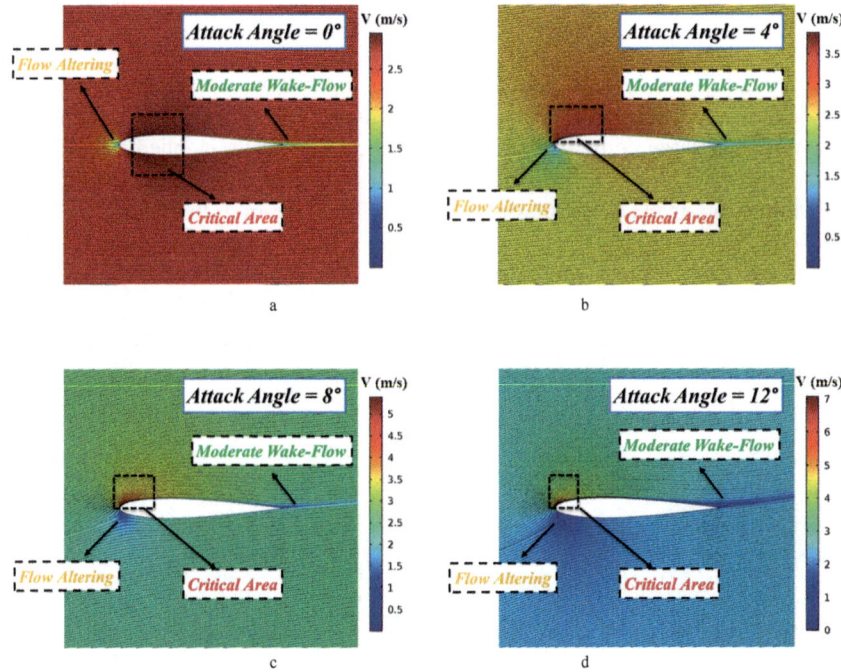

Fig. 5. Flow structure around bio-inspired bridge attachment with various attack angles

$$F_d = \frac{1}{8}C_D\pi\rho_f d_p^2 |u - U|(u - U) \tag{10}$$

where F_d is the drag force acting on particle i; C_D is the drag coefficient; ρ_f is the fluid density; d_p is the particle diameter; u, U are the fluid velocity and particle velocity.

The distribution of shear stress near the boundary of the bridge attachment with representative attack angles of 0°, 4°, 8° and 12° has been shown in Fig. 6. For different bed sediments, the critical shear stress varies dramatically. To calculate the scour range and depth, the relationship between the shear stress and critical shear stress can be utilized. It can be found that with the increase of attack angle, the maximum shear stress generated by the coming water gets larger from 19 Pa (0°) to 126 Pa (12°). For each of the model, the shear stress along the upstream boundary drops faster than along the downstream boundary. The critical area becomes smaller and with larger shear stress when the attack angle is larger. This means the erosion will occur in the critical area within a short time period, and the slope of scour hole in this area will be large. When the critical shear stress in the sediments around the pier is about 20, the erosion will be different in every case. In the first case, erosion will not occur; in the second case, only the area in the front of the attachment will be eroded; in the third and fourth case, half of the area around the attachment will be eroded. In addition, the upstream point, where flow firstly contacts the structure, witnesses the rapid drop of

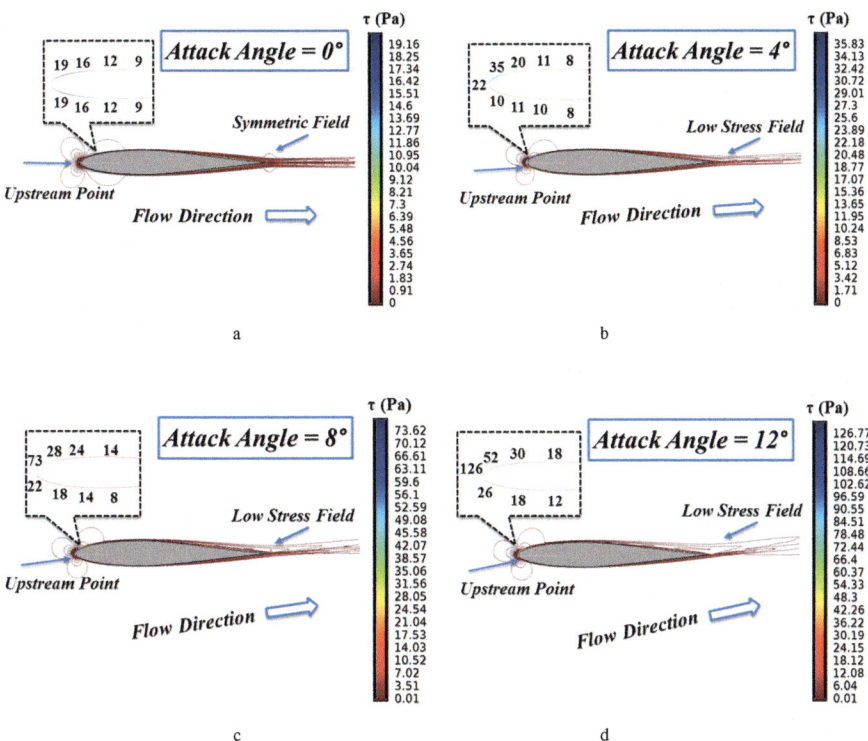

Fig. 6. Shear stress results around bio-inspired bridge attachment with various attack angles

shear stress. In this area, the shear stress and flow velocity drop dramatically, which means soil located in this area will be well protected during the scour process. This is the reason why fishes will swing towards the flow direction to save their energy during the daily life. In the next step, the bridge attachment can be designed to be self-adaptive in its interaction with the flow.

4   Conclusions

A finite element model is employed for simulating scour around a bio-inspired bridge attachment. The free surface model and the flow-structure model are coupled in this study. The influence of attack angle has been carried out with this model. The following conclusions are obtained:

1. The flow-structure interaction can be simulated and captured in the numerical simulation. With the change of flow direction, the flow field together with the distribution of shear stress around the bio-inspired bridge attachment is analyzed. When the attack angle gets larger, the interaction between flow and structure will become more serious.

2. Bio-inspired bridge attachment can alert the coming water to mitigate the power of the flow. This makes it hard to form the horseshoe vortex and downflow water. With its streamlined outline, the bridge attachment leads to a slight wake flow, which makes the fluid-solid interaction moderate.

3. The area around the upstream point in the attachment is with slow flow velocity and small shear stress. In the next step, the bridge attachment can be designed rotatable to be self-adaptive to the flow direction.

4. The flow structure and shear stress field around bridge attachment without attack angle are symmetric. While the scour-related parameters along the downstream boundary of the attachment decrease slower than those along upstream boundary when the flow comes with an attack angle. We should pay more attention to the back of the attachment during the scour monitoring.

Acknowledgements. This work was supported by the National Key R&D Program of China (Grant No. 2016YFC0800200), the National Natural Science Foundation of China (Grant No. 41172246), and the Fundamental Research Funds for the Central Universities of China, and China Scholarship Council (partial support via US National Science Foundation 0900401). Financial support from these organizations is gratefully acknowledged.

References

1. Melville, B.W., Coleman, S.E.: Bridge Scour. Water Resources Publication (2000)
2. Deng, L., Cai, C.S.: Bridge scour: Prediction, modeling, monitoring, and countermeasures—Review. Pract. Periodical Struct. Des. Constr. **15**(2), 125–134 (2009)
3. Breusers, H.N.C., Nicollet, G., Shen, H.W.: Local scour around cylindrical piers. J. Hydraul. Res. **15**(3), 211–252 (1977)
4. Wang, C., Yu, X., Liang, F.: A review of bridge scour: mechanism, estimation, monitoring and countermeasures. Nat. Hazards **87**, 1–26 (2017)
5. Chiew, Y.M.: Scour protection at bridge piers. J. Hydraul. Eng. **118**(9), 1260–1269 (1992)
6. Wang, C., Liang, F., Yu, X.: Experimental and numerical investigations on the performance of sacrificial piles in reducing local scour around pile groups. Nat. Hazards **85**, 1–19 (2017)
7. Heidarpour, M., Afzalimehr, H., Izadinia, E.: Reduction of local scour around bridge pier groups using collars. Int. J. Sedim. Res. **25**(4), 411–422 (2010)
8. Zarrati, A.R., Gholami, H., Mashahir, M.B.: Application of collar to control scouring around rectangular bridge piers. J. Hydraul. Res. **42**(1), 97–103 (2004)
9. Kumar, V., Rangaraju, K.G., Vittal, N.: Reduction of local scour around bridge piers using slot and collar. J. Hydraul. Eng. **125**(12), 1302–1305 (1999)
10. Gris, R.B.: Sheath for reducing local scour in bridge piers. In: Scour and Erosion, pp. 987–996 (2010)
11. Tao, J., Li, J.: Streamlining of bridge piers as scour countermeasures: effects of curvature of vertical profiles. Transp. Res. Rec. J. Transp. Res. Board **2521**, 172–182 (2015)
12. Menter, F.R.: Two-equation eddy-viscosity turbulence models for engineering applications. AIAA J. **32**(8), 1598–1605 (1994)
13. White, C.M.: The equilibrium of grains on the bed of a stream. In: Proceedings of the Royal Society of London, London, Series A, No. 958, vol. 174, pp. 322–338 (1940)

14. Shields, A.: Application of similarity principles and turbulence research to bed-load movement. Hydrodynamics Laboratory Publ. No. 167, California Institute of Technology, Pasadena, Calif (1936)
15. Briaud, J.L.: Case histories in soil and rock erosion: woodrow wilson bridge, Brazos River Meander, Normandy Cliffs, and New Orleans Levees. J. Geotech. Geoenviron. Eng. **134** (10), 1425–1447 (2008)
16. Melville, B. W.: Local scour at bridge sites. Report No. 117, School of Engineering, University of Auckland, Auckland, New Zealand (1975)
17. Wen, C.Y., Yu, Y.H.: Mechanics of Fluidization. Chem. Eng. Prog. Symp. Ser. **62**, 100–111 (1966)

Simulating the Caisson Penetration in Sand Using a Critical State Based Soil Model

Zhuang Jin[1], Zhen-Yu Yin[1,2,3(✉)], Ze-Xiang Wu[1],
and Panagiotis Kotronis[1]

[1] Ecole Centrale de Nantes, Université de Nantes, CNRS,
Institut de Recherche en Génie Civil et Mécanique (GeM),
1 rue de la Nöe, 44321 Nantes, France
zhenyu.yin@gmail.com
[2] Department of Geotechnical Engineering,
College of Civil Engineering, Tongji University, Shanghai, China
[3] Key Laboratory of Geotechnical and Underground Engineering of Ministry
of Education, Tongji University, Shanghai 200092, People's Republic of China

Abstract. The penetration of a caisson redistributes the soil stress state and influences the interactions between the foundation structure and the soil. However, few finite element studies on caisson foundations take the influence of installation into account. In this paper, a numerical simulation considering large deformations is presented to analyze the behavior of a caisson foundation during the installation phase. A newly developed elastoplastic model incorporating nonlinear elasticity, a plastic hardening law and the critical state concept is adopted and implemented into the finite element code ABAQUS. To solve the large deformation problem, the finite element method is combined with the Smoothed Particle Hydrodynamics method (SPH). The penetration area of the soil is simulated using SPH particles while classical finite elements are used elsewhere. Results demonstrate that the SPH method coupled with a critical state based model is a useful tool to analyze the behavior of caisson foundations, their failure modes and the soil-bucket interactions.

Keywords: Sand · Caisson foundation · Large deformation · SPH
Critical state

1 Introduction

A caisson is a closed-top steel tube, which is first lowered to the seafloor allowing to penetrate under its own weight, and then push to full depth with suction force producing by pumping water out of its interior. Recently, caissons have been widely used for different types of constructions, such as gravity platform jackets, jack-ups, offshore wind turbines, subsea systems and seabed protection structures. For an optimum design, a better understanding of the performance of caisson foundations is therefore necessary.

In recent years, a number of 2D and 3D numerical studies have been performed to research the caisson bearing capacity under different loading combinations and drainage conditions [1–6]. In all these studies, elastoplastic models were used for the soil,

© Springer Nature Singapore Pte Ltd. 2018
T. Qiu et al. (Eds.): GSIC 2018, *Proceedings of GeoShanghai 2018 International Conference: Advances in Soil Dynamics and Foundation Engineering*, pp. 740–748, 2018.
https://doi.org/10.1007/978-981-13-0131-5_80

however the influence of the installation process was ignored while considering the bearing capacity of caisson foundation. The simulation of caisson installation is a large deformation problem, which could lead to large mesh distortions in the high strain concentration zones around the skirt in the tradition finite element method [7, 8].

In this article, numerical simulations are presented of a caisson penetrating in sand. To deal with large deformations during the installation phase, a combined Lagrangian - Smoothed Particle Hydrodynamics method (SPH) is adopted and a critical state based soil model is used for the soil. The numerical results are compared with laboratory experiments and the ability of the combined Lagrangian-SPH method to reproduce the caisson installation and the soil stress state redistribution is discussed.

2 Introduction of the SIMSAND Model

2.1 Mathematical Description

In order to simulate the installation process of caisson foundation, a recently developed critical state based model, the SIMSAND model [9–14] is adopted hereafter. Under the framework of elastoplasticity, the strain rate is decomposed into an elastic and a plastic strain rate part:

$$\dot{\varepsilon}_{ij} = \dot{\varepsilon}^e_{ij} + \dot{\varepsilon}^p_{ij} \tag{1}$$

where $\dot{\varepsilon}_{ij}$ denotes the strain rate tensor and the superscripts e and p represent the elastic and plastic component respectively. The subscripts i and j vary from 1 to 3.

The nonlinear elastic behaviour is assumed isotropic with a bulk modulus K having the same form as the shear modulus:

$$\dot{\varepsilon}^e_v = \frac{\dot{p}'}{K}, \ \dot{\varepsilon}^e_d = \frac{\dot{q}}{3G} \tag{2}$$

$$\begin{cases} K = K_0 \cdot p_{at} \dfrac{(2.97 - e)^2}{(1+e)} \left(\dfrac{p'}{p_{at}}\right)^n \\[2mm] G = G_0 \cdot p_{at} \dfrac{(2.97 - e)^2}{(1+e)} \left(\dfrac{p'}{p_{at}}\right)^n \end{cases} \tag{3}$$

where G_0, K_0 and n are elastic constant parameters; p_{at} is the atmospheric pressure ($p_{at} = 101.3$ kPa). G_0 can be obtained as $G_0 = 3K_0(1-2v)/2(1 + v)$, where v is the Poisson's ratio for the case where the bulk modulus is determined from the isotropic compression curve.

The plastic strain rate is expressed as:

$$\dot{\varepsilon}^p_{ij} = d\lambda \frac{\partial g}{\partial \sigma'_{ij}} \tag{4}$$

The yield surface for shear sliding is expressed as:

$$f = \frac{q}{p'} - \frac{M_p \varepsilon_d^p}{k_p + \varepsilon_d^p} = 0 \tag{5}$$

where q is the deviatoric stress; k_p the plastic shear modulus; M_p the stress ratio corresponding to the peak strength, determined by the peak friction angle f_p ($M_p = 6\sin(f_p)/(3-\sin(f_p))$ in compression); ε_d^p the deviatoric plastic strain. The yield surface of SIMSAND model is plotted in the $p' - q$ and $e - p'$ planes in Fig. 1.

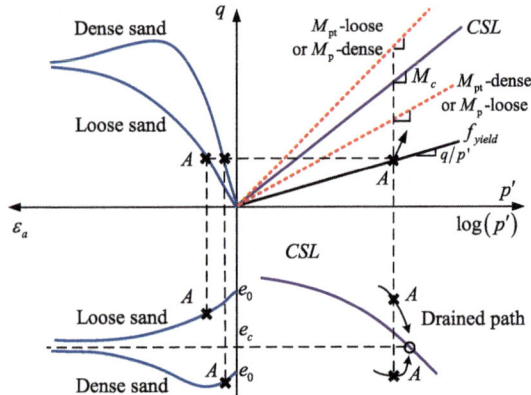

Fig. 1. Principle of the SIMPSAND critical state model

The model follows a non-associated flow rule where the derivatives of the potential function g have the following form:

$$\frac{\partial g}{\partial \sigma_{ij}} = \frac{\partial g}{\partial p'_{ij}} \frac{\partial p'}{\partial \sigma_{ij}} + \frac{\partial g}{\partial s_{ij}} \frac{\partial s_{ij}}{\partial \sigma_{ij}} \text{ with } \frac{\partial g}{\partial p'} = A_d \left(M_{pt} - \frac{q}{p'} \right); \\ \frac{\partial g}{\partial s_{ij}} = \{1\ 1\ 1\ 1\ 1\ 1\} \tag{6}$$

where A_d is a stress-dilatancy parameter; $M_{pt} = 6\sin(f_{pt})/(3-\sin(f_{pt}))$ in compression where f_{pt} is the phase transformation friction angle. The double indices ij is simplified to be $1\hat{=}11$, $2\hat{=}22$, $3\hat{=}33$, $4\hat{=}12$, $5\hat{=}23$, $6\hat{=}31$.

A nonlinear formulation of the Critical State Line (CSL) adopted:

$$e_c = e_{ref} \exp \left[-\lambda \left(\frac{p'}{p_{ref}} \right)^{\xi} \right] \tag{7}$$

and the effects of soil density and interlocking are introduced:

$$\tan \phi_p = \left(\frac{e_c}{e} \right) \tan \phi_u; \ \tan \phi_{pt} = \left(\frac{e_c}{e} \right)^{-1} \tan \phi_\mu \tag{8}$$

where e_c is the critical void ratio; e_{ref} the initial critical void ratio at $p' = 0$; $p_{ref} = 100$ kPa (fixed); λ and ζ are the parameters controlling the nonlinear shape of CSL in the $\ln p'$-e plane; ϕ_u the friction angle at critical state. The Lode angle dependent strength and the stress-dilatancy are taken into account by using the transformed stress method introduced by Yao et al. [15].

2.2 Model Parameters

The SIMSAND model parameters are divided in three groups: (1) elastic parameters (K_0, n and v), (2) critical state line related parameters (e_{ref}, λ, ζ and ϕ_u), and (3) plastic shear-sliding parameters (k_p and A_d). The calibration of the model parameters can be carried out using optimization methods [11, 12, 16, 17].

According to the three drained triaxial tests on Baskarp No.15 sand [18], the model parameters were identified as follows (Table 1).

Table 1. Model parameters for the Baskarp No.15 sand

Groups	Elastic			CSL				Shear-sliding	
Symbols	K_0 (kPa)	n	v	e_{ref}	λ	ζ	ϕ_u (°)	k_p	A_d
Values	344	0.58	0.25	0.71	0.022	0.71	33.8	0.0027	1.6

3 Numerical Modelling

3.1 Experimental Campaign

In order to validate the numerical model, a laboratory test on the installation of a caisson foundation in sand was selected [19]. The experimental set-up consists of a sand box (1600 mm × 1600 mm × 1150 mm), a loading frame and a hinged beam. A system of steel cables and pulleys induces a static (monotonic and sinusoidal) loading to the foundation through an electric motor drive placed on the hinged beam. The load, set by means of three weight hangers, is transferred to the foundation through a vertical beam bolted on the bucket lid. The foundation is instrumented with three LVDTs (Linear Variable Differential Transformer) and two load cells [20]. The testing set up follows the features of a similar testing apparatus [21]. Cone Penetration Tests (CPT) were carried out to assess the soil parameters. The caisson foundation is made of steel, has an outer diameter of 300 mm, a lid thickness of 11.5 mm, a skirt length of 300 mm and a skirt thickness of 1.5 mm.

3.2 Lagrangian and Combined Lagrangian - SPH Model

The ABAQUS software was adopted for the simulations. The finite element soil domain has side lengths of 1600 mm and a depth of 1150 mm. On the outer and bottom faces, the translational degrees of freedom were constrained. The plate thickness used for the skirt and lid of the bucket is 2 mm and 10 mm respectively. The diameter (D) of the bucket foundation and the skirt length (d) are equal to 300 mm. Following Foglia et al. [19], the density of the bucket is equal to 7800 kg/m^3, with the

Young modulus 200 GPa and the Poisson ratio 0.3. The soil parameters have been presented in the previous section.

One disadvantage of the meshless methods (e.g. SPH) over the Lagrangian models is their computational demand. The SPH method is also less accurate under small deformations. For this reason, only a part of the soil domain was modeled with the SPH method while a Lagrangian model was adopted for the rest. The "Tie Constraint" ABAQUS command was adopted, a command that "ties" two separate surfaces so that no relative motion exists; it allows fusing two domains even though their meshes are not identical. By using the "Tie Constraint" command, the SPH particles are thus "tied to" the Lagrangian domain. The friction coefficient of the interface between the caisson and the soil is 0.35.

In the combined Lagrangian - SPH model, only the portion of the soil experiencing the largest deformations was modeled with SPH particles (Fig. 2(a)). The SPH domain is 800 mm at each side and 1150 mm deep. For the CPT simulation presented however, each side of the SPH domain was reduced to 400 mm and the depth to 600 mm in order to reduce the calculation time, shown in Fig. 2(b).

3.3   Simulation Sequences

One CPT and one penetration test under pure vertical load were simulated. The applied load Vconst is 241 N which includes the buoyant self-weight of the bucket and the weight of the measuring system mounted on the foundation.

The combined Lagrangian - SPH model of Figs. 2(a) and (b) were used to simulate the caisson penetration test and the CPT and respectively. For all simulations, the dynamic explicit method was adopted. The spatial discretization parameters of the numerical models are given in Table 2.

Table 2. Spatial discretization parameters

Simulation	L (mm)	W (mm)	D (mm)	N° of SPH particles	N° of FE elements
CPT	400	400	600	23814	78304
Caisson penetration	1600	1600	1200	53361	147456

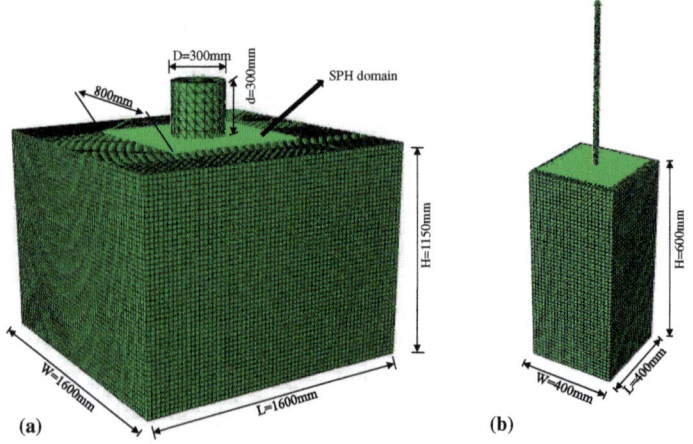

Fig. 2. Combined Lagrangian-SPH model for (a) caisson penetration and (b) CPT

3.4 Simulation Results for CPT

In order to validate the chosen material parameters and the numerical method, the CPT simulation is first presented. During the simulation, the cone velocity was kept constant and equal to 5 mm/s [18]. The comparison between experimental and numerical results is shown in Fig. 3, where the four CPT experimental results were taken from [19]. The chosen material parameters are found acceptable, as the numerical simulation curve agrees with the four experimental CPT results.

The fields of deviatoric plastic strain (PEEQ), deviatoric stress (S Mises, Pa) and mean effective stress (S Pressure, Pa) are plotted in Fig. 4 for the CPT. Reasonable distributions are found within a domain much smaller than the domain of the SPH particles.

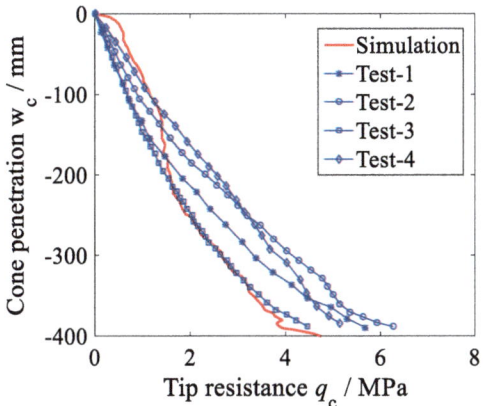

Fig. 3. CPT: experimental vs. numerical results

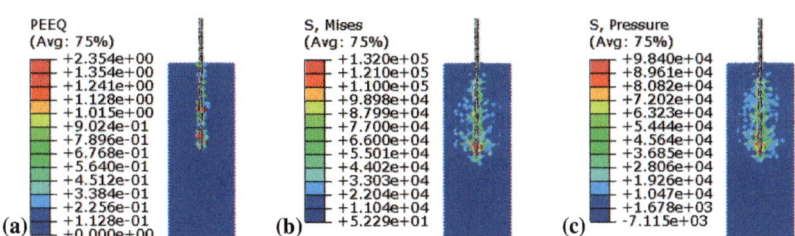

Fig. 4. Contour of (a) plastic deviatoric strain (PEEQ), (b) deviatoric stress (S, Mises, Pa) and (c) Mean stress (S Pressure, Pa) for CPT

3.5 Simulation Results for Caisson Penetration

The caisson foundation penetration calculation is presented hereafter. The caisson was forced into the sand at a constant velocity of 30 mm/s. Figure 5 shows the normalized vertical displacement versus vertical force and Fig. 6 the vertical stresses at the end of penetration. Comparison of the experimental with the numerical results clearly shows the capacity of the combined Lagrangian - SPH model to capture extremely large deformations.

The deviatoric plastic strain (SDV18), deviatoric stress (S Mises, kPa) and mean effective stress (S Pressure, kPa) fields are plotted in Fig. 6 for the pure vertical loading test. All results show reasonable distributions again within an area (large deformation domain in both vertical and horizontal directions) much smaller than the SPH particles domain.

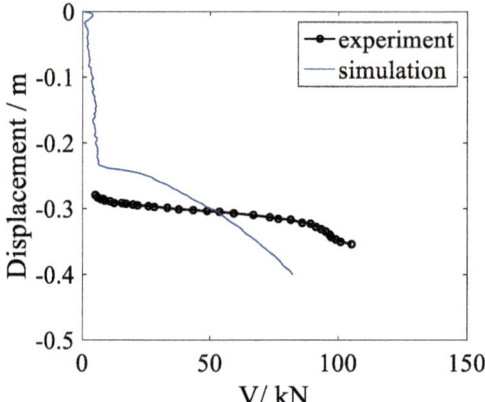

Fig. 5. Caisson penetration: experiment vs. numerical results

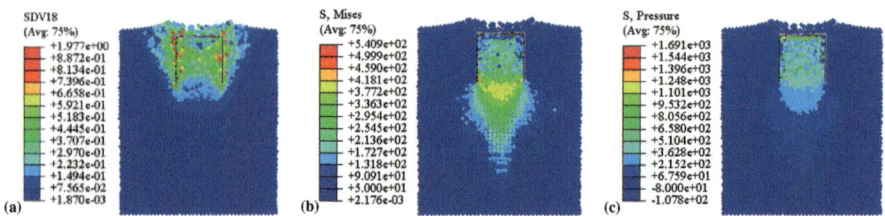

Fig. 6. Contour of (a) plastic deviatoric strain (SDV18), (b) deviatoric stress (S, Mises, kPa) and (c) Mean stress (S Pressure, kPa) for the pure vertical loading test

4 Conclusions

The Smoothed Particle Hydrodynamics method (SPH) is adopted in this article together with the SIMSAND critical state model to simulate the caisson penetration in dry sand. The material parameters are determined from the literature. A CPT test simulation was first carried out to validate the numerical method. The caisson-sand interactions during the penetration phase were then investigated. Results demonstrate that the combined Lagrangian–SPH model using a critical state based advanced model is able to reproduce the penetration of a caisson foundation in sand.

Acknowledgment. This research project is financially supported by National Natural Science Foundation of China (41372285 and 51579179).

References

1. Sukumaran, B., McCarron, W., Jeanjean, P., Abouseeda, H.: Efficient finite element techniques for limit analysis of suction caissons under lateral loads. Comput. Geotech. **24**, 89–107 (1999)
2. El-Gharbawy, S., Olson, R.: Modeling of suction caisson foundations. In: The Tenth International Offshore and Polar Engineering Conference, pp. 670–677. International Society of Offshore and Polar Engineers (1999)
3. Deng, W., Carter, J.: A theoretical study of the vertical uplift capacity of suction caissons. Int. J. Offshore Polar Eng. **12**, 342–349 (2002)
4. Liu, M., Yang, M., Wang, H.: Bearing behavior of wide-shallow bucket foundation for offshore wind turbines in drained silty sand. Ocean Eng. **82**, 169–179 (2014)
5. Li, D., Zhang, Y., Feng, L., Gao, Y.: Capacity of modified suction caissons in marine sand under static horizontal loading. Ocean Eng. **102**, 1–16 (2015)
6. Penzes, B., Jensen, M., Zania, V.: Suction caissons subjected to monotonic combined loading. In: The 17th Nordic Geotechnical Meeting (2016)
7. Vásquez, L.F.G., Maniar, D.R., Tassoulas, J.L.: Installation and axial pullout of suction caissons: numerical modeling. J. Geotech. Geoenviron. Eng. **136**, 1137–1147 (2010)
8. Li, Y., Yang, S., Yu, S.: Numerical simulation of installation process and uplift resistance for an integrated suction foundation in deep ocean. China Ocean Eng. **30**, 33–46 (2016)
9. Yin, Z.-Y., Xu, Q., Hicher, P.-Y.: A simple critical-state-based double-yield-surface model for clay behavior under complex loading. Acta Geotech. **8**, 509–523 (2013)
10. Jin, Y.-F., Wu, Z.-X., Yin, Z.-Y., Shen, J.S.: Estimation of critical state-related formula in advanced constitutive modeling of granular material. Acta Geotech. **12**, 1–23 (2017)
11. Yin, Z.-Y., Jin, Y.-F., Shen, S.-L., Huang, H.-W.: An efficient optimization method for identifying parameters of soft structured clay by an enhanced genetic algorithm and elastic–viscoplastic model. Acta Geotech. **12**, 849–867 (2017)
12. Jin, Y.-F., Yin, Z.-Y., Shen, S.-L., Hicher, P.-Y.: Selection of sand models and identification of parameters using an enhanced genetic algorithm. Int. J. Numer. Anal. Methods Geomech. **40**, 1219–1240 (2016)
13. Jin, Y.-F., Yin, Z.-Y., Shen, S.-L., Hicher, P.-Y.: Investigation into MOGA for identifying parameters of a critical-state-based sand model and parameters correlation by factor analysis. Acta Geotech. **11**, 1131–1145 (2016)
14. Wu, Z.-X., Yin, Z.-Y., Jin, Y.-F., Geng, X.-Y.: A straightforward procedure of parameters determination for sand: a bridge from critical state based constitutive modelling to finite element analysis. Eur. J. Environ. Civ. Eng., 1–23 (2017)
15. Yao, Y., Lu, D., Zhou, A., Zou, B.: Generalized non-linear strength theory and transformed stress space. Sci. China Ser. E Technol. Sci. **47**, 691–709 (2004)
16. Jin, Y.-F., Yin, Z.-Y., Shen, S.-L., Zhang, D.-M.: A new hybrid real-coded genetic algorithm and its application to parameters identification of soils. Inverse Prob. Sci. Eng. **25**, 1343–1366 (2017)
17. Yin, Z.-Y., Jin, Y.-F., Huang, H.-W., Shen, S.-L.: Evolutionary polynomial regression based modelling of clay compressibility using an enhanced hybrid real-coded genetic algorithm. Eng. Geol. **210**, 158–167 (2016)

18. Foglia, A., Ibsen, L.B.: Laboratory experiments of bucket foundations under cyclic loading. Department of Civil Engineering, Aalborg University (2014)
19. Foglia, A., Gottardi, G., Govoni, L., Ibsen, L.B.: Modelling the drained response of bucket foundations for offshore wind turbines under general monotonic and cyclic loading. Appl. Ocean Res. **52**, 80–91 (2015)
20. Foglia, A., Ibsen, L.B., Nicolai, G., Andersen, L.V.: Observations on bucket foundations under cyclic loading in dense saturated sand. In: The International Conference of Physical Modelling in Geotechnics, pp. 667–673. CRC Press LLC (2014)
21. Leblanc, C., Byrne, B., Houlsby, G.: Response of stiff piles to random two-way lateral loading. Géotechnique **60**, 715–721 (2010)

Modelling of Spudcan Foundation Penetrations Using an Improved Hypoplastic Model for Soft Clays

Jan Jerman[1(✉)], David Mašín[1], Raffaele Ragni[2], and Britta Bienen[2]

[1] Faculty of Science, Charles University in Prague,
Albertov 3, 128 43 Prague, Czech Republic
jan.jerman@natur.cuni.cz
[2] Centre for Offshore Foundation Systems and ARC,
CoE for Geotechnical Science and Engineering, University of Western Australia,
35 Stirling Hwy, Crawley, Perth, WA 6009, Australia

Abstract. In this paper, a new hypoplastic model for K_0 consolidated soft clays is presented. The enhanced model is characterized by a small number of parameters, adding two parameters to the original model, whilst significantly improving model predictions of undrained stress paths of K_0 consolidated soils. However, the calibration procedure remains simple – model can be calibrated using K_0 consolidated triaxial tests and oedometer only. It is demonstrated that the model predictions compare well with element test experimental data. The validation of the model involves large scale boundary value problem simulations incorporating very large deformations, in this case retrospective simulations of centrifuge test of spudcan installation. The enhanced model shows better performance compared to a parent hypoplastic model.

Keywords: Hypoplasticity · Constitutive modeling · Spudcan
Numerical modelling

1 Introduction

Energy infrastructure is increasingly located in areas with soft soil deposits. Typical example is offshore geotechnical engineering with two most prominent applications – (1) hydrocarbon exploration and production, (2) offshore wind turbines and wind farms. The inevitable progression from shallow to deeper waters led to change of soil types with softer and more clayey soils in deep water in contrast with prevalent sandy soils in shallow waters [1].

The majority of numerical studies in this field of geotechnical engineering have, to date, relied on relatively simple models for soil behavior. Most widely used constitutive models are basic elasto-plastic models (see [2]). Advanced constitutive models for soil have been developed, but their use in complex boundary value problem simulations of offshore structures is scarce. Ragni et al. [3] employed the hypoplastic model developed by Mašín [4] to investigate consolidation around jack-up foundations simulated within the framework of large deformation finite element analysis (RITSS strategy, [5]). The performance of the model implementation was validated through

© Springer Nature Singapore Pte Ltd. 2018
T. Qiu et al. (Eds.): GSIC 2018, *Proceedings of GeoShanghai 2018 International Conference: Advances in Soil Dynamics and Foundation Engineering*, pp. 749–756, 2018.
https://doi.org/10.1007/978-981-13-0131-5_81

retrospective simulation of centrifuge test data, carried out on Laminaria carbonate silty clay recovered from the northwest shelf of Australia. These analyses studied the effects of consolidation generated by set-up during spudcan penetration. However, a significant drawback of the simulations was disagreement between the predicted soil behavior on the boundary value problem and laboratory experiments. This fact limited the possibility for class A predictions of spudcan penetration, since the model required a calibration using both the element and centrifuge test results to predict the spudcan load-penetration curve well.

2 Constitutive Model Formulation

The approach chosen in this paper is based on hypoplasticity, which offers a different approach to model non-linearity than conventional elasto-plastic models. The model is characterized by a single equation non-linear in stretching **D**, thus allowing to model non-linearity in more straightforward way compared to elasto-plasticity [6]. Even without the notion of "yield surface", hypoplastic models can predict such features of soil behavior as critical state, non-linearity in large and small strains, stiffness dependency on loading direction, etc.

2.1 Hypoplastic Model for Soft Clays

The parent model used by Ragni et al. [3] is the hypoplastic model for clays proposed by Mašín [4]. This model was used as a base model for further developments described in the present paper. A general formulation of the hypoplastic model may be written as [7]:

$$\dot{\boldsymbol{\sigma}} = f_s \mathcal{L} : \boldsymbol{D} + f_s f_d \boldsymbol{N} \, \|\boldsymbol{D}\| \tag{1}$$

where $\dot{\boldsymbol{\sigma}}$ is the objective stress-rate tensor, $\boldsymbol{D}$ is the Euler stretching tensor, $\mathcal{L}$ and $\boldsymbol{N}$ are the fourth and second order constitutive tensors, respectively, and f_s and f_d the barotropy and pyknotropy factors. From the Eq. (1) it can be seen that the directional stiffness is modelled by a single equation, where combination of $\mathcal{L}$ and $\boldsymbol{N}$ governs both the directional stiffness and the strength. The normal compression line is described by the following equation:

$$ln(1 + e) = N - \lambda^* \ln\left(\frac{p}{pr}\right) \tag{2}$$

where parameters N and λ^* define the position and shape of the isotropic normal compression line [8]. To model a soil with such complex behavior as exhibited by Laminaria carbonate silty clay, it was also necessary to enhance the model by adding sensitivity framework developed by Mašín [9] in a model for structured clays, where the size of the state boundary surface p_e is multiplied by sensitivity s. The normal compression line for the enhanced model thus reads

$$ln(1+e) = N - \lambda^* \ln\left(\frac{p}{pr}\right) + \lambda^* \ln(s) \qquad (3)$$

2.2   Model Improvements

A shortcoming of the original model used by Ragni et al. [3] is an incorrect prediction of stress paths of anisotropically consolidated soils, as shown in Fig. 1b, where especially the initial stiffness is significantly underestimated. Modification of the model is based on procedure proposed by Mašín and Herle [10], where the tensor $\mathcal{L}$ is made bilinear in D, so the predictions resemble an elasto-plastic model with different tangent stiffness in loading and unloading – denoted as D-dependent L approach, abbreviated as DdepL hereafter.

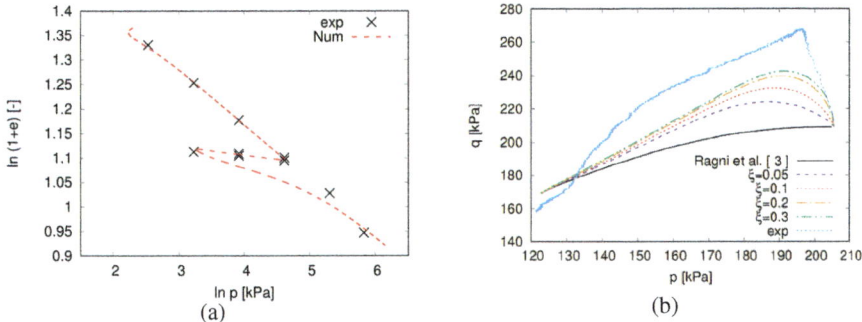

Fig. 1. Comparison of model predictions and experimental results. (a) Oedometric test, (b) stress paths of anisotropically consolidated compression triaxial tests for various values of ξ.

The rate form of the new model is defined as:

$$\dot{\sigma} = f_s \mathcal{L} : D + f_s f_d N D_n \qquad (4)$$

where $D_n = \| D \|$ of the original model is replaced by

$$D_n = w_y \|D\| + (1 - w_y)|d : D| \qquad (5)$$

The weight factor is defined as $w_y = F_{m_{sk'}}^{\xi}$, where ξ is a new model parameter. $F_{m_{sk}}$ is a Matsuoka-Nakai factor:

$$F_{m_{skew}} = \frac{9I_{3sk} + I_{1sk}I_{2sk}}{I_{3sk} + I_{1sk}I_{2sk}} \qquad (6)$$

The changes described above are sufficient for altering the stress paths for iso-topically consolidated soils, however, the procedure is not adequate for anisotopically consolidated soil, such as the soil of interest in the simulations – Laminaria carbonate silty clay. For this purpose, $\mathcal{L}, w_y$ and N (used to find $\mathcal{L}^D$) as well as stress invariants used in Eq. (5) needs to be calculated from a skewed stress space defined by Gajo and Muir Wood [11] as:

$$\boldsymbol{\sigma}_{sk} = \boldsymbol{\sigma} + \frac{1}{3} tr(\boldsymbol{\sigma})\boldsymbol{\gamma} \qquad (7)$$

where γ is a deviatoric second-order symmetric tensor defining skewing. When γ is null, the model reduces to model by Mašín and Herle [10] (defined using [8] as base model). The tensor γ is defined as

$$\gamma = \left[-\frac{2}{3} tg\gamma, \frac{1}{3} tg\gamma, \frac{1}{3} tg\gamma, 0, 0, 0 \right]^T \qquad (8)$$

where scalar γ stands for inclination of the skewed stress space with respect to p-axis in q-p plane. By skewing of the stress space and using the skewed stress for stress invariants and components of stiffness matrix, the stress path becomes perpendicular to p-axis at the stress deviator specified by the parameter γ.

The model can be rewritten in the form

$$\dot{\boldsymbol{\sigma}} = f_s \left(\mathcal{L}^D : \boldsymbol{D} + w_y f_d \boldsymbol{N} \|\boldsymbol{D}\| \right) \qquad (9)$$

Where the tensor $\mathcal{L}^D$ is dependent on the direction of stretching with respect to $\boldsymbol{d}$ via the equation

$$\mathcal{L}^D = \begin{cases} \mathcal{L} + f_d (1 - w_y)(\boldsymbol{N} \otimes \boldsymbol{d}) \text{ for } (\boldsymbol{d} : \boldsymbol{D}) > 0 \\ \mathcal{L} - f_d (1 - w_y)(\boldsymbol{N} \otimes \boldsymbol{d}) \text{ for } (\boldsymbol{d} : \boldsymbol{D}) \leq 0 \end{cases} \qquad (10)$$

For further inspection into implementation of $\mathcal{L}^D$ tensor and DdepL the reader is referred to the model by Mašín and Herle [10].

2.3 Parameter Calibration

The soil used in element tests and centrifuge experiments is reconstituted Laminaria carbonate silty clay recovered from the northwest shelf of Australia. Throughout the paper, '*Exp*' refers to experimental results and '*Num*' refers to numerical results by the enhanced model.

The basic clay hypoplastic model requires 5 material parameters equivalent to the Modified Cam clay model. The parameters are obtainable from standard element tests – oedometric and triaxial tests. However, the definitions of individual parameters are not identical to the Modified Cam clay model due to non-linear foundations of hypoplasticity, even though their physical meaning is similar.

In the following model calibration, the parameters of basic hypoplastic model [4] and sensitivity parameters from structured model by Mašín [9] are taken over from the work of Ragni et al. [3], where calibration of the original model and the model with meta-stable structure are thoroughly described. Sensitivity framework [9] adds sensitivity s as additional state variable and 3 more material parameters - k specifies the rate of structure degradation, A controls the influence of shear strains on structure degradation and the parameter s_f quantifies stable elements of structure. The enhanced model adds two parameters (ξ and γ) to the original model used by Ragni et al. [3]. These parameters can be obtained from K_0 consolidated samples. The parameter γ was obtained by following assumptions: the φ_{cr} of Laminaria was evaluated by Ragni et al. [3] to be 34°, which is equivalent to K_0 of 0.44 (using Jaky formula $1 - \sin(\varphi)$). From the value of K_0 and thus the mobilized q/p ratio, the equivalent $\gamma \approx 42°$ can directly be calculated. Following a parametric study of influence of various γ values, slightly higher value of $\gamma = 45°$ has been used in calculations. For calibration of parameter ξ extensive parametric study has been performed. Figure 1 shows stress paths of anisotropically consolidated compression triaxial tests for various values of ξ. The value of $\xi = 0.2$ has been evaluated as the closest representation of experimental results. Oedometric tests were re-calibrated using new values of ξ and γ to obtain similar results to calibration of Ragni et al. [3] by setting N and OCR to equivalent values, see Table 1 and Fig. 1.

Table 1. Model parameters for Laminaria carbonate silty clay used in the model. Values in brackets show minimum and maximum values used in numerical simulations.

Basic hypoplastic model for clays				
CS friction angle (φ)	NCL slope (λ)	Slope of URL (κ)	NCL for p = 1 kPa (N)	Shear modulus parameter (ν)
34°	0.113	0.013	1.721 (1.697 – 1.728)	0.1
Hypoplastic model for clays with meta-stable structure				
Initial sensitivity (S_{ini})	Final sensitivity (S_f)	Sensitivity degradation rate (k)	Parameter A	
2.9	1.0	0.05	0.2	
Soft clay model				
Rotation in q/p plane (γ)	Weighting factor (ξ)			
45°	0.2 (0–0.3)			
Other parameters				
OCR	K_0			
1.519 (1.205 – 1.616)	0.44			

T-bar tests were performed [12] to obtain sensitivity parameters, oedometer and anisotropically (K_0) consolidated triaxial tests were carried out for basic hypoplastic and enhanced hypoplastic soft clay models.

3   Model Evaluation and Numerical Model Results

Validation of the numerical model involved large scale boundary value problem incorporating very large deformations, namely retrospective simulations of centrifuge model data on spudcan installation performed by Bienen et al. [13]. The performance of the enhanced model has been evaluated against existing experimental data and their representation using parent hypoplastic model from Ragni et al. [3].

Remeshing and Interpolation Technique with Small Strain (RITSS) strategy [5] has been used as the large deformation finite element approach in the ABAQUS software. Identical numerical model as used by Ragni et al. [3] has been employed – the same geometry, mesh and boundary conditions were used to allow for consistency in evaluating performance of the enhanced and parent models' calibration.

To investigate the effect of parameter ξ, values ranging from $\xi = 0$ (equivalent to model by Ragni et al. [3]) to $\xi = 0.3$ (obtained from element test calibration) were used in the simulations. Figure 2 shows results of numerical simulations compared with

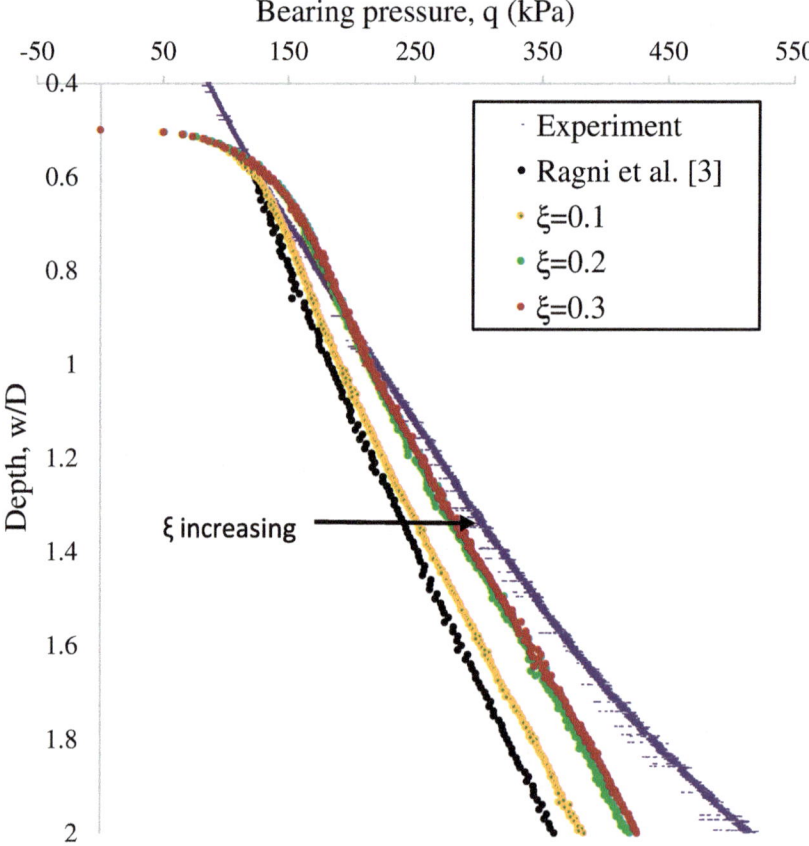

Fig. 2. Comparison of model numerical predictions and centrifuge experimental data. Vertical profiles of spudcan bearing capacity for various values of parameter ξ.

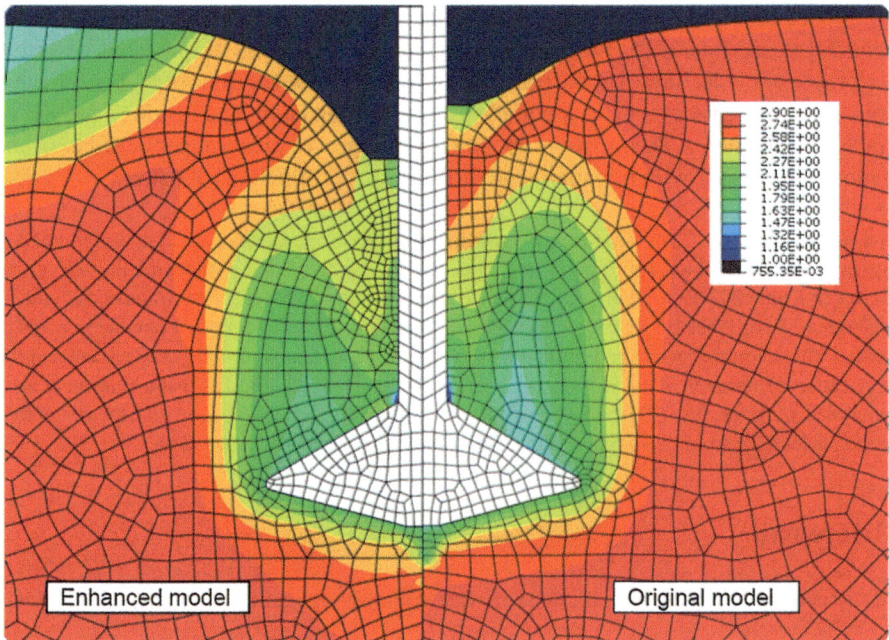

Fig. 3. Sensitivity *s* contour plots after the spudcan penetration to a depth of 1.2 w/D. Original simulation (RHS), enhanced model (LHS).

centrifuge data, vertical profiles of spudcan bearing capacity are plotted for various values of parameter ξ, showing increase of bearing capacity with increase of parameter ξ. The enhanced model predicts the bearing capacity better compared to the original model, which is agreement with improved predictions of element tests. Figure 3 compares sensitivity contour plots for simulation by Ragni et al. [3] and for the enhanced model, also showing the geometry and finite element mesh of the model.

In the original work by Ragni et al. [3] the prediction of bearing capacity was improved by adding parameter O_c defining position of critical state line. As spudcan penetration could not be reproduced correctly with O_c calibrated using element test data, it was adjusted using simulations of spudcan load-penetration curve. This apparent discrepancy between element and spudcan test results is resolved in the improved hypoplastic model.

4 Summary and Conclusions

A modification of a hypoplastic model for clays [4] is presented in the paper, aimed at improving the performance of the model for K_0 consolidated soils. Fundamental features of the model formulation were defined, followed by evaluation of performance of the model in element tests and the retrospective simulations of centrifuge model with respect to the original hypoplastic model. The model adds two additional parameters, which can be obtained from K_0 normally consolidated triaxial tests.

Model predictions compare well with element test experimental data as well as centrifuge experiments of spudcan penetration on Laminaria soil. It is demonstrated that the proposed model improves the prediction of undrained stress paths of K_0 consolidated soils as well as predictions in a complex offshore foundation engineering problem of spudcan penetration. The proposed modification improves practical applicability of the hypoplastic model by improving its predictions of soil initially in K_0 states.

Acknowledgment. This work forms part of the activities of the Centre for Offshore Foundation Systems (COFS), which is currently supported as one of the primary nodes of the Australian Research Council (ARC) Centre of Excellence for Geotechnical Science and Engineering and as a Centre of Excellence by the Lloyd's Register Foundation. Lloyd's Register Foundation helps to protect life and property by supporting engineering-related education, public engagement and the application of research. This support is gratefully acknowledged. The first author has also been supported by grant No. 1075516 of the Charles University Grant Agency. The second author received funding from grant No. 15-059355 of the Czech Science Foundation.

References

1. Randolph, M., Gourvenec, S.: Offshore Geotechnical Engineering. CRC Press, Boca Raton (2011)
2. Hossain, M.S., Hu, Y., Randolph, M.F., White, D.J.: Limiting cavity depth for spudcan foundations penetrating clay. Géotechnique **55**(9), 679–690 (2005)
3. Ragni, R., Wang, D., Mašín, D., Bienen, B., Cassidy, M.J., Stanier, A.S.: Numerical modelling of the effects of consolidation on jack-up spudcan penetration. Comput. Geotech. **78**, 25–37 (2016)
4. Mašín, D.: Clay hypoplasticity model including stiffness anisotropy. Géotechnique **64**(3), 232–238 (2014)
5. Hu, Y., Randolph, M.F.: A practical numerical approach for large deformation problems in soil. Int. J. Numer. Anal. Methods Geomech. **22**(5), 327–350 (1998)
6. Kolymbas, D.: An outline of hypoplasticity. Arch. Appl. Mech. **61**(3), 143–151 (1991)
7. Gudehus, G.: A comprehensive constitutive equation for granular materials. Soils Found. **36**(1), 1–12 (1996)
8. Mašín, D.: A hypoplastic constitutive model for clays. Int. J. Numer. Anal. Methods Geomech. **29**(4), 311–336 (2005)
9. Mašín, D.: A hypoplastic constitutive model for clays with meta-stable structure. Canad. Geotech. J. **44**(3), 363–375 (2007)
10. Mašín, D., Herle, I.: Improvement of a hypoplastic model to predict clay behaviour under undrained conditions. Acta Geotech. **2**(4), 261–268 (2007)
11. Gajo, A., Muir Wood, D.: A new approach to anisotropic, bounding surface plasticity: general formulation and simulations of natural and reconstituted clay behaviour. Int. J. Numer. Anal. Methods Geomech. **25**(3), 207–241 (2001)
12. Randolph, M.F., Hefer, P.A., Geise, J.M., Watson, P.G.: Improved seabed strength profiling using T-bar penetrometer. In: Proceedings of International Conference on Offshore Site Investigation and Foundation Behaviour - "New Frontiers", pp. 221–235. Society for Underwater Technology, London (1998)
13. Bienen, B., Ragni, R., Cassidy, M.J., Stanier, S.A.: Effects of consolidation under a penetrating footing in carbonate silty clay. J. Geotech. Geoenviron. Eng. **141**(9), 04015040 (2015)

Experimental Analysis on Interfacial Shear Strength in Penetration of Super-Long Piles in Ocean Engineering

Shu Lin[1], Shuwang Yan[1], Zhaolin Jia[2,3(✉)], and Ruiqing Lang[1]

[1] School of Civil Engineering, Tianjin University, Tianjin, China
[2] Key Laboratory of Soft Soil Engineering Character and Engineering Environment of Tianjin, Tianjin, China
jzhaolin@126.com
[3] Post-Doctoral Research Center, Huadian Heavy Industries Co., Ltd., Beijing, China

Abstract. In ocean engineering, the large-diameter and super-long pile can be seen as a frictional pile because of its long length, indicating that the lateral friction plays a significant role in the bearing capacity of the pile. According to in situ measured data, the lateral friction of the pile tends to be lower than the friction obtained by calculating the parameters of soils in the penetration process of super-long piles. This is caused by large deformation shearing during pile driving and might lead to a reduction of the pile-bearing capacity, which may render engineering projects unstable and unsafe. To study the lateral friction of piles in ocean engineering, large-scale direct shear tests and ring-shear tests were performed to investigate the interfacial shearing behavior of pile-soil during the pile-driving procedure. Three kinds of soil samples, including sand, silty clay and clay, were used in the tests to study the interfacial shearing behavior of soils of various textures. A reduction coefficient was proposed to determine the interfacial strength by referring to the soil strength. Based on the test results and previous works by other researchers, a discussion of the analysis of the factors that influence the residual strength of the soil is provided.. Advice for engineering projects is put forward in accordance with the analytic conclusions in this paper.

Keywords: Ocean engineering · Super-long pile · Shearing test
Interfacial strength · Residual strength

1 Introduction

With the development of ocean engineering, large-diameter and super-long pile has become one of the popular foundation structures of offshore platforms. The bearing capacity of the pile is now one of the hot research topics. According to engineering experience, pile-running or repelling hammering may occur during construction if the bearing capacity of the pile was misestimated in the design stage. For the super-long pile, the lateral friction contributes much to the bearing capacity, and it will significantly affect the stability of the pile foundations.

© Springer Nature Singapore Pte Ltd. 2018
T. Qiu et al. (Eds.): GSIC 2018, *Proceedings of GeoShanghai 2018 International Conference: Advances in Soil Dynamics and Foundation Engineering*, pp. 757–765, 2018.
https://doi.org/10.1007/978-981-13-0131-5_82

Based on several engineering projects, the penetration of the pile can decrease the lateral friction because of large deformation shearing that can weaken the soil and lead to a decrease in its strength. Therefore, shearing tests can be applied to investigate the lateral friction of the pile in terms of the interfacial shear strength. Studies of the interfacial shear strength between soils and structures have been carried out by many researchers [1–13]. Direct shear tests and ring shear tests were performed to investigate the shearing behavior between the soil and the structures made of various materials [1–8]. The influence factors of the shearing behavior, including cyclic loads, diameters of soil particles, and roughness of interface, were also discussed [7–9]. Nevertheless, the previous works were more concentrated on the soil on land, and there is little work that is reported on marine soil.

In this paper, large-scale direct shear tests and ring shear tests were conducted with various marine soils to investigate the behavior of shear strength change under large deformation shearing in ocean engineering. Based on the test results, a discussion is provided concerning the characteristics of friction between the soil and the pile. A reduction coefficient was introduced to describe the interfacial shearing characteristics between the soils and structures. The peak shear strength and the residual strength of the soils are also discussed, as are the influence factors under various loads.

2   Test Equipment and Methods

2.1   Soil Specimens

Four kinds of soil samples, including fine sand, medium sand, coarse sand and clay, were chosen in the shear tests. The samples were gathered from the Bohai Sea in China, and the basic physical properties of the specimens are shown in Tables 1 and 2.

Table 1. Basic physical properties of sand samples.

Sand	Particle size (mm)			Coefficient of uniformity	Coefficient of curvature	Dry density (g/cm^3)	
	d_{60}	d_{10}	d_{30}	C_u	C_c	Max	Min
Fine	0.152	0.080	0.103	1.90	0.870	1.72	1.45
Medium	0.4	0.15	0.25	1.6	1.042	1.75	1.48
Coarse (S3)	1.02	0.51	0.71	1.92	1.009	1.85	1.52

Table 2. Basic physical properties of cohesive soil samples.

Soil	Depth (m)	Water content w (%)	Unit weight γ (kN/m^3)	Void ratio e	Plasticity index	Saturation ratio (%)
S1	1.0–1.5	53.1	18.0	1.877	19	98.8
S2	1.5–2.0	46.1	17.9	1.767	20	97.7
S4	22.7–22.9	50.0	17.9	1.650	14	95.7
S5	1.0– 2.0	53.1	18.0	1.877	19	98.8

The sand samples were prepared by controlling the relative density of the soil with compaction, and gradually saturated from the bottom upwards by dropping the water slowly. For the clay, the samples were consolidated to the selected strength by vacuum preloading after being evenly stirred. The selected strength lay within the range of 5 to 20 kPa corresponding to the actual conditions in ocean engineering.

2.2 Experiment Apparatuses

Revised Large-Scale Direct Shear Apparatus. A revised large-scale direct shear apparatus was designed in accordance with the working principle of an ordinary direct shear apparatus. It consisted of a top box, bottom box, loading equipment and a testing system (Fig. 1a).

The detachable rectangular shear box, which had inside dimensions of 152 mm by 152 mm and a height of 50 mm, was composed of a top box and a bottom box. A 2-mm-wide gap between the two halves of the shear box was designed to reduce the effect of the interfacial friction of the boxes and to improve the accuracy of the tests.

Compared with the conventional direct shear apparatus, the large-scale direct shear apparatus boasts some advantages, including large scale, large shear movement and low instrumental errors. Moreover, it is suitable for the test considering the interface roughness and the simulation of actual shear conditions.

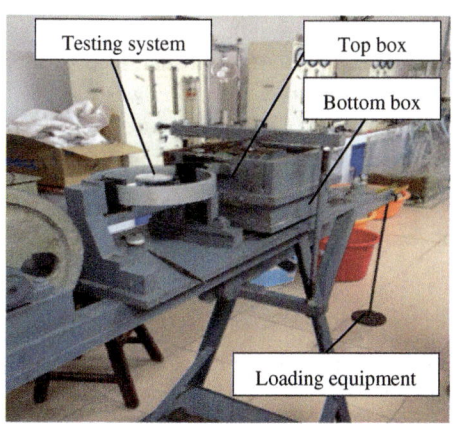
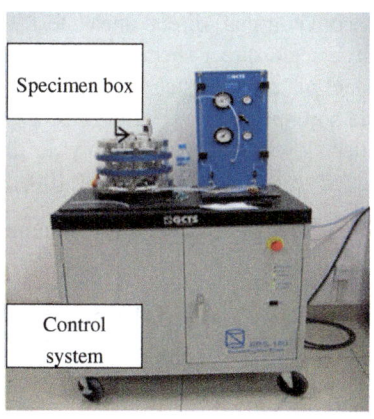

(a) Large-scale direct shear apparatus   (b) Ring shear apparatus

Fig. 1. Test apparatuses

Ring Shear Apparatus. The apparatus used in the ring shear test was SRS-150 dynamical ring shear apparatus (Fig. 1b) manufactured by GCTS in the USA. It is a type of Bromhead ring shear apparatus that has minimized the instrumental errors by its design. The samples were ring-shaped with an inner diameter of 100 mm, outer diameter of 150 mm and a height of 25 mm. The shear displacement can be much greater in the ring shear test than in others, which makes it easier to measure the residual strength.

2.3   Experiment Approaches

Direct Shear Apparatus

Definition of Reduction Coefficient. Interfacial characteristics refer to the slip-resistant characteristics of the interface when there is relative movement between the pile and soil, which can be denoted by

$$R_m = \tau_{pf}/\tau_{sf} \tag{1}$$

where R_m is the reduction coefficient, τ_{pf} is the interfacial shear strength between the pile and soil, and τ_{sf} is soil shear strength. The reduction coefficient should be different for sand and clay and therefore must be investigated by tests.

The structure material in this paper was a steel plate taken from Tianjin Harbor. The parameter of surface roughness R_z, defined as the maximum height of the surface profile of the plate, was 40–100 μm, which was consistent with the actual condition.

Test Approach. The soil shear strength as well as the interfacial strength between the structure (e.g., pile) and the soil was measured simultaneously in the tests so that the reduction coefficient could be obtained. The shear strength of the soil was measured by a conventional direct shear test and the interfacial strength was measured by a large-scale direct shear test. In the large-scale direct shear test, the interface material (steel plate) was placed in the upper shear box, and the lower shear box was filled with the soil specimen. To minimize the errors, the shear test was carried out immediately after slowly and gently applying the normal loads. Each kind of specimen was subjected to the normal loads of 100, 200, 300 and 400 kPa.

Ring Shear Tests. The test schemes are shown in Table 3.

Table 3. Experiment schemes of ring-shear tests.

Specimen number	Approach	Shear rate (mm/min)	Normal stress (kPa)	Consolidation time (h)
S3	Multistage shearing	1.0	100, 200, 300, 400	24
S4	Single-stage shearing	1.0	100, 200, 300, 400	24
S5	Single-stage shearing	1.0	200, 300, 400	24

For the coarse sand (S3), multistage shearing was adopted to perform the tests. Multistage shearing refers to the shear under staged consolidation pressures, which keeps the consistency of the specimen and shortens the testing time. To minimize the unstable vertical settlement due to the gradual increase of the normal loads, consolidation was carried out before increasing the normal loads to the next level.

For the silty clay and the clay, multistage loading could make the specimen squeeze out which could lead to a large error in the test. Therefore, single-staged shear was taken to perform the tests of specimens S4 and S5. Single-stage shearing is the shear test under a certain consolidation load.

3   Test Results

3.1   Direct Shear Tests

Test Results of Sand. For the relationships between the relative density and the reduction coefficient, the variation tendency was different for three kinds of sand. When the relative density increases, both the interfacial strength and the soil strength improve. If the interfacial strength increases faster than the soil strength, the relationship will show an increased variation tendency. A decreased variation tendency, however, may also appear when the interfacial strength increases slower than the soil strength. In the tests, the relationship showed an increased variation tendency for the fine and the medium sand, but it showed a decrease for the coarse sand. For the coarse sand, the soil particles interlocked with each other more tightly and the soil volume decreased under the shearing, which led to the relatively greater increase in the soil strength. The interfacial strength, however, increased slowly because the stiffness of the steel plate was great, and the sand particle was almost incompressible, which made the relationship between the relative density and the reduction coefficient show a decrease.

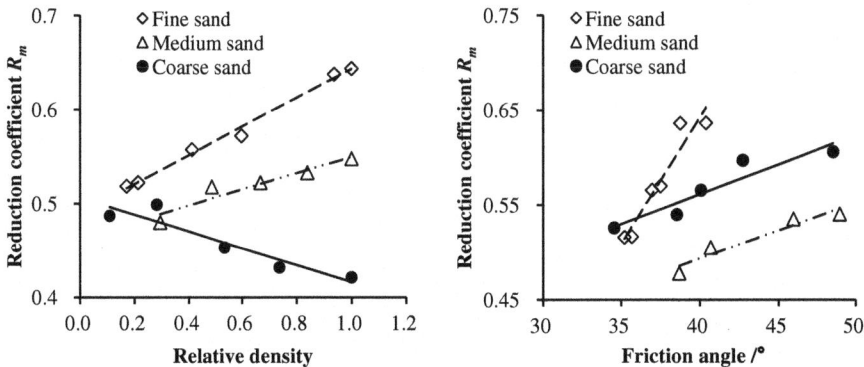

Fig. 2. Results of direct shear tests on sand

As shown in Fig. 2, the reduction coefficient increased with the increase in the friction angle. For different sands, the variety of shear strength was different, but all the reduction coefficients ranged in value between 0.45 and 0.65. Therefore, the reduction coefficient can be chosen between 0.45 and 0.65 according to strength parameters when the steel piles work within sand strata in engineering projects.

Test Results of Clay. Similar to the results in other studies [10], the interfacial strength between the steel plate and the clay was lower than the clay strength, which indicated that the failure plane was more likely to appear at the interface rather than inside the soil during the pile-driving procedure. According to the test results, when the clay strength lay within 5 to 20 kPa, the reduction coefficient of the clay varied from 0.3 to 1.0 (Fig. 3). As shown in Fig. 3, the coefficient was increasing with the development of the undrained soil strength, and approached 1.0 when the undrained strength is approximately 20 kPa. If the undrained strength developed beyond 20 kPa, the coefficient would be 1.0, indicating that the failure could be thought to occur inside the soil.

The reduction coefficient obtained by the design methods in the API codes is 1.0 for clay, which indicates that it is preferable to assume that the failure occurs inside the soil. It is unsafe for estimating the resistance of pile driving using these methods. Pile running may occur during the pile-driving procedure because of the overestimation of the skin friction as well as the pile-driving energy. Therefore, more reduction should be taken into account for the skin friction of pile based on the API codes when it is working in low-strength clay.

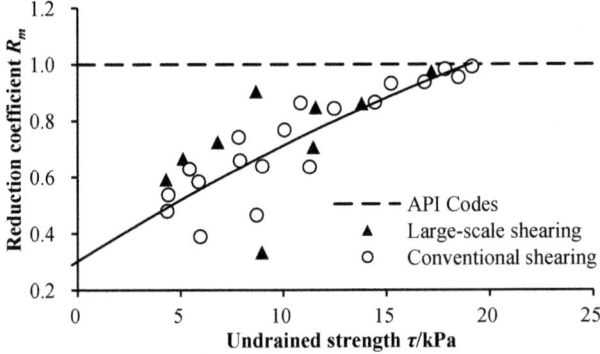

Fig. 3. Results of direct shear tests on clay

3.2   Ring Shear Tests

The relation between the shear stress and the shear strain under the multistage loading and the results of silty clay and clay are shown in Fig. 4. The peak strength and the residual strength of the specimens in each stage are summarized in Table 4.

According to the results, in each stage of shearing, the shear strength would fall to a certain stable value, seen as the residual strength, after it reached the peak value. The normally consolidated coarse sand showed the residual strength when the strain reached approximately 20 mm; the residual strength of silty clay appeared at larger strain than 20 mm, and it was shown at a much larger shear strain, approximately 30 mm, for the clay. The curves of the strength ratios are summarized in Fig. 4d. It starts from 0.833 to 0.989 for coarse sand and approaches 1.0 with increasing normal stress, indicating that the residual strength would be closer to the peak one under a greater normal stress. The strength ratios of silty clay still increased with the development of the normal stress, but they were more even than those of the sand. The ratio of clay varied from 0.68 to 0.81 with the normal stress from 200 to 400 kPa, indicating that the strength of clay decreased more under large deformation shearing.

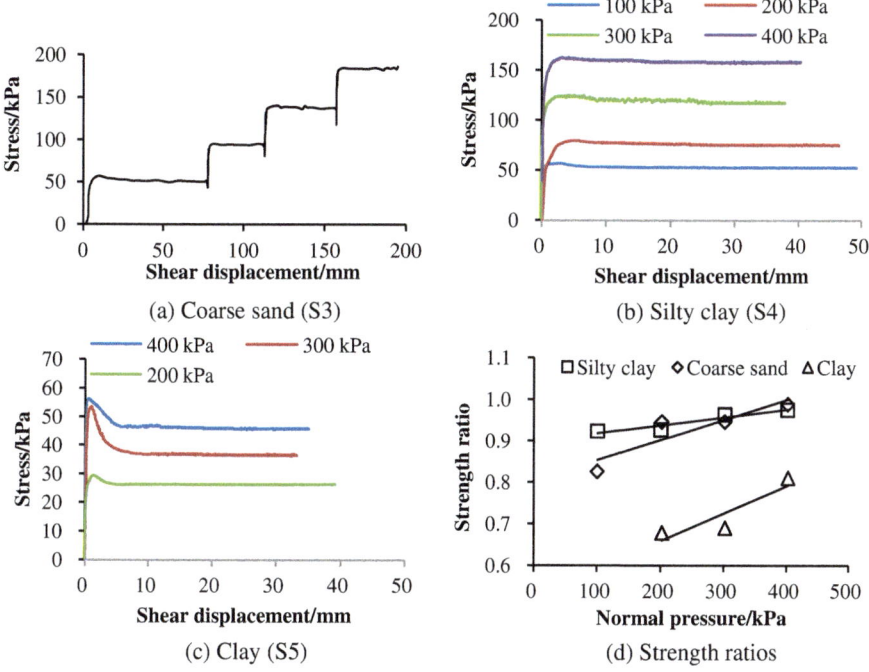

Fig. 4. Results of ring shear tests

Table 4. Strength of specimen in the ring shear test.

Specimen	Strength	Normal stress			
		100 kPa	200 kPa	300 kPa	400 kPa
S3	Peak strength/kPa	58.8	96.9	140.8	186.4
	Residual strength/kPa	49.0	92.2	134.0	184.5
	Ratio	0.833	0.951	0.952	0.989
S4	Peak strength/kPa	56.5	80.0	124.2	161.3
	Residual strength/kPa	52.1	74.2	119.4	157.6
	Ratio	0.922	0.928	0.960	0.977
S5	Peak strength/kPa	/	28.8	53.9	55.8
	Residual strength/kPa	/	19.7	37	45.7
	Ratio	/	0.684	0.686	0.819

In accordance with the test results, the shear ratios were obviously different for different soils. Skempton [7] noted that it was the content of cohesive particles that contributed much to the discrepancy among the strength ratios of soils. The residual strength was close to the peak value when the cohesive particle content was relatively low (lower than 20%), similar to sand and some silty clay soils. When the content was beyond 40%, the residual strength decreased more, and the decrease become greater with improving content of cohesive particles. The strength of cohesive soil decreases more than that of non-cohesive soil under large deformation shearing.

4   Conclusions

During pile driving in ocean engineering, the estimation of lateral friction was significant for a determination of the bearing capacity. The interfacial characteristics between the soil and steel plate and the residual strength of marine soils were respectively investigated through a large-scale direct shear test and ring shear test. A reduction coefficient was proposed to reflect the interfacial shear characteristics, and the interfacial strength could be deduced by referring to the reduction coefficient and the soil strength. According to the tests, the reduction coefficient was 0.45–0.65 for sand and 0.3–1.0 for clay. Moreover, the residual strength is much lower than the peak strength if the content of the cohesive particles is relatively high, and it is closer to the peak value when the cohesive particles take up less content in the soil. The results could be valuable to engineering projects.

References

1. Potyondy, J.G.: Skin friction between various soils and construction materials. Géotechnique 11(4), 339–353 (1961)
2. Lings, M.L., Dietz, M.S.: The peak strength of sand-steel interfaces and the role of dilation. Soils Found. Tokyo 45(6), 1–14 (2005)
3. Mortara, G., Mangiola, A., Ghionna, V.N.: Cyclic shear stress degradation and post-cyclic behaviour from sand–steel interface direct shear tests. Can. Geotech. J. 44(7), 739–752 (2007)
4. Zhou, K., Cheng, Y., Huang, X.: Experimental study on direct shear of different pile-soil interface. Subgrade Eng. 05, 93–95+99 (2011)
5. Wijewickreme, D., Amarasinghe, R., Eid, H.: Macro-scale direct shear test device for assessing soil-solid interface friction under low effective normal stresses. Geotech. Test. J. 37(1), 121–138 (2014)
6. Ebrahimian, B., Bauer, E.: Investigation of direct shear interface test using micro-polar continuum approach. In: Bifurcation and Degradation of Geomaterials in the New Millennium, pp. 143–148 (2015)
7. Cambio, D., Kang X., Ge L.: Characterization of Crushed Rock-Concrete Interface Behavior through the Parallel Gradation Technique, pp. 209–216. Geotechnical Special Publication (2011)
8. Kang, X., Cambio, D., Ge, L.: Effect of parallel gradations on crushed rock-concrete interface behaviors. J. Test. Eval. 40(1), 119–126 (2012)

9. Skempton, A.W.: Residual strength of clays in landslides, folded strata and the laboratory. Géotechnique **35**(1), 3–18 (1985)
10. Yin, Y., Li, F.: The shear properties of the contact surface between steel and coastal plain soil. Soil Eng. And Foundation **06**, 123–125 (2014)
11. Watry, S.M., Lade, P.V.: Residual shear strengths of bentonites on Palos Verdes Peninsula, California. Geo-Denver **289**, 323–342 (2014)
12. Wang, S., et al.: Shear behaviors of saturated loess in naturally drained ring-shear tests, pp. 19–27. Springer (2015)
13. Hong, Y., et al.: Shear strength of silty clay in Dalian by ring shear tests. J. Jilin Univ. **46**(5), 1475–1481 (2016). (Earth Science Edition)

Numerical Analysis of Current-Induced Local Scour Under a Vibrating Pipeline

Qi Zhang[1,2], Xiang-Lian Zhou[1,2(✉)], and Jian-Hua Wang[3]

[1] State Key Laboratory of Ocean Engineering, Shanghai Jiao Tong University,
Shanghai, People's Republic of China
zhouxl@sjtu.edu.cn
[2] Collaborative Innovation Center for Advanced Ship and Deep-Sea
Exploration (CISSE), Shanghai, People's Republic of China
[3] Centre for Marine Geotechnical Engineering, Shanghai Jiao Tong University,
Shanghai, People's Republic of China

Abstract. The dynamic interaction between the pipeline vibration and the local scour is investigated numerically. The *RNG k-ε* turbulence model and a sediment scour model are adopted to calculate the local scour below the vibrating pipeline. The vibrating pipeline is fully coupled with the fluid flow. The present numerical model is verified with the previous experimental results and shows good agreement. The coupling effects between the pipeline vibration and local scour are investigated numerically. The numerical results reveal that the maximum scour depth and the shape and scale of the scour hole are closely related to the amplitude of pipeline vibration. The vortices that shed behind the pipeline are influenced by the scour profile. The amplitude of pipeline vibration is influenced by the changes of scour depth.

Keywords: Local scour · Pipeline · Vortex-induced vibration · Current

1 Introduction

Submarine pipelines have been widely used in transporting the oil and gas from offshore to land. Submarine pipelines are in the complex hydrological environment, which may lead to local scour below the pipeline. Once the scour depth is deep enough, the free spanning pipelines may appear. As a result, the vortex-induced vibration (VIV) of the free spanning pipeline may occur if the length of the free spanning pipeline is long enough, which is directly related with the stability and fatigue of the pipeline.

In the past few decades, many researchers investigated the local scour below a fixed pipeline through experimental method [1, 2]. The mechanisms of local scour and the maximum scour depth under different experimental conditions were investigated in detail. Some empirical formulas were proposed to calculate the maximum scour depth. The fluid, pipeline, and seabed characteristics, which influence the local scour result significantly, were investigated and discussed [3–5]. In recent years, numerical method was widely used to study the local scour under the pipeline. Using different turbulence theories, researchers developed several two-dimensional and three-dimensional numerical models to investigate the local scour below the pipeline [6–8].

© Springer Nature Singapore Pte Ltd. 2018
T. Qiu et al. (Eds.): GSIC 2018, *Proceedings of GeoShanghai 2018 International Conference: Advances in Soil Dynamics and Foundation Engineering*, pp. 766–773, 2018.
https://doi.org/10.1007/978-981-13-0131-5_83

The VIV of pipeline was also investigated by the previous researchers in the last decades. It is known that the vortices that shed behind the free spanning pipeline may make the pipeline vibrate. On the other hand, the flow and pressure field around the pipeline are affected by the vibrating pipeline [9]. A large number of previous researches have focused on the mechanisms of VIV of pipeline through experimental and numerical method [10–12]. The effects of several fluid and pipeline factors, such as the pipeline diameter, pipeline mass ratio, current velocity, and Reynolds number, etc. on the pipeline vibration were also discussed [13, 14].

It can be seen that most of the previous researches mainly either focus on the local scour or the VIV of pipeline, studies on the dynamic interaction between the VIV of pipeline and local scour are scarce. Only a few researchers have studied the coupling effect between pipeline vibration and local scour [15–17]. Therefore, it is essential to understand the dynamic interaction between VIV of pipeline and local scour.

2 Numerical Model

In this research, both the fluid domain, seabed and the vibrating pipeline are integrated in a coupled model. The *RNG k-ε* turbulence model and a sediment scour model are adopted to simulate the fluid flow and local scour. The present numerical model solves the continuity equation and Navier-Stokes equations given below:

$$\frac{\partial \langle u_i A_i \rangle}{\partial x_i} = 0 \tag{1}$$

$$
\begin{aligned}
\frac{\partial u}{\partial t} + \frac{1}{V_F}\left(uA_x \frac{\partial u}{\partial x} + vA_y \frac{\partial u}{\partial y} + wA_z \frac{\partial u}{\partial z} \right) &= -\frac{1}{\rho}\frac{\partial p}{\partial x} + G_x + f_x \\
\frac{\partial v}{\partial t} + \frac{1}{V_F}\left(uA_x \frac{\partial v}{\partial x} + vA_y \frac{\partial v}{\partial y} + wA_z \frac{\partial v}{\partial z} \right) &= -\frac{1}{\rho}\frac{\partial p}{\partial y} + G_y + f_y \\
\frac{\partial w}{\partial t} + \frac{1}{V_F}\left(uA_x \frac{\partial w}{\partial x} + vA_y \frac{\partial w}{\partial y} + wA_z \frac{\partial w}{\partial z} \right) &= -\frac{1}{\rho}\frac{\partial p}{\partial z} + G_z + f_z
\end{aligned}
\tag{2}
$$

where x_i is the Cartesian coordinate, A_i is the area fraction, V_F is volume fraction, u_i (u, v and w) is the velocity, ρ is fluid density, p is the average hydrodynamic pressure, G_i is body acceleration, f_i is viscous acceleration ($i = x$, y, z).

The sediment scour model, which is fully coupled with the fluid domain, is adopted to simulate the local scour below the pipeline. By predicting the erosion, advection and deposition of the sediment, the sediment scour model describe the whole sediment transport processes [18]. The entrainment lift velocity of sediment $\mathbf{u}_{lift,i}$ and bed-load transport velocity $u_{bedload,i}$ can be written as [19, 20]:

$$\mathbf{u}_{lift,i} = \alpha_i \mathbf{n}_s d_*^{0.3} \left(\theta_i - \theta_{cr,i} \right)^{1.5} \sqrt{\frac{\|g\| d_i \left(\rho_i - \rho_f \right)}{\rho_f}} \tag{3}$$

$$u_{bedload,i} = \frac{q_{b,i}}{\delta_i c_{b,i} f_b} \qquad (4)$$

in which α_i is the entrainment parameter, $\mathbf{n}_s$ is the outward pointing normal to the packed bed interface, d_* is dimensionless diameter of sediment, θ_i is the local Shields parameter, $\theta_{cr,i}$ is the critical Shields parameter, $\|g\|$ is the magnitude of the acceleration of gravity, d_i is the diameter, ρ_i is the density of the sediment species i, ρ_f is the fluid density, f_b is the critical packing fraction of the sediment, $c_{b,i}$ is the volume fraction of species i in the bed material, $q_{b,i}$ is the volumetric bed-load transport rat, δ_i is the bed-load thickness.

The vibrating pipeline is fully coupled with the fluid flow. The governing equation of the pipeline vibration motion can be written as:

$$m\frac{\partial^2 X_z}{\partial t^2} + c\frac{\partial X_z}{\partial t} + kX_z = F_z \qquad (5)$$

where m is the pipeline mass, X_z is the pipeline displacement in the transverse direction (z-direction) with respect to the static balance position, c is the structural damping factor, k is the spring constant, and F_z is the transverse direction forces on pipeline.

The sketch of the numerical model is illustrated in Fig. 1. The total length of the computational domain is $50D$ ($25D$ upstream of the pipeline and $25D$ downstream of the pipeline. The fluid domain, seabed and the vibrating pipeline are contained in the integrated model. The pipeline is only allowed to vibrate in the z-direction. The boundary conditions of the numerical model are as follows: The top boundary of the model is considered to be the standard atmospheric pressure. A velocity vector is specified at the inlet boundary. At the inlet boundary, a velocity vector is specified to simulate the steady current. The Outflow boundary is used at the outlet boundary to avoid the reflection of flow. No-slip boundary is used at the pipeline surface.

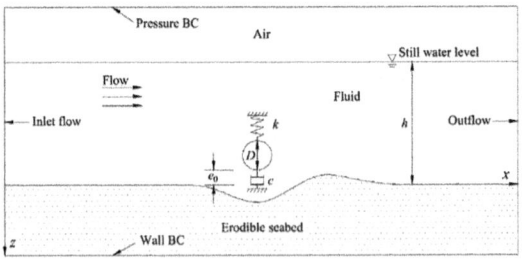

Fig. 1. Sketch of the numerical model.

3   Model Validation

In this research, both the sediment scour model and the vibrating pipeline model are verified with the previous experimental results in this research, respectively. Firstly, a validation of the local scour between the present model and the previous experimental result and numerical result are carried out to verify the accuracy of the sediment scour model [2, 21]. The test conditions are as follows: the current velocity is $V = 0.35$ m/s, the water depth is $h = 0.35$ m, the mean diameter of sand gran is $d_{50} = 0.036$ mm, the pipeline diameter is $D = 0.1$ m, the critical Shield number is 0.048, and the sediment density is $\rho = 2650$ kg/m^3.

Figure 2a shows the comparison of the scour profiles at the equilibrium stage between the present model and the previous results. It can be seen that the computed local scour profile results agree well with the previous experimental and numerical results. The numerical scour profile is slight deeper than the experimental result at the upstream of the pipeline but agrees well with the numerical result. The maximum scour depth are almost the same among these results. The result indicates the present numerical model is capable to study the sediment scour.

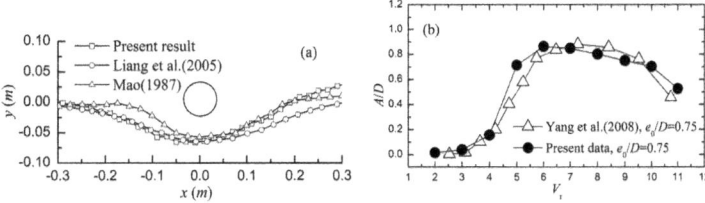

Fig. 2. Comparisons of the: (a) sediment scour model, and (b) pipeline vibration model.

The pipeline vibration model is also validated with the previous experimental results [16]. The test conditions are as follows: $D = 0.032$ m, $h = 0.3$ m, $V = 0.255$ m/s, $d_{50} = 0.38$ mm, the gap-to-diameter ratio is $e_0/D = 0.75$, the mass ratio of the pipeline is $m^* = 3.87$. Figure 2b shows the trends of the pipeline vibration amplitudes (A/D) versus the reduced velocity (V_r) between the present model and the experimental result [16]. It can be seen that the numerical result agrees well with the experimental result. The maximum pipeline vibration amplitude equals to 0.85D for both the numerical and experimental result when $V_r = 6.5$. Therefore, the results show that the present numerical model has enough accuracy to investigate the VIV of pipeline and local scour.

4   Numerical Results and Discussion

Numerical simulations were carried out to investigate the interaction between the VIV of pipeline and local scour. The water, pipeline, and seabed parameters of the numerical model are listed in Table 1. Figure 3 illustrates the comparison of the scour profiles at three instants of the scour process between the local scour under a fixed pipeline and a vibrating pipeline. It can be seen that the effects of VIV of pipeline on the local scour process are significant. The maximum scour depth and the scour hole width under the vibrating pipeline is much larger than under the fixed pipeline through the whole scour process.

Table 1. Table captions should be placed above the tables.

Water depth	0.4 m	Current velocity	0.3
Mass density of fluid	1000 kg/m^3	Pipeline diameter	0.032 m
Mass ratio of pipeline	1.0	Natural frequency in water	3.448
Initial gap-to-diameter ratio	0.5	Mean diameter of sand grain	0.36 mm
Mass density of sand grain	2650 kg/m^3	Critical Shields number	0.039

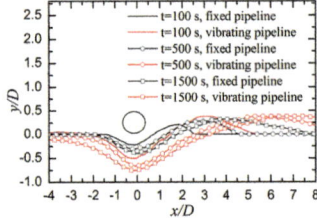

Fig. 3. Comparison of the scour profiles at three instants of the scour process between the local scour under a fixed pipeline and a vibrating pipeline.

At the initial stage of scour, the scour depth under the vibrating pipeline is slightly deeper than that below the fixed pipeline. Then, the scour hole below the pipeline extends gradually in both cases, and the sediment deposition mound is washed to the downstream of the pipeline gradually. The scouring velocity under the vibrating pipeline is much faster than that under the fixed pipeline. And the sediment deposition mound at the downstream of the pipeline grows much wider when considering the pipeline vibration. As the scour depth and the scour hole scale develop gradually, the effects of the wake vortex and the flow field on the local scour become weak. Finally, the scour profiles will no longer change. It can be seen that the maximum scour depth under the vibrating pipeline equals to $0.76D$ at the time instant t = 1500 s, which is much larger than that under the fixed pipeline with the value of $0.4D$. Also, the maximum scour hole width under the vibrating pipeline is about 7.5D, which is 1.88 times as wide as that under the fixed pipeline with the value of about 4D. The results indicate that the effects of the pipeline vibration on the local scour are significant.

Figure 4 shows the time history of maximum scour depth under the vibrating pipeline and the fixed pipeline. The maximum scour depth increases quickly at the initial stage of scour and then slows down gradually in both cases. After 1000 s, the maximum scour depth under the fixed pipeline almost has no changes, but the scour depth under the vibrating pipeline still increases slightly.

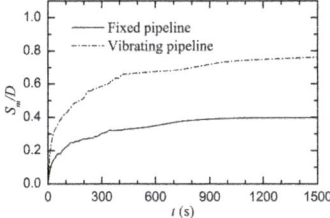

Fig. 4. Comparison of time development of scour depth under fixed pipeline and vibrating pipeline.

Figure 5 shows the instantaneous contours of vorticity magnitude around the vibrating pipeline upon the seabed at four instants in one vibration cycle. It can be seen that two single signed vortices released behind the pipeline per cycle of shedding. When the pipeline begins to move downward from the top maximum position, the vortex behind the upper side of pipeline is strong enough and begins to shed. As the pipeline moves downward, the vortex behind the upper side of pipeline moves to downstream at the angle of approximately 20° with respect to the horizontal direction and becomes round-shaped and weaker in strength gradually. It can be seen that the scour profile changes the shape and moving direction of the vortex that sheds behind the pipeline significantly. The vortex that sheds behind the lower side of pipeline becomes elongated and weaker quickly after it sheds behind the pipeline. The results indicate that the interaction between local scour and vortex shedding is significant. The vortices that shed behind the upper and lower sides of pipeline are significant different in shape and strength, which is closely related to the existence of the sediment deposition mound at the downstream of the pipeline.

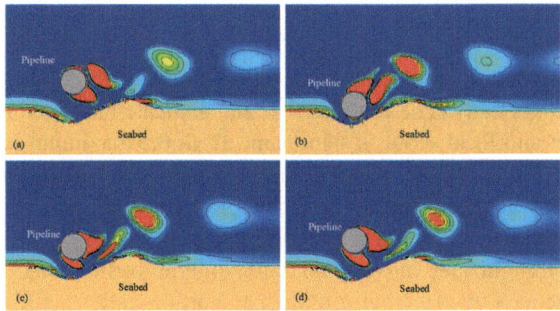

Fig. 5. Instantaneous contours of vorticity magnitude around the vibrating pipeline upon the seabed.

Figure 6a, b show the time history of vortex-induced pipeline vibration displacement upon the sandy seabed. The amplitude of pipeline vibration increases gradually and reaches to the value of about $0.55D$ with respect to the pipeline initial location, as shown in Fig. 6a. After 500 s, the amplitude of pipeline vibration is almost unchanged with the value of 0.52D. It can be seen that the effects of the boundary condition decreases as the scour hole expands and the scour depth increases gradually.

Figure 6c, d shows the time history of vortex-induced pipeline vibration displacement upon three different boundary conditions: rigid boundary, and wall-free condition. The amplitude of the pipeline vibration upon the rigid boundary equals to 0.47D, which is small than that upon the seabed with the value of 0.55D. It is worth noticing that the upward amplitude of the pipeline vibration with respect to the initial location are larger than the downward amplitude under the rigid boundary conditions, which mainly due to the effects of the rigid boundary, as shown in Fig. 6c. The average amplitude of the pipeline vibration equals to 0.05D with respect to the pipeline initial location under wall-free condition, as shown in Fig. 6d. This is due to the absence of boundary effects, the vortices that shed behind the pipeline are weak in strength to induce the pipeline vibrates. The results show that the effects of local scour on the pipeline vibration are significant.

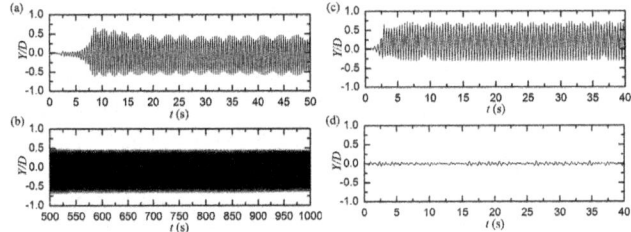

Fig. 6. Time history of vortex-induced pipeline vibration displacement.

5   Conclusions

In this study, the dynamic interaction between the pipeline vibration and the local scour is investigated numerically. The calculated results indicate that the interaction between pipeline vibration and local scour is significant. Both the maximum scour depth, scour hole scale, and the scale of the sediment deposition mound at the downstream of the pipeline increase when considering the pipeline vibration. The effect of boundary effects on the amplitude of pipeline vibration is significant.

The shape, strength and moving direction of the vortices that shed behind the pipeline are influenced by the scour profile. The vortices that shed from the upper side of pipeline move to downstream at the angle of approximately 20° with respect to the horizontal direction and become round-shaped. The vortices that shed from the bottom side of pipeline are influenced by the shapes of scour hole, which become elongated and weaker quickly after the vortices shed behind the pipeline.

References

1. Sumer, B.M., Jensen, H.R., Mao, Y., Fredsøe, J.: Effect of Lee-Wake on scour below pipelines in current. J. Waterw. Port Coast. Ocean Eng. **114**, 599–614 (1988)
2. Mao, Y.: The interaction between a pipeline and an erodible bed. Series Paper Technical University of Denmark (1987)
3. Sumer, B.M., Fredsøe, J.: The Mechanics of Scour in the Marine Environment. World Scientific, New Jersey (2002)
4. Sumer, B.M., Truelsen, C., Sichmann, T., Fredsøe, J.: Onset of scour below pipelines and self-burial. Coast. Eng. **42**, 313–335 (2001)
5. Sumer, B.M., Fredsøe, J.: Scour below Pipelines in Waves. J. Waterw. Port Coast. Ocean Eng. **116**, 307–323 (1990)
6. Mirmohammadi, A., Ketabdari, M.J.: Numerical simulation of wave scouring beneath marine pipeline using smoothed particle hydrodynamics. Int. J. Sedim. Res. **26**, 331–342 (2011)
7. Brørs, B.: Numerical modeling of flow and scour at pipelines. J. Hydraul. Eng. **125**(5), 511–523 (1999)
8. Li, F., Cheng, L.: Numerical model for local scour under offshore pipelines. J. Hydraul. Eng. **125**(4), 400–406 (1999)
9. Williamson, C.H.K., Govardhan, R.: Vortex-induced vibrations. Annu. Rev. Fluid Mech. **36**, 413–455 (2004)
10. Sumer, B.M., Fredsoe, J.: Review on vibrations of marine pipelines. Int. J. Offshore Polar Eng. **5**, 81–90 (1995)
11. Blevins, R.D., Coughran, C.S.: Experimental investigation of vortex-induced vibration in one and two dimensions with variable mass, damping, and reynolds number. J. Fluids Eng. **131**, 101202 (2009)
12. Williamson, C.H.K., Govardhan, R.: Vortex-induced vibrations. Ann. Rev. Fluid Mech. **36**, 413–455 (2004)
13. Zhao, M., Cheng, L.: Numerical simulation of two-degree-of-freedom vortex-induced vibration of a circular cylinder close to a plane boundary. J. Fluids Struct. **27**, 1097–1110 (2011)
14. Guilmineau, E., Queutey, P.: Numerical simulation of vortex-induced vibration of a circular cylinder with low mass-damping in a turbulent flow. J. Fluids Struct. **19**, 449–466 (2004)
15. Gao, F.P., Yang, B., Wu, Y.X., Yan, S.M.: Steady current induced seabed scour around a vibrating pipeline. Appl. Ocean Res. **28**, 291–298 (2006)
16. Yang, B., Gao, F.P., Jeng, D.S., Wu, Y.X.: Experimental study of vortex-induced vibrations of a pipeline near an erodible sandy seabed. Ocean Eng. **35**, 301–309 (2008)
17. Zhao, M., Cheng, L.: Numerical investigation of local scour below a vibrating pipeline under steady currents. Coast. Eng. **57**, 397–406 (2010)
18. Mastbergen, D.R., Van Den Berg, J.H.: Breaching in fine sands and the generation of sustained turbidity currents in submarine canyons. Sedimentology **50**, 625–637 (2003)
19. Zhang, Q., Zhou, X.L., Wang, J.H.: Numerical investigation of local scour around three adjacent piles with different arrangements under current. Ocean Eng. **142**, 625–638 (2017)
20. Van Rijn, L.: Sediment transport, part I: bed load transport. J. Hydraul. Eng. **110**, 1431–1456 (1984)
21. Liang, D., Cheng, L., Li, F.: Numerical modeling of flow and scour below a pipeline in currents: part II Scour simulation. Coast. Eng. **52**, 43–62 (2005)

Simulation and Prototype Measurements Analysis of Loading and Overflow Process of TSHD Hopper

Yuchi Hao[1(✉)], Runli Tao[1], Zheng Han[2], Chaozhe Yuan[1], and Lu Zhang[1]

[1] CCCC National Engineering Research Center of Dredging Technology and Equipment, Shanghai, China
haoyuchi@cccc-drc.com
[2] CHEC Dredging Co., Ltd., Shanghai, China

Abstract. In loading process of TSHD (Trailing Suction Hopper Dredgers), overflow is preferred if possible. The prediction of load and overflow losses is important for dredging production and environmental reasons. In this paper, based on Navier-Stokes equations and non-cohesive sediment transport empirical formulas, a three-dimensional numerical model was presented to simulate the loading and overflow process of TSHD. The model was compared with the typical Camp model. And then, a full-scale numerical model of TSHD XinHaihu4 was presented, the model was verified in a prototype situation by comparing with the measurements which were carried out onboard XinHaihu4 in DaFeng Waterway Engineering. Then a case with 4000 s loading time was simulated using the model of XinHaihu4, the loading process was analyzed to estimate the production and total overflow losses according to the results.

Keywords: TSHD · Hopper loading · Overflow loss · Deposition and erosion

1 Introduction

A trailing suction hopper dredger (TSHD) is a sea-going vessel that is equipped with suction pipes, loaded by pumping a sand-water mixture in hopper [1]. During the loading stage, a part of the dredged sand would not settle in the hopper but would be transported with the overflow discharge overboard, this is called overflow loss. To achieve an accurate prediction of production, improve the loading efficiency and reduce the dredging time, it is important to have knowledge of the sedimentation and scour characteristics of the dredged sediment during loading.

Currently, the sediment transport studies on the loading process are as follows: Vlasblom and Miedema's theoretical research about hopper sedimentation based on Camp model [2, 3], Ooijens added dynamics calculation in Camp model, and measured the velocity and concentration in the loading process by large-scale model experiments [4]. Cee van Rhee proposed a simplified two-dimensional CFD (computational fluid dynamics) model for loading process. In these studies, some studies estimated the overflow loss based on the sedimentation theory, some used the CFD method to simulate the flow process in the overflow phase to get the simulated production [6–8].

© Springer Nature Singapore Pte Ltd. 2018
T. Qiu et al. (Eds.): GSIC 2018, *Proceedings of GeoShanghai 2018 International Conference: Advances in Soil Dynamics and Foundation Engineering*, pp. 774–782, 2018.
https://doi.org/10.1007/978-981-13-0131-5_84

The method based on the Camp model is simple and quick, and it can get a satisfied production and overflow loss estimation, but it cannot reflect the effect of hopper geometry on loading process. Van Rhee's two-dimensional CFD model used a simple rectangular hopper, and it cannot reflect the hopper geometry, the type of inflow and overflow, and the sedimentation characteristic of particle sizes.

In this paper, a three-dimensional numerical model of hopper for simulating the loading process of TSHD was presented. The Navier-Stokes equations were used to simulate the flow in hopper. The sedimentation process was simulated by sediment empirical formulas. The simulation results were compared with those of the typical Camp model. On this basis, a prototype hopper numerical model of XinHaihu4 was presented, the full-scale validation of model was done by comparing with the prototype measurements in DaFeng Waterway dredging project. Then a 4000 s loading time case was simulated. Based on the results, the loading process of XinHaihu4 was analyzed. In addition, the production and total overflow losses are also estimated.

This model can simulate the sediment deposition and erosion process, and analyze the sedimentation characteristics of multi-particle size groups. These results can be used for new hopper design and TSHD production calculation.

2 Loading Process Description

Dredging process of TSHD can be composed of three phases:

Phase1: Start dredging until the sand-water mixture reached the overflow height. In this phase, all the sediment can be considered to stay in the hopper, suspended or deposited.

Phase2: Continue loading with overflow. In this phase, some sediment will settle, and others will be transported with the overflow discharge overboard. The deposition ratio depends on the inflow rate, energy dissipation, hopper geometry, overflow facilities, and particle size distribution. The increase of deposited sediment leads to sand bed rising, with the result that the flow section of the hopper decreases, which leads to the increase of flow velocity, and it causes the sediment re-erosion on top of the bed.

Phase3: In general, when the overflow sediment concentration reached the inflow sediment concentration, the loading process is over. Ignoring the subtle geometry that has little effect on the loading process, the simplified model of hopper is shown in Fig. 1. Due to the high velocity of inflow, the hydrodynamic of loading process is complex. The sediment movements in transport flow are randomness. The accurate simulation of sediment transport is the key to hopper loading simulation. Simulation of the loading process requires water flow equations and sediment empirical formulas.

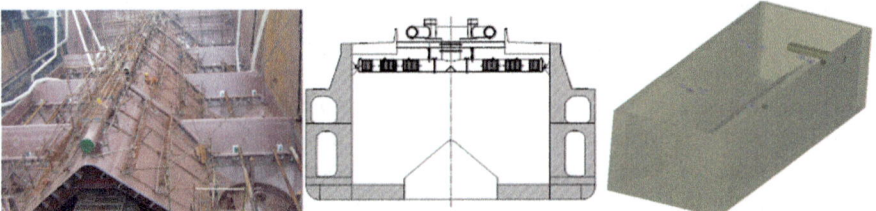

Fig. 1. Model of hopper

3   Basic Equations of the Model

3.1   Equations of Water Flow

For incompressible flow, the continuity and momentum equations are respectively:

$$V_F \frac{\partial \rho}{\partial t} + \frac{\partial}{\partial x}(\rho u A_x) + R \frac{\partial}{\partial y}(\rho v A_y) + \frac{\partial}{\partial z}(\rho w A_z) + \xi \frac{\rho u A_x}{x} = R_{DIF} + R_{SOR} \qquad (1)$$

$$\frac{\partial \rho u_i}{\partial t} + \frac{\partial}{\partial x_j}(\rho u_j u_i) = -\frac{\partial P}{\partial x_i} + G_i + \frac{\partial}{\partial x_j}\left(\rho v_{eff}\frac{\partial u_i}{\partial x_j}\right) - \frac{\partial}{\partial x_j}\left[r\left(u_{s,i}u_{s,j}\right)\right] \qquad (2)$$

Where ρ is the fluid density, u is the velocity, and v_{eff} is *effective viscosity*. The velocity components (u, v, w) correspond to the corresponding values of (x, y, z) coordinate system or (r, θ, z) coordinate system respectively. (A_x, A_y, A_z) is the area fraction of free flow in three directions. G_i (G_x, G_y, G_z) is the body acceleration, u_s is the slip velocity, r is the volume fraction.

3.2   Equations of Sediment Movements

The following types of sediment movements are considered in the simulation:

(1) Suspension and deposition: in the inflow section the incoming mixture move upward into suspension and settled under gravity;
(2) Scour: With the increase of settlement, the flow section area in hopper decreases, resulting in an increase in flow velocity and it causes the scour of settled sediment and re-suspension;
(3) Transport: the sediment transports along the bed, which leads to the changes of sea bed.

The suspended sediment convected and diffused with the fluid. Ignoring the interaction between the sediment particles in the transportation process, the suspended sediment transportation continuity equation is [6]:

$$\frac{\partial c_{s,i}}{\partial t} + \nabla \cdot (\bar{u} c_{s,i}) = 0 \qquad (3)$$

Where $c_{s,i}$ is the concentration of suspended sediment, $\bar{u}$ is the average velocity of sand-water mixture. Shields have conducted critical initiation experiments on all kinds of sediment particles. The Shields curve of the dimensionless critical initiation shear stress with the particle Reynolds number is obtained by these measured data. As the sand bed increases, the flow area decreases, so the velocity increases. When it reaches the critical velocity, the scour of settled sediment will occur, and the experimental formulas are used to calculate this process. According to the documents, the Mastbergen and Von den Berg empirical formula is chosen:

$$u_{lift,i} = \alpha_i n_s d_*^{0.3} (\theta_i - \theta_{cr,i})^{1.5} \sqrt{\frac{\|g\| d_{s,i}(\rho_{s,i} - \rho_f)}{\rho_f}} \tag{4}$$

$$d_* = d_{50} \left[\frac{\rho_f(\rho_{s,i} - \rho_f)\|g\|}{\mu^2}\right]^{\frac{1}{3}} \tag{5}$$

Shields number θ_i at the bed surface is calculated according to shear stress τ:

$$\theta_i = \frac{\tau}{\|g\| d_{s,i}(\rho_{s,i} - \rho_f)} \tag{6}$$

The critical Shields number is the crucial parameter to accurately simulate scour and initiation [9], which is commonly determined by experiments. If lack of experimental data, the Shields-Rouse equation is used to calculate the critical Shields number for the particles larger than 170 μm.

The large size particles rolling or jumping along the sedimentary surface and forms the bed-load transport. And it is calculated by Meyer, Peter and Muller equation, Φ_i is the bed-load transport rate:

$$\Phi_i = \beta_i(\theta_i - \theta''_{cr,i})^{1.5} \tag{7}$$

4 Comparison of the Model with Camp

4.1 Introduction

The Camp model calculations was published by Miedema [5]. In order to compare with this typical model, a rectangular hopper used whose size is $79.2 \times 22.4 \times 12.2$ m, and the capacity is about 21579 m^3. D_{50} of the sand is 0.4 mm. Table 1 shows the simulation case parameters.

Table 1. Loading process parameters

Operating parameters	Flow [m^3/sec]	Inflow density [kg/m^3]	Initial water level [m]
Value	14	1300	11.2

4.2   Comparison

According to the results in Fig. 2, the optimal loading time is suggested as about 6200 s. At this time, the total load of the dredger is about 32000t, and the overflow loss is about 10150t. The loading curve and the overflow loss curve is shown in Fig. 2. These results are basically consistent with the Camp model results [5].

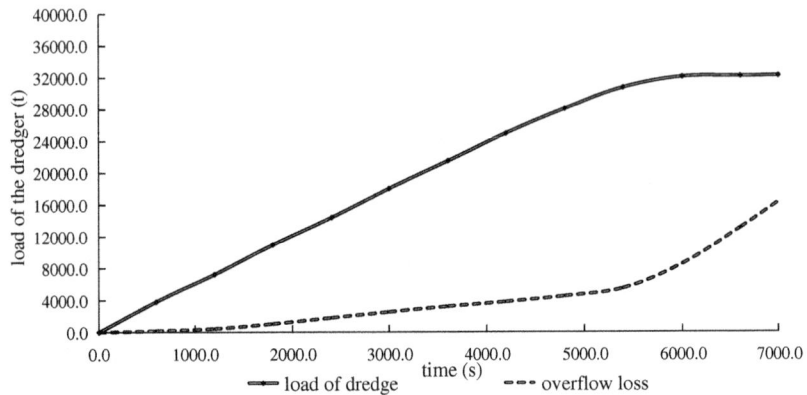

Fig. 2. The calculation results of load and overflow loss

5   Simulation of Prototype Loading Process

A three-dimensional full-scale numerical model of TSHD XinHaihu4 was presented based on the numerical model. To test the model in a prototype situation, the loading data in Dafeng waterway dredging project was measured onboard the TSHD Xinhaihu4. The measuring time is 1200 s, and the measuring date is Aug 23 2016, the dredging area is in JiangSu province China.

The inflow sediment concentration and flow rate mentioned in Table 2 are the mean values of the measured data. The calculation results is shown in Fig. 3. The calculation results were compared with the measured data of total load and dry soild (sediment finally loaded in the hopper). The simulation results are basically consistent with the measured data of the loading process of XinHaihu4.

Table 2. Prototype loading process simulation parameters

Operating parameters	Flow [m³/sec]	Inflow density [kg/m³]	Overflow height [m]	Initial water level [m]	d_{50} [mm]
Value	10.2	1185	11.4	8.2	0.17

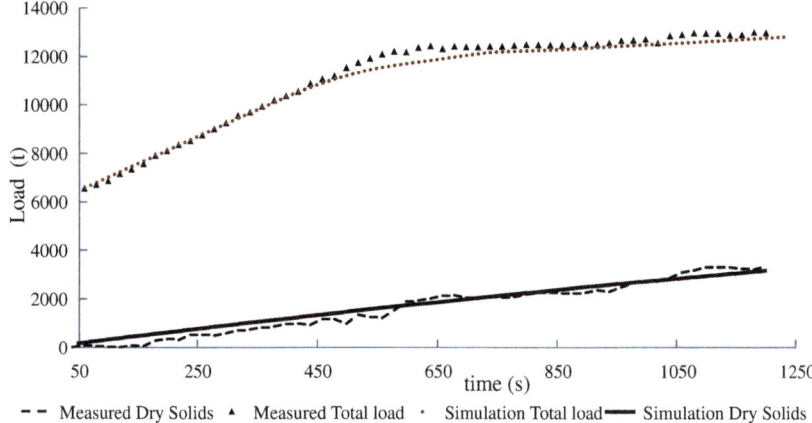

Fig. 3. Loading process verification of XinHaihu 4

Figure 4 shows the flow field in hopper. The inflow of XinHaihu 4 is simulated by this three-dimensional model, combined with the simulations of triangular chamber and bulkhead, and the flow field distribution in the hopper is simulated in detail.

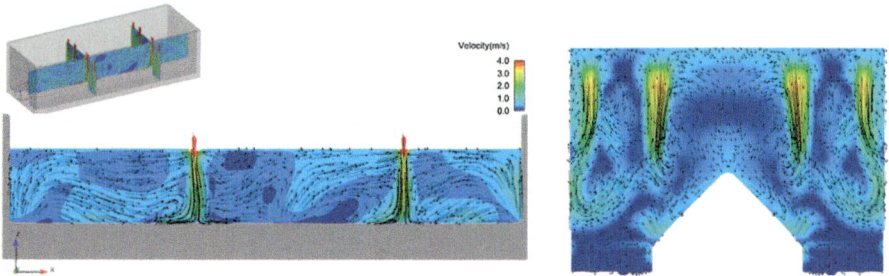

Fig. 4. The flow field distribution of loading process

The inflow section is set up with bi-directional inflow ports, which results in the formation of segmental circumfluence lengthwise in the hopper. Especially in the middle of the two sets of inflow ports, two circumfluences are formed, and the inflow energy can be effectively eliminated, which is beneficial for sand depositing rapidly.

Figure 5 shows the sediment deposition simulation in hopper of XinHaihu 4. The sediment settled in the hopper bottom is not much during the loading of 1200 s, most of the sediment loaded was suspended in the hopper.

Based on the prototype verification, a case with 4000 s loading time was simulated using the model of XinHaihu4, the loading process was analyzed to estimate the production and total overflow losses:

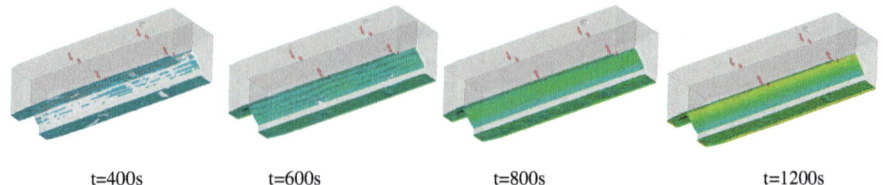

| t=400s | t=600s | t=800s | t=1200s |

Fig. 5. Local sedimentation in the loading process of XinHaihu 4

According to the calculation results, the overflow begins at about 370 s, and the overflow sediment concentration increases gradually, the growth rate (the slope of overflow sediment curve) significantly increases after about 750 s; the sediment content reaches 180 kg/m^3 at 1350 s, and then reaches about $280 \sim 290$ kg/m^3 at 2800 s, and the inflow sediment content is 300 kg/m^3. In Figs. 6 and 7, for loading process after about 2700 s $\sim$ 3000 s, the slope of the overflow sediment content curve becomes small, it means that the growth rate of overflow loss decreased, and the sediment content in overflow was about $280 \sim 290$ kg/m^3. According to the total load curve and accumulative overflow loss curve, after about 2700 s $\sim$ 3000 s, the total loading growth is limited, but the overflow loss increased significantly, indicating that the loading efficiency is low.

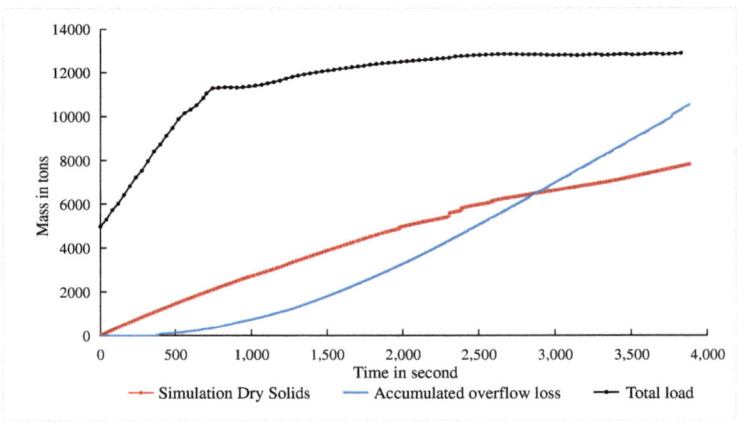

Fig. 6. Simulation result for loading process

The simulation and analysis for loading process give a suggestion of loading time. In TSHD dredging engineering, the reasonable loading time should be chosen by taking into account of actual dredging progress, the position of spoil site and environment requirements.

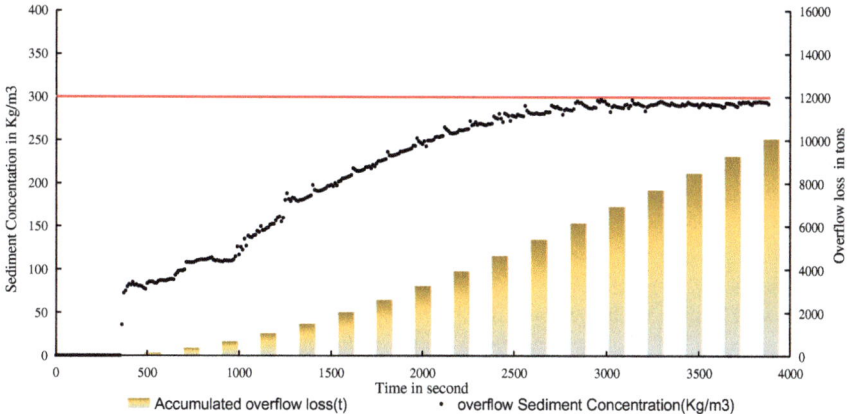

Fig. 7. Overflow loss calculation for loading process

6   Conclusions and Suggestions

(1) In this paper, based on Navier-Stokes equations and non-cohesive sediment transport empirical formulas, a three-dimensional numerical model was presented to simulate the loading and overflow process of TSHD. The model includes all important phenomena of the loading process.

(2) The model were compared with the typical Camp model, the two models give the same magnitude for total load and overflow loss.

(3) A full-scale numerical model of TSHD XinHaihu4 was presented, the model was verified in a prototype by comparing with the measurements which were carried out onboard XinHaihu4 in DaFeng Waterway dredging project.

(4) Based on the prototype verification, a case with 4000 s loading time was simulated using the model of XinHaihu4, the loading process was analyzed to estimate the production and total overflow losses according to the results, and suggestions about reasonable loading time are put forward to improve the loading efficiency.

(5) In order to save the calculation time of loading and overflow, the unimportant detailed geometry of the hopper is simplified. The inflow conditions and particle composition are also simplified. In the future study, it is suggested that further research of sediment model should be studied and relevant research should be carried out with full particle size groups.

(6) In this paper, we mainly study the loading process of non-cohesive sediment. Due to the limited calculation ability and time, the interaction between particles is simulated by parameter, and overflow loss is predicted by this model. For the future research on sediment simulation, the physical model experiments are proposed for sediment parameters test.

Acknowledgments. This work is supported by the National Key Research and Development Plan "Deepwater Channel Coordinated Regulation and Sedimentation Reduction Technology and Demonstration affected by runoff and tidal current" (Project No. 2016YFC0402107).

References

1. Van Rhee, C.: Modelling the sedimentation process in a trailing suction hopper dredger. Terra et Aqua, 86 (2002)
2. Camp, T.R.: Study of rational design of settling tanks. Sen. Works J. 742 (1936)
3. Miedema, S.A., Vlasblom, W.J.: Theory for Hopper Sedimentation. In: 29th Annual Texas A&M Dredging Seminar, New Orleans, (1996)
4. Ooijens, S.C.: Adding dynamics to the camp model for the calculation of overflow losses. Terra et Aqua **76**, 12–21 (1999)
5. Miedema, S.A., Van Rhee, C.: A Sensitivity Analysis on the effects of dimensions and geometry of trailing suction hopper dredges. Wodcon Orlando, USA (2007)
6. Miedema, S.A.: An analytical approach to the sedimentation progress in trailing suction hopper dredgers. Terra et Aqua **112**, 15–25 (2008)
7. Miedema, S.A.: An Analytical Method to Determine Scour. WEDA XXVIII & A&M 39, St. Louis, USA (2008)
8. Miedema, S.A.: The effect of the bed rise velocity on the sedimentation process in hopper dredges. J. Dredg. Eng. **10**(1), 10–31 (2009)
9. Vanoni V.A.: Sedimentation engineering. The Society (1975)

Novel Methods for Characterizing Pipe Walking Behaviours in Situ in Deep Water

Yue Yan[(✉)]

State Key Lab of Hydraulic Engineering Simulation and Safety,
Tianjin University, Tianjin, China
yueyan_geo@126.com

Abstract. This paper presents an overview of the most recent developments in geotechnical analyses with respect to the axial pipe 'walking' behaviour in deep water. This paper has followed a broad review of the offshore geotechnics as the current industry practice – OTC papers in 2006 and SUT paper in 2007, and state of the art paper at the 1st and 2nd Int. Symp. On Frontiers in Offshore Geotechnics. This paper has emphasized the major advances made in 2008 to 2016, with literatures emerged in 2009 to 2015. Novel framework to integrating data from multiple equipment and techniques during the project design phase are reviewed, and also the operating feedback. The tests results calibrated from these sophisticated modelling are presented. Physical modelling and analysis tools have advanced in parallel. It has now been able to replicate the interface shear behaviour in terms of effective stress at the low levels of consolidation stresses; and the implemented critical state theory can reveal the complex consolidation behaviours of an advancing pipe. The paper provides a basis for integrating the characterization data for the potential new site investigation tool, identifying the key areas for the ongoing research.

Keywords: Pipe walking · New equipment · Design methods
Axial pipe-soil interactions · Physical modeling · Analyses tools

1 Introduction

Over the last few years, the SAFEBUCK JIP has been addressing the axial pipe-soil interactions, and has collated a large range of data and identified the key parameters. These investigations have included in-situ measurement of soil properties [1], supplemented by two campaigns using Fugro's SMARTPIPE®; onshore model testing was undertaken at the Norwegian Geotechnical Institute (NGI) and laboratory tests involving standard and advanced low stress interface shear testing carried out at the University of Texas [2, 3], Cambridge University [4, 5], and UWA [6]. This large database has shown that the axial resistance, expressed as a proportion of the submerged pipe weight, can vary by an order of magnitude, depending on the rate of axial movement and cumulative time. The fundamental properties of the pipe-soil interface in terms of an effective stress strength envelope or the current undrained strength of the soil beneath the pipe have been omitted [7].

© Springer Nature Singapore Pte Ltd. 2018
T. Qiu et al. (Eds.): GSIC 2018, *Proceedings of GeoShanghai 2018 International Conference: Advances in Soil Dynamics and Foundation Engineering*, pp. 783–791, 2018.
https://doi.org/10.1007/978-981-13-0131-5_85

An important research study that has filled these voids has been conducted by White et al. [8], who proposed a robust framework for assessing the variation of friction factor. The framework uses concepts from critical state soil mechanics [9] to describe the transition of strength from undrained to drained behaviour, and these strength are also linked to the roughness and stress level effects. Hill et al. [1] has taken this framework a step further and provided the new assessment through the database that can synthesize these results into design recommendations. This opens the potential to confidently extrapolate from the particular conditions in the lab and field tests to the particular conditions relevant for the design of each pipe. Randolph et al. [10] have cast the variation of the axial friction factor in a more rigorous framework, and have proposed a theoretical basis to capture the various aspects of pipeline-soil interaction through analytical expressions. The model is based on the 'critical state' pore pressure generation model, while continuous shearing response incorporates excess pore pressure generation due to damage. Rate effects resulting in strength enhancement have been accounted for throughout the whole operation stages. This was the first time for this t-z framework to be constructed, providing insight into how the pipe response is affected by key parameters such as soil consolidation characteristics, and the pipe diameter relative to the sliding rate.

This paper provides a review of the development and associated challenges for the assessment of axial pipe-soil interaction. It discusses the most recent reports focusing on the newly developed axial pipe-soil framework, collating the data from a range of testing techniques for axial pipe-soil interaction. It provides a basis for integrating the characterization data for the potential new site investigation tool, identifying the key areas for the future research.

2 New Framework Interpretation of the Axial Sliding Response

2.1 New Framework for Axial Pipe-Soil Interaction

White et al. [6] presented a new design framework for axial pipe-soil interaction based on critical state soil mechanics, and encompassing the influences of stress level, pipeline roughness and shearing rate, including drainage and viscous effects. Different component of the design approach can be linked to fundamental soil properties that can be determined from element tests, as well as from model pipe tests in the laboratory or in situ. This section discusses in more details the experimental tests involved in quantifying components of the design framework.

The framework is illustrated schematically in Fig. 1, showing (1) a curved effective stress failure envelope, with a viscous effect, (2) a drained-undrained transition, associated with pore pressure generation, (3) an interface 'efficiency' linked to roughness and (4) a 'wedging effect due to curved pipe surface.

The fundamental strength criterion of the pipe-soil interface is an effective stress failure envelope. This is curved at low stresses leading to reducing drained friction factors with increasing stress level (illustrated by the red plane of Fig. 1). The results from either tilt table tests or shear box tests with very slow sliding velocity have been shown in Figs. 2a and b [6, 7], illustrating the curvature of the failure envelope for each soil type.

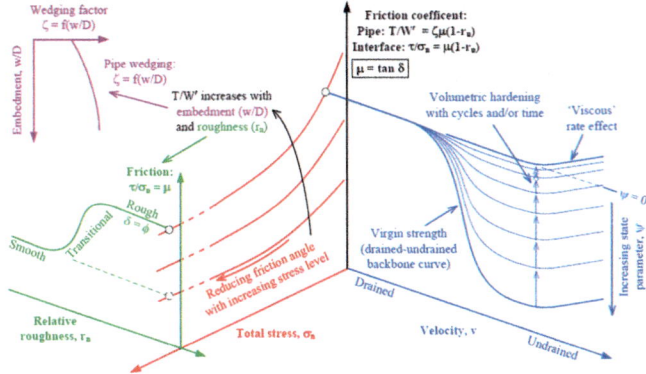

Fig. 1. Mechanisms affecting axial pipe-soil interaction [6]

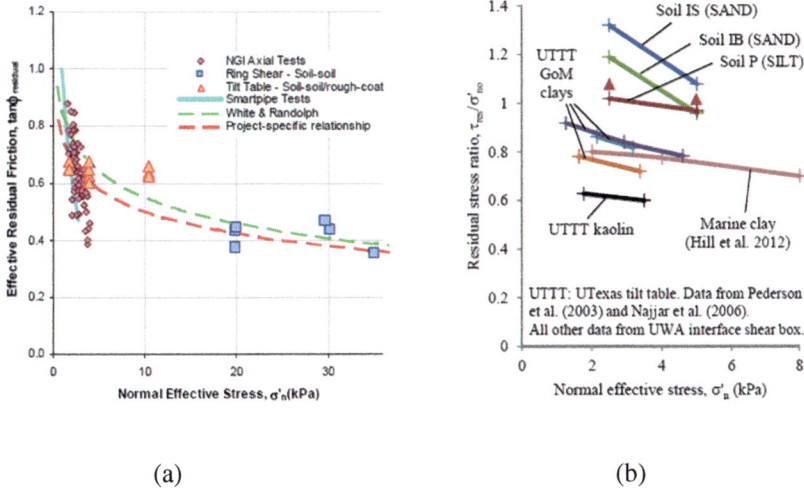

Fig. 2. Variation in drained strength with effective stress level: (a) collated by Bruton et al. [7]; (b) White et al. [6].

At faster rates of shearing, partially drained conditions prevail, and positive or negative excess pore pressure will be generated, depending on the OCR value of the soil (as illustrated in the blue curve in Fig. 1). Both dilatant and contractile undrained stress path are shown, relative to a drained path, with an allowance for the effect of the interface roughness (green lines). The corresponding variation in resistance with sliding velocity is evident, for soils with varying over-consolidation ratios (OCR) and varying degrees of contractancy. These S-shaped profiles can be fitted by a logarithmic sigmoid function [11]. Higher shearing rates also lead to viscous effects, which can be fitted by a hyperbolic sine function [12]. This behaviour has been investigated on the set of interface shear box testing, as shown in Fig. 3.

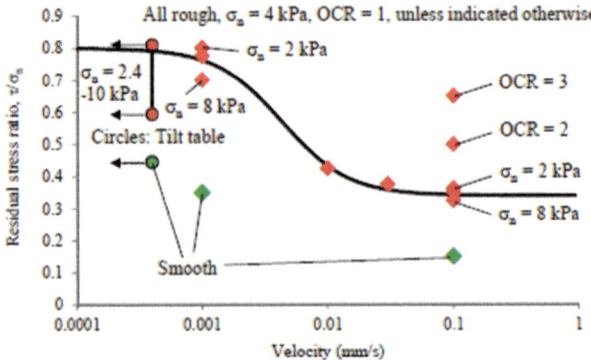

Fig. 3. Effect of velocity on interface shear strength (Hill et al. [1])

The roughness of the interface affects the sliding resistance by two mechanisms. First, the roughness influences the interface friction angle, and therefore the effective strength failure envelope (as shown in the green line in Fig. 1). Previous researchers [13, 14] have found that the interface friction varied with the surface roughness up to a critical value, which indicates that the failure will occur at whichever circumferential surface offers the least resistance. For the smooth pipe, failure may occur on the interface where pipe-soil shearing is easier than soil-soil shearing. Other published data has shown a correlation for the efficiency of the interface – proportion of the internal soil strength that is mobilised – with the relative roughness, $r_n = R_a/D_{50}$ [15], which is evident across the full range of particles from clay-sized to gravel sized. However, significant scatter is evident among different soils – perhaps due to the particle shape and grading effects, so site specific testing is essential for design. Secondly, the interface roughness also affects the level of excess pore pressure generated during sliding, as illustrated by data from shear box tests that incorporated pore pressure measurement [14].

The last parameter that will controlled the shearing behaviour is the 'wedging effect' (as shown in the purple plane in Fig. 1), which quantifies the higher total normal force around the pipe surface compared to the pipe submerged weight. Elastic solutions [16, 17] have shown that the wedging effect can create a 10–20% enhancement of the axial resistance for typical pipeline embedment.

2.2   Modelling the Axial Soil Resistance on Deep Water Pipelines

Randolph et al. [10] have presented an analytical framework for quantifying the magnitude of limiting drained and 'rate-dependent' undrained frictional resistance which underpins the proposed axial friction model.

The framework has been couched in terms of effective stresses, using critical state concepts to quantify the magnitude of excess pore pressure generated locally around the pipeline. Solutions for consolidation are introduced to quantify the duration of excess pore pressure. Two additional aspects of the behavior are added to match the observed velocity dependency of axial resistance, that is the (a) a damage term, leading to contractive volumetric strain at the interface, which will lead to reduction in friction as the velocity is increased due to the generation of excess pore pressure at the interface; and (b) strain rate dependency of the mobilised soil strength, which will give increasing friction with increasing velocity. The resulting variation of the normalised friction resistance has been validated using data from interface shear box testing, representing a planar idealisation of the same behavior, and from model pipe tests.

This approach captured the overall pipeline response incorporating the different facets of behavior in Fig. 4 for a range of different normalised velocity. The initial shearing response is based on the critical state pore pressure generation model, while the continuous shearing response following failure incorporates excess pore pressure generation due to damage. Enhancement by rate effects captures both pre- and post-yield stages, according to the Herschel-Bulkley model. The steady state friction factor ($\tau_{\text{steadystate}}/q$) can be expressed as Eq. 1.

$$\frac{\tau_{\text{steadystate}}}{q} = \mu_y \left[1 + \eta \left(\frac{v/D}{v_{\text{ref}}/D} \right)^{\beta} \right] \left\{ 1 - \frac{1}{1 + 0.24/[(\alpha/\lambda^*)T_{50}vD/c_v]} \right\} \qquad (1)$$

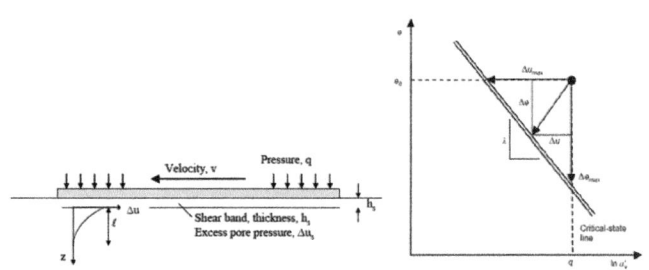

Fig. 4. Intelligent model for axial sliding pipeline (Randolph et al. [10])

Here, this equation has included three components: (1) the friction factor terms – μ_y is the yield stress (at close to zero strain rate), while at the reference shear strain rate, the effective friction ratio is $\mu_y(1 + \eta)$; T_{50} is the time factor at which the friction resistance is midway between the undrained and drained limits;

(2) the damage effects terms – α is proportional constant between the potential volumetric strain rate, and $\lambda^* = \lambda/(1 + e_0)$, the range of vD/c_v corresponds to ratios of v/v_{ref} from 2.5 down to 0.025.

(3) the strain rate terms – the normalised velocity, v/D represents the nominal shear strain rate, typical exponent value β lies in the range 0.05 to 0.15, with a tendency to increase with plasticity index. Here, v_{ref}/D has been taken as 10^{-3} s^{-1}, and the reference shear strain rate is 20%/hour (which is typical for a laboratory simple shear test).

The rate of excess pore pressure generation, for different sliding velocities follows a common response for the mobilization stage, was modeled as a hyperbolic response. During the post-yield stage they show an initial rise in resistance due to consolidation effect have compensated the new generated excess pore pressure before reducing to a steady state resistance according to the balance between the rates of pore pressure generation and dissipation.

To calibrate the proposed model, some parameter fitting are made in the back analysis of model pipe testing performed on soft marine clay reported by White et al. [8], as shown in Table 1 and Fig. 5. The back analysis has focused on the 15 sweeps of the total 65 sweeps that were performed after a pause period sufficient for full dissipation of the excess pore pressure from previous sweeps.

Table 1. Parameters used in back-analysis of model tests [10]

Back-analyses cases				
Parameters	A	B	C	D
c_v (m^2/year)	10			
Viscous parameter	$\eta = 0.25$, $\beta = 0.1$			
Friction parameter	$\mu_y(1 + \eta) = 0.7$			
Mobilization model	$R_f = 0.95$, $G_0/\tau_f = 30$			
Sliding consolidation model	$T_{50} = 0.05$, $m = 0.5$			
Soil compressibility	0.1			
$\Delta u_{max}/q$	0.5	0.1		
α	0.005	0.01		0.03
Pipe invert embedment	0.11	0.12	0.17	0.24

The curve fitting shows a reasonable fitting with the experimental data, and variations in the input parameter shows the soil state changes during the testing: (1) The critical state parameter $\Delta u_{max}/q$ is decreasing with preceding failure and reconsolidation cycles, which suggests the adjustment reflect the soil state moving closer to the critical state after cycles of failure and reconsolidation; (2) the change in fitting damage parameter corresponds to the variation of over consolidation when the pipe weight was increased by a factor of 2.65, and reduced back to the original weight.

This back analysis shows the advantage of proposed model and some complexities of the real behavior. It could be understood that repeated episodes of failure and reconsolidation leads to changes in soil state, and the resulting tendency for pore pressure generation. This behavior is observed in penetrometer tests [18] and soil element tests [19], and is usually tackled in practice by quantifying changes in strength based on trends observed in a suite of element tests – with varying degrees of drained pre-shearing for example [20]. The present model captures the equivalent behavior during interface shearing using parameters that can be similarly tuned, and the model could potentially be extended to include direct calculation of the changing soil state and damage potential as consolidation occurs.

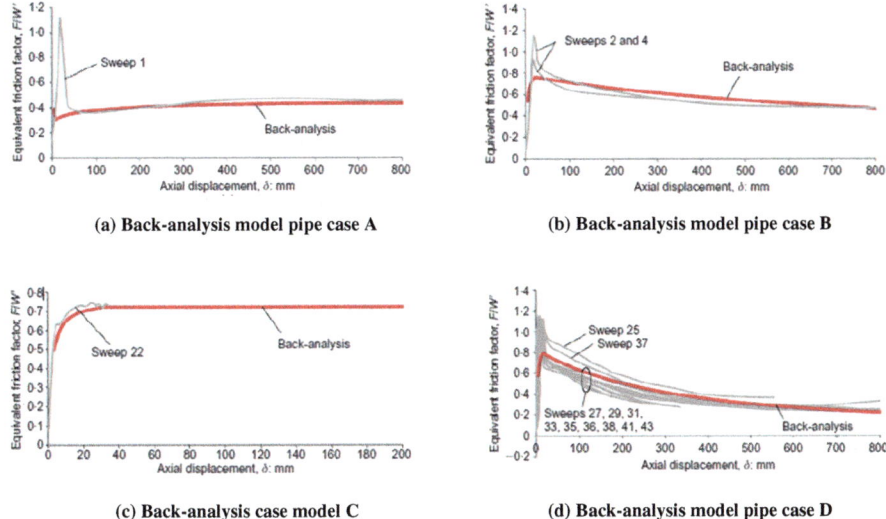

(a) Back-analysis model pipe case A

(b) Back-analysis model pipe case B

(c) Back-analysis case model C

(d) Back-analysis model pipe case D

Fig. 5. Back-analysis of model pipe test on soft marine clay (Randolph et al. [10])

3 Summary and Recommendation for the New Framework Used for Design

This paper has given a comprehensive review of the recent developments in analysis techniques for the axial pipe-soil interaction design. The common theme is the relative magnitude of the drained and undrained soil strength, and the evolutions of these strengths, and importance of recognizing the widely varying rates of shearing involved in the pipe-soil processes. Drawing on concepts from critical state soil mechanics and encompassing the influence of stress level, pipeline roughness and shearing rate, the new analysis framework offers an improvement in design practice.

To best utilize our observations from the different testing, it is necessary to integrate it in a systematic way, with a consistent theoretical framework. This has allowed the different datasets to be cross-checked against each other in a detailed way.

First, it should be understood that velocities and magnitudes of pipe movement during operation event can span undrained and partially-drained regions for many natural marine clays. As a consequence, both drained and undrained conditions and linkage between these should be considered. The lower bound axial resistance (assuming soil is contractile) is usually based on the normally consolidated strength ratio $(s_u/\sigma'_V)_{nc}$. Refinement of this value – ideally upward is a key aim when planning and interpreting soil characterization studies and pipe-soil model tests.

Second, soil remolding will occur during undrained penetration, with partial strength recovered during reconsolidation, so that the initial strength during undrained sliding is generally lower than the in situ strength at the given depth. In model tests, the pre consolidated surcharge pressure may control the undrained sliding resistance, whereas in the field it is more likely to be controlled by the consolidation stresses imposed by the pipe itself.

Third, the two frameworks shown in these two sections have been both based on the critical state framework, which captures the main mechanisms observed in the experiments, including the drained, undrained limiting resistance, wedging, rate effects and effective stress failure envelope with viscous effects. Each component can be identified and isolated, allowing more confident extrapolation of interpretation methods to different soil and pipe conditions.

Fourth, the latter framework has provided a rigorous 't-z' model, which quantifies the **essentially time-varying** friction. These analytical and numerical solutions have led to (a) a closed formed consolidation expression, rather than velocity; and (b) a speculative damage effects that captures observation of 'velocity-dependent' steady state.

Fifth, as the project progresses, two parallel activities are required: (1) **improved analyses:** incorporate in the analysis other effects: wedging, over consolidation, partial or full drainage that will only raise the estimate of the friction; (2) **improved parameter selection:** project specific characterisation data to reduce the soil property uncertainty.

References

1. Hill, A.J., White, D.J., Bruton, D.A.S., Langford, T.E., Meryer, V., Jewell, R.J., Ballard, J.-C.: New datasets and improved practice for practice for assessment of axial pipe-soil interaction. In: Proceedings of International Conference on Offshore Site Investigation and Geotechnics. SUT, London (2012)
2. Najjar, S.S., Gilbert, R.B., Liedtke, E.A., McCarron, B.: Tilt table test for interface shear resistance between flowlines and soils. In: Proceedings of Conference on Offshore Mechanics and Arctic Engineering, Cancun, Mexico (2003)
3. Pedersen, R.C., Olson, R.E., Rauch, A.F.: Shear and interface strength of clay at very low effective stress. Geotech. Test. J. **26**(6), 1–8 (2003)
4. Bolton, M.D., Ganesan, S.A., White, D.J.: Greater Plutonio Development: axial pipe-soil interaction using the Cam-shear device. Cambridge University Technical Services, Report No. SC-CUTS-0605-R00 to Boreas Consultants and BP (confidential) (2006)
5. White, D.J., Randolph, M.F.: Seabed characterisation and models for pipeline-soil interaction. Int. J. Offshore Polar Engng **17**(3), 193–204 (2007)
6. White, D.J., Campbell, M.E., Boylan, N.P., Bransby, M.F.: A new framework for axial pipe-soil interaction illustrated by shear box tests on carbonate soils. In: Proceedings of International Conference on Offshore Site Investigation and Geotechnics. SUT, London (2012)
7. Bruton, D.A.S., White, D.J., Langford, T.E., Hill, A.J.: Techniques for the assessment of pipe-soil interaction forces for future deepwater developments. In: Proceedings of Offshore Technology Conference, Houston, OTC 20096 (2009)
8. White, D.J., Ganesan, S.A., Bolton, M.D., Bruton, D.A.S., Ballard, J.-C., Langford, T.E.: SAFEBUCK JIP - observations of axial pipe-soil interaction from testing on soft natural clays. In: Proceedings of Offshore Technology Conference, Houston, OTC 21249 (2011)
9. Schofield, A., Wroth, C.P.: Critical State Soil Mechanics. Blackie Academic, Glasgow (1968)

10. Randolph, M.F., White, D.J., Yan, Y.: Modelling the axial soil resistance on deep water pipelines. Géotechnique **62**(9), 837–846 (2012)
11. House, A.R., Oliphant, J.R.M.S., Randolph, M.F.: Evaluating the coefficient of consolidation using penetration tests. Int. J. Phys. Model. Geotech. **1**(3), 17–25 (2001)
12. Randolph, M.F., Hope, S.N.: Effect of cone velocity on cone resistance and excess pore pressure. In: Proceedings of the Engineering Practice and Performance of Soft Deposits, Osaka, pp. 147–152 (2004)
13. Uesugi, M., Kishida, H.: Influenctial factors of friction between steel and dry sands. Soils Found. **26**(2), 33–46 (1986)
14. Tsubakihara, Y., Kishida, H., Nishiyama, T.: Frictional behaviour between normally consolidated clay and steel by two direct shear type apparatuses. Soils Found. **33**(2), 1–13 (1993)
15. Subba Rao, K.S., Allam, M.M., Robinson, R.G.: Interfacial friction between sands and solid surfaces. Geotech. Eng. **131**(2), 75–82 (1998). Proceedings of Institution of Civil Engineers
16. Gourvenec, S., White, D.J.: Elastic solutions for consolidation around seabed pipelines. In: Proceedings of Offshore Technology Conference, Houston, OTC 20554 (2010)
17. Krost, K., Gourvenec, S.M., White, D.J.: Consolidation around partially embedded seabed pipelines. Géotechnique **61**(2), 167–173 (2011)
18. White, D.J., Hodder, M.S.: A simple model for the effect on soft strength of episodes of remoulding and reconsolidation. Can. Geotech. J. **47**(7), 821–826 (2010)
19. O'Reilly, M.P., Brown, S.F., Overy, R.F.: Cyclic loading of silty clay with drainage periods. ASCE J. Geotech. Eng. **117**(2), 354–362 (1991)
20. Andersen, K.H.: Bearing capacity under cyclic loading - offshore, along the coast, and on land. Can. Geotech. J. **46**(5), 513–535 (2009). The 21st Bjerrum Lecture presented in Oslo, 23 November 2007

Experimental Set-up of the Toroid and Hemi-ball Penetrometers

Yue Yan[(✉)]

State Key Lab of Hydraulic Engineering Simulation and Safety,
Tianjin University, Tianjin, China
yueyan_geo@126.com

Abstract. The toroid and hemiball penetrometers described in this paper are new site investigation tools that have been developed to measure pipe-soil interaction parameters in situ, at shallow depth in the seabed. This paper reports an experimental program using the new in situ penetrometers, and also a conventional pipe segment, which was conducted at the University of Western Australia. The testing campaign included tests carried out under 1 g and 25 g centrifuge conditions. The 1 g testing were performed on the basis that self-weight effects are essentially negligible for the sliding failure that accompanies axial pipe movements or torsion. The centrifuge toroid testing is considered to provide the most reliable and valuable data to support the deployment of the new penetrometers. The testing was developed to provide data that spans the ranges of toroid/pipe in-service behaviour – including toroid/pipe laying, consolidation after overloading is applied and toroid/pipe axial movement. The focus of the testing program described in this paper was on the further examination of the functionality of the new penetrometer, and to illustrate how much shallow penetrometers could be used for assessing the interaction between pipeline and the seabed.

Keywords: Toroid penetrometer · Hemi-ball penetrometer · 1 g testing
Centrifuge testing · Axial displacement

1 Introduction

1.1 Backgrounds

The toroid and hemi-ball penetrometers described here are new site investigation tools that have been developed to measure pipe-soil interaction parameters in situ, at shallow depth in the seabed. The devices investigated here are scale models of the potential field tools. Each device comprises a model toroid or hemi-ball which can be driven in the vertical direction (for penetration) and then in a torsional mode to assess parameters for axial pipe-soil interaction. An overview of the devices and analysis of their response during vertical penetration and torsion was given in the previous numerical studies [1–4]. This paper reports an experimental programme using the new in situ penetrometers, and also a conventional pipe segment, which was conducted at the University of Western Australia. The testing campaign included tests carried out under 1 g and 25 g centrifuge conditions.

© Springer Nature Singapore Pte Ltd. 2018
T. Qiu et al. (Eds.): GSIC 2018, *Proceedings of GeoShanghai 2018 International Conference: Advances in Soil Dynamics and Foundation Engineering*, pp. 792–801, 2018.
https://doi.org/10.1007/978-981-13-0131-5_86

The tests under 1 g conditions were performed on the basis that self-weight effects are essentially negligible for the sliding failure that accompanies axial pipe movements or torsion. It is possible to conduct meaningful small scale tests outside the centrifuge environment, without incurring significant scaling issues, provided self-weight stresses are unimportant. The basic test logic of 1 g testing programme was to perform the axial sliding (i.e. torsion for the toroid and hemi-ball) tests for the three devices with similar drainage conditions during the period of deployment. The previous numerical investigations assessed the relative shearing velocity, shearing distance and elapsed time for each device to allow direct correlations between the three devices.

The centrifuge toroid testing is considered to provide the most reliable and valuable data to support the deployment of this new penetrometer. In total six toroid tests (referred to as C-T-01–C-T-06) and two complementary pipe tests (C-P-01 and C-P-02) were performed in one strongbox. The centrifuge testing was developed to provide data that spans the ranges of toroid/pipe in-service behaviour – including the toroid/pipe laying, consolidation after overloading is applied and the toroid/pipe axial movement. The focus of the centrifuge testing was on the further examination of the functionality of the new penetrometer, and to illustrate how such shallow penetrometers could be used for assessing the interaction between pipelines and the seabed.

1.2   Aims and Objectives

The purpose of this paper is to describe the experimental techniques for the new toroid and hemi-ball penetrometers under 25 g centrifuge and 1 g laboratory floor conditions. The technical details of the model tests performed are outlined, and the facilities, experimental devices, procedures and program are described.

2   Experimental Facilities

2.1   Centrifuge and 1 g Modelling

The technique of geotechnical centrifuge testing is a well-established modelling method which reliably models the behaviour of geotechnical materials at small scale. Comprehensive descriptions of centrifuge modelling and scaling principles are detailed in Taylor [5] and Wood [6]. The self-weight of the soil is scaled up by the ratio of centrifuge acceleration to Earth's gravity, allowing for the simulation of realistic stress levels, hence the strength and stiffness of soil. The centrifuge testing therefore faithfully replicates the correct modelling of strength ratio, $su/\gamma'D$ to reliably assess soil mechanisms, particularly where self-weight effects are important. Model testing on the laboratory floor at standard gravity (1 g) may be misleading for problems such as uplift of foundations or anchors where the self-weight stresses are important.

The self-weight effects may be ignored during axial (or torsional) shearing of pipe segments or the model penetrometers. As such, laboratory floor tests at 1 g were also undertaken. The equipment and data acquisition systems for the 1 g testing are not described separately, since they were identical to those for the 25 g centrifuge testing.

A complete description of the UWA beam centrifuge, as commissioned in 1989, can be found in Randolph et al. [7]. The centrifuge is an Accutronic Model 661 geotechnical centrifuge. It has a swinging platform radius of 1.8 m with a nominal working radius (R_e) of 1.55 m, and is rated at a capacity of 40 g-tonnes (which equates to the testing packages of a maximum payload of 200 kg tested at an acceleration level of 200 g).

2.2   Robotic Actuation and Control

The testing described in this chapter involves manipulating of a pipe segment, using a 2-directional actuator; and a toroid or hemi-ball penetrometer using the rotary actuator clamped within the existing 2-direction actuator [8]. The actuator is mounted on top of the strongbox, in which the soil sample is contained, and allows vertical, axial (horizontal for the pipe and torsional for hemi-ball/toroid) movements to be imposed on the pipe or penetrometer. An overview of the UWA actuators is shown in Fig. 1.

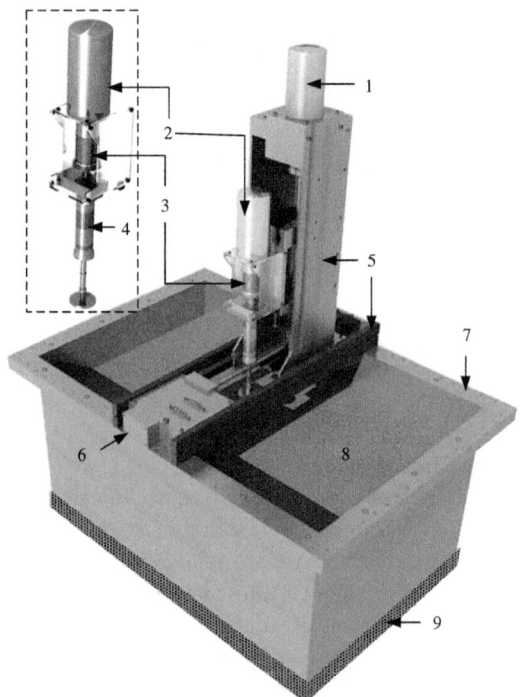

Fig. 1. Rotary actuator at 1 g testing trolley (1-Encoder, 2-Motor, gearbox & tachometer, 3-Miniature slip-ring, 4-Lead shaft, 5-Encoder, 6-Motor & Encoder, 7-strongbox, 8-clay sample, 9-drainage layer)

The vertical and horizontal motion is provided by two DC servo-motors driving vertical and horizontal leadscrews. The actuator can travel at a displacement rate up to

3.1 mm/s over strokes of 260 mm and 180 mm in the vertical and horizontal directions respectively. The torsional motion control is provided by a DC servomotor, running through a 1:100 gear ratio to provide up to 10 Nm of torque at speeds ranging from 0.2 deg/s to 360 deg/s.

The actuators are controlled by software written in-house at UWA using a Labview interface [9]. This primary software runs on an in-flight computer mounted on the centrifuge, and serves as a server to a master computer in the centrifuge control room, with communication via an Ethernet link to the in-flight PC. The software is operated by the operator in the control room via the Remote Desktop link to the in-flight computer. Load- and displacement-controlled motions are achieved by means of a software feedback loop using real time output feedback. Complex loading sequences, including varying control modes on various axes, have now become feasible with the development of the sophisticated motion control systems (PACS) [9].

2.3 Data Acquisition

The UWA geotechnical laboratory has been equipped with a novel high-speed wireless data acquisition system (WDAS) developed at UWA [10]. The WDAS system consists of up to 8 separate miniature units mounted on the centrifuge baskets, communicating with the control room via the Ethernet link going through the optic fibre slip ring. Throughout each test, the following data were recorded: time, centrifuge acceleration level, vertical pipe/toroid/hemi-ball displacement, toroid/hemi-ball torsional or pipe axial displacement, vertical force, torsional force on toroid/hemi-ball or axial force on pipe, and PPT readings.

2.4 Instrumented Loading Arm

Two loading legs were used in the experiments (Fig. 2).

1. Aluminum loading arm that fits into the rotary actuator: The loading arm was designed and manufactured by the workshop at UWA, and was initially used for the testing of combined loading on a jack-up footing on sands [11]. The loading arm consists of axial, bending, and torsion strain gauges. The loading leg is placed in a loading-rig frame that allows the leg to be driven in the rotational, horizontal and vertical directions independently. The outer shaft is fixed to the motor head piece and prevented from twisting, while the inner shaft, which extends down to where the penetrometer is attached, is allowed to twist.

2. Aluminum loading arm that fits into the two-directional actuator: The loading arm is instrumented with bending strain gauges at two locations. Using an electrical comparator, these gauges indicate the horizontal load acting on the pipe (parallel to the orientations of the strain gauges) attached to the loading arm.

   In the 1 g testing, both loading arms were equipped with an S-shaped load cell to measure the vertical force during penetration and subsequent axial or torsional shearing. It was designed and manufactured by the UWA workshop technicians so

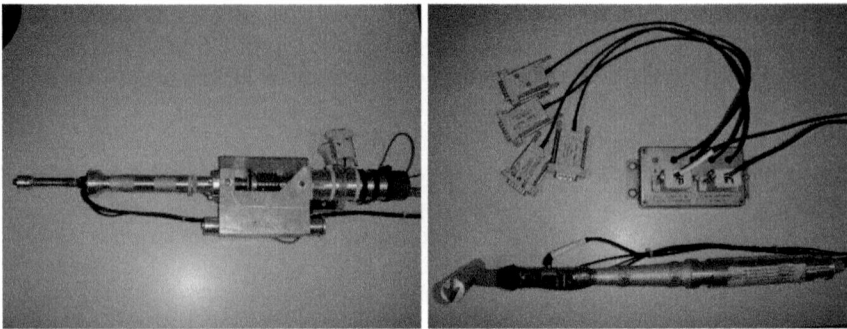

Fig. 2. Loading leg

as to ensure a low sensitivity to bending loads. A specific bending compensation unit (BCU) was used to eliminate the bending induced cross talk on the vertical loading signal. The new coupled loading arm ensured vertical load remained constant at a low magnitude (10–20 N) when the pipe or penetrometer was translated axially.

2.5   Model Toroid, Hemi-Ball and Pipe

The model instrumented toroid and hemi-ball (Fig. 3) used in the study were manufactured from aluminum, with an outside diameter (d) of 80 mm, and a lever arm (L) of 32 mm for the toroid, and a wall thickness (t) of 5.5 mm for the hemi-ball. The model pipe was also manufactured from aluminum, with a diameter of 20 mm, and a length of 120 mm. The surface was chosen to be relatively smooth, which was achieved by leaving the aluminum surface untouched.

For the centrifuge testing with an acceleration of 25 g, the model represents a prototype toroid of 0.4 m in diameter, and a prototype pipe of 0.5 m in diameter. The toroid and pipe dimensions are summarized in Table 1, at both model and prototype scales.

Fig. 3. Model toroid, hemi-ball and pipe in 1 g deployment condition

Table 1. Toroid, pipe dimension (model/prototype scale at 25 g)

Parameters	Toroid	Pipe
Lever arm, L (mm)	32/800	–
Diameter, L (mm)	16/400	20/500
Penetration velocity (mm/s)	0.1	0.125
Range of interface shearing rates* (expressed as v_{axial}, mm/s)	$v_{axial} = \omega L$ 0.001–1^	v_{axial} 0.01–1^
Interface shearing length (mm)	±100/±2500	±30/±750

*Varied during tests. ω: angular velocity (radian/s).
^Velocity shown only in model scale units.

3 Preparation of Clay Specimens

3.1 Preparation of Kaolin Slurry

The clay samples were prepared from the standard kaolin clay used extensively at UWA, and characterised by Stewart [12]. The kaolin clay powder was mixed to homogeneous slurry in a conventional barrel mixer at a water content of 120% (twice the liquid limit). A thorough mixing period normally took twelve hours to complete, with periodic intervention to monitor progress and manually break down the large clogs. The strongbox was prepared with a sand drain at the base (around 20 mm thick), overlain by a permeable drainage mat. The soil slurry was then poured manually into the strongbox, resulting in a slurry sample 270 mm high.

3.2 Preparation of Kaolin Clay for 1 g and Centrifuge Spinning

1 g-Strongbox A and B: For the two samples prepared for 1 g testing (Box B and Box C) the slurry was consolidated under a consolidation press. The target pressure of 50 kPa was applied to the top of the sample following the initial one day consolidation under 10 kPa. The final consolidation pressure was held for 8 days to minimise subsequent creep and secondary compression effects. The sample was then relaxed to atmosphere pressure under an additional 70 mm head of surface water, producing a backpressure for negative increments in pore pressure to prevent cavitation during swelling of the soil sample. The soil sample was maintained under ambient atmospheric pressure for 5 days for complete swelling, before being moved to the 1 g trolley area for testing.

C-Strongbox A: The slurry was first consolidated at 1 g under a consolidation press following an identical procedure as for Box A. The sample was then moved to the centrifuge and reconsolidated at a centrifuge acceleration of 100 g over 70 h. The test series commenced after five hours following reduction of the centrifuge acceleration to the test level of 25 g.

4 Testing Program

4.1 Testing Program in 1 g Conditions

The testing procedure for both 1 g and high g conditions follows the procedure shown in Table 2.

Table 2. Overview of 1 g and high testing procedure

Parameters	Description	Details
Embedment	Static device embedment	w/D = 0.25 during 1 g testing
Overloading	Overloading applied to $\sigma_{n0} = V/A_{nom} = 4$ kPa	Overloaded to the specified loading, and wait for consolidation
Consolidation	Yes	1 g-Strongbox A; C-Strongbox A
	No	1 g-Strongbox B
Episodes of axial sliding + reconsolidation	Backbone curve – harden from slowest value upwards	1 g-Strongbox A; C-Strongbox A 0.01–1^
	Backbone curve – harden from fastest value downwards	1 g-Strongbox B
Uplift	Static device uplift	

$A_{nom} = 2\pi LD$ for toroid, $A_{nom} = L_0 D$ for pipe, and $A_{nom} = \pi D^2/4$ for ball

Two toroid tests, two hemi-ball tests and two pipe tests were performed in Box A and Box B at 1 g. The testing involved undrained laying and varying scenarios of axial movements under the same normalised device weights V/A_{nom} of 4 kPa. During these tests, the testing velocity, displacement and intermittent reconsolidation time were carefully chosen so that good agreement of drainage conditions for three devices could be achieved. These tests may be grouped into two categories as follows, featuring the two different sequences of cyclic motions.

- Increasing velocity scenario (3 tests) with intermittent reconsolidation: These are labelled as 1 g-A-01, with A being the denotation for device (T for toroid, B for hemi-ball; and P for pipe).
- Decreasing velocity scenario (3 tests) with no intermittent reconsolidation: These are labelled as 1 g-A-02, with A being the denotation for device (T for toroid, B for hemi-ball; and P for pipe).

Following the justification described in Yan (2013), the specified shearing time (t), velocity (Ω, v) and distance (ϕ, s) for a ball (with $D_{ball} = 80$ mm, $r_{eff} = 23.1$ mm), and a toroid (with $D_{toroid} = 16$ mm) and a pipe (with $D_{pipe} = 20$ mm), for a start embedment ratio w/D of 0.25, shall be related in the following proportions, to mobilize comparable levels of consolidation:

$$\frac{t_{toroid}}{t_{ball}} = \frac{16t_{pipe}}{25t_{ball}} = 0.2 \tag{1}$$

$$\frac{\Omega_{toroid}}{\Omega_{ball}} = 0.25$$

$$\frac{\Omega_{toroid}L}{v_{pipe}} = 1.25$$

$$\frac{\phi_{toroid}}{\phi_{ball}} = 0.05$$

$$\frac{\phi_{toroid}L}{s_{ball}} = 0.8$$

4.2 Testing Program in Centrifuge Conditions

A total of six toroid tests and two cyclic axially-sliding pipe tests were performed in Box C at an acceleration level of 25 g. These are labelled as C-T-n (for toroid) and C-P-n (for pipe), with n being the number of the test. Each test involved cyclic axial movement of the model device (toroid or pipe) under varying device weights (i.e., applied vertical loads).

These tests included undrained pipe/toroid laying, overloading the device to represent a realistic pipeline operating weight, and consolidation followed by large longitudinal cyclic motion. Different values of weights after overloading were used during the test program to simulate light and medium weight pipes. These correspond to operative weights of submerged pipe/toroid bearing pressures, V/A_{nom} of 1.5–6 kPa.

5 Interpretation Methods of Model Test Results

5.1 Data Correction

Prior to all the toroid tests, the toroid was subjected to a cyclic torsional movement under displacement control before it touched the soil sample surface but whilst suspended in the water and with all sensors wired up. This indicated the modest torque correction due to the drag effect from the trailing wires of the pore pressure sensors. It also provided a correction for the pore pressure transducers (termed as "PPT") readings that arises from the curved water surface created by the off-centre position within the centrifuge during torsional movements (as illustrated in Fig. 4).

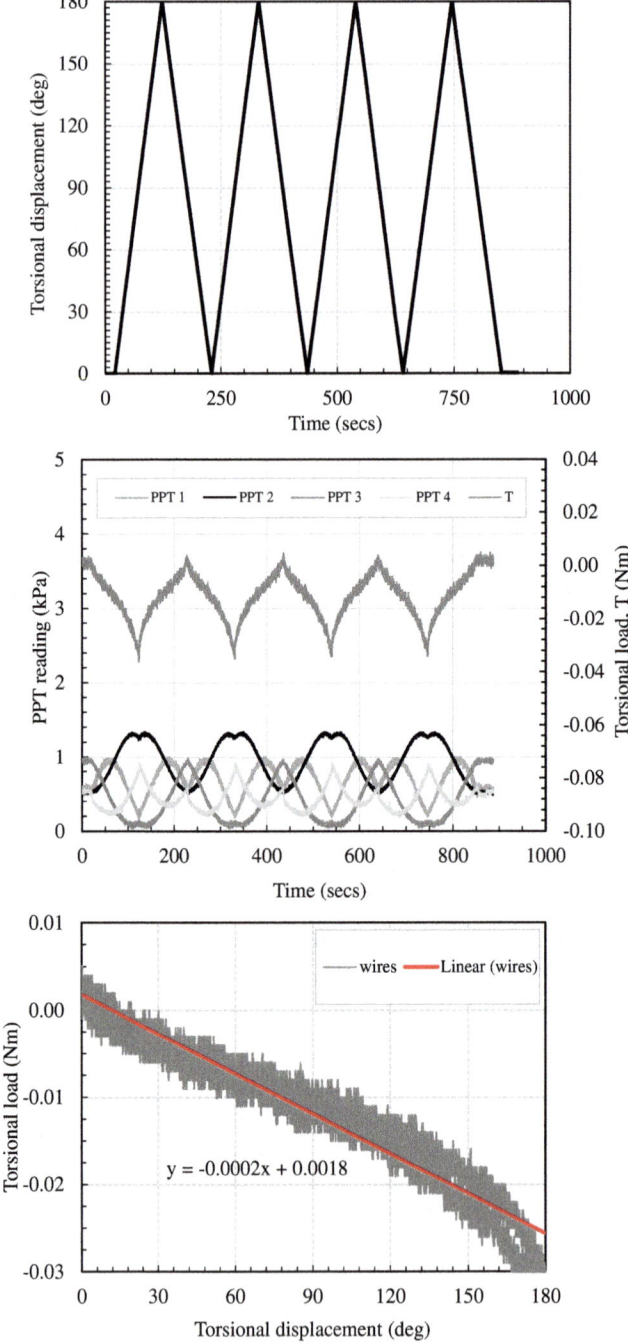

Fig. 4. Example PPT and wire corrections: C-T-02

5.2 Presentation of Results

Full interpretation of the testing is provided in the separate paper. The results are presented either at prototype scale or in a dimensionless form, including interpretation in the following forms:

1. Toroid/pipe/hemi-ball trajectory: settlement during device axial displacement.
2. Time history of vertical and axial displacements in prototype units.
3. Axial resistance – displacement curves in normalised dimensions.
4. Time history of excess pore pressures: generation of excess pore pressure during shearing and its subsequent dissipation.

These data are further interpreted in an effective stress framework. The processes of undrained penetration and equalisation are compared with results from the coupled FE analyses presented in Yan [4].

References

1. Yan, Y., White, D.J., Randolph, M.F.: Investigation into novel shallow penetrometers for fine-grained soils. In: Proceedings of 2nd International Symposium on Frontiers in Offshore Geotechnics, pp. 321–326 (2010)
2. Yan, Y., White, D.J., Randolph, M.F.: Penetration resistance and stiffness factors for hemispherical and toroidal penetrometers in uniform clay. Int. J. Geomech. ASCE **11**(4), 263–275 (2011)
3. Yan, Y., White, D.J., Randolph, M.F.: Cyclic consolidation and axial friction for seabed pipelines. Géotechnique Lett. **4**, 165–169 (2014)
4. Yan, Y.: Novel methods for characterizing pipe-soil interaction forces in situ in deep water. Ph.D. thesis, The Univesity of Western Australia (2013)
5. Taylor, R.N.: Geotechnical Centrifuge Technology. Blackie Academic Press, Glasgow (1995)
6. Wood, D.M.: Geotechnical Modelling. Spon Press/Taylor & Francis, New York (2004)
7. Randolph, M.F., Jewell, R.J., Stone, K.J.L., Brown, T.A.: Establishing a new centrifuge facility. In: Proceedings of the International Conference on Centrifuge Modelling – Centrifuge, vol. 91, Broulder, Colorado (1991)
8. Watson, P.G.: Performance of skirted foundations for offshore structures. Ph.D. thesis, The University of Western Australia (1999)
9. De Catania, S., Breen, J., Gaudin, C., White, D.J.: Development of a multiple-axis actuator control system. In: Proceedings of 2nd International Symposium on Frontiers in Offshore Geotechnics, Perth, Australia (2010)
10. Gaudin, C., White, D.J., Boylan, N., Breen, J., Brown, T.A., De Catania, S., Hontion, P.: A wireless data acquisition system for centrifuge model testing. Measur. Sci. Technol. **20**, 095709 (2009)
11. Cheong, J.: Physical testing of jack-up footings on sand subjected to torsion. Final year undergraduate thesis, Final year undergraduate thesis, The University of Western Australia (2002)
12. Stewart, D.P.: Lateral loading of piled bridge abutments due to embankment construction. Ph.D. thesis, The University of Western Australia (1992)

Centrifugal Model Test Study of Five-Pile Foundations for Offshore Wind Turbine

Zhaoxia Tong[1(✉)], Yang Chen[1], Zhaokun Zuo[2], and Jingyan Feng[1]

[1] Beihang University, Beijing, China
tongzx@buaa.edu.cn
[2] China Academy of Railway Sciences, Beijing, China

Abstract. The deformation and stability of pile foundation for offshore wind turbine have obtained much attention in recent years. Three series of centrifuge model tests are conducted on the five-pile foundations subjected to monotonic loading and cyclic loading. The displacement development of piles, the distribution of pile moment and the soil deformation around the piles are discussed. It is found that the horizontal displacement ratio between each pile maintains stable at the monotonic loading. With the increasing of cyclic loading, the horizontal displacement of each pile accumulates progressively, yet the accumulation rate decreases. The bending moment of pile along the buried depth increases firstly and then decreases. The maximum bending moment happens at about 20% of the pile buried depth. The marked horizontal displacement happens in top 40% region of soil along pile buried depth. Compared to the soil in the active zone, high horizontal displacement and large zone are produced for the soil in the passive zone.

Keywords: Centrifugal model test · Offshore wind turbine · Pile foundation
Deformation

1 Introduction

As a new form of energy, wind is clean and environmentally friendly [1, 2]. China has huge and promising wind power, especially offshore wind power. The development of offshore wind power is not only an efficient use of wind energy resources but also an important choice for the sustainable development in China [3]. However, the offshore wind farm experiences the complex cyclic loadings such as wind loading and wave loading. Due to the high requirements on the upper structure of the wind turbine, the foundation design and construction of the offshore wind turbine face big challenge [4].

Pile foundation is the main type of foundation for offshore wind turbine. The deformation and stability of pile foundation under ocean conditions have been paid much attention in the recent years. Matlock [5] carried out field observations for the five-pile and 10-pile foundations on a soft clay subgrade. He found that the variation of the shear force and the bending moment between each pile in one pile group is small when the horizontal load increases monotonically. Under the cyclic load, the degradation of the soil resistance always begins when each pile reaches the same deformation regardless of the difference in the number and space of the group piles. Zhu [6, 7] carried

© Springer Nature Singapore Pte Ltd. 2018
T. Qiu et al. (Eds.): GSIC 2018, *Proceedings of GeoShanghai 2018 International Conference:
Advances in Soil Dynamics and Foundation Engineering*, pp. 802–809, 2018.
https://doi.org/10.1007/978-981-13-0131-5_87

out large scale model test on the single pile foundation. Based on the distribution of the displacement and bending moment along depth before and after pile yielding, he presented a horizontal static and cyclic weakening p-y curve analysis model for large diameter single pile foundation in sandy soil subgrade.

This paper carries out centrifugal model tests for five-pile foundation under complex loadings. The deformation and the distribution of the bending moment along pile together with the soil response around the piles are studied.

2 Test Configurations

2.1 Test Equipment and Model

The experiments are conducted using a TH-50g-tons geotechnical centrifuge from Institute of Geotechnical Engineering, Tsinghua University. Its effective radius is 2 m and the maximum centrifugal acceleration is 250 g.

The five-pile model used in the tests was manufactured according to the pile foundation scheme for Rudong, Nantong wind farm project with the ratio of 1: 100. The parameters for model pile and prototype pile are shown in Table 1. Figure 1 shows schematic view of the five-pile foundation model. Due to the symmetries of the model and loads, a semi-model was used. The model box is aluminous, with internal space of 600 mm (length) × 200 mm (width) × 535 mm (height). One side of the box is a plexiglass plate, through which we can measure deformation of pile and the soil around the pile by non-contact displacement measurement system.

Table 1. Parameters for model pile and prototype pile

	Scale	Diameter/m	Pile length/m	Wall thickness/m	Stiffness/N·m	Pile space/m
Model piles	1:100	0.02	0.35	0.001	566.919	0.12
Prototype piles		2	35	0.03	1.89×10^{10}	12

The soils used in the tests are Fujian standard sands with particle size ranged from 0.25 to 0.5 mm. The moisture content of the soils were 5% with its dry density of 1.55 g/cm^3. To ensure the uniformity of the soil subgrade, the sands were filled into the model box in 5 layers with the same falling height. Each layer of sands are controlled to be 80 mm thick after compaction.

Four steps were conducted to install the pile foundation after the preparation of soil subgrade. Firstly, according to the spacing and layout of the five pile foundation, a circular steel pipe with the same diameter as the model pile was punched into the soil subgrade with the depth of 290 mm and then removed. Thus the holes for the model piles were made. Secondly, the prefabricated pile foundations are slowly inserted into the holes until the depth of 290 mm. Thirdly, the plexiglass plate is removed, and the mark points spacing about 150 mm on both sides of the pile model used to measure the displacement of the soil are placed. Finally, the plexiglass plate is reassembled with glass glue coating. After the solidification of glass glue, the deaired water was added from the bottom of model box to suturing the soil subgrade. To ensure the full saturation of soil subgrade, the water table was controlled to be 2 cm higher than the mud line.

The loading equipment used in the experiments is a hydraulic loading system and can be controlled by computer to apply monotonic load and harmonic load. The loading range is from 0 to 10 kN. The horizontal displacement of each pile and truss is measured by laser displacement sensor, and the pile moment is measured by strain gauge as Fig. 1 shown.

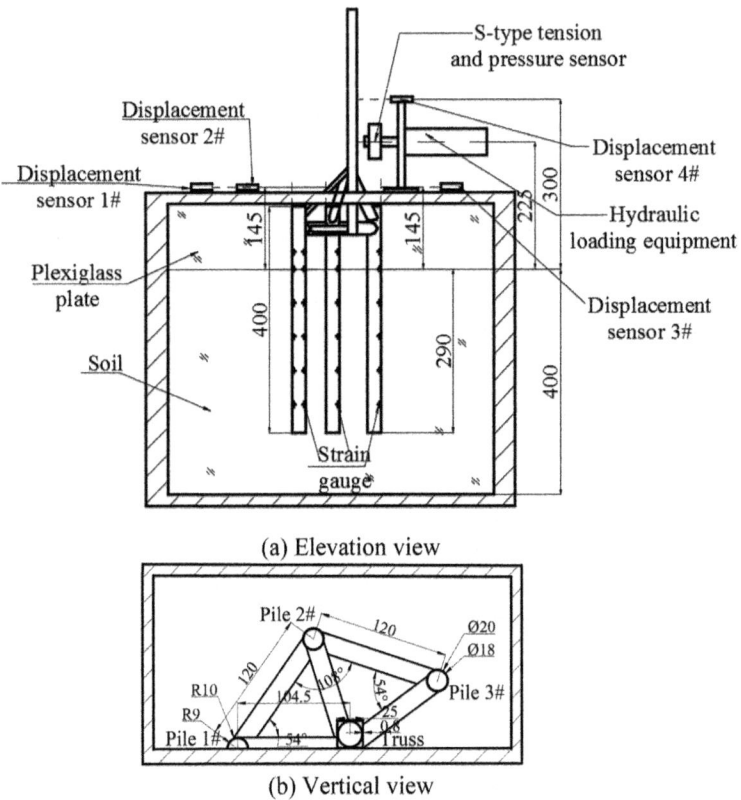

(a) Elevation view

(b) Vertical view

Fig. 1. Schematic view of the five-pile foundation model (unit: mm).

2.2 Experiment Scheme

In order to simulate the complex loadings experienced by the pile foundations, the experiments schemes are designed as Table 2 shown. The loads applied on the truss are exerted with three steps. The first step is the monotonic loading. The second and third step is cyclic loading with different average value and amplitude. All experiments are carried out under 50 g acceleration. Considering the wind common frequency of 0.02 Hz, the cyclic loading frequency is also controlled to be 1 Hz aiming to ensure the frequency similarity between the test and wind loadings.

Table 2. Experiment plan

Experiment	Loading conditions		
	Step 1/Monotonically loading to-	Step 2/Cyclic loading	Step 3/Cyclic loading
W1	500 N	Amplitude: 50 N Average value: 500 N	Amplitude: 180 N Average value: 340 N
W2	700 N	Amplitude: 100 N Average value: 550 N	Amplitude: 200 N Average value: 400 N
W3	700 N	Amplitude: 200 N Average value: 600 N	Amplitude: 100 N Average value: 680 N

3 Results

The displacements of piles, the distribution of pile moment and the soil deformation around the piles are discussed as follows.

3.1 Development of Pile Displacement

Figure 2 shows the truss and pile displacement in the first monotonic loading step for Test W3. It can be seen that each pile generates almost the same horizontal displacement when the loading is lower than 200 N. As the loading increases, the piles 1# and 2# generate the largest and smallest horizontal displacement respectively. The displacement and displacement ratio for each pile are given in Table 3. It can be seen that the displacement ratio for each pile has minor change when the loading increases to 700 N. The piles of 1#, 2# and 3# account for almost 44%, 26% and 30% respectively. It can be inferred that the pile group rotates around the position of pile 2# during the process of the monotonic loading.

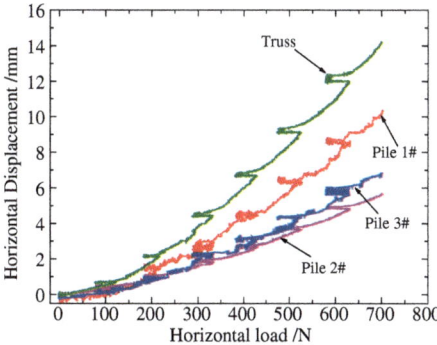

Fig. 2. Development of pile and truss displacement in step 1 for Test W3

Table 3. Horizontal pile displacement in monotonic loading step

Horizontal load/N	Displacement/mm			Displacement percentage/%		
	Pile 1#	Pile 2#	Pile 3#	Pile 1#	Pile 2#	Pile 3#
300	2.55	1.64	1.98	41.33	26.58	32.09
400	4.16	2.36	2.82	44.54	25.27	30.19
500	5.87	3.47	3.75	44.84	26.51	28.65
600	7.81	4.45	5.32	44.43	25.31	30.26
700	10.4	5.71	6.66	45.67	25.08	29.25

The displacement accumulation of the truss under two cyclic loading steps of Test W1 is shown in Fig. 3. It can be seen that under cyclic loading, the displacement of the truss increases steadily accompanied by cyclic vibration. However, with the increasing of the cyclic loading, its accumulation rate tends to decrease. The higher loading amplitude produces higher deformation amplitude.

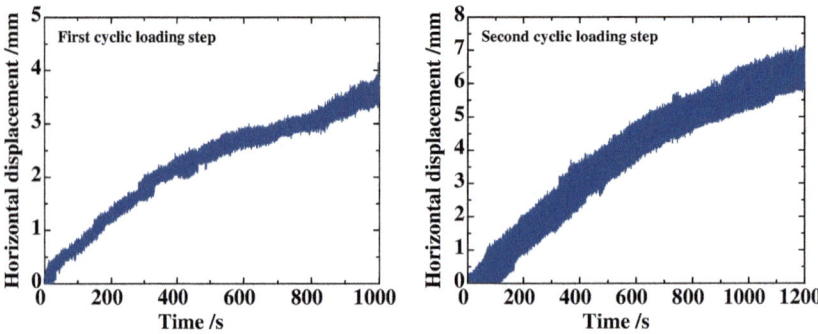

Fig. 3. Displacement accumulation curve of truss under cyclic loading

3.2 Distribution of Pile Moment

The distribution of bending moment along the buried depth for each pile is shown in Fig. 4 when the monotonic loading reaches 100 N, 200 N, 300 N, 400 N and 500 N for Test W2. It can be seen that along the buried pile depth (from top to bottom), the bending moment increases firstly and then decreases. The maximum bending moment happens at about 20% of the pile buried depth. Meanwhile, the loading has minor effects on the position of the maximum bending moment.

Table 4 shows the maximum bending moment and percentage of each pile. It can be seen that the pile 2# support the maximum bending moment, and the pile 1# supports the minimum bending moments when the loading reaches 500 N. As the loading increases, the bending moment percentage of piles 1# and 2# tends to increase, yet pile 3# tends to decrease. As the loading increases to 500 N, the bending moment percentage of piles 1# is still lower than 20%, the pile 2# and 3# account for over 80%.

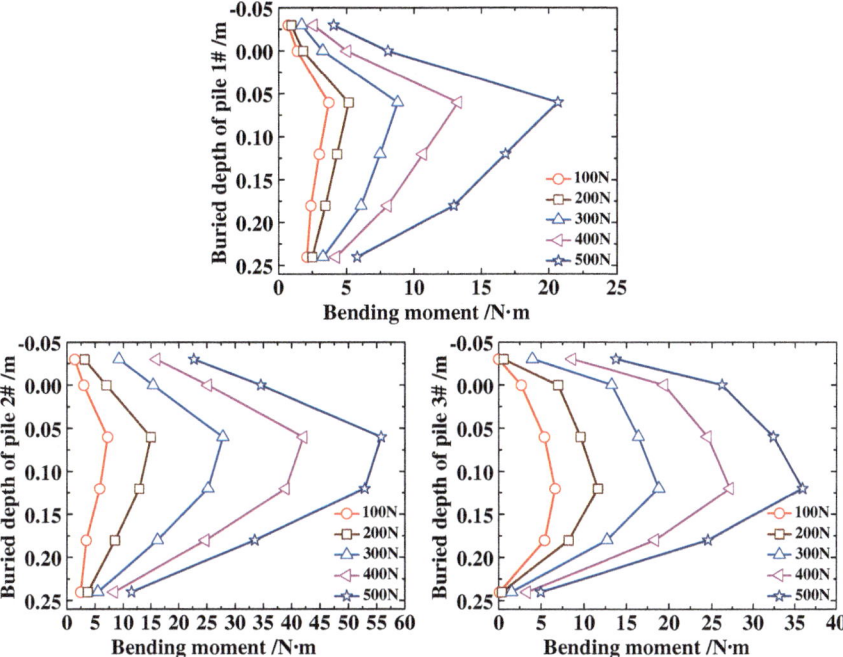

Fig. 4. The bending moment of pile in monotonic loading step in Test W2

Table 4. The maximum bending moment in monotonic loading step

Load/N	Maximum bending moment/N·m			Percentage/%		
	Pile 1#	Pile 2#	Pile 3#	Pile 1#	Pile 2#	Pile 3#
100	3.59	9.67	10.13	15.35	41.34	43.31
200	5.12	14.52	14.07	15.19	43.07	41.74
300	8.81	24.31	20.31	16.49	45.5	38.01
400	13.14	36.23	27.83	17.02	46.93	36.05
500	20.65	54.37	37.15	18.41	48.47	33.12

3.3   Soil Response Around Pile

Figure 5 shows the horizontal displacement vector of the soil around the pile #1 in the monotonic loading step for Test W1. It can be seen that the displacement vector of the passive soil in the left side of the pile pointed to the downward direction, indicating that this part of soil generates the horizontal and vertical displacement. Larger horizontal displacement is produced at the upper part. The marked horizontal displacement happens in top 40% region along pile buried depth. The displacement of the active soil in the right side of the pile is smaller than that in the passive soil. The marked horizontal displacement almost happens in top 20% region along pile buried depth.

Figure 6 shows the horizontal displacement contour when monotonically loading to 500 N. It is clear that higher horizontal displacement are produced at the passive zone. The soil region to generate horizontal displacement is larger in the passive zone than that in the active zone.

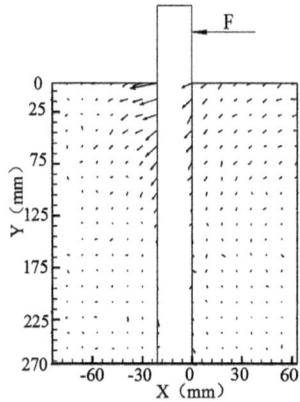

Fig. 5. Displacement vector diagram

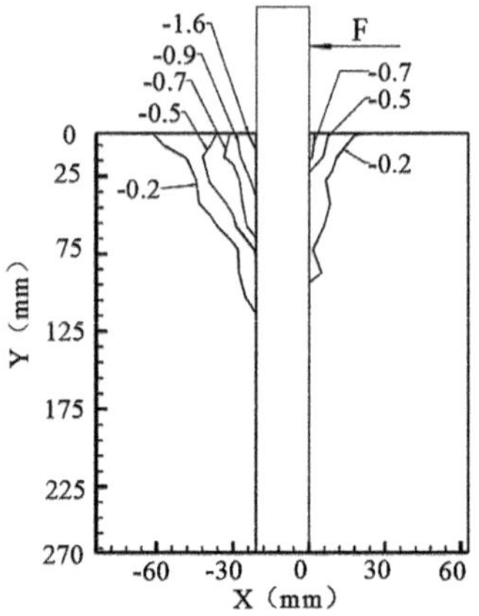

Fig. 6. Horizontal displacement contour

4   Conclusion

Three series of centrifuge model tests are conducted on the five-pile foundations for the offshore wind turbine. The displacement development of piles, the distribution of pile moment and the soil deformation around the piles are discussed. The main conclusion can be drawn as follow.

(1) The horizontal displacement ratio between each pile maintains stable at the monotonic loading. With the cyclic loading at the second and third step, the horizontal displacement of each pile accumulates progressively, yet the accumulation rate decreases.

(2) The bending moment of pile along the buried depth increases firstly and then decreases. The maximum bending moment happens at about 20% of the pile buried depth. The loading has minor effects on the position of the maximum bending moment.

(3) The marked horizontal displacement happens in top 40% region along pile buried depth. Compared to the soil in the active zone, high horizontal displacement and large zone are produced for the soil in the passive zone.

Acknowledgements. The present study is financially supported by the National Basic Research Program of China (973 Program) (contract number 2014CB047003) and the National Natural Science Foundation of China (contract number 10902005).

References

1. Jiang, N.: Development status and prospect of deep sea wind power generation technology. J. Circ. Syst. **3**(1), 21–24 (2015)
2. Zhao, W., Liu, Y., Wang, W.: Current situation and development of marine renewable energy power generation. Smart Grid **3**(6), 493–499 (2015)
3. Qi, H., Shen, D., Zhuang, Y., et al.: Analysis report of large scale wind power industry development in China in 2013–2014. In: Wind Energy Industry, pp. 7–26 (2014)
4. Liu, L., Song, J.: Discussion on China offshore wind resource and wind turbine selection. In: Proceedings of the Annual Conference of China Society of Environmental Science, pp. 6964–6968 (2014)
5. Hudson, M., Wayne, B.I., Allen, E.K., Dewaine, B.: Filed tests of the lateral-load behavior of pile groups in soft clay. In: Offshore Technology Conference, 3871-MS (1980)
6. Zhu, B., Zhu, R., Luo, J., et al.: Model test study on horizontal large deflection characteristics of high piled foundation. Chin. J. Geotech. Eng. **32**(4), 521–530 (2010)
7. Zhu, B., Xiong, G., Liu, J., et al.: Centrifuge modelling of a large-diameter single pile under lateral loads in sand. Chin. J. Geotech. Eng. **35**(10), 1807–1815 (2013)

Sample Disturbance of Onsøy Clay Due to Gas Exsolution

Shaoli Yang[1(✉)], Tom Lunne[1], and Gulin Yetginer[2]

[1] Norwegian Geotechnical Institute, 0805 Oslo, Norway
shaoli.yang@ngi.no
[2] Statoil ASA, 4313 Stavanger, Norway

Abstract. High quality block clay samples at Onsøy were obtained by using the Sherbrooke sampler. Pore water with various degrees of gas dissolved were introduced to soil specimen under a back pressure simulating 1500 m water depth. Then total stresses were released, resulting in gas coming out of solution before trimming samples for advanced tests. Results from both Triaxial compression tests (CAUC) and Direct simple shear tests (DSS) showed that sample disturbance is increasing due to gas exsolution up to $\eta = 20\%$ and there is less increase in effects on undrained shear strength for $\eta = 20$ to 67%. $\Delta e/e_i$ and anisotropy ratio indicate that sample disturbance due to gas exsolution has more influence on DSS and CRSC tests than on CAUC tests. CRSC tests showed that preconsolidation stress is decreasing with increasing degree of gas saturation. Tube sampling disturbance of onsøy clay from previous study is discussed and compared with disturbance due to gas coming out of solution.

Keywords: Gas exsolution · Undrained shear strength · Sample disturbance
Degree of gas saturation

1 Introduction

For deep water soil investigations, large stress relief happens when samples are brought to deck. If there is gas dissolved in the pore water of the samples, the gas will expand and come out of solution during the sampling process. As a consequence, cracks may be created. Sample disturbance or development of cracks due to gas exsolution and expansion were reported in deep water site and this has been discussed by Lunne et al. 2001 [1] and Yang et al. 2017 [2]. Effect of sample disturbance is an important issue in geotechnical engineering and this has been studied for many years (Bjerrum 1973 [3]; Lunne et al. 2006 [4]; Hight and Leroueil 2003 [5]; Ladd and Degroot 2003 [6]; Prasad et al. 2007 [7]; Rocchia et al. 2013 [8]).

However influence of the large stress relief process on the behaviour of the clays as measured by laboratory tests is not well known. Normally, samples taken from deep water are transported to onshore laboratory and some advanced tests will be carried out in order to evaluate in-situ undrained shear strength of clays. Laboratory tests on samples disturbed due to stress relief may not provide representative soil parameters for the soil condition in situ. Clays may have different degrees of gas saturation in situ.

© Springer Nature Singapore Pte Ltd. 2018
T. Qiu et al. (Eds.): GSIC 2018, *Proceedings of GeoShanghai 2018 International Conference: Advances in Soil Dynamics and Foundation Engineering*, pp. 810–818, 2018.
https://doi.org/10.1007/978-981-13-0131-5_88

Lunne et al. (2001) [1] carried out a comprehensive set of laboratory triaxial and oedometer tests on Lierstranda and Bothkennar clays where deep water sampling of clay with various amounts of gas dissolved in the pore water was simulated. In addition, tube sampling strains were also simulated in that work. Equipment and procedures developed in the previous study were used in the present testing programme described in this paper. See also description in Sect. 2. Triaxial compression tests (CAUC), direct simple shear tests (DSS), oedometer tests (CRSC) and SHANSEP tests were carried out in this study. Results on CAUC and SHANSEP tests were reported in Yang et al. (2017) [2]. Results from DSS tests are presented in this paper and compared with the CAUC tests for obtaining anisotropy ratios. Results from CRSC tests are also interpreted and discussed.

2 Tested Clay, Test Program and Procedures

2.1 Description of Clay Tested

Two high quality samples taken using the Sherbrooke block sampler (Lefebvre and Poulin 1979 [9]) at depth intervals 8.7 to 9.05 m and 9.9 to 10.25 m in Onsøy were selected. Onsøy is one of NGI's main soft clay test sites and is described by Lunne et al. (2003) [10]. Range in classification parameters are listed in Table 1 for the two block samples.

Table 1. Index parameters for the two block samples used for the laboratory testing.

Sample No.	Depth, m	Water content, %	Soil unit weight, kN/m³	Plasticity index, %	% particles < 2μ	Activity*	OCR
7	8.70–9.06	67.1–69.0	15.52–15.90	49.4	49.8	0.99	1.38
10	9.90–10.17	63.4–68.5	15.73–16.01	42.7–48.0	64.4–68.4	0.66–0.68	1.33

Note: Activity = Ip/% particles < 2μ.

The overconsolidation ratio is based on CRSC tests and Casagrande interpretation method, and the overconsolidation is caused by aging.

NGI (2000) [11] did some tests with degree of gas saturation of 0% and 20% on Onsøy block samples by using the same testing procedures as used in this study, and the results from these tests are utilized when relevant in this study.

2.2 Testing Program

CAUC, DSS, CRSC and SHANSEP tests, with various degree of gas saturation (η) of 0%, 6%, 20% and 67% were carried out in this study. The tests discussed in this study is shown in Table 2.

Table 2. Tests discussed in this paper

Tests	Degree of gas saturation			
	0%	6%	20%	67%
CAUC**	1	1	1	1
DSS	1	1	1	1
CRSC	1		1	1
DSS*	2		2	
CRSC*	2		2	

**Results for CAUC tests were presented in Yang et al. (2017) [2].
*Tests performed in 2000 at NGI (NGI 2000 [11]).

In addition to the advanced tests, X ray scanning was performed for all samples before and after stress relief in order to observe the structure of the clays and check if any cracks were created after stress relief. This study will discuss results from DSS and CRSC tests (Table 2).

2.3 Test Procedures for Specimens with Gas

In order to simulate sampling from 1500 m water depth of clay with gas dissolved in the pore water, the scheme used by Lunne et al. (2001) [1]. as summarized below was used.

(a) Start with a block sample. Trim a sample with diameter = 120 mm and height = 61 mm and place it in the big oedometer cell with dry filter disks and subject it to a vertical stress of about 0.1 times of σ'_{v0}. This is about 6.3 kPa for the tested Onsøy clay and is about equal to the swelling pressure for the clay tested. Then flush filter disks with water with the same salinity (about 30 g/l) as the pore water of the clay.

(b) Apply a back pressure high enough to achieve a certain percentage of gas saturation at 1500 m water depth. For example, if a gas saturation equal to 20% is wanted, apply back pressure, u_b, of 0.2 × 15 MPa = 3 MPa and percolate with gas saturated water done at this pressure as described below.

(c) Increase vertical effective stress to 0.7 times pc'.

(d) Percolate sample with pore water fully saturated with methane gas until about 4 pore volumes have passed through the sample.

(e) Increase the back pressure (i.e. the pore pressure) to 15 MPa (corresponding to 1500 m water depth). The degree of gas saturation of the pore water, η, is then

$$\frac{u_b \ (\text{in MPa})}{15} \times 100\%$$

This procedure was used to prepare samples from Onsøy with η = 0, 6, 20 and 67%.

(f) Reduce the effective vertical stress and then the back pressure to zero to simulate stress relief when bringing sample from 1500 water depth to deck level of the ship.

In real case, both effect of mechanical disturbance due to tube sampling and gas exsolution during the total stress relief exist. Only the effect of gas exsolution is discussed in this study. The combined disturbance due to tube sampling and stress relief was studied by Lunne et al. (2001) [1].

Specimens were trimmed from the large oedometer sample to DSS and CRSC specimens with area of 20 cm². The reason for using small specimens for this gas study is to reduce the amount of time it takes to percolate water through the specimen. For Onsøy the required time was about 4 weeks.

All DSS specimens were consolidated to the in-situ vertical effective stress. Sample quality is discussed in Sect. 3. CRSC tests were carried out using procedures given by Sandbækken et al. (1986) [12].

3   Test Result

3.1   DSS Tests

Figures 1 and 2 show stress strain and stress path curves for DSS tests, and Fig. 1 gives normalized strength ratios for the DSS tests. It seems there is a peak strength for η = 0% and 6%, the peak is not significant for η = 20 and 67%. The figures indicate that sample disturbance is increasing due to gas exsolution up to η = 20% and there is less increase in effects for η = 20 to 67% (Fig. 3). In addition, normalized strength ratios from the CAUC tests (Yang et al. 2017 [2]) are given in Fig. 3. When degree of gas saturation is changing from 0% to 67%, the decrease of normalized shear strength is about 8% for CAUC tests, while the decrease of normalized strength is about 24% for DSS tests.

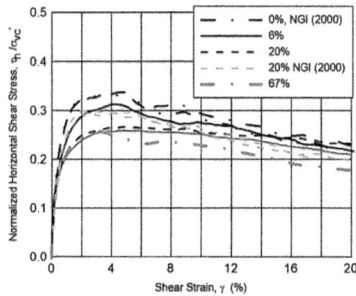

Fig. 1. Normalized stress versus shear strain from DSS tests (left)

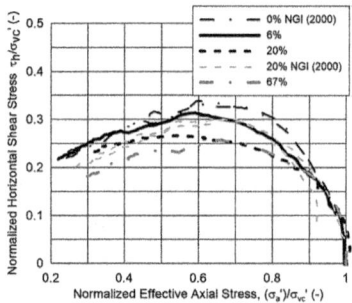

Fig. 2. Normalized stress path from DSS tests (right)

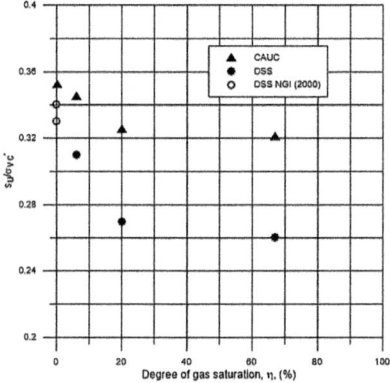

Fig. 3. Normalized strength versus degree of gas saturation (left)

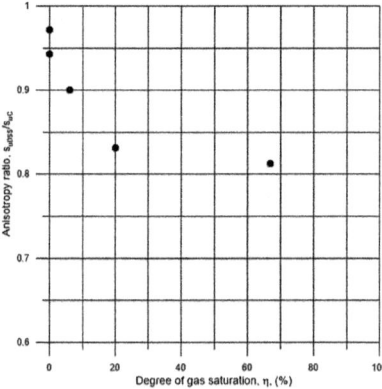

Fig. 4. Anisotropy ratio versus degree of gas saturation (right).

Anisotropy ratios in Fig. 4 shows that sample disturbance due to gas exsolution has more influence on DSS tests than on CAUC tests. This is discussed in details in Sect. 3.3.

3.2 CRSC Tests

Figure 5 shows that with an increasing in degree of gas saturation, preconsolidation stress is decreased. However for high degree of gas saturation, preconsolidation stress based on Casagrande method is not reliable due to poor sample quality. Test for 0% degree of gas saturation is not within the trend, and this may be due to the natural variation in the samples.

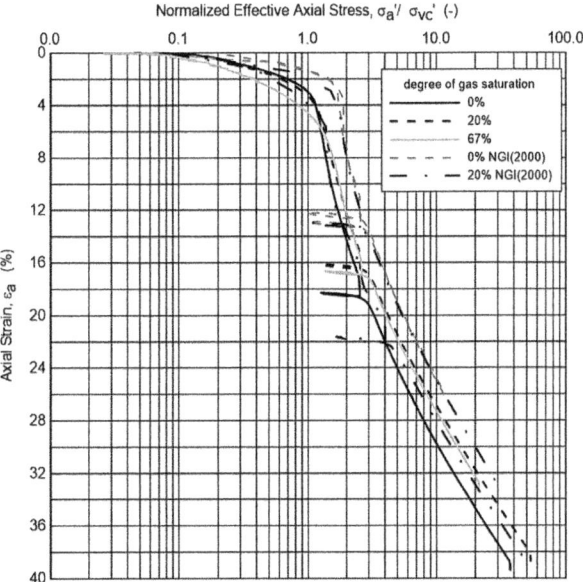

Fig. 5. Normalized stress versus axial strain from CRSC tests

3.3 Sample Quality

Cracks were observed after percolation for the big oedometer specimens, especially for samples with degrees of gas saturation of 20% and 67% (Fig. 6). X-ray pictures show that cracks are increasing with increase of degree of gas saturation (Fig. 7). All these pictures indicate that sample disturbance increasing with degree of gas saturation.

Fig. 6. Cracks observed after percolation and stress relief

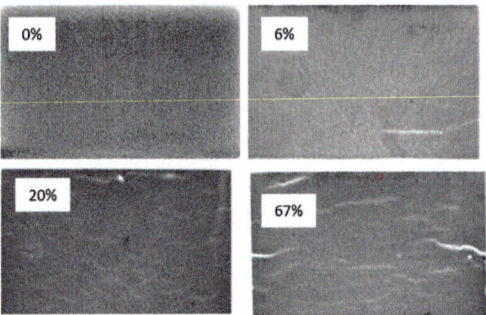

Fig. 7. Cracks observed by x-ray investigation after percolation and stress relief

Sample quality based on criteria in Table 3 is shown in Fig. 8. One DSS test showed poor sample quality for 0% degree of gas saturation, and this test is excluded from the discussion since this sample was not within the expected trend. In general, the higher the degree of gas saturation, the more disturbance for the specimen. CAUC tests performed on Onsøy clay (Lunne et al. 2006 [14]) from 54 mm and 75 mm tube and block samples showed that there is a significant effects of tube sample disturbance. Tube sample disturbance is much larger than disturbance from gas coming out of solution when the results from CAUC tests are compared. Figure 8 indicates that more disturbance can be observed for CRSC and DSS tests compared with CAUC tests. This may be due to specimen trimming method and initial contact conditions. In addition, most of the cracks developed are in horizontal direction which may contribute to low shear strength from DSS tests. In real case disturbance will be even more when tube sampling is combined with stress relief (Lunne et al. 2001 [1]).

Table 3. Criteria for evaluation of sample disturbance as quantified by the value of $\Delta e/e_i$ (Lunne et al. 1998 [13])

OCR	Sample quality category			
	Very good to excellent (1)	Good to fair (2)	Poor (3)	Very poor (4)
1–2	<0.04	0.04–0.07	0.07–0.14	>0.14
2–4	<0.03	0.03–0.05	0.05–0.10	>0.10

Note: The description of sample quality refers to use of the sample for measurement of mechanical properties.

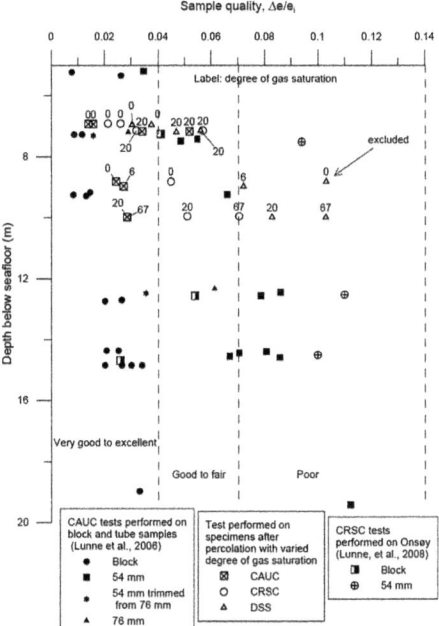

Fig. 8. Sample quality from tests with varied degree of gas saturation and tests on block and tube samples.

4 Conclusion

Cracks were created after percolation of gas saturated water through onsøy specimen and stress relief. Most of cracks developed are in horizontal direction. At deep water site, cracks could be developed when samples with dissolved gas in pore water are taken from the seabed to deck. Sample disturbance of tested Onsøy clay is increasing with increase of degree of gas saturation. The increase is less significant when degree of gas saturation is more than 20%. $\Delta e/e_i$ and anisotropy ratio indicate that sample disturbance due to gas exsolution has more influence on DSS and CRSC tests than on CAUC tests. CRSC tests show a decreasing of preconsolidation stress with increasing of degree of gas saturation.

Acknowledgment. The authors would like to acknowledge constructive discussions with NGI colleagues and the support from the Norwegian Deepwater Programme - Seabed Project, as represented by Statoil. The Norwegian Research Council has also given financial support through the Norwegian Geotest Site Project where Onsøy is one of the sites being developed as part a large infrastructure facility for future research related to geotechnical engineering.

References

1. Lunne, T., Berre, T., Strandvik, S., Andersen, K.H., Tjelta, T.I.: Deepwater sample disturbance due to stress relief. In: Proceedings OTRC Conference, April 2001, pp. 64–85 (2001)
2. Yang, S.L., Lunne, T., Andersen, K.H., Tjelta, G.: Experimental study on effect of stress relief and gas exsolution on sample quality. In: Proceedings on Smarter Solutions for Future Offshore Development, London, pp. 332–337 (2017)
3. Bjerrum, L.: Problems of soil mechanics and construction on soft clay. In: proceedings of the 8th International Conference on Soil Mechanics and Foundation Engineering, Moscow, vol. 3, pp. 111–159. A.A. Balkema, Rotterdam (1973)
4. Lunne, T., Berre, T., Andersen, K.H., Strandvik, S., Sjursen, M.: Effects of sample disturbance and consolidation procedures on measured shear strength of soft marine Norwegian clays. Can. Geotech. J. **43**, 726–750 (2006)
5. Hight, D.W., and Leroueil, S. 2003. Characterization of soils for engineering purposes. In: Tan, T.S., Phoon, K.K., Hight, D.W., Leroueil, S. (eds.) Proceedings of the International Workshop Characterization and Engineering Properties of Natural Solid, Singapore, 2–4 December 2002, vol. 2, pp. 149–162. A.A. Balkema, Rotterdam (2002)
6. Ladd, C.C., DeGroot, D.J.: Recommended practice for soft ground site characterization: arthur casagrande lecture. In: Culligan, P.J., Einstein, H.H., Whittle, A.J. (eds.) Soil and Rock America 2003: Proceedings of the 12th Panamerican Conference on Soil Mechanics and Geotechnical Engineering, Cambridge, 22–26 June 2003, vol. 1, pp. 3–57. VGE Verlag Gluckauf, Essen (2003)
7. Prasad, K.N., Triveni, S., Schanz, T., Nagaraj, T.S.: Sample disturbance in soft and sensitive clays: analysis and assessment. Mar. Georesour. Geotechnol. **25**, 181–197 (2007)
8. Rocchia, G., Vaciagoa, G., Fontanab, M., Da Prata, M.: Understanding sampling disturbance and behaviour of structured clays through constitutive modelling. Soils Found. **53**(2), 315–334 (2013)
9. Lefebvre, G., Poulin, C.: A new method of sampling in sensitive clay. CGJ **16**, 226–233 (1979)
10. Lunne, T., Long, M., Forsberg, C.F.: Characterisation and engineering properties of Onsøy clay. In: Tan, T.S., et al. (ed.) Characterisation and Engineering Properties of Natural Soils, Balkema 2003. Proceedings of the International Workshop, Singapore 2002, vol. 1, pp. 395–427 (2003)
11. NGI: Studying the effect of gas on soil properties measured on deepwater samples. Results of laboratory tests for evaluation of disturbance due to gas. NGI project 20001411-3. Final report dated on 5 March 2002 (2000)
12. Sandbækken, G., Berre, T., Lacasse, S.: Oedometer testing at the Norwegian Geotechnical Institute. Consolidation of Soils: Testing and Evaluation: A Symposium. Fort Lauderdale, Fla. 1985. American Society for Testing and Materials, vol. 892, pp. 329–353. Special Technical Publication (1986). Also Published in: Norwegian Geotechnical Institute, Oslo. Publication, vol. 168 (1987)
13. Lunne, T., Berre, T., Strandvik, S.: Sample disturbance effects in deep water soil investigations. In: Offshore Site Investigation and Foundation Behaviour 1998, New Frontiers, SUT Conference London 1998, pp. 199–220 (1998)
14. Lunne, T., Berre, T., Andersen, K.H., Strandvik, S., Sjursen, S.: Effects of sample disturbance and consolidation procedures on measured shear strength of soft marine Norwegian clays. Can. Geotech. J. **43**, 726–750 (2006)

In-situ Tests on Behaviour of Breakwater with Bucket Foundation Under Wave Loading

Yunfei Guan[✉], Yongyong Cao, Yi Tang, and Ning Zhang

Department of Geotechnical Engineering, Nanjing Hydraulic Research Institute,
Nanjing 210024, China
gyfnhri@163.com

Abstract. The bucket foundation can be used as the foundation of the break-water built on clay. In this paper in-situ tests are performed to study the behaviour of the bucket foundation using the newly developed automatic monitoring system. The test data shows the variations of stresses and pore water pressure with time. The total stress and pore pressure change periodically with time. The amplitude of variation is almost constant for the total stress and decreases slightly for the pore pressure. During the consolidation process, the effective stress increases with time while the excess pore water pressure dissipates gradually. The profiles of earth pressure on the inner and external surface of the bucket walls are also shown.

Keywords: Bucket foundation · In-situ test · Earth pressure profile

1 Introduction

A new bucket foundation is developed for the vertical breakwater built in deep clay layer under wave loading. The breakwater consists of several bucket structures built in the silty clay layer. Each structure consists of a bucket at the bottom and two barrels at the top (Fig. 1). The length and width of the bottom bucket is 30 m and 20 m, respectively. The inner space is divided into 9 compartments by the clapboards. The thickness of the wall and the clapboard is 0.4 m and 0.3 m, respectively. Two upper barrels are built at the top of the bottom bucket along the shorter axis with a diameter of 8.9 m and a thickness of 0.4 m. A wave wall is built at the top of the upper barrels with a height of 10.6 m. The whole structure is transited to predetermined location by the floating dock after being built. It is sunk by draining water and applying suction to the bottom bucket. The breakwater is used to reduce the intensity of wave action and prevent the sediment siltation.

Compared to those of the conventional breakwater, the advantages of the break-water with bucket foundation are less material demanding and convenient and economic construction. After being sunk, the stress and deformation of this structure are relatively complex under long-term cyclic wave loading.

Tjelta [1] performed experiments on the sinking of the bucket foundation in the center part of the North Sea. They presented test data of the friction on the wall of the bucket, the earth pressure and the pore water pressure. Achmus et al. [2] studied the

© Springer Nature Singapore Pte Ltd. 2018
T. Qiu et al. (Eds.): GSIC 2018, *Proceedings of GeoShanghai 2018 International Conference: Advances in Soil Dynamics and Foundation Engineering*, pp. 819–828, 2018.
https://doi.org/10.1007/978-981-13-0131-5_89

mechanical behaviour and bearing capacity of the suction bucket foundation in sand based on the Mohr-Coulomb constitutive model using ABAQUS. Attentions are paid to the relationship among the horizontal loading, moment and the rotation and bearing capacity of the bucket foundation for different heights of loading. They also studied the earth pressure profiles on the inner and external surface of the bucket walls under the limit condition and the effects of depth and diameter on the bearing capacity of the bucket.

Fig. 1. Breakwater with multi-compartment bucket foundation.

Cao and Wu [3] performed finite element analysis on the stability of the bucket foundation under wave loading. It was shown that the largest shear stress was found around the interface of the bottom of the bucket and the soil when the largest design wave loading was applied. The total strain was shown to be spherically distributed around the port side toe of the bucket.

The stress and deformation behaviours of different types of bucket foundations have been investigated using model tests and numerical modelling [4, 5]. It is difficult to model different layers of the foundation soil and to assume the boundary conditions, and thus these studies may not be sufficiently accurate. The in-situ tests focus on the sinking process under suction. However, the interaction between buckets and soil is relatively complex and may not be constant under surcharge and wave loading after being sunk. It is important to perform in-situ monitoring during the service period of the bucket foundation.

In this study in-situ tests are performed to study the behaviour of the bucket foundation using the newly developed automatic monitoring system. The variations of the earth pressure on the wall of the bucket and the pore water pressure during the service period under long-term cyclic wave loading are analyzed. The earth pressure profiles on the inner and external surface of the bucket walls are presented. This study provides test data and analytical methods for the design of the bucket foundations.

2   Overview of the In-situ Tests

The depth of water in the site is 5 m. The top of the foundation soil is the clay layer with a thickness of 7 to 10 m. The soil is of low strength and high sensitivity. The vertical breakwater with bucket foundation is used here.

The in-situ tests can be divided into three periods according to the construction method and the behaviour. They are sinking period with suction (interaction between bucket and soil under suction), operation period of the breakwater (interaction between bucket and soil under wave loading) and the operation period of the bucket foundation (the refilled earth pressure on the port side of the bucket).

This study analyzes the data of the in-situ tests on the breakwater with bucket foundation. The main contents of the tests are listed as follows:

(1) Tests on the wave loading on the upper barrels.
(2) Tests on the interaction between bucket and soil. The earth pressure and pore water pressure on the interface of the bucket and soil are measured. This includes the horizontal and vertical earth pressures and pore water pressure on the walls of the bottom buckets and the clipboards and the vertical earth pressure and pore water pressure on the inner surface of the roof of the bottom buckets.
(3) Tests on the internal force of the bucket foundation. The location of the large internal force is determined based on the bucket foundation model tests and numerical analyses. The stress of the steel bars and the strain of the concrete on the walls of the buckets, the roof of the bottom buckets and the key locations in the connecting walls under different conditions are measured.
(4) Tests on the displacement and deformation of the bucket foundation. The horizontal displacement, settlement and rotation of the structure are measured at the typical test points at the top of the buckets.

The measuring instruments are placed at different locations when building the buckets and used to measuring the internal and external forces of the structure. The displacement and deformation of the buckets during the operation period are measured using the inclinometers and the GNSS deformation monitoring system. The inclination of the buckets is measured by the inclinometers placed on the inner surface of the wall at each end of the longer axis of the upper and bottom buckets. The earth pressure and pore water pressure on the inner and external surface of the bottom bucket wall are measured using the vibrating wire earth pressure cells (VWEs) and vibrating wire piezometers (VWPs), respectively [6]. Figure 2 illustrates the locations of the VWEs and the VWPs. The steel bar gauges and the concrete strain gauges are used to measure the internal force. They are placed at the possible locations for the large internal force under loading.

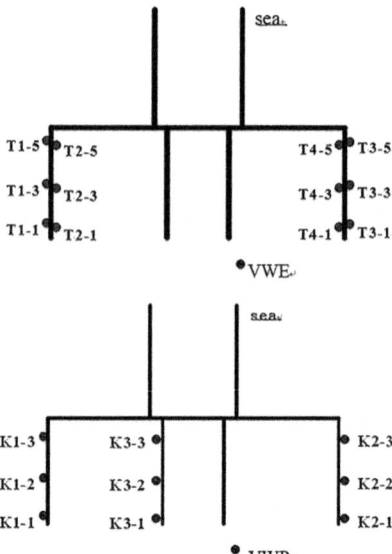

Fig. 2. Illustration of locations of VWEs and VWPs.

3   Monitoring of Earth Pressures on Buckets During Operation Period

The earth pressure on the walls of buckets is the main influence factor of the deformation and stability of the buckets. Figure 3 shows the variations of the total stresses at the locations of T1 and T2 on the port side and T3 and T4 on the seaside during 6 months after the buckets being sunk.

It can be seen that the variations of the total stresses on the bucket walls are sinusoidal at both port side and sea side. The total stresses change with sea level. There are largest and smallest values for every day and the average values are used when interpreting the data. For the total stress at the same phase point of different cycles, the measured values are almost the same. It means that the sum of the effective stress and the excess pore water pressure is constant and changes in the amplitude of the total stress are slight when the effect of tides is ignored.

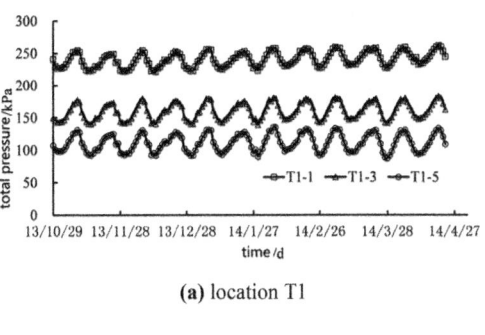

(a) location T1

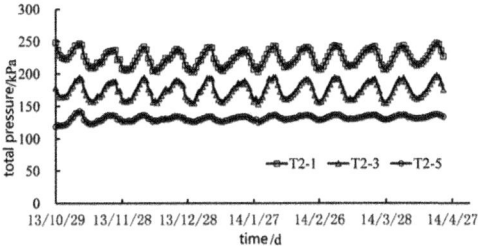

(b) location T2

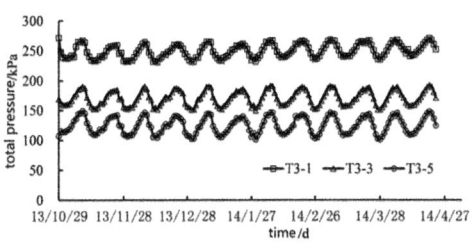

(c) location T3

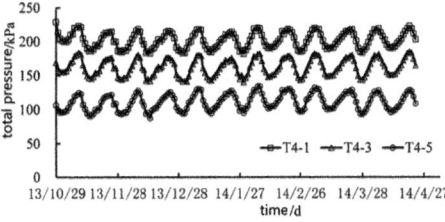

(d) location T4

Fig. 3. Variations of total stresses during operation period

4   Behaviour of Buckets During Operation Period

4.1   Analysis of Effective Stress and Pore Water Pressure

Stresses in the soil consist of effective stress and pore pressure. For the saturated soil, the pore pressure means the pore water pressure. The total stress as the sum of the effective stress and the pore water pressure is measured by the VWEs, while the pore water pressure is measured by the VWPs. The difference between the values measured by the VWEs and the VWPs is the effective stress. Figure 4 shows the variations of effective stresses on the external surface of the bucket walls at the locations of T1 and T3 at the port side and the seaside, respectively.

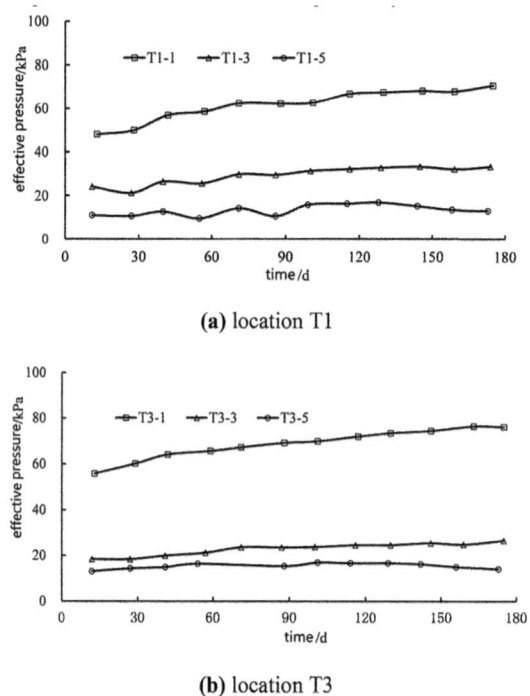

(a) location T1

(b) location T3

Fig. 4. Variations of effective stresses during operation period.

The pore pressures consist of the hydrostatic pore pressure and the excess pore pressure. The excess pore pressure can be obtained by deducting the hydrostatic pore pressure from the pore pressure measured by VWPs. Figure 5 shows the variations of the excess pore pressures on the external surface of the bucket walls at the locations of K1 and K2 at the port side and the seaside, respectively.

It can be found that the effective stresses increase at most measuring points at locations of T1 and T2. The rate of increase is the largest at T1-1 and T3-1 and the smallest at T1-3 and T3-3. The effective stresses increase from 48 kPa to 68 kPa

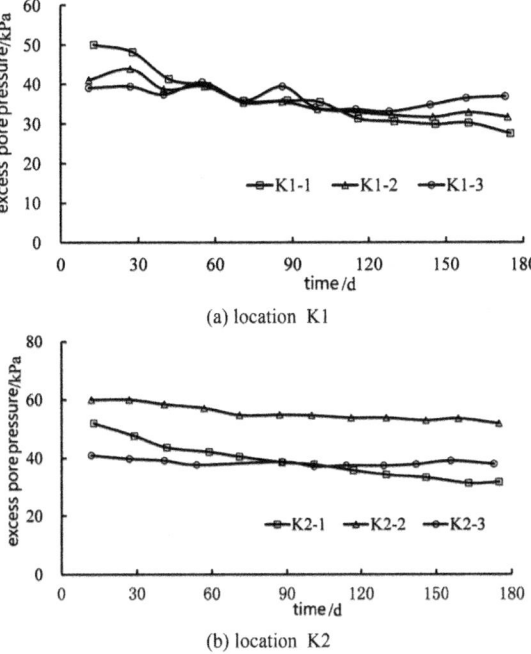

(a) location K1

(b) location K2

Fig. 5. Variations of excess pore pressures during operation period.

and from 46 kPa to 65 kPa at T1-1 and T3-1, respectively. The measured excess pore pressures at K1 and K2 decrease with time. According to the effective stress principle, under a certain static loading, the excess pore pressure dissipates and the effective stress increases during the consolidation process of the saturated soils. The variations of the excess pore pressure and effective stress shown in Fig. 4 to Fig. 5 present the consolidation process of the foundation soil. Additionally, the excess pore pressures at the top and bottom of the bucket walls are larger than those in the middle. This means the disturbing of the soil is greater at the top and bottom when the buckets are being sunk and it agrees with the tests results. The pore water pressures at the top and bottom generally decrease while those in the middle are almost constant.

Based on the data of the earth pressure and pore water pressure, the total stress changes periodically with time and the amplitude of variation is almost constant. The same periodically variations are also found for the pore water pressure and the amplitude decreases slightly with time. The effective stress and the excess pore pressure increases and decreases with time, respectively. This means the foundation soil consolidates gradually. It can be found that the variations of pore water pressure and earth pressure are reasonable. There is no obvious displacement observed. The breakwater with multi-compartments bucket foundation can be considered to be stable.

4.2   Analysis of Earth Pressures on Bucket Wall

The buckets are stable during the operation period due to the interaction between the wall and the bottom of the buckets and the foundation soil. However, there are inclination and displacement of the buckets under the wind and wave loadings. To study the interaction between the buckets and the soil, the analyses pf the variations of the earth pressure at T3 and T4 on the buckets walls after the buckets being sunk are performed.

Figure 6 shows the earth pressure profile of on the external surface of buckets walls at the sea side. The earth pressure on the bucket wall is mainly influenced by the depth of the bottom bucket and the displacement of the bucket. When the bottom bucket is rigid and static, the earth pressure on the wall is the static earth pressure. Figure 6 shows the static earth pressure on the bucket wall. The measured earth pressures for the depth of 0 m to 7.02 m are smaller than the static earth pressures due to the wind and wave loadings. This means for this range of depth the buckets move or rotate away from the soil at the seaside and thus the earth pressure is active earth pressure. The measured earth pressures for the depth of 7.02 m to 10.02 m are greater than the static earth pressures. This means for this range of depth the buckets move or rotate towards the soil at the seaside and thus the earth pressure is passive earth pressure.

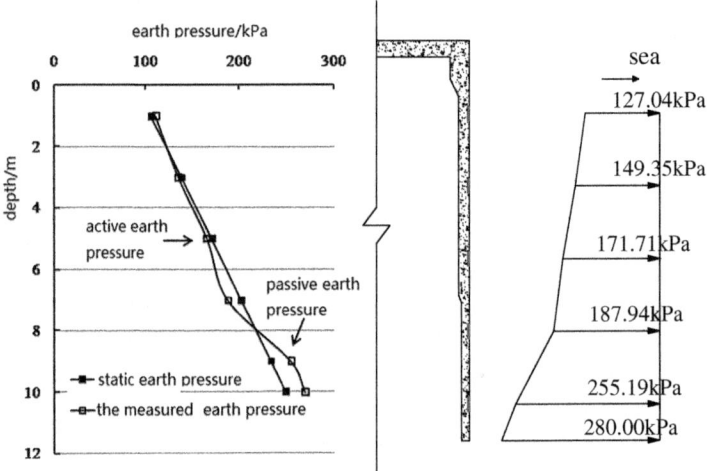

Fig. 6. Earth pressure profile of inner surface of buckets walls at seaside.

Figure 7 shows the earth pressure profile of on the inner surface of buckets walls at the seaside. It can be seen that the variation of earth pressure with depth are opposite to that at T3. It matches with the typical earth pressure profiles at two sides of the retaining wall. For the depth of 0 m to 5.02 m the bucket is under passive pressure which is greater than the static earth pressure. For the depth of 5.02 m to 10.02 m the bucket is under active pressure which is smaller than the static earth pressure. The earth pressure profiles show that the bucket foundation moves and rotates towards the port side, which agrees well with the data measured by the inclinometers and the GNSS deformation monitoring system.

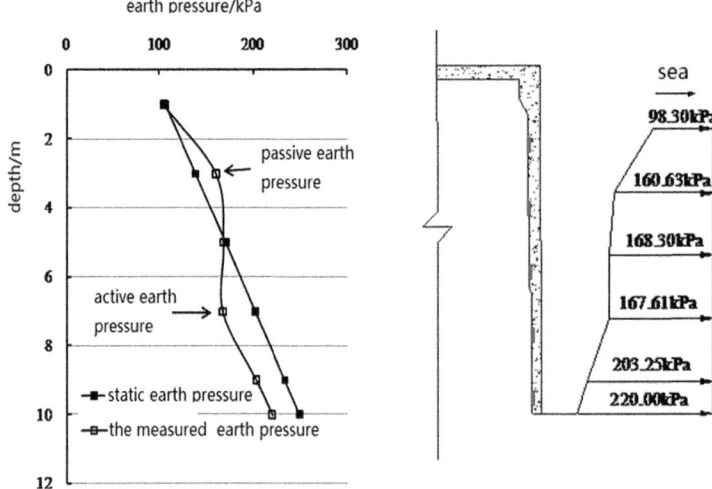

Fig. 7. Earth pressure profile of inner surface of buckets walls at seaside.

5 Conclusions

In this study the in-situ tests are performed on the breaking water with bucket foundation during the operation period. The main conclusions are drawn:

(1) During the operation period the total stress on the bucket foundation changes periodically with time. The effective stress increases with time while the excess pore water pressure dissipates. The foundation soil consolidates gradually.

(2) The breakwater with bucket foundation moves and rotates towards the port side under wave loading. The earth pressure is active for the upper part of the external surface of the bucket wall and passive for the lower part. The distribution is opposite for the inner surface.

References

1. Tjelta, T.I.: Geotechnical experience from the installation of the Europipe jacket with bucket foundations. In: Proceedings of Offshore Technology Conference (1995)
2. Achmus, M., Akdag, C.T., Thieken, K.: Load-bearing behavior of suction bucket foundations in sand. Appl. Ocean Res. **43**, 157–165 (2013)
3. Cao, Y., Wu, Y.: Numerical analysis on stability of the multi-compartment single-bucket structure during wave loading. China Harbour Engineering 02, 11-152016. (in Chinese)
4. De Groot, M.B., Bolton, M.D., Foray, P.: Physics of liquefaction phenomena around marine structures. J. Waterw. Port Coast. Ocean Eng. **132**(4), 227–243 (2006)

5. Wang, J., Li, C., Moran, K.: Cyclic undrained behavior of soft clays and cyclic bearing capacity of a single bucket foundation. In: Proceedings of 15th International Offshore and Polar Engineering Conference, Seoul, Korea, pp. 377–383 (2005)
6. Cao, Y., Cai, Z., Guan, Y.: Stability tests for new bucket-based breakwater driven by negative pressure. Port Waterw. Eng. **07**, 41–45 (2014). (in Chinese)

The Effect of Truss Rigidity
on the Deformation of Three-Pile Foundation

Zhao Xue[✉], Zhaoxia Tong, and Jinyan Feng

School of Transportation Science and Engineering, Beihang University,
Beijing 100191, China
Xuezhao93@buaa.edu.cn

Abstract. The pile foundation is the main foundation form of offshore wind turbines now. This paper uses the finite element software ABAQUS to simulate the effect of the truss rigidity on the stress and deformation of the three-pile foundation under horizontal loading. The truss rigidity is varied by changing the value of the Young modulus, the diameter and the thickness of the truss. The results show that the deformation of the three piles are not exactly the same. The moment and displacement of the pile in the same plane with the applied load are larger than those of other piles. When the truss is more rigid, the moment of each pile approaches to be more balanced. Meanwhile, with the increasing of the truss rigidity, the overall deformation can be reduced, and the stability of the foundation will be improved.

Keywords: Pile foundation · Truss rigidity · Offshore wind turbine
Deformation · Horizontal loading

1 Background

Wind energy is a kind of clean and renewable energy. It is the key point of our country's planning and development in 12th Five-Year [1].Wind power generation has the advantages of large reserves, high development efficiency, little environmental pollution and no occupation of land, hence it has broad prospects for development [2].

Pile foundation is a basic form widely used for offshore wind turbines [3]. In the experimental research of the current study on pile foundation of offshore wind turbine, Matlock et al. [4] carried out field tests on horizontal loading on the pile foundation. He also analyzed the stress and deformation characteristics and the influencing factors of the bearing capacity of the pile-soil system. Zhang and Chen [5], Zhu et al. [6] analyzed the displacement and deformation response of the single pile and group-pile system under cyclic loading through the centrifuge model test. Rong et al. [7] developed a centrifugal field dynamic loading equipment to simulate the deformation and bearing characteristics of the pile foundation under different loads. In addition to the experimental research, Liu et al. [8] have carried out the numerical simulation, and analyzed the response characteristics between the pile and foundation soil under different loads.

The environment of offshore wind field is complicated. The foundation is subject to complicated wind, wave loads, floating objects, collision and so on, so the requirements of the displacement of the wind turbine are very strict. Through numerical simulation

© Springer Nature Singapore Pte Ltd. 2018
T. Qiu et al. (Eds.): GSIC 2018, *Proceedings of GeoShanghai 2018 International Conference: Advances in Soil Dynamics and Foundation Engineering*, pp. 829–836, 2018.
https://doi.org/10.1007/978-981-13-0131-5_90

analysis of two-dimensional pile and soil system under cyclic loading, Zuo and Tong et al. [9] found that the loading of the first cycle had a significant influence on the truss. In order to analyze the main characteristics of the deformation of the group piles foundation, it is necessary to study on the rigidity of the truss which has great influence on the stress and deformation distribution of each pile. In this paper, numerical simulation has been used to analyze the three-pile foundation under horizontal gradient loads. The influence of the Young modulus, the diameter and the thickness of the truss on the stress and deformation of the three-pile foundation is emphatically analyzed. It is hoped to be beneficial for the study on the stress and deformation of the offshore wind turbine pile foundation under complex loads.

2   Numerical Model

2.1   Model Introduction

The finite element software ABAQUS was used to simulate the three pile foundation. The three-dimensional calculation model and the cross section is shown in Figs. 1 and 2. The length and width of the computational model are 110 m. The height is 70 m. The pile uses hollow round tube. The outer diameter of the pile is 2.5 m and the wall thickness is 30 mm. The three piles are distributed in an equilateral triangle, in which the spacing between the piles is 16 m. The pile length is 35 m, and the embedded depth is 29 m. Above the pile foundation is a truss with a height of 16 m, and the horizontal load is applied on the top of the truss. In order to transfer the displacement and rotation better, the binding mode between the pile and the truss is Coupling. In order to make comparison on the centrifuge model test of Zuo [10], we connect displacement sensor A, B and T on each pile and truss. Its length is 8.5 m and the connection mode is Tie. The contact mode between the pile and the surrounding soil uses Hard Contact in the normal direction and the tangent direction uses the friction mode. The friction coefficient is 0.5 according to the experience. In the finite element calculation of this paper, both the foundation soil and the pile use solid elements. The truss and sensor are simulated by B31 unit. The truss and the pile foundation use the linear elastic model, and the foundation soil uses Mohr-Coulomb constitutive model. The material parameters of the model are shown in Table 1. The boundary of the model is set as following: the three directions of the bottom boundary unit are fixed and the side boundary element nodes are restrained only in the horizontal direction. The number of units in the whole model is 25250 and the number of nodes is 28327.

As shown in Fig. 2. The entire model is symmetrically distributed along the x-z plane. The pile along the force F in the x-z plane is named as the 1# pile, and the other two are denoted as 2# pile and 3# pile. This paper takes the 1# and 2# pile for example to analyze the stress and deformation of the pile body.

Figure 3 shows the data processed according to the similarity ratio. Since the proposed centrifuge model test is applied in a graded manner due to the load, the line of experiment is not smooth. However the numerical simulation results including the displacement of the sensor T on the top of the truss and the horizontal displacement of the sensor A, B on the top of the 1# pile, 2# pile are close to the centrifuge model test results on the variation law and numerical value. It can be seen that the numerical simulation can give a good prediction for the deformation of three-pile foundation.

Fig. 1. Three dimensional sketch map of pile foundation

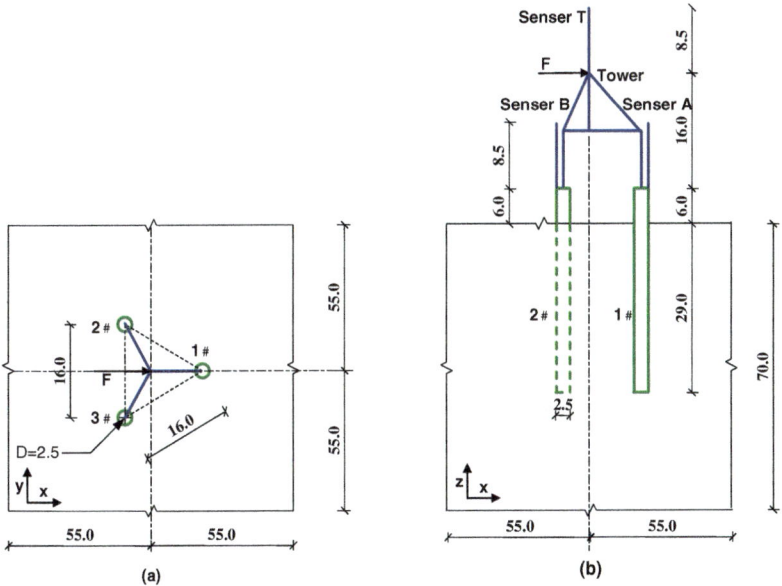

Fig. 2. Schematic view of pile foundation (unit: m). (a) vertical view; (b) elevation view.

Table 1. Parameters used in the model

Element type	Mass density γ (kg/m^3)	Modulus of elasticity E (MPa)	Poisson ratio ν	Friction angle φ (°)	Cohesion c (kPa)	Permeability coefficient k (m/s)
Soil	2800	5.4	0.3	20	10	1E-6
Pile and truss	8000	200000	0.25	–	–	–

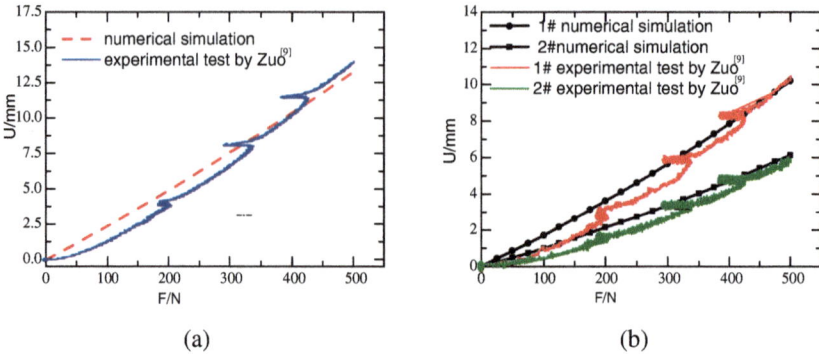

Fig. 3. The horizontal displacement comparison between the numerical simulation and the experimental test by Zuo [9]. (a) relation curve between the horizontal displacement at the top of the sensor T and the load; (b) relation curve between the horizontal displacement of the sensor of the 1# pile or the 2# pile and the load.

3   Simulation Results Discussed

In order to study the effect of the truss rigidity on the stress and deformation characteristics of the pile foundation, the rigidity of the truss is controlled by changing the Young modulus of the truss, the diameter and the thickness of the tube of the truss.

3.1   Influence of the Young Modulus of the Truss on the Three-Pile Foundation

First of all, the influence of the modulus of elasticity of the truss material on the horizontal displacement at the top of the truss (Fig. 4 (a)) is studied. The Young modulus of the truss is chosen as 12.5 GPa, 25 GPa, 50 GPa,100 GPa and 200 GPa. The horizontal displacement at the top of the truss decreases gradually. When the modulus of elasticity is 12.5 GPa, the horizontal displacement is the maximum value of 1.8 m. The distribution of the moment of the 1# pile and the 2# pile along with the soil depth is shown in Fig. 4 (b) and (c). As the Young modulus of the truss increases from 12.5 GPa to 200 GPa, the maximum bending moment of the 1# pile along the pile decreases by about 5000 kN·m, while the 2# pile decreases by about 1000 kN·m. The maximum bending moment of the two piles exists in the mud below the 1/4 pile. This characteristics is similar to the centrifuge model test of Lu et al. [11]. It shows that the model can reasonably reflect the deformation and stress characteristics of the pile foundation under horizontal load.

Figure 5 (a) and (b) show the change curves of the displacement of 1# pile and 2# pile with the soildepth. The top of the pile has the maximum displacement. With the increasing of the Young modulus of the truss, the maximum horizontal displacement of the 1# pile decreases by nearly 0.25 m, while the 2# pile has a minor change on displacement.

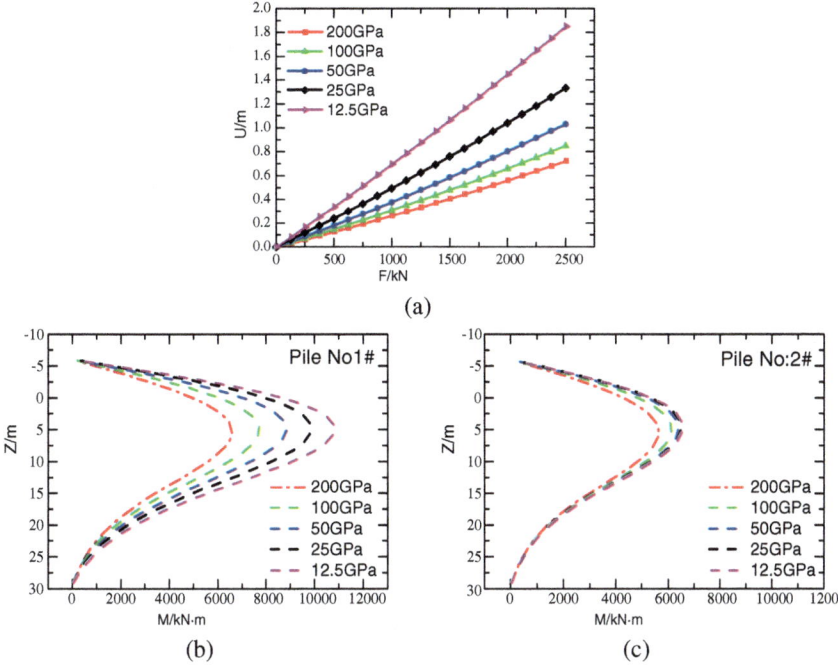

Fig. 4. The effect of the Young modulus of truss (a) the relationship between the horizontal displacement at the top of the column varies and the load; (b) and (c) the effect of the Young modulus of truss on the distribution of bending moment along the depth.

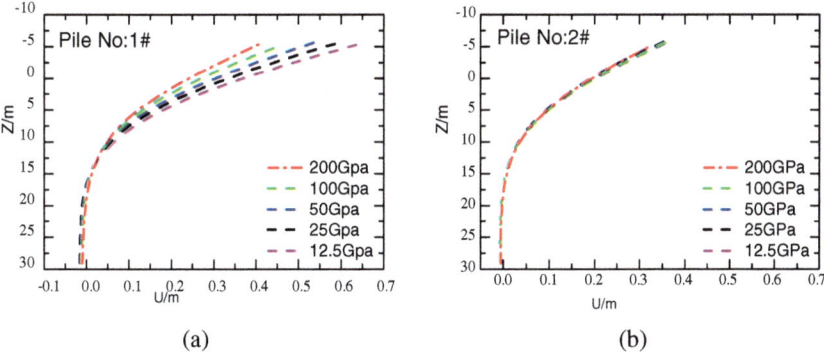

Fig. 5. The effect of the Young modulus of truss on the distribution of horizontal displacement along the depth.

Figure 6 (a) and (b) reflect that the maximum bending moment, Mmax, and maximum horizontal displacement, Umax, of the 1# pile are larger than those of the 2# pile. With the increasing of the Young modulus of the truss, the Mmax and the Umax of the 1# pile and 2# pile are getting closer. Mmax of the 1# pile decreases by 4500 kN·m, while Umax of the 1# pile decreases by about 0.4 m. The rate of change gets smaller and smaller. As shown in picture, the bending moment and deformation of 2# pile approach to 1# pile. The increase of the Young modulus of the truss will contribute to the improvement of the utilization ratio of the 2# pile.

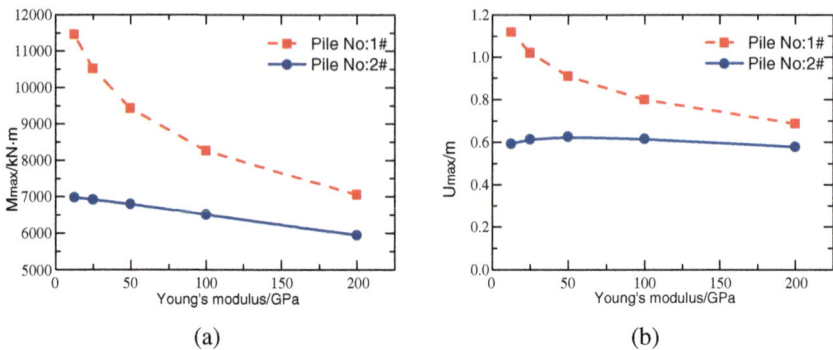

(a)   (b)

Fig. 6. The effect of the Young modulus of truss on the bending moment and deformation. (a) relation curve between maximum bending moment and Young modulus; (b) relationship between horizontal displacement and Young modulus.

3.2   Influence of Truss Size on Three-Pile Foundation

In addition to changing the modulus of elasticity of the truss material, the diameter D and the thickness T of the truss are changed to regulate the rigidity of the truss. Figure 7 (a), (b) reflect that the relation between the maximum bending moment, the maximum horizontal displacement and D. When D varies from 0.8 m, 1.6 m, 2.4 m, 3.2 m to 4 m, the maximum bending moment and the displacement of the two piles decreases. The Mmax of 1# pile body decreases by about 7500 kN·m and the 2# pile decreases by about 3000 kN·m. The Umax of 1# pile decreases by about 0.6 m and the 2# pile decreases by about 0.2 m. The following conclusions can be drawn from the picture: as D is smaller than 2.50 m, the change is faster. When diameter D is more than 2.50 m, the rate of change has slowed down.

Figure 8 reflects the relation between the stress and the deformation of the two piles when the thickness T of the truss is chosen as 30 mm, 60 mm, 90 mm, 120 mm and 150 mm. The regulation is the same as the increasing of the Young modulus and the diameter D of the truss. The maximum bending moment of 1# pile and the displacement of the top of 1# pile both decrease. The Mmax decreases by about 2500 kN·m, and the Umax decreases by about 0.2 m. The bending moment and deformation of 2# pile and 1# pile are getting closer.

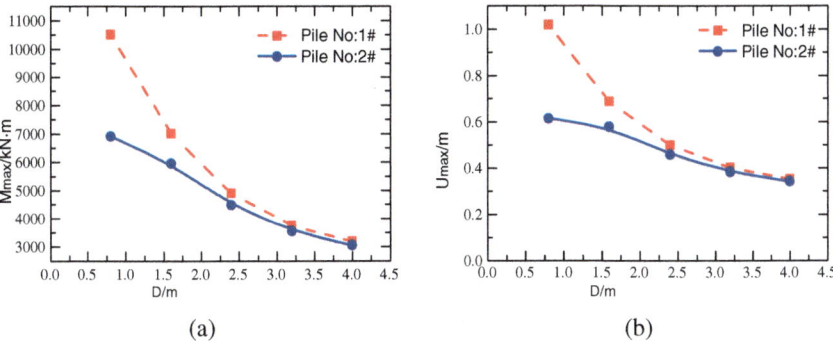

Fig. 7. The effect of the diameter D of truss on the bending moment and deformation. (a) relation curve of the maximum bending moment of pile body and diameter; (b) relationship of horizontal displacement of the top of the pile and the diameter.

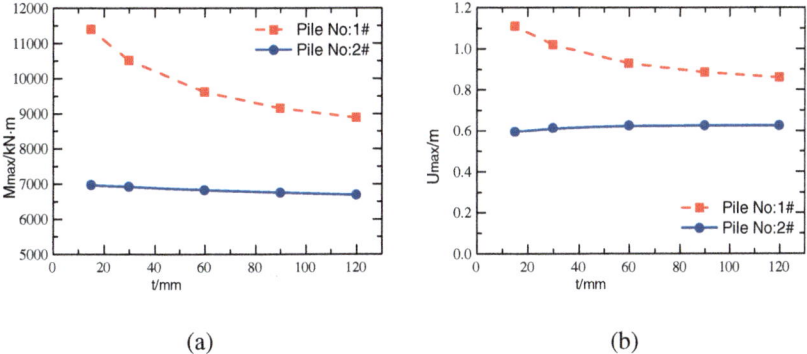

Fig. 8. The effect of the thickness T of truss on the bending moment and deformation. (a) relationship between maximum bending moment of pile and thickness of the truss (b) relationship between horizontal displacement of pile head and truss thickness.

4   Conclusion

This paper analyzes three pile foundation of offshore wind turbines under monotonic horizontal load by finite element software, and discusses the influence of Young modulus of the truss, the diameter and the thickness of the truss on bending moment and deformation of pile body. The following conclusions can be drawn:

(1) When the three piles are subjected to horizontal loads, the stress and deformation of the three piles are not exactly the same. The moment and displacement of the pile in the same plane with the load are larger than those of the other piles.

(2) The bending moment increases first and then decreases along the buried depth. The maximum moment occurred on the 1/4 of the pile in the mud. The horizontal displacement of the pile decreases gradually along the buried depth, and the bottom of the pile is barely changed.

(3) When increasing the truss stiffness, such as using materials with large modulus of elasticity, increasing the diameter and the thickness of the material, the stress of each pile can be more balanced and the material can be used more efficiently.

Acknowledgements. The present study is financially supported by the National Basic Research Program of China (973 Program) (contract number 2014CB047003) and the National Natural Science Foundation of China (contract number 10902005).

References

1. Liang, C., Jia, T., Chen, X.: Status quo and developing trend of wind energy. J. Shanghai Dianji Univ. **1**, 73–77 (2009)
2. Water Conservancy and Hydropower Planning and Design Institute. FD 003-2007 Wind turbine foundation design requirements (Trial) [S]. Beijing: China Water Conservancy and Hydropower Press (2007)
3. Wu, Z., Wang, F.: Blower foundation type in marine wind electric field and calculation method. Port Waterway Eng. **10**, 249–258 (2008)
4. Matlock, H., Wayne, B.I., Allen, E.K., Dewaine, B.:Filed tests of the lateral-load behavior of pile groups in soft clay. In: Offshore Technology Conference, 3871-MS (1980)
5. Zhang, L., Chen, Z.: Study of the model test of laterality loaded piles in cohesive soils. Chin. J. Geotech. Eng. **5**, 40–50 (1990)
6. Zhu, F., Fang, P., Huang, H.: Research on super-long pile in soft clay. Chin. J. Geotech. Eng. **25**(1), 76–79 (2003)
7. Rong, B., Zhang, G., Wang, F.: Wind motor in centrifugal field is complex pile load test equipment development. Rock soil Mech. **32**(5), 1596–1600 (2011)
8. Liu, C.: The Experimental Study and Numerical Analysis of the Mechanical Response of Offshore Wind Turbine Foundation Soil. Tsinghua University, Beijing (2014). (in Chinese)
9. Zuo, Z., Tong, Z.: Stress characteristics of soil around the pile of a monopole foundation in offshore wind turbines. China Earthquake Engineering Journal **36**(3), 549–554 (2014)
10. Zuo, Z.,Tong, Z.:Analysis of deformation and stability of offshore fan foundation under complex cyclic loading
11. Lu, W., Zhang, G., Wang, A.: Bearing behavior of multiple piles for offshore wind driven generator. Ocean Eng. **129**, 538–548 (2016)

Author Index

© Springer Nature Singapore Pte Ltd. 2018

837

T. Qiu et al. (Eds.): GSIC 2018, *Proceedings of GeoShanghai 2018 International Conference: Advances in Soil Dynamics and Foundation Engineering*, pp. 837–840, 2018.
https://doi.org/10.1007/978-981-13-0131-5